ESO ASTROPHYSICS SYMPOSIA
European Southern Observatory

Series Editor: Bruno Leibundgut

Springer
Berlin
Heidelberg
New York
Barcelona
Hong Kong
London
Milan
Paris
Tokyo

Physics and Astronomy   ONLINE LIBRARY

http://www.springer.de/phys/

ESO ASTROPHYSICS SYMPOSIA
European Southern Observatory

Series Editor: Bruno Leibundgut

Series homepage – http://www.springer.de/phys/books/eso/

M. Gilfanov R. Sunyaev E. Churazov (Eds.)

Lighthouses of the Universe:
The Most Luminous Celestial Objects and Their Use for Cosmology

Proceedings of the MPA/ESO/MPE/USM Joint Astronomy Conference Held in Garching, Germany, 6–10 August 2001

Springer

Volume Editors

Marat Gilfanov
Rashid Sunyaev
Eugene Churazov
Max-Planck-Institut für Astrophysik
Karl-Schwarzschild-Strasse 1
85741 Garching, Germany

Series Editor

Bruno Leibundgut
European Southern Observatory
Karl-Schwarzschild-Strasse 2
85748 Garching, Germany

Library of Congress Cataloging-in-Publication Data applied for.

Die Deutsche Bibliothek - CIP-Einheitsaufnahme

Lighthouses of the universe : the most luminous celestical objects and their
use for cosmology ; proceedings of the MPA ESO MPE USM Joint Astronomy
Conference, held in Garching, Germany, 6 - 10 August 2001 / M. Gilfanov ...
(ed.). - Berlin ; Heidelberg ; New York ; Barcelona ; Hong Kong ; London ;
Milan ; Paris ; Tokyo : Springer, 2002
    (ESO astrophysics symposia)
    (Physics and astronomy online library)
    ISBN 3-540-43769-X

ISBN 3-540-43769-X Springer-Verlag Berlin Heidelberg New York

Springer-Verlag Berlin Heidelberg New York
a member of BertelsmannSpringer Science+Business Media GmbH

http://www.springer.de

© Springer-Verlag Berlin Heidelberg 2002
Printed in Germany

Typesetting: Camera-ready by the authors/editors
Cover design: Erich Kirchner, Heidelberg

Printed on acid-free paper   SPIN: 10856495   55/3141/du - 5 4 3 2 1 0

Preface

During August 6–10, 2001 Garching hosted the conference "Lighthouses of the Universe", organized jointly by the Max-Planck-Institut für Astrophysik, Max-Planck-Institut für extraterrestrische Physik, European Southern Observatory and Universitäts-Sternwarte München.

The conference aim was to discuss all aspects of the most luminous objects in the Universe – its lighthouses. Broad theoretical reviews were complemented by fascinating results from the orbital observatories HST, Chandra and XMM-Newton, and from ground-based facilities such as Keck and VLT. We were gratified by the attendance of the meeting, the excellent reviews and contributed talks, the many lively discussions and poster sessions, and especially by the friendly and productive atmosphere over the five conference days. Very encouraging were the positive remarks about the meeting received both during and after the conference. Most importantly, we have been fortunate to receive excellent texts from essentially all of our invited and contributing speakers, and additionally from many poster papers. We would like to thank all of the participants for helping to achieve this.

Originally, we did not plan to publish any conference photos. However, by the end of the meeting we had discovered a number of "lively" pictures taken both during the conference and various associated events by our diploma students (with special thanks to Christian Haydn). The camera was typically aimed at the students and their friends, but on more than one occasion several renowned astrophysicists happened (by chance) to be in the background. We decided to include the best of these pictures in the book. They are by no means intended to be a complete account of the conference or a photo-gallery of its participants. We apologize to all of you who were not lucky enough to be in the right place at the right time and who did not appear in the photos we have in our possession. Of course, we also apologize to those who did happen to be in the right place at the right time.

Finally, we would like to use this opportunity to thank all those who helped with the planning and logistical support of the meeting. The meeting would not be possible without the help and cooperation of many people, among them the members of the local organizing and scientific advisory committees, the conference secretarial support of Kate O'Shea, Cornelia Rickl, Maria Depner and Gabriele Kratschmann, and other personnel from the Garching Research Campus. We express our sincere gratitude to all of them.

Garching, January 2002

Marat Gilfanov
Rashid Sunyaev
Eugene Churazov

Lighthouses of the Universe:

Contents

XIV   Contents

Part I

Clusters of Galaxies

Tales within Tales and Cutoffs within Cutoffs: What Sets the Mass Scale for Galaxies?

Paul L. Schechter

Massachusetts Institute of Technology, Cambridge MA 02139, USA

Abstract.

1. Does the mass function for clusters of galaxies cut off exponentially?
2. Does the luminosity function for galaxies cut off exponentially?
3. Is the dependence of virial velocity on galaxy luminosity a power law?
4. Does the velocity function for galaxies cut off exponentially?

1 Introduction

The luminosities of cosmic lighthouses are limited by a variety of physical processes. The Eddington limit immediately comes to mind, but like other such limits it is a function of mass. The question of limiting luminosities quickly becomes one of limiting masses.

At least four classes of cosmic lighthouses are on the program for this meeting and each has a different typical mass associated with it. It is my non-expert impression that for both stars and AGN we have some understanding of the physics that limits their masses. I will argue that the mass scale for clusters of galaxies is readily explained, but that our understanding (or at least my understanding) of the mass scale for galaxies is very incomplete.

2 The Mass Function for Clusters of Galaxies

The mass scale for clusters of galaxies is set by cosmological parameters which we take to be no different in character than the fundamental microscopic constants. The reigning paradigm is that clusters of galaxies grow by gravitational instability from small density perturbations in an otherwise uniform universe.

We start by asking how large a perturbation is needed at the time of recombination to collapse into an object which has just virialized, and find that we need a density perturbation with a fractional amplitude of roughly 0.001. We then ask how fluctuations in the underlying dark matter distribution manifest themselves as observed fluctuations in the cosmic microwave background. We find the corresponding fluctuations in the CMB on an angular scale which gives a mass typical of of clusters of galaxies.

The typical mass for clusters of galaxies might manifest itself in many different ways: it might be the mean of a Gaussian, the break in two power laws, a cutoff in a single power law or something else. The typical mass sets the scale but it remains to determine the shape of the mass function. For the purpose of the present meeting, we are especially interested the details of its high mass tail.

I did some work on the mass function for clusters as part of my PhD thesis. Gunn and Gott [8] just had done their well known work on the spherical collapse model for clusters. It seemed to me that since we knew that there were galaxies, whatever perturbations gave rise to galaxies would, on a larger scale, give rise to clusters. Starting, for the sake of argument, with the assumption that galaxies were distributed in Poisson fashion I arrived at an expression for the mass function for clusters of galaxies that consisted of a power law with an exponential cutoff.

There were problems with the argument and they had me worried. I had equated two quantities that I knew where not the same thing. The nature of the swindle is more obvious if one describes the quantities in words:

$$\begin{pmatrix} \text{fraction of} \\ \text{masses } M \\ \text{with } \delta > \delta_c \end{pmatrix} \approx \begin{pmatrix} \text{fraction of Universe} \\ \text{collapsed in objects} \\ \text{with masses } > M \end{pmatrix} \quad . \tag{1}$$

This *ansatz* made it possible to derive a mass function, but I was very nervous about the slope at the low mass end. An argument that gave the same slope in any number of dimensions couldn't possibly be right. I struggled with this for the better part of a year. Some time thereafter Bill Press joined the effort. He generalized my Poisson result to a power-law spectrum of arbitrary index n, emphasizing the self-similarity of the process and carried out what, by Ed Bertschinger's [2] reckoning, appears to have been the first cosmological N-body simulations (with N = 1000) for the purpose of checking our results. We published what we had [18] and never did find a way to avoid the swindle.

Our mass function,

$$n(M) = \frac{1}{2\sqrt{\pi}} \left(1 + \frac{n}{3} \right) \frac{\bar{\rho}}{M^2} \left(\frac{M}{M*} \right)^{3+n} \exp \left[- \left(\frac{M}{M*} \right)^{(3+n)/3} \right] \quad , \tag{2}$$

has undergone a number of extensions and improvements. A fudge factor has been successfully explained, and the approach has been "extended" (see Lacey and Cole [11] and the review by Schuecker *et al.* [20]) to include conditional merger probabilities looking backward and forward in time. It has also been adapted to non-spherical collapse [22]. Its deficiencies have been investigated by N-body experiment, leading several groups to suggest modifications [21,9]. Pierluigi Monaco writes [15]:

> The history of the mass function theory is reviewed in Monaco (1998) [14] but it can effectively be summarized in a sentence: there is a simple, effective and wrong way to describe the cosmological mass function. Wrong of course, does not refer to the results but to the whole procedure.

Not only did we get the calculation wrong, we had the input physics wrong as well. We assumed that the matter that was clustering was strictly baryonic. At recombination baryonic density perturbations have roughly the same amplitude as the CMB fluctuations, which we now know were much too small to produce today's clusters of galaxies. What we didn't appreciate was that non-baryonic dark matter perturbations, uncoupled

to the photons and the baryons, could have been growing while the baryonic perturbations were locked into the photon fluid. Clusters of galaxies are fundamentally self-
gravitating, pressure supported spheroids of dark, non-baryonic matter. Their baryons
comprise only a small fraction of their mass. And the galaxies from which they take
their name include only a small fraction of the baryons.

These shortcomings notwithstanding, many investigators have found the Press-
Schechter recipe to be an acceptable first cut at the mass function for clusters of galaxies. The answer to our first question appears to be "yes" – the mass function has an
exponential cutoff.

3   The Luminosity Function for Galaxies

3.1   The Shape of the Luminosity Function

Hubble, Zwicky and George Abell all carried out studies of the luminosity function
for galaxies, and and all of them found that there is a characteristic luminosity. They
differed considerably, however, on the shape of the luminosity function. Hubble found
it to be roughly Gaussian and Abell found the cumulative luminosity function to be a
broken power law.

In one of the opening volleys of the science wars, Abraham Maslow wrote [13]
wrote:

> I suppose it is tempting, if the only tool you have is a hammer, to treat
> everything as if it were a nail.

It's true. My hammer was the power law with an exponential cutoff and I hammered on
the luminosity function for galaxies [19].

Many investigators have wielded this hammer since, and some have found it unsatisfactory. But it would still seem to be of some use. In their analysis of the SDSS
commissioning data, Blanton *et al.* find, to their evident surprise, that a power law with
an exponential cutoff fits their data better than an alternative, non-parametric model [3].
The answer to the second question would also appear to be "yes."

3.2   Implications for the Mass Function

From the outset it seemed somewhat inconsistent, perhaps even hypocritical, to use
roughly the same functional form for both the mass function for galaxies and the luminosity function for galaxies. One might argue that the same process which gives rise to
the cluster mass function gives rise to the galaxy mass function at some earlier time,
but this argument is flawed. The idea of self-similar condensation was that the objects
seen at one epoch would merge to form larger objects. How could they merge and yet
survive? We would be eating our cake and having it too. If the same process were to
work twice, an early epoch of structure had to be frozen in such a way that it could
survive subsequent mergers. This question has been addressed repeatedly, both theoretically and experimentally with N-body simulations. Only recently have the latter begun
to give reliable answers. We discuss them in greater detail below, but for the moment
the important conclusion is that a small but significant fraction of the mass, perhaps
10%, survives in what look like galaxies.

4 The Faber-Jackson and Tully-Fisher Relations

In papers published within a year of each other, Faber and Jackson [7] (henceforth FJ) and Tully and Fisher [25] (henceforth TF) found, respectively, that elliptical and spiral galaxies exhibited a similar scaling between internal velocities (velocity dispersion and circular velocity) and galaxy luminosity. They found luminosity varied roughly as the fourth power of observed velocity, with a scatter about the mean relation of 30-40%. The FJ relation has since been refined to include surface brightness by Dressler *et al.* [6] and Djorgovski and Davis [5] giving what the latter call the "fundamental plane."

Bernardi *et al.* [1] recently completed a massive study of the fundamental plane finding, among other things, that it is, very nearly, a plane. When they compute the mean velocity dispersion at fixed absolute magnitude and correct for evolution they see little if any deviation from a power law – the FJ relation holds. The answer to our third question is another "yes".

5 The Velocity Function for Galaxies

5.1 Why Not the Mass Function?

As our subject is the mass scale for galaxies, the distribution of virial velocities – the "velocity function" – may seem a diversion, deflecting us from our goal. Alas it is nearly impossible to measure masses for galaxies. Mass measurements combine a velocity and a scale length measurement. While one can measure scale lengths for the baryons one sees in galaxies – stars or gas – there is every reason to believe that this is not representative of the dark matter. Baryonic collapse is dissipational while dark matter is non-dissipational. Differential collapse by roughly a factor of ten is not unreasonable. We exhaust the supply of stars and gas with which to measure galaxy measure rotation curves and velocity dispersions before we reach the outer limits of the dark matter.

By contrast, the velocities we measure in the baryonic matter *are* representative of the dark matter. The virial theorem, in its simplest incarnation, tells us that the time averaged squared (virial) velocity is given by

$$< v^2 >= \left\langle \vec{r} \cdot \frac{\partial \Phi}{\vec{\partial r}} \right\rangle \quad , \tag{3}$$

where Φ is the gravitational potential. The dark matter and the stars orbit in the same potential, and would have the same virial velocity on the same orbit. Since galaxies appear to be nearly isothermal, the righthand side of the above equation is not a strong function of position, and it is not unreasonable to take the virial velocity for the baryonic matter as a proxy for that of the dark matter.

A similar circumstance holds for studies of clusters of galaxies, which sometimes use "temperature functions" rather than mass functions. Masses for clusters are similarly dependent upon details of how the scale length is measured. By contrast measurement of the temperature of X-ray emitting gas is relatively straightforward.

5.2   The Observations

At first it would seem a straightforward matter to compute a "velocity function" from the luminosity function and either the FJ or TF relation, especially if the luminosity-velocity relations were indeed power laws. For example, if if the luminosity function cut off as $exp(-L/L*)$ and luminosity varied as v^4, one would have velocity varying as $exp[-(v/v*)^4]$. If the luminosity-velocity relation saturated at large velocities, the cutoff would be sharper.

But there are complications. Bernardi et al. [1] point out that in carrying out the above calculation, one must take into account the spread in virial velocity at fixed luminosity. Moreover, there is a potential problem [10] in covariances of the luminosity function and the velocity-luminosity relation with hidden variables. Pahre et al. (as reported by Kochanek [10]) have taken such covariances into account. Their velocity function, as shown in Kochanek's figures 6,7 and 9, cuts off exceedingly sharply.

The separations of multiply imaged quasars vary as the square of the virial velocity. Kochanek [10] has computed a velocity function from these, and while the statistics are meager, it is consistent with an exponential cutoff. Taking all of this into consideration, the answer to our fourth question, as determined from observations, is "most probably."

5.3   The Simulations

Determining the merger histories for present day galaxies from observations is a daunting problem, even for our own Milky Way. Luckily the hierarchical condensation of dark matter can be simulated with N-body experiments, at least in principal. But for many years N-body practitioners were frustrated by their simulations – they formed beautiful clusters but there were no galaxies. Previous generations of structure merged to form the present generation. Sometime around 1986 this phenomenon was given a name by Jerry Ostriker – overmerging. In the last two or three years several groups (notably those based in New Mexico, Seattle and Garching) appear to have traced the overmerging problem to the numerical necessity of softening the $1/r$ gravitational potential. It now seems that substructure can survive to the present epoch.

This is a major accomplishment, and cause for celebration, but we shouldn't forget that while some substructure survives, most of it is destroyed [24]. The astronomical jargon used to describe this is still in flux, but the substructures which survive (identified with today's galaxies) are called sub-halos and the larger structures in which they are embedded (identified with today's clusters) are called halos. To order of magnitude, it appears that 10% of the dark matter remains more closely associated with sub-halos rather than the larger halos. [1]

---

[1] This raises an interesting question question of physics which, while not strictly applicable to the universe in which we live, may nonetheless help to understand it. The paper by Bill Press and myself made much of the self-similarity of the growth of structure. This depended first, upon the power-law nature of the perturbation spectrum and second, on the assumption of an Einstein-de Sitter cosmology. Today we believe that the perturbation spectrum at recombination is the product of a number of competing processes that would have destroyed the power-law nature of any input spectrum. But we can still ask what would happen in a hypothetical Einstein-de Sitter universe in which the input spectrum *was* a power law. Would self-similarity

The next step after identifying sub-halos is to construct a mass function. Some investigators choose to construct a velocity function as well, either for the sake of comparison with observation or because it is less sensitive to the details of the algorithm used to identify sub-halos.

While the results are relatively new and the details may change as the experiments improve, two important conclusions can already be drawn. First, it appears that the mass function for sub-halos is very much steeper than the luminosity function for galaxies. Second, there is at best weak evidence for an exponential cutoff in the masses of sub-halos – the mass functions look very much like power laws.

The same holds true for velocity functions. They look like power laws without any obvious cutoff [4,24]. The answer to our fourth question, as determined from N-body simulations, is "evidently not." In an odd way this is reassuring – it is not obvious what would introduce either a mass scale or velocity scale for the sub-halos, so it is just as well that we don't see one.

We face a serious dilemma. Observation gives a positive answer to our fourth question and N-body experiment gives a negative answer.

6   The Baryonic Mass Function for Galaxies

As discussed in the previous section, the dark matter mass function for galaxies is perversely difficult to measure. By comparison the baryonic mass function is fairly easy – indeed to first order it is the luminosity function scaled by a mass-to-light ratio.

The N-body experiments in which sub-halos form within halos involve only dark matter particles – they interact only by non-dissipative, mutual gravitation. By contrast, luminosity functions, both optical (for galaxies) and X-ray (for clusters), describe baryonic matter, which can radiate away its energy. All manner of physics can cause dark matter and baryonic matter, however uniformly distributed at first, to become separated over the course of time. There is every reason to think that mass function for baryons will therefore be quite different from the mass function for sub-halos (and halos).

A variety of "gastrophysical" mechanisms have been invoked to explain the differences between the observed (baryonic) luminosity function for galaxies and sub-halo mass function derived from N-body experiments. At the low mass end, baryons must either be ejected from sub-halos or something must prevent them from collapsing along with the collapse of dark matter. Somerville [23] has used the word "squelching" to describe the suppression of dwarf galaxy formation by the re-ionization of the IGM. There may be some disagreement about the physics but it's an apt word to describe the phenomenon.

Whatever causes the flattening of the mass function at the low mass end, something else must cause a steepening at the high mass end. White and Rees [27] proposed that the cooling time for baryonic gas, assumed to be uniformly distributed in a dark matter potential well, increases with increasing mass to the point at which the baryons no

---

still hold? In particular, would the fraction of mass in sub-halos always be the same percentage of the total mass at all times? Or would we expect to find sub-sub-halos inside sub-halos and so forth? I suspect that this is the case, although I doubt anyone has the code or the perseverance to carry this out.

longer have time to cool and condense. It should be remembered, however, that White and Rees went to considerable lengths to emphasize that substructure would be destroyed. If substructure is not destroyed, then the gas associated with this substructure would be clumpy and would have much shorter cooling times.

Whatever the detailed physics, the underlying idea of White and Rees would seem to be that sub-halos with large internal velocities exist, but we don't see them because their baryons could not and have not condensed.

A half-dozen groups are engaged in a heroic enterprise that goes by the name "semi-analytic modeling." They attempt to form galaxies in dark matter halos, taking care to include the gravitational condensation of halos and the gas-dynamical condensation of baryons within these halos. They include star formation, energy feedback, chemical enrichment and obscuration by dust. All of these are accomplished by a series of more-or-less realistic recipes. More recently some of these modelers have substituted high resolution N-body simulations for analytically derived halo histories. These models have many free parameters – enough to fit almost any set of constraints. Yet with all these parameters, the modelers still have difficulty in cutting off the bright end of the luminosity function. The adopted values of the parameters seem extreme, and in some cases the assumptions seem *ad-hoc*.

The TF and FJ relations place strong constraints on the recipes. One sort of recipe might have the fraction of the baryons which condense within a halo decrease as the halo velocity dispersion increases. A second sort might have the condensation of baryons in a halo be an all-or-nothing affair, but with baryon condensation increasingly unlikely with increasing halo dispersion. Recipes of the first sort might run afoul of the velocity-luminosity constraints, producing halos with large velocities but relatively low luminosities.

I suspect that something on the order of the solution proposed by White and Rees will ultimately prove successful in inhibiting the collapse of baryons into the largest halos, thereby producing a baryonic mass cutoff for galaxies. But for the moment the modelers have not yet persuaded each other; one wonders whether they have even persuaded themselves.

For the sake of argument let's grant that something squelches the condensation of baryons into low mass halos that something else inhibits the condensation of baryons into high mass halos. This brings us back to the velocity function constructed from gravitational lensing. If baryonic infall had no effect on the structure of sub-halos, the distribution of lens separations could be derived from the velocity function determined from N-body simulations and one ought not to see a cutoff. But White and Kochanek [26] have argued that baryonic infall *does* alter the gravitational potentials of halos, particularly at the relatively small radii sampled by lensed quasars, so even the lensed quasar separation distribution may exhibit a cutoff, or at least a kink. While this might solve our problem, it should be noted (as they do) that their model reproduces the observed distribution of lens separations only if the baryonic mass fraction in galaxies is larger than is usually assumed.

7 cD Galaxies: Sub-Halos or Halos?

No discussion of galaxies as lighthouses would be complete without mention of cD galaxies. It is helpful to make a distinction between those which are instantaneously the brightest and those which are the brightest when averaged over cosmic timescales. In their contributions to these proceedings LONGAIR and CESARSKY discuss sub-mm sources and ULIRGs, which are instantaneously the brightest galaxies in the universe. Averaged over time this honor may belong, instead, to the cD galaxies found at the centers of many (but not all) clusters of galaxies.

The "cD" classification originated in a paper by Matthews, Morgan and Schmidt [12] (henceforth MMS). The lowercase "c" comes from Morgan's classification for supergiant stars and the uppercase D stands for diffuse. With NGC 6166 in Abell 2199 as the archetype [16], these galaxies had large, low surface brightness halos. They frequently seemed to harbor radio sources.

The designation cD is sometimes taken to stand for the words central and dominant. While this was not the original meaning of MMS, there is more than a bit of truth in it. Such galaxies are indeed found preferentially at the centers of clusters of galaxies and they tend to be the dominant galaxy. Oegerle and Hill [17] have found that they also tend to lie at the center of the velocity distribution for the cluster in which they reside.

There is an old but still interesting question as to whether the cD galaxies are members of the same population as the other galaxies in a cluster, or are in some way special. If the luminosities of galaxies really do cut off exponentially, there's no explaining cD's.

There are other peculiarities associated with cD's as well – their velocity dispersion profiles rise rather than fall as a function of radius. This is consistent with their extended envelopes and even with a flattening of the slopes of their light profiles. Moreover they seem to have an excess of globular clusters.

The same N-body experiments which give rise to interesting substructure also seem to produce a condensation at the center of each cluster which has the density of a galaxy. The special position of galaxies at the centers of cluster potentials might make them different. Today they seem largely to be star piles, the piling up of fragments of previously formed galaxies.

At earlier times cD's may have been special in a different way. The galaxies at the centers of potentials are the natural sinks for cooling baryons, and one expects baryons to collect there – one would think in disks. In the semi-analytic models the galaxy at the center of a halo occupies a unique position – it is the only galaxy which accretes baryons. In such a model a dark matter halo collects baryons only until it is subsumed into a larger halo in which it is no longer in the privileged central position. The other galaxies in the halo are called satellites to distinguish them from the central galaxy. But if they have any baryons, then at one time they must have been at the center of the cluster. In short, every galaxy must have at one time been a central dominant galaxy, though not a cD galaxy in the sense of MMS.

8 Conclusion

The answers to the four questions posed in the abstract are yes, yes, yes and yes-and-no. The observed velocity function for galaxies appears to cut off exponentially while the velocity function for N-body sub-halos appears not to. There is as yet no detailed physical model which cuts off the condensation of baryons into dark matter halos in a manner which conforms to the observations.

References

1. Bernardi, M. *et al.* 2001 (astro-ph/0110344)
2. Bertschinger, E. 1998, ARAA, **36**, 599
3. Blanton, M. R. *et al.* 2001, AJ, **121**, 2358
4. Bullock, J. S., Kravtsov, A. V., & Weinberg, D. H. 2001, ApJ, **548**, 33
5. Djorgovski, S. & Davis, M. 1987, ApJ, **313**, 59
6. Dressler, A., Lynden-Bell, D., Burstein, D., Davies, R. L., Faber, S. M., Terlevich, R., & Wegner, G. 1987, ApJ, **313**, 42
7. Faber, S. M. & Jackson, R. E. 1976, ApJ, **204**, 668
8. Gunn, J. E. & Gott, J. R. I. 1972, ApJ, **176**, 1
9. Jenkins, A., Frenk, C. S., White, S. D. M., Colberg, J. M., Cole, S., Evrard, A. E., Couchman, H. M. P., & Yoshida, N. 2001, MNRAS, **321**, 372
10. Kochanek, C. S. 2001, in *The Dark Universe meeting at STScI, April 2-5, 2001* M. Livio, ed., Cambridge University Press, (astro-ph/0108160)
11. Lacey, C. & Cole, S. 1993, MNRAS, **262**, 627
12. Matthews, T. A., Morgan, W. W., & Schmidt, M. 1964, ApJ, **140**, 35
13. Maslow, A. H. 1966, *The Psychology of Science; A Reconnaissance* (New York: Harper, Row & Publishers), p. 15
14. Monaco, P. 1998, Fundamentals of Cosmic Physics, **19**, 157
15. Monaco, P. 1998, ASP Conf. Ser. 146: *The Young Universe: Galaxy Formation and Evolution at Intermediate and High Redshift*, 318
16. Morgan, W. W. & Rountree Lesh, J. 1965, ApJ, **142**, 1364
17. Oegerle, W. R. & Hill, J. M. 1994, AJ, **107**, 857
18. Press, W. H. & Schechter, P. 1974, ApJ, **193**, 437
19. Schechter, P. 1976, ApJ, **203**, 297
20. Schuecker, P., Böhringer, H., Arzner, K., & Reiprich, T. H. 2001, A&A, **370**, 715
21. Sheth, R. K. & Tormen, G. 1999, MNRAS, **308**, 119
22. Sheth, R. K., Mo, H. J., & Tormen, G. 2001, MNRAS, **323,** 1
23. Somerville, R. S. 2001 (astro-ph/017507)
24. Springel, V., White, S. D. M., Tormen, G., & Kauffmann, G. 2001, MNRAS, **328**, 726
25. Tully, R. B. & Fisher, J. R. 1977, A&A, **54**, 661
26. White, M. & Kochanek, C. S. 2001, ApJ, **560**, 539
27. White, S. D. M. & Rees, M. J. 1978, MNRAS, **183**, 341

Galaxy Clusters: Cosmic High-Energy Laboratories to Study the Structure of Our Universe

Hans Böhringer

Max-Planck-Institut für Extraterrestrische Physik, D-85748 Garching, Germany

Abstract. This contribution illustrates the study of galaxy clusters as astrophysical laboratories as well as probes for the large-scale structure of the Universe. Using the REFLEX Cluster Survey, the measurement of the statistics of the large-scale structure on scales up to 500 h^{-1} Mpc is illustrated. The results clearly favour a low density Universe.

Clusters constitute, in addition, well defined astrophysical laboratory environments in which some very interesting large-scale phenomena can be studied. As an illustration we show some spectacular new XMM X-ray spectroscopic results on the thermal structure of cooling flows and the interaction effects of AGN with this hot intracluster medium. The X-ray observations with XMM-Newton show a lack of spectral evidence for large amounts of cooling and condensing gas in the centers of galaxy clusters believed to harbour strong cooling flows. To explain these findings we consider the heating of the core regions of clusters by jets from a central AGN. We find that the power output the AGN jets is well sufficient. The requirements such a heating model has to fulfill are explored and we find a very promising scenario of self-regulated Bondi accretion of the central black hole.

1 Introduction

Galaxy clusters with masses from about 10^{14} to over 10^{15} M$_\odot$ are the largest clearly defined building blocks of our universe. Their formation and evolution is tightly connected to the evolution of the large-scale structure of our Universe as a whole. Clusters are therefore ideal probes for the study of the large-scale matter distribution. Due to the hot, intracluster plasma trapped in the potential well of galaxy clusters, which can take on temperatures of several keV (several 10 Million degrees), galaxy clusters are the most luminous X-ray emitters in the Universe next only to quasars. The hot gas and its X-ray emission is a good tracer of the gravitational potential of the clusters and thus allows us to obtain mass estimates of the clusters, to study their morphology and dynamical state, and to detect clusters as gravitationally bound entities out to very large distances (Sarazin 1986). Systematic studies have shown that clusters have within a first order description, a quite standardized, self-similar appearance. This is reflected observationally in quite narrow correlations of observable parameters, like e.g. the temperature - X-ray luminosity relation. The for our application most important relation is that of mass and X-ray luminosity. Since, for the construction of a cosmologically interesting cluster sample, we are interested to collect all the clusters above a certain mass limit, this relation makes it possible to select the most massive clusters essentially by their X-ray luminosity.

The first detailed mass-X-ray luminosity relation compiled from X-ray observations is shown in Fig.1 (Reiprich & Böhringer 1999, 2001). From these results a cluster

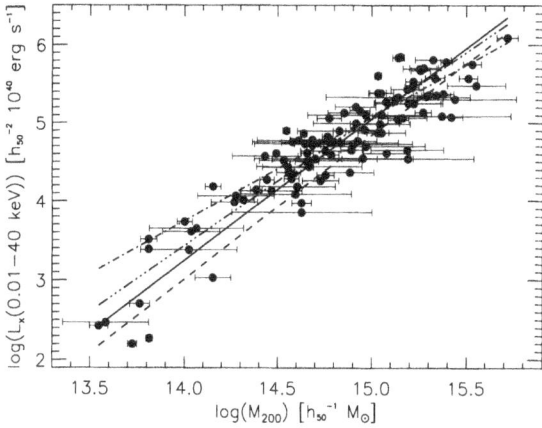

Fig. 1. Mass-X-ray luminosity relation determined from the 106 brightest galaxy clusters found in the ROSAT All-Sky Survey by Reiprich & Böhringer (1999, 2001)

mass function and its integral, the matter density in the Universe bound in clusters was derived. In Fig. 2 we show this cumulative cluster mass density function, yielding $\Omega_{clusters} \sim 0.02$. Thus for the currently most popular value for the matter density $\Omega_m \sim 0.3$ about 6% of the matter in the Universe is bound in clusters with a mass larger than $6.4 \cdot 10^{13} h_{50}^{-1} \, M_\odot$.

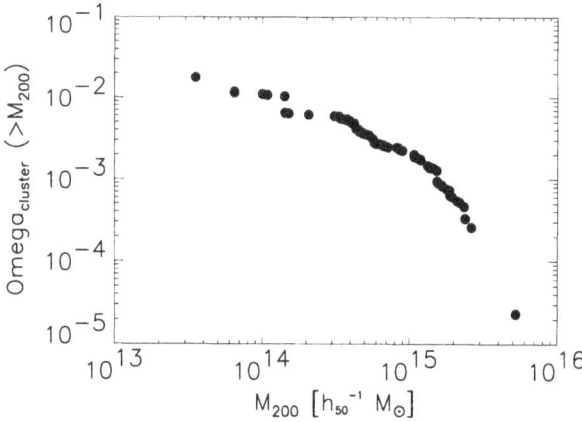

Fig. 2. Cumulative fraction of the mass density bound in clusters in the Universe compared to the critical density (Reiprich & Böhringer 2001)

2 Measurement of the Large-Scale Structure

Based on the ROSAT All-Sky X-ray Survey - so far the only all-sky survey conducted with an X-ray telescope - we exploit the above mentioned tight X-ray luminosity-mass relation with X-ray selected cluster redshift surveys to study the large-scale structure of the Universe. Fig.3 shows the sky distribution of the brightest X-ray galaxy clusters identified so far in the ROSAT All-Sky Survey within two projects: the northern NO-RAS Survey (Böhringer et al. 2000) and the southern REFLEX Survey (Böhringer et al. 2001a). The latter project is currently more complete comprising a sample of 452 clusters and therefore the following results are based on this project. Most of the redshifts for this survey have been obtained within the frame of an ESO key program (Böhringer et al. 1998, Guzzo et al. 1999) .

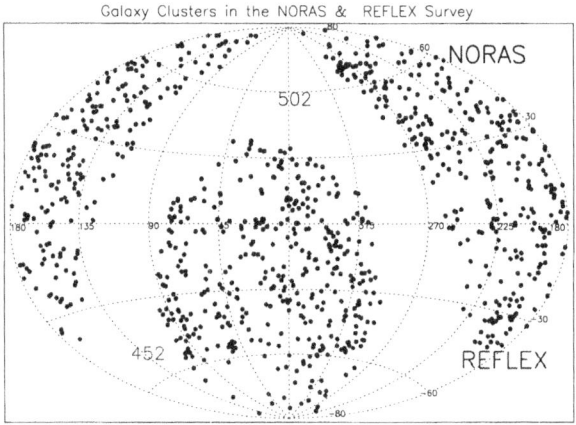

Fig. 3. Sky distribution of the brightest galaxy clusters found in the ROSAT All-Sky Survey investigated in the NORAS and REFLEX redshift surveys.

The basic census of the REFLEX cluster survey is the X-ray luminosity function shown in Fig. 4 (Böhringer et al. 2002). The essential goal of this survey is the assesment of the statistics of the large-scale structure. The most fundamental statistical description of the spatial structure is based on the second moments on the distribution, characterized either by the two-point-correlation function or its Fourier transform, the density fluctuation power spectrum. The two-point correlation function of the clusters in the REFLEX sample has been derived by Collins et al. (2000). The results show a power law shaped correlation function with a slope of 1.83, a correlation length of $18.8h_{100}^{-1}$ Mpc and a possible zero crossing at $\sim 45h_{100}^{-1}$ Mpc. The density fluctuation power spectrum (Fig. 5) has been determined by Schuecker et al. (2001a). The power spectrum is characterized by a power law at large values of the wave vector, k, with a slope of $\propto k^{-2}$ for $k \leq 0.1h$ Mpc^{-1} and a maximum around $k \sim 0.03h$ Mpc^{-1} (corresponding to a wavelength of about $200h^{-1}$ Mpc). This maximum reflects the size of the horizon when the Universe featured equal energy density in radiation and matter and is a sensitive measure of the mean density of the Universe, Ω_0. For standard open and flat Cold Dark Matter models (OCDM and ΛCDM) we find the following constraints $h\Omega_0 = 0.12$ to 0.26 (Schuecker et al. 2001a).

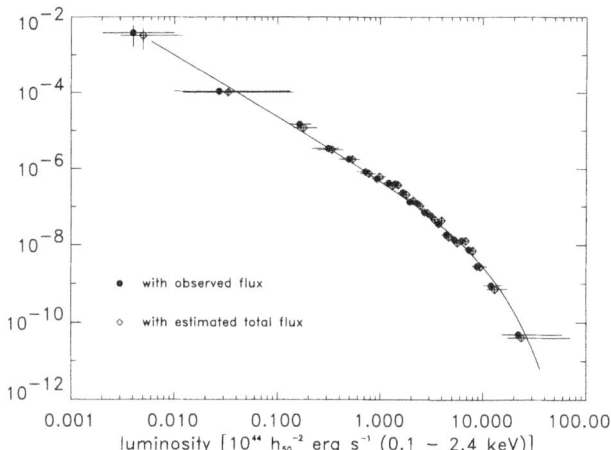

Fig. 4. X-ray luminosity function of the REFLEX cluster Survey (Böhringer et al. 2002). The two series of data point refer to the calculation of the luminosities from the X-ray fluxes as observed and corrected for the missing flux in the outer very low surface brightness regions, respectively.

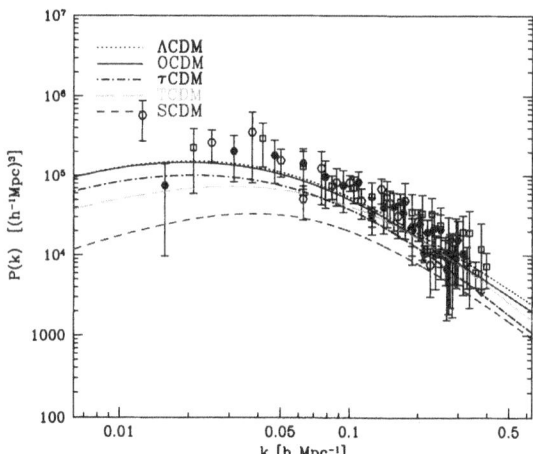

Fig. 5. Power spectra of the density fluctuations in the REFLEX cluster sample together with predictions from various popular cosmological models taken from the literature. For details see Schuecker et al. (2001a).

3   Cosmological Tests with Galaxy Clusters

In addition to the above mentioned large-scale structure test for cosmological models several further studies yield information on cosmological parameters. These include the modelling of the mass, temperature, and luminosity functions (e.g. Reiprich & Böhringer 2001, Ikebe et al. 2001), the study of the mass-to-light ratio in connection with galaxy biasing Carlberg et al. (1996; which yields $\Omega_0 \sim 0.24 \pm 0.1$), the baryon mass fraction in clusters, and the statistics of merging clusters as observed in the RE-FLEX cluster sample (Schuecker et al. 2001b). All these different cosmological studies using clusters point towards a low density universe. These results can be compared to the results obtained from observations of the cosmic microwave anisotropies (e.g. De Bernardis et al. 2000, 2001) and of the study of distant SN Ia (e.g. Perlmutter et al. 1999). While these two investigations provide combined constraints that encircle a region in the model parameter space spanned by the cosmological parameters Ω_0 and Ω_Λ around values of $\Omega_0 = 0.3$ and $\Omega_\Lambda = 0.7$, the galaxy cluster results provide a different cut through this parameter space crossing the other two results at their intersection. That is, the cluster results provide at present (without the inclusion of an investigation of a very large redshift range) no significant constraints on the Ω_Λ-parameter, but allow values of for $\Omega_0 = 0.2 - 0.4$ in the range consistent with the combined cosmic microwave and SN Ia data. Thus, the evidence for a low density universe is solidifying.

4   Cluster Cooling Cores

X-ray imaging observations have shown that the X-ray emitting, hot gas in a large fraction of all galaxy clusters reaches high enough densities in the cluster centers that the cooling time of the gas falls below the Hubble time, and gas may cool and condense in the absence of a suitable fine-tuned heating source (e.g. Silk 1976, Fabian & Nulsen 1977). From the detailed analysis of surface brightness profiles of X-ray images of clusters obtained with the *Einstein*, *EXOSAT*, and *ROSAT* observatories, the detailed, self-consistent scenario of inhomogeneous, comoving cooling flows emerged (e.g. Nulsen 1986, Thomas, Fabian, & Nulsen 1987, Fabian 1994). The main assumptions on which the cooling flow model is based and some important implications are: (i) Each radial zone in the cooling flow region comprises different plasma phases covering a wide range of temperatures. The consequence of this temperature distribution is that gas will cool to low temperature and condense over a wide range of radii. (ii) The gas features an inflow in which all phases with different temperature move with the same flow speed. (iii) There is no energy exchange between the different phases, between material at different radii, and no heating.

Now the first analysis of high resolution X-ray spectra and imaging spectroscopy obtained with *XMM-Newton* has shown to our surprise that the spectra show no signatures of cooler phases of the cooling flow gas below an intermediate temperature (e.g. Peterson et al. 2001, Tamura et al. 2001) and local isothermality in the cooling flow region (e.g. Böhringer et al. 2001b, Matsushita et al. 2001, Molendi & Pizzolato 2001) in conflict with the inhomogeneous cooling flow model. Here, we discuss these new spectroscopic results and their implications and point out the way to a new possible model for this phenomenon. The results are mostly based on the detailed observations of the M87 X-ray halo. A detailed description of this study is provided by Böhringer et al. (2001c).

5   Spectroscopic Diagnostics of Cluster Cooling Cores

XMM Reflection Grating Spectrometer (RGS) observations of several cooling core regions show signatures of different temperature phases ranging approximately from the hot virial temperature of the cluster to a lower limiting temperature, T_{low}. Clearly observable spectroscopic features of even lower temperature gas expected for a cooling flow model are not observed. A1835 with a bulk temperature of about 8.3 keV has T_{low} around 2.7 keV (Peterson et al. 2001) and similar results have been derived for A1795 (Tamura et al. 2001). These results are very well confirmed by *XMM* observations with the energy sensitive imaging devices, EPN and EMOS, providing spectral information across the entire cooling core region, yielding the result that (for M87, A1795, and A1835) single temperature models provide a better representation of the data than cooling flow models (Böhringer et al. 2001a, Molendi & Pizzolato 2001) also implying the lack of low temperature components. The very detailed analysis of M87 by Matsushita et al. (2001 and the contribution to this workshop) has shown that the temperature structure is well described locally by a single temperature over most of the cooling core region, except for the regions of the radio lobes and the very center ($r \leq 1$ arcmin, ~ 5 kpc).

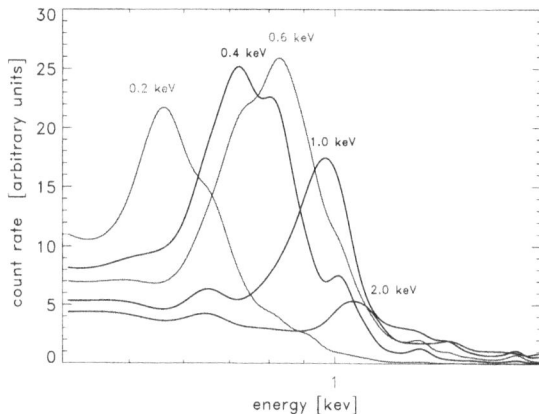

Fig. 6. The Fe L-line complex in X-ray spectra as a function of the plasma temperature for a metallicity value of 0.7 solar. The simulations show the appearance of the spectra as seen with the XMM EPN. The emission measure was kept fixed when the temperature was varied.

Among the spectroscopic signatures which are sensitive to the plasma temperature in the relevant temperature range, the complex of iron L-shell lines is most important. Fig. 6 shows simulated X-ray spectra as predicted for the XMM EPN instrument in the spectral region around the Fe L-shell lines for a single-temperature plasma at various temperatures from 0.4 to 2.0 keV and 0.7 solar metallicity. There is a very obvious shift in the location of the peak making this feature an excellent thermometer. For a cooling flow with a broad range of temperatures one expects a composite of several of the relatively narrow line blend features, resulting in a quite broad peak. Fig. 7a shows for example the deprojected spectrum of the M87 halo plasma for the radial range 1 - 2 arcmin (outside the inner radio lobes) and a fit of a cooling flow model with a mass deposition rate slightly less than 1 $M_\odot$ yr^{-1} as expected for this radial range from the analysis of the surface brightness profile (e.g. Matsushita et al. 2001). It is evident that the peak in the cooling flow model is much broader than the observed spectral feature. For comparison Fig. 7b shows the same spectrum fitted by a cooling flow model where a temperature of 2 keV was chosen for the maximum temperature and a suitable lower temperature cut-off (1.44 keV) was determined by the fit. The very narrow temperature interval (almost isothermality) is well consistent with the narrow peak. A similar result is obtained for other clusters, e.g. A1795 as shown in Fig. 7d.

Since this diagnostics of the temperature structure is essentially based on the observation of metal lines, an inhomogeneous distribution of the metal abundances in the cluster ICM and a resulting suppression of line emission at low temperatures was suggested as a possible way to reconcile the above findings with the standard cooling flow model by Fabian et al. (2001 and contribution in these proceedings). As shown by Böhringer et al. (2001c) such a scenario will still result in a relatively broad Fe L-line feature and does not solve the problem in this case of M87.

6   Internal Absorption

Another possible attempt to obtain consistency is to allow the absorption parameter in the fit to adjust freely. This is demonstrated in Fig. 7c with the same observed spectrum where the best fitting absorption column density is selected in such a way by the fit that the absorption edge limits the extent of the Fe-L line feature towards lower energies. This is actually the general finding with ASCA observations showing two possible options for the interpretation of the spectra of cluster core regions: (1) an interpretation of the results in form of an inhomogeneous cooling flow model which than necessarily includes an internal absorption component (e.g. Allen 2000, Allen et al. 2001), or (2) an explanation of the spectra in terms of a two-temperature component model (e.g. Ikebe et al. 1999, Makishima 2001) where the hot component is roughly equivalent to the hot bulk temperature of the clusters and the cool component corresponds approximately to T_{low}. Thus for the cooling flow interpretation to work and to produce a sharp Fe-L line feature as observed, the absorption edge has to appear at the right energy and therefore values for the absorption column of typically around $3 \cdot 10^{21}$ cm^{-2} are needed (e.g. Allen 2000 and Allen et al. 2001 who find values in the range $1.5 - 5 \cdot 10^{21}$ cm^{-2}).

It is therefore important to perform an independent test on the presence of absorbing material in the cluster cores. Thanks to *CHANDRA* and *XMM-Newton* we can now use central cluster AGN as independent light sources for probing. Using the nucleus and jet of M87 (with *XMM*, see Fig. 3) and the nucleus of NGC1275 (with *CHANDRA*)

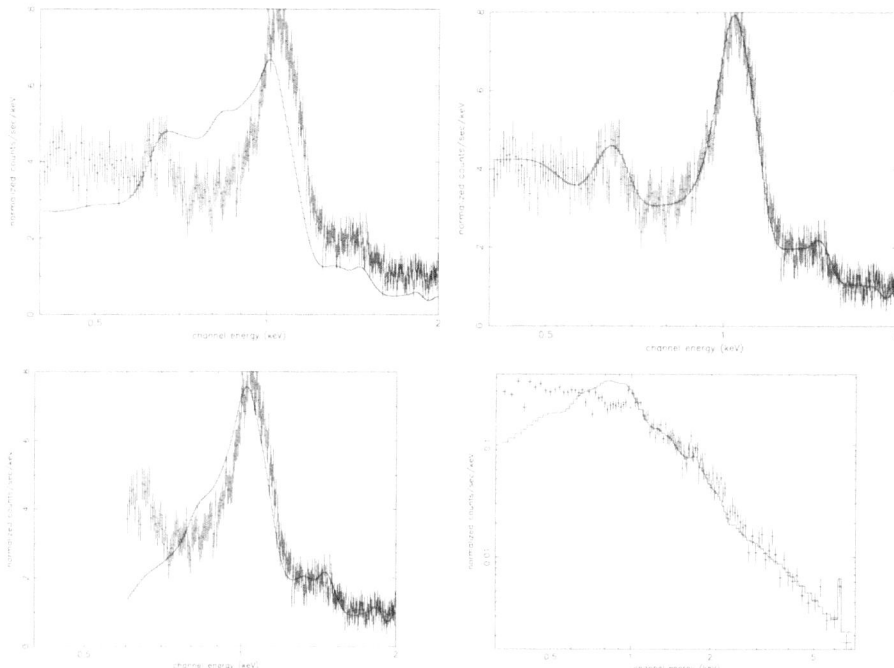

Fig. 7. XMM EPN spectrum of the M87 X-ray halo in the radial range $R = 1 - 2$ arcmin. The spectrum has been fitted with a cooling flow model with a best fitting mass deposition rate of 0.96 $M_\odot$ yr^{-1} and a fixed absorption column density of $1.8 \cdot 10^{20}$ cm^{-2}, the galactic value, and a parameter for T_{low} of 0.01 keV. (**b - upper right**): same spectrum fitted by a cooling flow spectrum artificially constraint to emission from the narrow temperature interval 1.44 - 2.0 keV, where T_{low} was treated as a free fitting parameter. (**c - lower left**): same spectrum fitted with a free parameter for the internal excess absorption. The spectrum was constraint to the energy interval 0.6 to 2.0 keV. (**d - lower right**): *XMM* EPN spectrum of A1795 fitted with a cooling flow model with the galactic value for absorption.

we find no signature of internal absorption. Thus at least for these two cases internal absorption is not observed.

7   Heating Model

In view of these difficulties of interpreting the observations with the standard cooling flow model, we may consider the possibility that the cooling and mass deposition rates are much smaller than previously thought, that is reduced by at least one order of magnitude. To decrease the mass condensation under energy conservation some form of heating is clearly necessary. Three forms of heat input into the cooling flow region have been discussed: (i) heating by the energy output of the central AGN (e.g. Pedlar et al. 1990, Tabor & Binney 1993, McNamara et al. 2000, (ii) heating by heat conduction from the hotter gas outside the cooling flow (e.g. Tucker & Rosner 1983, Bertschinger

& Meiksin 1986), and (iii) heating by magnetic fields, basically through some form of reconnection (e.g. Soker & Sarazin 1990, Makishima et al. 2001). The latter two processes depend on poorly known plasma physical conditions and are thus more speculative. The energy output of the central AGN, however, can be determined as shown below.

A heating scenario can only successfully explain the observations if among others the two most important requirements are met: (i) The energy input has to provide sufficient heating to balance the cooling flow losses, that is about 10^{60} to 10^{61} erg in 10 Gyr or on average about $3 \cdot 10^{43} - 3 \cdot 10^{44}$ erg s^{-1}, and (ii) The energy input has to be fine-tuned. Too much heating would result in an outflow from the central region and the central regions would be less dense than observed. Too little heat will not reduce the cooling flow by a large factor. Therefore the heating process has to be self-regulated: mass deposition triggers the heating process and the heating process reduces the mass deposition.

Further constraints are discussed by Böhringer et al. (2001c). The total energy input into the ICM by the relativistic jets of the central AGN can be estimated by the interaction effect of the jets with the ICM by means of the scenario described in Churazov et al. (2000). It relies on a comparison of the inflation and buoyant rise time of the bubbles of relativistic plasma which are observed e.g. in the case of NGC 1275 (Böhringer et al. 1993, Fabian et al. 2000). The estimated total energy output is for three examples, M87: $1.2 \cdot 10^{44}$ erg s^{-1}, Perseus: $1 \cdot 10^{45}$ erg s^{-1}, and Hydra A: $2 \cdot 10^{45}$ erg s^{-1}. These values for the energy input have to be compared with the energy loss in the cooling flow, which is of the order of 10^{43} erg s^{-1} for M87 and about 10^{44} erg s^{-1} for Perseus. Thus in these cases the energy input is larger than the radiation losses in the cooling flow for at least about the last 10^8 yr. We have, however, evidence that this energy input continued for a longer time with evidence given by the outer radio halo around M87 with an outer radius of 35 - 40 kpc (e.g. Kassim et al. 1993, Rottmann et al. 1996). Owen et al. (2000) give a detailed physical account of the halo and model the energy input into it. They estimate the total current energy content in the halo in form of relativistic

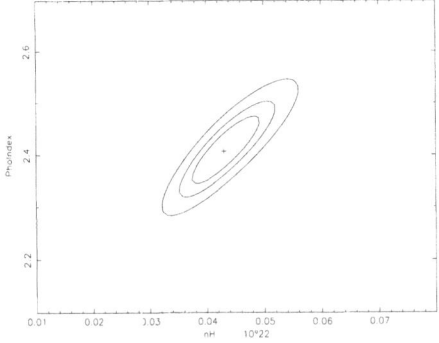

Fig. 8. Constraints on the shape of the *XMM EPN*-spectrum of the nucleus of M87. The lines show the 1, 2, and 3σ confidence intervals for the combined fit of the slope (photon index) of the power law spectrum and the value for the absorbing column density, n_H, in units of 10^{22} cm^{-2}.

plasma to $3 \cdot 10^{59}$ erg and the power input for a lifetime of about 10^8 years, which is also close to the lifetime of the synchrotron emitting electrons, to the order of 10^{44} erg s^{-1}, consistent with our estimate. The very characteristic sharp outer boundary of the outer radio halo of M87, noted by Owen et al. (2000), has the important implications, that this could not have been produced by magnetic field advection in a cooling flow.

Thus, we find a radio structure providing evidence for a power input from the central AGN into the halo region of the order of about ten times the radiative energy loss rate over at least about 10^8 years (for M87). The energy input could therefore balance the heating for at least about 10^9 years. The observation of active AGN in the centers of cooling flows is a very common phenomenon. E.g. Ball et al. (1993) find in a systematic VLA study of the radio properties of cD galaxies in cluster centers, that 71% of the cooling flow clusters have radio loud cDs compared to 23% of the non-cooling flow cluster cDs. Therefore we can safely assume that the current episode of activity was not the only one in the life of M87 and its cooling flow.

The mechanism for a fine-tuned heating of the cooling flow region should most probably be searched for in a feeding mechanism of the AGN by the cooling flow gas. The most simple physical situation would be given if simple Bondi type of accretion from the inner cooling core region would roughly provide the order of magnitude of the power output that is observed and required. Using the classical formula for spherical accretion from a hot gas by Bondi (1952) we can obtain a very rough estimate for this number. For the proton density near the M87 nucleus ($r \leq 15$ arcsec) of about 0.1 cm^{-3}, a temperature of about 10^7 K (e.g. Matsushita et al. 2001), and a black hole mass of $3 \cdot 10^9$ $M_{\odot}$ (e.g. Ford et al. 1994) we find a mass accretion rate of about 0.01 $M_{\odot}$ yr^{-1} and an energy output of about $7 \cdot 10^{43}$ erg s^{-1}, where we have assumed the canonical value of 0.1 for the ratio of the rest mass accretion rate to the energy output. The corresponding accretion radius is about 50 pc (~ 0.6 arcsec). This accretion rate is more than a factor of 1000 below the Eddington value and thus no reduction effects of the spherical accretion rate by radiation pressure has to be expected. Small changes in the temperature and density structure in the inner cooling core region will directly have an effect on the accretion rate. Therefore we have all the best prospects for building a successful self-regulated AGN-feeding and cooling flow-heating model.

8   Conclusions

Several observational constraints have let us to the conclusion that the mass deposition rates in galaxy cluster cooling cores are not as high as previously predicted. The new X-ray spectroscopic observations with a lack of spectral signatures for the coolest gas phases expected for cooling flows and the lower mass deposition rates indicated at other wavelength bands than X-rays are more consistent with mass deposition rates reduced by one or two orders of magnitude below the previously derived values. This can, however, only be achieved if the gas in the cooling flow region is heated. The most promising heating model is a self-regulated heating model powered by the large energy output of the central AGN in most cooling flows. for cooling flows and the lower mass deposition rates indicated at Most of the guidance and the support of the heating model proposed here (based on concepts developed in Churazov et al. 2000, 2001) is taken from the detailed observations of a cooling core region in the halo of M87 and to a smaller part from the observations in the Perseus cluster. These observations show that the central AGN produces sufficient heat for the energy balance of the cooling flow, that the most fundamental and classical accretion process originally proposed by Bondi (1952) provides an elegant way of devising a self-regulated model of AGN heating of the cooling flow, and that most of the further requirements that have to be met by a heating model to be consistent with the observations can most probably be fulfilled. Since these ideas are mostly developed to match the conditions in M87, it is important to extent such detailed studies to most other nearby cooling flow clusters.

In this new perspective the cooling cores of galaxy clusters become the sites where most of the energy output of the central cluster AGN is finally dissipated. Strong cooling flows should therefore be the locations of AGN with the largest mass accretion rates. While in the case of M87 with a possible current mass accretion rate of about 0.01 $M_\odot$ y^{-1} the mass addition to the black hole (with an estimated mass of about $3 \cdot 10^9$ $M_\odot$) is a smaller fraction of the total mass, the mass build-up may become very important for the formation of massive black holes in the most massive cooling flows, where mass accretion rates above 0.1 $M_\odot$ y^{-1} become important over cosmological times.

I like to thank the ROSAT team, the ESO key program team, the NORAS team, and in particular Peter Schuecker, Luigi Guzzo, Chris Collins, Kyoko Matsushita, Eugene Churazov, Yasushi Ikebe, Thomas Reiprich, and Yasuo Tanaka for the pleasant and fruitful collaboration. I like to thank in particular Rashid Sunyaev for the organization of this very magnificent conference!

References

1. Allen, S.W., 2000, MNRAS, **315**, 269
2. Allen, S.W., Fabian, A.C., Johnstone, R.M., et al., 2001, MNRAS, **322**, 589
3. Ball, R., Burns, J.O., Loken, C., 1993, AJ, **105**, 53
4. Bertschinger, E. & Meiksin, A., 1986, ApJ, **306**, L1
5. Böhringer, H., Voges, W., Fabian, A.C. Edge, A.C., and Neumann, D., 1993, MNRAS, **264**, L25
6. Böhringer, H., Guzzo, L., Collins, C.A., et al. 1998, The Messenger, No. **94**, 21
7. Böhringer, H., Voges, W., Huchra, J.P., et al., 2000, ApJS, **129**, 435
8. Böhringer, H., Schuecker, P., Guzzo, L., et al., 2001a, A&A, **369**, 826
9. Böhringer, H., Belsole, E., Kennea, J., et al., 2001b, A&A, **365**, L181
10. Böhringer, H., Matsushita, K., Ikebe, Y., et al. 2001c, A&A, in press
11. Böhringer, H., Collins, C.A., Guzzo, L., 2002, ApJ, **566**, 1
12. Bondi, H., 1952, MNRAS, **112**, 195
13. Carlberg, R.G., Yee, H.K.C., Ellingson, E., et al. 1996, ApJ, **462**, 32
14. Collins, C.A., Guzzo, L., Böhringer, H., 2000, MNRAS, **319**, 939
15. Churazov, E., Forman, W., Jones, C., Böhringer, H., 2000, A&A, **356**, 788
16. Churazov, E., Brüggen, M., Kaiser, C.R., Böhringer, H., & Forman, W., 2001, ApJ, **554**, 261
17. De Bernardis et al., 2000, Nature, **404**, 955
18. De Bernardis et al., 2001, astro-ph/0105296
19. Fabian, A.C., 1994, ARA&A, **32**, 277
20. Fabian, A. C., Sanders, J. S., Ettori, S., et al. 2000, MNRAS **318**, 65
21. Fabian, A.C., Mushotzky, R.F., Nulsen, P.E.J., Peterson, J.R., 2001, MNRAS, **321**, 20

22. Ford, H.C., et al. 1994, ApJ, **435**, L27
23. Guzzo, L., Böhringer, H., Schuecker, P., et al., 1999, The Messenger, No. **95**, 27
24. Ikebe, Y., Makishima, K., Fukazawa, Y., et al., 1999, ApJ, **525**, 58
25. Ikebe, Y., Reiprich, T.H., Böhringer, H., Tanaka, Y., 2001, A&A, in press
26. Kassim, N., Perley, R.A., Erickson, W.C., Dwarakanath, K.S., 1993, AJ, **106**, 2218
27. Makishima, K., Ezawa, H., Fukazawa, Y., et al., 2001, PASJ, **53**, 401
28. Matsushita, K., Belsole, E., Finoguenov, A., Böhringer, H., 2001, A&A, in press
29. McNamara, B., Wise, M., Nulsen, P.E.J., et al. 2000, ApJ, **534**, L135
30. Molendi, S. & Pizzolato, F., 2001, ApJ, 560, 194
31. Nulsen, P.E.J., 1986, MNRAS, **221**, 377
32. Owen, F.N., Eilek, J.A., Kassim, N.E., 2000, ApJ, **543**, 611
33. Pedlar, A., Ghataure, H. S., Davies, R. D., et al., 1990, MNRAS, **246**, 477
34. Perlmutter, N., 1999, ApJ, **517**, 565
35. Peterson, J.R., Paerels, F.B.S., Kaastra, J.S., et al., 2001, A&A, **365**, L104
36. Reiprich, T.H. & Böhringer, H., 1999, Astron. Nachr., **320**, 296
37. Reiprich, T.H. & Böhringer, H., 2001, ApJ, submitted
38. Rottmann, H., Mack, K.-H., Klein, U., Wielebinski, R., 1996, A&A, **309**, L9
39. Sarazin, C.L., 1986, Rev. Mod. Phys., **58**, 1
40. Schuecker, P., Böhringer, H., Guzzo, et al., 2001a, A&A, **368**, 86
41. Schuecker, P., Böhringer, H., Reiprich, T.H., Ferreti, L., 2001b, A&A, 378, 408
42. Silk, J., 1976, ApJ., **208**, 646
43. Soker, N. & Sarazin, C.L., 1990, ApJ, **348**, 73
44. Tabor, G. & Binney, J., 1993, MNRAS, **263**, 323
45. Tamura, T., Kaastra, J.S., Peterson, J.R., et al., 2001, A&A, **365**, L87
46. Thomas, P.A., Fabian, A.C., & Nulsen, P.E.J., 1987, MNRAS, **228**, 973
47. Tucker, W.H. & Rosner, R., 1983, ApJ, **267**, 547

Cooling Flows in Clusters of Galaxies

Andrew C. Fabian

Institute of Astronomy, Madingley Road, Cambridge CB3 0HA, UK

Abstract. The gas temperature in the cores of many clusters of galaxies drops inward by about a factor of three or more within the central 100 kpc radius. The radiative cooling time drops over the same region from 5 or more Gyr down to about 10^8 yr. Although it would seem that cooling has taken place, XMM and Chandra spectra show no evidence for strong mass cooling rates of gas below 1–2 keV. Chandra images show holes coincident with radio lobes and cold fronts indicating that the core regions are complex. The observational situation is reviewed here and ways in which continued cooling may be hidden are discussed, togther with the implications for any heat source which balances radiative cooling.

1 Introduction

The gas density within the central 100 kpc or so of the centre of most clusters of galaxies is high enough that the radiative cooling time of the gas is less than 10^{10} yr. The cooling time drops further at smaller radii, suggesting that in the absence of any balancing heat source much of the gas in the central regions is cooling out of the hot intracluster medium. In order to maintain the pressure required to support the weight of the overlying gas, a slow, subsonic inflow known as a cooling flow develops.

X-ray observations made before Chandra and XMM-Newton were broadly consistent with the cooling flow picture (see 23 for a review and 42 for an opposing view), although several issues remained unresolved. The first issue was the observed X-ray surface brightness profile, which was not as peaked as expected from a homogeneous flow. Instead a multiphase gas was assumed, dropping cold gas over a range of radii. The second was the fate of the cooled gas. At the rates of 100s to more than $1000\,\mathrm{M_\odot\,yr^{-1}}$ found in some clusters, the central galaxies should be very bright and blue if the cooled gas forms stars with a normal intial-mass-function. In many cases they do have excess blue light indicative of massive star formation [36, 1, 13, 15], but at rates which are a factor of 10 to 100 times lower than the X-ray deduced mass cooling rate. It has been argued [46] that there is no significant sink in terms of cold gas clouds. A third issue involved the shape of the soft X-ray spectrum, which was inconsistent with a simple cooling flow. Absorption intrinsic to the flow was found to be a possible explanation [2, 3]. A final, major, issue was whether the neglect of heating is justified. The effect of gravitational heating as the gas flows was taken into account, but the effects pf any central radio source, which pumps energy into the surrounding gas via jets, together with disturbances due to subclusters plunging into the core every few Gyr were not included due to a lack of quantitative information. Heat flow due to thermal conduction was also generally assumed negligible.

The situation with cluster cooling flows has been clarified over the past year, particularly by the high spatial resolution imaging of Chandra and the high spectral resolution of the XMM-Newton Reflection Grating Spectrometer (RGS). Chandra images show much detail in the cores of clusters, with bubbles from radio sources [41, 24] and cold fronts [40, 57] seen. RGS spectra [50, 54, 38] confirm the presence of a range of temperatures in cooling flow clusters but fail to show evidence for gas cooling below 1–2 keV. Simply put, the data are consistent with gas cooling at a high rate to about one third of the mean temperature beyond 100 kpc but then vanishing.

At about the same time, the evidence for both warm [35,17, 20] and cold [19] molecular gas at the centres of cooling flows clusters has become widespread. In some extreme cases there may be over 10^{11} $M_\odot$ of cold gas [19]. The presence of dust in these regions is also widespread, as demonstrated by the Balmer decrement in the optical/UV nebulosities commonly seen (e.g. 32, 15), dustlanes, and submm and IR detections [18, 3, 34]. It is therefore possible that more star formation, and in particular cold gas clouds, may be found in and around central cluster galaxies (see also 29 for a discussion of the properties of very cold gas clouds). There has also been the intriguing detection of OVI emission from A2597 with FUSE [47]. Lastly, recent numerical simulations of evolving cluster which include radiative cooling of the gas predict cooling flows (e.g. 48).

At face value the X-ray data tempt many to assume that some form of heating balances cooling and so dismiss cooling flows altogther. That ignores the how, why and what of the heating, which remains unsolved, although several candidates have been identified [56, 6, 16, 10, 11]. Some form of feedback is probably required to prevent all of the gas from being heated up. If feedback does occur we have a good chance to observe how it works, since the region is spatially resolved and optically thin. The process is of wide importance, since it provides the upper mass limit for galaxies (in simulations of the galaxy luminosity function, [39] switch off cooling in massive galaxies).

My own view is to treat it as an intriguing astrophysical puzzle whch can be tackled observationally. Heating from radio sources and infalling subclusters must occur at some level, but whether it can balance radiative cooling over the required spatial scales to better than a factor of a few is not yet clear. Cooling probably does account for the observed star formation and cold gas clouds. A major remaining issue is whether the mass cooling rates are reduced from the earlier X-ray deduced rates by a factor of a few, ten or a hundred. We may be witnessing a nearby example of the kind of feedback processes common in galaxy formation; in particular, one in which accretion onto the central black hole and the resultant kinetic energy release play a major role.

2   Chandra Results

Chandra images show structure in cluster cores. The X-ray emission is steeply peaked into the centres of many clusters but there are holes and fronts in the peak. Markevitch et al [40] found sharply defined cold fronts on A2142, across which the pressure is continuous yet the temperature changes by a factor of about 2. Ettori & Fabian [21] note that thermal conduction must be heavily suppressed in order that such sharp features can last long enough to be common. The fronts probably indicate that the gas of subclusters does not readily mix with the existing intracluster medium, presumably because they are separate magnetic structures. Also the cores of infalling subclusters may not be strongly shocked in decelerating into the core (see [44, 22]).

2.1   The Perseus Cluster

Holes in the X-ray surface brightness are seen to coincide with some radio lobes. The best examples are in the Perseus cluster, and were first seen with ROSAT [7]. Chandra shows that they have bright rims of X-ray *cool* gas [7]. This is contrary to the work of [33] who predicted that the rims would signify shocks. Other holes coincident with radio lobes are found in Hydra A [41, 16] and many other clusters. The puzzling aspect if radio sources are heating the cooling gas is that in all cases reported the *coolest* gas seen is that closest to the radio lobes. Of course there is much energy going into the lobes, but the energy from the PdV work expended in forming the holes can propagate away as sound waves, and the relativistic energy stored in the bubbles can be lifted away and out of the immediate core by buoyancy.

Provided that the filling factor of the holes by relativistic plasma is high, then the jet power required to make the holes in the Perseus cluster is considerable at about $10^{45} \, \mathrm{erg \, s^{-1}}$ [27].

2.2   A1795, A2199 and the Centaurus Cluster

The Chandra image of A1795 [25] shows an 80 kpc long soft X-ray feature coincident with an $H\alpha$ filament found by [14]. It is plausibly a cooling wake trailing behind the central galaxy, which is at the head of the filament. The galaxy is moving around in the core of the cluster at a few hundred $\mathrm{km \, s^{-1}}$. There is no evidence from the temperatures of the gas that this motion has heated the gas significantly. The central galaxy in A2199 may also be oscillating, as deduced from the unusual morphology of its radio source [12]. Again, the coolest gas appears to be close to both the radio source and the central galaxy [37].

The Centaurus cluster shows a cool, plume-like feature which appears about 20 kpc long [52]. The gas temperature drops from about 3.5 keV down to about 1 keV at the centre, with evidence that more than one phase is present there. Interestingly, the known strong abundance gradient in iron peaks at about 1.5 times Solar at about 15 kpc before dropping back down to about 0.5 near the centre. This limits the amount of widespread mixing or convection that can have occurred within the core.

In summary, the high-spatial-resolution Chandra data are revealing that cluster cores are complex with a combination of holes due to radio lobes as well as filaments, plumes and cold fronts. All cores studied so far where the radiative cooling time of the gas is a few Gyr show significant central temperature drops.

3   XMM-Newton Results

The most striking results have come from the RGS data which show little evidence for gas cooling below 1–2 keV [50, 53, 38]. Emission lines from FeXX and XVII, for

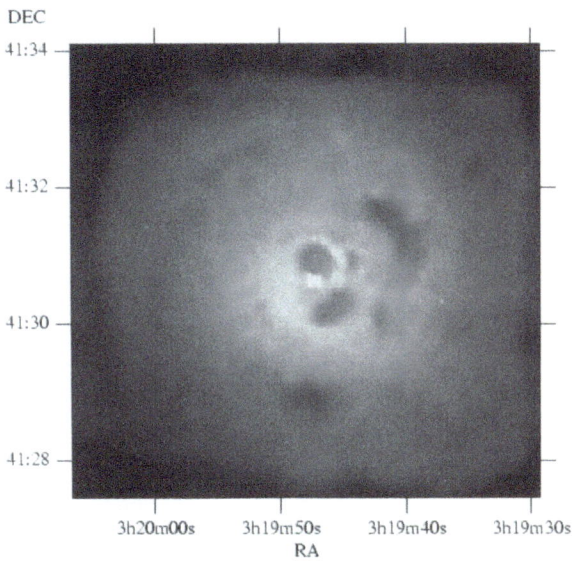

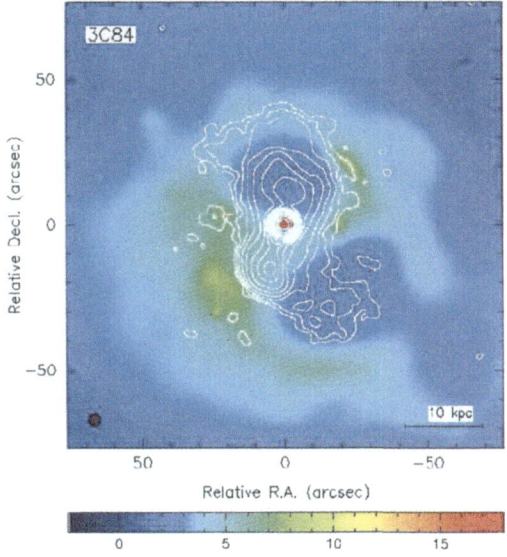

Fig. 1. (Top) Adaptively-smoothed 0.5–7 keV ACIS-S X-ray image of the centre of the Perseus cluster. (Bottom) Radio image (1.4 GHz restored with a 5 arcsec beam, produced by G. Taylor; see [24], overlaid on an adaptively smoothed 0.5–7 keV X-ray map.

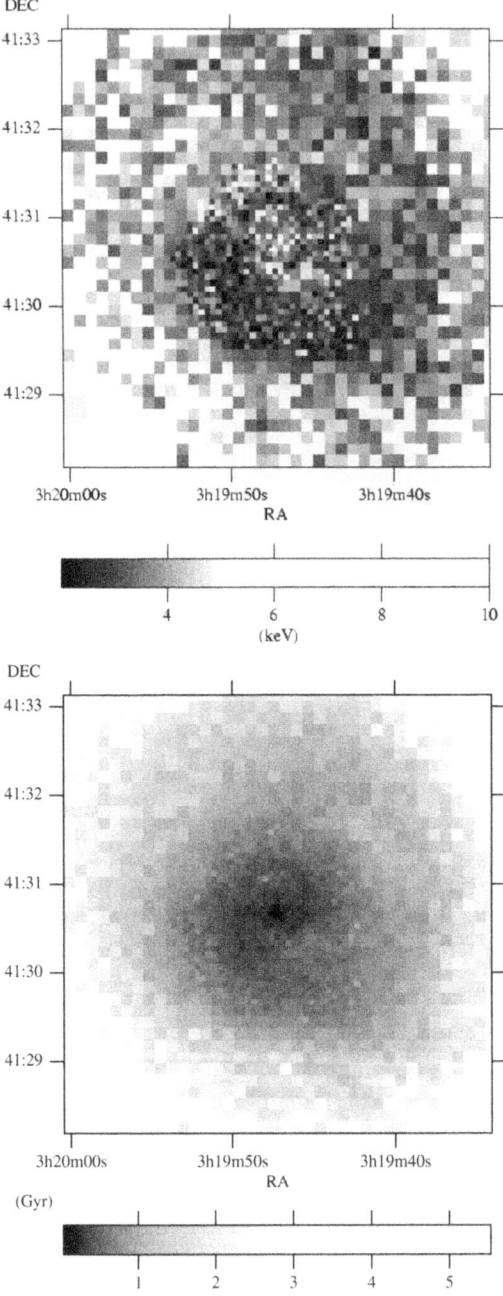

Fig. 2. (Top) Temperature and (Bottom) radiative cooling time maps produced from the X-ray colour ratios [24]. Note that the coolest gas ($T \sim 2.5 \, \text{keV}$) with the shortest cooling time ($\sim 0.3 \, \text{Gyr}$) lies in the rim around the N lobe and in the E bright blob. Single-phase gas has been assumed for the analysis.

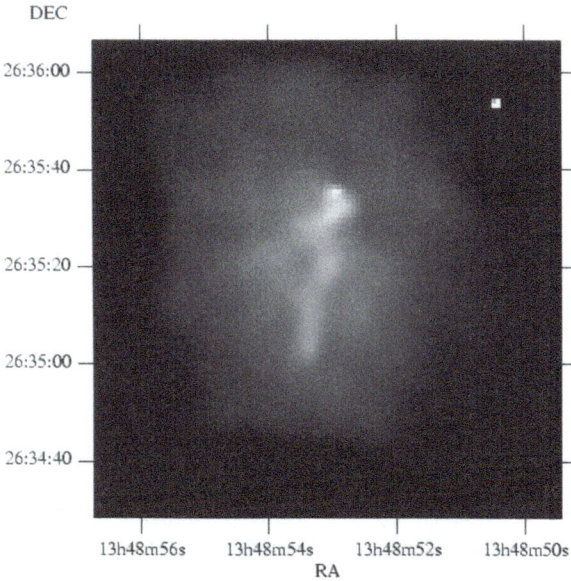

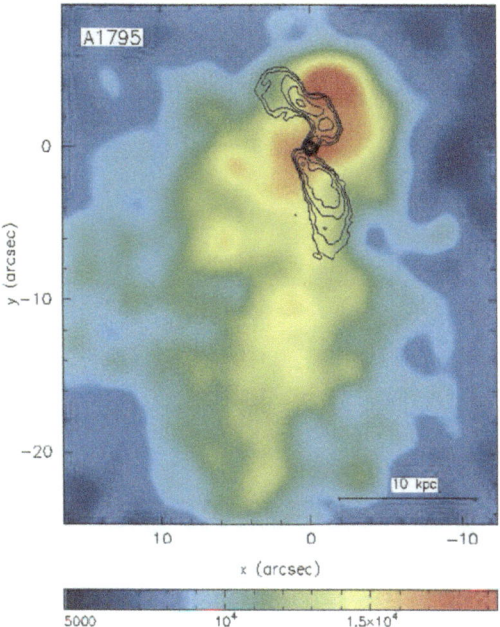

Fig. 3. (Top) Adaptively-smoothed X-ray image of the centre of A1795 [25]. (Bottom) Overlay of the 3.6 cm radio emission [31] on the X-ray image.

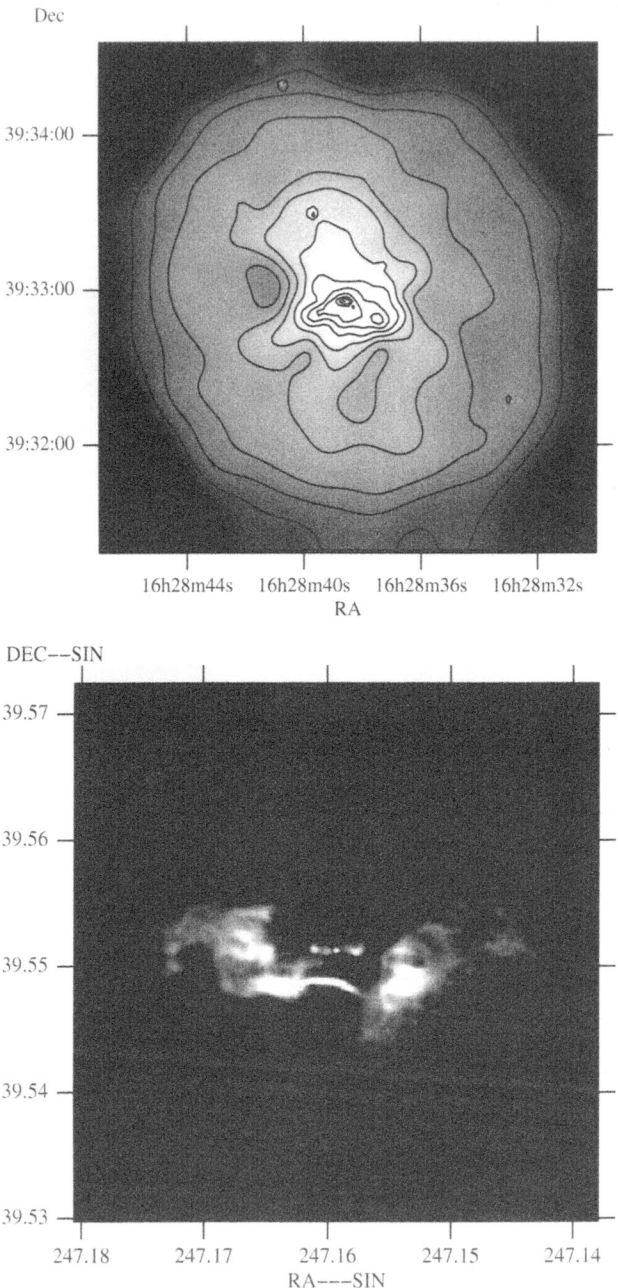

Fig. 4. (Top) Adaptively-smoothed 0.5–3 keV X-ray image of the core of A2199 (the contours are logarithmic); (Bottom) 1.7 GHz radio image [30]. See [37] for analysis of the X-ray image).the X-ray surface brightness drops at the position of the outer radio lobes.

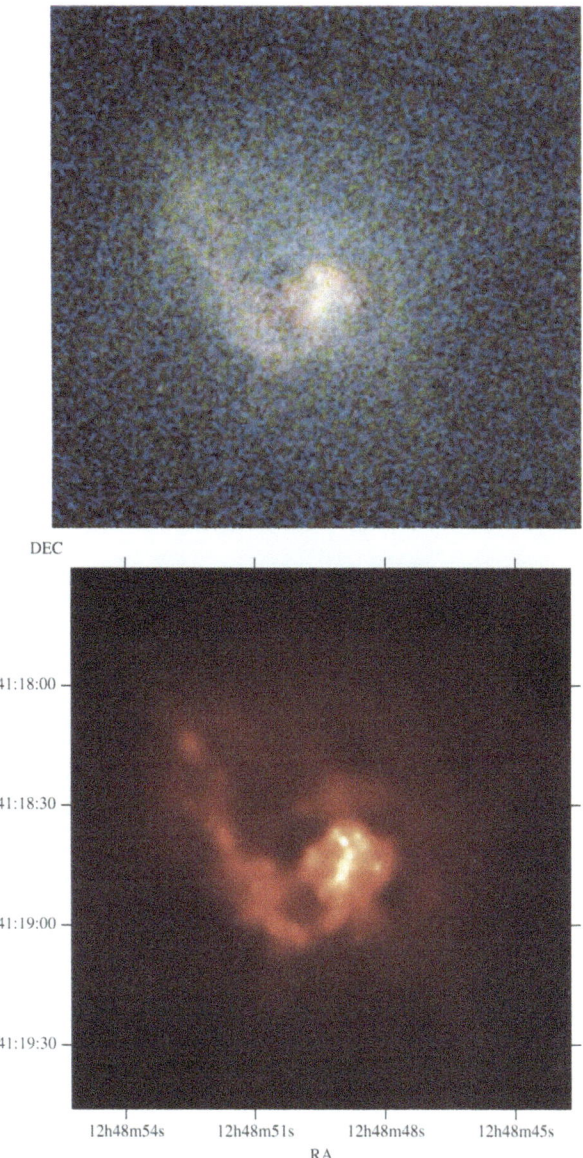

Fig. 5. (Top) Raw colour image and (Bottom) adaptively-smoothed 0.5–3 keV X-ray image of the core of the Centaurus cluster [52]. The radio source in the central cluster galaxy, NGC 4696, has a complex cap-like structure, 30 arcsec wide, fitting around the head of the X-ray plume-like feature [55].

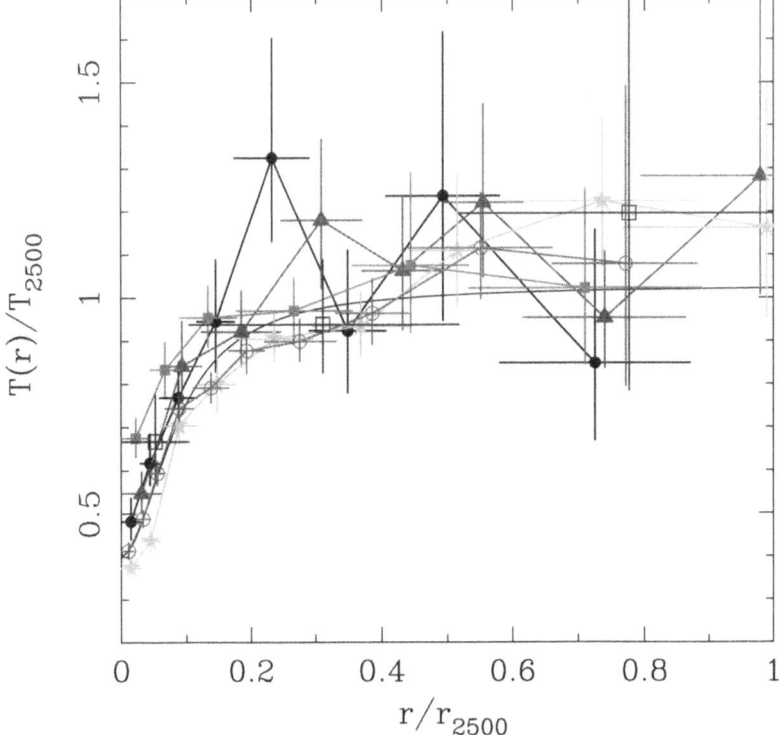

Fig. 6. Chandra temperature profiles for six massive clusters: PKS0745-191, A2390, A1835, MS2137-2353, RXJ1347-1145 and 3C295, from [4]. Note the temperature drop within $\sim$ $0.2r_{2500} \sim 120$ kpc.

example, should be bright and easily seen if the mass cooling rate is high, but they are absent. EPIC CCD spectra (e.g. [9, 43]) confirm this result.

Various explanations have been given [50, 26, 28]. The gas may be cooling and yet appear to vanish when it reaches say 2 keV. Clouds of cold gas may photoelectrically absorb the soft X-rays, or the gas may have become dense enough to separate from the flow and mix in with surrounding hotter gas [45]. Alternatively it may mix in with colder gas (for example that associated with the optical filaments or the molecular gas – note that this also explains why the filaments are so bright; [28]).

Another possibility is that the metals in the gas are not uniformly mixed in, but have a bimodal distribution [26]. Gas in which ten per cent has a metallicity of 3 times solar and 90 per cent has zero metallicity has the same spectrum as gas at 0.3 solar if cooling is unimportant. When it does cool however, the metal emission lines cool only 10 per cent of the gas and so are much reduced as compared with the situation if they were responsible for cooling all the gas.

If heating is the explanation, then it cannot just be some low level of heat which stops the gas at 1–2 keV, since it would then accumulate at that temperature, contrary

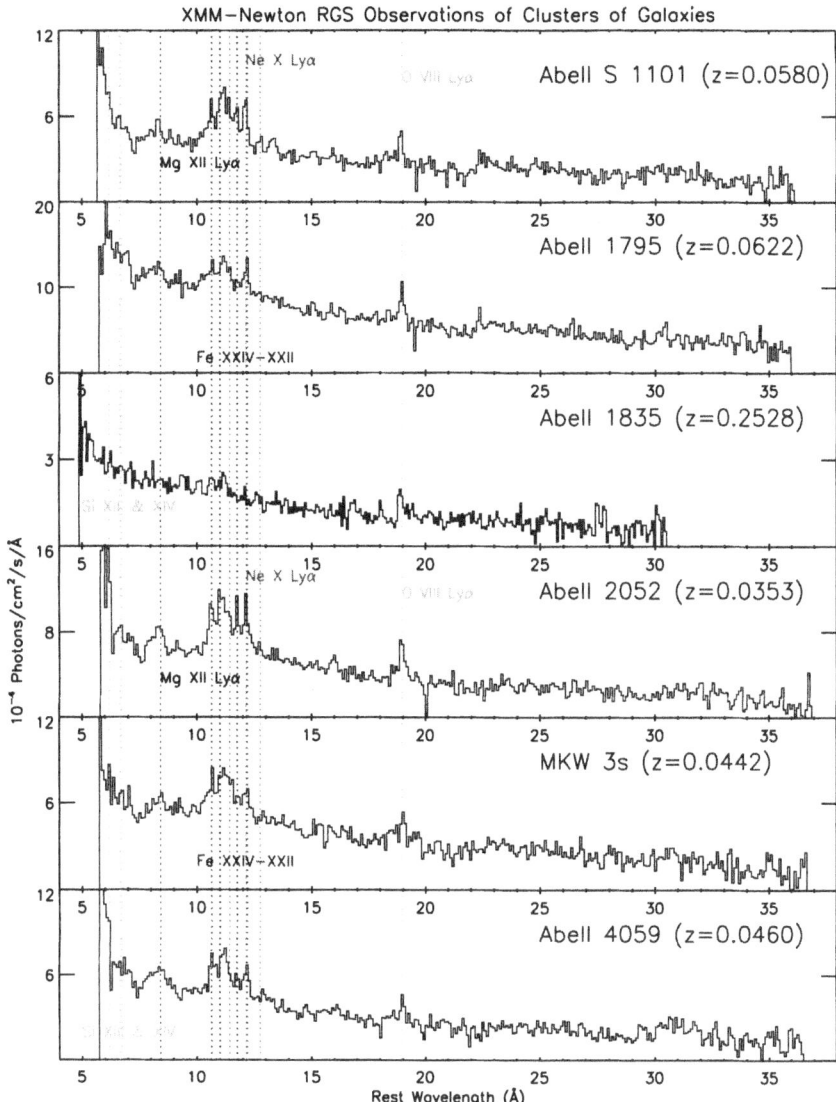

Fig. 7. RGS spectra of 6 cooling flow clusters, kindly provided by J. Peterson and J. Kaastra. Emission lines between 10-13 A indicate the presence of cooler gas in these clusters (at 1–3 keV) but the lack of lines between 13 and 18 A shows that gas is not radiatively cooling below 1–2 keV at high rates in any simple unobscured manner.

to observation. It has to halt the cooling over the full range of temperatures, and thus radii. How this can happen is a puzzle. If the radio source is responsible, then it may be intermittent. Maybe we do not see the heating phase, which is short lived. However the power required to stem the flow during the heating phase then goes up to high values. There may not be any problem in the Virgo cluster around M87 where the energy rquirements are relatively small, but in a massive cluster like A1835 [50, 52] the necessary power may exceed 10^{46} erg s^{-1} [26].

4   Discussion

The central 100 kpc radius region in most clusters has a radiative cooling time shorter than 5 Gyr (Peres et al 1997) and many have central cooling times of only 10^8 yr. The gas temperature drops by a factor of 3 or more over this radius range. It seems plausible that the temperature drop and short radiative cooling times are related and that the low temperatures are caused by radiative cooling.

It is then a puzzle as to why gas which has cooled by a factor of three, and for which the radiative cooling time has reduced by a factor of ten or more, is not seen to cool further.

There are two obvious solutions, both of which have some difficulties. The first solution is that the gas does cool but either the soft X-ray emission is absorbed or the cooling is non-radiative and due, say, to mixing. The problem of the fate of cooled gas then remains. The second solution is that some heating balances cooling. The problem here is that the heat has to balance cooling over a wide range of radii and a wide range of timescales. Also observations of radio lobes which are a likely source of heat indicate that they coincide with the coolest gas in cluster cores.

The answer may be more complex, with the major temperature drop being due to a combination of in situ radiative cooling and gas introduced from dense cooling subclusters. Heat from the kinetic energy of infalling subclusters, and turbulence, continues to be dissipated throughout the core, reducing the age of any steady central cooling region to only a few Gyr. An intermittent central radio source powered by accretion from the intracluster medium heats and churns up some of the coolest gas at the centre. Radiatively-cooling clumps (possibly metal rich) fall out of the mean slow inflow once their temperature drops to below one third of the outer temperature and rapidly mix with cooler gas clouds closer to the centre. The mixture would have a temperature of about 10^5 K (as in mixing layers; Begelman & Fabian 1990), and rapidly lose its thermal energy by UV emission. Most of this would be absorbed by neighbouring cold gas and dust, to be reradiated as optical/UV line emission and infrared dust emission [28]. Massive star formation take place in cooled clumps and spreads dust into the surrounding gas, further enhancing cooling via infrared emission.

5   Acknowledgements

I am grateful to my many collaborators, in particular Steve Allen, Carolin Crawford, Stefano Ettori, Roderick Johnstone, Jeremy Sanders, Robert Schmidt and Greg Taylor for help and discussions, and the organisers for creating a timely and memorable meeting. The Royal Society is thanked for support.

References

1. Allen S. W., 1995, MNRAS, 276, 947
2. Allen S.W., Fabian A.C., 1997, MNRAS, 286, 583
3. Allen S. W., Fabian A. C., Johnstone R. M., Arnaud K. A., Nulsen P. E. J., 2001, MNRAS, 322, 589
4. Allen SW, Schmidt RW, Fabian AC 2001, MNRAS, 328, L37
5. Begelman MC, Fabian AC, 1990, MNRAS, 244, 26
6. Binney J., Tabor G., 1995, MNRAS, 276, 663
7. Böhringer H., Voges W., Fabian A.C., Edge A.C., Neumann D.M., 1993, MNRAS, 264, L25
8. Böhringer H., Nulsen P.E.J., Braun R., Fabian A.C., 1995, MNRAS, 274, L67
9. Böhringer H. et al 2000, A&A, 365, L181
10. Böhringer H., et al 2001, A&A, in press, astro-ph/0111112
11. Brüggen M Kaiser CR Churazov E Ensslin TA 2001, MNRAS in press, astro-ph/0108486
12. Burns JO, Schwendeman O, White RA, 1983, ApJ, 271, 575
13. Cardiel N., Gorgas J, Arago-Salamanca A., 1998, MNRAS, 298, 977
14. Cowie L.L., Hu E.M., Jenkins E.B., York D.G., 1983 ApJ, 272, 29
15. Crawford C.S., Allen S.W., Ebeling H., Edge A.C., Fabian A.C., 1999, MNRAS, 306, 875
16. David L, Nulsen P.E.J.,McNamara B.R., Forman, W., Jones C., Robertson B., Wise M., 2001, ApJ, 557, 546
17. Donahue M., Mack J., Voit G. M., Sparks W., Elston R., Maloney P. R., ApJ, 2000, 545, 670
18. Edge A. C., Ivison R. J., Smail I., Blain A. W., Kneib J.-P., 1999, MNRAS, 306, 599
19. Edge A. C., 2001, MNRAS, 328, 762
20. Edge A. C., et al 2001, MNRAS, submitted
21. Ettori S., Fabian A.C., 2000, MNRAS, 317, L57
22. Fabian A. C., Daines S. J., 1991, MNRAS, 252, 17
23. Fabian A.C., 1994, ARAA, 32, 277
24. Fabian A.C., et al 2000a, MNRAS, 318, L65
25. Fabian A.C., et al 2001d, MNRAS, 321, L33
26. Fabian A.C., Mushotzky R.F., Nulsen P.E.J., Peterson J., 2001a MNRAS, 321, L20
27. Fabian A.C., Celotti A Blundell KM Kassim NE Perley RA 2001b, MNRAS submitted, astro-ph/0111418
28. Fabian A.C., et al 2001c, MNRAS submitted
29. Ferland G., Johnstone RM Fabian AC 2001 MNRAS submitted
30. Giovannini G., Cotton W.D., Feretti L., Lara L., Venturi T., 1998, ApJ, 493, 632
31. Ge J.P., Owen F.N., 1993, AJ, 105, 778
32. Heckman T. M., Baum S. A., van Breugel W. J. M., McCarthy P., 1989, ApJ, 338, 48
33. Heinz S., Reynolds C.S., Begelman M.C., 1998, ApJ, 501, 126
34. Irwin JA Stil M Bridges TJ 2001, MNRAS, 328, 359
35. Jaffe W., Bremer M. N., 1997, MNRAS, 284, L1
36. Johnstone R. M., Fabian A. C., Nulsen P. E. J., 1987, MNRAS, 224, 75
37. Johnstone RM Fabian AC Allen SW Sanders JS 2002 MNRAS submitted
38. Kaastra J., et al, 2001, A&A 365, L99
39. Kauffmann G., Guiderdoni B., White S. D. M., 1994, 267, 981
40. Markevitch M. et al, 2000, ApJ, 541, 542
41. McNamara B. et al, 2000a, ApJ, 534, L135

42. Makishima K. et al, 2001, PASJ, 53, 401
43. Molendi S., Pizzolato F., 2001, ApJ, 560, 194
44. Motl PM Burns JO Loken C Norman ML 2001, BAAS, 198, 9202
45. Norman C., Meiksin A., 1996, ApJ, 468, 97
46. O'Dea C. et al 1994, ApJ, 422, 467
47. Oegerle W. R., Cowie L., Davidsen A., Hu E., Hutchings J., Murphy E., Sembach K., Woodgate B., 2001, ApJ, 560, 187
48. Pearce FP, Thomas PA Couchman HMP, Edge AC, 2000, MNRAS, 317, 1029
49. Peres C. B., Fabian A. C., Edge A. C., Allen S. W., Johnstone R. M., White D. A., 1998, MNRAS, 298, 416
50. Peterson J.A. et al 2001, A&A, 365, L104
51. Rosner R., Tucker W., 1989, ApJ, 338, 761
52. Sanders J, Fabian AC, 2001, MNRAS submitted, astro-ph/0109336
53. Schmidt R.W., Allen S.W., Fabian A.C., MNRAS, 327, 1057
54. Tamura T. et al 2001, A&A, 365, L87
55. Taylor GB Fabian AC Allen SW, 2001, MNRAS submitted
56. Tucker W.H., Rosner R., 1983, 267, 547
57. Vikhlinin A., Markevitch M., Murray S.S., 2000, ApJ, 549, L47

Bubble-Heated Cooling Flows

Eugene Churazov[1,2], Hans Böhringer[3], Marcus Brüggen[1,4], William Forman[5],
Christine Jones[5], Christian Kaiser[6], and Rashid Sunyaev[1,2]

[1] MPI für Astrophysik, Karl-Schwarzschild-Strasse 1, 85740 Garching, Germany
[2] Space Research Institute (IKI), Profsoyuznaya 84/32, Moscow 117810, Russia
[3] MPI für Extraterrestrische Physik, P.O.Box 1603, 85740 Garching, Germany
[4] International University Bremen, Campus Ring 1, 28759 Bremen, Germany
[5] Harvard-Smithsonian Center for Astrophysics, 60 Garden St., Cambridge, MA 02138
[6] Department of Physics & Astronomy, University of Southampton, University Road,
Southampton SO17 1BJ, UK

Abstract. We speculate on the possibility that radiative energy losses of the gas in cooling flows
are balanced by the (mechanical) energy input from a central supermassive black hole. We argue
that the efficiency of the mechanical energy dissipation can be high even in the absence of strong
shocks. A qualitative picture of a cooling flow heated by buoyant bubbles is outlined.

1 Evidence for Bubbles in the Cooling Flows

The radiative cooling time of gas in the central parts of rich galaxy clusters is much
shorter than the Hubble time. Without an external energy source, this gas must cool
below X–ray temperatures forming so-called "cooling flows" (see [17] for a review).
The fate of the gas, after it has cooled, is not clear and so far no solid evidence for a
massive repository of cold gas has been found. While observations do show cooler gas
in the central regions of many clusters the most recent XMM and Chandra data suggest
that either gas does not cool below $\sim 1 - 3$ keV or it is somehow hidden from us (e.g.
[19,7]). These new data revived interest in various sources of energy capable to balance
the radiation losses. One possibility, discussed for two decades, is that energy produced
by AGN activity in the massive galaxy at the center of a cooling flow balances radiative
losses of the X–ray emitting gas (see e.g. [3,14]).

According to radio observations [10], a large fraction ($\sim$71%) of cD galaxies at
the centers of cluster cooling flows shows evidence for radio activity. This fraction is
much higher than in non-cooling flow clusters and suggests a possible link between the
presence of cooling gas and the activity of a central supermassive black hole. Given
that $\sim 10^8$–$10^{10} M_\odot$ black holes are likely to reside in these galaxies, one can imagine
that these supermassive black holes, accreting matter at a rate well below the critical
Eddington value, could provide the necessary amount of energy to offset cooling. Since
the gas in cooling flows is optically thin (at least in the continuum) the heating efficiency
of the gas by AGN radiation (see [14]) is low. A more efficient source of heat could
be mechanical power injected by AGN into cluster gas. In fact for some of the radio
sources in the cooling flows, well-collimated jets are observed and (model dependent)
estimates of the total mechanical power of the jets are large (e.g. [32,31]) and sufficient
to compensate radiative losses of the whole cooling flow region.

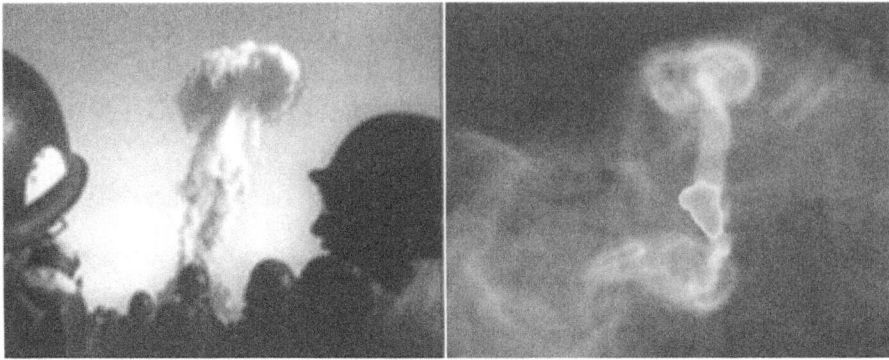

Fig. 1. On the left: a mushroom created in the Earth atmosphere by a powerful explosion. On the right: a mushroom-like structure in the 90 cm radio map of M87 (from [31]). The original image (part of it) has been rotate 90 degrees clockwise for easier comparison with the left figure.

Models involving gas heating by an AGN have been extensively discussed in the literature (e.g. [34,3,33,16,12,13,23]). In the discussion below, we present a qualitative picture based particularly on the ROSAT, Chandra and XMM-Newton data on M87 and NGC 1275 [5–8,18,11,12]. Other examples where radio lobes inflated by a central AGN are interacting with the cluster gas include e.g. Hydra-A [28,16,30], A2597 [29], A2052 [4] etc. This interaction is common in groups and early type galaxies, but we focus our discussion on clusters.

ROSAT observations provided first clear evidence [5] for a "mechanical" interaction of the radio lobes around NGC 1275 and X–ray emitting gas – holes in the X–ray surface brightness distribution to the North and South from the nucleus were found to coincide with the bright radio emitting regions, suggesting that the cosmic ray pressure is at least comparable to that of the hot intracluster gas. Subsequently, many more cases were found where the X-ray emitting gas seems to be displaced by the radio emitting regions e.g. [28,21,4]. While interaction of the radio and X–ray emitting regions certainly takes place in cooling flows there is no widespread evidence (but see [23]) for direct heating of the cooling flow gas by strong shocks. In the best studied cases like NGC 1275 the bright shells surrounding radio lobes are in fact cooler [18] than the ambient gas – contrary to the expectations in the shock heating scenario (e.g. [22]).

In another well studied case of M87 instead of a clear anticorrelation of the radio and X–ray emitting plasmas (as in many other sources) the features in the X–ray surface brightness distribution were found to resemble the morphology of some prominent features in the outer radio lobes [6,7,2]. This makes the case of M87 especially interesting. A new radio map of M87 at 90 cm [31] showed the structure of radio emitting plasma with unprecedented clarity and revealed many peculiar features, in particular torus–like features to the East and to the West from the nucleus. One can note striking similarity of these features with some evolutionary stages of hot buoyant bubbles formed by a powerful (e.g. nuclear) explosion in the Earth's atmosphere (Fig.1). The transformation of the initially spherical bubble into a torus is a common and well known property of buoyant bubbles lacking strong surface tension. Morphologically similar structures resembling "mushrooms" appear in Rayleigh-Taylor unstable configurations: as the fluid rises through the ambient medium, Kelvin-Helmholtz instabilities create the torus-like head of the "mushroom". The observed correlations of X–ray and radio structures in M87 are then explained as due to cold gas captured by the rising bubbles and uplifted to large distances from the central source. One can further exploit this similarity to bi-uld a qualitative model of a cooling flow (outlined in [8,12,13]) where buoyant bubbles inflated by an AGN serves as a major source of energy balancing radiative cooling.

2   Heating by Bubbles

We assume that most AGN power goes into an outflow and bubbles of hot, high entropy gas are created near the center of the cluster. We further assume that direct microscopic heating (e.g. heating of the cluster gas by the high entropy gas in bubbles via thermal conduction) is inefficient. We are then left with the "mechanical" energy exchange between the bubbles and the cluster gas. The most important question here is what fraction of energy goes into sound waves which can escape the cooling flow region thus reducing the heating efficiency.

2.1   Active, Early Stage of Bubble Formation

At early stages of a bubble's (radio lobes) formation, shocks are likely to occur. If the energy injection occurs in a small region of undisturbed cluster gas then strong shocks will be sent into the medium. Strong shocks are efficient in energy dissipation and there would be direct heating of the gas as it passed through any strong shock. In a spherically symmetric cluster atmosphere, with a density profile typical of a cooling flow ($\rho \sim 1/r$), the strength of the shock decreases and it evolves into a sound (compression) wave which can carry away some part of the shock energy. Strong shocks leave behind relatively compact regions of high entropy gas which contain significant fractions of the energy. The numerical value of this energy fraction depends on the gas adiabatic index and the radial profile of the ambient gas. However, an analogy with the Sedov solution for a strong explosion suggests that the fraction of the shock energy deposited is "of the order of several times 10%" [35].

2.2   The Late, Passive Stage of Bubble Evolution

Let us now consider the latest stages of bubble evolution when any strong shock has already moved away. From simulations of rising bubbles (e.g. [12,9]), one can note that the characteristic dimensionless numbers like the Mach number $M = v/c_s$ or Froude number $F = v/\sqrt{gL}$ (where v is the characteristic velocity, c_s is the sound velocity, g is the gravitational acceleration and L is the characteristic length) are not very far from unity. This precludes an immediate conclusion on the character of generated disturbances and on the fate of the deposited energy. On the other hand the same argument suggests that while generation of sound waves certainly occurs during all stages of the lobe evolution, it is unlikely that a bulk of energy is channeled into sound waves and that these waves carry away most of the energy (unless very special conditions are created).

Buoyancy drives the bubbles upward with a subsonic velocity and the bubbles maintain approximate pressure equilibrium with the ambient gas. During their upward motion, the bubbles dissipate their energy when crossing roughly a few scale heights as is clear from considering the behavior of the total enthalpy of the rising bubble during adiabatic expansion. The enthalpy is given by:

$$E = \frac{\gamma}{\gamma - 1} PV = E_0 \left(\frac{P}{P_0} \right)^{1-1/\gamma}, \tag{1}$$

where E_0 is the initial enthalpy of the bubble, P_0 and P are the initial and final pressure in the bubble respectively, γ is the adiabatic index of the hot gas inside the bubble. As an example, for $\gamma = 4/3$ and a pressure profile derived from the XMM-Newton data for M87 [7,27] half of the energy is lost by a bubble after reaching a distance of ~ 20 kpc from the its origin. Part of the energy goes into the PdV work done by the expanding bubble on the cluster gas. The remainder of the energy is extracted from the bubble by the ambient gas through drag exerted on the rising bubble (we assume that the mass density of the bubble's contents is very low and we neglect the potential energy associated with it).

An approximate estimate of how much energy goes into sound waves during the initial phase of the bubble's rise can be obtained assuming that the bubble instantly accelerates to a terminal velocity v and then maintains this velocity. For a solid sphere, Landau & Lifshitz [24] give the energy emitted in sound waves as $E \sim \frac{\pi}{3} \rho R^3 v^2$, where ρ is the ambient gas density and R is the radius of the sphere. This energy is less than a quarter of the internal energy of the bubble $E_i \sim \frac{1}{(\gamma-1)\gamma} \frac{4}{3} \pi \rho R^3 c_s^2$, where c_s is the sound speed of the gas outside the bubble. After the initial acceleration, the motion of the bubble is subsonic and steady (as seen in the simulations) and we do not expect that the drag is predominantly due to the generation of sound waves and weak shocks. Actual values of the drag estimated from simulations [12] also support this expectation. A large fraction of energy goes instead into the kinetic and potential energy of the entrained gas, turbulence in the wake of the bubbles, and possibly into internal gravity waves.

Given the steep entropy profiles in typical cooling flow clusters (the temperature is rising with radius while the density is deacreasing), large masses of entrained gas (even if accelerated to a fraction of the sound velocity) cannot travel a long distance with the rising bubbles (see Fig.10 in [12]) and therefore kinetic energy can not efficiently "leak" outside the cooling flow region with the entrained gas.

Internal gravity waves are trapped within the inner region because the buoyancy frequency is a decreasing function of radius in clusters (e.g. [1,25]). The buoyancy (Brunt-Väisälä) frequency N can be written as

$$N^2 = \frac{g}{\gamma} \frac{d \ln s}{d r}, \tag{2}$$

where $s = P/\rho^\gamma$ is the gas entropy, r is the radius, g is the gravitational acceleration, and P is the pressure. For M87, the buoyancy frequency can be estimated directly using the gas temperature and density profiles derived from the XMM-Newton data [7,27] under the assumption that the gas is in hydrostatic equilibrium. These estimates give

$N \sim 1.2\,10^{-7}, 3.5\,10^{-8}, 1\,10^{-8}$ rad/y at radii of 2, 10 and 50 kpc respectively. Thus, the frequency of the internal waves excited in the inner region is larger than the buoyancy frequency in the outer parts and the waves are reflected back into the central region.

Finally, we note that for subsonic turbulence (created in the wake of a rising bubble), the ratio of the rates of generation of sound waves and dissipation is [24]

$$\frac{\varepsilon_s}{\varepsilon_d} \sim \left(\frac{v}{c_s}\right)^5 \ll 1, \tag{3}$$

where v is a characteristic velocity and c_s is the sound speed.

Thus after the gas passes through a phase of complex rearrangements and mixing, a significant fraction of the initial bubble energy is dissipated within the cooling flow region. Although sound waves are inevitably generated during this process, they can carry away only a fraction of the energy comparable to that which is dissipated.

The bottom line is that the fraction of mechanical energy deposited by an AGN and dissipated *within the cooling flow region* is high.

3   Sketch of the Cooling Flow Region

Based on the above discussion, it is possible to outline the basic properties of a cooling flow heated by buoyant bubbles, which are quasi-continuously generated by an AGN. Bubbles are created in the innermost region and are assumed to be filled with a hot, radio emitting plasma. The size of a typical bubble is set by injection power, gravity, and parameters of the cluster gas in such a way that the lifetime of a bubble due to buoyancy is comparable with the expansion time of the bubble due to injection power. For M87 and NGC1275 this corresponds to a size in the range of few – 10 kpc. The bubbles have much larger entropy than the ambient gas and they rise through the whole cooling flow region. Perhaps the most important process operating in the innermost region is not a net heating, but entrainment which continuously removes the low entropy gas from the central region [12,30]. If bubbles are created initially along some preferred direction, then bubble induced gas motion may resemble the motion of air in the room with a heater: the higher entropy gas rises along one direction while lower entropy gas accretes along others. This means that the radial entropy profile of a cooling flow does not necessarily develop a flat core. Instead along most radial directions the entropy is an increasing function of radius as in the standard cooling flow picture.

When crossing a few scale heights, the bubbles i) lose their energy through drag forces and PdV work and ii) decrease in radio brightness significantly. At this stage, bubbles can be identified as X–ray surface brightness "holes" that are not associated with strong radio emission [11,18,29]. The dissipation of energy extracted from the bubbles is distributed spatially with no obvious "heating" features e.g. shock waves. As seen from the simulations of the bubbles in M87, the velocity with which the bubbles rise is smaller than the sound velocity of the cooling flow gas, but not by a large factor. Therefore the bubbles pass through the cooling flow during a time interval only slightly longer than a sound crossing time t_s. This is at least an order of magnitude shorter than a typical gas cooling time t_c. An assumption that radiative losses of the gas are compensated by the (efficient) dissipation of the bubbles energy then implies that the total volume of the bubbles created during one cooling time of the gas is approximately equal to the cooling flow volume (or in other words the total energy of the bubbles is equal to the total energy of the cooling flow). Therefore a volume filling factor of the bubbles within the cooling flow region is approximately $t_s/t_c \ll 1$.

Thermal instabilities can take place on a background of extensive convective motions. Overdense (cool) lumps of thermal gas, entrained by the rising bubbles or falling back to the central region may produce filamentary structures, stretching radially from the galaxy. One possible example of such a filament may be found in A1795 [15,20] (but see also [26]).

4   Conclusions

ROSAT, Chandra and XMM-Newton have identified many nearby cooling flow clusters where radio lobes inflated by a central AGN are interacting with the cluster gas. Buoyancy force drives the lobes through the cluster gas similarly to the rise of bubbles of hot gas created by a strong explosion in the Earth atmosphere. We argue that even in the absence of strong shocks a significant fraction of the energy of radio lobes is dissipated in the cooling flow region. This opens a possibility to consider a model of a cooling flow where mechanical energy deposition by rising bubbles is able to offset radiative losses of the cluster gas.

Although our discussion has focussed on the interaction between bubbles and gas in rich clusters, many of the same phenomena must occur in compact groups and early type galaxies. These systems also have bright, active galaxies near the centers of hot, gaseous halos that also have short cooling times in their cores. Furthermore, several groups and early type galaxies show evidence for the presence and interaction of relatvistic bubbles with the gaseous atmospheres. Hence, we expect that the basic ideas describing the importance of mechanical power produced by a massive black hole can also be important for the energetics of the hot, X-ray emitting gas in groups and early-type galaxies.

5   Acknowledgements

WRF and CJF acknowledges support from NASA contract NAS8-39073.

References

1. Balbus, S. A. & Soker, N. 1990, ApJ, 357, 353
2. Belsole, E. et al. 2001, A&A, 365, L188
3. Binney, J. & Tabor, G. 1995, MNRAS, 276, 663
4. Blanton, E. L., Sarazin, C. L., McNamara, B. R., & Wise, M. W. 2001, ApJL, 558, L15
5. Böhringer, H., Voges, W., Fabian, A. C., Edge, A. C. and Neumann, D. M. 1993, MNRAS, 264, L25
6. Böhringer, H., Nulsen, P. E. J., Braun, R. and Fabian, A. C. 1995, MNRAS, 274, L67
7. Böhringer, H. et al. 2001, A&A, 365, L181
8. Böhringer et al., 2002, A&A, accepted

9. Brüggen et al., 2002, MNRAS, accepted
10. Burns, J. O. 1990, AJ, 99, 14
11. Churazov, E., Forman, W., Jones, C., & Böhringer, H. 2000, A&A, 356, 788
12. Churazov, E., Brüggen, M., Kaiser, C. R., Böhringer, H., & Forman, W. 2001, ApJ, 554, 261
13. Churazov, E., Sunyaev, R., Forman, W, Böhringer, H. 2002, MNRAS, accepted
14. Ciotti, L. & Ostriker, J. P. 2001, ApJ, 551, 131
15. Cowie, L. L., Hu, E. M., Jenkins, E. B. and York, D. G. 1983, ApJ, 272, 29
16. David, L. P., Nulsen, P. E. J., McNamara, B. R., Forman, W., Jones, C., Ponman, T., Robertson, B., & Wise, M. 2001, ApJ, 557, 546
17. Fabian, A. C. 1994, ARA&A, 32, 277
18. Fabian, A. C. et al. 2000, MNRAS, 318, L65
19. Fabian, A. C., Mushotzky, R. F., Nulsen, P. E. J., & Peterson, J. R. 2001a, MNRAS, 321, L20
20. Fabian, A. C., Sanders, J. S., Ettori, S., Taylor, G. B., Allen, S. W., Crawford, C. S., Iwasawa, K., & Johnstone, R. M. 2001b, MNRAS, 321, L33
21. Finoguenov, A. & Jones, C. 2001, ApJL, 547, L107
22. Heinz, S., Reynolds, C. S., & Begelman, M. C. 1998, ApJ, 501, 126
23. Jones C. et al., 2001, ApJl, submitted (astro-ph/0108114)
24. Landau, L.D., Lifshitz, E.M., 1986, Hydrodynamics (in russian), Moscow, Nauka
25. Lufkin, E. A., Balbus, S. A., & Hawley, J. F. 1995, ApJ, 446, 529
26. Markevitch, M., Vikhlinin, A., & Mazzotta, P. 2001, ApJL, 562, L153
27. Matsushita, K. et al., 2002, A&A, accepted
28. McNamara, B. R. et al. 2000, ApJl, 534, L135
29. McNamara, B. R. et al. 2001, ApJ, accepted, (astro-ph/0110554).
30. Nulsen, P. E. J. et al., 2002, ApJ, accepted (astro-ph/0110523)
31. Owen, F. N., Eilek, J. A., & Kassim, N. E. 2000, ApJ, 543, 611
32. Pedlar, A., Ghataure, H. S., Davies, R. D., Harrison, B. A., Perley, R., Crane, P. C., & Unger, S. W. 1990, MNRAS, 246, 477
33. Soker, N., White, R. E., David, L. P., & McNamara, B. R. 2001, ApJ, 549, 832
34. Tucker, W. H. & Rosner, R. 1983, ApJ, 267, 547
35. Zeldovich, Ya.B., Raizer, Yu.P. 1966, Physics of Shock Waves and High-Temperature Hydrodynamic Phenomena, Academic Press

Cooling and Heating the ICM
with High-Resolution Simulations

Stefano Borgani[1] and Fabio Governato[2]

[1] INFN, c/o Dipartimento di Astronomia dell'Università, via Tiepolo 11, I-34131 Trieste, Italy
[2] Osservatorio Astronomico di Brera, via Brera 28, I-20131, Milano, Italy

Abstract. We present results from high–resolution Tree+SPH simulations of galaxy clusters and groups, aimed at studying the effect of non–gravitational heating and of radiative cooling on the physics of the intra–cluster medium (ICM). We simulate three systems, having emission–weighted virial temperature T_{ew} $\simeq$0.6,1 and 3 keV, with spatial resolution better than 1% of the virial radius. We consider the effect of different prescriptions for non–gravitational ICM heating, such as energy feedback from type-II supaernovae (SN), as predicted by semi–analytical models (SAMs) of galaxy formation, and two different minimum entropy floors, $S_{fl} = 50$ and 100 keV cm^2, imposed at $z = 3$. Simulations with only gravitational heating nicely reproduce predictions from self–similar ICM models on the luminosity–temperature relation. We use observational results on the relation between bolometric luminosity and temperature, to constrain the amount of extra–heating required. We find that setting the entropy floor to $S_{fl} = 50$ keV cm^2, which corresponds to an extra heating energy of about 1 keV per particle, reproduces observational results. When cooling is included, a too large gas fraction, $\simeq 50\%$, goes to the cold phase in the absence of any extra–heating. This fraction is only slightly decreased by heating the gas with $\sim 1/3$ keV/part, modulated in redshift according to the star–formation rate predicted by SAMs.

1 Introduction

The past generation of X–ray satellites (e.g., ROSAT, ASCA and Beppo-SAX) provided during the last decade invaluable information about the physics of this diffuse intra–cluster medium (ICM). With the advent of the Chandra–AXAF and Newton–XMM satellites, the level of details at which the ICM physics can be described, is undergoing an order–of–magnitude improvement, both in spatial and in energy resolution.

From the theoretical viewpoint, the first attempt to model the thermodynamical properties of the ICM assumed them to be entirely determined by gravitational processes [1]. Since gravity alone does not introduce characteristic scales, this model predicted the gas within clusters of different mass to behave in a self–similar way. Under the assumption of hydrostatic equilibrium and X–ray emissivity dominated by bremsstrahlung, the self–similar model gives the scaling $L_X \propto T^2(1+z)^{3/2}$ between X–ray luminosity and gas virial temperature[1]. In a similar way, if one characterizes the ICM entropy through the quantity $S = T/n_e^{2/3}$ (n_e: electron number density; [2]), then self–similar scaling implies $S \propto T(1+z)^2$. These predictions have been found to be at variance with respect to several observations facts: (a) $L_X \propto T^{\sim 3}$ is measured for

---

[1] Strictly speaking, this scaling holds for an Einstein–De-Sitter Universe with power–law shape for the power spectrum of density perturbations.

$T > 2$ keV clusters [3–5], with a further steepening for $T \lesssim 1$ keV groups [6], and no evidence of evolution out to $z \sim 1$ [7]; (b) the entropy at one–tenth of the cluster virial radius tends to a constant value, $S \sim 100$ keV cm^2 for $T \lesssim 2$ keV, with self–similar scaling recovered for hot, $T \gtrsim 6$ keV, systems [2,8].

A possible interpretation for such discrepancies requires self–similarity to be broken by non–gravitational gas heating [9,9–12]. This extra heating would place gas on a higher adiabat, prevent it from reaching high central densities and, therefore, suppress the X–ray luminosity. As for the astrophysical source for this heating, the two most credible hypothesis are based on energy feedback from supernovae (SN) [13–15] or from AGN activity [16,17]. A further possibility, that has been advocated to break self–similarity, is represented by radiative cooling, which would selectively remove low–entropy gas in central regions thus allowing higher–entropy gas to flow in from cluster external regions, while suppressing X–ray emissivity as a consequence of the lower density of hot diffuse gas [18]. Different groups have run numerical simulations aimed at understanding in details the effect of non–gravitational heating [19,20] and cooling [21,22].

We present here results from our undergoing project of running of high–resolution simulations of a few clusters and groups, using different schemes for injecting non–gravitational energy feedback into the ICM. We describe here the effect of extra–heating on the ICM entropy and X–ray luminosity. Results of simulations are compared to observational results. Further details about the simulations and their analysis are presented in two separate papers [23,24].

2   The Simulations

We use PKDGRAV, a parallel, multistepping treecode with periodic boundary conditions, and an SPH treatment of hydrodynamic forces to re–simulate at high resolution three halos taken from a cosmological box (100 Mpc aside) of a ΛCDM Universe, with $\Omega_m = 0.3$, $\Omega_\Lambda = 0.7$, $\sigma_8 = 1$, $H_0 = 70$ km s^{-1}Mpc^{-1} and $f_{bar} = 0.13$ for the baryon fraction. The main characteristics of the three halos are listed in Table 1. In the following we will refer to the three simulated structures as Virgo cluster, Fornax group and Hickson group. In order to have the effect of numerical resolution under control, we also realized a lower–resolution replica of each simulation, by degrading the mass resolution by a factor 2^3 for the Virgo clusters and by a factor 1.5^3 for the groups, by also rescaling the softening parameter according to the mean interparticle distance.

The first scheme for non–gravitational heating is based on setting a minimum entropy value at some pre–collapse redshift [19,11,20]. For gas with local electron number density n_e and temperature T, we define the entropy as $S = T/n_e^{2/3}$. At $z = 3$, we select all the gas particles with overdensity $\delta_g > 5$, and, after assuming a minimum floor entropy, S_{fl}, each gas particle having $s_i < S_{fl}$ is assigned an extra thermal energy, so as to bring its entropy to the floor value. We choose two values for this entropy floor, $S_{fl} = 50$ and 100 keV cm^2. We estimate the amount of energy injected in the ICM in these pre–heating schemes by selecting at $z = 0$ all the gas particles within the virial radius of the simulated cluster and tracing them back to $z = 3$. We find that $S_{fl} = 50$ keV cm^2 amounts to an average extra heating energy of $E_h = \frac{3}{2}T_h \simeq 1.4$ keV/part for

Table 1. Characteristics of the simulated systems. Column 2: total virial mass ($10^{13} M_\odot$); Column 3: virial radius (Mpc); Column 4: Emission–weighted virial temperature (keV) for the runs including only gravitational heating. Column 5: mass of gas particles ($10^8 M_\odot$). Column 6: Plummer–equivalent force softening parameter (kpc).

Run	M_{vir}	R_{vir}	T_{ew}	m_{gas}	ϵ
Virgo	30.4	1.75	2.07	2.21	7.5
Fornax	5.91	1.01	0.95	0.65	5.0
Hickson	2.49	0.76	0.60	0.65	5.0

the two groups and $E_h \simeq 0.9$ keV/part for the Virgo cluster. Such values are twice as large for $S_{\mathrm{fl}} = 100$ keV cm^2.

As for the pre–heating from SN feedback [17], we resort to semi–analytical modelling of galaxy formation to estimate the star–formation rate within halos having the same mass as the simulated ones (see ref.[25] for a detailed description of this computation). We assume a low feedback parameter so as to reproduce both the local B-band luminosity function and the Tully–Fisher relation [26]. The resulting star–formation rates are used to derive the heating of the intergalactic medium as a function of redshift, as provided by type II SN. During the cluster evolution, the energy released by SN is shared among all the gas particles having $\delta_g \geq 50$. Under the assumption that all the energy released by SN is thermalized into the ICM, this scheme dumps a total amount of about 0.35 keV/part extra energy per particle.

Finally, we also run a low–resolution version of the Hickson group with radiative cooling, both without any extra heating and by including the SN feedback, in order to verify to what extent this energy source can contrast the problem of overcooling [28].

3   Results

We provide in Figure 1 a qualitative description of the effect of non–gravitational heating on the Virgo–like cluster by showing the map of the gas density and entropy for gravitational heating only and when setting an entropy threshold. In the absence of extra–heating (left panels) the high resolution achieved in our simulation reveals a wealth of substructures in the gas distribution and in the entropy pattern. The main structure of the cluster shows a high–density, low–entropy core, surrounded by regions of progressively higher entropy associated to recently accreted gas. Small halos merging into the cluster main body are able to keep their low–entropy structure for a few crossing time scales, before their gas is stripped. As a consequence, sharp structures arise well inside the virial region, with entropy discontinuities and tail of gas stripped from the merging subhalos by the effect of the ram pressure. This picture, dramatically changes as the gas receives non–gravitational heating: the gas is placed on a higher adiabat, is no longer able to accrete inside the small–mass halos and, therefore, accretion shocks are switched off. However, while the small–scale features are washed out, a halo

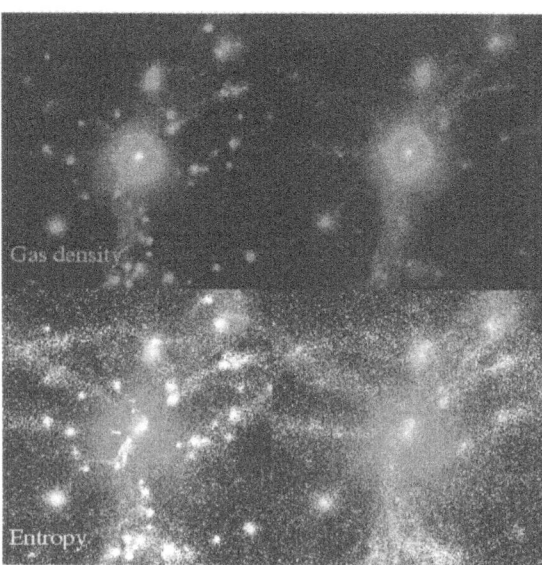

Fig. 1. Maps of gas density and entropy for the Virgo cluster at $z = 0$, within a box of 12.5 Mpc. Left panels: gravitational heating only. Right panels: entropy floor $S_{\mathrm{thr}} = 50$ keV cm^2 imposed at $z = 3$. Brighter regions indicate gas with higher density and with lower entropy. The dashed line is the analytical prediction by [29], based on gravitational heating only.

of high–entropy, recently shocked gas still surrounds the cluster main body. We refer to ref.[23] for a discussion on the effect of pre–heating on the ICM entropy pattern.

The left panel of Figure 2 shows the effect of resolution on the relation between X–ray luminosity, computed for a pure bremsstrahlung emissivity, and mass–weighted temperature. Our results show that the analytical prediction for an underlying NFW DM profile, computed for the simulated cosmology, is reached only when the ICM is resolved with $\sim 5 \times 10^4$ gas particles within the virial radius. Quite remarkably, as this resolution is attained, simulation results agree very well with analytical predictions. This plot also shows the effect of emission from metal lines. This contribution actually goes in the "wrong" direction, in that it relatively increases L_X for smaller systems, thus further increasing the discrepancy with respect to the observed L_X–T relation.

The right panel of Fig. 2 shows the effect of non–gravitational heating on the L_X–T relation and compare it to observational results for nearby clusters and groups. The effect of extra heating manifests itself through a suppression of the X–ray luminosity, which is the consequence of the decreased gas density in the central cluster regions. Quite remarkably, heating with about one-third of keV per particle, with redshift modulation as predicted by our SN model, has a too small effect. Both heating recipes, based on setting an entropy floor at $z = 3$, have a much larger effect, while still leaving T_{ew} almost unchanged. It turns out that ~ 1 keV per particle of extra energy is required to steepen the L_X–T relation to the observed slope. This result is consistent with that derived to account for the excess entropy in central cluster regions [23].

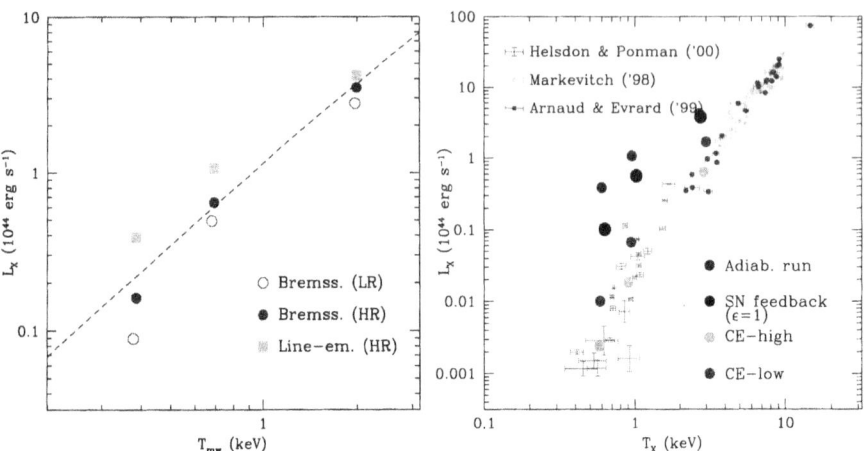

Fig. 2. The luminosity–temperature relation. Left panel: the relation between L_X and the mass–weighted temperature, T_{mw}, for bremsstrahlung emissivity (circles) and by accounting for metal–line emissivity through a Raymond–Smith spectrum with metallicity $Z = 0.3 Z_\odot$ (squares). Filled and open circles are for the high– and low–resolution runs, respectively. Right panel: the comparison between observational data on the local L_X–T_{ew} relation and results from the simulations, for the different pre-heating schemes, as indicated in the labels.

Which are the implications of this result on the astrophysical sources responsible for non gravitational heating of the inter–galactic medium? Although our results suggest that the majority of the required energy budget can not be supplied by type II SN, the final answer to the above question requires a better understanding and a more accurate treatment of several physical processes [27].

Radiative cooling has been advocated as a possible solution to the problem of ICM excess entropy and slope of the L_X–T relation [18]. Gas undergoing cooling in central cluster regions is converted into collisionless stars. This has the twofold effect of de-creasing the amount of diffuse gas contributing to the X–ray emission and to decrease the pressure support. Therefore, strongly shocked gas in the cluster outskirts starts flow-ing inside, thus increasing the central entropy level. This is explicitly shown in the left panel of Figure 3, where we plot the entropy profile for the Hickson group simulations, with and without cooling. Although an explicit algorithm of star formation has not been included in this simulation, we discard in the computation of the X–ray emissivity all those gas particles having $T < 3 \times 10^4$ K and overdensity $\delta > 10^3$. When cooling is switched on, the gas left in the hot phase increases its entropy, which reaches a level consistent with that indicated by data (~ 150 keV cm^2; [2]). However, since the cool-ing time scale is much shorter than the dynamical time scale in central cluster regions, a large fraction of gas is converted into cold stars [28]. This is demonstrated in the right panel of Fig.3, which shows that $\simeq 50\%$ of gas is in a cold phase, at a temperature $T < 10^4$ K. This result is at variance with observations which, instead, indicate that only a small fraction, $\sim 10\%$, of cluster baryons resides in the cold phase. This calls for the presence of a feedback mechanism, which were able to prevent the "cooling crisis"

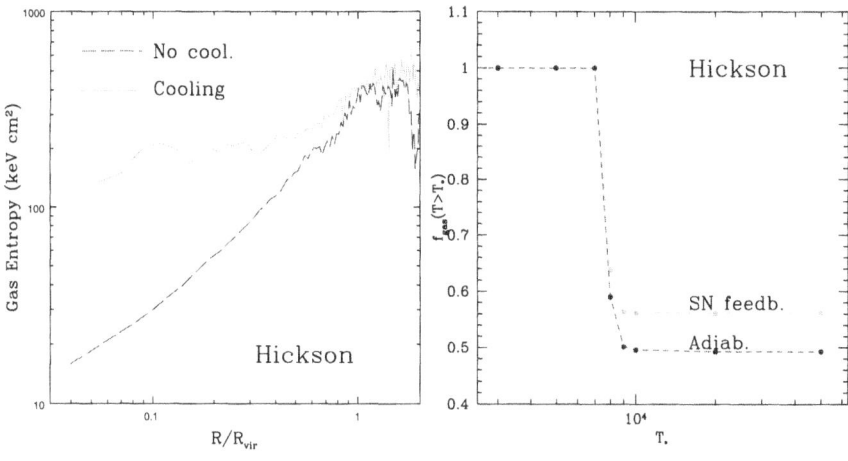

Fig. 3. Left panel: the entropy profile of the Hickson group, without extra heating, for the runs with and without cooling (solid and dashed lines, respectively). Right panel: the fraction of gas particles above a given temperature T_*, for the cooling runs, without extra–heating and with the SN feedback scheme (lower and upper lines, respectively).

[28]. Our way to include feedback energy from SN is admittedly oversimplified. However, it shows that heating with $\sim 1/3$ keV/particle, released with a realistic redshift modulation and targetted to gas particles above the virial overdensity, is not enough to alleviate much overcooling.

It is clear that the solution to this problem will require cooling and feedback from some heating source to be included in realistic simulations in a self–consistent way. To this purpose, a detailed understanding of the process galaxy formation and evolution, and of the diffusion of energy feedback into the ICM are mandatory. In this respect, the improvement of observational data on the abundance and spatial distribution of heavy elements from Chandra and Newton-XMM satellites will shed light on the interplay between ICM physics and the history of star formation in clusters.

Acknowledgments

We wish to thank all the people that is contributing to this project. They are: J. Wadsley, N. Menci, P. Tozzi, L. Tornatore, G. Lake, T. Quinn and J. Stadel. The simulations have been realized at CINECA (Bologna) and ARSC (Fairbanks) supercomputing centers.

References

1. N. Kaiser: MNRAS **222**, 323 (1986)
2. T.J. Ponman, D.B. Cannon, F.J. Navarro: Nature **397**, 135 (1999)
3. S.W. Allen, A.C. Fabian: MNRAS **297**, 63 (1998)

4. K.A. Arnaud, A.E. Evrard: MNRAS **305**, 631 (1999)
5. M. Markevitch: Astrophys. J. **504**, 27 (1998)
6. S.F. Helsdon, T.J. Ponman: MNRAS **315**, 356 (2000)
7. S. Borgani, P. Rosati, et al.: Astrophys. J., in press, preprint astro–ph/0106428 (2001)
8. E.J. Lloyd-Davies, T.J. Ponman, D.B. Cannon: MNRAS **315**, 689 (2000)
9. A. Cavaliere, N. Menci, P. Tozzi: Astrophys. J. **501**, 493 (1998)
10. M.L. Balogh, A. Babul, D.R. Patton: MNRAS **307**, 463 (1998)
11. P. Tozzi, C. Norman: Astrophys. J. **546**, 63 (2001)
12. F. Brighenti, W.G. Mathews: Astrophys. J. **553**, 103 (2001)
13. M. Loewenstein: Astrophys. J. **532**, 17 (2000)
14. R.G. Bower, A.J. Benson, C.L. Bough, S. Cole, C.S. Frenk, C.G. Lacey: MNRAS, **325**, 497 (2001)
15. N. Menci, A. Cavaliere: MNRAS **311**, 50 (2000)
16. P. Valageas, J. Silk: Astron. Astrophys. **350**, 725 (1999)
17. K.K.S. Wu, A.C. Fabian, P.E.J. Nulsen: MNRAS **318**, 889 (2000)
18. G.L. Bryan: Astrophys. J. **544**, L1 (2000)
19. J.F. Navarro, C.S. Frenk, S.D.M. White: MNRAS **275**, 720 (1995)
20. J.J. Bialek, A.E. Evrard, J.J. Mohr: ApJ **555**, 597 (2000)
21. F.R. Pearce, P.A. Thomas, H.M.P. Couchman, A.C. Edge: MNRAS **317**, 1029 (2000)
22. O. Muanwong, P.A. Thomas, S.T. Kay, F.R. Pearce, H.M.P. Couchman: Astrophys. J. **552**, L27 (2001)
23. S. Borgani, F. Governato, et al.: Astrophys. J. **559**, L71 (2001)
24. F. Governato, S. Borgani, et al.: in preparation (2001)
25. F. Poli, E. Giallongo, N. Menci, S. D'Odorico, A. Fontana: Astrophys. J. **527**, 662 (1999)
26. S. Cole, C. Lacey, C. Baugh, C.S. Frenk: MNRAS, **318**, 168 (2000)
27. A. Pipino, F. Matteucci, S. Borgani, A. Biviano: Astron. Astrophys., submitted (2001)
28. M.L. Balogh, F.R. Pearce, R.G. Bower, S.T. Kay: MNRAS **326**, 1228 (2001)
29. V.R. Eke, J.F. Navarro, C.S. Frenk: Astrophys. J. **503**, 569 (1998)

Chandra Observations of the Components of Clusters, Groups, and Galaxies and Their Interactions

William Forman[1], Christine Jones[1], Maxim Markevitch[1,3], Alexey Vikhlinin[1,3], and Eugene Churazov[2,3]

[1] Smithsonian Astrophysical Observatory, 60 Garden St. Cambridge, MA USA
[2] Max-Planck-Institut für Astrophysik, Karl-Schwarzschild-Strasse 1, 85741 Garching, Germany
[3] Space Research Institute (IKI), Profsoyuznaya 84/32, Moscow 117810, Russia

Abstract. We discuss two themes from Chandra observations of galaxies, groups, and clusters. First, we review the merging process as seen through the high angular resolution of Chandra. We present examples of sharp, edge-like surface brightness structures "cold fronts", the boundaries of the remaining cores of merger components and the Chandra observations of CL0657, the first clear example of a strong cluster merger shock. In addition to reviewing already published work, we present observations of the cold front around the elliptical galaxy NGC1404 which is infalling into the Fornax cluster. Second, we review the effects of relativistic, radio-emitting plasmas or "bubbles", inflated by active galactic nuclei, on the hot X-ray emitting gaseous atmospheres in galaxies and clusters. We review published work and also discuss the unusual X-ray structures surrounding the galaxies NGC4636 and NGC507.

1 Introduction

With its first images, the Einstein Observatory changed our view of clusters and galaxies. Clusters, rather than being dynamically old, relaxed systems showed extensive substructure reflecting complex gravitational potentials with "double" clusters merging on times scales of $\sim 10^9$ yrs [31,32,19]. Luminous elliptical galaxies, rather than being gas poor, were instead found to be gas rich systems with *hot* coronae having masses up to $\sim 10^9$ $M_\odot$ [21]. ROSAT and ASCA continued the revolution with studies of cluster merging and substructure including Coma, A2256, A754, Cygnus A, Centaurus, A1367,and Virgo, to mention just a few [4,5,1,51,29,30,28,52,7,37,10,46,35].

The high angular resolution provided by Chandra has again brought us new views of old friends – early type galaxies and clusters of galaxies – that we thought we knew pretty well. We are familiar with the ingredients, galaxies, radio emitting plasma, hot gas, and dark matter. The recipe is simple – *mix vigorously*. With these simple instructions we find new and unexpected phenomena in the Chandra observations.

2 A New Aspect of Cluster Mergers – Cold Fronts

For many years clusters were thought to be dynamically relaxed systems evolving slowly after an initial, short-lived episode of violent relaxation. However, in a prescient paper, Gunn & Gott argued that, while the dynamical timescale for the Coma cluster, the prototype of a relaxed cluster, was comfortably less than the Hubble time, other less dense clusters had dynamical timescales comparable to or longer than the age of the Universe [26]. Gunn & Gott concluded that "The present is the epoch of cluster formation". With the launch of the Einstein Observatory came the ability to "image" the gravitational potential around clusters. Many papers in the 1980's exploited the imaging capability of the Einstein Observatory and showed the rich and complex structure of galaxy clusters.

The X-ray observations supported the now prevalent idea that structure in the Universe has grown through gravitational amplification of small scale instabilities or hierarchical clustering. At one extreme, some clusters grow, in their final phase, through mergers of nearly equal mass components. Such mergers can be spectacular events involving kinetic energies as large as $\sim 10^{64}$ ergs, the most energetic events since the Big Bang. More common are smaller mergers and accretion of material from large scale filaments. An example showing the relationship between large scale structure and cluster merging is seen in the ROSAT image of A85 where small groups are detected, infalling along a filament into the main cluster. The central cluster galaxy, the bright cluster galaxies, an X-ray filament and nearby groups and clusters all show a common alignment at a position angle of about 160° extending from 100 kpc (the outer isophotes of the central cD galaxy) to 25 Mpc (the alignment of nearby clusters) [12]. Such common alignments over a wide range of scales are expected if clusters form through accretion of matter from filaments [49].

Chandra's high angular resolution has further illuminated the merging process and the complexity of the X-ray emitting intracluster medium (ICM). Prior to the launch of Chandra, sharp gas density discontinuities had been observed in the ROSAT images of A2142 and A3667 [37]. Since both clusters exhibited characteristics of major mergers, these features were expected to be shock fronts. However, the first Chandra observations showed that these were not shocks, but a new kind of structure – cold fronts [36]. Their study has provided new and detailed insights into the physics of the ICM [53,54].

2.1 Cold Fronts in Cluster Mergers

The first cold front observed by Chandra was in the hot ($kT \sim 9$ keV), X-ray-luminous cluster A2142 ($z = 0.089$). Two bright elliptical galaxies, whose velocities differ by 1840 km s^{-1} lie near the cluster center and are aligned in the general direction of the X-ray brightness elongation [36]. A2142 exhibits two sharp surface brightness edges – one lies $\sim 3'$ northwest of the cluster center (seen earlier in the ROSAT image) and a second lies $\sim 1'$ south of the center [36]. The gas temperature distributions across the edges show sharp and significant *increases* as the surface brightness (gas density) *decreases* [36]. The gas density changes across the edges compensate, within the uncertainties, for the temperature increases so that the gas pressures across the edges are consistent with being equal. One possible origin of the A2142 structures is that they arise from the merger of two inter-penetrating systems whose dense cores have survived the merger process [36]. We observe A2142 as it would appear after the shock fronts have passed by each of the dense cores. The outer, lower density gas has been shock heated, but the dense cores remain "cold" [36]. Each sharp edge is then a boundary of a ram pressure-stripped subcluster core.

A particularly beautiful example of a cold front is seen in the Chandra observation of the Fornax cluster. In Fig. 1a, we see gas bound to the infalling bright elliptical galaxy NGC1404 as it approaches the cluster center (to the northwest). The image clearly shows the sharp edge of the surface brightness discontinuity, shaped by the ram pressure of the cluster gas. The temperature map (Fig. 1b) confirms that the infalling cloud is cold compared to the hotter Fornax ICM.

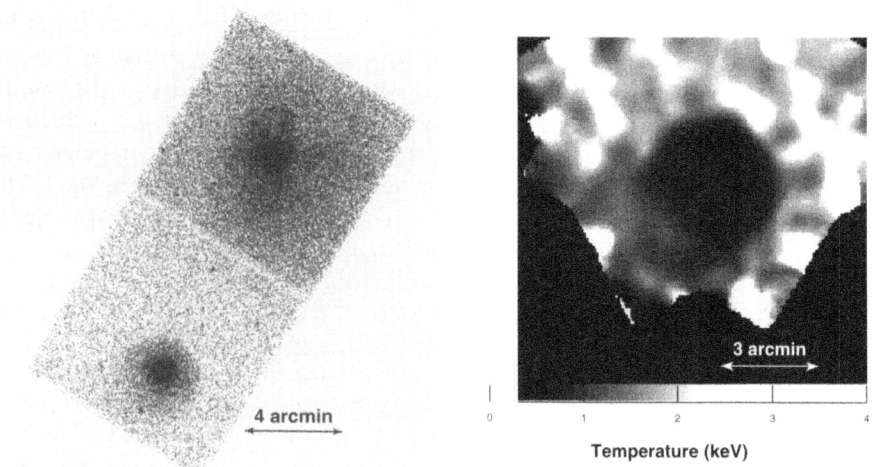

Fig. 1. The ACIS observation of NGC1404 and NGC1399. (**a**) shows the 0.5–2.0 keV band image of the Fornax cluster. The gas filled dark halo surrounding NGC1404 is at the lower left (southeast) while the cluster core, dominated by the halo surrounding NGC1399 lies at the upper right (northwest). (**b**) The temperature map of the Fornax region. The cold core surrounding NGC1404 has a temperature of less $\sim$ 1 keV while the surrounding gas has a temperature of $\gtrsim$ 1.5 keV.

Another example of a galaxy infalling into a cluster potential is M86. X-ray emission from M86 was first observed with the Einstein Observatory and subsequently by ROSAT. Its unusual "plume" was explained as a ram pressure stripped galactic corona produced as M86 crosses the core of the Virgo cluster at supersonic velocity [20,14,56,45]. The Chandra image and temperature map confirm a cool ($\sim$ 0.6 keV) corona embedded in the hot ($\sim$ 3 keV) Virgo ICM [22]. Based on the difference between the velocity of M86 and the Virgo cluster core, about $\sim$1500 km sec^{-1}, M86 must be crossing the Virgo core at supersonic velocity. However, we see no clear cold front as in NGC1404 (for the image and temperature map of M86 see Fig. 6 and Fig. 7 in Forman et al. [22]). The most likely explanation is that M86 is moving nearly in the line of sight, directly towards us, and the resulting cold front (and possible shocks) are difficult to see in projection.

2.2 Cluster Physics and Edges

A3667, a moderately distant cluster ($z = 0.055$), also was expected to exhibit a shock front based on its ROSAT image [37]. However, as with A2142, the sharp feature is the boundary of a dense cold cloud, a merger remnant [53,54]. As Vikhlinin et al. showed, the edge is accurately modeled as a spheroid (see Fig. 2a) [53]. From the surface brightness profile, converted to gas density, and precise gas temperatures, the gas pressure on both sides of the cold front can be accurately calculated. The difference between the two pressures is a measure of the ram pressure of the ICM on the moving cold front. Hence, the precise measurement of the gas parameters yields the cloud velocity. The factor of two difference in pressures between the free streaming region and the region immediately inside the cold front yields a Mach number of the cloud of 1 ± 0.2 (1430 ± 290 km s^{-1}) [53].

In addition to the edge, a weak shock is detected. The distance between the cold front and the weak shock ($\sim$350 kpc) and the observed gas density jump at the shock (a factor of 1.1-1.2) yield the shock's propagation velocity, $\sim$1600 km s^{-1}, which is consistent with that derived independently from the pressure jump across the cold front [53].

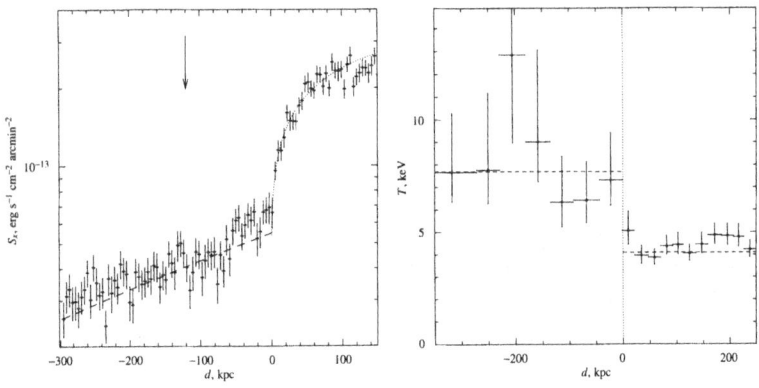

Fig. 2. (a) The surface brightness profile of A3667 extracted in elliptical regions across the cold front. The sharp "edge" is clearly seen. The dashed line is the ROSAT PSPC fit to the outer surface brightness distribution and agrees well with the Chandra observation. The dotted curve is a fit to a spheroid with a sharp boundary. The vertical arrow indicates the position of the weak shock. As discussed by Vikhlinin et al., the excess at distances of 0-50 kpc in front of the edge represents gas that accumulates in the stagnation region [53]. **(b)** The temperature profile across the cold front. The temperature *increases* from $\sim$ 4 keV to $\sim$ 8 keV across the front.

The A3667 observation provides important information on the efficiency of transport processes in clusters. As the surface brightness profile shows (see Fig. 2), the density "edge" is very sharp. Quantitatively, Vikhlinin et al. found that the width of the front was less than 3.5'' (5 kpc). This sharp edge requires that transport processes across the edge be suppressed, presumably by magnetic fields. Without such suppression, the edge should be broader since the relevant Coulomb mean free path for electrons is about 13 kpc, several times the width of the cold front [53]. Furthermore, Vikhlinin et al. observed that the cold front appears sharp only over a sector of about $\pm 30°$ centered on the direction of motion, while at larger angles, the sharp boundary disappears [54]. The disappearance can be explained by the onset of Kelvin-Helmholtz instabilities, as the ambient ICM gas flows past the moving cold front. To explain the observed extent of the sharp boundary, the instability must be partially suppressed, e.g., by a magnetic field parallel to the boundary with a strength of $7 - 16\mu G$. Such a parallel magnetic field may be drawn out by the flow along the front. This measured value of the magnetic field in the cold front implies that the pressure from magnetic fields is small (only 10-20% of the thermal pressure) and, hence, supports the accuracy of cluster gravitating mass estimates derived from X-ray measurements that assume that the X-ray emitting gas is in hydrostatic equilibrium and is supported by thermal pressure [54].

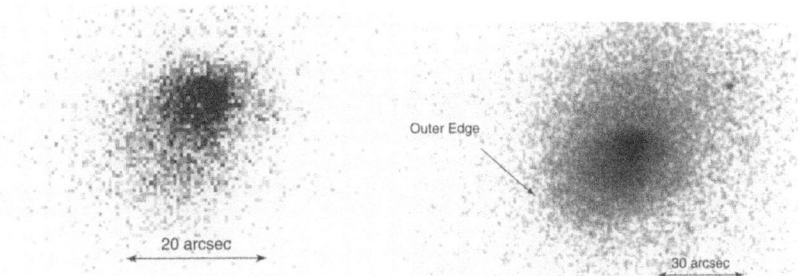

Fig. 3. Multiple X-ray surface brightness edges in ZW3146 ($z = 0.296$). (**a**) The 0.5-2.0 keV image of the central region of ZW3146 shows the two inner edges at $3''$ and $8''$. (**b**) The right panel shows the edge at $35''$.

2.3   ZW3146 – Multiple Cold Fronts

ZW3146 is a moderately distant ($z = 0.2906$; 5.74 kpc per arcsec) cluster with a remarkably high mass deposition rate that was estimated to exceed 1000 $M_\odot$ yr^{-1} [13]. The Chandra image further demonstrates the remarkable nature of this cluster – on scales from $3''$ to $30''$ ($\sim$ 20 kpc to 170 kpc), three separate X-ray surface brightness edges are detected (see Fig. 3 and Forman et al. [22]). At the smallest radii, two edges are seen to the northwest and north of the center (see Fig.3a). The first, at a radius of $\sim 3''$ (17 kpc), spans an angle of nearly 180° with a surface brightness drop of almost a factor of 2. The second edge, at a radius of $\sim 8''$ (45 kpc) spans only 90° but has a surface brightness drop of almost a factor of 4. The third edge (see Fig. 3b) lies to the southeast, about $35''$ (200 kpc) from the cluster center, has a decrease of about a factor of 2, and, as with the first edge, extends over an angle of almost 180°.

   The variety of morphologies and scales exhibited by these sharp edges or cold fronts is quite remarkable. Possibly the edges may arise from moving cold gas clouds that are the remnants of merger activity as observed in A2142 and A3667 or as oscillations (or "sloshing") of the cool gas at the center of the cluster potential as observed in A1795 [38]. The extremely regular morphology of ZW3146 on large linear scales seems to exclude a recent merger and, hence, "sloshing" of the gas seems the more likely explanation for the observed edges. High resolution, large scale structure simulations show that dense halos, formed at very early epochs, would not be disrupted as clusters collapse [24,25]. While most of the dark matter halos, having galaxy size masses, are associated with the sites of galaxy formation, the larger mass halos also may survive (without their gas) or may have fallen into the cluster only recently. Hence, we might expect to find a range of halo mass distributions moving within the cluster potential. We speculate that, as these halos move, the varying gravitational potential could accelerate the cool dense gas that has accumulated in the cluster core and could produce the "sloshing" needed to give rise to the multiple surface brightness edges observed in some clusters. Simulations are needed to confirm such a possibility.

2.4   CL0657 – A Prototypical Cluster Shock Front

CL0657 ($z = 0.296$) was discovered by Tucker et al. as part of a search for "failed" clusters, clusters that were X-ray bright but had few, if any, optical galaxies [47]. From ASCA observations, this cluster was found to have a remarkably hot gas temperature of about 17 keV, making it the hottest cluster known [48].

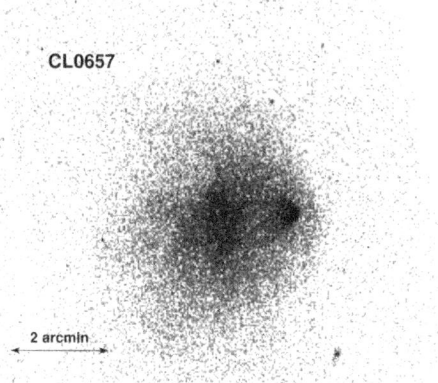

Fig. 4. The Chandra image of the cluster CL0657. The cluster exhibits the classic properties of a supersonic merger – a dense (cold) "bullet" traversing the hot cluster with a leading shock front (Mach cone). The gas parameters across the front imply the cold core is traversing the cluster at a supersonic velocity with a Mach number of $M \sim 2 - 3$. The disrupted core of the cluster can be seen to the east of the "bullet".

The Chandra image of CL0657 shows the classic properties of a supersonic merger (see Markevitch et al. for a detailed discussion of this cluster [39]). We see a dense (cold) core moving to the west after having traversed, and disrupted, the core of the main cluster. Leading the cold, dense core is a density discontinuity which appears as a shock front (Mach cone) and is confirmed by the spectral data to be hotter to the east (trailing the shock), unlike the cold fronts discussed above (or the eastern boundary of the bullet which also is a cold front). The detailed gas density parameters confirm that the "bullet" is moving to the west with a velocity of 3000–4000 km sec^{-1}, approximately 2-3 times the sound speed of the ambient gas. CL0657 is the first clear example of a relatively strong shock arising from cluster mergers.

The Chandra observation confirms the unusually high mean cluster temperature of 14-15 keV, but shows regions with temperatures as high as 20 keV. The unrelaxed nature of CL0657 urges caution in the use of high temperatures of a few extreme clusters to derive cosmological constraints. Since the present mean temperature is likely to differ from what it will become after the cluster achieves hydrostatic equilibrium, the derived cluster mass would be in error and would result in incorrect cosmological constraints.

3   The Radio – X-ray Connection – or Bubbles, Bubbles Everywhere

Prior to the launch of Chandra, ROSAT observations of NGC1275 and M87 provided hints of complex interactions between radio emitting plasmas ejected from AGN within the nuclei of dominant, central cluster galaxies [2,3,8]. With the launch of Chandra, the interaction between the radio emitting plasma and the hot intracluster medium (ICM) has been observed in many systems and now can be studied in detail.

3.1   Hot Plasma Bubbles in Cluster and Galaxy Atmospheres

One of the first, and clearest, examples of the effect of plasma bubbles on the hot intra-cluster medium was found in the Perseus cluster around the bright active, central galaxy NGC1275 (3C84). First studied in ROSAT images[2], the radio emitting cavities to the north and south of NGC1275 are clearly seen in the Chandra images with bright X-ray emitting rims surrounding the cavities that coincide with the inner radio lobes [15]. For NGC1275/Perseus, the radio lobes are in approximate pressure equilibrium with the ambient, denser and cooler gas and the bright X-ray rims surrounding the cavities are softer than the ambient gas. The central galaxy in the Hydra A cluster also harbors X-ray cavities associated with radio lobes that also show no evidence for shock heating [42]. Both sets of radio bubbles, being of lower density than the ambient gas, must be buoyant.

The Chandra images of Perseus/NGC1275 also suggest the presence of older bubbles produced by earlier outbursts [15]. These older bubbles appear as X-ray surface brightness "holes", but unlike the inner bubbles, these outer holes show no detectable radio emission, suggesting that the synchrotron emitting electrons may have decayed away leaving a heated, plasma bubble (see Fabian et al. who recently reported low frequency radio spurs extending towards the outer bubbles in NGC1275, consistent with this scenario [16]). Such bubbles, with no attendant radio emission, are seen by Chandra in the galaxy groups HCG62 and MKW3s [55,41].

The examples of bubbles described above concentrate on those around central dominant cluster galaxies. However, bubbles, and their effects are seen in more common early type galaxies. For example, in the E1 galaxy M84 (NGC4374), Chandra observed an unusual X-ray morphology which is explained by the effect of the radio lobes on the hot gas [17]. The X-ray emission appears $\mathcal{H}$-shaped, with a bar extending east-west with two nearly parallel filaments perpendicular to this bar. The complex X-ray surface brightness distribution arises from the presence of two radio lobes (approximately north and south of the galaxy) that produce two low density regions surrounded by higher density X-ray filaments. As with Perseus/NGC1275 and Hydra A, the filaments, defining the $\mathcal{H}$-shaped emission, have gas temperatures comparable to the gas in the central and outer regions of the galaxy and hence argue against any strong shock heating of the galaxy atmosphere by the radio plasma. By deriving the gas density surrounding the radio lobes, Finoguenov & Jones were able to calculate the strength of the magnetic field using the observed Faraday rotation. They inferred a line-of-sight magnetic field of 0.8μ Gauss [17].

3.2 Evolution of Buoyant Plasma Bubbles in Hot Gaseous Atmospheres

The 327 MHz high resolution, high dynamic range radio map of M87 shows a well-defined torus-like eastern bubble and a less well-defined western bubble, both of which are connected to the central emission by a column, and two very faint almost circular emission regions northeast and southwest of the center [43]. The correlation between X–ray and radio emitting features has been remarked by several authors [18,3,27].

Motivated by the similarity in appearance between M87 and hot bubbles rising in a gaseous atmosphere, Churazov et al. developed a simple model of the M87 bubbles which is generally applicable to the many bubble-like systems seen in the Chandra observations [9]. An initial buoyant, spherical bubble transforms into a torus as it rises through the galaxy or cluster atmosphere. By entraining cool gas as it rises, it exhibits a characteristic "mushroom" appearance, similar to an atmospheric nuclear explosion. This may qualitatively explain the correlation of the radio and X–ray emitting plasmas and naturally accounts for the thermal nature of the X-ray emission associated with the rising torus [3]. Finally, in the last evolutionary phase of an atmospheric explosion, the bubble reaches a height at which the ambient gas density equals that of the bubble. The bubble then expands to form a thin layer (a "pancake"). The large low surface brightness features in the M87 radio map could be just such pancakes – the final evolutionary phase of the bubbles. In the simulations performed by Churazov et al. the buoyant bubbles behaved as expected and did produce the features observed in both X-rays and radio for M87. Although the exact form of the rising bubbles was sensitive to initial conditions, the toroidal structures were a common feature. Ambient gas was uplifted in the cluster atmosphere reducing the effects of gas cooling and flowing to the center and producing the "stem" of the mushroom that is brighter than the surrounding regions [9].

3.3 Explosive Cavities

NGC4636 is one of the nearest and most X-ray luminous "normal" elliptical galaxies ($L_X \sim 2 \times 10^{41}$ ergs s^{-1}). The first X-ray imaging observations of NGC4636 from Einstein showed that, like other luminous elliptical galaxies, NGC4636 was surrounded by an extensive hot gas corona [21].

The Chandra observation of NGC4636 shows a new phenomenon – shocks produced by nuclear outbursts (see Jones et al. for details of the Chandra observation of NGC4636 [33]). The high angular resolution Chandra image (see Fig. 5) shows symmetric, $\sim$ 8 kpc long, arm-like features in the X-ray halo surrounding NGC4636. The leading edges of these features are sharp and are accompanied by temperature increases of $\sim 30\%$, as expected from shocks propagating in a galaxy atmosphere.

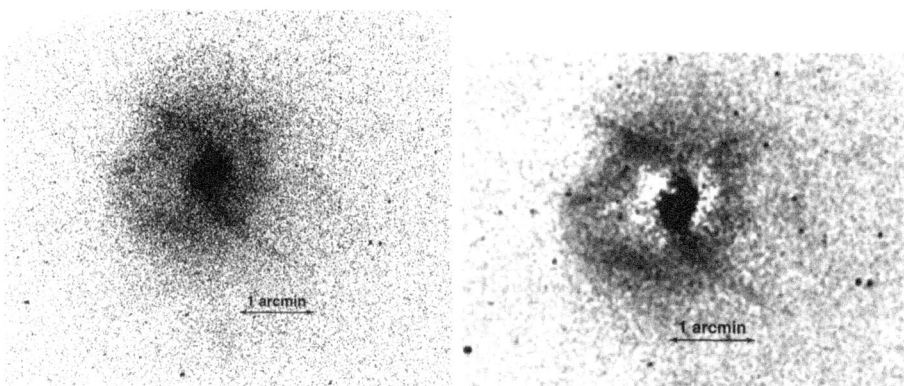

Fig. 5. (**a**) shows the 0.5-2.0 keV ACIS-S image of NGC4636 at full resolution (1 pixel = 0.492″). (**b**) shows the emission after an azimuthally symmetric model describing the galaxy corona has been subtracted. The remaining emission was smoothed with a two pixel Gaussian. Shocks from a nuclear outburst could produce the brighter arm-like structures, while the additional features could arise from other outbursts.

Although the sharpness of the edges of the NE and SW arms appears similar to the sharp edges found along "fronts" in clusters (see discussion and references above), the cluster "fronts" are cold, while those in NGC4636 are hot. Also, while the presence of sharp fronts suggests the possibility of an ongoing merger, the east-west symmetry of the halo structures, the similarity of this structure to that seen around radio lobes, as well as the lack of a disturbed morphology in the stellar core or in the stellar velocities suggest an outburst from the nucleus as the underlying cause. In particular, the bright SW arm, the fainter NW arm and the bright NE arm can be produced by the projected edges of two paraboloidal shock fronts expanding about an east – west axis through the nucleus. A shock model is also consistent with the evacuated cavities to the east and west of the central region.

The size, symmetry, and gas density and temperature profiles of the shocks are consistent with a nuclear outburt of energy $\sim 6 \times 10^{56}$ ergs having occurred about $\sim 3 \times 10^6$ years ago. It is tempting to suggest that these outbursts are part of a cycle in which cooling gas fuels nuclear outbursts that periodically reheat the cooling gas. Such outbursts if sufficiently frequent could prevent the accumulation of significant amounts of cooled gas in the galaxy center.

3.4   NGC507 – the Central Galaxy in a Group

NGC507 is the central galaxy in a nearby ($z = 0.016$) group that has been studied extensively in X-rays [34,40,6,23]. The galaxy is the site of a weak B2 radio source (luminosity $\sim 10^{37}$ ergs s^{-1}) [11]. The Chandra X-ray image, shown in Fig. 6a, covers only the central, high surface brightness emission of the group around NGC507. The

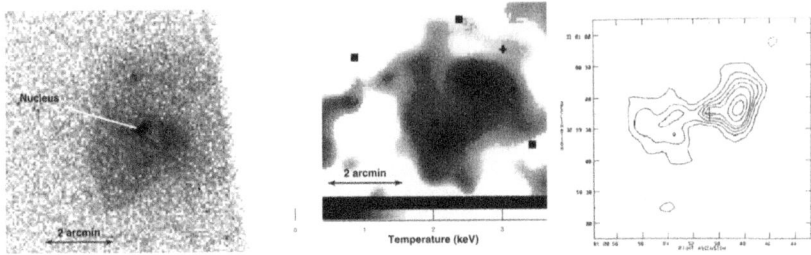

Fig. 6. (a) The 0.5-2.0 keV surface brightness distribution of NGC507. **(b)** The temperature map of the central region of NGC507. The galaxy center is cool as are the region to the west, the north-south edge to the east, and the edge to the south running from northwest to southeast. **(c)** The VLA radio map showing the central point source, a jet emanating to the west and two radio lobes [11]. The depression in the X-ray surface brightness to the west of the galaxy peak coincides with the western radio lobe.

0.5-2.0 keV surface brightness distribution shows sharp edges to the southwest, southeast and north, reminiscent of those in the clusters A2142 and A3667. In addition to the edges, there are two X-ray peaks. The first, to the east, coincides with the nucleus of NGC507. A second peak, 1′ to the west has no optical counterpart. However, comparing the X-ray and the radio map (Fig. 6a, c) shows that the western radio lobe lies precisely in the surface brightness trough between the nucleus and the peak to the west. Thus, it seems likely that the radio lobe, probably a buoyant bubble, has displaced X-ray emitting gas generating a trough in the X-ray surface brightness distribution.

The origin of the peculiar sharp surface brightness discontinuities around NGC507 is unclear. The bright emission is well fit by a thermal model with gas temperatures near 1 keV, consistent with the mean ASCA temperature of 1.10 ± 0.05 keV [40]. The emission from the central region is resolved and hence the contribution from a central AGN is relatively small (see Forman et al. [22] for additional discussion of NGC507). Perhaps the X-ray surface brightness features arise from motion of NGC507 and its dark halo within the larger group potential as suggested for the multiple edges in clusters [38].

4   Conclusions

We did not expect the rich variety of new structures seen in the Chandra high angular resolution observations of clusters and early type galaxies. Instead of confirming our prejudices, Chandra has brought us a wealth of new information on the interaction of radio sources with the hot gas in both galaxy and cluster atmospheres. We see "edges" in many systems with hot and cold gas in close proximity and have been able to extract important new parameters of the ICM from their study. We have only barely begun to digest the import of the Chandra cluster and galaxy observations. We can only expect the unexpected as Chandra observations continue and as our understanding of how best to use this new observatory matures.

We acknowledge support from NASA contract NAS8 39073, NASA grants NAG5-3065 and NAG5-6749 and the Smithsonian Institution.

References

1. H. Bohringer, U. Briel, R. Schwarz, W. Voges, G. Hartner, J. Trumper: Nature **368**, 828 (1994)
2. H. Bohringer et al.: MNRAS **264**, L25 (1993)
3. H. Bohringer, P. Nulsen, R. Braun, A. Fabian: MNRAS **274**, L67 (1995)
4. U. Briel et al.: A&A, **246**, L10 (1991)
5. U. Briel & J. Henry: A&A, **259**, L31 (1992)
6. D. Buote & A. Fabian: MNRAS, **296**, 977 (1998)
7. E. Churazov, et al.: ApJ, **520**, 105 (1999)
8. E. Churazov, W. Forman, C. Jones, H. Bohringer: A&A, **356**, 788 (2000a)
9. E. Churazov et al.: ApJ **554**, 261 (2001)
10. R. Donnelly et al.: ApJ **500**, 138 (1998)
11. H. R. de Ruiter, P. Parma, P., C. Fanti, R. Fanti: A&AS, **65**, 111 (1986)
12. F. Durret, F. et al.: A&A, **335**, 41 (1998)
13. A. Edge et al.: MNRAS, **270**, L1 (1994)
14. A. Fabian, J. Schwarz, W. Forman: MNRAS **192**, 135 (1980)
15. A. Fabian et al.: MNRAS **318**, L65 (2000)
16. A. C. Fabian, A. Celotti, K.M. Blundell, N.E. Kassim, R.A. Perley: MNRAS in press, astro-ph/0111418 (2001)
17. A. Finoguenov, C. Jones: ApJL **547**, L107 (2001)
18. E. Feigelson, P. Wood, E. Schreier, D. Harris, M. Reid: ApJ **312**, 101 (1987)
19. W. Forman, J. Bechtold, W. Blair, R. Giacconi, L. van Speybroeck, C. Jones: ApJL **243**, 133 (1981)
20. W. Forman, J. Schwarz, C. Jones, W. Liller, A. Fabian: ApJL **234**, 27 (1979)
21. W. Forman, C. Jones, W. Tucker: ApJ **293**, 102 (1985)
22. W. Forman, M. Markevitch, C. Jones, A. Vikhlinin, E. Churazov, astro-ph/0110087 (2001)
23. Y. Fukazawa: PASJ, **50**, 187 (1998)
24. S. Ghigna, B. Moore, F. Governato, G. Lake, T. Guinn, J. Stadel: MNRAS, **300**, 146 (1998)
25. S. Ghigna, B. Moore, F. Governato, G. Lake, T. Guinn, J. Stadel: ApJ, **544**, 616 (2000)
26. J. Gunn, & R. Gott: ApJ **176**, 1 (1972)
27. D. Harris, F. N. Owen, J. A. Biretta, W. Junor: Diffuse Thermal and Relativistic Plasma in Galaxy Clusters eds. H.Böhringer, L.Feretti, P. Schuecker, MPE Report **271**, 111 (1999)
28. M. Henriksen & M. Markevitch: ApJ **466**, L79 (1996)
29. J. P. Henry & U. Briel: ApJL **443**, 9 (1995)
30. H. Honda et al.: ApJL **473**, 71 (1996)
31. C. Jones et al.: ApJL **234**, 21 (1979)
32. C. Jones & W. Forman: ApJ **276**, 38 (1984)
33. C. Jones, W. Forman, A. Vikhlinin, M. Markevitch, L. David, A. Warmflash, S. Murray, P. E. J. Nulsen: submitted to **ApJL**, astro-ph/0108114 (2001)
34. D. Kim & G. Fabbiano: ApJ **441**, 182 (1995)
35. K. Makishima et al.: PASJ, **53**, 401 (2001)
36. M. Markevitch et al.: ApJ **541**, 542 (2000)
37. M. Markevitch, C. Sarazin, A. Vikhlinin: ApJ **521**, 526 (1999)
38. M. Markevitch, A. Vikhlinin, P. Mazzotta: ApJL **562**, L153 (2001)
39. M. Markevitch et al.: ApJL in press, astro-ph/0110468 (2001)
40. H. Matsumoto et al.: ApJ **482**, 133 (1997)
41. P. Mazzotta et al.: ApJL in press, astro-ph/0107557 (2001)
42. B. McNamara et al.: ApJL **534**, 135 (2000)
43. F. Owen, J. Eilek, N. Kassim: ApJ **543**, 611 (2000)
44. T. Ponman et al.: Nature, **369**, 462 (1994)

45. F. Rangarajan, D. White, H. Ebeling, A. Fabian: MNRAS, **277**, 1047 (1995)
46. S. Schindler, B. Binggeli, H. Bohringer: A&A **343**, 420 (1999)
47. W. Tucker, H. Tananbaum, & R. Remillard: ApJ **444**, 532 (1995)
48. W. Tucker et al.: ApJ **496**, L5 (1998)
49. M. Van Haarlem & R. Van de Weygaert: ApJ **418**, 544 (1993)
50. A. Vikhlinin et al.: ApJL **520**, 1 (1999)
51. A. Vikhlinin, W. Forman, C. Jones: ApJ **435**, 162 (1994)
52. A. Vikhlinin, W. Forman, C. Jones: ApJL **474**, 7 (1997)
53. A. Vikhlinin, M. Markevitch, S. Murray: ApJ **551** 160 (2000a)
54. A. Vikhlinin, M. Markevitch, S. Murray: ApJ **549**, L47 (2000b)
55. J. Vrtilek et al. , in preparation and IAP 2000 Workshop (2001)
56. D. White, A. Fabian, W. Forman, C. Jones, C. Stern: ApJ **375**, 35 (1991)

XMM-Newton Observation of M87 and Its X-Ray Halo

Kyoko Matsushita[1], Alexis Finoguenov[1], and Hans Böhringer[1]

Max-Planck-Institut für Extraterrestrial Physik, D-85748 Garching, Germany

Abstract. We report the results of detailed analysis of the X-ray emitting plasma halo of M 87, the cD galaxy of the Virgo Cluster. The data indicate that the intracluster medium has a single phase structure locally, except for the regions associated to the the radio lobe structures. The single-phase nature of the intracluster medium conflicts with the standard cooling flow model which is based on a multi phase temperature structure. In addition the signature of gas cooling below 0.8 keV to zero temperature is not observed as expected for a cooling flow.

The abundance profiles of O, Si, S and Fe are derived from the deprojected analysis. The consistency of Si and Fe abundance and a flatter profile of O indicate that ICM is dominated by supernova Ia ejecta where Si abundance is larger than Fe abundance in the solar unit.

1 Introduction

In cores of many clusters, X-ray imaging data show a highly peaked surface brightness profile (Fabian 1994 for a review). The radiative cooling time in these region is much less than a Hubble time. Without a heating process, the gas cools to low temperature and results in a "cooling flow". (Fabian 1994 for a review). The mass flow rate, $\dot{M}$, that is deduced in the standard cooling flow model, is approximately proportional to the radius. This implies that matter is deposited throughout the entire cooling flow region. It also implies that the gas in the cooling flow zone is "multi-phase" on scales small.

ASCA and ROSAT observations confirmed the existence of cooler gas in the cores of cooling flow clusters as expected (e.g. Allen & Fabian 1994, Ikebe et al. 1999, Ikebe 2001). However, without assuming excess absorption, the values of $\dot{M}_S$ are systematically lower than $\dot{M}_I$ (e.g. Ikebe et al. 1999; Makishima et al. 2001). The central temperatures of some cD galaxies obtained with the ROSAT PSPC are close to those obtained from normal elliptical galaxies with the same stellar velocity dispersion (Matsushita 2001). Recently, it was discovered with the RGS instrument onboard XMM-Newton that there is little X-ray emission from a component with a temperature below a certain lower cutoff value that differs from object to object (e.g. Tamura et al. 2001; Kaastra et al. 2001). In addition to the XMM RGS results mentioned above, the XMM-Newton observatory offers the possibility to perform a very detailed spectral and spatial analysis of the cooling flow area. This is in particular important since the RGS spectrum taken at the very central part of the intracluster medium (ICM).

M 87 is the cD galaxy of the nearest rich cluster of galaxies, the Virgo Cluster. It is suggested to have a "cooling flow" with $\dot{M}$ of about $10M_\odot\ \mathrm{yr}^{-1}$ (Stewart et al. 1984). M 87 also hosts a central AGN. The radio emission is complex and there are strong two lobe structures. An enhancement of the X-ray emission around the lobes was discovered with the EINSTEIN HRI (Feigelson et al. 1987) and ROSAT PSPC (Böhringer et al. 1995)

1.1 Observation

M87 was observed with XMM-Newton on June 19th, 2000. The effective exposure of the EPN and the EMOS detector are 30ks and 40 ks for the EPN and the EMOS, respectively, The thin filter is used for both detectors. We have excluded regions with excess emission related to the jet and radio lobes (Belsole et al. 2001) and accumulated spectra in annular regions, although these regions are not fully excluded within 2'. Deprojected spectra are calculated by subtracting the contribution from outer shell regions assuming the ICM is spherically symmetric as done by Nulsen and Böhringer (1995).

2   Temperature Structure

When the regions associated to the radio structures (Belsole et al. 2001) are excluded, the single temperature MEKAL model gives reasonable fits at $R > 2'$. The temperature profile has a positive temperature gradient, ~ 1.7 keV at $R = 1$' and ~ 2.5 keV at $R=15'$ (Figure 1). Within $R < 2'$, the two temperature model gives better fits than the single temperature model. The temperatures of the cooler and hotter component show nearly almost constant values, at ~ 1 keV and 1.7 keV, respectively (Figure 1). The spatial distribution of the cooler component indicates that it relates to the region corresponding to the radio lobes (see also Belsole et al. 2001).

The central stellar velocity dispersion of M 87 is 350 km/s (Whitemore et al. 1985). The assumption of $\beta_{spec}=1$ implies a temperature of 0.8 keV, which agrees well with the observed central ICM temperature of M 87, which implies that the central region of the intracluster medium is equivalent to a virialized interstellar medium in M 87. Therefore, as indicated in Matsushita (2001) the central temperature closely reflects the gravitational potential depth of the central galaxy, rather than the existence of a cooling flow.

Because there are some problems in the determination of the temperature through X-ray spectral fitting such as resonant line scattering and Fe-L atomic data, we compare the temperatures obtained from the Fe-L energy band, and the continuum spectra, and S line ratios (Figure 2). At R> 1' temperature derived from continuum, that from ratio of He-like and H-like lines of S, of given deprojected spectrum agree within 10%. The temperature obtained from Fe-L band also agrees well with that derived from the continuum. These results strongly indicate that at least at $R > 1'$, the ICM at a given radius is dominated by only a single temperature component.

3   Application of the Cooling Flow Model

With the RGS we can observe only the very central core of cooling flow clusters. In contrast, using the CCD detectors, we can accumulate spectra within the entire cooling flow region. At $R = 4'$, the derived cooling time is only a few Gyr and $\dot{M}_I$ is 8 $M_\odot y^{-1}$,

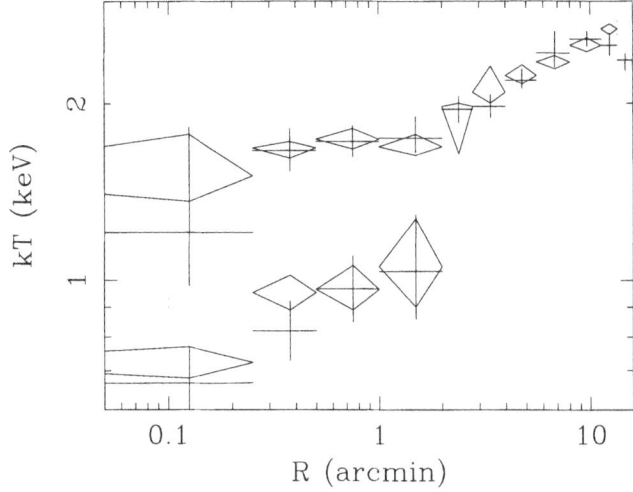

Fig. 1. Deprojected radial profile of the temperature obtained with EMOS (diamonds) and EPN (crosses) by fitting the spectra with the single temperature model ($R > 2'$) and the two temperature MEKAL model ($R < 2'$).

respectively. The net effect of resonance scattering is negligible for the radiative output in this integrated circular region. Without any heating, the gas should cool down to 0 keV, emitting X-rays. Therefore, the spectral cooling rate, $\dot{M}_S$, within the radius should be $8 M_\odot y^{-1}$.

We fitted the deprojected spectrum within $R < 4'$ with a cooling flow model using the MEKAL code modified by photoelectric absorption. As in other cooling flow clusters observed by RGS, in order to fit the deprojected spectra within $R < 4'$ with a cooling flow model, we need a cut-off temperature, lowT, which is found to be 1.4 keV. Figure 3 shows the EPN spectrum within $R < 4'$, fitted with the cooling flow model. We also plotted in the figure, other cooling flow model with lowT to 0.1 keV, fitted only to the data in the spectral range above 1.2 keV. The difference in the continuum level between 0.2 and 0.5 keV is due to the instrumental low-energy tail of Fe-L and O-K lines. The component cooling radiatively to 0 keV should have emitted strong Fe-L lines between the 0.6 to 1.0 keV, which are not seen in the spectrum. The upper limit on spectral cooling flow component is only 0.8 $M_\odot y^{-1}$, which is a factor of 10 smaller than the expected value from the standard cooling flow model at the same radius. Fabian et al. (2001) made a proposal to solve the cooling flow problem, that metals in ICM are not uniformly distributed. In this suggestion, a metal poor part of the gas cools without emitting lines and a metal rich part cools rapidly. In this way, total strength of Fe-L line emission should be reduced. Figure 3 also shows an example of a bimodal metal model which is sum of a metal poor and a metal rich component. Because of the dependence of the cooling function on metallicity, the total Fe-L spectrum depends on the metallicity distribution. The metal rich component also emit strong Fe-L lines below 0.9 keV, although its strength is slightly reduced, but the combined effect is small. Any multi-abundance model cannot explain the observed Fe-L profiles. In order to explain it, we

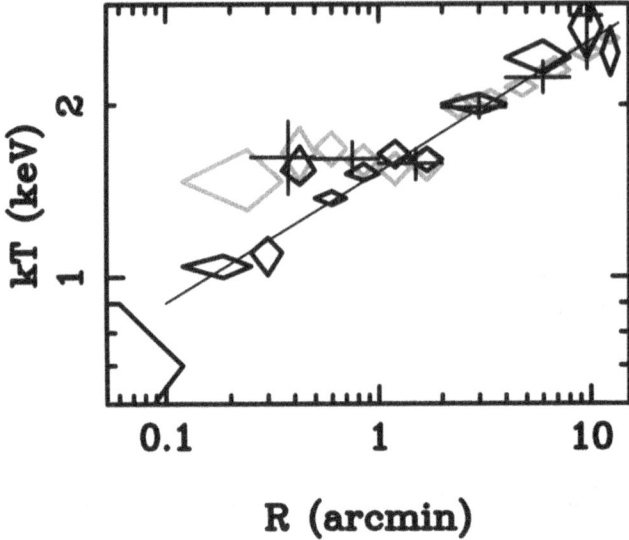

Fig. 2. Deprojected radial profile of the temperature obtained with EMOS, by fitting the spectra of 0.7–1.3 keV (black diamonds), above 1.6 keV (gray diamonds) with a MEKAL model. The results obtained from the ratio of S lines are also shown (black crosses). The solid line corresponds to the best fit regression line for the MEKAL model using the whole energy band of the EMOS.

needs a sharp cut off in the temperature distribution. Applying intrinsic absorption to the cooling flow component do not solve the problem (Böhringer et al. 2002)

4   Problem in the Standard Cooling Flow Model

The single phase nature of the ICM is inconsistent with the the standard cooling flow model, which indicates that the ICM is multi-temperature on small scales, since mass is thought to be deposited within the whole cooling flow region. In addition, the cooling matter should emit strong Fe-L lines below 0.9 keV. We cannot detect a spectral cooling flow component from the whole field of view of the detector and the ICM at any given radius is dominated by a single temperature component. The upper limit of $\dot{M}_S$ is an order of magnitude smaller than $\dot{M}_I$.

The resonant line scattering may reduce the strengths of some lines. However, all the Fe-L lines below 0.9 keV, which include both resonance lines and non-resonance lines, are suppressed from the expected values. In addition, the spectral cooling flow component is also small where the resonant line scattering is not effective. Therefore, the resonant line scattering is not the cause to suppress the Fe-L lines below 0.9 keV.

We conclude that the scenario for the dense gas with short cooling time in the centers of clusters needs to be revised. Probably heating processes that can substantially reduced the mass deposition have to be reconsidered, like the energy input of the central AGN.

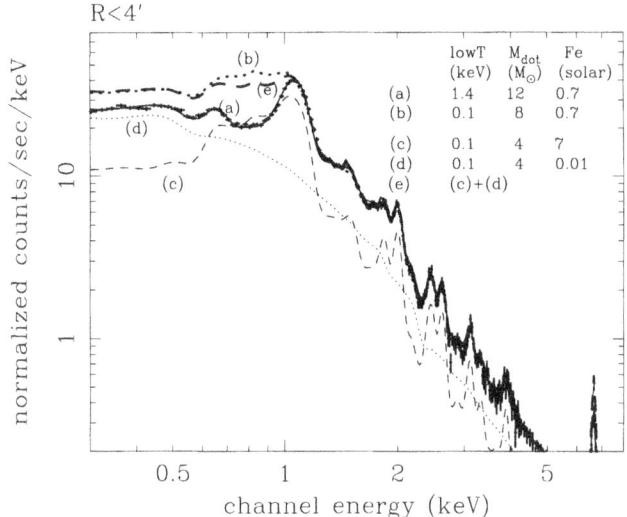

Fig. 3. Deprojected spectra within $R < 4'$ fitted with a cooling flow model with a cut-off temperature, lowT=1.44 keV (solid line, a). The bold dotted line (b) shows the model with cooling to 0.1 keV. The thin dashed line (c) and the dotted line (d) corresponds to the model with a model with 10 times higher metallicity and a model with low metal abundance, respectively. The bold dashed line (e) is the sum of the high and low abundance models.

5   Abundance Structure

Figure 4 shows abundance profiles obtained from the deprojected analysis. We adopt for the solar iron abundance the 'photospheric' value, Fe/H $= 4.68 \times 10^{-5}$ by number (Anders and Grevesse 1989). Within $2'$, we also presented the results of two-temperature MEKAL model, since there is a small amount of a cooler component with a temperature of 1 keV associated to the radio lobes. The two temperature MEKAL fit gives significantly larger abundances for all the elements than the single temperature MEKAL fit, although the contributions of the cooler component are small. For example, at R=0.5-1', including of 10 % of the cooler component in units of emission measure changes Fe abundances by a factor 2. As a result, the central abundance drop which is seen in the single MEKAL fit is not seen in the result of the two MEKAL fit.

As derived from projected spectra (Böhringer et al. 2001; Finoguenov et al. 2001), obtained abundances of Fe, Si and S show a strong negative gradient. Abundance profile of Si and Fe are consistent to be each other, although absolute values of Si is a factor of 1.7 larger than Fe. S abundance is similar with Si at center but lower than Si at outer regions. In contrast, O have a flatter profile, and values of O are a factor of 2 smaller than those of Si. For Kα lines of Hydrogen like O and Si, resonant line scattering may effective within 0.5 '. In order to explain the Si and Fe abundances consistency, we need that present SN Ia products have larger Si abundances than Fe. This results may reflects SN Ia in old stellar system are sub-luminous, due to small amount of Ni^{56}.

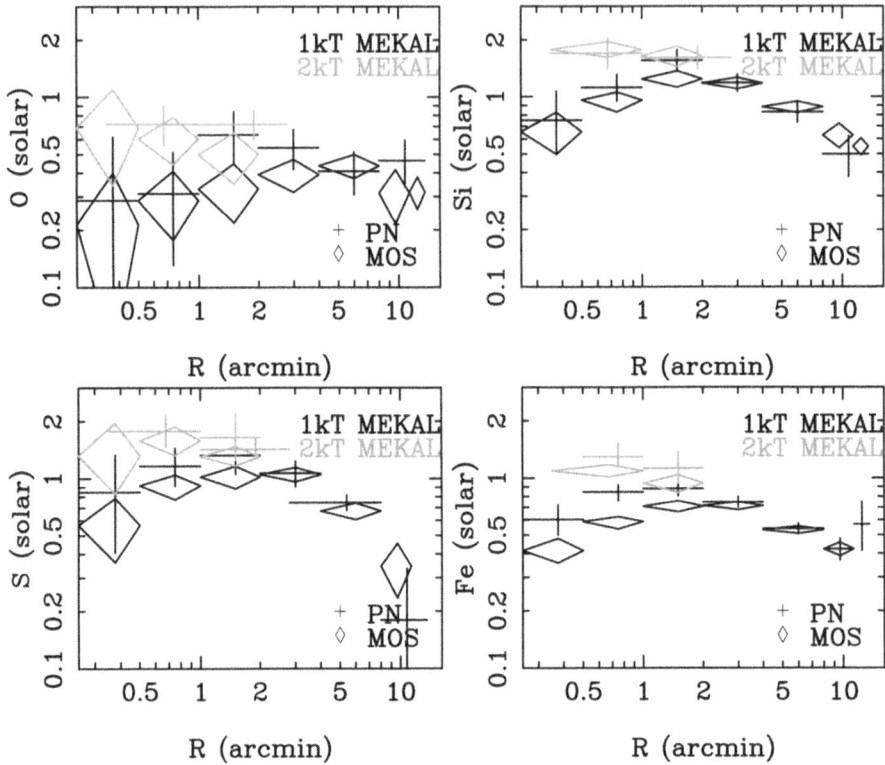

Fig. 4. Deprojected abundance profile of O, Si, S and Fe derived from the single temperature (black) and the two temperature (gray) MEKAL model.

References

1. Allen S. W., & Fabian A. C., MNRAS, 269, 409, (1994)
2. Belsole E., et al. A&A, 365, L188 (2001)
3. Böhringer H., Nulsen P.E.J., Braun R., & Fabian A.C., MNRAS, 274, 67, (1995)
4. Böhringer et al. A&A in press, astro-ph/0111112, (2002)
5. Fabian A. C., ARA&A, 32,277 (1994)
6. Fabian A.C., Mushotzky R.F., Nulsen P.E.J., & Peterson J.R., 2001, MNRAS, 321, 20
7. Feigelson E.D. et al., 312, 101 (1987)
8. Finoguenov A., et al. 2001, A&A in press
9. Ikebe Y., et al., ApJ, 525, 581 (1991)
10. Kaastra J.S. et al, A&A, 365, L99 (2001)
11. Makishima K., Ezawa H., Fukazawa Y., Honda H., Ikebe Y., Kamae T., Kikuchi K., Matsushita K., Nakazawa K., Ohashi T., Takahashi T., Tamura T., & Xu H., PASJ, 53, 401 (2001)
12. Matsushita K. , ApJ, 547, 693 (2001)
13. Stewart G., Canizares C.R., Fabian A.C., & Nulsen P.E.J., ApJ, 278, 536 (1984)
14. Tamura T., et al., A&A, 365, L87 (2001)
15. Whitemore B.C., McElroy D.B., & Tonry J.L., ApJS, 59, 1 (1985)

First Results from the ESO Distant Cluster Survey

Douglas Clowe[1] and Luc Simard[2]

[1] IAEF, Universität Bonn, Auf dem Hügel 71, 53121 Bonn, Germany
[2] Steward Observatory, 933 N. Cheery Ave., University of Arizona, Tucson, AZ, 83721, USA

Abstract. We present initial results from the ESO Distant Cluster Survey. The survey, a two year ESO large program, has had most of the optical imaging and about half of the infrared imaging completed on the sample of 20 high-redshift clusters. From the optical imaging we perform a weak lensing analysis and obtain measurements of the masses of the clusters. We also perform a bulge-disk decomposition on the cluster galaxies, and find that, with two notable exceptions, the higher mass clusters have a higher fraction of bulge dominated galaxies.

1 Introduction

The ESO Distant Cluster Survey (EDisCS) is a study of 10 $z \sim 0.5$ and 10 $z \sim 0.8$ clusters of galaxies. The clusters were selected from the Las Campanas Cluster Survey[1]. The clusters were chosen from an initial sample of 13 $z \sim 0.5$ and 17 $z \sim 0.8$ using a criteria of a minimum overdensity of red galaxies in a narrow color range in $V - I$ or $R - I$ colors respectively. The goal of the survey is to provide an unbiased and systematically selected sample of clusters and cluster galaxies which can be compared to existing low redshift surveys.

The survey is composed of three phases. In the first phase, the cluster candidates are observed in two filters, V and I for $z \sim 0.5$ candidates and R and I for $z \sim 0.8$ candidates, on the VLT for twenty minutes in each passband. The second part of the survey consists of obtaining deep optical and infrared images of each of the twenty selected clusters, with the goal of obtaining photometry on cluster galaxies four magnitudes fainter than the brightest cluster galaxy (BCG) with errors in each passband smaller than 0.1 magnitudes. During this segment we expect to get 45 minutes integration in B, V, and I on the VLT with FORS2 and 3 hours in K' on the NTT with SOFI for the $z \sim 0.5$ clusters, and 2 hours integration in V, R, and I on the VLT with FORS2 and 6 hours in J and K' on the NTT with SOFI for the $z \sim 0.8$ sample. During the final segment we will acquire optical spectroscopy on ~ 50 cluster galaxies per cluster with the VLT/FORS2 with sufficient signal-to-noise to obtain both stellar population and kinematic information on each galaxy.

Currently, we have finished with phase one of the survey and are $\sim 85\%$ complete with the phase two optical imaging and $\sim 55\%$ complete with the infrared observations. The remaining optical imaging should be completed in early 2002 and the infrared imaging in early summer of 2002. The third phase spectroscopic observations are scheduled to be taken in the spring and early summer of 2002. In the following sections we present some initial results using the optical imaging portion of the survey. Section two contains weak lensing analyzes of the clusters and section three contains (insert short summary of bulge/disk decomposition here).

2 Weak Lensing

Weak lensing is a study in which one measures the ellipticities of background galaxies and looks for a statistical deviation from an isotropic ellipticity distribution. This has the advantages over other methods to measure cluster masses that one gets a direct measure of the mass with no assumptions regarding the dynamical state of the cluster, and that, with current technology, the maximum radius at which one can measure the signal is limited primarily by the size of the detectors.

Objects were detected and had their sizes, magnitudes, and second moments of the surface brightness measured using the IMCAT software package written by Nick Kaiser. Ellipticities were created from the measured second moments and were corrected for PSF anisotropy and circular smearing using the KSB techniques [2]. The corrected ellipticities can then be used as a direct estimator of the reduced shear, g, at the location of the galaxy. For the $z \sim 0.8$ cluster sample, only those galaxies with $R - I < 1$, $I > 22$, and a signal-to-noise in the detection greater than 10 were used to estimate the shear, which resulted in a typical galaxy density of ~ 44 galaxies/sq. arcmin. For the $z \sim 0.5$ cluster sample, only $V - I < 1.5$, $I > 21.5$, and a signal-to-noise in the detection greater than 10 were used, resulting in a typical galaxy density of ~ 25 galaxies/sq. arcmin.

The resulting cluster shears were then analyzed with two methods. The first method was using an inversion algorithm which uses the fact that both the shear and the mass surface density are second derivates of the surface potential to create a two dimensional image of the mass [3]. This surface density distribution, however, can only be determined to within an unknown additive constant. The second method bins the shear radially around a chosen center of mass, the brightest cluster galaxy in this case. The resulting shear profile can then be fit with various mass profiles. In Table 1 we give the best fit singular isothermal sphere velocity dispersion to the shear profiles for each cluster along with the significance of this fit from a 0 km/s model and the reduced χ^2 of the fit. In Fig. 1 we present six of the massmaps of the clusters.

Table 1. weak lensing results

\multicolumn{4}{c}{$z \sim 0.5$ clusters}	\multicolumn{4}{c}{$z \sim 0.8$ clusters}
cluster name	σ (km/s)	S/N	χ^2	cluster name	σ (km/s)	S/N	χ^2
cl1018-1211	960	4.7	0.80	cl1037-1243	860	2.6	0.80
cl1059-1253	1010	4.8	0.99	cl1040-1155	630	1.3	1.04
cl1119-1129	400	0.7	0.54	cl1054-1146	850	2.3	1.14
cl1202-1224	670	2.3	0.85	cl1054-1245	930	2.8	0.85
cl1232-1250	1030	5.7	0.74	cl1103-1245	640	1.1	1.04
cl1238-1144	300	0.4	0.80	cl1122-1136	700	1.4	1.00
cl1301-1139	730	2.9	0.77	cl1138-1133	730	1.9	0.89
cl1353-1137	690	2.4	0.74	cl1216-1201	810	2.1	0.85
cl1411-1148	320	0.5	0.78	cl1227-1138	580	1.0	0.86
cl1420-1236	600	1.8	0.83	cl1354-1230	940	2.9	1.00

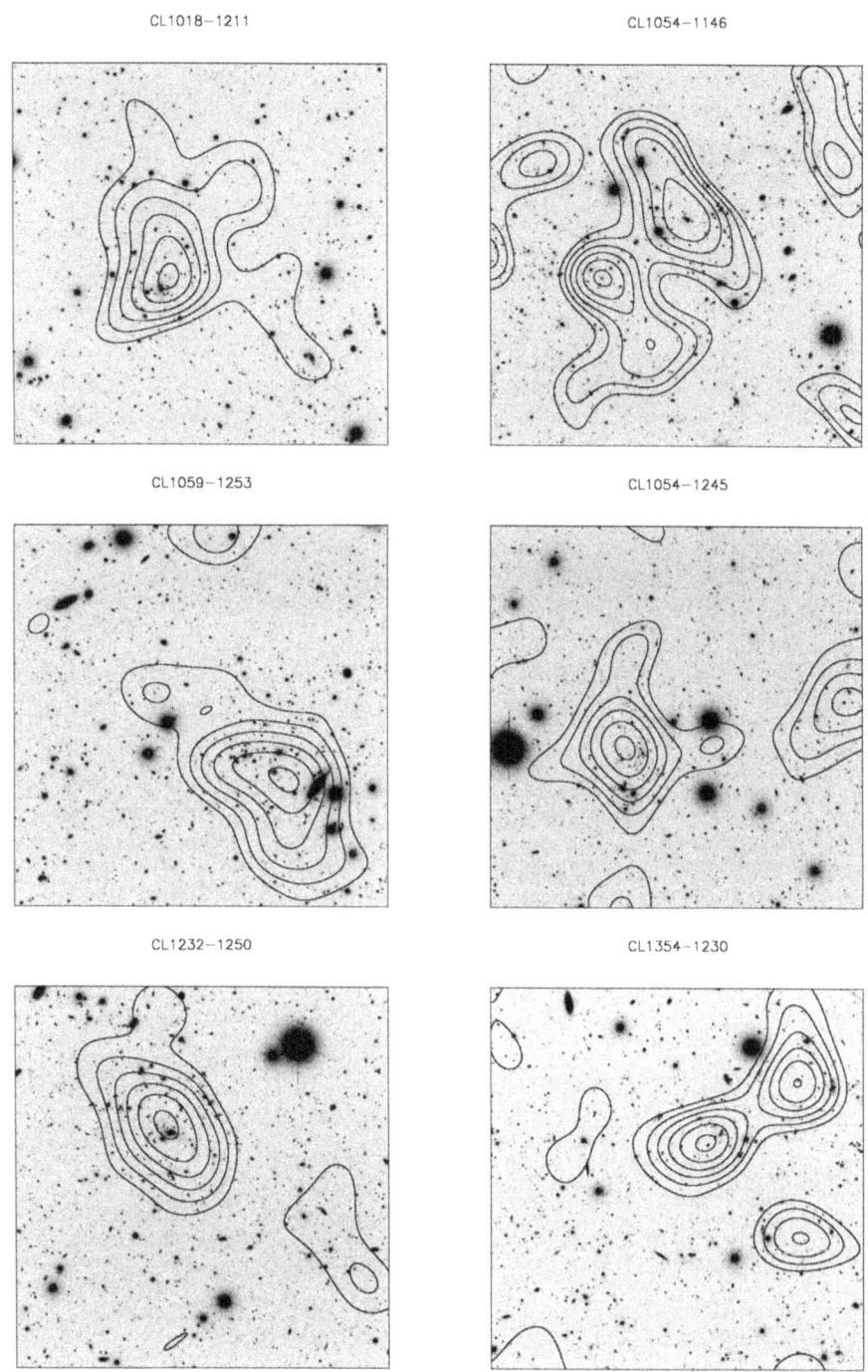

Fig. 1. Above are shown, in contours overlaying the *I*-band image, 6 of the massmaps resulting from weak lensing analysis of the EDisCS cluster sample.

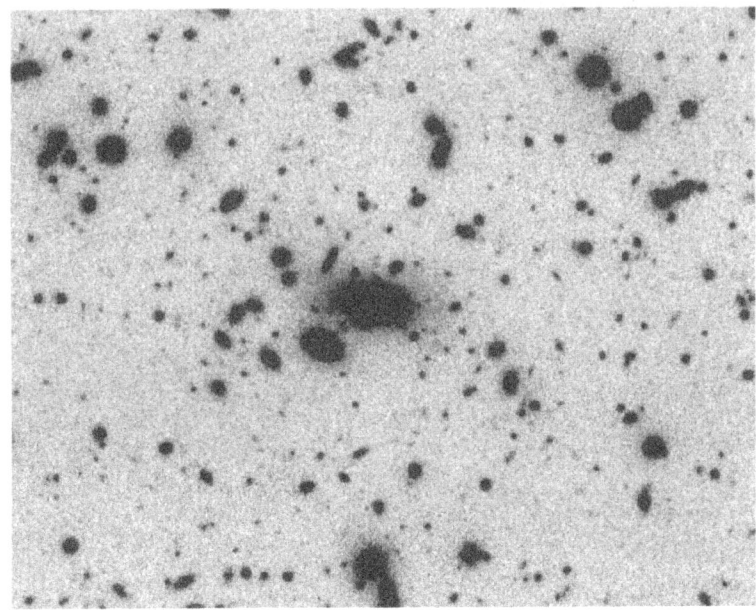

Fig. 2. FORS2 I-band image of the core of EDisCS cluster CL1232-1250 at $z = 0.53$. Image scale = 0".20 pixel^{-1}, total exposure time = 45 minutes, and seeing = 0".52 FWHM.

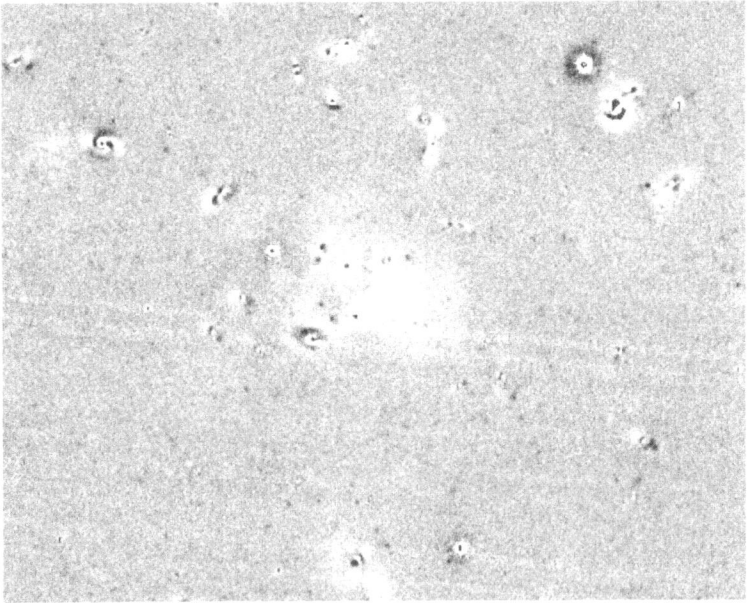

Fig. 3. FORS2 I-band residual image of the core of EDisCS cluster CL1232-1250 after processing through the GIM2D bulge+disk fitting pipeline. Note the wealth of residual structures that can be used to quantify the effects of cluster environment on galaxy morphology.

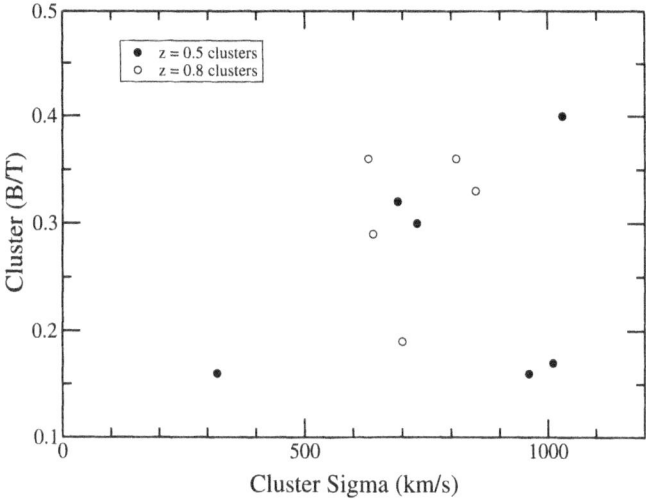

Fig. 4. Total fraction (B/T) of I-band light in bulges versus cluster σ measured from the weak-lensing maps.

The results shown in table 1 have several large systematic errors. The velocity dispersions are calculated assuming that the clusters are at a redshift of either 0.5 or 0.8 and that the background galaxies lie in a sheet at $z = 1.2$. The mass estimates will increase with increasing cluster redshift and decrease with increasing background galaxy redshift. Further, the velocity dispersions have been increased by 25% over that measured to correct for the expected number of foreground stars and galaxies and cluster galaxies in the background galaxy catalog used to measure the shear. Once the survey is complete, we will be able to determine the cluster redshifts spectroscopically and have photometric redshift measurements for all the objects. We thus expect to remove most of the systematic errors and also increase the signal-to-noise of the lensing detections.

3   Bulge+Disk Decomposition

The morphologies of galaxies in all 20 EDisCS clusters and the surrounding field are being quantitatively characterized using full two-dimensional, PSF-convolved bulge+disk surface brightness models. We used a *completely automated* pipeline based on the GIM2D software package (Simard et al. 2001). The pipelines consists of the following steps: (1) Source detection, deblending and extraction using SExtractor, (2) Construction of a spatially-varying Point-Spread-Function using DAOPHOT and the list of objects with high SExtractor stellarity values, (3) Creation of a PSF for each object to be fit, (4) Bulge+disk fitting with GIM2D, (5) Creation of residual images and computation of residual image asymmetry indices, and (6) Creation of structural parameter catalogs from the GIM2D output log files. The fact that our morphological classifications are performed without any human intervention is essential to a meaningful characterization of the systematic biases in our classifications and of the selection function of our sample. Our initial sets of fits are performed independently in each of our filters using the I-band as a reference for object detection and pixel masks. The next phase will fit our galaxies simultaneously in all filters to fully leverage all the information our images.

The 2D galaxy surface brightness model used by GIM2D has a maximum of twelve parameters: the total flux F, the bulge fraction B/T ($\equiv 0$ for pure disk systems), the bulge semi-major axis effective radius r_e, the bulge ellipticity e ($e \equiv 1 - b/a$, $b \equiv$ semi-minor axis, $a \equiv$ semi-major axis), the bulge position angle of the major axis ϕ_b on the image (clockwise, y-axis $\equiv 0$), the disk semi-major axis exponential scale length r_d (also denoted h in the literature), the disk inclination i (face-on $\equiv 0$), the disk position angle ϕ_d on the image, the dx and dy offsets of the model center with respect to the initial position provided by SExtractor, the background residual level db, and the Sérsic index n (we use $n = 4$ in all our bulge+disk fits). The position angles ϕ_b and ϕ_d were not forced to be equal for two reasons: (1) a large difference between these position angles is a signature of barred galaxies, and (2) some observed galaxies do have *bona fide* bulges that are not quite aligned with the disk position angle. Galaxy types are primarily defined by their bulge fraction, and we subdivide our samples in three broad categories: disks ($B/T \leq 0.3$), intermediate ($0.3 < B/T \leq 0.6$) and ellipticals ($B/T > 0.6$). Our resulting structural catalogs contain a wealth of information that can be used to compare the morphological mixtures, scaling relations (e. g. disk luminosity-size) and bulge/disk colors of galaxy populations in our distant clusters with the predictions of sophisticated numerical simulations.

Eleven (6 at $z = 0.5$ and 5 at $z = 0.8$) of the 20 EDisCS clusters have been processed in all three optical bands through our GIM2D bulge+disk fitting pipeline. Figures 2 and 3 shows the core of the EDisCS cluster CL1232-1250 at $z = 0.53$ before and after processing through the GIM2D pipeline. Figure 4 shows the total fraction of light residing in bulges (the "cluster bulge fraction") as a function of cluster velocity dispersion as measured from the weak-lensing map. The cluster bulge fraction was calculated using galaxies with redshifts within ± 0.2 from the cluster redshift and with clustercentric distances less than 1 Mpc ($H_0 = 70$, $\Omega_m = 0.3$, $\Omega_\Lambda = 0.7$). The solid circles are the six $z = 0.5$ clusters, and the open circles are the five $z = 0.8$ clusters. Apart from the two clusters with $B/T \leq 0.2$ and $\sigma \geq 950$ km s^{-1}, there appears to be a trend of larger B/T with larger σ. More massive clusters would appear to be more efficient at producing bulges. The B/T values of $z = 0.8$ clusters are not systematically different from those of the $z = 0.5$ clusters at a fixed σ. Though our results are still preliminary, it is interesting to note that some clusters with $\sigma \sim 1000$ km s^{-1} can have low B/T values. The most massive clusters could therefore contain a wide range of galaxy populations.

References

1. A. Gonzales, D. Zaritsky, J. Delcanton, A. Nelson: ApJS in press (2001)
2. N. Kaiser, G. Squires, T. Broadhurst: ApJ **449**, 460 (1995)
3. N. Kaiser, G. Squires: ApJ **404**, 441 (1993)

A 400 Square Degrees Serendipitous *ROSAT* PSPC Cluster Survey.

Rodion Burenin[1,2], Alexey Vikhlinin[1,2], Mikhail Pavlinsky[1],
Nail Sakhibullin[3], Ilfan Bikmaev[3], Zeki Aslan[4], and Irek Khamitov[4]

[1] Space Research Institute, Moscow, Russia
[2] Harvard-Smithsonian Center for Astrophysics, Cambridge MA, USA
[3] Kazan State University, Academy of Sciences of Tatarstan, Kazan, Russia
[4] TUBITAK National Observatory, Turkey

Abstract. Our new survey is specially designed to detect a large number of X-ray luminous galaxy clusters out to redshift $z \approx 1$. We describe the construction and current status of the survey, as well as the results of recent observations of our cluster candidates on Russian 1.5-m telescope AZT-22 at the TUBITAK National Observatory (Turkey). To demonstrate a potential of the survey we obtain constraints on the evolution of most luminous clusters using a small subsample, which was best observed so far.

1 Introduction

The most X-ray luminous and massive galaxy clusters are most important for studying cosmology. Theory of hierarchical structure formation predict their evolution to be faster and more sensitive to the cosmological parameters.

Study of the evolution of these clusters is hampered by their poor sampling at high redshifts, $z > 0.3$, in the existing X-ray cluster surveys. Large area surveys (EMSS [1,2], surveys based on RASS, e.g. [3]) have too low sensitivities to detect a large number of high redshift clusters. Deep *ROSAT* surveys (e.g. RDCS [4], WARPS [5]) have too low sky coverage to detect these rare systems at $z < 1$. At higher redshifts the optical followup observations are very difficult even with largest telescopes.

CfA-*ROSAT* 160 square degrees X-ray cluster survey [6] (hereinafter 160 deg^2) is better suited for these studies, since it has large sky coverage while keeping a good sensitivity. One of its important results is the confirmation of the negative evolution of luminous clusters [7,8] originally reported by EMSS [2]. However, even this survey was not designed specially to detect luminous clusters. In particular, it excluded a number of *ROSAT* fields which were not well suited to detect X-ray faint clusters. The goal of our new survey is to increase the coverage for *luminous*, $L_{44} > 3$ ($\equiv L/10^{44}$erg s^{-1},0.5–2 keV) clusters at $z < 0.8$, i.e. for clusters with fluxes $f_{-13} > 1$ ($\equiv f/10^{-13}$erg s^{-1}cm^{-2}) by relaxing field selection criteria of 160 deg^2.

2 The Survey

As compared to 160 deg^2 the sky coverage of our new survey is increased by including observations at lower Galactic latitudes $|b| > 25°$, with lower exposures, using

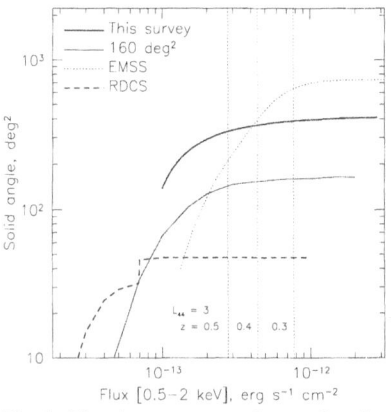

Fig. 1. The sky coverage of as a function of flux.

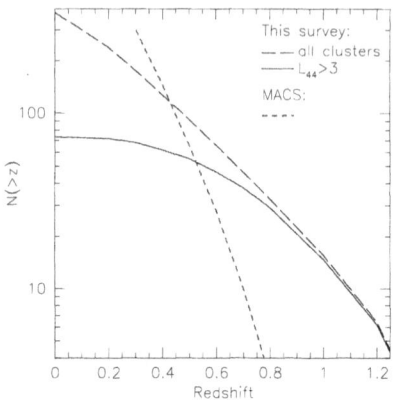

Fig. 2. The expected number of clusters above limiting redshift (no evolution).

pointings to extended targets. We also used new fields which appeared in the *ROSAT* archive since 1997. The total number of *ROSAT* PSPC fields used in our survey is 1632 compared to 647 in 160 deg^2.

We detect clusters using their angular extent in the inner $17'.5$ of the *ROSAT* PSPC field where the angular resolution is good. The cluster detection procedure is fully automated and is essentially the same as that in 160 deg^2. We detected 384 cluster candidates with fluxes $f_{-13} > 1$, of them 334 were identified with galaxy clusters. The majority of unidentified candidates are the objects with no CCD data and DSS counterparts, and most likely are distant, $z > 0.5$, clusters.

The sky coverage of our survey as a function of flux is shown in Fig. 1 in comparison with other X-ray clusters surveys. It is significantly larger at the fluxes of interest. We also compare our survey with the ongoing MAssive Cluster Survey (MACS) [9]. In Fig. 2, we present the no-evolution predictions for the redshift distribution of clusters in two surveys (MACS data were taken from [9]). A significantly larger number of high-L_x clusters at redshifts $z > 0.5$ is expected in our survey compared to MACS. However, it should be emphasized that these two surveys are very different in design and should not be compared directly.

3 Observations on AZT-22 Telescope

CCD imaging of our cluster candidates is necessary to obtain photometric redshifts estimates, and optical identifications for unidentified clusters. We have started the program of imaging observations on the Russian 1.5-m telescope AZT-22 at the TUBITAK National Observatory (Turkey). Our program allocated 11 photometric nights in February-June, 2001. Good photometry has been obtained for 38 clusters. Our photometric redshift estimates use the correlation between the magnitude of the brightest cluster galaxy and redshift, calibrated in 160 deg^2 [6]. Standard stars observations show that the accuracy of photometric redshifts from the AZT-22 data is similar to that obtained on other telescopes in the 160 deg^2 survey, $\Delta z = ^{+0.04}_{-0.07}$.

4   Evolution of the Most Luminous Clusters

The optical observations of our clusters are in the very preliminary which prevents a detailed study of cluster evolution. However, to demonstrate a potential of our survey we obtain constraints on the evolution of the high-L_x clusters in the best-observed part of our survey. Currently it consist in clusters with fluxes $f_{-13} > 3$ detected in 806 fields at $\delta > -20°$ in α range $8^h - 16^h30^m$. This subsample contains 60 clusters. Of them, 44 has known redshifts: 28 come from the 160 deg^2 survey and 16 more are listed in NED. Ten clusters with unknown redshifts were observed by AZT-22, and have photometric redshift estimates. Inspection of NED and DSS plates for the remaining 6 clusters shows that 3 of them are in fact poor groups at $z = 0.1$–0.2, and 3 are Abell clusters, which are unlikely to be at $z > 0.4$ according to $m_{10} - z$ relation [10].

There are no clusters with measured redshifts $z > 0.4$ and luminosities $L_{44} > 3$ in this subsample, and only two clusters with upper bounds on their redshifts $z < 0.467$, < 0.448 and on their luminosities $L_{44} < 3.99$, < 3.32, inferred from their photometric redshift estimates. Assuming no evolution of the cluster X-ray luminosity function, in this part of our survey we expect to find 9.8 clusters with $L_{44} > 3$ and $f_{-13} > 3$ at $z > 0.4$. Given that we observe at most 2 such clusters, the no-evolution hypothesis is rejected at the 99.7% confidence level.

Note that the analysis above uses the 160 deg^2 survey data, and therefore is not completely independent. However, it shows that we already reach a significance for detection of cluster evolution at a level comparable to that in EMSS [2], 160 deg^2 [7,8], and RDCS [11]. When our survey is completed we will be able to detect a negative evolution with a $4 - 6\sigma$ significance.

This work was supported by NASA grant NAG 5-9217 and partially by RBRF grants 00-02-17124 and 01-02-06300.

References

1. Stocke J. T., Morris S. L., Gioia I. M., et al.: ApJS, **76**, 813 (1991).
2. Henry J. P., Gioia I. M., Maccacaro T., et al.: ApJ, **386**, 408 (1992).
3. Ebeling H., Edge A. C., Allen S. W., et al.: MNRAS, **318**, 333 (2000).
4. Rosati P. et al.: ApJL, **492**, L21 (1998).
5. Jones L. R. et al.: ApJ, **495**, 100 (1998).
6. Vikhlinin A., McNamara B. R., Forman W., Jones C., Quintana H., Hornstrup A.: ApJ, **502**, 558 (160 deg^2, 1998).
7. Vikhlinin A., McNamara B. R., Forman W., et al.: ApJL, **498**, L21 (1998).
8. Vikhlinin, A., McNamara B. R., Quintana H., et al.: 'Distant cluster X-ray luminosity function derived from the 160 square degrees *ROSAT* survey'. In: *Large Scale Structure in the X-ray Universe, 0-22 September 1999 Workshop, Santorini, Greece* ed. by Plionis M. & Georgantopoulos I. (Atlantisciences, Paris, France, 1999) p.31.
9. Ebeling H., Edge A. C., Henry J. P.: ApJ, **553**, 668 (2001).
10. Ebeling H., Voges W., Bohringer H.: MNRAS, **281**, 799 (1996).
11. Rosati P., et al.: astro-ph/0001119.

Magnetic Fields in Clusters of Galaxies

Federica Govoni[1,2], Luigina Feretti[2], Gabriele Giovannini[1,3], Daniele Dallacasa[1,2], and Gregory B. Taylor[4]

[1] Istituto di Radioastronomia CNR, Bologna, Italy
[2] Dip. di Astronomia Univ. di Bologna, Italy
[3] Dip. di Fisica Univ. di Bologna, Italy
[4] National Radio Astronomy Observatory, Socorro, NM, USA

1 Introduction

Clusters of galaxies are the largest virialized structures in the Universe. In the last years the presence of a intra-cluster magnetic field has been unambiguously proven although its strength and structure are still poorly known. To investigate the presence and structure of large scale magnetic fields we are studying the rotation measure (RM) of radio galaxies and radio halos in clusters of galaxies. Here we present new results on the magnetic field estimate through RM determination in A2255 and the equipartition magnetic field derived in 4 clusters with an extended radio halo.

1.1 Magnetic Fields in A2255

A2255 (z=0.0806) is known to contain a radio halo and several radio sources located at various distances from the cluster center (see Fig. 1). Here we investigate the strength and the structure of the magnetic field in Abell 2255 by means of the study of the polarized emission from the radio galaxies A2255B, A2255E located within the cluster. Feretti et al. (1997) estimated an equipartition magnetic field strength of 0.5 μG from the study of the extended halo source. Following the definition: $\Psi_\lambda = \Psi_{int} + RM\lambda^2$, where Ψ_{int} is the intrinsic position angle and Ψ_λ is the position angle at wavelength λ, we derived the RM image by fitting the polarization angle images obtained at two wavelengths within 6 cm band and at two wavelengths within 3.6 cm band using the Very Large Array (VLA). In the Table 1, along with the distance from the cluster center, we report the maximum absolute value, the mean, and the σ of the RM values for the two sources. Assuming a simple model for the cluster magnetic field (Felten et al. 1996), the derived RMs can reasonably be explained by the presence of cluster magnetic fields of about 5 μG with a scale length of about 10 kpc (H_0=50 Km/sec Mpc).

1.2 Radio Halos

Radio halos are wide diffuse radio sources located in the cluster center with no obvious connection to the galaxy population of the clusters. They have always been considered very rare phenomenon. However, recently, thanks to the improved sensitivity of the radio telescopes and to the existence of deep radio surveys, the number of known halos has increased.

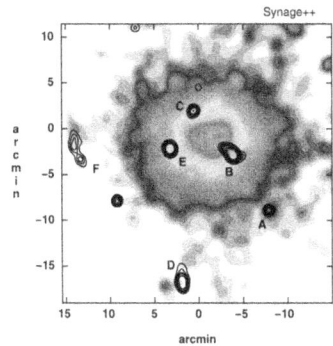

Fig. 1. The 1.4 GHz radio image obtained from the NVSS (contour) superimposed to the X-Ray image of Abell 2255 taken from Rosat PSPC archive. The radio contour levels are 5, 10, 20, 30, 50, and 70 mJy/beam. The radio image has an angular resolution of $45''$. The Rosat image has an angular resolution of $25''$ (FWHM) and has been smoothed with a Gaussian of $\sigma = 30''$. The sensitivity of the NVSS image is not enough to show the halo source.

Table 1. Rotation Measure

Name	Dist. center ($arcmin$)	$|\mathrm{RM}_{max}|$ (rad/m^2)	$< RM >$ (rad/m^2)	σ_{RM} (rad/m^2)
A2255B	3.2	-225	-89	196
A2255E	5.0	$+176$	94	212

The presence of a possible radio halo in the clusters A520, A773, A2254, A2744 was suggested by Giovannini et al. (1999). Thanks to deep radio observations we confirm the presence of a radio halo in all of these clusters. Fig. 2 shows a radio image obtained with the VLA at 1.4GHz with a resolution of $15''$ of the halo source in A520. In the Table 2 we give some parameters of the radio emission. All the radio halos are powerful and extended, moreover the radio halo in A2744 is one of the largest known so far. The equipartition magnetic fields are of the order of about 1 μG except in A773 where the presence of a very steep spectral index $\alpha \simeq 2.8$ ($S(\nu) \propto \nu^{-\alpha}$) leads to a magnetic field of about 4 μG.

2   Discussion

The presence of intra-cluster magnetic fields is well established. We can obtain measures of their strength and structure through RM studies or estimating the equipartition magnetic field in clusters with a radio halo. The magnetic field measured through the study of the RM in radio sources belonging to the cluster is in general higher than the magnetic field calculated from the equipartition in the radio halos. Some explanations

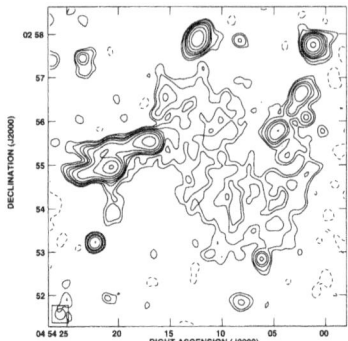

Fig. 2. Radio image at 1.4 GHz of the halo source in A520, the FWHM is 15″.

Table 2. Radio parameters. Col. 1 Cluster name; Col. 2 Redshift; Col. 3 Flux density; Col. 4 Largest Linear Size of the halo; Col. 5 Radio Power; Col. 6 Equipartition magnetic field.

Name	z	S_{1400} (mJy)	LLS (Mpc)	P_{1400} (Watt/Hz)	H_{eq} (μG)
A520	0.199	34.4	1.4	6.4×10^{24}	1.3
A773	0.217	12.6	1.6	2.8×10^{24}	4.3
A2254	0.308	33.7	1.2	4.9×10^{24}	0.7
A2744	0.178	57.1	2.3	2.6×10^{25}	0.8

for the discrepancy obtained using different methods of analysis have been proposed in literature (Carilli & Taylor 2002 and the references therein) however this is still an open problem, and it will be investigated in more detail in the future.

References

1. C.L. Carilli, G.B. Taylor : ARA&A, (2002) in press
2. J.E. Felten: 1996, Clusters, Lensing, and the Future of the Universe, ASP Conference Series, Vol. 88, Eds. V.Trimble and A.Reisenegger, p. 271
3. L. Feretti, H. Böhringer, G. Giovannini, D. Neumann: A&A **317**, 432 (1997)
4. G. Giovannini, M. Tordi, L. Feretti : New Ast. **4**, 141 (1999)

X-Ray Statistical Properties of the Central Cool Component in Clusters of Galaxies

Yasushi Ikebe

Max-Planck-Institut für extraterrestrische Physik, Postfach 1312, 85741 Garching, Germany

Abstract. Central cool gas component that is often observed from a well-relaxed cluster system has long been interpreted as a consequence of "Cooling Flow" (CF), radiative cooling followed by inflow of Intra-Cluster Medium (ICM). However, recent XMM-Newton spectroscopy has shown no signatures of cooler gas phases below certain temperatures in typical CF clusters (A1795, Tamura et al. 2001; A1835, Peterson et al. 2001). This contradicts the conventional CF model or at least requires a major revision of the model. In order to investigate statistical properties of the central cool component, we performed systematic analysis of *ASCA* data on 85 clusters. We found that 1) temperature of the central cool component strongly depends on the temperature of the main ICM, 2) the cool component is selectively found around a brightest cluster galaxy (BCG) that coincide with the X-ray peak position, and 3) the luminosity-temperature (L–T) relation of the cool component shows nice agreement with the L–T relation of the main ICM. Together with the previous observational fact that, in some of the "CF" clusters, the total gravitating mass is clustering in two distinct spatial scales, a main cluster component and a second small-scale system, we conclude that the central cool component is associated with the second small-scale self-gravitating system that is immersed in the host cluster, and the cool component temperature reflects the gravitational potential depth.

1 Sample and ASCA Analysis

For the study we used brightest 85 clusters observed with *ASCA* whose redshifts distribute from 0.004 to 0.201 with a median of 0.046. From each GIS and SIS instrument, a spectrum over a central region of $2'$ radius and one outside $2'$ radius were accumulated. We fitted each set of *ASCA* spectra with the two-temperature model (2T model), in which a hot component is filling entire cluster region, while in the central region a cool component is allowed to coexist with the hot component forming two-phase plasma. We then obtained X-ray luminosity in 0.1-2.4 keV band and temperature of cool and hot component, L_c, T_c, L_h, and T_h. Full results are given in Ikebe et al. (2002).

We have tested the validity of the 2T model using the latest X-ray data of Abell 1795 taken with XMM-Newton, and found that the 2T modeling is valid to the most advanced X-ray data and usable to characterize the property of the central cool component and the surrounding ICM.

2 Results and Discussions

Based on the results from the 2T model fitting to the *ASCA* data of the 85 selected clusters, we investigated various correlations. In the study, we classified the cluster sample into three groups, which are Strong Cool Component (SCC), X-ray Dominant

(XD) and non-X-ray Dominant (nXD). An SCC cluster is defined as one showing a very strong (statistically very significant) cool component in the *ASCA* spectrum. Among the rest of the clusters, one in which the X-ray peak coincides with the Brightest Cluster Galaxy (BCG) is classified as an XD cluster, while nXD is defined as one without such a galaxy. All the SCC clusters turned out to be XDs, too.

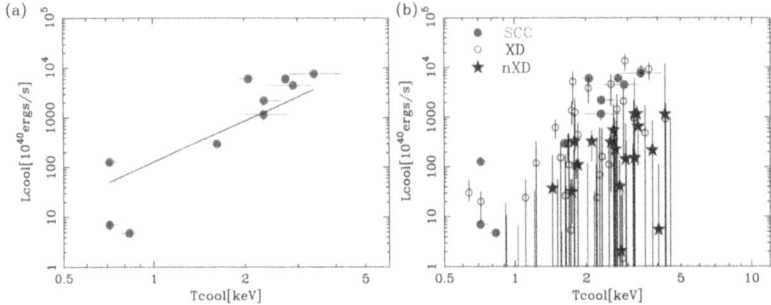

Fig. 1. (a) The L_c–T_c relation for the SCC clusters. The line is the best-fit power-law function to the L_h–T_h relation illustrated in Fig. 3. (b) The L_c–T_c relation for all the sample clusters. Filled circles, open circles, and filled stars specify SCC, XD, and nXD clusters, respectively.

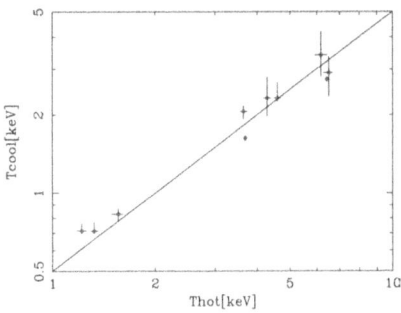

Fig. 2. T_c–T_h relation for the SCC clusters. The line represents $T_c = T_h/2$.

For the SCC clusters, we determined the temperatures and emission measures of each cool and hot component. Figure 1(a) shows the cool component luminosity as a function of the cool component temperature for the SCC clusters. A positive correlation is clearly seen. Figure 2 shows the correlation between the two temperatures of the hot and cool component for the SCC clusters. Surprisingly, the ratios between the two temperatures, T_c/T_h, is virtually constant and the relation, $T_c = T_h/2$ well represents the correlation.

For clusters that do not show very strong cool component, i.e. non-SCC clusters, we fixed T_c at the half value of each mean temperature derived with a single-temperature model fitting, according to the $T_c = T_h/2$ relation. The luminosity of the cool component thus estimated for the non-SCC clusters is illustrated in Fig. 1(b). The results of the SCC clusters are also overlayed. SCC and XD clusters exhibit systematically more luminous cool component than nXD clusters. Actually, many of nXD clusters give only upper limit to L_c.

We obtained the $L - T$ relation of the hot component, which is illustrated in Fig. 3(a). Unlike for the cool component, the three classes do not show any systematic difference. It is then clearly shown that the X-ray characteristics of the XD and nXD clusters are segregated mainly by the central cool component. To compare with the $L - T$ relation of the cool component would be very interesting. In the Fig. 3(b), the $L_c - T_c$ relation for the SCC clusters shown in Fig. 1 is overlayed on the $L_h - T_h$ relation. A surprising agreement is clearly seen. The best fit power-law function to the $L_h - T_h$ relation is compared with the SCC $L_c - T_c$ relation in Fig. 1. The similarity of the two $L - T$ relation suggests that both the cool and hot components are related to individual gravitational bound objects. Therefore, the cool component can be naturally interpreted as ICM filling a self-gravitating system whose size is comparable to a giant elliptical galaxy or a group of galaxies, which is immersed in the host cluster.

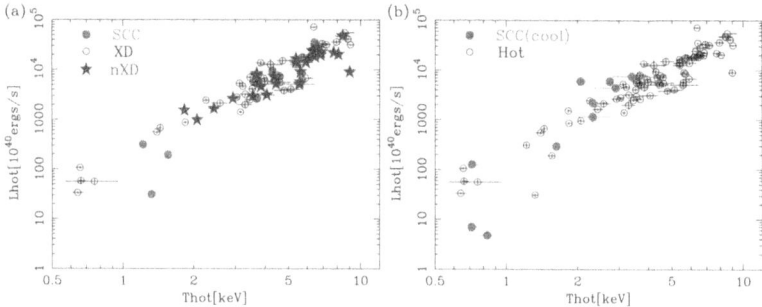

Fig. 3. (a) The $L_h - T_h$ relation for all the sample clusters. Filled circles, open circles, and filled stars specify SCC, XD, and nXD clusters, respectively. (b) The $L_c - T_c$ relation for the SCC clusters (filled circles) is overlayed on the $L_h - T_h$ relation (open circles).

References

1. Ikebe et al. 2002, A&A, in press
2. Tamura et al. 2001, A&A, 365, L87
3. Peterson et al. 2001, A&A, 365, L104

Constraining Cosmological Models with the Brightest Galaxy Clusters in the X-Ray Sky

Thomas H. Reiprich[1,2] and Hans Böhringer[1]

[1] Max-Planck-Institut für extraterrestrische Physik, P.O. Box 1312, 85741 Garching, Germany
[2] University of Virginia, P.O. Box 3818, Charlottesville, VA 22903-0818, USA

Abstract. A new X-ray selected and X-ray flux-limited galaxy cluster sample has been constructed. Based on the ROSAT All-Sky Survey the 63 brightest clusters with galactic latitude $|b_{II}| \geq 20$ deg and flux $f_X(0.1 - 2.4 \, \text{keV}) \geq 2 \times 10^{-11}$ ergs s^{-1} cm^{-2} have been compiled. Gravitational masses have been determined utilizing intracluster gas density profiles, derived mainly from ROSAT PSPC pointed observations, and gas temperatures, as published mainly from ASCA observations, assuming hydrostatic equilibrium. Using this sample the galaxy cluster mass function has been determined and used to constrain the mean cosmic matter density and the amplitude of mass fluctuations. Comparison to Press–Schechter type model mass functions in the framework of Cold Dark Matter cosmological models yields the constraints $\Omega_m = 0.12^{+0.06}_{-0.04}$ and $\sigma_8 = 0.96^{+0.15}_{-0.12}$ (90 % c.l.). Various sources of possible systematic uncertainties have been tested. Adding all identified systematic uncertainties to the statistical uncertainty in a worst case fashion results in an upper limit $\Omega_m < 0.31$. For comparison to previous results a relation $\sigma_8 = 0.43 \, \Omega_m^{-0.38}$ has been derived. The results obtained here are compared to results from measurements of distant supernovae and cosmic microwave background (CMB) fluctuations.

The galaxy cluster mass function measures the cluster number density as a function of gravitational mass. Theoretical prescriptions have been derived to predict the cluster mass function for specific cosmological models, e.g., [6,2]. A comparison between observed and predicted mass functions allows to put constraints on the mean cosmic matter density, Ω_m, and the amplitude of mass fluctuations within spheres of radius 8 Mpc, σ_8.

In this work the first mass function for X-ray selected galaxy clusters based on the ROSAT All-Sky Survey has been constructed using individually determined cluster masses. The observed mass function for this sample, *HIFLUGCS* (the HIghest X-ray FLUx Galaxy Cluster Sample) [7], has been quantitatively compared to predictions using a standard χ^2 procedure with Ω_m and σ_8 as free parameters. The minimum and statistical error ellipses for some standard confidence levels (c.l.) are given in Fig. 1. The tight constraints obtained show that with *HIFLUGCS* we can go beyond determining an Ω_m–σ_8 relation and put limits on Ω_m and σ_8 individually. It is found that

$$\Omega_m = 0.12^{+0.06}_{-0.04} \quad \text{and} \quad \sigma_8 = 0.96^{+0.15}_{-0.12} \tag{1}$$

(90 % c.l. statistical uncertainty for two interesting parameters), indicating a relatively low value for the density parameter. The large covered mass range, the specific region in Ω_m–σ_8 parameter space, and the assumption of CDM cosmological models with given Hubble constant, H_0, baryon density, Ω_b, primordial power spectral index, n, and CMB temperature, T_0, allow to derive these tight constraints from a local cluster sample. For

comparison for a given σ_8 value the Ω_m value which minimizes χ^2 is calculated. These pairs can then roughly be described by a straight line in log space given by

$$\sigma_8 = 0.43 \, \Omega_m^{-0.38} \, . \tag{2}$$

In Fig. 1 we also plot the best fit model mass functions for given $\Omega_m = 0.5$ and $\Omega_m = 1.0$ and one notes immediately that these value pairs give a poorer description of the shape of the mass function.

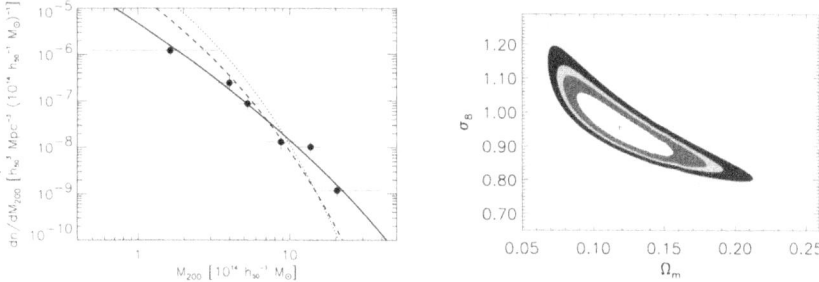

Fig. 1. Left: *HIFLUGCS* mass function compared to the best fit model mass function with $\Omega_m = 0.12$ and $\sigma_8 = 0.96$ (solid line). Also shown are the best fit model mass functions for fixed $\Omega_m = 0.5$ ($\Rightarrow \sigma_8 = 0.56$, dashed line) and $\Omega_m = 1.0$ ($\Rightarrow \sigma_8 = 0.43$, dotted line). Right: Statistical confidence contours. The cross indicates the position of the minimum. Ellipses indicate the 68 %, 90 %, 95 %, and 99 % confidence levels for two interesting parameters.

The quoted error ranges have been calculated from the χ^2 procedure. However, also a large number of possible systematic uncertainties has been quantified (see [7] for details). Almost all identified systematics give rise to smaller uncertainties than the statistical uncertainties. However, it is not impossible that several systematics combined have a significant effect. In order to determine the highest Ω_m value allowed by a conspiracy of all identified uncertainties, all positive $\Delta\Omega_m$ values have been added to the positive statistical uncertainty for Ω_m and this sum has been added to the Ω_m value which minimizes χ^2. In this worst case procedure the upper limit $\Omega_m < 0.31$ has been found.

How do the results of this work fit into the general picture? How do they compare to constraints derived from completely independent measurements und what conclusions can be drawn from a comparison? Among the various methods that have been applied, e.g., [4], currently two more methods seem especially encouraging for constraining the relevant parameters, e.g., [1]. One approach uses measurements of temperature fluctuations in the CMB, another distant supernovae as standard candles. The constraints (95 % c.l.) achieved recently by these two methods, e.g., [3,5] are compared in an Ω_m–Ω_Λ diagram to the results obtained here in Fig. 2, where Ω_Λ is the normalized cosmological constant. The first noteworthy aspect is that these independent approaches nicely complement each other in the way they constrain different regions in parameter space.

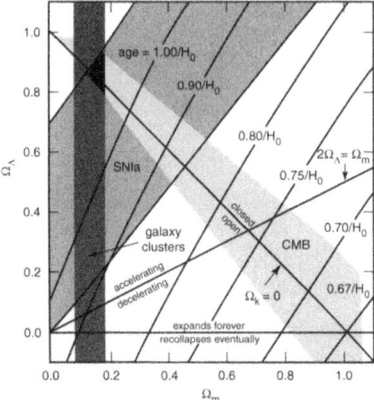

Fig. 2. Observational constraints on two fundamental cosmological parameters. The area labeled galaxy clusters shows the statistical constraints obtained in this work. Note that the worst case upper limit for the galaxy cluster result including all identified systematic uncertainties is given by $\Omega_m < 0.31$. (Basic Figure adopted from [1,3].)

Secondly it is found that all three error ranges overlap; there is an area for which all measurements are consistent. This consistency is very encouraging.

What are the implications of the fact that the concordance is confined to a specific region? In Fig. 2 some additional lines are drawn that separate special regions. The region of consistency is very close to the line $\Omega_m + \Omega_\Lambda = 1$, i.e. $\Omega_k = 0$. This indicates that the universe may have a flat geometry. Furthermore the region is located clearly within a parameter region where the universe will never collapse again, but expand infinitely ($\Omega_m \leq 1 \wedge \Omega_\Lambda \geq 0$). Moreover the region lies well above the line that separates the states in which the expansion of the universe is decelerating or accelerating ($2\,\Omega_\Lambda = \Omega_m$). This implies that not only will the universe expand forever but it will do so ever faster.

References

1. N. A. Bahcall et al.: Sci, **284**, 1481 (1999)
2. J. R. Bond et al.: ApJ, **379**, 440 (1991)
3. P. de Bernardis et al.: Nat, **404**, 955 (2000)
4. P. J. E. Peebles: PASP, **111**, 274 (1999)
5. S. Perlmutter et al.: ApJ, **517** (1999)
6. W. H. Press and P. Schechter: ApJ, **187**, 425 (1974)
7. T. H. Reiprich and H. Böhringer: ApJ, submitted

During the Conference

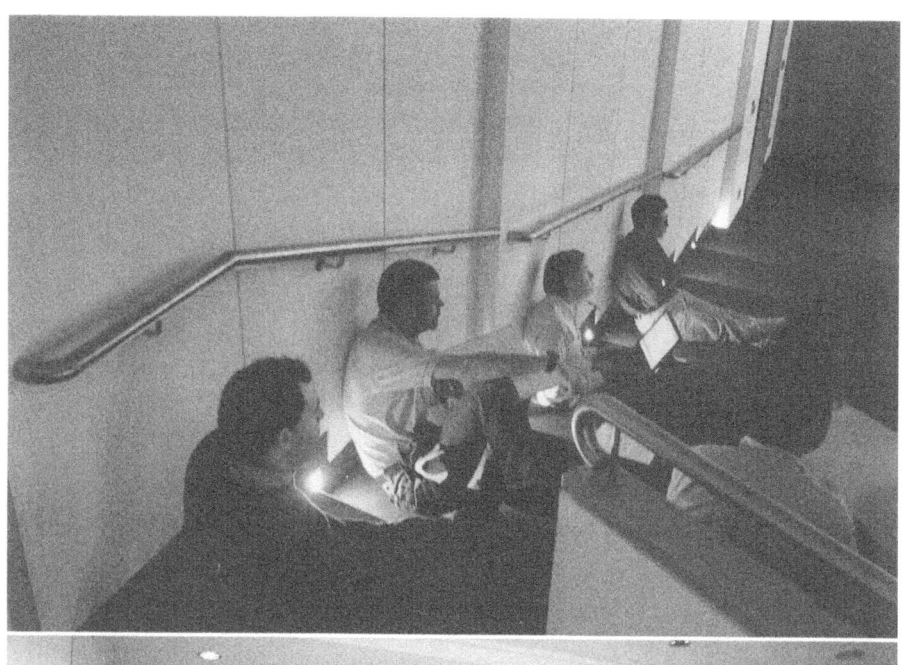

Part II

Brightest Galaxies

Far-Infrared and Submillimetre Lighthouses

Malcolm Longair[1] and Andrew Blain[2]

[1] Cavendish Laboratory, Madingley Road, Cambridge CB3 0HE
[2] Department of Astronomy, California Institute of Technology, Pasadena, CA91125.

Abstract. Evidence is discussed which strongly suggests that the sources discovered in deep submillimetre (submm) surveys form a major population of distant dusty galaxies. These sources can account for the background radiation observed in these wavebands. Three sources are identified with certainty with distant galaxies with measured redshifts. They are all hyperluminous, and are likely to be massive galaxies in which a significant fraction of their stellar populations has already been formed. It is a challenge to account for such a large population of high-redshift luminous galaxies. The relation of these observations to the problem of accounting for the evolution of the cosmic star formation rate is discussed.

1   Introduction

Over ten years ago, a convincing case could be made that the submm and far-infrared (IR) wavebands would prove to be of fundamental significance for astrophysical cosmology, in particular, in detecting the earliest generations of star-forming galaxies. A very small number of us had that vision even earlier – this occasion cannot be allowed to pass without showing the spectrum of the background radiation published by Rashid Sunyaev and one of us (MSL) in 1971. In our review of the background radiation in all wavebands, we guessed how much background radiation might be radiated in the far-IR and submm wavebands on the basis of the few observations of star forming regions then observed in the far-IR waveband (Fig. 1). It turns out that we guessed almost exactly the 'right answer' on the basis of essentially no observational data.

One of the great astronomical achievements of the last five years has been the detection of a very large population of faint sub-millimetre sources, far in excess of even our most optimistic predictions. In addition, estimates of the background radiation due to discrete sources have also been made in these wavebands from observations with the FIRAS instrument of COBE. In this brief review, the following topics will be discussed: (i) our present understanding of the deep counts of submm sources, (ii) the problems of determining their redshift distribution, (iii) the question of whether or not these observations are consistent with the favoured hierarchical clustering picture of galaxy formation, and (iv) a number of puzzles which arise from these studies. Throughout this review, we will rely heavily upon the many papers on different aspect of these studies carried out in collaboration with Ian Smail, Rob Ivison and Jean-Paul Kneib over the period 1997-2001.

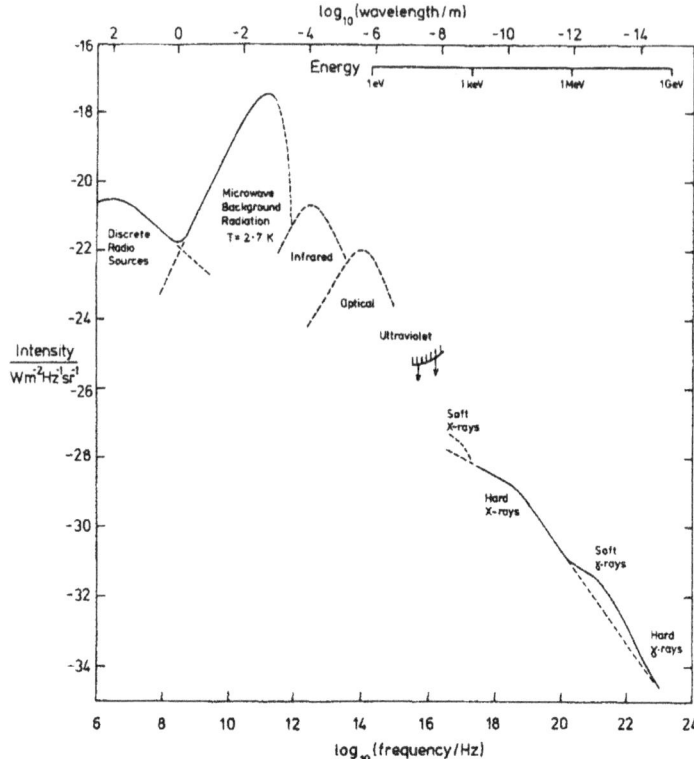

Fig. 1. The spectrum of the extragalactic background radiation as it was known in 1969 (Longair and Sunyaev 1971). The figure still provides a good impression of the overall spectral energy distribution of the background radiation from radio frequencies to γ-ray energies. Regions of the spectrum in which only theoretical estimates were available at that time are shown by dashed lines.

2   The Deep Submillimetre Sky

The radiation process which dominates astrophysical cosmology in the submm waveband is the emission of dust grains with temperatures typically in the range 20 to 80 K. The dust is optically thin in the long wavelength Rayleigh-Jeans region of the spectrum, resulting in a spectrum which can be approximated by $S_\nu \propto \nu^{3-4}$. As a result, the K-corrections for distant dusty galaxies are so large and negative that, for redshifts $1 \leq z \leq 10$, the flux density-redshift relation is more or less independent of redshift (Fig. 2). One of us (MSL) carried out early analytic calculations of this effect in 1990 and presented them to the JCMT Board when it was agonising over whether or not the SCUBA instrument for the JCMT should be funded. This argument played its part in persuading the Board to be bold and approve the funding of SCUBA, which has been such an enormous success and essential for the astrophysics described in this paper.

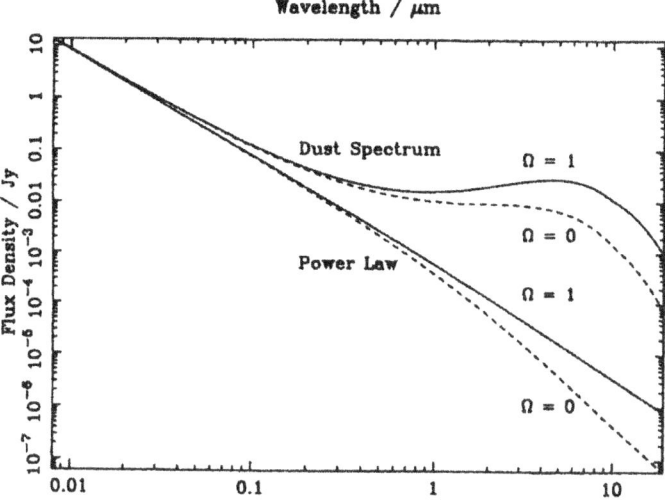

Fig. 2. The flux-density redshift relation for a standard dust spectrum at a temperature of 45 K for world models with $\Omega = 1$ and 0. For comparison, the same relations are shown for a population of sources with power-law spectra $S_\nu \propto \nu^{-1}$. (Blain and Longair 1993).

Some useful rules of thumb are as follows. For observations made at a wavelength of 850 μm, which for many reasons is the most effective wavelength for carrying out the deepest observations with ground-based submm telescopes, an ultraluminous far-IR galaxy with $L_{\mathrm{FIR}} = 3 \times 10^{12}\,L_\odot$, similar to Arp 220, has a star-formation rate of about $300\,M_\odot$ year^{-1} and flux density $S \geq 3$ mJy at 850 μm out to $z = 10$.

The hero of the story is undoubtedly the SCUBA submm camera, which was commissioned on the JCMT in 1997 (Holland et al. 1999). In the first deep SCUBA observations, Smail, Ivison and Blain (1997) discovered evidence for a large population of very faint submm sources, in excess of even our most optimistic predictions (Blain and Longair 1993). These pioneering observations have been followed up by many other deep surveys, including observations of the Hubble Deep Field. These observations are very challenging, typically only one or two faint submm sources being observed in each SCUBA image in long integrations. Smail *et al.* (2001) have reviewed in detail the number counts of faint submm sources from all these surveys. They are self-consistent, and the existence of a large population of faint sub-millimetre sources is beyond doubt. We will concentrate upon the *SCUBA Lens Survey* of Smail *et al.* (2001). There are three important advantages in concentrating upon this set of observations. First of all, the numbers of sources are known to be complete within well defined flux density limits. Second, the use of gravitational lensing enables the survey to be extended by about a factor of two fainter than would be possible without gravitational lensing. Third, because the clusters have been studied in great detail by the Hubble Space Telescope and other instruments such as the VLA. As a result, the cluster potentials are very well defined and accurate source sensitivity maps can be created for each cluster.

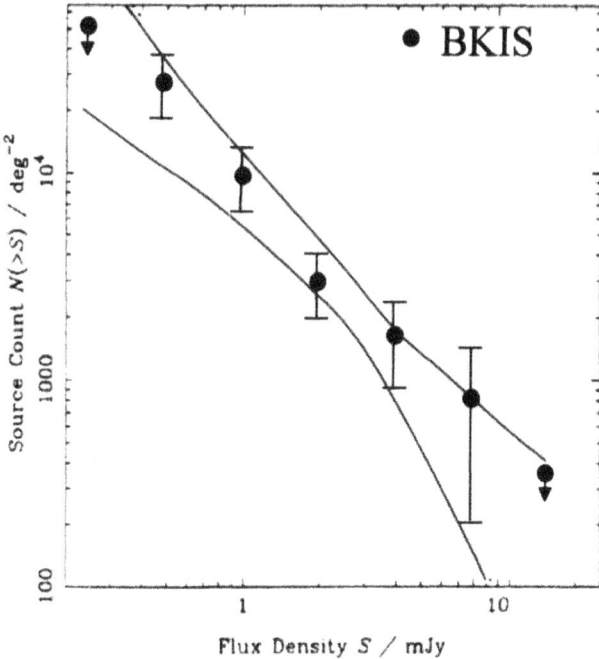

Fig. 3. Deep 850-μm integral number counts from the SCUBA Lens Survey. As integral counts, the numbers of sources in each flux density bin are not independent (Blain *et al.* 1999a). Other deep surveys are in agreement with these results (Smail *et al.* 2001).

The cumulative number counts of faint submm sources are shown in Fig. 3 for the 15 background sources detected in the fields of 7 rich clusters in the Smail *et al.* (2001) survey. Two additional sources in the survey are certainly associated with central cluster galaxies. As the counts are cumulative, $N(\geq S)$, the numbers in each flux density bin are not independent. Only three of the 15 sources have been securely identified and redshifts measured. Part of the reason for the difficulties in making secure identifications is that the 15-arcsec beam of the JCMT means that the positional information from SCUBA is not adequate on its own for identifying faint sources. The SCUBA data need to be supplemented by radio and millimetre interferometric observations, as well as deep optical and near-IR imaging. It is revealing to summarise the available data on the three convincing identifications. Note the enormous amount of observational effort with many different telescopes which has been necessary to reach even this modest level of completeness.

- **SMM J02399-0136**
    $z = 2.80$
    $L_{\rm FIR} \sim 10^{13}\,{\rm L_\odot}$ corresponding to a star formation rate of about $10^3\,{\rm M_\odot\,yr^{-1}}$.
    Interacting/merging pair of galaxies.
    UV spectrum has broad lines ($\Delta v \sim 1000\,{\rm km\,s^{-1}}$) suggesting a Seyfert-2 galaxy.

Strong CO emission observed, indicating at least $M_{H_2} = 2 \times 10^{11}$ M$_\odot$. Could make a further 10^{11} M$_\odot$ of stars in next 10^8 yr.

The system already has $L \approx 6L^*$ optically.

Seems to involve both an AGN and a starburst. Detected by *Chandra* (Bautz *et al.* 2000).

See Ivison *et al.* (1998); Frayer *et al.* (1998); Vernet & Cimatti (2002).

- **SMM J14011-0252**

  $z = 2.56$

  $L_{\rm FIR} \sim 6 \times 10^{12}\, L_\odot$ corresponding to a star formation rate of about 600 $M_\odot$ year^{-1}. Interacting/merging pair of galaxies.

  No evidence of AGN. No detection with *Chandra* (Fabian *et al.* 2000)

  Strong narrow CO observed indicating $M_{H_2} \sim 10^{11}$ M$_\odot$.

  See Ivison *et al.* (2000, 2001); Frayer *et al.* (1999).

- **SMM J02399-0134**

  $z = 1.06$

  $L_{\rm FIR} \sim 6.5 \times 10^{12}\, L_\odot$ corresponding to a star formation rate of about 650 $M_\odot$ year^{-1}. Ring galaxy.

  Optical observations suggest a Seyfert 1.5-2 galaxy. Detected by *Chandra* (Bautz *et al.* 2000).

  See Soucail *et al.* (1999); Kneib *et al.* (2002).

In these cases, the galaxies are interacting or peculiar and there is evidence for the presence of an active nucleus in two cases. What makes these really convincing associations has been the discovery of CO emission at the observed redshift of the identification. The large amounts of molecular hydrogen present in these sources is a strong indication that the emission is likely to be associated with active star formation.

The host galaxies of the 12 submm sources without secure redshifts are much fainter (Smail *et al.* 2001). A strong indication that they are also very distant galaxies is provided by deep VLA observations (Smail *et al.* 2000). There is known to be a strong correlation between the radio and far-IR luminosities of *IRAS* galaxies. Since the continuum spectra of these galaxies are very different in the radio and submm wavebands, Carilli & Yun (1999, 2000) showed that the ratio of flux densities at 350 and 1.4 GHz can be used to estimate their redshifts. Fig. 4 shows their predicted relation, as well as the 350/1.4 GHz spectral indices for a sample of far-IR sources with known redshifts. It can be seen that crude redshift information can be obtained for the unidentified sources from Fig. 4. Fortunately, since the cluster fields have been surveyed in the radio waveband with the VLA, it has been possible to make estimates of the redshift ranges in which they are likely to lie.

The submm–radio spectral indices for the three galaxies discussed in some detail above are in reasonable agreement with the correlation in Fig. 4. From the available data on their 1.4-GHz flux densities, the other 12 sources are all expected to have redshifts greater than 1. The data would be consistent with the mean redshift being about 2 or 3, but this needs to be confirmed by further detailed studies. It is noteworthy that all three certain sources have 850-μm flux densities greater than 10 mJy and so these are among the brightest sources found in the SCUBA Lens Survey.

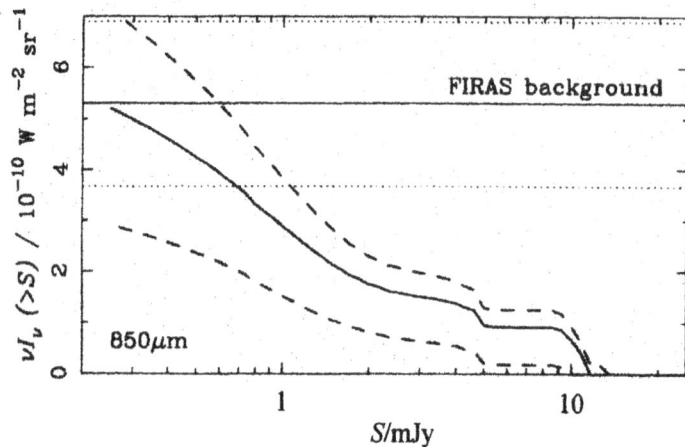

Fig. 4. Mean spectral indices between 1.4 and 350 GHz for 17 far-IR galaxies studied by Carilli and Yun. The dotted lines indicate $\pm 1\sigma$ estimates for the predicted redshifts. (Carilli and Yun 2000)

Therefore, the fifteen sources have far-IR luminosities which place them among the most luminous far-IR galaxies known. The origin of the dust far-IR luminosity is not certain, there being evidence in two of the three identified sources for both intense star-formation activity and the presence of an active galactic nucleus. The statistics are too sparse to come to any definite conclusion, but the results would be consistent with the admixture found by Veilleux *et al.* (1999) in their sample of luminous low-redshift far-IR sources.

3   The Origin of the Far-Infared Background

One of the important aspects of these observations is the contribution which these sources make to the background radiation as determined by the FIRAS instrument of COBE (Puget *et al.* 1996; Fixsen *et al.* 1998). Fig. 5 shows the results of the analysis of Smail *et al.* (2001), comparing the background due to the integrated number counts with the FIRAS estimates of the background at 850 μm. It is striking that more than half of the background is accounted for by sources with flux densities $S_{850\,\mu m} \geq 1$ mJy.

The inference is that most of the background radiation must be associated with sources which are within a factor of 10 in luminosity of the galaxies already identified in the SCUBA Lens Survey. Interpreted literally, this would mean that the sources which are responsible for the background radiation have luminosities $L \geq 10^{12}\,L_{\odot}$. This is something of a surprise, since the luminosity function of *IRAS* galaxies is continous down to galaxies of much lower far-IR luminosity. To express this another way, the background would be dominated by luminous far-IR galaxies at large redshifts and the evolution of less luminous sources must be significantly less than these sources. The situation is similar to that encountered in the counts of radio galaxies and quasars which

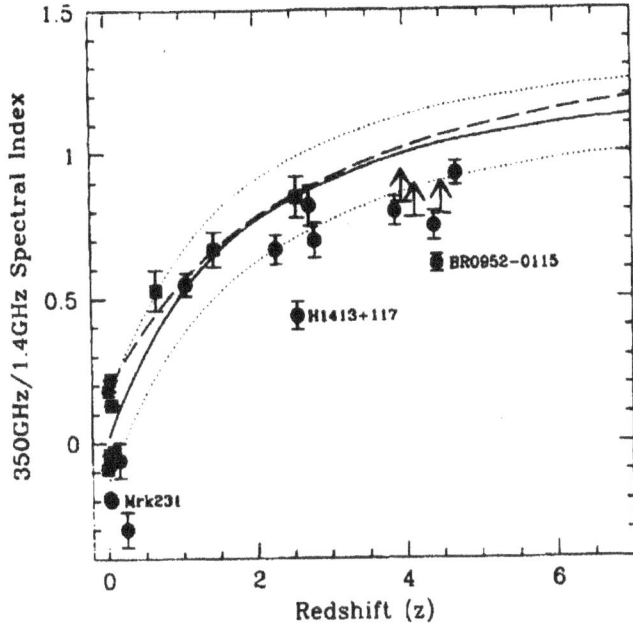

Fig. 5. The contribution of discrete sources to the far-IR background radiation as a function of limiting flux density compared with the background inferred from observations made with the FIRAS instrument of COBE. (Smail *et al.* 2001.)

can be roughly explained by some form of luminosity evolution. The puzzle is all the greater in that, according to the favoured hierarchical clustering picture, massive galaxies should build up their stellar populations by coalescence and therefore they should be less luminous at large redshifts (for detailed estimates, see Efstathiou and Rees (1988), Efstathiou (1995) and White (this volume)). On the contrary, the IR background seems to be dominated by very luminous galaxies, which are presumably massive and already well-formed at large redshifts.

4   The Cosmic Star Formation Rate

In his PhD dissertation, Allon Jameson showed how the equations of cosmic chemical evolution can be combined with the Press-Schechter formalism to produce a rather simple result which indicates clearly where the key physics lies in understanding the cosmic evolution of the chemical elements and the cosmic star formation rate (Jameson 1999, Longair 2000).

Although Paul Schechter described at this meeting his concerns about the Press–Schechter formalism for the formation of large-scale structures such as galaxies, Simon White showed that it is not such a bad approximation after all – it has the great advantage of providing an analytic tool for studying aspects of galaxy formation by hierarchical clustering. The evolution of the Press–Schechter mass function,

$$N(M) = \frac{\bar{\varrho}}{\sqrt{\pi}} \frac{\gamma}{M^2} \left(\frac{M}{M^*}\right)^{\gamma/2} \exp\left[-\left(\frac{M}{M^*}\right)^{\gamma}\right],$$

with cosmic epoch is controlled by the value of the reference mass $M^*(z) = M^*(0)[\delta(z)/\delta(0)]^{2/\gamma}$, above which the mass spectrum cuts off exponentially. $\delta(z)$ is the function which describes the growth of linear perturbations in an expanding Universe and $\gamma = 1 + (n/3)$, where n is the spectral index of the power spectrum of initial density perturbations, $P(k) = k^n$ (Press and Schechter 1974; for a simple derivation of these results, see Longair 1998). We had previously derived a simple analytic expression for the merger rate of galaxies from the Press–Schechter mass function (Blain & Longair 1993a,b; Blain et al. 1999b).

In order to derive an expression for the luminosity density per unit comoving volume due to star formation during mergers, assumptions need to be made about the amount of energy liberated by star formation activity. In our 1993 approach, it was assumed that a fixed fraction x of the total masses of the merging galaxies M was converted into stars, liberating $0.007xMc^2$ of energy, and resulting in the formation of a mass xM of heavy elements. It is more realistic to allow x to vary as a function of redshift, $x(z)$ and so we write $x(z) = x_0\varepsilon_*(z)$, introducing a *star formation efficiency* $\varepsilon_*(z)$, normalized to unity at $z = 0$. The luminosity density at redshift z is then,

$$\rho_L(z) = 0.007x_0\varepsilon_*(z)c^2 \int_0^\infty M\dot{N}_{\text{form}}\,dM = 0.007\bar{\rho}\varepsilon_*(z)c^2\frac{2}{\gamma}\beta\frac{\dot{\delta}(z)}{\delta(z)}, \qquad (1)$$

where $\beta = \phi x_0/\sqrt{\alpha}$; ϕ and α are constants with values close to unity. In an Einstein–de Sitter model $\dot{\delta}(z)/\delta(z) = H_0(1 + z)^{3/2}$.

We adopt the simplest form of the equations of cosmic chemical evolution given by Pei & Fall (1995) for the closed box model, which can be expressed as $\dot{\Omega}_* + \dot{\Omega}_g = 0$, in which the density parameters Ω_* and Ω_g refer to the stellar and gaseous components of the galaxies respectively. For the closed box model, $\Omega_* + \Omega_g = \Omega_{g\infty}$, where $\Omega_{g\infty}$ is the initial density parameter of primordial gas. At large redshifts, all the baryons were in the form of gas and there were no stars.

Some assumption must be adopted about the dependence of the star formation efficiency ε_* upon the density of gas and metals. It is reasonable to assume that it is proportional to the product of the average density of gas Ω_g, representing the availability of fuel for star formation, and the average density of heavy elements synthesized by a given epoch Ω_m – the latter represents the efficiency of cooling, mediated by the optically thin emission from dust and atomic fine-structure transitions of heavy elements (Si, O, N and C) in dense molecular clouds. Under these assumptions, we can write $\varepsilon_*(z) = k\Omega_g\Omega_m$, where k is a constant. This is a Schmidt law of star formation (Kennicutt 1998).

The result is an analytic expression for the star formation efficiency

$$\varepsilon_* = a\,\text{sech}^2[b\,J(z) - \cosh^{-1}\sqrt{a}]. \qquad (2)$$

where the constants a and b have the values, $a = \Omega^2_{g\infty} kQ/4$ and $b = \beta k \Omega_{g\infty}/\gamma$, while the function $J(z)$ is

$$J(z) = \int_0^z \frac{\dot\delta}{\delta} \frac{dr}{dz'} \frac{1}{c(1+z')}\, dz'.$$

In an Einstein–de Sitter world model, $dr/dz = (c/H_0)(1+z)^{-3/2}$, $\delta \propto (1+z)^{-1}$ and so $J = \ln(1+z)$. The expression (2) is then analytic,

$$\varepsilon_* = a\, \mathrm{sech}^2[b\ln(1+z) - \cosh^{-1}\sqrt{a}].$$

These simple formulae completely define the star formation rate as a function of epoch. The magnitude of a determines the height of the peak of the star formation efficiency as compared with its value at $z = 0$, while the ratio of a and b determines the redshift of the peak. The maximum occurs when the argument of the sech^2 function is zero, that is at,

$$z = \exp\left(\frac{\cosh^{-1}\sqrt{a}}{b}\right) - 1,$$

for an Einstein-de Sitter model. In this model $\varepsilon_* \propto (1+z)^{-2b}$ at large redshifts and $\varepsilon_* = (1+z)^{2b\sqrt{1-a^{-1}}}$ at small redshifts. The star formation rate for the Einstein-de Sitter model, normalised to its value at the present epoch, is therefore,

$$\dot\Omega_* = (1+z)^{3/2} a\, \mathrm{sech}^2[b\ln(1+z) - \cosh^{-1}\sqrt{a}]. \tag{3}$$

In Fig. 6, examples of the variations of the star formation rate with cosmic epoch have been derived by fitting the expression (3) to observations of the background spectrum derived by Fixsen *et al.* (1998) and the 60 μm IRAS number counts (Bertin *et al.* 1997) for different assumed temperatures of the emitting dust grains. For comparison, various direct estimates of the star formation rate are shown. The derived curves provide a satisafactory envelope for the observations, recalling that the vertical normalisation is arbitrary.

Once expression (3) for the global star formation rate has been established, it is straightforward to work out the variation of the density of metals, the gas density, the metallicity and the build up of the heavy elements in the intergalactic gas as functions of redshift. We have shown that the resulting expressions can provide a good account of the observed variations of these quantities with cosmic epoch.

How seriously should we take these models? The same type of formalism can be made to work for a wide variety of cosmological models. The key feature of the calculations is that the formation of stars takes place through the conversion of gas in the interstellar medium into stars during galaxy mergers – that process essentially guarantees that there will be a maximum in the star formation rate at redshifts about 2–3. In the present calculations, one of the most important features concerns the assumptions made about the dependence of the efficiency of the star formation rate upon density and metallicity and that must be present in all similar approaches. Nonetheless, I would recommend the sech^2 formula (3) for the rate of star formation to you.

Is this picture relevant to what is observed in the number counts of faint submm sources? The good news is that, taking together the information on the far-IR background, the IRAS luminosity function and the counts, the sech^2 function provides a

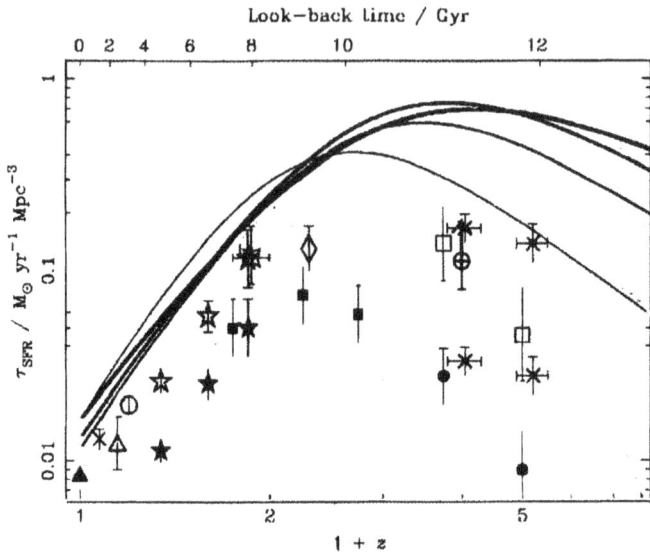

Fig. 6. Star formation histories derived for the best-fitting models with dust temperatures $T_d =$ 35, 40, 45 and 50 K (solid lines in order of increasing thickness). The vertical normalization of the curves is arbitrary.

very good fit to all the data and is an economical means of describing how the star formation rate has evolved with cosmic epoch. There is plenty of flexibility for accommodating the different behaviours which have been claimed by different authors.

There is however a puzzle. The inference would be that most of the star and element formation is associated with those galaxies which dominate the number counts of faint submm sources – we have argued that the latter are associated with massive hyperluminous galaxies. A literal interpretation of these results would therefore be that most of the star and element formation activity is associated with massive star-forming galaxies at large redshifts. To express it another way, it would not be a coincidence that the number density of far-IR galaxies and the cosmic star-formation rate both attain a maximum at $z \sim 2 - 3$.

A great deal remains to be unravelled. The message of this paper is that we need to be able to pin down the redshift distribution for the faint sub-millimetre sources as well as obtaining good observations for much larger and deeper samples of these sources. One of the key instruments for achieving this is the ALMA submm array. Our view is that ALMA is the key to unravelling many of these puzzles and mysteries.

References

1. Bautz M. et al., 2000, ApJ, 543, L119
2. Bertin E., Dennefeld M., Moshir M., 1997, A&A, 323, 685
3. Blain A. W., Kneib J.-P., Smail I., Iviosn R. J., 1999a, ApJ, 512, L87
4. Blain A. W., Jameson A., Smail I., Longair M. S., Kneib J.-P., Ivison R. J., 1999b, MNRAS, 309, 715
5. Blain A. W., Longair M. S., 1993a, MNRAS, 264, 509
6. Blain A. W., Longair M. S., 1993b, MNRAS, 265, L21
7. Blain, A.W., Longair, M.S. 1996, MNRAS, 279, 847.
8. Carilli, C. L., Yun, M. S., 1999, ApJ, 513, L13.
9. Carilli, C. L., Yun, M. S., 2000, ApJ, 530, 618.
10. Efstathiou, G. (1995). In *Galaxies in the Young Universe*, (eds. H. Hippelein, K. Meisenheimer and H.-J. Röser), 299. Berlin: Springer-Verlag.
11. Efstathiou, G. and Rees, M.J. 1988, MNRAS, 230, 5P.
12. Fabian A. C., et al. 2000, MNRAS, 315, L8.
13. Fixsen D. J., Dwek E., Mather J. C., Bennett C. L., Shafer R. A., 1998, ApJ, 508, 123
14. Frayer D. T. et al., 1998, ApJ, 506, L7.
15. Frayer D. T., et al., 1999, ApJ, 514, L13.
16. Holland W. S., et al., 1999, MNRAS, 303, 657.
17. Ivison R. J. et al., 1998, MNRAS, 298, 583.
18. Ivison R. J. et al., 2000, MNRAS, 315, 209.
19. Ivison R. J. et al., 2001, ApJL, in press (astro-ph/0110085)
20. Jameson, A., 1999, PhD Dissertation, University of Cambridge.
21. Kennicutt R. C., 1998, ApJ, 498, 541
22. Kneib J.-P., et al., 2002, A&A in preparation.
23. Longair, M.S. 1998. *Galaxy Formation*, Berlin: Springer-Verlag.
24. Longair, M.S. 2000. In *26th International Cosmic Ray Conference, Salt Lake City*, (eds. B.L. Dingus, D.B. Kieda and M.H. Salamon), 3. New York: AIP Conference Proceedings 516.
25. Longair, M.S. and Sunyaev, R.A. 1971, *Uspekhi Fiz. Nauk.*, 105, 41. [English translation: *Soviet Physics Uspekhi*, 14, 569.]
26. Pei, Y.C., Fall, S.M. 1995, ApJ, 454, 6
27. Press W. H., Schechter P., 1974, ApJ, 187, 425
28. Puget, J-L., et al., 1996, A&A, 308, L5
29. Smail I., Ivison R. J., Blain A. W., ApJ, 1997, 490, L5.
30. Smail I. et al., 2000, MNRAS, 528, 612.
31. Smail, I., Ivison, R., Blain, A.W. and Kneib, J.-P., 2001, MNRAS, submitted.
32. Soucail, G. et al., 1999, A&A, 343, L70.
33. Veilleux, S., Sanders, D. B., Kim, D.-C., 1997, ApJ, 484, 92.
34. Vernet, J. and Cimatti, A., 2002, A&A, in press.

The Most Luminous Galaxies

I. Félix Mirabel[1,2]

[1] CEA/DSM/DAPNIA, Service d'Astrophysique, 91191 Gif/Yvette. France
[2] IAFE/CONICET. cc 67, suc 28. Ciudad Universitaria. 1428 Buenos Aires. Argentina

Abstract. Ultraluminous galaxies in the local universe ($z \leq 0.2$) emit the bulk of their energy in the mid and far-infrared. The multiwavelength approach to these objects has shown that they are advanced mergers of gas-rich spiral galaxies. Galaxy-galaxy collisions took place on all cosmological time-scales, and nearby mergers serve as local analogs to gain insight into the physical processes that lead to the formation and trans-formation of galaxies in the more distant universe. Here I review multiwavelength observations –with particular emphasis on recent results obtained with ISO– of mergers of massive galaxies driving the formation of: 1) luminous infrared galaxies, 2) elliptical galaxy cores, 3) luminous dust-enshrouded extranuclear starbursts, 4) symbiotic galaxies that host AGNs, and 5) tidal dwarf galaxies. The most important implication for studies on the formation of galaxies at early cosmological timescales is that the distant analogs to the local ultraluminous infrared galaxies are invisible in the ultraviolet and optical wavelength rest-frames and should be detected as sub-millimeter sources with no optical counterparts.

1 Luminous Galaxies

One of the most important discoveries from extragalactic observations at mid- and far-infrared wavelengths has been the identification of a class of "Luminous Infrared Galaxies" (LIGs), objects that emit more energy in the infrared ($\sim$ 5–500 μm) than at all other wavelengths combined (see [18] for a comprhensive review). The first all-sky survey at far-infrared wavelengths carried out in 1983 by the *Infrared Astronomical Satellite* (*IRAS*) resulted in the detection of tens of thousands of galaxies, the vast majority of which were too faint to have been included in previous optical catalogs. It is now clear that part of the reason for the large number of detections is the fact that the majority of the most luminous galaxies in the Universe are extremely dusty. Previous assumptions, based primarily on optical observations, about the relative distributions of different types of luminous galaxies—e.g. starbursts, Seyferts, and quasi-stellar objects (QSOs)—need to be revised.

Galaxies bolometrically more luminous than $\sim 4\,L^*$ (i.e. $L_{bol} \geq 10^{11}\,L_\odot$) appear to be heavily obscured by dust. Although luminous infrared galaxies (hearafter LIGs: $L_{ir} > 10^{11}\,L_\odot$) are relatively rare objects, reasonable assumptions about the lifetime of the infrared phase suggest that a substantial fraction of all galaxies with $L_B > 10^{10}L_\odot$ pass through such a stage of intense infrared emission [23].

A comparison of the luminosity function of infrared bright galaxies with other classes of extragalactic objects in the local universe is shown in Figure 1. At luminosities below $10^{11}\,L_\odot$, *IRAS* observations confirm that the majority of optically selected objects are relatively weak far-infrared emitters. Surveys of Markarian galaxies confirm that both Markarian starbursts and Seyferts have properties (e.g. f_{60}/f_{100} and L_{ir}/L_B

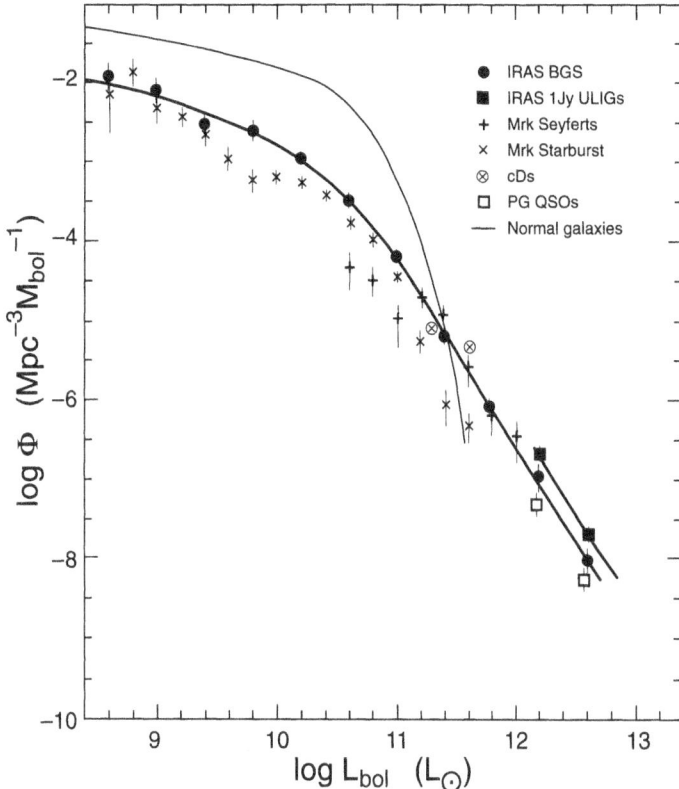

Fig. 1. Galaxy luminosity function of Infrared Galaxies compared with other extragalactic objects in the local universe. Among the most luminous galaxies ($L_{bol} > 10^{11.5}$ L$_\odot$), infrared galaxies selected from the IRAS survey outnumber optically selected Seyferts and quasars. For references see [18].

ratios) closer to infrared selected samples as does the subclass of optically selected interacting galaxies. However because the most luminous galaxies are enshrouded in dust, relatively few objects in optically selected samples are found with $L_{ir} > 10^{11.5}$ L$_\odot$.

The high luminosity tail of the infrared galaxy luminosity function is clearly in excess of what is expected from the Schechter function. For $L_{bol} = 10^{11} - 10^{12}$ L$_\odot$, LIGs are as numerous as Markarian Seyferts and ~ 3 times more numerous than Markarian starbursts. Ultraluminous infrared galaxies (hereafter ULIGs: $L_{ir} > 10^{12}$ L$_\odot$) appear to be ~ 2 times more numerous than optically selected QSOs, the only other previously known population of objects with comparable bolometric luminosities.

Although LIGs comprise the dominant population of extragalactic objects at $L_{bol} > 10^{11}$ L$_\odot$, they are still relatively rare. For example, Figure 1 suggests that only one object with $L_{ir} > 10^{12}$ L$_\odot$ will be found out to a redshift of ~ 0.033, and indeed, Arp 220 ($z = 0.018$) is the only ULIG within this volume. The total infrared luminosity from

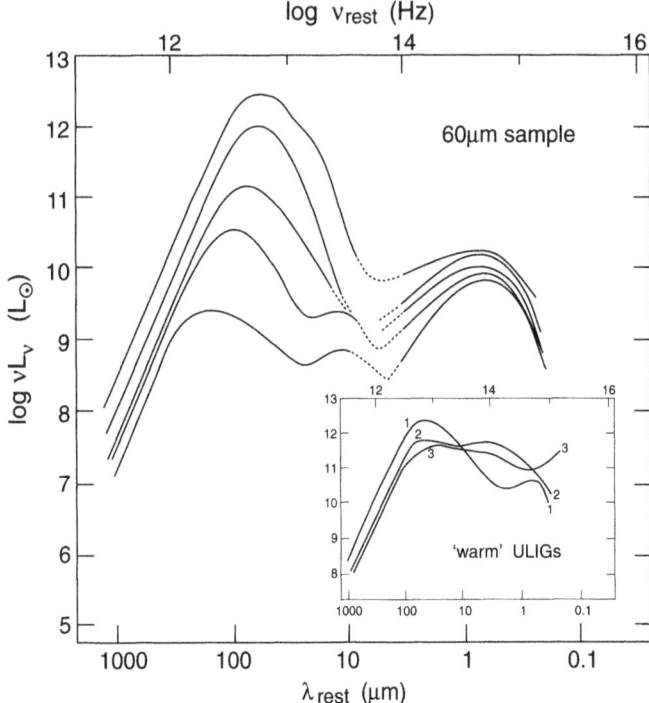

Fig. 2. Variation of the mean Spectral Energy Distribution (from submillimeter to UV wave-lengths) with increasing L_{ir} for a 60 μm sample of infrared galaxies. (Insert) Examples of the subset ($\sim$ 15%) of ULIGs with "warm" infrared color ($f_{25}/f_{60} > 0.3$). Three objects (1—the powerful Wolf-Rayet galaxy IRAS 01002–2238, 2—the "infrared QSO" IRAS 07598+6508, 3—the optically selected QSO I Zw 1) are shown in the inset. For references see [18].

LIGs in the *IRAS* Bright Galaxy Survey (BGS) is only $\sim$ 6% of the infrared emission in the local Universe [24].

There are preliminary indications that ULIGs have been more numerous in the past. Comparison of the space density of nearby ULIGs with the more distant population provides evidence for possible strong evolution in the luminosity function at the highest infrared luminosities. Assuming pure density evolution of the form $\Phi(z) \propto (1 + z)^n$, [8] found $n \sim 7 \pm 3$ for a complete flux-limited sample of ULIGs.

The infrared properties for the complete *IRAS* Bright Galaxy Sample have been summarized and combined with optical data to determine the relative luminosity output from galaxies in the local Universe at wavelengths ~ 0.1–$1000\,\mu m$ [24]. Figure 2 illustrates how the shape of the mean spectral energy distribution (SED) varies for galaxies with increasing total infrared luminosity. Systematic variations are observed in the mean infrared colors; the ratio f_{60}/f_{100} increases while f_{12}/f_{25} decreases with increasing infrared luminosity. Figure 2 also illustrates that the observed range of over

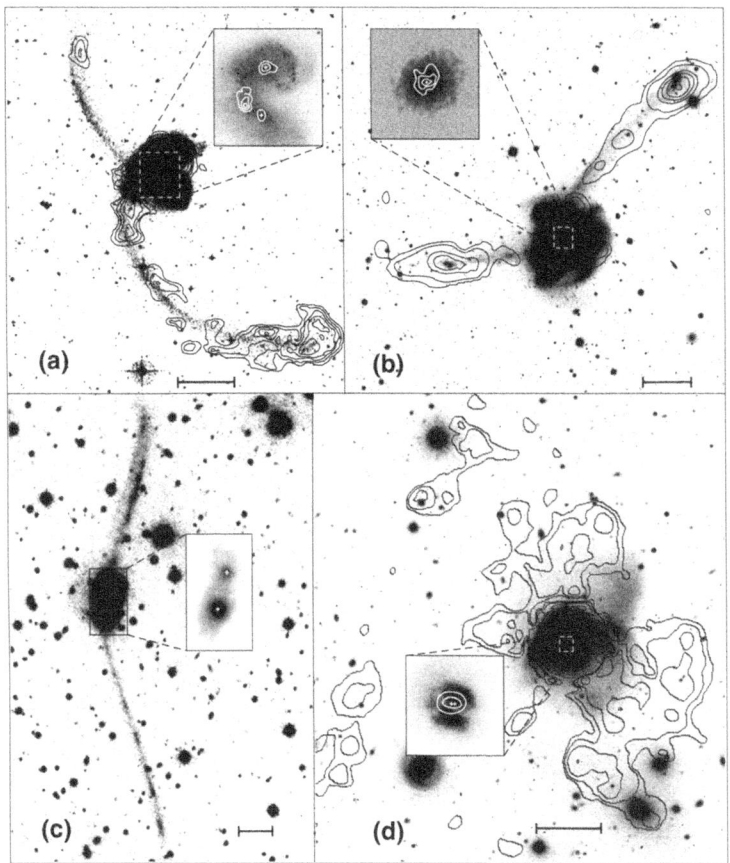

Fig. 3. Well-studied mergers: *(a)* NGC 4038/39 (Arp 244 = "The Antennae"); *(b)* NGC 7252 (Arp 226 = "Atoms for Peace"); *(c)* IRAS 19254–7245 ("The Super Antennae"); *(d)* IC 4553/54 (Arp 220). The two at the top are LIGs whereas the two at the bottom are ULIGs. Contours of HI 21-cm line column density *(black)* are superimposed on deep optical (*r*-band) images. Inserts show a more detailed view in the *K*-band (2.2 μm) of the nuclear regions of NGC 4038/39, NGC 7252, and IRAS 19254–7245, and in the *r*-band (0.65 μm) of Arp 220. White contours represent the CO(1→0) line integrated intensity as measured by the OVRO millimeter-wave interferometer. No HI or CO interferometer data are available for the southern hemisphere object IRAS 19254–7245. The scale bar represents 10 kpc.

3 orders of magnitude in L_{ir} for infrared-selected galaxies is accompanied by less than a factor of 3–4 change in the optical luminosity.

[20] showed that a small but significant fraction of ULIGs, those with "warm" ($f_{25}/f_{60} > 0.3$) infrared colors, have SEDs with mid-infrared emission ($\sim$5–40 μm) over an order of magnitude stronger than the larger fraction of "cooler" ULIGs. These warm galaxies (Figure 2 insert), which appear to span a wide range of classes of extragalactic objects including powerful radio galaxies (PRGs: $L_{408MHz} \geq 10^{25} W Hz^{-1}$) and

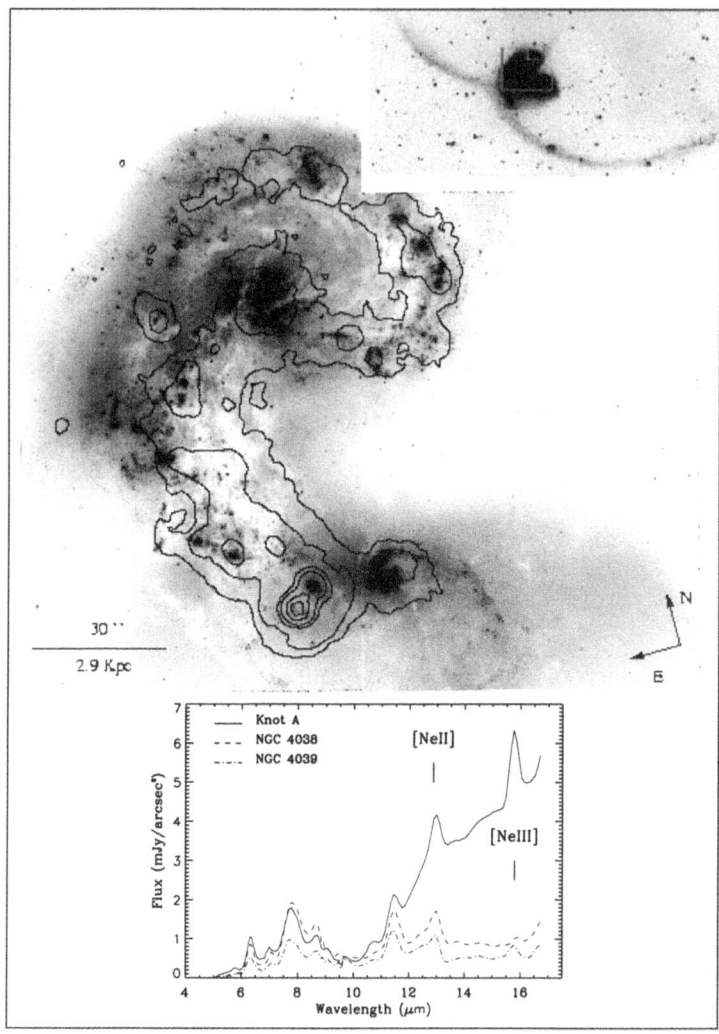

Fig. 4. The upper figure from [16] shows a superposition of the mid-infrared (12 -17 μ, countours) image of the Antennae galaxies obtained with the Infrared Space Observatory, on the composite optical image with V (5252 Å) and I (8269 Å) filters recovered from the Hubble Space Telescope archive . About half of the mid-infrared emission from the gas and dust that is being heated by recently formed massive stars comes from an off-nuclear region that is clearly displaced from the most prominent dark lanes seen in the optical. The brightest mid-infrared emission comes from a region that is relativelly inconspicuous at optical wavelengths. The ISOCAM image was made with a 1.5″ pixel field of view. Contours are 0.4, 1, 3, 5, 10, and 15 mJy. The lower figure shows the spectrum of the brightest mid-infrared knot and of the nuclei of NGC 4038 and NGC 4039. The rise of the continuum above 10 μm and strong NeIII line emission observed in the brightest mid-infrared knot indicate that the most massive stars in this system of interacting galaxies are being formed in that optically obscured region, still enshrouded in large quantities of gas and dust.

optically selected QSOs, have been used as evidence for an evolutionary connection between ULIGs and QSOs (e.g. [19,20]).

There is a strong correlation between the broad band colors (from optical to far-infrared) and morphological type [18]. In particular, the fraction of objects that are interacting/merger systems appears to increase systematically with increasing infrared luminosity. The imaging surveys of objects in the local universe [19,12] have shown that the fraction of strongly interacting/merger systems increases from $\sim 10\%$ at $log\ (L_{ir}/L_{\odot})$ = 10.5–11 to $\sim 100\%$ at $log\ (L_{ir}/L_{\odot}) > 12$. In pannel (c) of Figure 3 is shown the "Super-antennae", which is the prototype of ULIG [14]. ISO observations [10] have shown that more than 98% of the mid-infrared flux from this object comes from the southern component which hosts a Seyfert 2 nucleus.

From the detailed studies of nearby ultraluminous infrared galaxies the following conclusions were reached. 1) They are mergers of evolved gas-rich giant spiral galaxies (e.g. Milky Way with Andromeda), and not "primival" galaxies. 2) To boost the luminosity above 10^{12} $L_{\odot}$ the nuclei must have approached at least 10 kpc, namely, they are advanced mergers. 3) Due to the gravitational impact the interstellar gas decouples from the stars and large amounts of interstellar matter fall at high rates to the central region. This is the condition to produce a nuclear starburst, and/or feed a supermassive black hole at super-Eddington accretion rates. To produce such large accretion rates, the gravitational potential wheels of massive buldges are needed.

A workshop on the question concerning the ultimate source of energy (starbursts versus AGN's) took place in Ringberg on October 1998. Below 2 10^{12} $L_{\odot}$ starbursts dominate the energy budget, but above 3 10^{12} $L_{\odot}$ AGN's seem to be always present and become an important source of energy. In this respect it is interesting to note that it is found with ISO that in the prototype Seyfert 2 galaxy NGC 1068, about 80% of the mid-infrared flux between 4 and 18 μm comes from the AGN [11].

A caveat for the subject of this conference is that the pre-encounter objects that merged at high redshifts must have been different from the metal-rich evolved galaxies merging at present. Another caveat is that ultraluminous IR galaxies at high redshifts may be very difficult to detect using the Lyman break technique. Due to the large amounts of dust in ultraluminous objects, very little or none continuum leaks out at ultraviolet wavelengths. Therefore, surveys with submillimeter arrays as ALMA will be needed to detect ultraluminous galaxies at high redshifts.

2   ISO Observation of Extranuclear Starbursts

The starbursts in ultraluminous galaxies take place in the nuclear region. One of the new findings with ISO is a class of very luminous dust-enshrouded extranuclear starbursts in nearby spiral-spiral mergers. When the pre-encounter galaxies do not have prominent buldges, namely, when the mergers are -for instance- two Sc galaxies, the most luminous starbursts may take place in extranuclear regions that are inconspicuous at optical wavelengths. These extranuclear starbursts have sizes $\leq$ 100 pc in radius and can produce up to 50% of the overall mid-infrared output from these systems. Furthermore, the analyses of the mid-infrared spectra indicate that the most massive stars in these systems are formed inside these optically invisible knots.

In Figure 4 is shown in contours the mid-infrared (12-17 μm) image of the Antennae galaxies obtained with ISO [16], superimposed on the optical image from HST. Below are shown representative spectra of the two nuclei and the brightest mid-infrared knot. It shows that the most massive stars are formed in an obscured knot of 50 pc radius, which produces about 15% of the total luminosity from the Antennae galaxies between 12.5 and 17 μm. A more extreme case is found in NGC 3690 [6], where it is observed an extranuclear region $\leq$100 pc in radius that radates $\sim$45% of the overall mid-infrared output from this system. If the fraction of far-infrared fluxes is the same as in the mid-infrared, such compact region produces a luminosity of 2 10^{11} L$_\odot$. Therefore, the luminosity of a few compact starburst knots of this type would be comparable to the total bolometric luminosity of a ULIG such as Arp 220 (Figure 3d).

The multiwavelength view of this nearby sample of prototype merging systems suggests caution in deriving scenarios of early evolution of galaxies at high redshift using only observations in the narrow rest-frame ultraviolet wavelength range [16]. Although the actual numbers of this type of systems may not be large, we must keep in mind that the most intense starbursts are enshrouded in dust and no ultraviolet light leaks out from these regions.

3   Symbiotic Galaxies

Giant radio galaxies are thought to be massive ellipticals powered by accretion of interstellar matter onto a supermassive black hole. Interactions with gas rich galaxies may provide the interstellar matter to feed the active galactic nucleus (AGN). To power radio lobes that extend up to distances of hundreds of kiloparsecs, gas has to be funneled from kiloparsec size scales down to the AGN at rates of $\sim$1 M$_\odot$ yr^{-1} during $\geq 10^8$ years. Therefore, large and massive quasi-stable structures of gas and dust should exist in the deep interior of the giant elliptical hosts of double lobe radio galaxies. Recent mid-infrared observations with ISO revealed for the first time a bisymmetric spiral structure with the dimensions of a small galaxy at the centre of Centaurus A [17]. The spiral was formed out of the tidal debris of accreted gas-rich object(s) and has a dust morphology that is remarkably similar to that found in barred spiral galaxies (see Figure 5). The observations of the closest AGN to Earth suggest that the dusty hosts of giant radio galaxies like CenA, are "symbiotic" galaxies composed of a barred spiral inside an elliptical, where the bar serves to funnel gas toward the AGN.

The barred spiral at the centre of CenA has dimensions comparable to that of the small Local Group galaxy Messier 33. It lies on a plane that is almost parallel to the minor axis of the giant elliptical. Whereas the spiral rotates with maximum radial velocities of ~ 250 km s^{-1}, the ellipsoidal stellar component seems to rotate slowly (maximum line-of-sight velocity is ~ 40 km s^{-1}) approximately perpendicular to the dust lane. The genesis, morphology, and dynamics of the spiral formed at the centre of CenA are determined by the gravitational potential of the elliptical, much as a usual spiral with its dark matter halo. On the other hand, the AGN that powers the radio jets is fed by gas funneled to the center via the bar structure of the spiral. The spatial co-existence and intimate association between these two distinct and dissimilar systems suggest that Cen A is the result from a cosmic symbiosis.

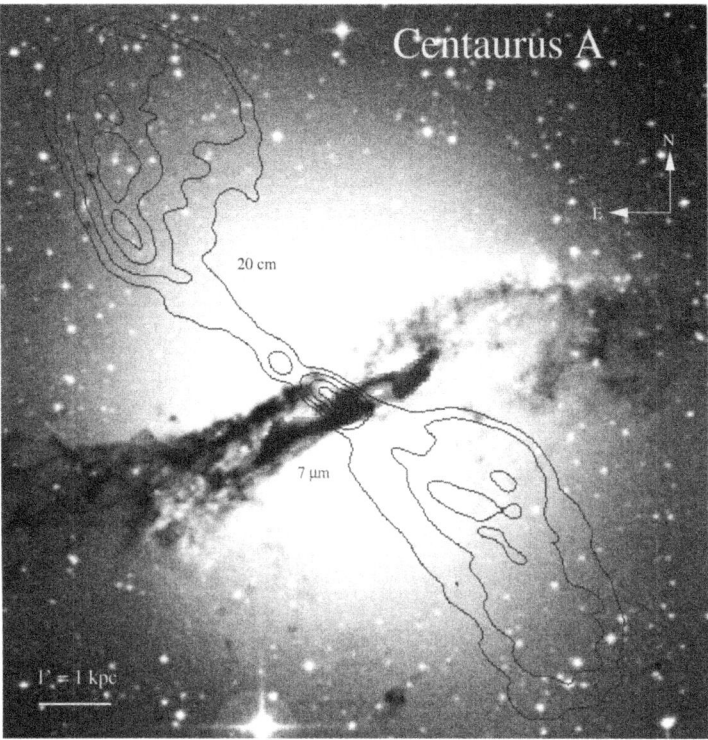

Fig. 5. The ISO 7 μm emission (dark structure; [17]) and VLA 20 cm continuum in contours [1], overlaid on an optical image from the Palomar Digital Sky Survey. The emission from dust with a bisymmetric morphology at the centre is about 10 times smaller than the overall size of the shell structure in the elliptical and lies on a plane that is almost parallel to the minor axis of its giant host. Whereas the gas associated to the spiral rotates with a maximum radial velocity of 250 km s^{-1}, the ellipsoidal stellar component rotates slowly approximately perpendicular to the dust lane [27]. The synchrotron radio jets shown in this figure correspond to the inner structure of a double lobe radio source that extends up to 5° ($\sim$ 300 kpc) on the sky. The jets are believed to be powered by a massive black hole located at the common dynamic center of the elliptical and spiral structures.

4   Formation of Ellipticals

In disk-disk collisions of galaxies, dynamical friction and subsequent relaxation may produce a mass distribution similar to that in classic elliptical galaxies. From the relative numbers of mergers and ellipticals in the New General Catalogue [26] estimated that a large fraction of ellipticals could be formed via merging. The first direct observational evidence for the transition from a disk-disk merger toward an elliptical was presented in the optical study of NGC 7252 by [22]. The brightness distribution over most of the main body of this galaxy which is shown in Figure 3c is closely approximated by a de Vacouleurs ($r^{-1/4}$) profile. However, NGC 7252 still contains large amounts of interstellar gas and exhibits a pair of prominent tidal tails (see Figure 3b); neither property is typical of ellipticals.

Near-infrared images are less affected by dust extinction and also provide a better probe of the older stellar population, which contains most of the disk mass and therefore determines the gravitational potential. K-band images of six mergers by [28] showed that the infrared radial brightness profiles for two LIGs—Arp 220 and NGC 2623— follow an $r^{-1/4}$ law over most of the observable disks. Among eight merger remnants, [25] found K-band brightness profiles for four objects that were well fitted by an $r^{-1/4}$ law over most of the observable disks. [8] finds a similar proportion ($\sim 50\%$) of ULIGs whose K-band profiles are well fit by a $r^{-1/4}$ law.

More recently, [9] have proposed that ULIGs are elliptical galaxies forming by merger-induced dissipative collapse. The extremely large central gas densities ($\sim 10^2$–$10^3 \, M_\odot \, pc^{-3}$) observed in many nearby ULIGs, and the large stellar velocity dispersions found in the nuclei of Arp 220 and NGC 6240 are comparable to the stellar densities and velocity dispersions respectively, in the central compact cores of ellipticals.

Despite the K-band and CO evidence that LIGs may be forming ellipticals, we still need to account for two important additional properties of ellipticals: 1. the large population of globular clusters in the extended halos of elliptical galaxies, which cannot be accounted for by the sum of globulars in two preexisting spirals, and   2. the need to remove the large amounts of cold gas and dust present in infrared-luminous mergers in order to approximate the relative gas-poor properties of ellipticals. These two issues have been discussed by [18].

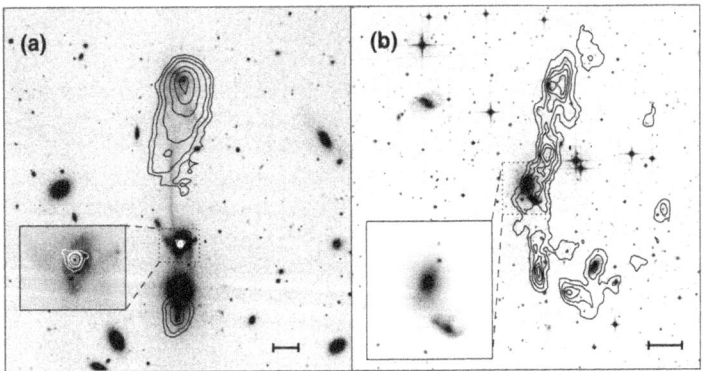

Fig. 6. *(a) NGC 3561A/B (Arp 105) from [3];   (b) NGC 5291A/B ("Sea shell") from [4]. Tidal dwarfs may have different morphologies: Blue compacts, Magellanic Irregulars, and Dwarf Irregulars. Contours of H I 21-cm line column density (black) are superimposed on deep optical (r-band) images. Inserts show a more detailed view in r-band of the spiral galaxy NGC 3561A [3], and of the interacting pair NGC 5291A/B. White contours represent the CO(1→0) line integrated intensity as measured by the IRAM millimeter-wave interferometer. CO emission has not been detected in NGC 5291A/B. The scale bar represents 20 kpc.*

5 Tidal Dwarf Galaxies

Collisions between giant disk galaxies may trigger the formation of dwarf galaxies. This idea, which was first proposed by [29] and later by [21], has received recent observational support [14,15,5,3]. Renewed interest in this phenomenon arose from the inspection of the optical images of ULIGs, which frequently exhibit patches of optically emitting material along the tidal tails (see Figure 3a–c). These objects appear to become bluer near the tips of the tails at the position of massive clouds of H I. These condensations have a wide range of absolute magnitudes, $M_V \sim -14$ to -19.2, and H I masses, $M(H I) \sim 5 \times 10^8$ to $6 \times 10^9 M_\odot$. [13] have shown that objects resembling irregular dwarfs, blue compacts, and irregulars of Magellanic type are formed in the tails. These small galaxies of tidal origin are likely to become detached systems, namely, isolated dwarf galaxies. Because the matter out of which they are formed has been removed from the outer parts of giant disk galaxies, the tidal dwarfs we observe forming today have a metallicity of about one third solar [2].

It is interesting that in these recycled galaxies of tidal origin there is—as in globular clusters—no compelling evidence for dark matter [13]. To find the true fraction of dwarf galaxies that may have been formed by processes similar to the tidal interactions we observe today between giant spiral galaxies, more extensive observations of interacting systems will be needed. A recent step forward is the statistical finding that perhaps as much as one half of the dwarf population in groups is the product of interactions among the parent galaxies [7].

Tidal dwarfs are formed not only during spiral-spiral mergers, but also in encounters of spirals with massive ellipticals in clusters of galaxies. In Figure 6 are shown the results from the multiwavelength study of Arp 105 and NGC 5291A/B which are in clusters of galaxies. In Arp 105, [3] find tidal dwarfs that resemble Magellanic Irregulars and a blue compact. In NGC 5291, about 10 tidal dwarfs of irregular morphology are found associated to the 200 kpc HI ring shown in Figure 6 [4].

6 Conclusions

1) Scenarios on the history of star formation that use only observations in the UV and optical rest-frames result in luminosity functions that are strongly biased in the high luminosity end.

2) The most luminous nuclear and off-nuclear starbursts are enshrouded in dust. In merging galaxies ISO revealed off-nuclear starburst knots with sizes ≤ 100 pc that produce bolometric luminosities of up to $2\ 10^{11}\ L_\odot$ (e.g. NGC 3690). A few of these starburst knots can produce the overall bolometric luminosity of an ultraluminous galaxy such as Arp 220.

3) The observation with ISO of the nearest AGN to Earth (Centaurus A) opens the general question on whether the hosts of giant radio galaxies are symbiotic galaxies composed of spirals at the centre of giant ellipticals.

4) Mergers of disks can produce metal-rich elliptical galaxy cores.

5) Collisions between giant disk galaxies trigger the formation of dwarf galaxies out tidal debris. A fraction of these re-cycled galaxies become detached systems with

diverse morphologies: blue compact dwarfs, dwarf irregulars, and irregulars of Magellanic type.

Acknowledgements: Most of the work review here was carried out in collaboration with D.B. Sanders, P-A. Duc, V. Charmandaris and O. Laurent.

References

1. J.J. Condon et al.: A&ASS **103**, 81 (1996)
2. P.-A. Duc: Genèse de galaxies naines dans les systèmes en interaction. PhD thesis, Univ. Paris (1995)
3. P.-A. Duc & I. F. Mirabel: A&A **289**, 83 (1994)
4. P.-A. Duc & I. F. Mirabel: A&A **333**, 813 (1998)
5. B. G. Elmegreen, M. Kaufman, M. Thomasson: ApJ **412**, 90 (1993)
6. P. Gallais et al.: The Universe as seen by ISO, ESA/SP-427 vol. 2, 881 (1999)
7. S. D. Hunsberger, J. C. Charlton, D. Zaritsky: ApJ **462**, 50 (1996)
8. D. C. Kim: The IRAS 1 Jy survey of ultraluminous infrared galaxies. PhD thesis, Univ. Hawaii (1995)
9. J. Kormendy, D. B. Sanders: ApJ **390**, L53 (1992)
10. O. Laurent, V. Charmandaris, I. F. Mirabel et al.: A&A **359**, 359 (2000)
11. E. Le Floc'h, I. F. Mirabel, O. Laurent et al.: A&A **367**, 487 (2001)
12. J. Melnick, I. F. Mirabel: A&A **231**, L19 (1990)
13. I. F. Mirabel, P.-A. Duc, H. Dottori: In *Dwarf Galaxies*, ed. G Meylan, P Prugniel, p. 371. Garching bei Munchen:ESO (1995)
14. I. F. Mirabel, D. Lutz, J. Maza: A&A **243**, 367 (1991)
15. I. F. Mirabel, H. Dottori, D. Lutz: A&A **256**, L19 (1992)
16. I. F. Mirabel, L. Vigroux, V. Charmandaris et al.: A&A **333**, L1 (1998)
17. I. F. Mirabel, O. Laurent, D. B. Sanders et al.: A&A **341**, 667 (1999)
18. D. B. Sanders, I. F. Mirabel: ARAA **34**, 749 (1996)
19. D. B. Sanders, B. T. Soifer, J. H. Elias, et al.: ApJ **325**, 74 (1998)
20. D. B. Sanders, B. T. Soifer, J. H. Elias, G. Neugebauer, K. Matthews: ApJ **328**, L35 (1998)
21. F. Schweizer: In *Structure and Properties of Nearby Galaxies*, ed. EM Berkhuijsen, R Wielebinski, p. 279. Dordrecht:Reidel (1978)
22. F. Schweizer: ApJ **252**, 455 (1982)
23. B. T. Soifer, D. B. Sanders, B. F. Madore, G. Neugebuer, et al.: ApJ **320**, 238 (1987)
24. B. T. Soifer, G. Neugebauer: AJ **101**, 354 (1991)
25. S. A. Stanford, H. A. Bushouse: ApJ **371**, 92 (1991)
26. A. Toomre: In *The Evolution of Galaxies and Stellar Populations*, ed. BM Tinsley, RB Larson, p. 401. New Haven, CT:Yale Univ. Obs. (1977)
27. A. Wilkinson et al.: MNRAS **218**, 297 (1986)
28. G. S. Wright, P. A. James, R. D. Joseph, I. S. McLean: Nature **344**, 417 (1990)
29. F. Zwicky: Ergeb. Exakten Naturwiss. **29**, 34 (1956)

Resolving the Cosmic Infrared Background with ISOCAM

Catherine Cesarsky[2], David Elbaz[1], Herve Aussel[3], Dario Fadda[5], Alberto Franceschini[4], Hector Flores[6], and Pierre Chanial[1]

[1] DAPNIA/Service d'Astrophysique, CEA/Saclay, 91191 Gif-sur-Yvette Cedex, France
[2] European Southern Observatory, Karl-Schwarzschild-Strasse 2, D-85748 Garching bei Muenchen, Germany
[3] Institute For Astronomy, 2680 Woodlawn Drive, Honolulu, Hawaii 96822, USA
[4] Dipartimento di Astronomia, Vicolo Osservatorio 2, I-35122 Padova, Italy
[5] Instituto de Astrofisica de Canarias, Via Lactea, S/N E38200, La Laguna (Tenerife), SPAIN
[6] Observatoire de Paris-Meudon, DAEC, F92195, Meudon Principal CEDEX, FRANCE

The discovery in the COBE all-sky maps of a bright isotropic background in the far-IR/sub-mm, of likely extragalactic origin (CIRB), with a proeminent hump at a wavelength of 140 μm, has been interpreted at first as the integrated emission by dust present in very distant galaxies (Puget et al. 1996; Hauser et al. 1998). What is the relation between the CIRB and the topic of this Conference, Lighthouses of the universe? In the infrared, the lighthouses are the LIGs ($L_{IR} \geq 10^{11}$ $L_\odot$ and the ULIGS ($L_{IR} \geq 10^{12}$ $L_\odot$). Discovered by IRAS, the LIGs and ULIGs appear to be mostly powered by star formation, although at least some also harbor an AGN. In the local universe, LIGs and ULIGs together emit less than 10% of the IR light, and less than 2% of the bolometric light. Is their contribution much more important at large distances? Are they ultimately responsible for the CIRB?

The energy content of the CIRB is very high: it represents at least half and perhaps two thirds of the overall cosmic background. It surpasses the X ray background by a factor of ~ 200 and the optical by a factor of order 2. In the local universe, only 30% of the light of the galaxies is absorbed by dust and reemitted in the infrared. Thus, in the distant universe the infrared radiation must be dominant. But how distant are the sources of the CIRB? To answer this question, it is necessary to resolve the CIRB and identify its sources.

The sensitivity and angular resolution achievable up to date in the FIR, in the region of the maximum of the CIRB distribution, are insufficient to resolve the background. With ISOPHOT, less than 10 % of the CIRB is resolved at 170 μm (Dole et al. 2001), and a preliminary follow up indicates that a large fraction of the detected galaxies are located nearby. With the modern arrays of bolometers, the wavelength region at 450 μm and 850 μm have been investigated (see presentation by M.Longair in these proceedings). At 850 μm, SCUBA has resolved 20% of the background and up to 60% in gravitationally lensed areas. However very few of the SCUBA sources have measured redshifts, given the width of the PSF and the faintness of the galaxies. It is thus not possible to reliably estimate the fraction of the emission at the peak of the CIRB due to the SCUBA galaxies.

Some inferences can be made on the distance of the CIRB sources if the infrared lighthouses of the distant universe are not too dissimilar from the local ones,and if indeed they are powered by starlight converted into IR emission by dust. The spectral energy distribution in the IR of local galaxies peaks above 60 μm and typically around 80 ± 20 μm. As a result, the distant galaxies responsible for the peak of the CIRB should be located at z below ~ 1.3 and present a redshift distribution peaked around z ~ 0.8.

Shortwards of the peak of the CIRB it is necessary to go down to 15 μm to encounter a spectral region where the depth of the surveys and the angular resolution are sufficient to unravel the background. This is because the camera on board of the Infrared Space Observatory (Kessler et al. 1996), ISOCAM (Cesarsky et al 1996), allowed to perform surveys with 1000 times better sensitivity and 60 times better resolution than IRAS. A wide range of surveys were performed with ISOCAM. ~ 1000 galaxies detected in the 15 μm surveys were used to produce number counts (i.e. surface density of galaxies as a function of flux density; see Elbaz et al. 1999). As it turns out, 15 μm is not a bad wavelength for attempting to detect CIRB sources. If they are at z ~ 0.85, the surveys detect radiation which was emitted at ~ 8 μm, just where the MIR emission of galaxies is most intense thanks to the preeminence in their spectrum of the broad bands usually denominated PAHs. Indeed, for the detection of starforming galaxies in the relevent range of redshifts ISOCAM easily surpasses not only IRAS, SCUBA or ISOPHOT but also the deep VLA radio surveys. The steep slope of the 15 μm counts below 1 mJy indicates the presence of an excess of faint sources by one order of magnitude in comparison with predictions assuming no evolution of the 15 μm luminosity function with redshift.

The ISOCAM sources are relatively easy to identify. The redshift distribution of the ISOCAM galaxies indeed peaks at z ~ 0.8, as found from the identifications performed from the ultradeep images in the HDF N field. Using the ISOCAM surveys, it is possible to set a lower limit to the integrated galaxy light at 15 μm of (2.4 ± 0.5) nW m^{-2} sr^{-1}, a result consistent with the upper limit of 5 nW m^{-2} sr^{-1} derived from blazar observations in gamma rays.

Elbaz et al (2002) find that the rest frame 6.75, 12 and 15 μm luminosities of local galaxies are correlated with their bolometric IR luminosities (8-1000 μm). These correlations allow to compute the FIR and the bolometric luminosity of the galaxies in the HDF N. For the sources for which they are available, the values obtained are very well correlated also with radio luminosities, as expected given the well known radio-FIR correlation. It also becomes possible to identify the ULIGs and LIGs contribution.

LIGs and ULIGs produce $\sim 60\%$ of the integrated galactic light at 15 μm; the comoving density of IR luminosity produced by the lighthouses was (70 ± 35) times larger at z ~ 1 than today, while in B or UV it has only decreased by a factor ~ 3.

To estimate the fraction of the CIRB emmited by the ISOCAM sources, it is necessary to evaluate the contribution to the MIR emission of active nuclei, as opposed to star formation. For this purpose, Fadda et al. (2002) analysed the correlation of deep X ray and MIR observation in two of the regions covered by ISOCAM surveys,the HDF N and an area of the Lockman Hole. Deep X ray surveys are available for both regions, the former by XMM-Newton, the latter by Chandra. It is noteworthy that 30% of the

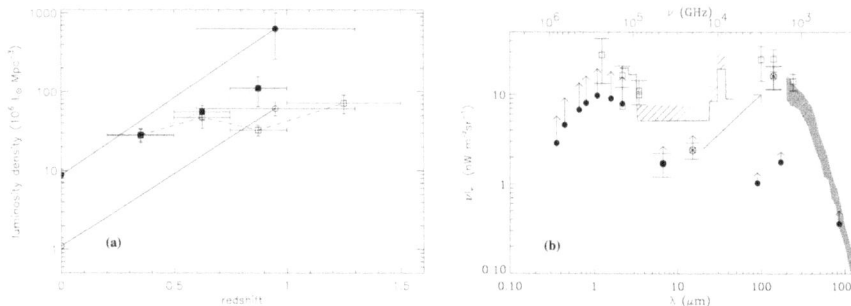

Fig. 1. (from Elbaz et al., 2002)**a)** FIR (filled circles), MIR-15 μm (open circles) and UV-2800Å luminosity density (in $L_\odot$ Mpc^{-3}) as a function of redshift. UV-2800Å: open squares rom Cowie, Songaila & Barger (1999); filled squares from Lilly et al. (1996). **b)** Integrated Galaxy Light (IGL, filled dots) and Extragalactic Background Light (EBL, open squares, grey area) from the UV to sub-millimeter. EBL measurements from COBE: 200-1500 μm EBL from COBE-FIRAS (grey area, Lagache et al. 1999), 1.25, 2.2, 3.5, 100, 140 μm EBL from COBE-DIRBE (open squares). IGL in the U,B,V,I,J,H,K bands from Madau & Pozzetti (2000). The upper end of the arrows indicate the revised values suggested by Bernstein et al. (2001, factor two higher). The estimate of the 15 μm IGL (2.4 $\pm$ 0.5 nW m^{-2} sr^{-1}) is marked with a surrounded star. 6.75 μm (ISOCAM-LW2 filter) IGL from Altieri et al. (1999, filled dot). Hatched upper limit from Mkn 501 (Stanev & Franceschini 1998). The arrow from 15 to 140 μm indicates the computation of the 140 μm IGL due to ISOCAM galaxies.

0.5-2 KeV sources, and up to 63% of the 5-10 KeV sources in the Lockman Hole are seen by ISOCAM, while only $\sim$ 10% of the sources show X ray emission. The X ray sources in this field are essentially interpreted as active nuclei. Such is not the case for the very deep X ray survey in HDF N, capable of detecting also starforming galaxies. Considering also results on the bright large area sample observed by ISOCAM and Beppo SAX in the ELAIS S 1 region, Fadda et al. (2002) arrive at an upper limit for the AGN contribution to the 15 μm background of (17 $\pm$ 2)%.

Assuming that AGNs have a spectrum like NGC 1068, Elbaz et al. (2002) have derived the contribution of ISOCAM galaxies to the peak of the CIRB at 140 μm: 16 $\pm$ 5 nW m^{-2} sr^{-1} (of which $\sim$ 4% is due to AGNs), as compared to the measured value of (25 $\pm$ 7) nW m^{-2} sr^{-1} (figure 1). Thus, the dusty galaxies unveiled by ISOCAM are mostly IR lighthouses, LIGs or ULIGs, and are responsible for the bulk of the IR background.

References

1. Altieri, B., Metcalfe, L., Kneib, J.P., et al. 1999, A&A 343L, 65
2. Aussel, H., Cesarsky, C.J., Elbaz, D., Starck, J.L. 1999, A&A 342, 313
3. Bernstein, R.A., Freedman, W. L., Madore, B.F. 2001, ApJ, accepted (astro-ph/0112193)
4. Cesarsky, C.J., Abergel, A., Agnèse, P., et al. 1996b, A&A 315, L32
5. Cowie, L.L., Songaila, A., Barger, A.J. 1999, ApJ 118, 603
6. Dole, H., Gispert, R., Lagache, G., et al. 2001, A&A 372, 364

7. Elbaz, D., Cesarsky, C.J., Fadda, D., et al. 1999, A&A 351, L37
8. Elbaz, D., Cesarsky, C.J., Chanial, P. et al. 2002, A&A, accepted (astro-ph/0201328)
9. Fadda, D., Flores, H., Hasinger, G., et al. 2001, A&A, accepted (astro-ph/0111412)
10. Hauser, M.G., Arendt, R.G., Kelsall, T., et al. 1998, ApJ 508, 25
11. Kessler, M., Steinz, J., Anderegg, M., et al. 1996, A&A 315, L27
12. Lagache, G., Abergel, A., Boulanger, F., Desert, F.-X., Puget, J.-L. 1999, A&A 344, 322
13. Lilly, S.J., Lefévre, O., Hammer, F., Crampton, D. 1996, ApJ 460, L1
14. Madau, P., Pozzetti, L. 2000, MNRAS 312, L9
15. Puget, J.L., Abergel, A., Bernard, J.P., et al. 1996, A&A 308, L5
16. Stanev, T., Franceschini, A. 1998, ApJ 494, L159

The Environment of Optically Very Luminous Galaxies

Alberto Cappi[1], Christophe Benoist[2], Luiz N. da Costa[3], and Sophie Maurogordato[4]

[1] Osservatorio Astronomico di Bologna, via Ranzani 1, I-40127, Bologna, Italy
[2] Observatoire de la Côte d'Azur, B4229, Le Mont–Gros, F–06304 Nice, France
[3] ESO, Karl-Schwarzschild-Strasse 2, D–85748, Garching bei München, Germany
[4] CERGA, CNRS, Observatoire de la Côte d'Azur, B4229, Le Mont–Gros, F–06304, Nice Cedex 4, France

Abstract. We discuss the main properties of Very Luminous Galaxies (VLGs, $M_B \leq -21$) and present a preliminary analysis of galaxy systems associated to them.

1 Properties of VLGs and Their Systems

The amplitude of the correlation function increases significantly as a function of galaxy luminosity, but only when $L > L^*$ (Benoist et al. 1996; Norberg et al. 2001): this bias is non linear (Benoist et al. 1999). The VLGs have the largest correlation amplitude, but most of them are not in clusters, and many are late type galaxies (Cappi et al. 1998; Giuricin et al. 2001). Visual inspection of DSS images has shown the presence of fainter companions and/or evidence of interaction around many VLGs (Cappi et al. 1998), but the reality of these associations requires spectroscopic confirmation. Understanding the nature of VLGs and of their associated systems is important from the point of view of galaxy formation and evolution, raising a number of interesting questions, concerning the mass and extent of the associated halos, the role of merging (spirals could undergo only minor episodes of merging), and the evolution of VLG systems. Notice that the groups including one or two VLGs and much fainter members are systems comparable to the Local Group (M31 is in fact a VLG), and they would not have been included in present group catalogs. For these reasons we are carrying out observations and collecting public data from recent surveys, such as the 2dF Galaxy Redshift Survey. Within the maximum distance of our SSRS2 VLGs volume–limited sample (~ 0.065), the 2dFGRS is volume–limited at $M = -17$. Therefore from the first public data of the 2dFGRS we can obtain redshifts of faint galaxies around some SSRS2 VLGs. In table 1 we present preliminary data for 4 systems associated to VLGs 61, 45, 48, 53: the first was observed at OHP, while for the others we retrieved the 2dF redshifts of all galaxies within a projected radius of 1.5 h^{-1} Mpc and within 1500 km/s from the VLG, and applied a 3–σ clipping algorithm to select group members.

Apparently the velocity dispersion of these systems is comparable, in the range 300–500 km/s (but we have found some hint of a dependence on the VLG morphological type). Late type dominated groups do not have significant X–ray emission (see the analysis of 12 poor groups by Zabludoff & Mulchaey 1998). Can a group dominated by one or two late–type VLGs evolve into an early–type dominated group through merging? The analysis we are carrying out (Cappi et al. 2002) will enable us to define the properties of VLG systems and possibly to answer this and other questions directly related to galaxy formation and evolution.

Fig. 1. VLG061.

Table 1. Data for a few VLG systems.

VLG Id.	Type	N_g	System redshift	Velocity dispersion
VLG045	S	20	0.0574 ± 0.0004	$465\,^{+100}_{-63}$
VLG048	Sbc	16	0.0585 ± 0.0004	$419\,^{+105}_{-63}$
VLG053	S0	32	0.0575 ± 0.0002	$356\,^{+57}_{-41}$
VLG061	SB(rs)c	7	0.0193 ± 0.0004	318^{+177}_{-67}
VLG083	E	31	0.0580 ± 0.0002	$319\,^{+53}_{-38}$

References

1. C. Benoist, A. Cappi, L.N. da Costa, S. Maurogordato, F. Bouchet, R. Schaeffer: ApJ **514**, 563 (1999)
2. C. Benoist, S. Maurogordato, L.N. da Costa, A. Cappi, R. Schaeffer: ApJ **472**, 452 (1996)
3. A. Cappi A., L.N. da Costa, C. Benoist, S. Maurogordato, P.S. Pellegrini: AJ **115**, 2250 (1998)
4. G. Giuricin, S. Samurovic, M. Girardi, M. Mezzetti, C. Marinoni: ApJ **554**, 857 (2001)
5. P. Norberg et al.: MNRAS, in press (astro-ph/0105500) (2001)
6. A.I. Zabludoff, J.S. Mulchaey: ApJ **496**, 39 (1998)

What Triggers Radio Galaxies?

Maria José Cruz and Katherine M. Blundell

Oxford University Astrophysics, Keble Road, Oxford, OX1 3RK, UK

Abstract. We are investigating possible triggering scenarios which may give rise to powerful jetted active galaxies such as classical double radio galaxies. Important clues come from studying those objects which have most recently been triggered. The newly discovered 'Youth-Redshift Degeneracy' for classical double radio galaxies means that the highest redshift objects in a flux-limited survey will be observed to be significantly younger (more recently triggered) than the more nearby objects. We are pursuing this investigation with a new sample specially filtered to favour the detection of high-redshift radio galaxies. We will present some preliminary results of a near-IR imaging campaign using the UFTI on the UKIRT.

1 Motivation and target objects

Radio galaxies can be extremely luminous at radio wavelengths (for example $\gtrsim 10^{28}\,\mathrm{WHz^{-1}sr^{-1}}$), and so they are detectable out to very high redshifts ($z \sim 4$). Classical double radio galaxies exhibit radio structures consisting of two lobes of emission which can extend over hundreds of kiloparsecs and which lie either side of their host galaxies. The plasma responsible for the lobe radiation arrives in the lobes only after being transported along the oppositely directed jets which emanate from the central black hole within the host galaxy [2].

The luminosity of the plasma at a particular radio frequency will not remain constant throughout the life of the radio galaxy. A number of lines of argument suggest that the luminosity decreases as the source gets older. One such example is the observation [1] that there are no sources which have very large linear sizes which are also highly luminous: the only highly luminous sources have short linear sizes suggesting that as sources age and grow in size, their luminosity must decrease. Consideration of the energetics of evolving radio sources [4] strongly supports the view that sources are at their most luminous when they are youngest.

Since for any flux limited survey the more distant sources selected will have to be the more intrinsically luminous, observed high-redshift radio galaxies will be inevitably *younger* than their nearby counterparts — this is called the "Youth-Redshift Degeneracy" [3]. Very high redshift sources, being inevitably young, are seen only a short time ($< 10^7$ years) after their jet-triggering event. This is much shorter than galaxy dynamical timescales. Thus the study of such young systems as these high redshift radio galaxies may give useful clues about the jet-triggering process. An important benefit of studying the host galaxies of radio galaxies is that observations are not affected by bright quasar nuclei.

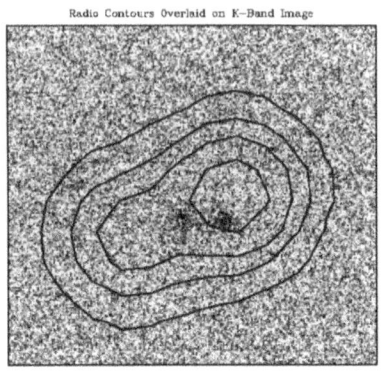

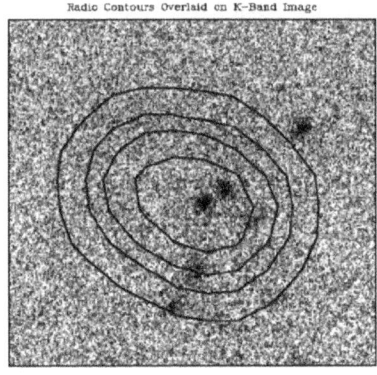

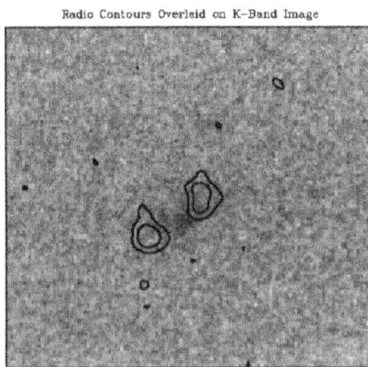

Fig. 1. Radio contours at 1.4 GHz (top images) and at 8.4 GHz (bottom image) are overlaid on infra-red K-Band images from the UFTI instrument on UKIRT. All these radio galaxies are at $z > 2$ and show evidence of multiple components of host galaxies.

2   Present Work

It has been suggested that powerful radio galaxies are triggered by galaxy-galaxy mergers [5]. We wish to test this hypothesis by examining very recently triggered (young) and luminous radio galaxies: as described above, high-redshift classical double radio galaxies are such objects.

We are using a new sample of radio galaxies specially filtered to preferentially select high redshift members. This sample, called 6C**, is a sample selected to be brighter than 1 Jy at 151 MHz and then filtered with criteria chosen to optimize the chances of finding radio galaxies at high redshifts. These criteria are based on two well understood characteristics of very distant radio galaxies: small angular size and steep radio spectral index. The sample, 6C**, excludes objects larger than 12" or whose spectral indices between 151 MHz and 1.4 GHz are flatter than 1.

Observations done in recent years with the VLA, the WHT and the UKIRT have resulted in a nearly complete dataset of high resolution radio maps, redshifts and near-infrared identifications respectively. From our preliminary analysis of a near-IR campaign of the 6C** sample using UFTI on the UKIRT we are looking for evidence of disturbed morphologies along with cases in which more than one host galaxy is identified with the radio galaxy. Some examples of this are shown in Figure 1.

3   Future Work

In the near future we will be doing a careful statistical analysis of the frequency of the disturbed and non-disturbed morphology occurrence in our sample of high redshift radio galaxies. In the long run we plan to take a more kinematical approach to this investigation and not just merely a morphological one. We wish to study the relative dynamics of the identified pairs of galaxies via spectroscopy.

Acknowledgments

MJC would like to thank Corpus Christi College, Oxford, for providing travel funds and would also like to acknowledge the financial support from Fundação para a Ciência e a Tecnologia (Portugal) under the grant SFRH/BD/1025/2000.

References

1. J.E. Baldwin: 'Evolutionary Tracks of Extended Radio Sources'. In: *Extragalactic Radio Sources*, ed. by D.S. Heeschen, C.M. Wade (Reidel, Dordrecht 1982) pp. 21-24
2. M.C. Begelman, R.D. Blandford, M.J. Rees: Rev. Mod. Phys. **56**, 280 (1984)
3. K.M. Blundell, S. Rawlings: Nature **399**, 330 (1999)
4. K.M. Blundell, S. Rawlings, C.J. Willott: AJ **117**, 677 (1999)
5. D.B. Sanders, B.T. Soifer, J.H. Elias, B.F. Madore, K. Matthews, G. Neugebauer, N.Z. Scoville: ApJ **325**, 74 (1988)

Evolution and Clustering of Giant Elliptical Galaxies

Steven Phillipps[1] and Nathan Horleston[1]

Department of Physics, University of Bristol, Bristol, UK.

Abstract. Number-magnitude counts of morphologically selected elliptical galaxies can place constraints on galaxy evolution and cosmological models. Similarly, the observed clustering of faint galaxies is dependent on the development of large scale structure in the universe (e.g. clustering evolution in a given cosmological model) and on the evolution of the galaxies. We are investigating whether mutually consistent results can be obtained from these separate considerations.

1 Number Counts

Counts of galaxies as a function of apparent magnitude are well known to be dependent both on the cosmological model (specifically the density parameter Ω_0 and the contribution to the curvature from the cosmological constant, Ω_Λ) and on the evolution of the type of galaxy being considered. Even in the (arguably simplest) case of giant elliptical galaxies, where we are moderately confident in our ability to account for the evolution of the stellar content, there is still strong degeneracy between the cosmological parameters and the amount of evolution via galaxy mergers (see [1] and [2] for discussion of counts in the Hubble Deep Fields).

However in the standard CDM hierarchical clustering scheme of structure formation, the amount of merging itself (at least, that of the galaxies' dark halos) is also dependent on the cosmological model parameters (e.g. [3], [4]). One might therefore hope that there exists a self consistent picture.

2 Clustering

Further, these same models will predict the level of clustering for any particular cosmology and this can provide a second observable for the consistency check. Linking the mass distribution to that of the galaxies obviously introduces a further complication, but we can use the most recent semi-analytic models of galaxy formation within CDM halos to make this step (see e.g. [5] and G. Kauffmann, this meeting). The evolution of clustering then becomes dependent on the type of galaxy being studied and is not simply related to the overall galaxy correlation function, not least because of the evolution of the galaxies within their halos (e.g., S. White, this meeting).

However, if we tailor our approach suitably, we can obtain measures of clustering as a function of redshift which we may be able to compare with the analytic predictions. For instance, we may wish to study the evolution of those galaxies with luminosities (representing masses) above a certain, possibly evolving, value. Alternatively we may wish to consider the N most luminous galaxies per fixed comoving volume at different redshifts.

This is relatively straightforward to do if we adopt the approach of Phillipps & Shanks ([6], [7]) which utilises a set of 'centre' galaxies with known redshifts. The (excess over random) number of galaxies seen in projection around such a galaxy (and brighter than a selectable absolute magnitude at the redshift of the centre galaxy) can be related to the galaxy correlation function

3 Theory

Consider the expected number of galaxies seen in a volume element δV a distance r from the observer O and subtending a solid angle $\delta\Omega$ at O (i.e. $\delta V = r^2\delta r\delta\Omega$ in the Euclidean case; it is straightforward to incorporate the real cosmological volume elements). If δV is at a distance y from a given galaxy G_1 whose distance r_1 is known, n is the mean number density of galaxies and ϕ is the selection function (effectively the fraction of galaxies bright enough to be detectable at a distance r), then by the definition of the spatial correlation function ξ,

$$\delta^2 n = n[1 + \xi(y)]\phi(r)r^2\delta r\delta\Omega \tag{1}$$

The expected number in $\delta\Omega$ is then simply the integral over all r,

$$\delta n = n\delta\Omega \int_0^\infty [1 + \xi(y)]\phi(r)r^2\,dr \tag{2}$$

At this point it would be possible to define a correlation function $e_1(\theta)$, say, in terms of the excess probability of finding a galaxy in $\delta\Omega$ an angle θ from G. However, in order to obtain ξ from e_1 we would need to know $\phi(r)$ for the galaxies of unknown distance – in general this could be entirely different from that of the known-redshift galaxies.

In the case where we have magnitude information for our field galaxies, however, a much more amenable statistic can be obtained. Consider the expected number of galaxies per unit solid angle a distance θ from G_1

$$E_1(\theta) = N + n \int \xi(y)\phi(r)r^2\,dr \tag{3}$$

where N is the mean surface density. Since the distance to G_1 is known we can find the number per unit area at G as a function of projected comoving separation, $s = r_1\theta$,

$$\Sigma'(s) = \frac{N}{r_1^2} + \frac{n}{r_1^2} \int \xi(y)\phi(r)r^2\,dr \tag{4}$$

In the case of a power law correlation function

$$\xi = By^{-\gamma} \tag{5}$$

this can be seen to reduce to

$$\Sigma'(s) = Nr_1^{-2} + n\phi(r_1)BGs^{1-\gamma} \tag{6}$$

where the contant $G = \sqrt{\pi}\Gamma(\gamma-1/2)/\Gamma(\gamma/2)$ and we have made the usual assumption that ξ is significantly non-zero only on scales much less than the depth of the survey, here r_1.

If we now note that $n\phi(r_1)$ is the number density of galaxies brighter than some limit, it is apparent that a useful move at this point is to take the same limiting absolute magnitude at r_1, M, for the counts around each centre galaxy G_i. In this case, taking the mean over many centre galaxies,

$$\Sigma(s) = \langle N(r_i)r_1^{-2}\rangle + \Phi(M)BGs^{1-\gamma} \tag{7}$$

where Φ is the galaxies' integral luminosity function. Note that N now depends on r_i as the limiting apparent magnitude for the counts will vary as $m_i = M + 25 + 5\log r_i$ (this is ignoring cosmological and k-corrections which have to be included in practice but this is straightforward). The first term on the rhs of the above equation can either be evaluated from the observed galaxy counts as a function of m_i and the known values of r_i, or simply obtained as the limiting value of Σ at large s, in which case it is appropriate to write

$$\Sigma(s) = \Sigma_\infty + \Phi(M)BGs^{1-\gamma} \tag{8}$$

4 Conclusion

As can be seen, from the above form it becomes relatively straightforward to calculate $\Sigma(s)$ on a galaxy by galaxy basis. Thus each 'centre' galaxy gives an estimate of the amplitude of $\xi(y, z)$, call it $B(z)$, at its particular redshift z. Note that the latter need only be modestly accurate (since we use no depth information in calculating separations), so photometric redshifts should be adequate. (And, of course, we do not need to know the individual redshifts of the projected neighbours at all, though we could utilise them to cut down on the poisson noise in the number of unrelated galaxies). Similarly, we do not need to account for redshift space distortion. Finally, it is simple to recalculate the statistics for any given cosmological model or limiting magnitude M (which may be chosen to evolve with z if required). Results of our first application of this technique – again to galaxies in the Hubble Deep Fields – will be presented elsewhere.

References

1. S.P. Driver, R.A. Windhorst, S. Phillipps, P.D. Bristow: ApJ, **461**, 525 (1996)
2. S. Phillipps, S.P. Driver et al.: MNRAS, **319**, 807 (2000)
3. C. Baugh, S. Cole, C.S. Frenk: MNRAS, **283**, 1361 (1996)
4. G. Kauffmann, S. Charlot: MNRAS, **297**, L23 (1998)
5. G. Kauffmann, J.M. Colberg, A. Diaferio, S.D.M. White: MNRAS, **307**, 529 (1999)
6. S. Phillipps, T. Shanks: MNRAS, **227**, 115 (1987)
7. S. Phillipps, T. Shanks: MNRAS, **229**, 621 (1987)

Bier und Brez'n session

Part III

Gamma-Ray Bursts as the Lighthouses

Novel Ways to Probe the Universe with Gamma-Ray Bursts and Quasars

Abraham Loeb

Astronomy Department, Harvard University
Cambridge, MA 02138, USA

Abstract. I consider novel ways by which Gamma-Ray Bursts (GRBs) and quasars can be used to probe the universe. Clues about how and when was the intergalactic medium ionized can be read off the UV emission spectrum of GRB explosions from the first generation of stars. The existence of intergalactic and galactic stars can be inferred from their gravitational microlensing effect on GRB afterglows. Prior to reionization, quasars should be surrounded by a halo of scattered Lyα radiation which probes the neutral intergalactic medium (IGM) around them. The situation is analogous to the appearance of a halo of scattered light around a street lamp which is embedded in a dense fog. Outflows from quasars magnetize the IGM at all redshifts. As a result, the shocks produced by converging flows during the formation of large scale structure in the IGM, accelerate electrons to relativistic energies and become visible in the radio regime through their synchrotron emission and in the γ–ray regime through their inverse–Compton scattering of the microwave background photons. During transient episodes of strong mergers, X-ray clusters should therefore appear as extended radio or γ–ray sources on the sky.

1 Novel Cosmological Studies with Gamma-Ray Bursts

1.1 Preface

Since their discovery four decades ago, quasars have been used as powerful lighthouses which probe the intervening universe out to high redshifts, $z \sim 6$ [42,20]. The spectra of almost all quasars show strong emission lines of metals, indicating super-solar enrichment of the emitting gas [40]. This implies that at least in the cores of galaxies, formation of massive stars and their evolution to supernovae preceded the observed quasar activity. If Gamma-Ray Bursts (GRBs) originate from the remnants of massive stars (such as neutron stars or black holes), as seems likely based on recent estimates of their energy output [89,24,23], then they should exist at least out to the same redshift as quasars. Although GRBs are transient events, their peak optical-UV flux can be as bright as that of quasars. Hence, GRBs promise to be as useful as quasars in probing the high–redshift universe.

Not much is known observationally about the universe in the redshift interval $z = 6$–30, when the first generation of galaxies condensed out of the primordial gas left over from the Big Bang (see reviews [3] and [56]). Observations of the microwave background anisotropies indicate that the cosmic gas became neutral at $z \sim 1000$ and remained so at least down to $z \sim 30$ (see, e.g. [87]). On the other hand, the existence of transmitted flux shortward of the Lyα resonance in the spectrum of the highest-redshift quasars and galaxies (see, for example, Figure 1), indicates that the intergalactic medium was reionized to a level better than 99.9999% by a redshift $z \sim 6$. This follows from the fact that the Lyα optical depth of the intergalactic medium at high-redshifts ($z \gg 1$) is [38],

$$\tau_\alpha = \frac{\pi e^2 f_\alpha \lambda_\alpha n_{HI}(z)}{m_e c H(z)} \approx 6.15 \times 10^5 x_{HI} \left(\frac{\Omega_b h}{0.03}\right) \left(\frac{\Omega_m}{0.3}\right)^{-1/2} \left(\frac{1+z}{10}\right)^{3/2}, \quad (1)$$

where $H \approx 100h$ km s^{-1} Mpc$^{-1} \Omega_m^{1/2}(1+z)^{3/2}$ is the Hubble parameter at the source redshift z, $f_\alpha = 0.4162$ and $\lambda_\alpha = 1216$Å are the oscillator strength and the wavelength of the Lyα transition; $n_{HI}(z)$ is the average intergalactic density of neutral hydrogen at the source redshift (assuming primordial abundances); Ω_m and Ω_b are the present-day density parameters of all matter and of baryons, respectively; and x_{HI} is the average fraction of neutral hydrogen. Modeling [20] of the *transmitted flux* in Figure 1 implies $\tau_\alpha < 0.5$ or $x_{HI} < 10^{-6}$, i.e., most of the low-density gas throughout the universe is ionized to a level of 99.999% at $z < 6$. However, there are some dark intervals in the spectrum which could be indicative of regions with a higher neutral fraction [14]. In fact, Becker et al. [4] reported the detection of an extended (> 300Å long) dark interval just shortward of the Lyα emission line in the spectrum of the newly discovered quasar SDSS 1030+0524 at $z = 6.28$. The suppression of the flux by a factor > 150 may indicate the first detection of the Gunn-Peterson trough, although caution is warranted since the inferred optical depth $\tau_\alpha > 5$ can be produced by a neutral fraction as small as $X_{HI} \sim 2 \times 10^{-4}$ according to equation (1).

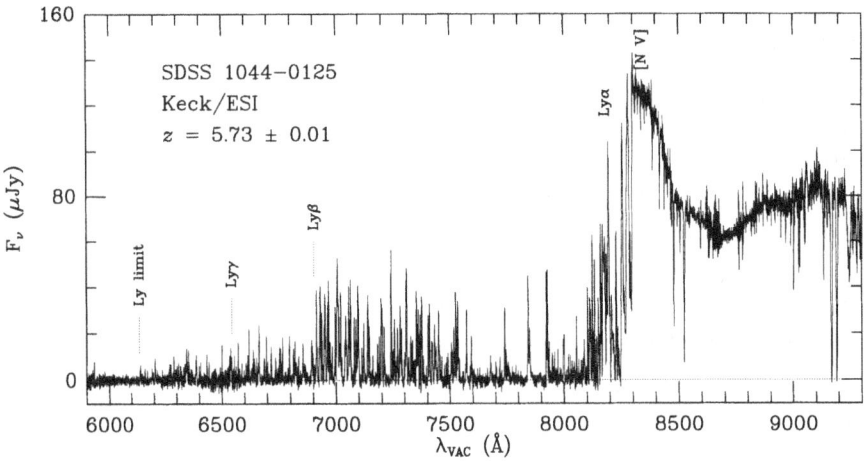

Fig. 1. Spectrum of the quasar SDSS1044-0125 at $z = 5.73$ (Djorgovski et al. 2001), originally discovered by the Sloan Digital Sky Survey (Fan et al. 2000).

The question: *how and when was the universe reionized?* defines a new frontier in observational cosmology [56]. The UV spectrum of GRB afterglows can be used to probe the ionization and thermal state of the intergalactic gas during the epoch of reionization, at redshifts $z \sim 7$–10 [64]. The stretching of the temporal evolution of GRB lightcurves by the cosmological redshift factor $(1 + z)$, makes it easier for an observer to react in time and measure a spectrum of their optical-UV emission at its peak.

Energy arguments suggest that reionization resulted from photoionization and not from collisional ionization [81,28]. The corresponding sources of UV photons were either stars or quasars. Recent simulations of the first generation of stars that formed out of the primordial metal–free gas indicate that these stars were likely massive [9,1]. If GRBs result from compact stellar remnants, such as black holes or neutron stars, then the fraction of all stars that lead to GRBs may have been higher at early cosmic times. This, however, is true only if the GRB phenomena is triggered on a time scale much shorter than the age of the universe at the corresponding redshift, which for $z \gg 1$ is $\sim 5.4 \times 10^8$ yr $(h/0.7)^{-1}(\Omega_m/0.3)^{-1/2}[(1 + z)/10]^{-3/2}$.

1.2 Properties of High-Redshift GRB Afterglows

Young (days to weeks old) GRBs outshine their host galaxies in the optical regime. In the standard hierarchical model of galaxy formation, the characteristic mass and hence optical luminosity of galaxies and quasars declines with increasing redshift [39,79,3]. Hence, GRBs should become easier to observe than galaxies or quasars at increasing redshift. Similarly to quasars, GRB afterglows possess broad-band spectra which extend into the rest-frame UV and can probe the ionization state and metallicity of the IGM out to the epoch when it was reionized at redshifts $z \sim 7$–10 [56]. Lamb & Reichart [54] have extrapolated the observed γ-ray and afterglow spectra of known GRBs to high redshifts and emphasized the important role that their detection might play in probing the IGM. Simple scaling of the long-wavelength spectra and temporal evolution of afterglows with redshift implies that at a fixed time lag after the GRB trigger in the observer's frame, there is only a mild change in the *observed* flux at infrared or radio wavelengths as the GRB redshift increases. Ciardi & Loeb [12] demonstrated this behavior using a detailed extrapolation of the GRB fireball solution into the non-relativistic regime (see the 2μm curves in Figure 2). Despite the strong increase of the luminosity distance with redshift, the observed flux for a given observed age is almost independent of redshift in part because of the special spectrum of GRB afterglows (see Figure 4), but mainly because afterglows are brighter at earlier times and a given observed time refers to an earlier intrinsic time in the source frame as the source redshift increases. The mild dependence of the long-wavelength ($\lambda_{\mathrm{obs}} > 1\mu$m) flux on redshift stands in contrast to other high-redshift sources such as galaxies or quasars, which fade rapidly with increasing redshift [39,79,3]. Hence, GRBs provide exceptional lighthouses for probing the universe at $z = 6$–30, at the epoch when the first stars had formed.

Assuming that the GRB rate is proportional to the star formation rate and that the characteristic energy output of GRBs is $\sim 10^{52}$ ergs, Ciardi & Loeb [12] predicted that there are always ~ 15 GRBs from redshifts $z > 5$ across the sky which are brighter than ~ 100 nJy at an observed wavelength of $\sim 2\mu$m. The infrared spectrum of these sources could be taken with future telescopes such as the *Next Generation Space Telescope* (planned for launch in 2009; see http://ngst.gsfc.nasa.gov/), as a follow-up on

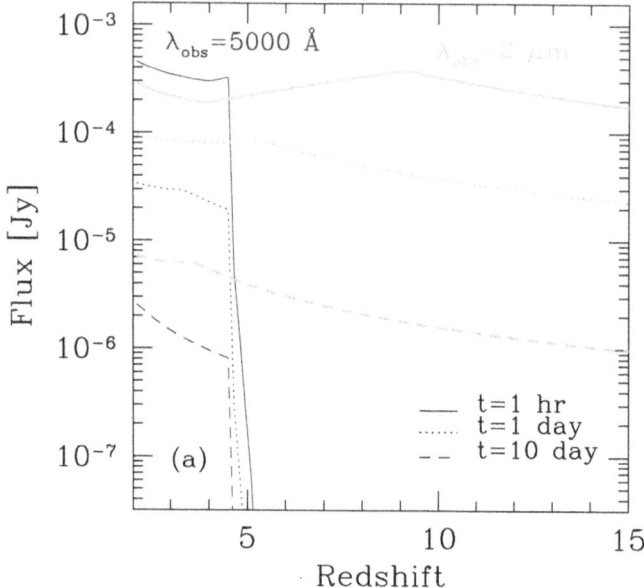

Fig. 2. Theoretical expectation for the observed afterglow flux of a GRB as a function of its redshift (from Ciardi & Loeb 2000). The curves refer to an observed wavelength of 5000 Å(thin lines) and 2μm (thick lines). Different line types refer to different observed times after the GRB trigger, namely 1 hour (solid line), 1 day (dotted) and 10 days (dashed). The 5000Å flux is strongly absorbed at $z > 4.5$ by intergalactic hydrogen. However, at infrared and radio wavelengths the observed afterglow flux shows only a mild dependence on the source redshift.

their early X-ray localization with the *Swift* satellite (planned for launch in 2003; see http://swift.sonoma.edu/).

The redshifts of GRB afterglows can be estimated photometrically from either the Lyman limit or Lyα troughs in their spectra. At low redshifts, the question of whether the Lyman limit or Lyα trough interpretation applies depends on the absorption properties of the host galaxy. If the GRB originates from within the disk of a star–forming galaxy, then the afterglow spectrum will likely show a damped Lyα trough. At $z > 6$ the Lyα trough would inevitably exist since the intergalactic Lyα opacity is $> 90\%$ (see Figure 13 in [79]). Interestingly, an absorption feature in the afterglow spectrum which is due to the neutral hydrogen within a molecular cloud or the disk of the host galaxy, is likely to be time-dependent due to the ionization caused by the UV illumination of the afterglow itself along the line-of-sight [69].

So far, there have been two claims for high-redshift GRBs. Fruchter [25] argued that the photometry of GRB 980329 is consistent with a Lyα trough due to IGM absorption at $z \sim 5$. Anderson et al. [2] inferred a redshift of $z = 4.5$ for GRB 000131 based on a crude optical spectrum that was taken by the VLT a few days after the GRB trigger. *These cases emphasize the need for a coordinated observing program that will alert 10-meter class telescopes to take a spectrum of an afterglow about a day after the GRB trigger, based on a photometric assessment (obtained with a smaller telecope using the Lyman limit or Lyα troughs) that the GRB may have originated at a high redshift.*

In the following two subsections, I illustrate the usefulness of GRB afterglows for cosmological studies through two examples.

DETERMINING THE REIONIZATION REDSHIFT

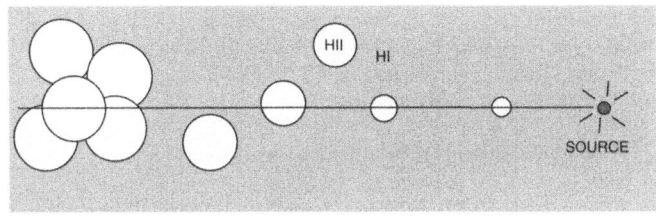

Spectrum

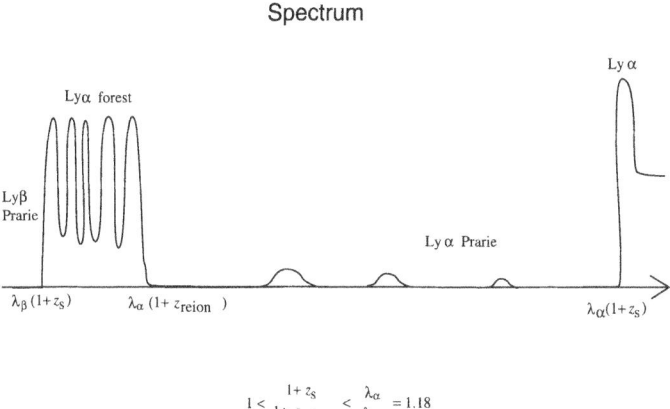

$$1 < \frac{1+z_s}{1+z_{reion}} < \frac{\lambda_\alpha}{\lambda_\beta} = 1.18$$

Fig. 3. Sketch of the expected spectrum of a source at a redshift z_s slightly above the reionization redshift z_{reion}. The transmitted flux due to HII bubbles in the pre-reionization era, and the Lyα forest in the post-reionization era, are exaggerated for illustration.

1.3   Probing the Reionization Epoch and Metallicity History of the IGM

The UV spectra of GRB afterglows can be used to measure the evolution of the neutral intergalactic hydrogen with redshift. Figure 3 illustrates schematically the expected absorption just beyond the reionization redshift. Resonant scattering suppresses the spectrum at all wavelengths corresponding to the Lyα resonance prior to reionization. Since the Lyα cross-section is very large, any transmitted flux prior to reionization reflects a large volume of ionized hydrogen along the line-of-sight. If the GRB is located at a redshift larger by $> 18\%$ than the reionization redshift, then the Lyα and the Lyβ troughs will overlap. Unlike quasars, GRBs do not ionize the IGM around them; their limited energy supply $\sim 10^{52}$ ergs [89,24,23] can ionize only $\sim 4 \times 10^5 M_\odot$ of neutral hydrogen within their host galaxy.

Quasar spectra indicate the existence of an IGM metallicity which is a fraction of a percent of the solar value [17]. The metals were likely dispersed into the IGM through outflows from galaxies, driven by either supernova or quasar winds [3,28]. Detection of metal absorption lines in the spectrum of GRB afterglows, produced either in the IGM or the host galaxy of the GRB, can help unravel the evolution of the IGM metallicity with redshift and its link to the evolution of galaxies. Detection of X-ray absorption by intergalactic metals can be used to establish the existence of the warm component of the IGM which has not been observed so far [22,43,69].

1.4 Cosmological Microlensing of Gamma-Ray Bursts

Loeb & Perna [57] noted the coincidence between the angular size of a solar-mass lens at a cosmological distance and the micro-arcsecond size of the image of a GRB afterglow. They therefore suggested that microlensing by stars can be used to resolve the photospheres of GRB fireballs at cosmological distances. (Alternative methods, such as radio scintillations, only provide a constraint on the radio afterglow image size [34,89] but do not reveal its detailed surface brightness distribution, because of uncertainties in the scattering properties of the Galactic interstellar medium.)

The fireball of a GRB afterglow is predicted to appear on the sky as a ring (in the optical band) or a disk (at low radio frequencies) which expands laterally at a superluminal speed, $\sim \Gamma c$, where $\Gamma \gg 1$ is the Lorentz factor of the relativistic blast wave which emits the afterglow radiation [88,76,68,36]. For a spherical explosion into a constant density medium (such as the interstellar medium), the physical radius of the afterglow image is of order the fireball radius over Γ, or more precisely [36]

$$R_s = 3.9 \times 10^{16} \left(\frac{E_{52}}{n_1} \right)^{1/8} \left(\frac{t_{\text{days}}}{1+z} \right)^{5/8} \text{ cm,} \qquad (2)$$

where E_{52} is the hydrodynamic energy output of the GRB explosion in units of 10^{52} ergs, n_1 is the ambient medium density in units of 1 cm^{-3}, and t_{days} is the observed time in days. At a cosmological redshift z, this radius of the GRB image occupies an angle $\theta_s = R_s/D_A$, where $D_A(z)$ is the angular diameter distance at the GRB redshift, z. For the typical cosmological distance, $D_A \sim 10^{28}$ cm, the angular size is of order a micro-arcsecond (μas). Coincidentally, this image size is comparable to the Einstein angle of a solar mass lens at a cosmological distance,

$$\theta_E = \left(\frac{4GM_{\text{lens}}}{c^2 D} \right)^{1/2} = 1.6 \left(\frac{M_{\text{lens}}}{1 M_\odot} \right)^{1/2} \left(\frac{D}{10^{28} \text{ cm}} \right)^{-1/2} \mu\text{as,} \qquad (3)$$

where M_{lens} is the lens mass, and $D \equiv D_{\text{os}} D_{\text{ol}}/D_{\text{ls}}$ is the ratio of the angular-diameter distances between the observer and the source, the observer and the lens, and the lens and the source [77]. Loeb & Perna [57] showed that because the ring expands laterally faster than the speed of light, the duration of the microlensing event is only a few days

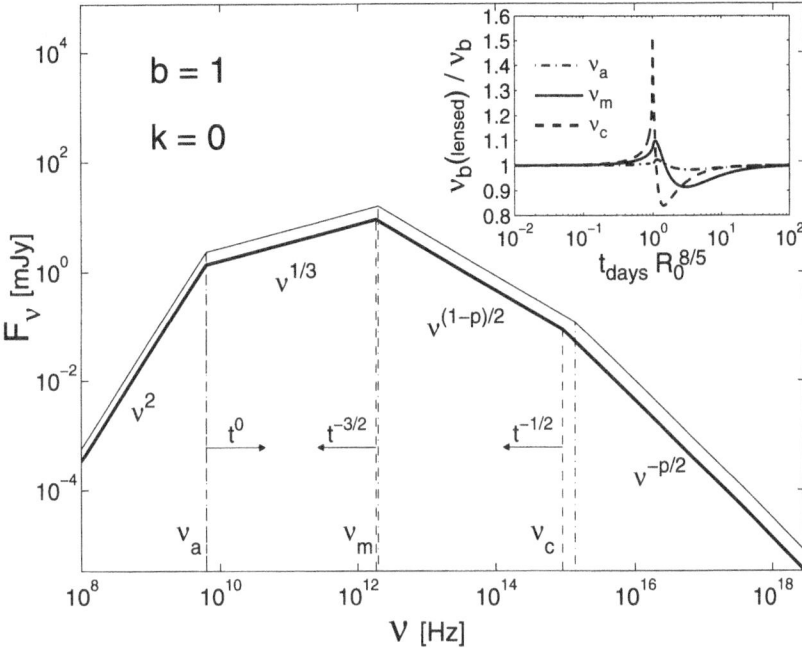

Fig. 4. A typical broken power-law spectrum of a GRB afterglow at a redshift $z = 1$ (from Granot & Loeb 2001). The observed flux density, F_ν, as a function of frequency, ν, is shown by the boldface solid line at an observed time $t_{\mathrm{days}} = 1$ for an explosion with a total energy output of 10^{52} ergs in a uniform interstellar medium ($k = 0$) with a hydrogen density of 1 cm^{-3}, and post-shock energy fractions in accelerated electrons and magnetic field of $\epsilon_e = 0.1$ and $\epsilon_B = 0.03$, respectively. The thin solid line shows the same spectrum when it is microlensed by an intervening star with an impact parameter equal to the Einstein angle and $R_0 \equiv [\theta_s(1 \text{ day})/\theta_E] = 1$. The insert shows the excess evolution of the break frequencies $\nu_b = \nu_a$, ν_m and ν_c (normalized by their unlensed values) due to microlensing.

rather than tens of years, as is the case for more typical astrophysical sources which move at a few hundred km s^{-1} or $\sim 10^{-3}c$.

The microlensing lightcurve goes through three phases: (i) constant magnification at early times, when the source is much smaller than the source-lens angular separation; (ii) peak magnification when the ring-like image of the GRB first intersects the lens center on the sky; and (iii) fading magnification as the source expands to larger radii.

Granot & Loeb [35] calculated the radial surface brightness profile (SBP) of the image of a Gamma-Ray-Burst (GRB) afterglow as a function of frequency and ambient medium properties, and inferred the corresponding magnification lightcurves due to microlensing by an intervening star. The afterglow spectrum consists of several power-law segments separated by breaks, as illustrated by Figure 4. The image profile changes considerably across each of the spectral breaks, as shown in Figure 5. It also depends on the power–law index, k, of the radial density profile of the ambient medium into which the GRB fireball propagates. Gaudi & Loeb [31] have shown that intensive monitoring

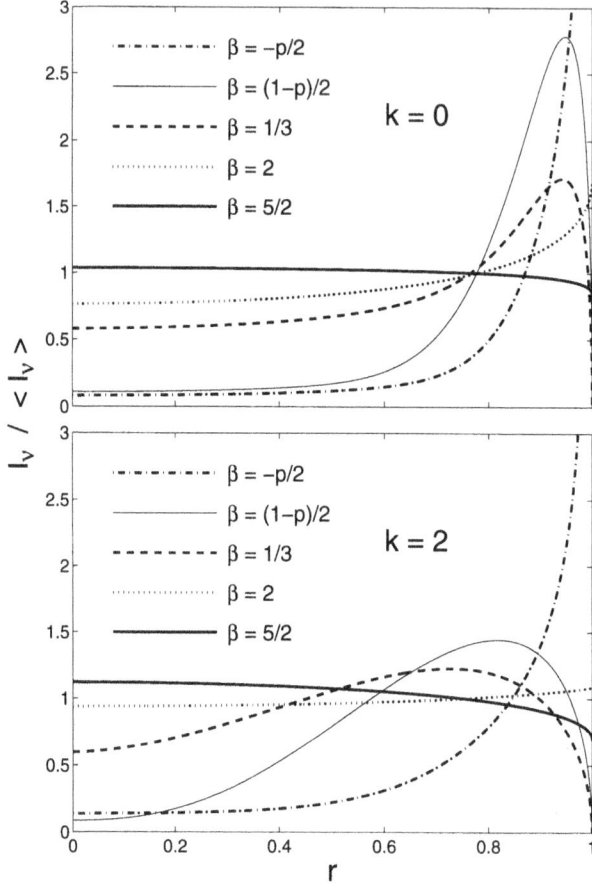

Fig. 5. The surface brightness, normalized by its average value, as a function of the normalized radius, r, from the center of a GRB afterglow image (where $r = 0$ at the center and $r = 1$ at the outer edge). The image profile changes considerably between different power-law segments of the afterglow spectrum, $F_\nu \propto \nu^\beta$ (see Figure 4). There is also a strong dependence on the power–law index of the radial density profile of the external medium around the source, $\rho \propto R^{-k}$ (taken from Granot & Loeb 2001).

of a microlensed afterglow lightcurve can be used to reconstruct the parameters of the fireball and its environment. The dependence of the afterglow image on frequency offers a fingerprint that can be used to identify a microlensing event and distinguish it from alternative interpretations. It can also be used to constrain the relativistic dynamics of the fireball and the properties of its gaseous environment. At the highest frequencies, the divergence of the surface brightness near the edge of the afterglow image ($r = 1$ in Figure 5) depends on the thickness of the emitting layer behind the relativistic shock front, which is affected by the length scale required for particle acceleration and magnetic field amplification behind the shock[60,37].

Ioka & Nakamura [44] considered the more complicated case where the explosion is collimated and centered around the viewing axis. More general orientations that violate circular symmetry need to be considered in the future.

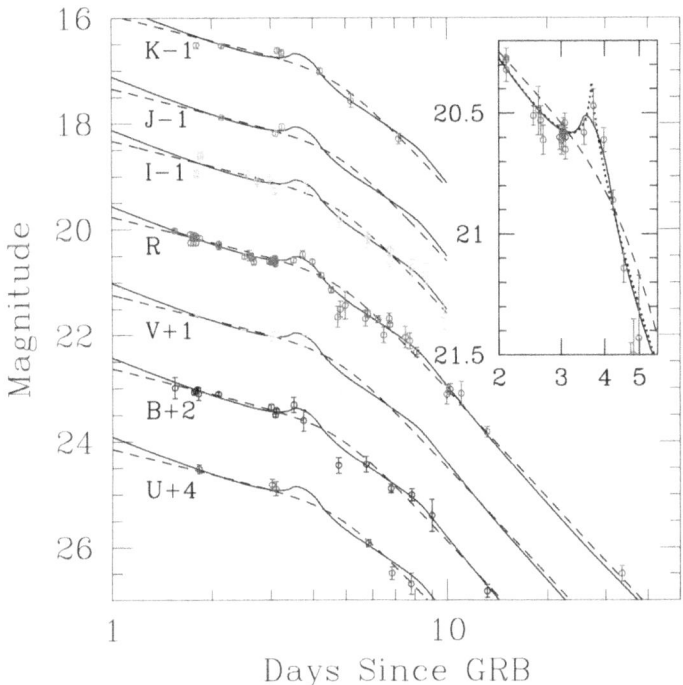

Fig. 6. $UBVRIJK$ photometry of GRB 000301C as a function of time in days from the GRB trigger (from Garnavich et al. 2000; Gaudi et al. 2001). Data points have been offset by the indicated amount for clarity. The dashed line is the best-fit smooth, double power-law lightcurve (with no lensing), while the solid line is the overall best-fit microlensing model, where the SBP has been determined from direct inversion. The inset shows the R-band data only. The dotted line is the best-fit microlensing model with theoretically calculated SBP, for $k = 0$ and $\nu > \nu_c$.

GRB 000301C Garnavich, Loeb, & Stanek [29] have reported the possible detection of a microlensing magnification feature in the optical-infrared light curve of GRB 000301C (see Figure 6). The achromatic transient feature is well fitted by a microlensing event of a $0.5M_\odot$ lens separated by an Einstein angle from the source center, and resembles the prediction of Loeb & Perna [57] for a ring-like source image with a narrow fractional width ($\sim 10\%$). Alternative interpretations relate the transient achromatic brightening to a higher density clump into which the fireball propagates [5], or to a refreshment of the decelerating shock either by a shell which catches up with it from behind or by continuous energy injection from the source [91]. However, the microlensing model has a smaller number of free parameters. If with better data, a future event will show the generic temporal and spectral characteristics of a microlensing event, then these alternative interpretations will be much less viable. A galaxy 2″ from GRB 000301C might be the host of the stellar lens, but current data provides only an upper-limit on its surface brightness at the GRB position. The existence of an intervening galaxy increases the probability for microlensing over that of a random line-of-sight.

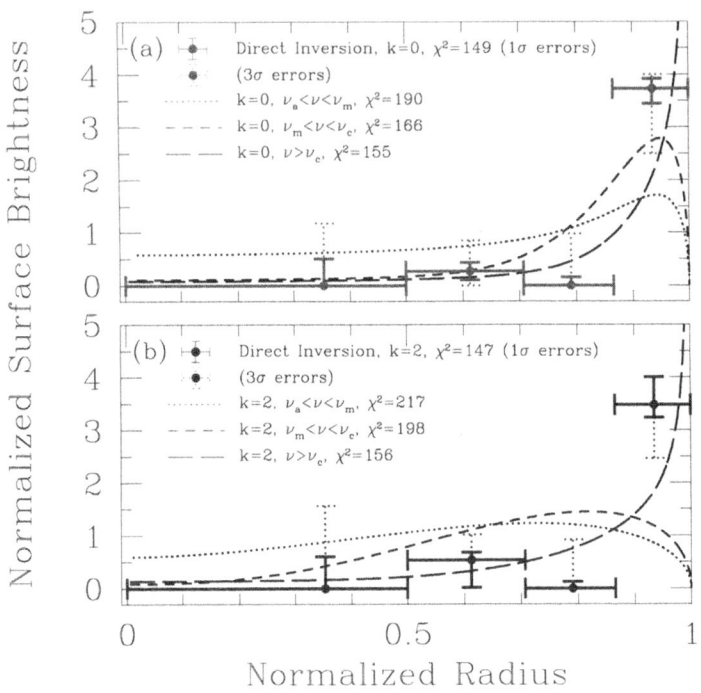

Fig. 7. Fitting GRB 000301C with different SBPs as a function of normalized radius (taken from Gaudi et al. 2001). The points are the SBPs determined from direct inversion, with 1σ errors (solid) and 3σ errors (dotted). The curves are theoretically calculated SBPs for various frequency regimes (see Figure 5). (a) Uniform external medium, $k = 0$. (b) Stellar wind environment, $k = 2$. The number of degrees of freedom is 92 for the direct inversion points and 89 for the curves.

Gaudi, Granot, & Loeb [32] have shown that direct inversion of the observed light curve for GRB 000301C yields a surface brightness profile (SBP) of the afterglow image which is strongly limb-brightened, as expected theoretically (see Figure 7).

Obviously, realistic lens systems could be more complicated due to the external shear of the host galaxy or a binary companion. Mao & Loeb [53] calculated the magnification light curves in these cases, and found that binary lenses may produce multiple peaks of magnification. They also demonstrated that *all* afterglows are likely to show variability at the level of a few percent about a year following the explosion, due to stars which are separated by tens of Einstein angles from the line-of-sight.

What is the probability for microlensing? If the lenses are not strongly clustered so that their cross-sections overlap on the sky, then the probability for having an intervening lens star at a projected angular separation θ from a source at a redshift $z \sim 2$ is $\sim 0.3\Omega_\star(\theta/\theta_E)^2$ [71,6,66,67], where $\Omega_\star$ is the cosmological density parameter of stars. The value of $\Omega_\star$ is bounded between the density of the luminous stars in galaxies and the total baryonic density as inferred from Big Bang nucleosynthesis, $7 \times 10^{-3} < \Omega_\star < 5 \times 10^{-2}$ [27]. Hence, *all* GRB afterglows should show evidence for events with $\theta \sim 30\theta_E$, for which microlensing provides a small perturbation to the light curve[53]. (This crude estimate ignores the need to subtract those stars which are embedded in the dense central regions of galaxies, where macrolensing dominates and the microlensing optical depth is of order unity.) However, only one out of roughly a hundred afterglows is expected to be strongly microlensed with an impact parameter smaller than the Einstein angle. Indeed, Koopmans & Wambsganss [48] have found that the 'a posteriori' probability for the observed microlensing event of GRB 000301C along a random line-of-sight is between 0.7–2.7% if 20-100% of the dark matter is in compact objects.

Microlensing events are rare but precious. Detailed monitoring of a few strong microlensing events among the hundreds of afterglows detected per year by the forthcoming Swift satellite, could be used to constrain the environment and the dynamics of relativistic GRB fireballs, as well as their magnetic structure and particle acceleration process.

2 Illumination of the Intergalactic Medium by Quasars

2.1 Lyα Halos

The absorption trough created in the spectrum of a distant sources due to neutral hydrogen diminishes down to undetectable flux levels as soon as the mean neutral fraction of hydrogen is larger than $\sim 10^{-4}$ (see equation 1). Once the transmitted flux reaches very low levels, it is difficult to infer whether the neutral fraction of hydrogen is as low as 0.01% or as high as 100%. A novel way to study the reionization phase transition relies on a direct detection of intergalactic neutral hydrogen. Loeb & Rybicki [58,75] have shown that the neutral gas prior to reionization can be probed through narrow-band imaging of embedded Lyα sources. The physical situation is analogous to the appearance of a halo of scattered light around a street lamp which is embedded in a dense fog. The "street lamp" in this metaphor is a high–redshift quasar which emits Lyα photons into the surrounding neutral hydrogen gas. The IGM scatters these photons and acts as a fog.

The radiation of the first galaxies is strongly absorbed shortward of their restframe Lyα wavelength by neutral hydrogen in the intervening IGM. However, the Lyα photons emitted by these sources are not eliminated but rather scatter until they redshift out of resonance and escape due to the Hubble expansion of the surrounding intergalactic hydrogen (see Fig. 8). Detection of the diffuse Lyα halos around high redshift sources would provide a unique tool for probing the neutral IGM before the epoch of reion-ization. Loeb & Rybicki [58,75] explored the above effect for a uniform, fully-neutral IGM in a pure Hubble flow. It is important to extend this analysis to more realistic cases of sources embedded in an inhomogeneous IGM, which is partially ionized by these sources [10]. It would be interesting to extract particular realizations of the perturbed IGM around massive galaxies from numerical simulations, and to apply a suitable ra-diative transfer code in propagating the Lyα photons from the embedded galaxies. Map-ping of the properties of the associated Lyα halos will allow to assess their detectability with future observations.

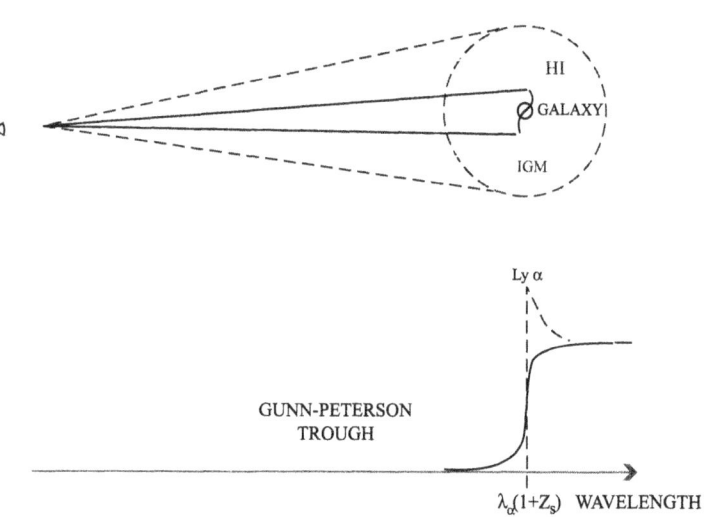

Fig. 8. *Loeb–Rybicki halos:* Scattering of Lyα line photons from a galaxy embedded in the neutral intergalactic medium prior to reionization. The line photons diffuse in frequency due to the Hub-ble expansion of the surrounding medium and eventually redshift out of resonance and escape to infinity. A distant observer sees a Lyα halo surrounding the source, along with a characteris-tically asymmetric line profile. The observed line should be broadened and redshifted by about one thousand km s^{-1} relative to other lines (such as H$_\alpha$) emitted by the galaxy.

2.2 Magnetization of the Intergalactic Medium by Quasar Outflows

Magnetic fields and cosmic rays play an important dynamical role in the interstellar medium of galaxies. However, they are often ignored in discussions of the intergalactic medium (IGM). *How significant is the pressure from these non-thermal components in the IGM? Has the magnetic field observed in collapsed objects, such as galaxies or clusters of galaxies, originated from the IGM?* The first question is particularly relevant for the reconstruction of the mass distribution of galaxy clusters from X-ray data.

Outflows from quasars inevitably pollute the IGM with magnetic fields. The short-lived activity of a quasar leaves behind an expanding magnetized bubble in the IGM. Furlanetto & Loeb [28] modelled the expansion of the remnant quasar bubbles and calculated their distribution as a function of size and magnetic field strength at different redshifts. They have found that generically by a redshift $z \sim 3$, about 5–20% of the IGM volume is filled by magnetic fields with an energy density $> 10\%$ of the mean thermal energy density of a photo-ionized IGM at $\sim 10^4$ K (see Figure 9). As massive galaxies and X-ray clusters condense out of the magnetized IGM, the adiabatic compression of the magnetic field could result in the field strength observed in these systems without a need for further dynamo amplification. The intergalactic magnetic field could also provide a non–thermal contribution to the pressure of the photo-ionized gas that may account for the claimed discrepancy between the simulated and observed Doppler width distributions of the Lyα forest at $z > 2$ [11,83]. However, the supplied magnetic energy is unlikely to be dynamically important today since the present-day IGM was heated by gravitationally-induced shocks to an average mass-weighted temperature of $\sim 3 \times 10^6$K [13], larger by two orders of magnitude than the magnetic energy input from quasars.

2.3 Particle Acceleration in Magnetized Intergalactic Shocks

Even though the present-day magnetic pressure may not be dynamically important, the magnetization of the IGM by quasar outflows has important consequences. Once magnetized, the shocks produced in the IGM during the formation of large-scale structure can accelerate a population of highly relativistic electrons by the Fermi mechanism, similarly to the collisionless shocks of supernova remnants. The accelerated electrons emit synchrotron radiation as they gyrate in the embedded magnetic fields and produce γ-ray radiation by inverse–Compton scattering the microwave background photons. This non-thermal radiation spans some ~ 20 orders of magnitude in photon energies from the radio to the TeV γ-ray regime. The radiation is expected to delineate the strong shocks formed in the cosmic web of large scale sheets and filaments in the IGM. The brightest emission originates from the strongest intergalactic shocks around galaxy clusters.

2.4 Non-Thermal Radio and Gamma-Ray Emission

More than a third of all X-ray clusters with luminosities $> 10^{45}$ erg s^{-1} possess diffuse radio halos [33]. Based on energy arguments and circumstantial evidence, these radio halos are believed to be caused by synchrotron emission from shock-accelerated electrons during the merger events of their host clusters ([41,86,21]; see [55] for references to alternative, less successful models). These highly-relativistic electrons cool primarily through inverse-Compton scattering off the microwave background. Since their cooling time is much shorter than the dynamical time of their host cluster, the radio emission is expected to last as long as the shock persists and continues to accelerate fresh electrons to relativistic energies. Intergalactic shocks also occur along the filaments and sheets that channel mass into the clusters. These structures, also traced by the galaxy distribution [16], are induced by gravity and form due to converging large-scale flows in the IGM.

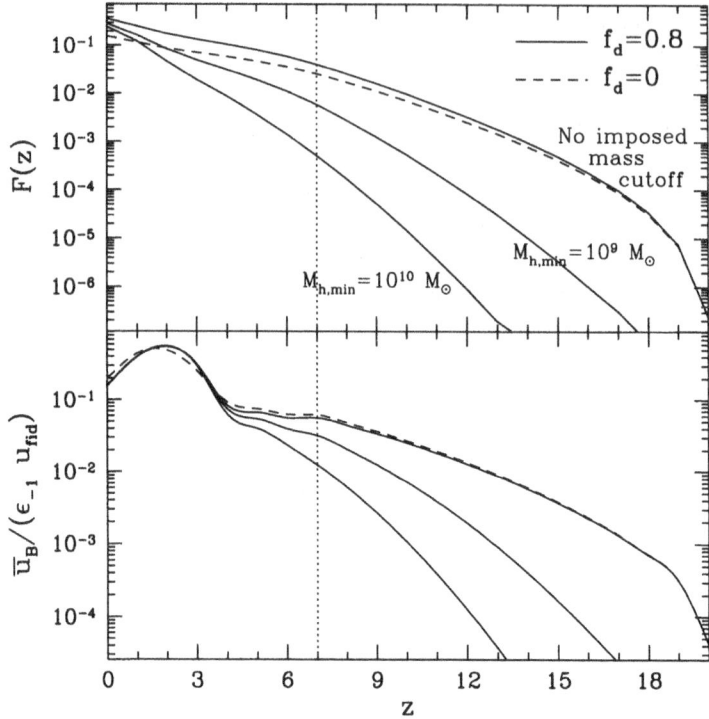

Fig. 9. *Upper Panel:* Volume filling fraction of magnetized bubbles $F(z)$, as a function of red-shift (taken from Furlanetto & Loeb 2001). *Lower Panel:* Ratio of magnetic energy density, $\bar{u}_B/\epsilon_{-1}$, to a fiducial thermal energy density $u_{fid} = 3n(z)kT_{IGM}$ of a photoionized IGM, where $T_{IGM} = 10^4$ K, as a function of redshift. The solid curves assume a different minimum mass for a galaxy out of which a quasar outflow may originate, $M_{h,min}$. Examples include a case where this minimum mass is determined by atomic cooling before reionization and by infall sup-pression afterward (top curve), and fixed-value cases with $M_{h,min} = 10^9 M_\odot$ (middle curve), and $M_{h,min} = 10^{10} M_\odot$ (bottom curve). The vertical dotted line indicates the assumed redshift of reionization, $z_r = 7$.

The intergalactic synchrotron emission contaminates the cosmic microwave back-ground (CMB) anisotropies – relic from the epoch of recombination, and as such needs to be considered in the design and analysis of anisotropy experiments at low frequen-cies. Previous estimates of synchrotron contamination of the cosmic anisotropies fo-cused on Galactic emission, which occurs primarily at low Galactic latitudes and large angular scales [82]. Waxman & Loeb [90] calculated the intergalactic synchrotron con-tribution to the fluctuations in the radio sky as a function of frequency and angular scale (see Figure 10). They assumed that most of the emission originates from the virializa-tion shocks around X-ray clusters, and used the Press–Schechter [72] mass function to describe the abundance of such clusters as a function of redshift (see also subsequent work in [26]). Although the synchrotron background amounts to only a small fraction of the CMB intensity, Waxman & Loeb [90] showed that its fluctuations could dom-

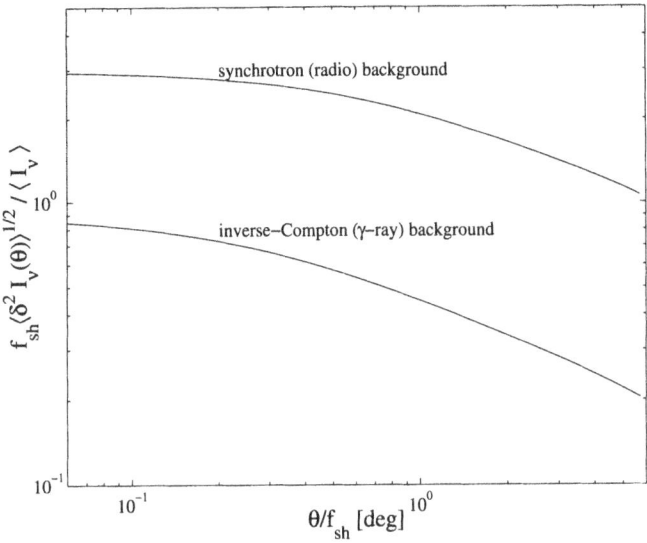

Fig. 10. Fractional intensity fluctuations, $(\langle I_\nu(0)I_\nu(\theta)\rangle - \langle I_\nu\rangle^2)^{1/2}/\langle I_\nu\rangle$, in synchrotron (radio) and inverse-Compton (γ-ray) background flux (taken from Waxman & Loeb 2000). The dimensionless coefficient, $f_{\rm sh}$, is of order unity. The ratio of synchrotron to CMB intensity is $\langle I_\nu^{\rm syn}\rangle/I_\nu^{\rm CMB} = 6 \times 10^{-6} f_{\rm sh}^{-2}(\xi_B/0.01)(\nu/10{\rm GHz})^{-3}$, where the post-shock magnetic energy fraction ξ_B is assumed to be of order 0.01, corresponding to a magnetic field strength of 0.1μG in the virialization shock of X-ray clusters. The post-shock energy fraction carried by relativistic electrons is assumed to be $\xi_e \sim 0.05$, similar to the value inferred in the shocks of supernova remnants.

inate over the primordial CMB fluctuations at low photon frequencies, $\nu < 10$ GHz. They found that radio emission from cluster shocks contributes a fluctuation amplitude of $\sim 40\mu$K $\times(\nu/10{\rm GHz})^{-3}$ to the CMB on angular scales between 1 and 0.1°, respectively. Interestingly, current anisotropy experiments are just sensitive to this level of fluctuations. Existing detections by CAT ($50 \pm 15\mu$K at 15 GHz on 0.2–0.5° scales) and OVRO ($56^{+8.5}_{-6.6}\mu$K at 20 GHz on 0.1–0.6° scales), as well as 95% upper limits ($< 40\mu$K on arcminute scales at 9-15 GHz by the ATCA, RYLE and VLA detectors) are consistent with our prediction.

Loeb & Waxman [59] have shown that a significat fraction of the diffuse γ–ray background [78] might have been generated by the shocks resulting from the formation of large-scale structure in the intergalactic medium. Similarly to the collisionless shocks of supernova remnants [7], these shocks produce a population of highly-relativistic electrons with a maximum Lorentz factor $> 10^7$ that scatter a small fraction of the microwave background photons in the present-day universe up to γ-ray energies, thereby providing the γ-ray background. The predicted flux agrees with the observed diffuse background over more than four decades in photon energy and is not sensitive to the precise magnetic field value, provided that the fraction of shock energy carried by relativistic electrons is $\xi_e \sim 0.05$ (a value consistent with that inferred for supernovae shocks) The same electrons that emit γ-rays by inverse-Compton scattering of microwave background photons, also produce synchrotron radiation in the radio band due to intergalactic magnetic fields. The existence of magnetic fields with an amplitude $> 0.1\mu G$, is inferred in cluster halos [47,30,74,45], and is also required for the Fermi acceleration of these electrons. The appearance of radio halos around young X-ray clusters is therefore a natural consequence and an important test of this model for the extragalactic γ-ray background. The combination of radio and γ-ray data can be used to calibrate ξ_e and determine the strength of the intergalactic magnetic field.

The assumed acceleration of electrons by collisionless shocks is similar to that observed in supernova remnants. Recent X-ray [49,50] and TeV [80,65] observations of the supernova remnants SN1006 and SNR RX J1713.7–3946 imply that electrons are accelerated in the remnant shocks up to an energy ~ 100 TeV, and are confined to the collisionless fluid by magnetic fields. These shocks have a velocity of order 10^3 km s^{-1}, similar to the velocity of the intergalactic shocks we consider here. Although the plasma density is very different in the two cases, the density may be scaled out of the problem by measuring time in units of ω_{pe}^{-1}, where ω_{pe} is the electron plasma frequency (shock characteristics may also depend on the pre-shock magnetic field, which introduces a dimensionless parameter into the problem, ω_{ce}/ω_{pe}, where ω_{ce} is the electron cyclotron frequency. However, for both supernova and IGM shocks, $\omega_{ce}/\omega_{pe} \ll 1$, suggesting a similar behavior in both cases). For supernova shocks, the inferred energy density in relativistic electrons constitutes 1–10% of the post-shock energy density in these remnants [18], a fraction consistent with the global ratio between the mean energy density of cosmic-ray electrons and the turbulent energy density in the interstellar medium of our galaxy.

The production of anisotropic backgrounds of radio and γ-ray radiation by strong intergalactic shocks is a natural consequence of structure formation in the Universe. The brightest emission originates from the shocks on Mpc scales around newly formed, massive X-ray clusters. The foreground synchrotron fluctuations might be comparable to the anisotropy signals detected by existing low-frequency microwave background experiments, and can be easily isolated through multi-frequency observations. Polarization anisotropy experiments could then constrain the coherence length of the intergalactic magnetic field.

Different astrophysical sources may contribute to the strength of the intergalactic magnetic fields; aside from radio sources [15,28] they include: the large scale shocks themselves [52], the first generation of stars [73] and supernova-driven winds from galaxies [51]. Different contributions may lead to different scalings of the magnetic field energy density with IGM gas parameters. The non-thermal γ-ray emission is only weakly dependent on the magnetic field strength [59]. However, the non-thermal radio emission is highly sensitive to this parameter and different models may lead to a different dependence of the radio emission on scale and redshift, which may have observable consequences (e.g. modifying the functional dependence of the correlation function shown in Figure 10).

Recent observations of the giant radio galaxy NGC 315 may constitute the first direct detection of collisionless large scale structure shocks, accelerating electrons to high energies [19]. Radio observations of this source are most naturally explained by the existence of a large scale shock produced by a flow converging towards an intersection of galaxy filaments [19].

Waxman & Loeb [90] predicted a fluctuation amplitude $> 40\%$ in the γ-ray background intensity on sub-degree scale, and the existence of extended, $> 1°$, γ-ray halos, associated with newly formed massive clusters. On scales larger than a degree the fluctuation amplitude declines and is well below the anisotropy limits from EGRET (see Figure 5 in [78]). Detection of the predicted signals will provide a calibration of the uncertain model parameter, ξ_e. The high-energy maps required to detect the predicted anisotropy signal will be made between 20 MeV and 300 GeV by the GLAST instrument (planned for launch in 2005; see http://glast.gsfc.nasa.gov/), which is expected to be more sensitive than EGRET by an order-of-magnitude [8]. The predicted γ-ray halos may constitute a significant fraction of the unidentified extra-Galactic EGRET sources [84,85]. However, since the angular extension of the brightest halos is large, a more careful analysis is required in order to assess the detectability of such halos by EGRET.

A future, dedicated, all-sky anisotropy experiment, operating at several frequencies below 10 GHz, would be able to map the fluctuations in the intergalactic synchrotron background. The resulting synchrotron map could then be cross-correlated with full-sky maps at hard X-ray or γ-ray energies to confirm its cosmic origin. Identification of the synchrotron fluctuations together with their counterpart inverse-Compton emission of hard X-rays or γ-rays by the same population of shock-accelerated electrons, can be used to empirically determine the strength and spatial distribution of the intergalactic magnetic field. Similarly, the correlation between radio and γ-ray halos may be detectable around individual X-ray clusters. Strong radio halos could be the best indicators for bright γ-ray clusters, which would provide the first obvious targets for GLAST.

Semi-analytic models [90] identify the intergalactic shocks with smooth spherical accretion of gas onto clusters, while in reality they result from asymmetric mergers as well as from converging flows in large scale sheets and filaments. The more realistic emission from these complex geometries can be best modeled through detailed hydrodynamic simulations [61,63,46]. Mergers of comparable mass clusters, for example, would tend to produce only mild shocks due to the prior heating of the shocked gas, leading to reduced non-thermal emission due to the steep power-law slope of the accelerated electrons.

In analogy with the collisionless shocks of supernova remnants, the intergalactic shocks are also expected to accelerate a hadronic cosmic-ray component that would acquire a fraction (~ 10–50%) of the post-shock energy more substantial than that carried by relativistic electrons (~ 1–10%, see [18]). Collisions of the hadronic cosmic-rays with IGM protons produce pions, π^0, which decay into γ-ray photons; however for typical parameters this radiation component is expected to be sub-dominant relative to the inverse-Compton production of γ-rays by the relativistic electrons [59,90,62]. The buoyancy of the relativistic cosmic–ray fluid may lead to instabilities that would eventually separate it from the IGM gas and allow it to expand into the large-scale voids that fill most of the volume in between the sheets and filaments of the IGM. For a relativistic fluid, the adiabatic decrease of the pressure with increasing volume ($p \propto V^{-4/3}$) is milder than for the IGM gas ($p \propto V^{-5/3}$). It is possible that the intergalactic voids are currently being dominated by cosmic–ray and magnetic field pressure which suppresses the formation of galaxies there.

Acknowledgements

I thank all my collaborators on the topics described in this brief review: Rennan Barkana, Benedetta Ciardi, Steve Furlanetto, Peter Garnavitch, Scott Gaudi, Jonathan Granot, Zoltan Haiman, Shude Mao, Rosalba Perna, George Rybicki, Kris Stanek, and Eli Waxman. This work was supported in part by NASA grants NAG 5-7039, 5-7768, and by NSF grants AST-9900877, AST-0071019.

References

1. T. Abel, G. Bryan, M. L. Norman: Ap. J., **540**, 39 (2000)
2. A. I. Anderson, et al.: A. & A., **364**, L54 (2000)
3. R. Barkana, A. Loeb: Physics Reports, **349**, 125 (2001)
4. R. H. Becker et al: Astr. J., **submitted**; astro-ph/0108097 (2001)
5. E. Berger, et al.: Ap. J., **545**, 56 (2000)
6. O. M. Blaes, R. L. Webster: Ap. J., **391**, L63 (1992)
7. R. Blandford, D. Eichler: Physics Reports, **154**, 1 (1987)
8. E. D. Bloom: Space Sci. Rev., **75**, 109 (1996)
9. V. Bromm, P. S. Coppi, R. B. Larson: **527**, 5 (1999)
10. V. Bromm, A. Loeb, G. Rybicki: in preparation (2001)
11. G.L. Bryan, M. Machacek, P. Anninos, M.L. Norman: Ap. J., **517**, 13 (1999)
12. B. Ciardi, A. Loeb: Ap. J., **540**, 687 (2000)
13. R. Dave, et al.: Ap. J., **552**, 473 (2001)
14. S. G. Djorgovski, S. M. Castro, D. Stern, A. Mahabal: Ap. J. L., **submitted**; astro-ph/0108069 (2001)
15. R. A. Daly, A. Loeb: Ap. J. **364**, 451 (1990)
16. A. G. Doroshkevich et al.: M.N.R.A.S., **283**, 1281 (1996)
17. S. L. Ellison, A. Songaila, J. Schaye, M. Pettini: A. J., **120**, 1175 (2000)
18. D. C. Ellison, P. Slane, B. M. Gaensler: Ap. J., **submitted**; astro-ph/0106257 (2001)
19. T. A. Ensslin et al.: Ap. J., **549**, L39 (2001)
20. X. Fan, et al.: A. J., **120**, 1167 (2000)
21. L. Feretti: to appear in Proc. of IAU Symp. # 199 on "The Universe at Low Radio Frequencies", Pune, India; astro-ph/0006379 (2000)
22. F. Fiore, F. Nicastro, S. Savaglio, L. Stella, M. Vietri: Ap. J., **544**, L7 (2000)
23. D. A. Frail, et al.: Nature, **submitted** (2001); astro-ph/0102282
24. D. L. Freedman, E. Waxman: Ap. J., **547**, 922 (2001)
25. A. Fruchter: Ap. J., **512**, L1 (1999)
26. Y. Fujita, C. L. Sarazin, Ap. J., **in press**; astro-ph/0108369 (2001)
27. M. Fukugita, C. J. Hogan, P. J. E. Peebles: Nature, **366**, 309 (1998)
28. S. R. Furlanetto, A. Loeb: Ap. J., **556**, 619 (2001)
29. P. M. Garnavich, A. Loeb, K. Z. Stanek: Ap. J., **544**, L11 (2000)
30. R. Fusco-Femiano et al.: Ap. J., **513**, L21 (1999)
31. B. S. Gaudi, A. Loeb: Ap. J., **558** (2001)
32. B. S. Gaudi, J. Granot, A. Loeb: Ap. J., **in press** (2001)
33. G. Giovannini, M. Tordi, L. Feretti: New Astronomy, **4**, 141 (1999)
34. J. Goodman: New Astronomy, **2**, 449 (1997)
35. J. Granot, A. Loeb: Ap. J., **551**, L63 (2001)

36. J. Granot, T. Piran, R. Sari: Ap. J., **513**, 679 (1999a); **527**, 236 (1999b)
37. A. Gruzinov, E. Waxman: Ap. J., **511**, 852 (1999)
38. J. E. Gunn, B. A. Peterson: Ap. J., **142**, 1633, (1965)
39. Z. Haiman, A. Loeb: Ap. J., **483**, 21 (1997); **503**, 505 (1998); **552**, 459 (2001)
40. F. Hamann, G. Ferland: A. R. A. & A. **37**, 487 (1999)
41. D. E. Harris, V. K. Kapahi, R. D. Ekers: Astr. &Ap. Suppl., **39**, 215 (1995)
42. F. D. A. Hartwick, D. Schade: A. R. A. & A., **28**, 437 (1990)
43. U. Hellsten, N. Y. Gnedin, J. Miralda-Escude: ApJ, **509**, 56 (1998)
44. K. Ioka, K., T. Nakamura: Ap. J., **in press**; astro-ph/0102028 (2001)
45. J. S. Kaastra, et al.: Ap. J., **519**, L119 (1999)
46. U. Keshet, E. Waxman, V. Springel, A. Loeb, L. Hernquist: Ap. J., in preparation (2001)
47. K.-T. Kim, P. P. Kronberg, G. Giovannini, T. Venturi: Nature, **341**, 720 (1989)
48. L. V. E. Koopmans, J. Wambsganss: M.N.R.A.S., **in press**; astro-ph/0011029 (2001)
49. K. Koyama et al.: Nature, **378**, 255 (1995)
50. K. Koyama et al.: PSAJ, **49**, L7 (1997)
51. P. P. Kronberg, H. Lesch, U. Hopp: Ap. J., **511**, 56 (1999)
52. R. M. Kulsrud, R. Cen, J. P. Ostriker, D. Ryu: Ap. J., **454**, 60 (1997)
53. S. Mao, A. Loeb: Ap. J., **547**, L97 (2001)
54. D. Q. Lamb, D. E. Reichart: Ap. J. **536**, 1 (2000)
55. H. Liang, R. W. Hunstead, M. Birkinshaw, P. Andreani: Ap. J., **544**, 686 (2000)
56. A. Loeb, R. Barkana: ARA&A, **in press** (2001)
57. A. Loeb, R. Perna: Ap. J., **495**, 597 (1998)
58. A. Loeb, & G. Rybicki: Ap. J., **524**, 527 (1999)
59. A. Loeb, E. Waxman: Nature, **405**, 156 (2000)
60. M. V. Medvedev, A. Loeb: Ap. J. **526**, 697 (1999)
61. F. Miniati, D. Ryu, H. Kang, T. W. Jones, R. Cen, J. P. Ostriker: Ap. J., **542**, 608 (2000)
62. F. Miniati, D. Ryu, H. Kang, T. W. Jones: Ap. J., **559**, 1 (2001)
63. F. Miniati, T. W. Jones, H. Kang, D. Ryu: Ap. J., **in press**; astro-ph/0108305 (2001)
64. J. Miralda-Escudé: Ap. J., **501**, 15 (1998)
65. H. Muraishi et al.: Astr. & Ap., **354**, L57 (2000)
66. R. Nemiroff: Ap.& SS, **259**, 309 (1998)
67. R. Nemiroff, J. P. Norris, J. T. Bonnell, G. F. Marani: Ap. J., **494**, L173 (1998)
68. A. Panaitescu, P. Mészáros: Ap. J., **493**, L31 (1998)
69. R. Perna, A. Loeb: Ap. J., **501**, 467 (1998)
70. R. Perna, & A. Loeb: Ap. J., **503**, L135 (1998)
71. W. H. Press, J. E. Gunn: Ap. J., **185**, 397 (1973)
72. W. H. Press, P. Schechter: Ap. J., **187**, 425 (1974)
73. M. J. Rees: QJRAS, **28**, 197 (1987)
74. Y. Rephaeli, D. Gruber, P. Blanco: Ap. J., **511**, L21 (1999)
75. G. Rybicki, & A. Loeb: Ap. J., **520**, L79 (1999)
76. R. Sari: Ap. J., **494**, L49 (1998)
77. P. Schneider, J. Ehlers, E. E. Falco: *Gravitational Lenses* (Springer, Heidelberg 1992)
78. P. Sreekumar et al.: Ap. J., **494**, 523 (1998)
79. D. Stern, H. Spinrad: P. A. S. P., **111**, 1475 (1999)
80. T. Tanimori et al.: Ap. J., **497**, L25 (1998)
81. M. Tegmark, J. Silk, A. Evrard: Ap. J., **417**, 54 (1993)
82. M. Tegmark, D.J. Eisenstein, W. Hu, A. de Oliveira-Costa: Ap. J., **530**, 133 (2000)
83. T. Theuns, A. Leonard, G. Efstathiou, F.R. Pearce, P.R. Thomas: M.N.R.A.S., **301**, 478 (1998)
84. T. Totani, T. Kitayama: Ap. J., **545**, 572 (2000)

85. W. Kawasaki, T. Totani: Ap. J., **submitted**; astro-ph/0108309 (2001)
86. P. Tribble: M.N.R.A.S., 263, 31 (1993)
87. X. Wang, M. Tegmark, M. Zaldarriaga: Phys. Rev. D, **submitted**; astro-ph/0105091 (2001)
88. E. Waxman: Ap. J., **491**, L19 (1997)
89. E. Waxman, S. R. Kulkrani, D. A. Frail: Ap. J., **497**, 288 (1998)
90. E. Waxman, A. Loeb: Ap. J., **545**, L11 (2000)
91. Z. Zhang, P. Meszaros: Ap. J., **552**, L35 (2001)

Gamma-Ray Bursts as a Probe of Cosmology

Donald Q. Lamb[1]

Department of Astronomy & Astrophysics, University of Chicago, 5640 South Ellis Avenue, Chicago, IL 60637

Abstract. We show that, if the long GRBs are produced by the collapse of massive stars, GRBs and their afterglows may provide a powerful probe of cosmology and the early universe.

1 Introduction

There is increasingly strong evidence that gamma-ray bursts (GRBs) are associated with star-forming galaxies [1,2,3,4] and occur near or in the star-forming regions of these galaxies [2,3,4,5,6]. These associations provide indirect evidence that at least the long GRBs detected by BeppoSAX are a result of the collapse of massive stars. The discovery of what appear to be supernova components in the afterglows of GRBs 970228 [7,8] and 980326 [9] provides tantalizing direct evidence that at least some GRBs are related to the deaths of massive stars, as predicted by the widely-discussed collapsar model of GRBs [10,11,12,13,14]. If GRBs are indeed related to the collapse of massive stars, one expects the GRB rate to be approximately proportional to the star-formation rate (SFR).

2 Detectability of GRBs and Their Afterglows

We have calculated the limiting redshifts detectable by BATSE and HETE-2, and by *Swift*, for the sixteen GRBs with well-established redshifts and published peak photon number fluxes. In doing so, we have used the peak photon number fluxes given in Table 1 of [15], taken a detection threshold of 0.2 ph s^{-1} for BATSE and HETE-2 and 0.04 ph s^{-1} for *Swift*, and set $H_0 = 65$ km s^{-1} Mpc^{-1}, $\Omega_m = 0.3$, and $\Omega_\Lambda = 0.7$ (other cosmologies give similar results). Figure 1 displays the results. This figure shows that BATSE and HETE-2 would be able to detect half of these GRBs out to a redshift $z = 20$ and 20% of them out to a redshift $z = 50$. *Swift* would be able to detect half of them out to redshifts $z = 70$, and 20% of them out to a redshift $z = 200$, although it is unlikely that GRBs occur at such extreme redshifts. Consequently, if GRBs occur at very high ($z > 5$) redshifts (VHRs), BATSE has probably already detected GRBs at these redshifts, and HETE-2 and *Swift* should detect them as well.

The soft X-ray, optical and infrared afterglows of GRBs are also detectable out to VHRs. The effects of distance and redshift tend to reduce the spectral flux in GRB afterglows in a given frequency band, but time dilation tends to increase it at a fixed time of observation after the GRB, since afterglow intensities tend to decrease with time. These effects combine to produce little or no decrease in the spectral energy flux

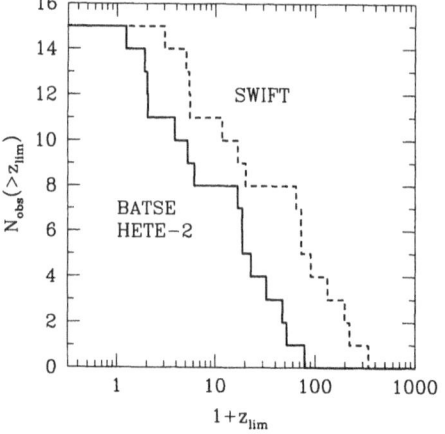
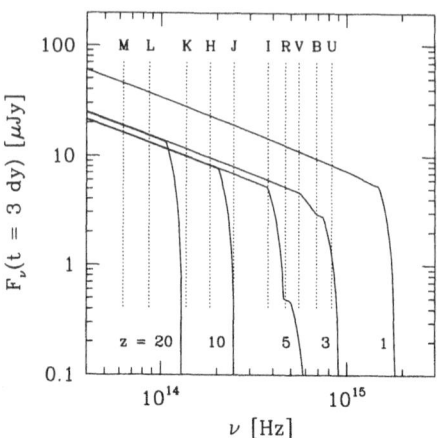

Fig. 1. Cumulative distributions of the limiting redshifts at which the 15 GRBs with well-determined redshifts and published peak photon number fluxes would be detectable by BATSE and HETE-2, and by *Swift*.

Fig. 2. The best-fit spectral flux distribution of the early afterglow of GRB 000131, as observed one day after the burst, after transforming it to various redshifts, and extinguishing it with a model of the Lyα forest.

F_ν of GRB afterglows in a given frequency band and at a fixed time of observation after the GRB with increasing redshift:

$$F_\nu(\nu, t) = \frac{L_\nu(\nu, t)}{4\pi D^2(z)(1 + z)^{1-a+b}}, \tag{1}$$

where $L_\nu \propto \nu^a t^b$ is the intrinsic spectral luminosity of the GRB afterglow, which we assume applies even at early times, and $D(z)$ is the comoving distance to the burst. Many afterglows fade like $b \approx -4/3$, which implies that $F_\nu(\nu, t) \propto D(z)^{-2}(1+z)^{-5/9}$ in the simplest afterglow model, where $a = 2b/3$ [16]. In addition, $D(z)$ increases very slowly with redshift at redshifts greater than a few. Consequently, there is little or no decrease in the spectral flux of GRB afterglows with increasing redshift beyond $z \approx 3$.

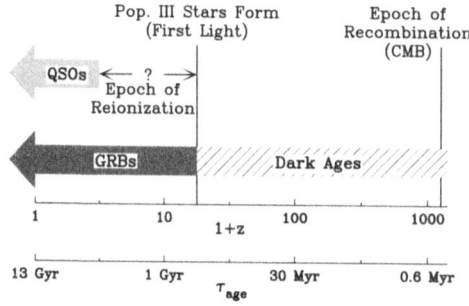

Fig. 3. Cosmological context of VHR GRBs. Shown are the epochs of recombination, first light, and re-ionization. Also shown are the ranges of redshifts corresponding to the "dark ages," and probed by QSOs and GRBs.

In fact, in the simplest afterglow model where $a = 2b/3$, if the afterglow declines more rapidly than $b \approx 1.7$, the spectral flux actually *increases* as one moves the burst to higher redshifts! An example of this is the afterglow of GRB 000131. Its peak flux F_{peak} was in the top 5% of all BATSE bursts and the break energy E_{break} in its spectrum was 164 keV, yet it occurred at a redshift $z = 4.50$. We have calculated the best-fit spectral flux distribution of the afterglow of GRB 000131 from [17], as observed three days after the burst, transformed to various redshifts. The transformation involves (1) dimming the afterglow, (2) redshifting its spectrum, (3) time dilating its light curve, and (4) extinguishing the spectrum using a model of the Lyα forest (for details, see [15]). Finally, we have convolved the transformed spectra with a top hat smearing function of width $\Delta\nu = 0.2\nu$. This models these spectra as they would be sampled photometrically, as opposed to spectroscopically; i.e., this transforms the model spectra into model spectral flux distributions.

Figure 2 shows the resulting spectral flux distribution. The spectral flux distribution of the afterglow is cut off by the Lyα forest at progressively lower frequencies as one moves out in redshift. Thus high redshift afterglows are characterized by an optical "dropout" [4], and VHR afterglows by a near infrared "dropout." We conclude that, if GRBs occur at very high redshifts, both they and their afterglows can be easily detected.

3 GRBs as a Probe of Cosmology and the Early Universe

Theoretical calculations show that the birth rate of Pop III stars produces a peak in the SFR in the universe at redshifts $16 \lesssim z \lesssim 20$, while the birth rate of Pop II stars produces a much larger and broader peak at redshifts $2 \lesssim z \lesssim 10$ [18,19,20]. Therefore one expects GRBs to occur out to at least $z \approx 10$ and possibly $z \approx 15 - 20$, redshifts that are far larger than those expected for the most distant quasars.

Figure 3 places GRBs in a cosmological context. At recombination, which occurs at redshift $z = 1100$, the universe becomes transparent. The cosmic background radiation originates at this redshift. Shortly afterwards, the temperature of the cosmic background radiation falls below 3000 K and the universe enters the "dark ages" during which there is no visible light in the universe. "First light," which occurs at $z \approx 20$, corresponds to the epoch when the first stars form. Ultraviolet radiation from these first stars and/or from the first active galactic nuclei re-ionizes the universe. Afterward, the universe is transparent in the ultraviolet.

QSOs are currently the most powerful probes of the high redshift universe. GRBs have several advantages relative to QSOs as probes of cosmology. First, GRBs are expected to occur out to $z \approx 20$, whereas QSOs occur out to only $z \approx 5$. Second, very high redshift GRB afterglows can be 100 - 1000 times brighter at early times than are high redshift QSOs. This makes possible very sensitive high dispersion spectroscopy of the metal absorption lines and the Lyman α forest in the spectrum of the afterglows. Third, no "proximity effect" on intergalactic distances scales is expected for GRBs and their afterglows, in contrast to QSOs. Thus GRBs may be relatively "clean" probes of the intergalactic medium, the Lyman α forest, and damped Lyman α clouds, even in the vicinity of the GRBs.

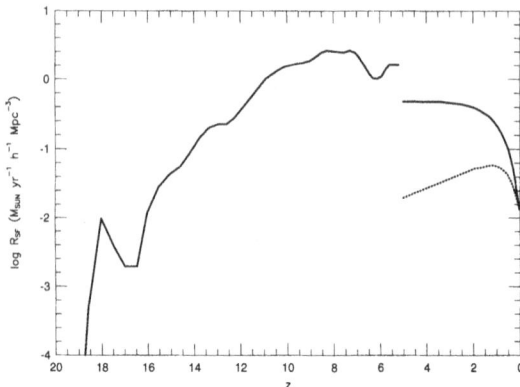

Fig. 4. The cosmic SFR R_{SF} as a function of redshift z. The solid curve at $z < 5$ is the SFR derived by [25]; the solid curve at $z \geq 5$ is the SFR calculated by [18] (the dip in this curve at $z \approx 6$ is an artifact of their numerical simulation). The dotted curve is the SFR derived by [24]. From [15].

The important cosmological questions that observations of GRBs and their afterglows may be able to address include the following:

• Information about the epoch of "first light" and the earliest generations of stars from merely the detection of GRBs at very high redshifts;

• Information about the growth of metallicity in the universe in the star-forming entities in which the bursts occur, in damped Lyman α clouds, and in the Lyman α forest from observations of the metal absorption line systems in the spectra of their afterglows;

• Information about the large-scale structure of the universe at VHRs from the clustering of the Lyman α forest lines and the metal absorption-line systems in the spectra of their afterglows; and

• Information about the epoch of re-ionization from the depth of the Lyman α break in the spectra of their afterglows.

Below we consider the first of these questions: the epoch of "first light" and the earliest generations of stars.

4   GRBs as a Probe of Star Formation

Observational estimates [21,22,23,24] indicate that the SFR in the universe was about 15 times larger at a redshift $z \approx 1$ than it is today. The data at higher redshifts from the Hubble Deep Field (HDF) in the north suggests a peak in the SFR at $z \approx 1 - 2$ [24], but the actual situation is highly uncertain.

In Figure 4, we have plotted the SFR versus redshift from a phenomenological fit [25] to the SFR derived from submillimeter, infrared, and UV data at redshifts $z < 5$, and from a numerical simulation by [18] at redshifts $z \geq 5$. The simulations done by [18] indicate that the SFR increases with increasing redshift until $z \approx 10$, at which point it levels off. The smaller peak in the SFR at $z \approx 18$ corresponds to the formation of Population III stars, brought on by cooling by molecular hydrogen. Since GRBs are detectable at these VHRs and their redshifts may be measurable from the absorption-line systems and the Lyα break in the afterglows [4], if the GRB rate is proportional to

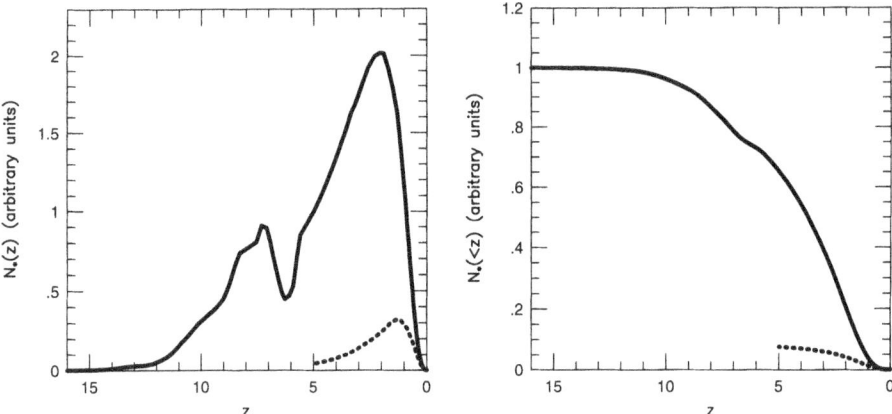

Fig. 5. Left panel: The number N_* of stars expected as a function of redshift z (i.e., the SFR from Figure 4, weighted by the differential comoving volume, and time-dilated) assuming that $\Omega_M = 0.3$ and $\Omega_\Lambda = 0.7$. Right panel: The cumulative distribution of the number N_* of stars expected as a function of redshift z. Note that $\approx 40\%$ of all stars have redshifts $z > 5$. The solid and dashed curves in both panels have the same meanings as in Figure 4. From [15].

the SFR, then GRBs could provide unique information about the star-formation history of the VHR universe.

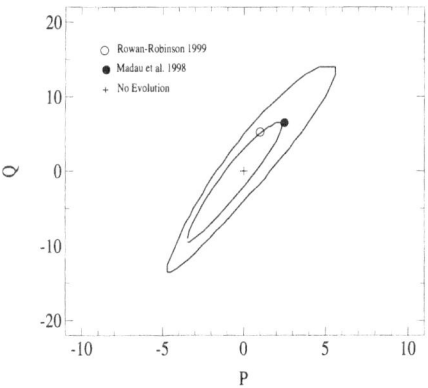

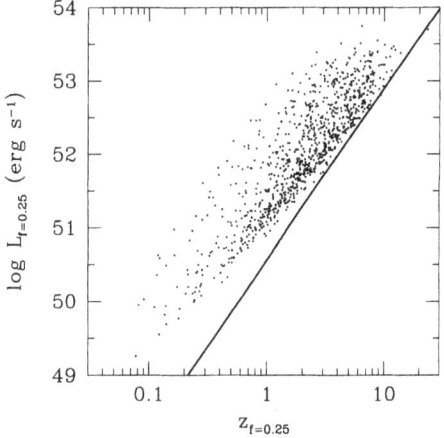

Fig. 6. Credible regions for the GRB rate parameters P and Q. The solid curves correspond to the 68% and 95% probability contours. Also shown are the (P,Q)-values corresponding to no space density evolution, the Madau et al. [24] SFR, and the Rowan-Robinson model fit to IR, optical and UV data [25]. From [26].

Fig. 7. The joint redshift and luminosity distribution of the qualitatively acceptable redshift distribution (see Figure 3 in [27]). The diagonal solid line shows the 10% detection threshold of BATSE. From [27].

We have calculated the expected number N_* of stars as a function of z assuming (1) that the GRB rate is proportional to the SFR[1], and (2) that the SFR is that given in Figure 4 (see [15] for details). The left panel of Figure 5 shows our results for $N_*(z)$ for an assumed cosmology $\Omega_M = 0.3$ and $\Omega_\Lambda = 0.7$ (other cosmologies give similar results). The solid curve corresponds to the star-formation rate in Figure 4; the dashed curve corresponds to the star-formation rate derived by [24]. Figure 5 shows that $N_*(z)$ peaks sharply at $z \approx 2$ and then drops off fairly rapidly at higher z, with a tail that extends out to $z \approx 12$. The rapid rise in $N_*(z)$ out to $z \approx 2$ is due to the rapidly increasing volume of space. The rapid decline beyond $z \approx 2$ is due almost completely to the "edge" in the spatial distribution produced by the cosmology. In essence, the sharp peak in $N_*(z)$ at $z \approx 2$ reflects the fact that the SFR we have taken is fairly broad in z, and consequently, the behavior of $N_*(z)$ is dominated by the behavior of the co-moving volume $dV(z)/dz$; i.e., the shape of $N_*(z)$ is due almost entirely to cosmology. The right panel in Figure 5 shows the cumulative distribution $N_*(> z)$ of the number of stars expected as a function of redshift z. The solid and dashed curves have the same meaning as in the upper panel. Figure 5 shows that for the particular SFR we have assumed, $\approx 40\%$ of all stars (and therefore of all GRBs) have redshifts $z > 5$.

5   Estimates of the GRB Rate

Is the GRB rate indeed proportional to the SFR (at least roughly)? We address this question in two ways. First, we consider the sample of GRBs with known redshifts. We adopt a Bayesian approach and calculate the likelihood of the data, assuming a very general model for the GRB rate and a power-law model for the intrinsic GRB photon luminosity distribution [26]. We fit the model jointly to the peak fluxes and redshifts of the 14 GRBs with known z, and the 7 GRBs for which there are constraints on z. Figure 6 shows our preliminary results. These suggest that the SFR models lie at about a 68% excursion from the best-fit GRB rate model. Thus we find that, despite the qualitative differences that exist between the observed GRB rate and estimates of the SFR in the universe, current data are consistent with the actual GRB rate being approximately proportional to the SFR when observational selection effects are taken into account.

Second, we use the variability of the lightcurves of long GRBs to estimate their distribution as a function of intrinsic photon luminosity and redshift [27]. Figure 7 shows the resulting joint redshift and luminosity distribution. This distribution suggests that the GRB rate continues to increase at very high redshifts, and that the intrinsic luminosities of GRBs evolve with redshift.

6   Conclusions

If the long GRBs are indeed produced by the collapse of massive stars, one expects GRBs to occur out to $z \approx 15 - 20$, redshifts that are far larger than those expected

---

[1] This may underestimate the GRB rate at VHRs since it is generally thought that the initial mass function will be tilted toward a greater fraction of massive stars at VHRs because of less efficient cooling due to the lower metallicity of the universe at these early times.

for the most distant QSOs. We have shown that both GRBs and their afterglows are easily detected out to these VHRs. GRBs can therefore give us information about the star-formation history of the universe, including the earliest generations of stars. The absorption-line systems and the Lyα forest visible in the spectra of GRB afterglows can be used to trace the evolution of metallicity in the universe, and to probe the large-scale structure of the universe at VHRs. Finally, measurement of the Lyα break in the spectra of GRB afterglows can be used to constrain, or possibly measure, the epoch at which re-ionization of the universe occurred.

References

1. Castander, F. J., & Lamb, D. Q. 1999, ApJ, **523**, 593
2. Fruchter, A. S., et al. 1999, ApJ, **516**, 683
3. Kulkarni, S. R., et al. 1998, Nature, **395**, 663
4. Fruchter, A. S. 1999, ApJ, **516**, 683
5. Sahu, K. C., et al. 1997, Nature, **387**, 476
6. Kulkarni, S. R., et al. 1999, Nature, **398**, 389
7. Reichart, D. E., 1999, ApJ, **521**, L111
8. Galama, T. J., et al. 2000, ApJ, **536**, 185
9. Bloom, J. S., et al. 1999, Nature, **401**, 453
10. Woosley, S. E. 1993, ApJ, **405**, 273
11. Woosley, S. E. 1996, in Gamma-Ray Bursts, eds. C. A. Meegan, R. D. Preece, & T. M. Koshut (New York: AIP), 520
12. Paczyński, B. 1998, ApJ, **494**, L45
13. MacFadyen, A. I., & Woosley, S. E. 1999, ApJ, **524**, 262
14. Wheeler, J. C., et al. 2000, ApJ, **537**, 810
15. Lamb, D. Q., & Reichart, D. E., 2000, ApJ, **536**, 1
16. Wijers, R. A. M. J., Rees, M. J., & Mészáros, P. 1997, MNRAS, **288**, L51
17. Andersen, M. I., et al. 2000, A&A, **364**, L54
18. Ostriker, J. P., & Gnedin, N. Y. 1996, ApJ, **472**, L63
19. Gnedin, N. Y., & Ostriker, J. P. 1997, ApJ, **486**, 581
20. Valageas, P., & Silk, J. 1999, A&A, **347**, 1
21. Gallego, J. 1995, ApJ, **455**, L1
22. Lilly, S. J., et al. 1996, ApJ, **460**, L1
23. Connolly, A. J. 1997, ApJ, **486**, L11
24. Madau, P., Pozzetti, L., & Dickinson, M. 1998, ApJ, **498**, 106
25. Rowan-Robinson, M. 1999, Ap&SS, **266**, 291
26. Weinberg, N., Graziani, C., Lamb, D. Q., and Reichart, D. E. 2001, in Proceedings of the Rome Workshop, in press (astro-ph/010759)
27. Reichart, D. E. and Lamb, D. Q. 2001, in Proceedings of the 20th Texas Symposium on Relativistic Astrophysics, in press (astro-ph/0103255)

Observational Tests of the Electro-Magnetic Black Hole Theory in Gamma Ray Bursts

Remo Ruffini[1]

ICRA - International Center for Relativistic Astrophysics and Physics Department, University of Rome "La Sapienza", I-00185 Rome, Italy.

Abstract. The Relative Space-Time Transformation (RSTT) Paradigm [44] and the Interpretation of the Burst Structure (IBS) Paradigm [45] are applied to the analysis of the structure of the burst and afterglow of Gamma-Ray Bursts within the theory based on the vacuum polarization process occurring in an Electro-Magnetic Black Hole, the EMBH theory. This framework is applied to the study of the GRB 991216 which is used as a prototype. The GRB-Supernova Time Sequence (GSTS) Paradigm, which introduces the concept of induced gravitational collapse in the Supernovae-GRB association [46], is announced and will be applied, within the EMBH theory, to GRB 980425 as a prototype in a forthcoming paper.

1 Introduction

I am very pleased to present here in Munich some observational tests of our Electro-Magnetic-Black-Hole theory, for short the EMBH theory, explaining some features of Gamma Ray Bursts, for short GRBs. The EMBH theory is rooted in discussions I had from 1971 to 1975 with Werner Heisenberg here in Munich, in Washington and Stanford.

GRBs are today promoting one of the most ample scientific effort in the entire field of science, both in the observational and theoretical domains. Following the discovery of the GRBs by the Vela satellites [58], the observations from the Compton satellite and BATSE had shown the isotropical distribution of the GRBs strongly suggesting a cosmological nature for their origin. It was still trough the data of BATSE that the existence of two families of bursts, the "short bursts" and the "long bursts" was presented, opening an intense scientific dialogue on their origin still active today, as we see in the talk of M. Schmidt in these proceedings.

An enormous momentum was gained in this field by the discovery of the afterglow phenomena by the BeppoSAX satellite and the optical identification of the GRBs sources at cosmological distances (see e.g. [9]). It has become apparent that fluxes of 10^{54} ergs/s are reached: during its peak emission the energy of a single GRB equals the energy emitted by all the stars of the Universe (see e.g. [48]).

From an observational point of view, an unprecedented campaign of observations is at work using the largest deployment of observational techniques from space with the satellites CGRO - BATSE, Beppo-SAX, Chandra, R-XTE, XMM-Newton, HETE-2, as well as the HST, and from the ground with optical KECK and the VLT and radio by the VLA observatories. Possibility of further examining correlation with detection of ultrahigh energy cosmic rays and neutrinos should be reachable in the near future thanks to developments of the AUGER and AMANDA experiments (see also [21]).

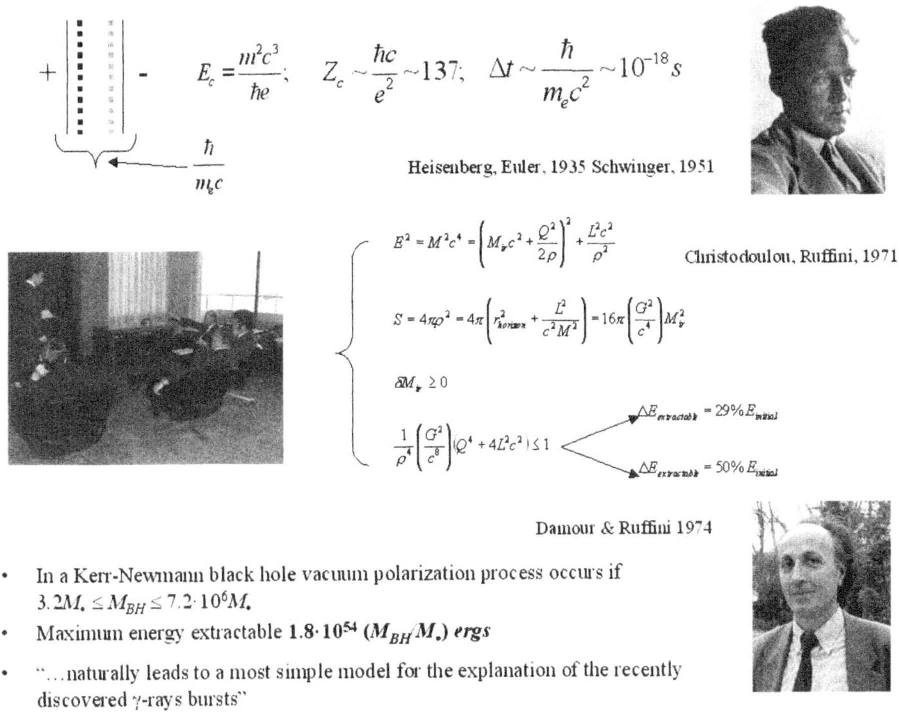

$$+\ \fbox{$\begin{matrix}\vdots\end{matrix}$}\ -\qquad E_c=\frac{m^2c^3}{\hbar e};\qquad Z_c\sim\frac{\hbar c}{e^2}\sim137;\qquad \Delta t\sim\frac{\hbar}{m_e c^2}\sim10^{-18}s$$

$$\underbrace{\qquad\qquad}_{\displaystyle\frac{\hbar}{m_e c}}$$

Heisenberg, Euler, 1935 Schwinger, 1951

$$E^2=M^2c^4=\left(M_\rho c^2+\frac{Q^2}{2\rho}\right)^2+\frac{L^2c^2}{\rho^2}$$

Christodoulou, Ruffini, 1971

$$S=4\pi\rho^2=4\pi\left(r_{korsun}^2+\frac{L^2}{c^2M^2}\right)=16\pi\left(\frac{G^2}{c^4}\right)M_\rho^2$$

$$\delta M_\rho\geq0$$

$$\frac{1}{\rho^4}\left(\frac{G^2}{c^8}\right)(Q^4+4L^2c^2)\leq1$$

$\Delta E_{extractable}=29\%\,E_{initial}$

$\Delta E_{extractable}=50\%\,E_{initial}$

Damour & Ruffini 1974

- In a Kerr-Newmann black hole vacuum polarization process occurs if
  $3.2M_\odot\leq M_{BH}\leq7.2\cdot10^6 M_\odot$.
- Maximum energy extractable $1.8\cdot10^{54}\,(M_{BH}/M_\odot)$ ergs
- "...naturally leads to a most simple model for the explanation of the recently discovered γ-rays bursts"

Fig. 1. Theoretical background of the EMBH model. The critical field against the breakdown of the vacuum is given next to a picture of Werner Heisenberg. The mass formula of EMBH is given next to the picture of the Ph.D. thesis discussion in Princeton by Demetrios Christodoulou, at the age of 19, in front of the committee with members: R. Ruffini (supervisor), J.A. Wheeler, E. Wigner and D. Wilkinson. The energetics of the vacuum polarization process of an EMBH is given next to the picture of T. Damour. These results were obtained, with R. Ruffini as supervisor, at Princeton University in 1974 during the preparation of the thesis of Doctorat d' Etat finally presented in Paris.

From a theoretical point of view, the GRBs offers comparable opportunities to develop new domains in yet untested directions of fundamental science. For the first time within the EMBH model, see Fig. 2, the opportunity exist to theoretically approach the following fundamental issues:

1. extreme relativistic hydrodynamic phenomena of an electron positron plasma expanding with sharply varying Lorentz gamma factors in the range 10^2 to 10^4. Analyze as well the very high collision of such expanding plasma with baryonic matter reaching intensities 10^{40} larger then the ones usually obtained in Earth based accelerators, see [50] and references therein,

2. the bulk process of vacuum polarization created by overcritical electromagnetic fields, in the sense of Heisenberg, Euler [22] and Schwinger [57]. This long sought quantum ultrarelativistic effect, not yet convincingly observed in heavy ion colli-

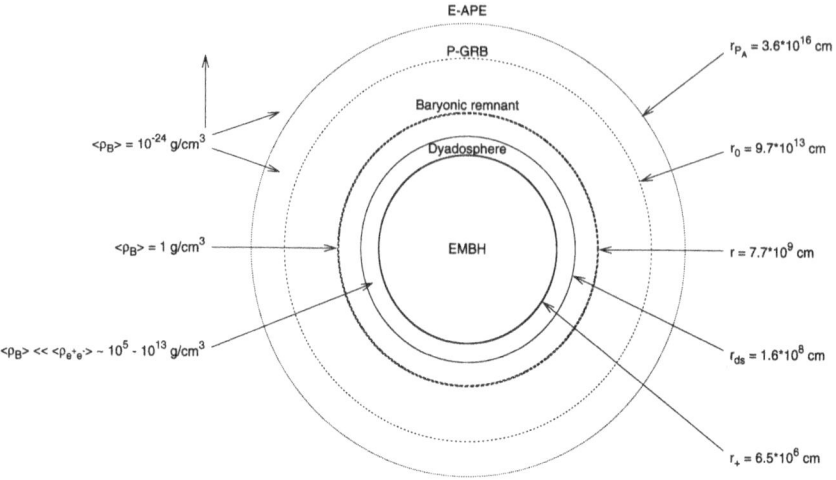

Fig. 2. Selected events of the expansion of the pulse in the EMBH model are represented, together with the distance from the EMBH in the laboratory frame at which they occur and with the density of the surrounding medium. The radial distances are represented in logarithmic scale.

sion on the Earth in Darmstadt, may indeed make its first appearance in the strong electromagnetic fields developed in astrophysical conditions during the process of gravitational collapse to a black hole, see [36,41,42],

3. a novel form of energy source: the extractable energy of a black hole, introduced by Christodoulou and Ruffini [7]. The enormous energies, released almost instantly in the observed GRBs, points to the possibility that for the first time we are witnessing the release of the extractable energy of an EMBH, during the process of gravitational collapse itself. We can compute and, if observationally confirmed, have the opportunity to study all general relativistic effects as the horizon of the Black hole is approached and is being formed, together with the associated ultrahigh energy quantum phenomena, see [6,49,53].

It is clear that in the approach to such a vast new field of research implying previously unobserved relativistic regimes it is not possible to proceed *as usual* adopting an uncritical comparison of observational data to theoretical models within the classical schemes of astronomy and astrophysics. Some insight to the new approach needed can be gained from past experiences in the interpretation of relativistic effects in high energy particle physics as well as from the explanation of some relativistic effects observed in the astrophysical domain. All those relativistic regimes are however much less extreme then the new ones encountered in GRBs.

There are at least three major new features in relativistic systems which have to be taken into due account:

1. Practically the totally of data in astronomical and astrophysical systems are acquired by using photons arrival time. It was Einstein [15] at the very initial steps of

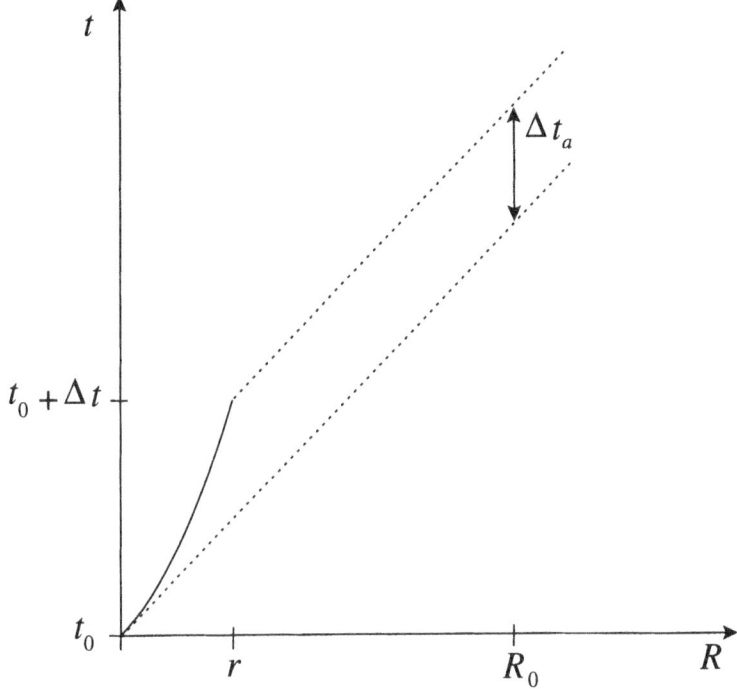

Fig. 3. This qualitative diagram illustrates the relation between the laboratory time interval Δt and the arrival time interval Δt_a for a pulse moving with velocity v in the laboratory time (solid line). We have indicated here the case where the motion of the source has a nonzero acceleration. The arrival time is measured using light signals emitted by the pulse (dotted lines). R_0 is the distance of the observer from the EMBH, t_0 is the laboratory time corresponding to the onset of the gravitational collapse, and r is the radius of the expanding pulse at a time $t = t_0 + \Delta t$. See also [44].

special relativity to caution on the use of such an arrival time analysis and to state that, when dealing with objects in motion, due care should be put in defining time synchronization procedure in order to construct the correct spacetime coordinate grid (see Fig. 3). It is not surprising that as soon as the first bulk motions relativistic effects were observed by radio and optical telescopes their interpretations within the classical framework of astrophysics led to the concept of "superluminal" motion. These observations refer to extragalactic radio sources, with Lorentz gamma factors ~ 6 [4] and to microquasars in our own galaxy with Lorentz gamma factor ~ 5 [28]. It has been recognized [38] that no "superluminal" motion exist as soon as the prescriptions indicated by Einstein are used in order to establish the correct spacetime grid for the astrophysics systems. In the present context of GRBs, where the Lorentz gamma factor can easily surpass 10^2, the direct application of classical concepts leads to enormous "superluminal" behaviors [50]. An approach based on classical arrival time considerations, as done sometime in current literature, completely subvert the causal relation in the observed astrophysical phenomenon.

2. One of the clear success of relativistic field theorists has been the understandings of the role of four momentum energy conservation laws in multiparticle collisions and decays such as the reaction: $n \rightarrow p + e + \bar{\nu}$. From the works of Pauli anf Fermi it became clear how in such processes, contrary to the case of classical mechanics, it is impossible to analyze a single term of the decay, the electron or the proton or the neutrino or the neutron, out of the context of the global point of view of the relativistic conservation of the total four momentum of the system, which involves the knowledge of the system during the entire decay process. These rules are routinely used by workers in high energy particle physics and have become part of their cultural background. If we apply these same rules to the case of the relativistic system of a GRB it is clear that it is just impossible to consider a part of the system, e.g. the afterglow, out of the general conservation laws and history of the *entire relativistic regime* of the system. The description of the sole afterglow, as has been done at times in the literature, could indeed be done within the framework of classical astronomy and astrophysics, but not in a relativistic astrophysics where the entire space-time grid necessary for the description of the afterglow depends on all the previous relativistic part of the worldline of the system.

3. The very lifetime of a phenomenon has not an absolute meaning, special and general relativity have shown. It depends both from inertial reference frame of the laboratory and of the observer and their relative motion. Such a phenomenon, generally expressed in the "twin paradox" has been extensively checked and confirmed to extreme high accuracy in elementary particle physics in the CERN experiments. This situation is much more extreme in GRBs due to the very large (in the range $10^2 - 10^4$) and time varying (on time scales ranging from fractions of seconds to months) gamma Lorentz factors between the Laboratory frame and the far away observer. Such an observer is moreover in the GRBs context further affected by the cosmological recession velocities of its local Lorentz frame.

These are some of the reasons why we have recently presented a basic Relative Space-Time Transformations RSTT paradigm [44] to be applied prior to the interpretation of GRBs data.

The first step is the establishment of the constitutive equations relating:

a) The comoving time of the pulse (τ)

b) The laboratory time (t)

c) The arrival time (t_a)

d) The arrival time at the detector (t_a^d)

The book-keeping of the four different times and corresponding space variables must be done carefully in order to keep the correct causal relation in the time sequence of the events involved.

The RSST paradigm reads: *"the necessary condition in order to interpret the GRB data is the knowledge of the entire worldline of the source from the moment of gravitational collapse. In order to meet this condition, given a proper theoretical description and the correct constitutive equations, it is sufficient to know the energy of the dyadosphere and the mass of the remnant of the progenitor star".*

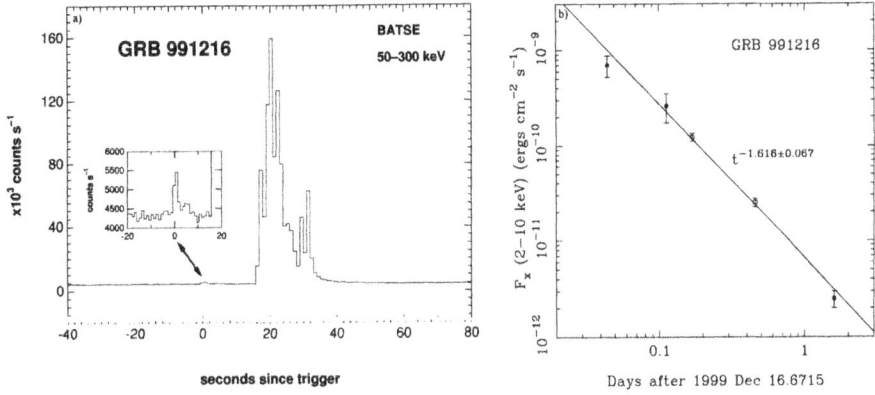

Fig. 4. a) The peak emission of GRB 991216 as seen by BATSE (reproduced from [1]); **b)** The afterglow emission of GRB 991216 as seen by XTE and Chandra (reproduced from [20]).

Clearly such an approach is at variance with the works in the current literature which mainly emphasize either some qualitative description of the sources or some quantitative phenomenological multiparameter fitting of the sole afterglow era.

- Many works in the literature have addressed the issue of the sources of GRBs. They include works on binary neutron stars mergers (see e.g. [14,29,24,25]), black hole - white dwarf [18] and black hole - neutron star binaries [30,27], Hypernovae (see [32]), failed supernovae or collapsars (see [63,23]), supranovae (see [60,61]). Only those based on binary neutron stars have reached the definition of detailed quantitative estimates of a model, but they present serious difficulties in the energetics, as well as in the explanation of "long bursts" (see [54,62]), and in the observed location of the GRBs' sources in star forming regions (see [5]). In the remaining cases was presented a sole qualitative analysis of the sources without addressing the overall problem from the source to the observations: the necessary details to formulate the equations of the dynamical evolution of the system are generally still missing.

Other works in the literature have mainly addressed the problem of fitting the data of the afterglow observations by phenomenological analysis. They are separated in two major classes:

- The "internal shock models", first introduced by [39], are by far the most popular ones having been developed in many different aspects, e.g. by [31,55,16,17]. The underlying assumption is that all the variabilities of GRBs in the range $\Delta t \sim 1$ ms up to the overall duration T of the order of 50 s are determined by a yet undetermined "inner engine". The difficulties of explaining the long time scale bursts by a single explosive event has evolved into a variety of assumptions on yet unspecified family of "inner engines" with a prolonged activity (see e.g. [34] and references therein).

- The "external shock models", also introduced by [26], are less popular today. There is the distinct possibility, within these models, that "GRBs' light curves are tomographic images of the density distribution of the medium surrounding the sources of GRBs" ([12], see also [11,13] and references therein). It is generally outlined that the structure of the burst does not depend directly from the "inner engine".

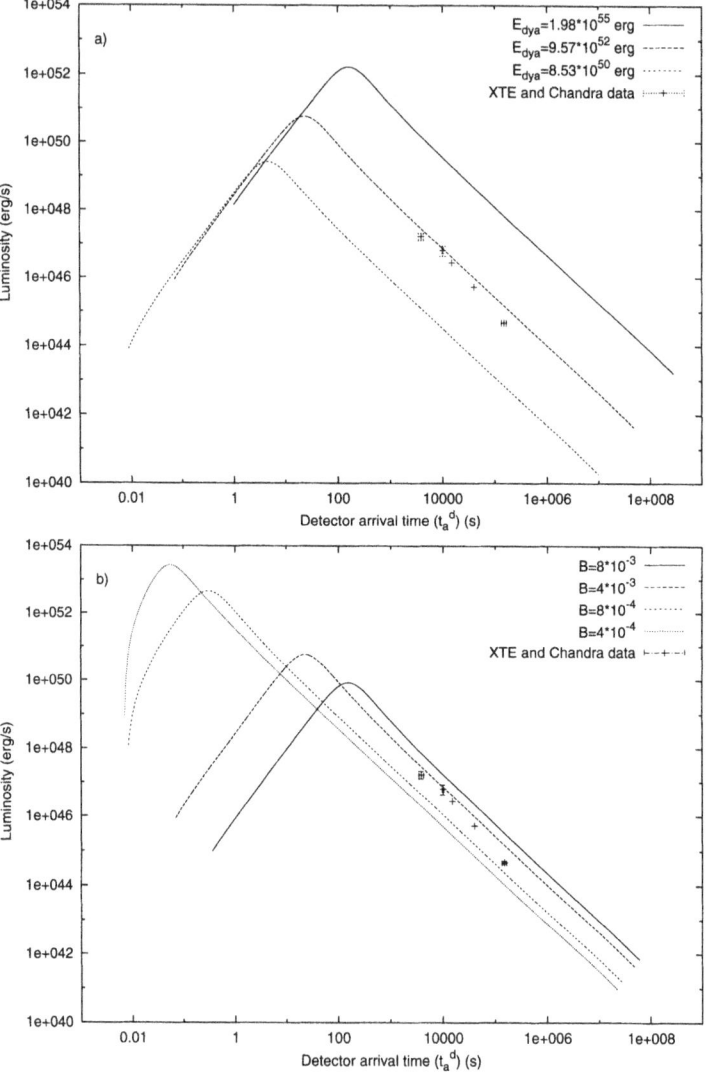

Fig. 5. a) Afterglow luminosity computed for an EMBH of $E_{dya} = 8.53 \times 10^{50}$ ergs, $E_{dya} = 9.57 \times 10^{52}$ ergs, $E_{dya} = 1.98 \times 10^{55}$ ergs and $B = 4 \times 10^{-3}$. b) for the $E_{dya} = 9.57 \times 10^{52}$, we give the afterglow luminosities corresponding respectively to $B = 4 \times 10^{-4}$, 8×10^{-4}, 4×10^{-3}, 8×10^{-3}.

All these works encounter the above mentioned difficulties, they present only a piecewise description of the GRB phenomenon and by neglecting the earlier phases, their space time grid is undefined and, as we have shown in detail in [50], results are reached at variance from the ones obtained in a complete and unitary description of the GRBs phenomenon. We outline in the following how such an unitary description naturally leads to new characteristic features both in the burst and afterglow of GRBs.

In a series of papers, we have developed the above mentioned EMBH model [40] which has the great advantage, despite its simplicity, that all eras of the model, following the process of gravitational collapse, are described by precise field equations which can then be numerically integrated. The three basic starting points are:

- the extractable energy of an EMBH introduced in [7],
- the vacuum polarization process *à la* Heisenberg-Euler-Schwinger [22,57] in the field of an EMBH, first computed in [10],
- the fact that vacuum polarization process can indeed be a realization of the reversible transformation of an EMBH introduced in [7].

These were the themes of discussions with Werner Heisenberg. He was supposed to inaugurate the 1975 Varenna Summer School [19] directed to these subjects if he had not died a few weeks earlier. In that school and in [10] the possibility that the process in [22,57] duly extended to the EMBH were at the very basis of the explanation of GRBs was advanced (see Fig. 1).

Following the Beppo SAX observations and the energetics requirements, we have returned to our EMBH model [40] and developed the dyadosphere concept [36]. The dynamics of the e^+e^--pairs and electromagnetic radiation of the plasma generated in the dyadosphere propagating away from the EMBH in a sharp Pairs-Electro-Magnetic pulse (the PEM pulse) has been studied by us and validated by the numerical codes at Livermore Lab [41]. The collision of the still optically thick PEM pulse with the baryonic matter of the remnant of the progenitor star has been again studied and validated by the Livermore Lab codes [42]. The further evolution of the sharp pulse of pairs-electromagnetic radiation and baryons (the PEMB pulse) further proceeds with increasing values of the Lorentz gamma factor until the condition of transparency is reached [2]. At this stage the Proper-Gamma Ray Burst (P-GRB) is emitted [45] and a pulse of Accelerated-Baryonic-Matter the (ABM pulse) is injected in the interstellar medium giving rise to the afterglow.

The interaction of the ABM-Pulse giving origin to the afterglow has been recently developed and presented in detail in [50]. We recall the minimum set of assumptions we have adopted:

1. the collision of the ABM pulse is assumed to occur with a constant homogeneous interstellar medium of number density $n_{ism} \sim 1cm^{-3}$. The energy emitted in the collision is assumed to be instantaneously radiated away (fully radiative condition). The description of the collision and emission process is done in spherical symmetry, taking only the radial approximation neglecting all the delay emission by scattered radiation.
2. special attention is given to numerically compute the power of the afterglow as a function of the arrival time using the correct constitutive equations for the space-time transformations in line with the RSTT paradigm.

3. finally some approximate solutions are adopted in order to obtain the determination of the power law indexes of the afterglow flux and compare and contrast them with the observational results as well as with the alternative results in the literature.

In [50] we have considered uniquely the above radial approximation in order to concentrate on the special role of the correct space time transformations in the RSST paradigm and to explicitly illustrate their impact on the determination of the power law index of the afterglow. This topic has been unduly neglected in the literature. We enter in a forthcoming papers both in the details of the role of beaming of the radiation and of the diffusion due to off axis emission [51,3].

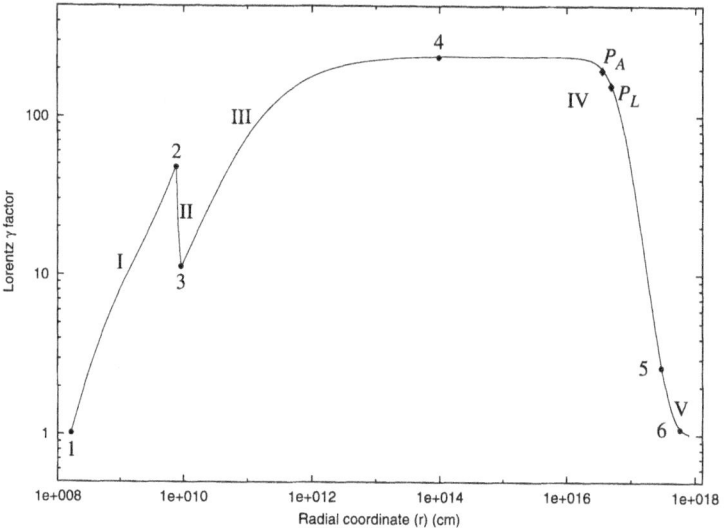

Fig. 6. The theoretically computed Lorentz gamma factor for the parameter values $E_{dya} = 9.57 \times 10^{52}$ erg, $B = 4 \times 10^{-3}$ is given as a function of the radial coordinate in the laboratory frame. The corresponding values in the comoving time, laboratory time and arrival time are given in [50]. The different eras, indicated by roman numerals, are illustrated in the text, while the points 1,2,3,4,5 mark the beginning and end of each of these eras. The points P_L and P_A mark the maximum of the afterglow flux, respectively in emission time and in arrival time (see [45]). The point 6 is the beginning of Phase D in era V. At point 4 the transparency condition is reached.

It is now clear after the observations of GRB 980425 that the afterglow phenomena can present, especially in the optical and radio wavelengths, features originating from phenomena spatially and causally distinct from the GRB phenomena. There is evidence in four different GRB systems, including GRB 991216 due to the observed emission in the iron lines and their shift, that a second component exist associated to the GRBs: a supernova. This supernova explosion follows in time the GRB emission and occurs at distances of $10^{16} - 10^{17}$ cm away from the location where the gravitotional collapse to the EMBH has occurred. The space time analysis of these events can be correctly performed when the RSTT paradigm is adopted [44,45]. These facts have motivated us to introduce the novel concept of induced supernovae [46]. In the current litterature the existence of these two different components has not been recognized and attempts have been made of fitting the data of the supernova, in the x-rays as well as in the optical and radio, as part of the afterglow within the framework of a multiparameters fitt. For the above mentioned reasons, such an approach adds further difficulties to the already critical situation of the current literature.

We have therefore confronted the theoretical results of the EMBH model with the data of GRB 991216 as a prototypical case (see Fig. 4). The reason of this choice are simply given:

1. This is one of the strongest GRBs in x-rays and is also quite general in the sense that shows relevant cosmological effects. It radiates mainly in X-rays and in γ-rays and less then 3% is emitted in optical and radio bands (see [20]). Also the emission of the supernova, inferred from the iron lines, is in this case weaker then the autentic GRB energy flux.
2. The excellent data obtained by BATSE on the burst [1] are complemented by the data on the afterglow acquired by the Chandra [35] and RXTE [8], and also superb data have been obtained from spectroscopy of the iron lines [35].
3. A very precise value for the slope of the energy emission during the afterglow as a function of time has been obtained: $n = -1.64$ [59] and $n = -1.616 \pm 0.067$ [20].

The comparison of the EMBH model to the data of the GRB 991216 and its afterglow has naturally led to a new paradigm for the interpretation of the burst structures (IBS paradigm) of GRBs [45]. The IBS paradigm reads: *"in GRBs we can distinguish an injector phase and a beam-target phase. The injector phase includes the process of gravitational collapse, the formation of the dyadosphere, as well as era I (the PEM pulse), era II (the engulfment of the baryonic matter of the remnant) and era III (the PEMB pulse). The injector phase terminates with the P-GRB emission. The beam-target phase addresses the interaction of the ABM pulse, namely the beam generated during the injection phase, with the ISM as the target. It gives rise to the E-APE and the decaying part of the afterglow"*.

We recall that:

a) The **injector phase** starts from the moment of gravitational collapse all the way to the emission of the proper GRB (the P-GRB) and encompasses the following eras:

- *The zeroth era*: the formation of the dyadosphere;
- *The era I*: the expansion of the PEM pulse;
- *The era II*: the interaction of the PEM pulse with the remnant left over by the collapse of the progenitor star;
- *The era III*: the further expansion of the PEMB pulse; The injector phase is concluded by the emission of the P-GRB and the ABM pulse, as the condition of transparency is reached.

b) The **beam-target phase**, in which the accelerated baryonic matter (ABM) generated in the injector phase collides with the interstellar medium (ISM), gives origin to the afterglow and encompasses the following eras:

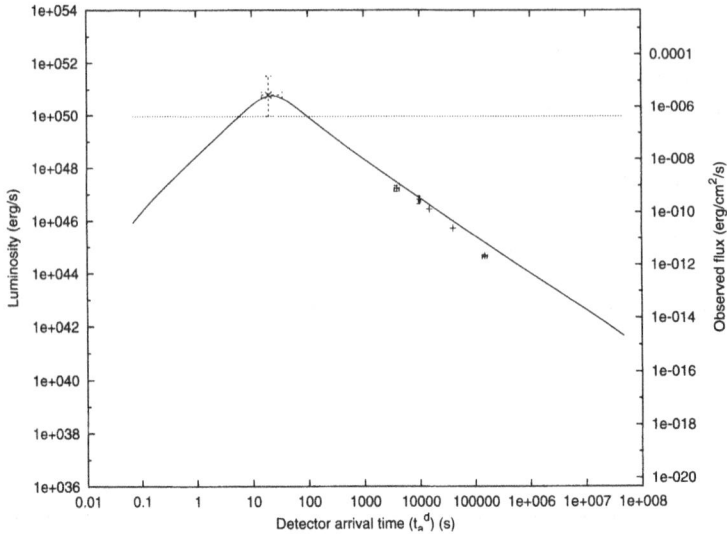

Fig. 7. Best fit of the afterglow data of Chandra, RXTE as well as of the range of variability of the BATSE data on the major burst, by a unique afterglow curve leading to the parameter values $E_{dya} = 9.57 \times 10^{52} erg$, $B = 4 \times 10^{-3}$. The horizontal dotted line indicates the BATSE noise threshold. On the left axis the luminosity is given in units of the energy emitted at the source, while the right axis gives the flux as received by the detectors.

- *The Era IV*: the ultra relativistic and relativistic regimes in the afterglow: the emitted flux first increases to reach a maximum value and then monotonically decrease following well defined power laws in the arrival time;
- *The Era V*: the approach to the non relativistic regimes in the afterglow, also describable by specific power laws in the arrival time;

Some qualitative representation of these eras as a function of the radial coordinate, in logarithmic scale are represented in Fig. 2.

The comparison of the EMBH theory to the data of GRB 991216 has allowed the determination of the only two free parameters of the theory: the energy of the Dyadosphere E_{dya} and the mass of the baryonic remnant left over by the collapse of the progenitor star, measured in units of the E_{dya}, defined by the dimensionless B parameter. Details are given in [50].

We have then obtained, for the first time, the complete history of the Lorentz gamma factor from the moment of gravitational collapse to the latest phases of the afterglow observations (see Fig. 6). We have as well determined the entire space time grid of the GRB 991216 by giving (see Table 1 in [50]) the radial coordinate of the GRBs phenomenon as a function of the four coordinate time variables. The extreme relativistic regimes at work in GRB 991216 lead to enormous superluminal behavior (up to $10^4 c$!) if the classical astrophysical concepts were adopted using the arrival time as the independent variable (see Table 1 in [50]). In turn this implies that any causal relation

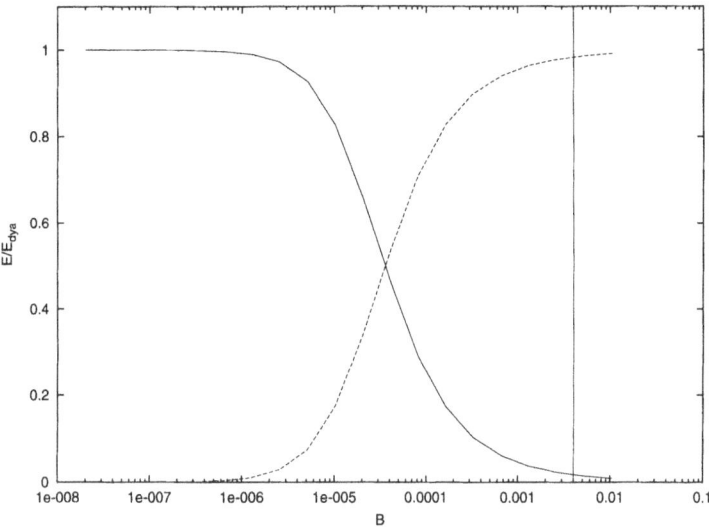

Fig. 8. Relative intensities of the E-APE (dashed line) and the P-GRB (solid line), as predicted by the EMBH model corresponding to the values of the parameters determined in Fig. 7, as a function of B. Details are given in [50]. The vertical line corresponds to the value $B = 4 \times 10^{-3}$.

based on classical astrophysics and the arrival time data, as often done in current GRBs literature, is incorrect.

We have just stressed how the analysis of the sole afterglow of GRB 991216 data, obtained by BATSE and the Chandra and RXTE satellites, has allowed to fix the only two free parameters of the EMBH theory. As a first byproduct of this analysis we can conclude, at variance with results in [20,33] pointing to a sharply collimated beamed emission in GRB 991216, that no evidence of beaming is found as a consequence of the perfect agreement between the observed slope of the afterglow and the theoretical value obtained within the EMBH model.

We can now proceed to acquire the predictions of the EMBH theory with reference to two fundamental quantities and their role within the GRBs structure: the P-GRB and the peak emission of the afterglow.

It soon appeared clear that the :

1. the so called "long bursts" observed by BATSE are actually not bursts at all. Once the proper space-time greed is given it is immediately clear that the long bursts are generated at distances of 4×10^{16} cm from the EMBH. The long burst coincides with the Extended-Afterglow-Peak-Emission, which we will call E-APE, of the afterglow: they were interpreted as bursts only due to the high threshold of the BATSE detectors see Fig. 7.
2. The time variability observed in simply due to the density inhomogeneities intrinsic in an interstellar cloud, as the ABM pulse impact on it [47], also see Fig. 10.
3. The previous two conclusions are based on the simplified pure radial description of the afterglow presented in [50]. The effects of angular scattering and spreading

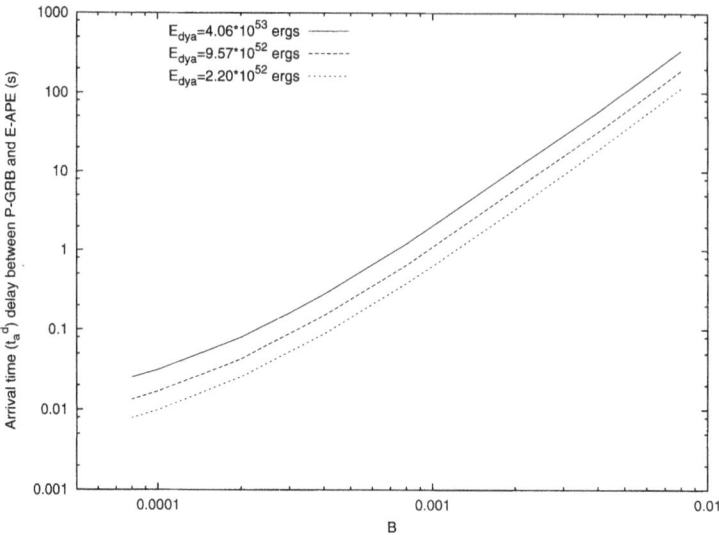

Fig. 9. Time delays between P-GRB and E-APE as a function of the B parameter for selected values of the E_{dya}.

in the signal has been considered [51,3]. This more complex approach leads to interesting new results, but does not affect the two above conclusions.

The hunt of the P-GRB then started. The interest in identifying it is mainly because some general relativistic and relativistic quantum field theory effects originating in the process of gravitational collapse during the formation of the EMBH are, in principle, encoded in the structure of the P-GRB.

Having the only two free parameter of the EMBH theory been fixed, there are two fundamental diagrams to be analyzed. The first, Fig. 9 relates the precise separation in time between the E-APE and the P-GRB, as a function of the amount of baryonic matter left over by the gravitational collapse of the progenitor star expressed by parameter B and for selected values of the E_{dya}. The second relates the intensities of the P-GRB to the E-APE, in units of the E_{dya}, to the parameter B, see Fig. 8. We stress that indeed this last diagram is an universal one, in the adopted variables. From these diagrams we can identify with the precision of *a few percent* in the intensity and with an approximation of *a few tenth of milliseconds* the P-GRB with the "precursor" in the BATSE data, (see Fig. 4 and Fig. 11).

Before concluding we would like to stress one final important consequence and prediction of the EMBH model:

1. the most general GRB is composed of the P-GRB and and the afterglow (see Fig. 11). The relative intensity of the P-GRB and the E-APE is a function of the B parameter.
2. for $B < 3.5 \times 10^{-5}$ the energy of the P-GRB is larger then the one of the E-APE and the energy of the dyadosphere is mainly emitted in what have been called

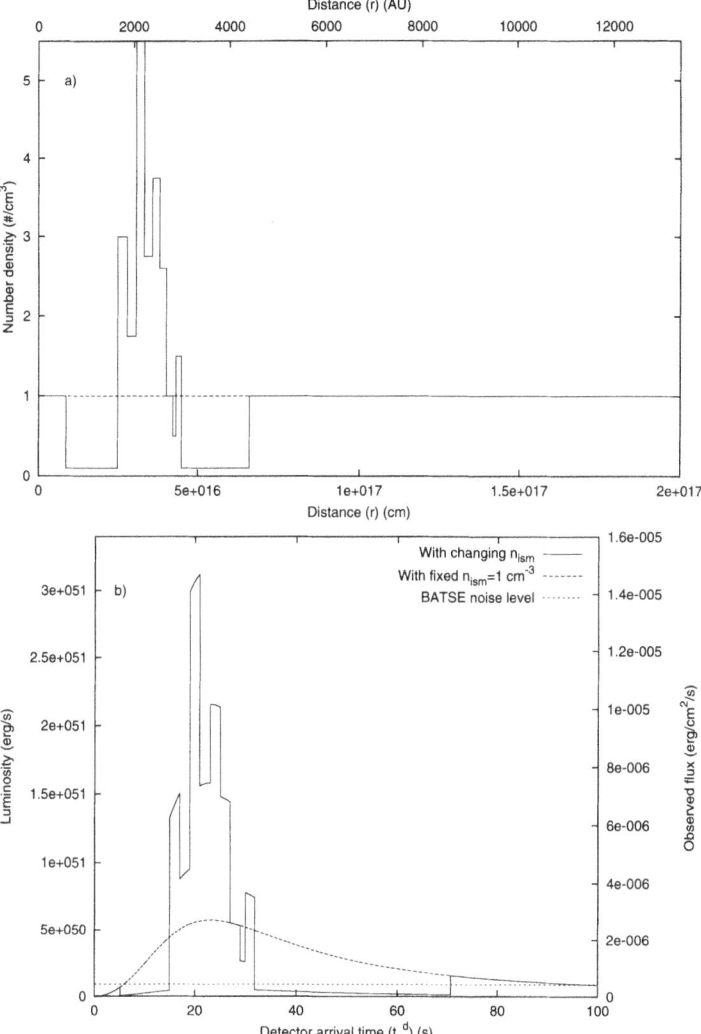

Fig. 10. a) The density contrast of the ISM cloud profile introduced in order to fit the observation of the burst of GRB 991216. The dashed line indicates the average uniform density $n = 1cm^{-3}$. b) Flux computed in the collision of the ABM pulse with an ISM cloud with the density profile given in a). The dashed line indicates the emission from an uniform ISM with $n = 1cm^{-3}$. The dotted line indicates the BATSE noise level. Note that 20 seconds in arrival time do corresponds to $\sim 5.0 \times 10^{16}$ cm in laboratory frame! Details are given in Table 1 of [50]. Compare and contrast these theoretical curves with the actual data reported in Fig. 4.

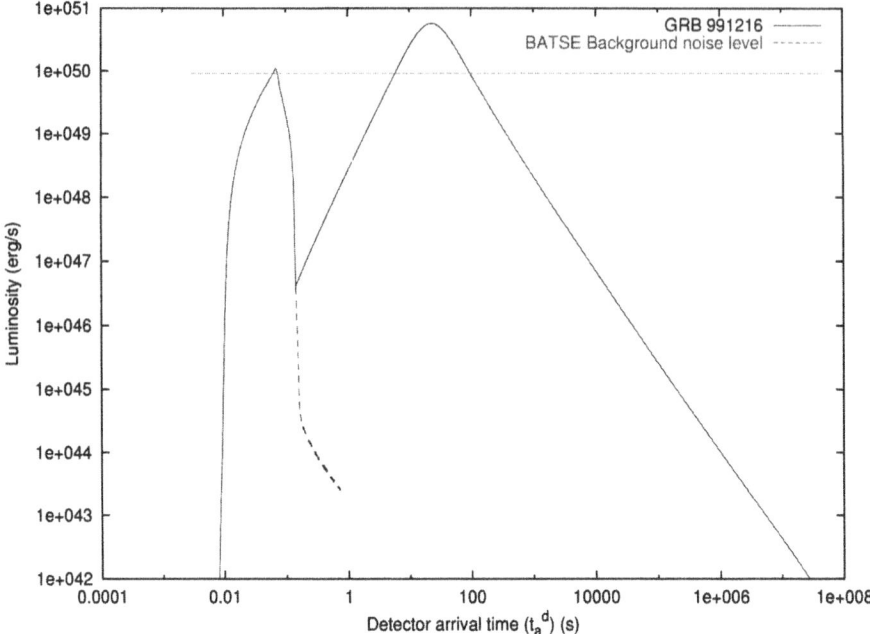

Fig. 11. Diagram showing the full picture of the model, with both P-GRB and E-APE and their relative intensities and time separation. The numerical values are presented and tabulated in [50].

the "short bursts". Their afterglow have been systematically lower then the BATSE threshold.

3. for $B > 3.5 \times 10^{-5}$ the energy of the E-APE predominates and the energy of the dyadosphere is mainly carried by the ABM pulse and emitted in the afterglow. The corresponding E-APE have been improperly called "long bursts".

It is interesting that this classification also explains at once the recently found conclusion that the distribution of short GRBs and long GRBs have essentially the same characteristic peak luminosity [56] and the fact that the short bursts are systematically harder then long bursts, see [50] for details.

Similarly the application of the RSTT and IBS paradigm has naturally conduced as to the concept of induced gravitational collapse [46], in order to explain the observed iron lines emission in the late phases of the afterglow of GRB 991216 and analogous effects in three additional GRBs showing clear correlation with supernovae. This is the topic of fortcoming publications [52].

The understanding of the role of P-GRB and E-APE in GRB 991216, the fact that both their absolute and relative intensities have been predicted within a few percent accuracy as well as that their arrival time has been computed with the precision of a few milliseconds, see [50], and Figs. 7,9,11, can be considered one of the major success of the EMBH theory.

New space missions have to be conceived to explore, on additional GRB sources, the theoretical predictions in the first 10^2 seconds of Fig. 11. This region has been left vastly unexplored by the BATSE data due to the high threshold. Current missions are exploring with great accuracy mainly the later phases of the afterglow. These observations as well as the ones of short bursts, which in the EMBH theory are P-GRB emissions, are indeed crucial, since indeed we re-hiterate: *all general relativistic and relativistic quantum field theory effects originating in the process of gravitational collapse during the formation of the EMBH are, in principle, encoded in the structure of the P-GRB* [6,49,53].

The long lasting debate, started in Princeton in 1971, of how an EMBH is formed has also by now been clarified in [43]. The needed charge segregation process occurs in the magnetosphere of a rotating magnetized star. The charged collapsing core, surrounded by an oppositely charged remnant, approach the EMBH final stages in ~ 30 seconds for a $10M_\odot$ progenitor star. The leading process of discharge of the EMBH is due to the vacuum polarization process in view of their very short time scale 10^{-19} seconds, see [37].

The EMBH theory can now be applied to all other GRB sources.

References

1. BATSE Rapid Burst Response, 1999, http://gammaray.msfc.nasa.gov/ kippen/batserbr/
2. Bianco, C.L., Ruffini, R., Xue, S.-S., 2001, A&A, 368, 377
3. Bianco, C.L., Chardonnet, P., Ruffini, R., Xue, S.-S., 2002, in preparation
4. Biretta, J.A., Sparks, W.B., Macchetto, F., 1999, ApJ, 520, 621
5. Bloom, J.S., Kulkarni, S.R., Djorgowski, S.G., 2000, submitted to Astron. Journ., astro-ph/0010176
6. Cherubini, C., Ruffini, R., Vitagliano, L., 2002, in preparation
7. Christodoulou, D., Ruffini, R., 1971, Phys. Rev. D, 4, 3552
8. Corbet, R., Smith, D.A., 2000, in "Rossi2000: Astrophysics with the Rossi X-ray Timing Explorer", Greenbelt, USA
9. Costa, E., invited talk in "IX Marcel Grossmann Meeting on General Relativity", V. Gurzadyan, R. Jantzen and R. Ruffini Ed., World Scientific, 2001, in press
10. Damour, T., Ruffini, R., 1975, Phys. Rev. Lett., 35, 463
11. Dermer, C.D., Chiang, J., Böttcher, M., 1999, ApJ, 513, 656
12. Dermer, C.D., Mitman, K.E., 1999, ApJ, 513, L5
13. Dermer, C.D., 2000, astro-ph/0005440
14. Eichler, D., Livio, M., Piran, T., Schramm, D.N., 1989, Nature, 340, 126
15. Einstein, A., 1905, Ann. Phys. (Germany), 17, 891
16. Fenimore, E.E., 1999, ApJ, 518, 375
17. Fenimore, E.E., Cooper C., Ramirez-Ruiz, E., Sumner, M.C., Yoshida, A., Namiki, M., 1999, ApJ, 512, 683
18. Fryer, C.L., Woosley, S.E., Herant, M., Davies, M.B., 1999, ApJ, 520, 650
19. Giacconi, R., Ruffini, R., Ed. and coauthors, *Physics and Astrophysics of Neutron Stars and Black Holes*, North Holland, Amsterdam, 1978
20. Halpern, J.P., Uglesich, R., Mirabal, N., Kassin, S., Thorstensen, J., Keel, W.C., Diercks, A., Bloom, J.S., Harrison, F., Mattox, J., Eracleous, M., 2000, ApJ, 543, 697
21. Halzen, F., "High energy neutrino astronomy" in "Weak Interactions and Neutrinos, Proceedings of the 17th International Workshop. Cape Town, South Africa", Edited by C. A. Dominguez and R. D. Viollier, World Scientific Publishers (Singapore, 2000), p.123

22. Heisenberg, W., Euler, H., 1935, Zeits. Phys., 98, 714
23. MacFadyen, A.I., Woosley, S.E., 1999, ApJ, 524, 262
24. Mészáros, P., Rees, M.J., 1992a, MNRAS, 257, 29p
25. Mészáros, P., Rees, M.J., 1992b, ApJ, 397, 570
26. Mészáros, P., Rees, M.J., 1993, ApJ, 405, 278
27. Mészáros, P., Rees, M.J., 1997, ApJ, 482, L29
28. Mirabel, I.F., Rodriguez, L.F., 1999, A.R.A.A., 37, 409
29. Narayan, R., Paczyński, B., Piran, T., 1992, ApJ, 395, L83
30. Paczyński, B., 1991, Acta Astronomica, 41, 257
31. Paczyński, B., Xu, G., 1994, ApJ, 427, 708
32. Paczyński, B., 1998, ApJ, 494, L45
33. Panaitescu, A., Kumar, P., 2001, ApJ, 560, L49
34. Piran, T., talk at 2000 Texas Meeting, see also astro-ph/0104134
35. Piro, L., et al., 2000, Science, 290, 955
36. Preparata, G., Ruffini, R., Xue, S.-S., 1998, A&A, 338, L87
37. Preparata, G., Ruffini, R., Xue, S.-S., 2002, in "Proceedings of The Seventh ICRA Network Workshop and The Seventh Korean-Italian Meeting on General Relativity".
38. Rees M.J., 1966, Nature 211, 468
39. Rees, M.J., Mészáros, P., 1994, ApJ, 430, L93
40. Ruffini, R., 1998, in "Black Holes and High Energy Astrophysics", Proceedings of the 49th Yamada Conference Ed. H. Sato and N. Sugiyama, Universal Ac. Press, Tokyo, 1998
41. Ruffini, R., Salmonson, J.D., Wilson, J.R., Xue, S.S., 1999, A&A, 350, 334, A&AS, 138, 511
42. Ruffini, R., Salmonson, J.D., Wilson, J.R., Xue, S.S., 2000, A&A, 359, 855
43. Ruffini R., 2002, in "Proceedings of the Ninth Marcel Grossmann Meeting on General Relativity", World Scientific, Singapore, in press
44. Ruffini, R., Bianco, C.L., Chardonnet, P., Fraschetti, F., Xue, S.-S., 2001, ApJ, 555, L107
45. Ruffini, R., Bianco, C.L., Chardonnet, P., Fraschetti, F., Xue, S.-S., 2001, ApJ, 555, L113
46. Ruffini, R., Bianco, C.L., Chardonnet, P., Fraschetti, F., Xue, S.-S., 2001, ApJ, 555, L117
47. Ruffini, R., Bianco, C.L., Chardonnet, P., Fraschetti, F., Xue, S.-S., 2001, Nuovo Cimento B, 116, 99
48. Ruffini, R., in "Fluctuating Paths and Fields - Dedicated to Hagen Kleinert on the Occasion of His 60th Birthday", Eds. W. Janke, A. Pelster, H.-J. Schmidt, and M. Bachmann, World Scientific, Singapore, 2001, p. 771
49. Ruffini, R., Vitagliano, L., 2002, submitted for pubblication
50. Ruffini, R., Bianco, C.L., Chardonnet, P., Fraschetti, F., Xue, S.-S., 2002, submitted for publication
51. Ruffini, R., Bianco, C.L., Chardonnet, P., Xue, S.-S., 2002, in preparation
52. Ruffini, R., Bianco, C.L., Chardonnet, P., Xue, S.-S., 2002, in preparation
53. Ruffini, R., Vitagliano, L., Xue, S.-S., 2002, in preparation
54. Salmonson, J.D., Wilson, J.R., Mathews, G.J, 2001, ApJ, 553, 471
55. Sari, R., Piran, T., 1997, ApJ, 485, 270
56. Schmidt, M., 2001, ApJ, 559, L79
57. Schwinger, J., 1951, Phys. Rev., 98, 714
58. Strong, I.B., in "Neutron Stars, Black Holes and Binary X-Ray Sources", Gursky, H. and Ruffini, R., editors, D. Reidel Publishing Company, 1975
59. Takeshima, T., Markwardt, C., Marshall, F., Giblin, T., Kippen, R.M., 1999, GCN Circ. 478 (http://gcn.gsfc.nasa.gov/gcn/gcn3/478.gcn3)
60. Vietri, M., Stella, L., 1998, ApJ, 507, L45
61. Vietri, M., Stella, L., 1999, ApJ, 527, L43
62. Wilson, J.R., Mathews, G.J., Marronetti, P., 1996, Phys. Rev. D, 54, 1317
63. Woosley, S.E., 1993, ApJ, 405, 273

Effect of the Pair-Annihilation on the Break Energy of GRB Spectra

Katsuaki Asano and Shiho Kobayashi

Department of Earth and Space Science, Osaka University,
Toyonaka 560-0043, Japan

1 Introduction

The widely accepted scenario for producing gamma-ray bursts (GRBs) and their after-glow is the dissipation of the kinetic energy of a relativistic flow by relativistic shocks. The rapid temporal variability requires that the GRB itself must arise from internal shocks within the flow, while the afterglow is due to the external shock produced as the flow is decelerated upon collision with the ambient medium. However, burst hardnesses are generally less than 1 MeV [1], and the apparent clustering of the break energies of GRB spectra in the 50keV-1MeV range is reported [2]. Recently Guetta, Spada & Waxman [3] showed by using a Monte-Carlo simulation that the Thomson optical depth due to $e^{\pm}$ pairs produced by synchrotron photons plays an important role in the internal shock model. They argued that the inclusion of the pair optical depth effects is essential in determining the break energy. In their study Guetta et al. use the standard inelastic collision approximation in which two colliding shells merge into a single uniform shell. However this method does not necessarily reproduce the internal energy in shocked shell and the break energy adequately, especially for a small relative Lorentz factor of two collisional shells. In this paper, we reexamine the pair optical depth effect using a simple analytic model. We show how the photosphere depends on the parameters of the internal shock model.

1.1 Internal Shocks

Consider equal mass shells with different Lorentz factors, $x \equiv \gamma_r/\gamma_s > 1$. The relative Lorentz factor is $\gamma \sim (x + 1/x)/2$. The rapid and the slower shells are denoted by the subscripts r and s, respectively. We assume that the widths of the shells L are comparable in the interstellar medium (ISM) rest frame. Under these assumptions, the ratio of the baryon number density is given by $n_r/n_s = 1/x$. In this paper, thermodynamical quantities, n and internal energy density e, are measured in the fluids' rest frame.

When the rapid shell catches up with the slower one at some radius r, the forward and reverse shocks form. The shock conditions and the equality of pressures along the contact discontinuity yield

$$\frac{(\gamma_F - 1)(4\gamma_F + 3)}{(\gamma_R - 1)(4\gamma_R + 3)} = \frac{n_r}{n_s}, \tag{1}$$

where γ_F and γ_R are Lorentz factors of the relative motion between the regions separated by the forward shock and by the reverse shock, respectively. Since the equality of velocities along the contact discontinuity gives $\gamma_R = \gamma_F \gamma - \sqrt{(\gamma_F^2 - 1)(\gamma^2 - 1)}$, the solution for γ_R depends only on one parameter x. The width of the shocked shell at the comoving frame l is $x\gamma_s L/(4\gamma_R + 3)$.

We now consider the synchrotron emission from the shocked shells. Since the rapid shell has larger energy and lower number density, the reverse shock emission is more luminous and harder. It dominates the forward shock emission. Therefore, we consider only the emission from the reverse shock. The shock accelerates electrons in the shell material into a power-law distribution: $N(\gamma_e) \propto \gamma_e^{-p}(\gamma_e \geq \gamma_{e,min})$. Assuming that constant fractions ϵ_B and ϵ_e of the internal energy go into the magnetic field and the electrons, respectively, one finds that the magnetic field and the typical random Lorentz factor of the electrons are given by $B^2 \propto \epsilon_B e$ and $\gamma_{e,min} \propto \epsilon_e(\gamma_R - 1)$, respectively.

In the fast cooling case, the observed characteristic photon energy is

$$\varepsilon_p^{obs} = 610\ \epsilon_e^2 F(x)\gamma_{s,2}\sqrt{\Sigma/\Delta}\ \ \text{keV}, \tag{2}$$

where $F(x) = x^{1/2}(\gamma_R - 1)^{5/2}(4\gamma_R + 3)^{1/2}(\gamma_R - \sqrt{\gamma_R^2 - 1})$. We have scaled the parameters as $\epsilon_{B,-1} = \epsilon_B/0.1$, $M_{48} = Mc^2/10^{48}$ ergs, $r_{13} = r/10^{13}$ cm, $L_7 = L/10^7$ cm, and $\gamma_{s,2} = \gamma_s/100$. We adopt $p = 2.5$ here and hereafter. Since some parameters appear in the following formulae in the same combinations, we have defined two variables, a surface density $\Sigma = M_{48}/r_{13}^2$ and an effective comoving shell width $\Delta = L_7\gamma_{s,2}/\epsilon_{B,-1}$ for convenience.

2 Photosphere

A fireball ejecta is initially optically thick due to the Thomson scattering by electrons associated with baryons in the ejecta. We can not observe the emission from it, until it comes out from the photosphere where the optical depth $\tau_T = 2\sigma_T M/4\pi r^2 m_p \sim 0.7\Sigma$ becomes unity. The maximum break energy E_M corresponds to ε_p^{obs} coming from photospheres ($\tau_T = 1$). This is a well-known result, but the estimate may undergo a significant change in the internal shock model if we take into account $e^{\pm}$ pairs produced by the synchrotron photons [3].

Using a simple approximation, we can analytically estimate the optical depth of the Thomson scattering due to the pairs as

$$\tau_\pm \sim 300\epsilon_e^3 G^2(x)\Sigma^{9/4}\Delta^{-1/4}, \tag{3}$$

where $G(x) = (\gamma_R - 1)^{13/8}(4\gamma_R + 3)^{1/8}/x^{1/8}$.

Since the characteristic energy ε_p^{obs} is proportional to $\Sigma^{1/2}$, a collision occurring with larger Σ radiates harder photons. However, we cannot observe collisions inside the photosphere. Then, the emission just above the photosphere is the hardest for given values of the parameters x, Δ and γ_s. Figure 1 depicts the maximum characteristic energy E_M as a function of x. In order to plot Figure 1, we numerically solved equation (1) for γ_R, obtained the maximum Σ by $\tau_\pm$ without the approximation, and substituted the maximum Σ for Σ in equation (2). We can see that the photon pair-creation effect makes E_M significantly smaller (solid lines), compared to the case without the effect (dotted lines).

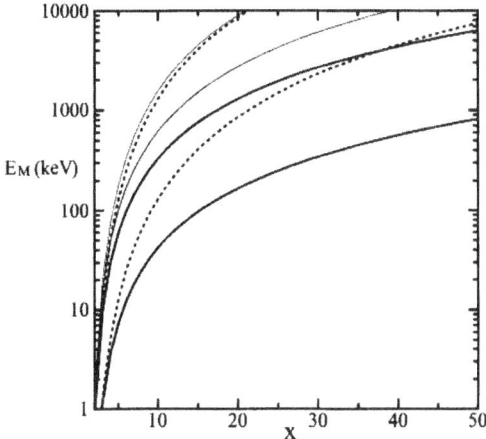

Fig. 1. Plot of E_M (solid line) against x. The dotted lines are values obtained without the process of pair-creation. The thick and thin lines are for equal-mass and equal-energy cases, respectively. The below thick lines are for $\Delta = 100$. The other lines are for $\Delta = 1$. Here we assume $\epsilon_e = 0.5$ and $\gamma_{s,2} = 1$.

In order to collide two shells outside the photosphere $\sim 10^{13}$ cm, we need $\gamma_s \geq 100$. On the other hand, the energy efficiency argument on the internal shock models requires a large dispersion of the initial Lorentz factors. However, the Lorentz factor should be smaller than few thousands in order to avoid abnormally large power of the central engine. Thus we suppose the typical Lorentz factor of the rapid shell, γ_r, is about 1000. Thus we assume the value of x is about 10-20. Then E_M is in the range of 300 keV - 1 MeV for $\Delta = 1$. The dependence on Δ of E_M is rather weak as $\propto \Delta^{-4/9}$. The value Δ might be determined by the size of the central engine. We expect small dispersion of the values of Δ.

If numerous internal shocks occur around the photosphere, the break energy may correspond to the highest ε_p^{obs}, namely E_M. Hence, the break energy should be determined by the photosphere and become few hundreds keV.

References

1. M.J. Harris, G.H.Share: ApJ **494**, 724 (1998)
2. R.D. Preece, M.S. Briggs et al.: ApJS **126**, 19 (2000)
3. D. Guetta, M. Spada, E. Waxman: ApJ **557**, 399 (2001)

The Thermal Precursors of Gamma-Ray Bursts

Frédéric Daigne

MPI für Astrophysik, Karl-Schwarzschild-Str. 1, D-81748 Garching bei München

Abstract. The prompt emission of GRBs probably comes from a highly relativistic wind which converts its kinetic energy into radiation via the formation of shocks within the wind itself. Such "internal shocks" can occur if the wind is generated with a highly non uniform distribution of Lorentz factor. For such a variable Lorentz factor, we estimate the expected thermal emission of the relativistic wind when it becomes transparent. We compare this emission to the emission produced by the internal shocks. In most cases we predict a rather bright thermal emission that could easily be detected. This favors wind acceleration mechanisms where the main energy reservoir is not under internal energy form (but magnetic ?). Such scenarios lead to the appearence of thermal X-ray precursors comparables to those observed by GINGA and WATCH/GRANAT.

1 The Photosphere of a Highly Relativistic Wind

The nature of the source responsible for the energy release leading to a GRB is not discussed here. We suppose that a relativistic wind carrying the energy has emerged with an average Lorentz factor $\bar{\Gamma} \geq 100$ and that the acceleration is complete at $r = r_{\mathrm{acc}}$, where the energy injection rate is $\dot{E}(t)$ and the initial Lorentz factor is $\Gamma(t)$, for $t = 0$ to $t = t_{\mathrm{w}}$. At $r = r_{\mathrm{acc}}$, the wind is still optically thick. Photons emitted by a layer successively cross all the layers emitted earlier by the source before escaping the relativistic wind. The optical depth $\tau(r)$ and the corresponding photosphere radius r_{ph} are approximatively given by

$$\tau(r) \simeq \frac{\kappa \dot{E}}{8\pi c^2 \Gamma^3 r} \text{ and } r_{\mathrm{ph}} \simeq \frac{\kappa \dot{E}}{8\pi c^3 \Gamma^2}, \tag{1}$$

where κ is the Thomson opacity. Fig. 1 shows r_{ph} for a wind lasting $t_{\mathrm{w}} = 10$ s with $\dot{E} = 10^{52}$ erg/s and an initial Lorentz factor also plotted in fig. 1.

2 Spectrum and Time Profile of the Photospheric Emission

In the framework of the fireball model (see e.g. Piran 1999), the luminosity and temperature of the photosphere are (the central black hole mass is $\mu_1 10\,\mathrm{M}_\odot$) :

$$L_{\mathrm{ph}} \simeq \dot{E} \left(\frac{r_{\mathrm{ph}}}{r_{\mathrm{acc}}}\right)^{-2/3} \text{ and } kT_{\mathrm{ph}} \simeq 1.3\,\mathrm{MeV}\ \dot{E}_{52}^{1/4} \mu_1^{-1/2} \left(\frac{r_{\mathrm{ph}}}{r_{\mathrm{acc}}}\right)^{-2/3}. \tag{2}$$

Fig. 1 shows L_{ph} and kT_{ph} in our example. The spectrum $dn^{\mathrm{ph}}(E)/dE$ and the count rate C_{12}^{ph} in any energy band $[E_1; E_2]$ are computed by assuming a Planck distribution for the photons. The result for our example is plotted in fig. 2.

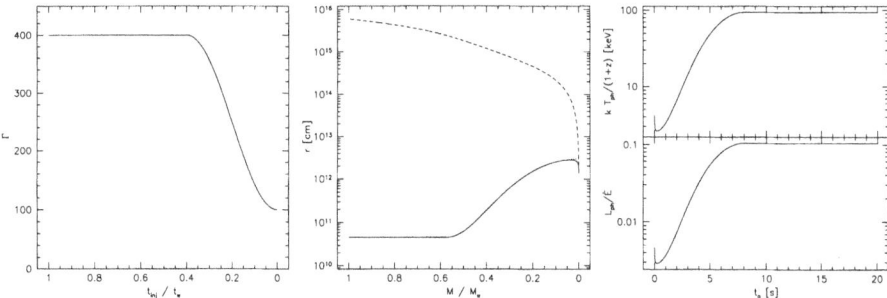

Fig. 1. *Left:* The initial Lorentz factor $\Gamma(t)$ as a function of the injection time $t_{\rm inj}/t_{\rm w}$. *Center:* The photosphere radius $r_{\rm ph}$ (solid line) as a function of the mass coordinate $M/M_{\rm w}$ of layers ($M_{\rm w}$ is the total mass). The dashed line shows the radius at which the photons emitted at the photosphere of each layer escape the wind. *Right:* The photospheric temperature and luminosity as a function of the arrival time $t_{\rm a}$ of photons. A redshift of $z = 1$ is assumed.

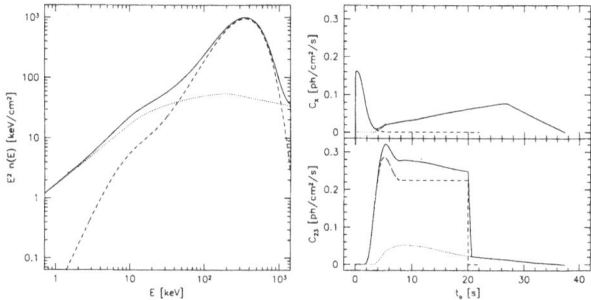

Fig. 2. *Left:* $E^2 n(E)$ is plotted as a function of the photon energy in keV (thick line). The spectrum is dominated by the photospheric thermal emission (dashed line). The non-thermal emission from the internal shocks is also plotted (thin line). *Right:* The count rate of the total (thick line), photospheric (dashed line) and internal shocks (thin line) emission are plotted as a function of the arrival time $t_{\rm a}$. *Top:* X-ray profile (3.5–8.5 keV); *Bottom:* γ-ray profile (50–300 keV). A redshift of $z = 1$ is assumed.

3   Spectrum and Time Profile of the Internal Shocks

The internal shock luminosity is $L_{\rm IS} = f_\gamma \dot{E}$, where the efficiency f_γ is the product of 3 terms : (i) the efficiency $f_{\rm d}$ of the kinetic to internal energy conversion behind the shock waves; (ii) the fraction $\alpha_{\rm e}$ of the internal energy which is injected in non-thermal electrons; (iii) the fraction $f_{\rm rad}$ of the electron energy which is radiated. The spectrum is modelized with the GRB-function (Band et al. 1993), whose 4 parameters are : the low and high energy slopes α and β, the peak energy $E_{\rm p}$ and the amplitude (related to $L_{\rm IS}$). This allows to compute the spectrum $dn^{\rm IS}(E)/dE$ and the count rate $C_{12}^{\rm IS}$. The result for our example is plotted in fig. 2. The internal shocks have be simulated using the model developped in Daigne & Mochkovitch 1998. We have adopted $\alpha = -1.0$, $\beta = -2.25$ and $E_{\rm p} = 200$ keV which are the typical values observed by Preece et al. (2000).

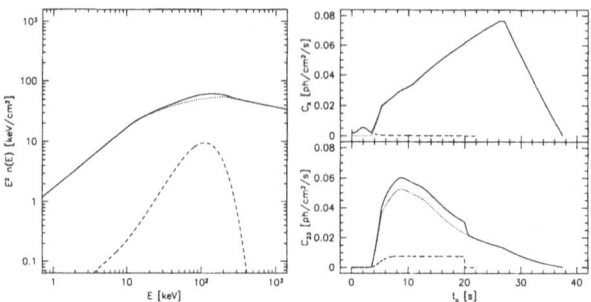

Fig. 3. Same as in fig. 2 when the photosphere is less hot and luminous ($\lambda = 0.01$).

4 Discussion

The ratio of the photospheric over internal shock count rates is given by :

$$R_{12} = \frac{C_{12}^{\mathrm{ph}}}{C_{12}^{\mathrm{IS}}} = 1.6 \; f_{\gamma\,0.1}^{-1} \; \dot{E}_{52}^{-1/4} \mu_1^{1/2} \frac{(1+z)E_\mathrm{p}}{200\,\mathrm{keV}} \frac{\mathcal{I}_{\mathrm{Band}}}{\mathcal{I}_{\mathrm{Planck}}} \frac{\int_{x_1}^{x_2} \frac{x^2 dx}{\exp x - 1}}{\int_{x_1}^{x_2} f_{\mathrm{Band}}(x) dx} \; . \tag{3}$$

As in our example, this ratio is usually greater than 1 in the X- and γ-ray bands so that the thermal emission dominates. This is in disagreement with the observations showing a non-thermal spectrum in this spectral range. Eq. 3 suggests 2 solutions : (i) **Increase of f_γ** : however, even with extreme values of $\alpha_\mathrm{e} \simeq 1$ and $f_{\mathrm{rad}} \simeq 1$, f_γ will always be limited by the dissipation efficiency which cannot exceed $f_\mathrm{d} \simeq 0.1 - 0.4$. This is insufficient to avoid a well detectable thermal emission. (ii) **Decrease of L_{ph} and kT_{ph}** : the only remaining solution to recover non-thermal bursts is to assume a less hot and luminous photosphere. We define λ as the fraction of the energy released by the source which is initially under internal energy form (the standard fireball model corresponds to $\lambda = 1$). Then the photospheric luminosity and temperature are multiplied respectively by λ and $\lambda^{3/4}$ so that the ratio R_{12} given by eq. 3 is now multiplied by a factor $\lambda^{3/4}$. Small values of λ can be expected, if, for instance, a large fraction of the energy released by the source is initially under magnetic form. Fig. 3 shows the new spectrum and profiles for our example adopting $\lambda = 0.01$. The photospheric emission is now only detectable in the X-ray band at the very beginning of the burst. This could correspond to the X-ray precursors observed by GINGA (Murakami et al. 1991) and WATCH/GRANAT (Sazonov et al. 1998) in several bursts. This last point, as well as the other results presented in this contribution, will be discussed in more details in a future paper (Daigne & Mochkovitch 2001).

References

1. Band, D. et al., 1993, ApJ, 413, 281.
2. Daigne, F. & Mochkovitch, R., 1998, MNRAS, 296, 275.
3. Murakami, T. et al., 1991, Nature, 350, 592.
4. Piran, T., 1999, Physics Reports, 314, 575.
5. Preece, R.D. et al., 2000, ApJS, 126, 19.
6. Sazonov et al., 1998, A&ASS, 129, 1.

The Optical Afterglow and Host Galaxy of GRB 000926

Johan P.U. Fynbo[1], Javier Gorosabel[2], Palle Møller[1], Jens Hjorth[3], Michael I. Andersen[4], Mathias P. Egholm[5], Brian L. Jensen[3], Holger Pedersen[3], Bjarne Thomsen[5], and Michael Weidinger[5]

[1] ESO, Garching, Germany
[2] DSRI, Copenhagen, Denmark
[3] Copenhagen Observatory, Denmark
[4] Division of Astronomy, Oulu University, Finland
[5] IFA, Århus, Denmark

Abstract. In this paper we illustrate with the case of GRB 000926 how Gamma Ray Bursts (GRBs) can be used as cosmological lighthouses to identify and study star forming galaxies at high redshifts. The optical afterglow of the burst was located with optical imaging at the Nordic Optical Telescope 20.7 hours after the burst. Rapid follow-up spectroscopy allowed the determination of the redshift of the burst and a measurement of the host galaxy HI-column density in front of the burst. With late-time narrow band Lyα as well as broad band imaging, we have studied the emission from the host galaxy and found that it is a strong Lyα emitter in a state of active star formation.

1 Introduction

Although the nature of the "central engines" of GRBs still are a subject of intense debate it is now well established that the majority of the long duration GRBs occur in star forming galaxies at cosmological redshifts (see Van Paradijs, Kouveliotou & Wijers 2000 for a review).

In this paper we focus on GRB 000926. This long duration burst was detected on September 26.9927 (UT) 2000 by three instruments (Ulysses, Konus and NEAR) in the Interplanetary Network (IPN, e.g. Klebsabel et al. 1982; Hurley et al. 2000), and localized to a 35 arcmin^2 error box which was circulated via the GRB Coordinates Network (GCN)[1] 20.3 hours after the burst (Hurley 2000). GRB 000926 was well studied over a wide range of the electromagnetic spectrum (Fynbo et al. 2001; Price et al. 2000, Piro et al. 2001; Harrison et al. 2001).

2 Identification of the Afterglow

Optical follow-up observations started at the Nordic Optical Telescope 17 minutes after the release of the IPN error box coordinates (Hurley 2000). In the main panel of Fig. 1, we show an R–band image of the Optical Afterglow (OA). The OA is marked with an arrow. The three small panels show the fading of the afterglow during the following days.

---

[1] http://gcn.gsfc.nasa.gov/gcn/

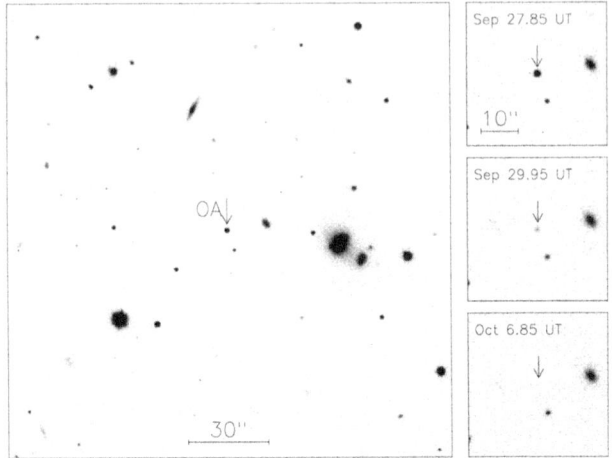

Fig. 1. *Left panel*: The R–band image of the OA taken 17 minutes after the release of the IPN error box coordinates. The OA is marked with an arrow. *Right panels*: Three smaller R–band images at three epochs showing the decline of the OA.

3   Spectroscopy

Optical spectra of the afterglow were obtained 21.7 hours and 44.4 hours after the burst. The combined spectrum is shown in Fig. 2. Seen are a number of strong metal absorption lines from which a redshift of z=2.0377 is determined. We detect no lines from intervening absorbers. In the blue end of the spectrum there is a damped absorption line due to neutral hydrogen from which we infer a HI column density of around 2×10^{21} cm^{-2}. The equivalent widths of the metal absorption lines are stronger than for any known Damped Lyα Absorber at similar redshifts (Møller et al. in preparation).

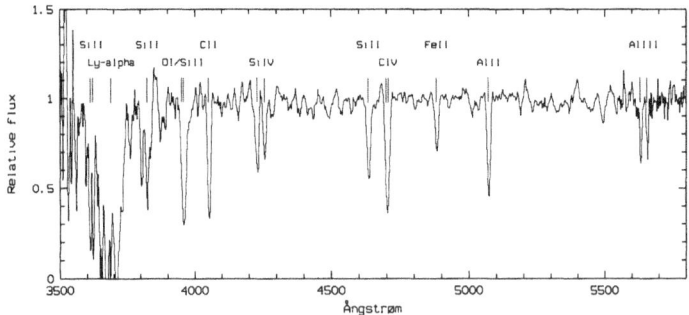

Fig. 2. The spectrum of GRB 000926 showing strong metal absorption lines as well as a damped Lyα absorption line at z=2.0377.

4   The Host Galaxy

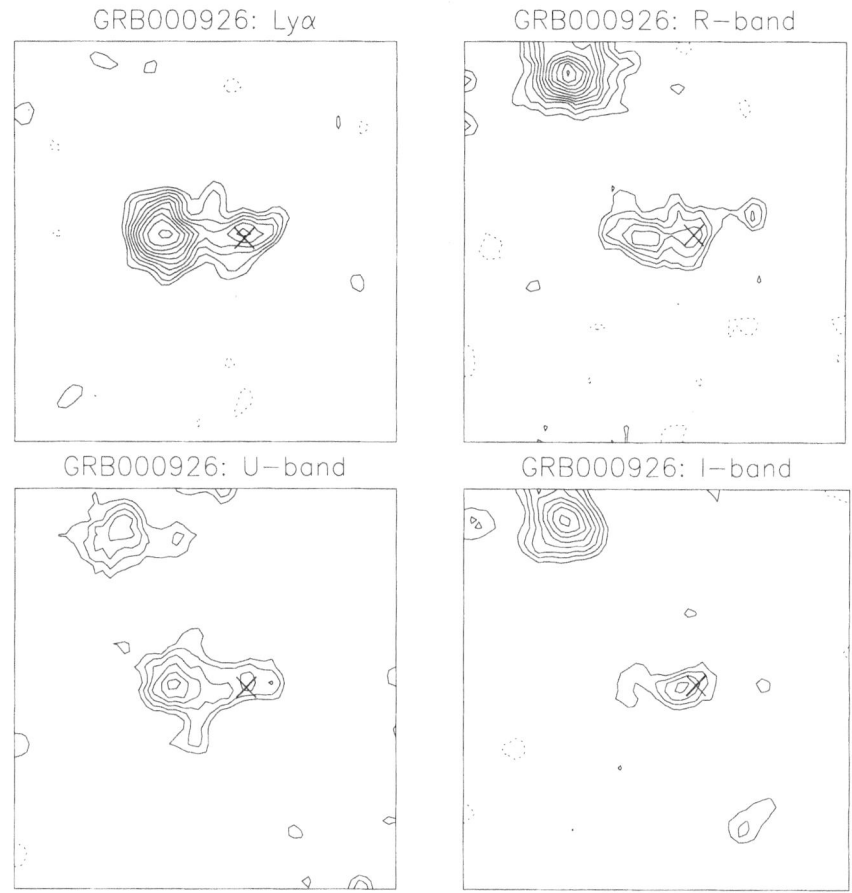

Fig. 3. The host galaxy of GRB 000926 as imaged in Lyα, U, R and I. The R-band image is taken from Fynbo et al. (2001). East is to the left and north is up. The size of the images is 10×10 arcsec^2. The position of the optical afterglow is indicated with an ×. The GRB occurred in the redder and fainter western part of the Lyα emitting region.

With deep R–band imaging obtained 1 month after the burst we detect the host galaxy as an extended object consisting of several compact knots. In order to study the host galaxy further, we obtained a special narrow (fwhm 45Å) band filter designed to cover Lyα at the GRB redshift. In May 2000 we obtained 12 hours of imaging in the narrow filter as well as 7 hours of imaging in the U filter and 3 hours of imaging in the I filter at the Nordic Optical Telescope. We detect the host galaxy in all bands (Fig. 3). The host galaxy is a relatively strong Lyα emitter (we detect one brighter Lyα emitter in the field). About 65% of the Lyα emission comes from the eastern knot of the host and the remaining 35% from the western knot in which the GRB occured. Although the signal-to-noise ratio is low, the western knot seems to be brightest in the I-band, and it must hence be redder than the other components. This indicates either more extinction or the presence of an older stellar population.

5 Summary

GRB 000926 occurred in a star forming galaxy at a redshift of z=2.0377. The optical spectrum of the afterglow shows strong metal absorption as well as a damped Lyα absorption from metal enriched gas in the host galaxy of the burst. The galaxy is a strong Lyα emitter consisting of at least two compact knots. The GRB occurred in the western and reddest of the knots.

Acknowledgments

The data presented here have been taken using ALFOSC, which is owned by the Instituto de Astrofisica de Andalucia (IAA) and operated at the Nordic Optical Telescope under agreement between IAA and the NBIfAFG of the Astronomical Observatory of Copenhagen. Nordic Optical Telescope is operated on the island of La Palma jointly by Denmark, Finland, Iceland, Norway, and Sweden, in the Spanish Observatorio del Roque de los Muchachos of the Instituto de Astrofisica de Canarias. MPE and MW acknowledge support from the ESO Directors Discretionary Fund. JG acknowledges the receipt of a Marie Curie Research Grant from the European Commission. This work was supported by the Danish Natural Science Research Council (SNF).

References

1. J.U. Fynbo, J. Gorosabel., T.H. Dall, et al.: A&A **373**, 796 (2001)
2. F.A. Harrison, S.A. Yost, R. Sari, et al.: ApJ **559**, 123 (2001)
3. K. Hurley, 2000, GCN 801 and 802
4. K. Hurley, T. Cline, E. Mazets, et al.: ApJL **534**, L23 (2000)
5. R. Klebsabel, W. Evans, D. Laros, et al.: ApJL **259**, L51 (1982)
6. J. van Paradijs, C. Kouveliotou, R.A.M.J. Wijers: ARA&A **38**, 379 (2000)
7. L. Piro, G. Garmire, M.R. Garcia, et al.: ApJ **558**, 442 (2001)
8. P.A. Price, F.A. Harrison, T.J. Galama, et al.: ApJL **549**, 7 (2000)

Peak Luminosity-Spectral Lag Relation Caused by the Viewing Angle of the Collimated Gamma-Ray Bursts

Kunihito Ioka[1] and Takashi Nakamura[1]

[1] Department of Earth and Space Science, Graduate School of Science, Osaka University, Toyonaka 560-0043, Japan
[2] Yukawa Institute for Theoretical Physics, Kyoto University, Kyoto 606-8502, Japan

Abstract. We compute the kinematical dependence of the peak luminosity, the pulse width and the spectral lag of the peak luminosity on the viewing angle θ_v of a jet. For appropriate model parameters we obtain the peak luminosity-spectral lag relation similar to the observed one including GRB980425. A bright (dim) peak with short (long) spectral lag corresponds to a jet with small (large) viewing angle. This suggests that the viewing angle of the jet might cause various relations in GRBs such as the peak luminosity-variability relation and the luminosity-width relation. Our model also suggests that X-ray rich GRBs (or X-ray flushes or Fast X-ray transients) are typical GRBs observed from large θ_v with large spectral lag and low variability.

Introduction: Afterglows of the GRBs are believed to be produced by the external shocks while the GRB itself is by the internal shocks. There is a growing evidence that some of afterglows are collimated with an opening half-angle of $\Delta\theta^{(a)} \sim 0.1$. However, the collimation half-angle $\Delta\theta$ of the internal shocks remains unknown since we can observe only the angular region inside the relativistic beaming half-angle $\sim \gamma^{-1}$ where γ is the Lorentz factor of the flow and is larger than ~ 100 to avoid the compactness problem. Since the angular size of a causally connected region is also $\sim \gamma^{-1}$, the minimum possible $\Delta\theta$ is $\sim \gamma^{-1} \leq 10^{-2} \ll \Delta\theta^{(a)}$, that is, $\Delta\theta$ can be much smaller than $\Delta\theta^{(a)}$. We here assume that the internal shocks consist of such sub-jets with the collimation half-angle of $\Delta\theta \ll \Delta\theta^{(a)}$. On the other hand, there are some possible relations between the peak luminosity, the spectral lag, the pulse width and the variability of GRBs. From six GRBs with known redshifts, Norris, Marani & Bonnell [1] found that the isotropic peak luminosity is inversely proportional to the spectral lag which is defined as the time lag of the peak luminosity between different energy bands. Fenimore & Ramirez-Ruiz [2] found that the luminosities of seven bursts with known redshifts are correlated with the variabilities of the bursts (see also [3]). If both relations are correct, there should be a relation between the variability and the spectral lag. Schaefer, Deng & Band [4] showed that this is the case by plotting the variability and the spectral lag of 112 GRBs. Here, we will show that depending on the viewing angle of a single jet to our line-of-sight, the luminosity-spectral lag/variability relation may arise kinematically [5].

Luminosity-lag/variability relation: The internal shock can be modeled by a collision of relativistic shells. A collision of two shells produces a single pulse light curve. Here we consider a single pulse. Let us consider an emitting thin shell that is confined to a cone of an opening half-angle $\Delta\theta$ and moves radially outward with the Lorentz factor of $\gamma = 1/\sqrt{1 - \beta^2}$. The emission is instantaneous at a time $t = t_0$ and a distance $r = r_0$ from the source. Let also the detector be at a distance D from the source

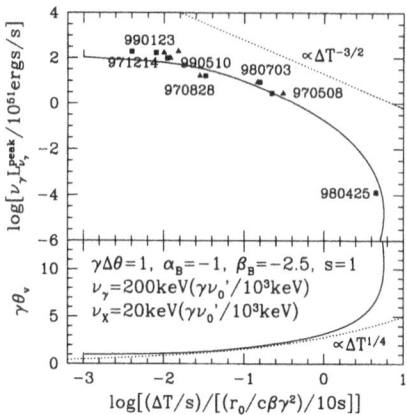

Fig. 1. The peak luminosity-spectral lag relation.

and θ_v be the angle that the axis of the emission cone makes with the axis connecting the detector with the source. Then, the flux of a single pulse at an observed time T and frequency ν is given by [5]

$$F_\nu(T) = \frac{2cA_0r_0\gamma^2}{D^2} \frac{\Delta\phi(T)f\left[\nu\gamma(1 - \beta\cos\theta(T))\right]}{\left[\gamma^2(1 - \beta\cos\theta(T))\right]^2}, \tag{1}$$

where $1 - \beta\cos\theta(T) = (c\beta/r_0)(T - T_0)$ and $T_0 = t_0 - r_0/c\beta$. For $\Delta\theta > \theta_v$ and $0 < \theta(T) \le \Delta\theta - \theta_v$, $\Delta\phi(T) = \pi$, otherwise $\Delta\phi(T) = \cos^{-1}\left[\frac{\cos\Delta\theta - \cos\theta(T)\cos\theta_v}{\sin\theta_v\sin\theta(T)}\right]$. For $\theta_v < \Delta\theta$, $\theta(T)$ varies from 0 to $\theta_v + \Delta\theta$ while from $\theta_v - \Delta\theta$ to $\theta_v + \Delta\theta$ for $\theta_v > \Delta\theta$. In order to have a spectral shape similar to the Band spectrum, we adopt the spectrum in the comoving frame as $f(\nu') = (\nu'/\nu_0')^{1+\alpha_B}[1 + (\nu'/\nu_0')^s]^{(\beta_B - \alpha_B)/s}$, where α_B (β_B) is the low (high) energy power law index, and s describes the smoothness of the transition between the high and low energy. $\alpha_B \sim -1$ and $\beta_B \sim -2.5$ are typical values. For $\theta_v \ll 1$ and $\Delta\theta \ll 1$, equation (1) depends on $\gamma\nu_0'$, $\gamma\theta_v$, $\gamma\Delta\theta$, $r_0/c\beta\gamma^2$, T_0, α_B, β_B, s and $2cA_0r_0\gamma^2/D^2$. To study the viewing angle dependence, we fix other parameters: $\gamma\Delta\theta = 1$, $\alpha_B = -1$, $\beta_B = -2.5$ and $s = 1$. The lab-frame frequency, the observed time and the flux per frequency are measured in units of $\gamma\nu_0'$, $r_0/c\beta\gamma^2$ and $2cA_0r_0\gamma^2/D^2$, respectively.

In the upper panel of Fig. 1, varying the viewing angle $\gamma\theta_v$ we show the isotropic peak luminosity at frequency $\nu_\gamma = 200\text{keV}(\gamma\nu_0'/10^3\text{keV})$ as a function of the spectral lag in an arbitrary vertical scale of $2cA_0r_0\gamma^2/D^2$. For simplicity, we define the spectral lag ΔT as the difference of the peak time between frequencies $\nu_\gamma = 200\text{keV}(\gamma\nu_0'/10^3\text{keV})$ and $\nu_X = 20\text{keV}(\gamma\nu_0'/10^3\text{keV})$. All points with $\Delta T \le 10^{-4}(r_0/c\beta\gamma^2)$ are plotted at $\Delta T = 10^{-4}(r_0/c\beta\gamma^2)$. In the lower panel of Fig. 1 the solid line shows the corresponding viewing angle $\gamma\theta_v$ as a function of ΔT. We adopt $r_0/c\beta\gamma^2 = 10$ s. The observed luminosity-spectral lag relations for seven bursts with know redshifts (from Table 1 of [1]) are also plotted for comparison. Surprisingly

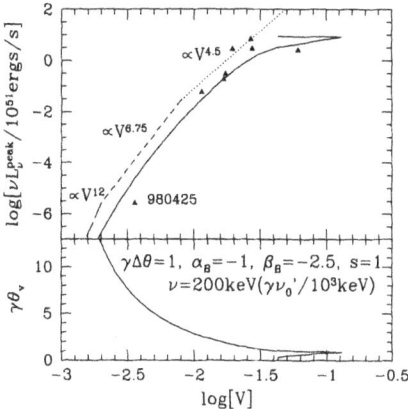

Fig. 2. The peak luminosity-variability relation.

enough, a simple sub-jet model happens to reproduce the observation quite well including GRB980425 which has the extremely dim luminosity and large spectral lag.

The sub-jets model might be compatible with the luminosity-variability relation [2–4]. In the upper panel of Fig. 2, we show the isotropic peak luminosity is shown as a function of the variability V by varying $\gamma\theta_v$ (see [5] for more details) together with the observation for eight bursts with known redshifts (from Table 1 of [2]). Surprisingly again, a sub-jet model happens to reproduce the observation quite well including GRB980425 which has the extremely low variability.

If we observe the jet from $\theta_v \sim$ several $\times \gamma^{-1}$ the maximum frequency is in the X-ray band while for $\theta_v \sim \gamma^{-1}$ it is in the γ-ray region so that depending on the viewing angle, GRBs may be observed as X-ray rich GRBs in our model [5,6]. The luminosity-width relation from equation (1) [5] is also consistent with the observation $\nu F_\nu^{peak} \propto W_{FWHM}^{-2.8}$ [7].

References

1. J. P. Norris, G. F. Marani, J. T. Bonnell: ApJ **534**, 248 (2000)
2. E. Fenimore, E. Ramirez-Ruiz: astro-ph/0004176
3. D. E. Reichart et al.: ApJ **552**, 57 (2001)
4. B. E. Schaefer, M. Deng, D. L. Band: astro-ph/0101461
5. K. Ioka, T. Nakamura: ApJ **554**, L163 (2001)
6. T. Nakamura: ApJ **534**, L159 (2000)
7. E. Ramirez-Ruiz, E. E. Fenimore: ApJ **539**, 712 (2000)

High Frequencies in Power Spectrum of Gamma-Ray Bursts

Alexei Pozanenko and Vladimir Loznikov

Space Research Institute, Moscow 117810, Russia

Abstract. Power spectra of Gamma-Ray Burst (GRB) were investigated in the frequency range $0.01 - 500\ Hz$ using BATSE data. Significant variability was found on the scale up to $100\ Hz$. We found that high frequency part ($5 - 100\ Hz$) of averaged power spectrum is approximated by power law with power law index ~ -2.2.

1   Data analysis and results

We analyzed bursts from BATSE catalog [1] using $64\ ms$ time profiles from the BATSE Large Area Detectors (DISCSC+PREB data types). Time profiles are the sum of Discriminator channels 2 and 3, which correspond approximately to $50 - 300\ keV$ energy range. Background subtraction is performed in each discriminator channel separately, using the interpolation of a fitted polynomial model. Time profile of this data type usually covers almost all burst starting 2 seconds before trigger time T_0 and spanning up to the end of the burst.

To extend the frequency range of PDS we use BATSE Time Tagged Events (TTE) data type. Original TTE data were re-binned into $1\ ms$ time bins. The TTE data cover early stage of the bursts, i.e. time period from 1 to 4 seconds around trigger time T_0.

We investigated only long bursts with $T_{90} > 2\ s$, because short bursts ($T_{90} < 2\ s$) may represent different population of GRB. (Bi-modal distribution of duration parameter T_{90} points out to possible different GRB populations.) Total amount of bursts is 1542 and 1056, and total accumulation time is about $200\ Ksec$ and $1.1\ Ksec$ for $64\ ms$ and TTE data, respectively. The different number of GRB for each data type are due to data availability (telemetry loss and algorithm of onboard software).

Before averaging Power Density Spectrum (PDS) of individual bursts Poisson noise is subtracted from each spectrum in accordance to non-paralizable model of dead time and dead time constant reported by BATSE team [2]. Each PDS has different amplitude and shape, which are due to different intrinsic variability and observed intensity of particular burst [3,4]. To provide equal weight for each PDS to be averaged we use normalization of the total power of bursts to the same value. Otherwise bright bursts would dominate in the averaged spectrum. The choice of normalization is appropriate for investigation of PDS continuum.

Averaged spectrum combined from both data types ($1\ ms$ and $64\ ms$) is presented in Fig. 1. Left cut off is associated with finite duration of bursts. Indeed, the logarithmic mean of long mode of T_{90} distribution is around 10 seconds. Power laws with different indices can approximate high frequency tail and central part of the spectrum.

For quantitative analysis we use Band's law previously used for GRB energy spectra approximation [5] $P_f \sim \nu^{-a} exp(-\nu/\nu_0)$, if $(b - a)\nu_0 \geq \nu$ and $P_f \sim \nu^{-b}[(b -$

$a)\nu_0]^{(b-a)} exp(b-a)$, if $(b-a)\nu_0 < \nu$, which is a modified broken power law with break at ν_0. Results of the PDS approximation in the frequency range $0.03 - 100\,Hz$ are presented in Table 1. Central part of the spectrum is approximated by power law with index about $a = -1.5$ ([6,7]).

2 Discussion

We calculated averaged PDS of BATSE Gamma-Ray Bursts with $T_{90} > 2\,s$ from 0.01 up to $500\,Hz$ frequency range taking into account the dead time model of BATSE detectors. Significant variability is observed up to $100\,Hz$. Due to statistical uncertainties the spectrum cannot be considered above $100\,Hz$. The resulting spectrum can be approximated by broken power law in the wide range of frequencies $0.03 - 100\,Hz$. The power law index at low frequencies is about -1.5, and in high frequency tail is about -2.2. One may suggest that high frequency part of the spectrum represents properties of photon emission processes, while power law below break describes properties of time profile structure including time profile envelope.

Behavior of the PDS below $\sim 1/2\,s = 0.5\,Hz$ results from both intrinsic variability of emission processes in the burst source and profile structure properties. Actually we investigate only long bursts with $T_{90} > 2\,s$. Power law with the same index observed in two decades for a frequency range $0.03 - 5\,Hz$. The lower limit of the range is well

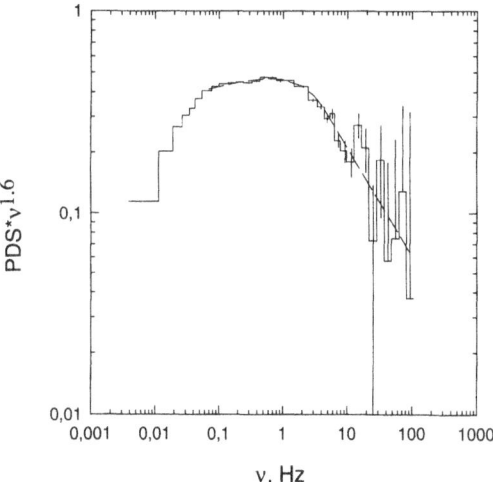

Fig. 1. Averaged Power Density Spectrum of BATSE gamma-gay bursts. Histogram is a spectrum combined from $64\,ms$ and $1\,ms$ data types. Dashed line is an approximation of the spectrum by a modified broken power law (Band's law, see text).

below of $0.5\ Hz$ and therefore one can conclude that in this frequency range profile structure variability dominates over emission variability.

If bi-modality of the duration distribution (T_{90}) results from profile structure of time profile, then one would expect specific time structure around $2\ s$ and consequently specific break in the power spectrum near $1/2\ Hz$. Absence of the break near $0.5\ Hz$ support the assumption of two populations of GRBs: short and long populations associated with the two modes of bi-modal distribution.

Investigation of the break position against burst intensity can help to understand cosmological properties of GRB. If the specific time scale associated with this break is an intrinsic property of GRB sources then it should evolve with redshift z as $(1+z)$. This investigation should include dependence of variability against of photon energy [8].

Table 1.

a	b	ν_0
		Hz
1.51 ± 0.02	2.15 ± 0.07	7.26 ± 1.57

References

1. C.A. Meegan et al.: Catalog of BATSE
2. R. Preece: private communication (2000)
3. B.M. Belli: Ap.J., 393, 266 (1992)
4. W.T. Giblin et al.: in AIP Conf.Proceedings 428, 241 (1998)
5. D. Band: Ap.J., 413, 281 (1993)
6. A.S. Pozanenko et al.: in AIP Conf.Proceedings 526, 220 (2000)
7. A.M. Beloborodov, B.E. Stern, R. Svensson: ApJ, 535, 158 (2000)
8. E.E. Fenimore et al.: Ap.J., 448, L101 (1995)

The Nature of the Host Galaxy of GRB 010222

Isabel Salamanca[1], Paul M. Vreeswijk[1], Lex Kaper[1], Evert Rol[1], Edward P.J. van den Heuvel[1], Nial Tanvir[2], and Sara Ellison[3]

[1] Astronomical Institute 'Anton Pannekoek', Univ. Amsterdam, The Netherlands
[2] Department of Physical Science, Univ. Hertfordshire, UK
[3] European Southern Observatory, Santiago, Chile

Abstract. The absorption lines in the spectrum of the afterglow of gamma-ray burst GRB 010222 indicate that its host galaxy, at $z = 1.48$, is an exceptionally strong damped Lyman-α (DLA) system. We present arguments that the host galaxies of other long-duration, bright GRBs likely fulfill the criteria of DLA systems as well.

1 Introduction

Gamma-ray bursts (GRBs) are the most powerful cosmic explosions. Their extreme luminosities provide the opportunity to study the properties of their distant host galaxies that otherwise would have remained undetected. With this information it becomes possible to find out whether GRBs preferentially occur in gas-rich, star-forming galaxies (which is expected if a GRB is caused by the collapse of a massive star forming a black hole). Furthermore, the absorption lines in GRB afterglow spectra can be used to, e.g., determine the metallicity of the interstellar medium in galaxies in the high-redshift universe.

GRB 010222 was discovered on February 22, 2001, at 7:23:30 UT by the *Beppo*SAX satellite ([11]), and is one of the brightest bursts ever recorded (fluence 9×10^{-5} erg cm^{-2} at 40-700 keV). At a redshift of $z = 1.48$ (see below), the isotropic energy output is 7.8×10^{53} erg, which makes it the third in luminosity of all GRBs with known redshift ([11], [17]). Here we report on spectroscopic observations of the afterglow of GRB 010222, obtained with the 4.2m *William Herschel Telescope* at the "El Roque de los Muchachos" observatory, La Palma, Spain.

2 The Host of GRB 010222: a Damped Lyman-α System

The optical spectrum of the V$\sim$18 afterglow of GRB 010222 includes many absorption lines; three absorption-line systems are present: at $z =$ 1.476, 1.156 and 0.927. The highest redshift system is identified as the host galaxy of GRB 010222 ([6], [9], [13]). The absorption lines are produced by resonance transitions of abundant ions like C IV, Si IV and Mg I. We measured their equivalent width (EW) and compared the values to those measured for DLA systems in the sample of Rao & Turnshek ([12]), see Fig. 2. These authors find that if EW(MgII 2796)>0.6 Å and EW(FeII 2600)>0.5, it has a 50% probability of being a DLA system. The host of GRB 010222 certainly fulfills these conditions ([9], [13]). Furthermore, the EW ratio is $\sim$1.5, very close to the average value of 1.45 deduced for DLAs ([12]). Finally, the line ratio of the Mg II doublet

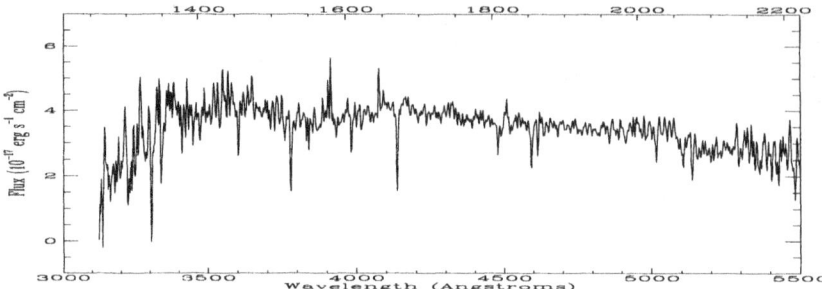

Fig. 1. Optical spectrum of the afterglow of GRB 010222, taken on February 23, 2001 (22.66 hours after the burst) with the double spectrograph ISIS at the 4.2m WHT. Here we show the blue arm. The spectral resolution is 3.3 Å.

in all the DLA systems in their sample is less than 1.5, as is the case for the host of GRB 010222.

That the host of GRB 010222 definitely is a DLA system, is based on the detection of the red wing of Lyman-α. We have deduced the column density of H I by fitting a Voigt profile to the red wing of Lyman-α. The value we obtain is N(H I) = (5 ± 2) $\times 10^{22}$ cm^{-2}. By definition, a damped Lyman-α system has N(H I)$> 2 \times 10^{20}$ cm^{-2}. The measured neutral hydrogen column density is consistent with that derived from the X-ray afterglow ([17]). The very high value of N(H I) ranks the host of GRB 010222 at the top of the column density distribution.

What can we say about the host galaxies of other GRBs? The EWs of the Mg II, Mg I and Fe II lines ([7], [10], [15], [16]) indicate a DLA nature for the host galaxies of GRB 990123, GRB 970708, GRB 990510, GRB 990712 and GRB 991216 as well (Fig. 2). Only in one other case, the host of GRB 000301C, observational evidence is presented for the direct detection of Lyman-α ([5]).

3   Discussion and Conclusion

From the analysis of the optical spectrum of the afterglow of GRB 010222, we conclude that the host galaxy is a DLA system, with the highest hydrogen column density known. The constant and strong mm and sub-mm emission (3.74 $\pm$ 0.53 mJy at 350 GHz) of GRB 010222 detected by Frail et al. ([3]) suggests that its host galaxy is very dusty and currently undergoing an intense burst of star formation (~ 500 M$_{\odot}$ yr^{-1}). In the optical and near-infrared it is a blue sub-L$_*$ galaxy. This is in good agreement with the accumulating observational evidence that suggests that GRB host galaxies are vigorously forming stars. The sample of intermediate-redshift DLA systems do not trace the *bulk* of star-forming galaxies, but rather a slowly evolving population of galaxies with a wide range of luminosities and morphological types (see [12] and references therein). However, there is no reason why a starburst cannot be a DLA. Furthermore, DLA systems having N(H I)$> 10^{22}$ cm^{-2} are very rare. This might be a selection effect: the sightlines towards QSOs probe galaxies that happen to be between us and the quasar. The

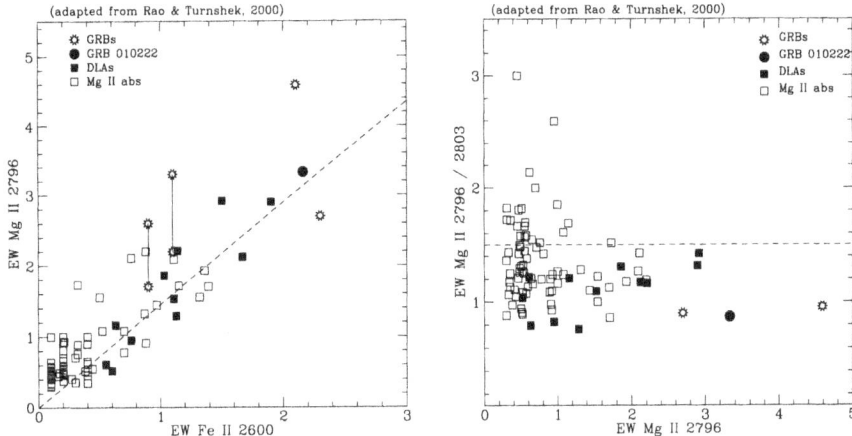

Fig. 2. All GRB host galaxies likely are damped Lyman-α systems. **Left:** The correlation between the equivalent widths of Mg II 2796 Å and Fe II 2600 Å for the Mg II absorbers can be extended to the GRB host galaxies. A vertical line is drawn in case the Mg II doublet lines are blended: the upper value is the total measured EW and the lower value is the EW divided by two. **Right:** The EW ratio of the Mg II doublet lines is less than 1.5 for all the DLA systems in the Rao & Turnshek sample, as well as for the three GRB hosts for which this information is available.

probability to randomly probe an area close to the dense nucleus of a galaxy is small. As GRBs occur close to the center of a galaxy ([4]), and their progenitors are likely embedded in gas-rich star-forming regions, one would predict that GRB host galaxies are associated with rather strong absorption-line systems. According to Schaye ([14]), however, at such high column densities the gas in the interstellar clouds is converted into molecules, becoming much more compact and short lived. These clouds have a much smaller chance of being intercepted by the line of sight towards QSOs.

GRBs provide a unique opportunity to study galaxies in the high-redshift universe, and they might probe a different population of galaxies as detected in QSO absorption lines.

References

1. A.C. Danks: Astrophysics and Space Science **269-270**, 639 (1999)
2. S.G. Djorgovski et al: Astrophys.J. **508**, L17 (1997)
3. D.A. Frail et al.: Astrophys.J accepted (2001) (astro-ph/0108436)
4. A.S. Fruchter et al.: in prep. (2001)
5. B.L. Jensen et al.: Astron.&Astrophys. **370**, 909 (2001)
6. S. Jha et al.: Astrophys.J **554**, L155 (2001)
7. S.R. Kulkarni et al.: Nature **398**, 389 (1999)
8. S.R. Kulkarni et al.: GCN Circ. 996 2001
9. N. Masetti et al.: Astron.&Astrophys. **374**, 382 (2001)
10. M.R. Metzger et al.: Nature **387**, 878 (1997)
11. L. Piro: GCN Circ. 959 (2001)

12. S.M. Rao & D.A. Turnshek: Astrophys.J. Supl. Ser. **130**, 1 (2000)
13. I. Salamanca et al.: in prep. (2001)
14. J. Schaye: Astrophys.J. submitted (2001) (astro-ph/0109280)
15. P.M. Vreeswijk et al.: Astrophys.J. **546**, 672 (2001)
16. P.M. Vreeswijk et al.: Astrophys.J. submitted (2001)
17. J.J.M. in 't Zand et al.: Astrophys.J. submitted (2001) (astro-ph/0104362)

On the Power Spectrum of Gamma Ray Bursts

Motoko Suzuki[1], Masahiro Morikawa[1], and Izumi Joichi[2]

[1] Department of Physics, Ochanomizu University
[2] School of Science and Engineering, Teikyo University

Abstract. Gamma ray bursts are known to have rapid time variation less than 64 milli-sec. The average of their power spectrum density (PSD) shows power law of index -1.67 through two decades. This is often argued in relation with the Kolmogorov law in turbulence; many self similar fluctuations are thought to yield the power law. We focus on the origin of the power law in PSD. We conclude that the origin is not the self similarity but a special form of the dominant shot.

1 Introduction

Gamma-ray bursts (GRBs) are commonly observed signals in the Universe. Their wave length rage $20 - 300 \mathrm{keV}$ with duration time scale about $10\,\mathrm{sec}$, the typical energy $10^{-5}\mathrm{erg/cm}^2$, and the frequency $1000/\mathrm{year}$. Despite the recent concentrated observations, the origin of GRBs has not yet been clarified.

GRBs often show rapid variability in their luminosity profile and therefore the Fourier transform technique is thought to be effective to reveal their intrinsic nature. Actually Beloborodov et al.[1] analyzed 214 light curves of long GRBs $T_{90} > 20\,\mathrm{sec}$. Averaging the power spectrum density (PSD), they obtained the power law: $I \propto f^{-1.67\pm0.02}$. Since the power index is very close to $5/3 = 1.66...$, this is often thought to imply the Kolmogorov spectrum of self similarity velocity fluctuations in turbulent medium.

However, the power law behavior in PSD does not necessarily imply the self similarity (fractal) structures[2]. At least we have three possible mechanisms which yield power law in PSDF 1. A special form of the shot such as $t^{-1/2}e^{-t}\theta\,(t)$ 2. Superposition of many self similar shots. 3. Levy-type random noise.

In this paper, we would like to identify which of the above mechanisms produces the power law in PSD of GRBs.

2 Power Spectrum Analysis

We use the data of Burst and Transient Source Experiment (BATSE) on Compton Gamma Ray Observatory (CGRO) with 64ms resolution[3]. The energy rage is $50 < \hbar\nu < 100$ (highest and lowest energy rages are uncertain). We subtract the background by the linear fitting and select the data with the peak count rate larger than 250 counts/ 64ms-bin to avoid noises. Finally 297 data set remain in our analysis. We individually Fourier transform them and calculate PSD for each data. Individual PSD of GRB data shows already power law. Therefore each PSD is fitted by the function $P\,(f) = Af^\alpha + B$.

The distribution of the power index α apparently shows the Gaussian distribution with the mean -1.76 and the dispersion 0.65[4]. The mean of the slope -1.76 is slightly different from the previous value -1.67 [1] that is the slope of the averaged data.

3   What Determines the Individual Power Law Index?

A single shot of the simple decay type a) $x\,(t)\;=\;ht^{-p}\exp\left[-t/\tau\right]\theta\,(t)$ has the PSD $P\,(f)\;=\;h^2\left[\Gamma\,(1-p)\right]^2\left[\left(\frac{1}{\tau}\right)^2+f^2\right]^{p-1}$. A single shot of the grow-and-decay (FRED) type b) $x\,(t)$ =(power law decay toward left)+(power law decay toward right) has the PSD $P\,(f)$ =(left PSD) + (right PSD) + (rapidly oscillating interference term). The superposition of several shots of the above type, for $x_k\,(t)\;=\;A_k x\,(t-a_k)$, $x_T\,(t)=\sum_{k=1}^{n} x_k\,(t)$ has the PSD

$P_T\,(f) = P\,(f)\left[\sum_{j=1}^{n} A_j^2 + \sum_{j>k}^{n} 2A_j A_k \cos\left(f\,(a_j - a_k)\right)\right].$

We decompose the luminosity curve in the following sequence: (1) Identify the peak of a luminosity curve. (2) Fit the data around the peak with the type a) or b)→skeleton and subtract it from the original data →1SD (3) Identify the peak of the subtracted data. (4) Repeat the procedure 2 → 2SD, c (5) Stop at 5SD.

The following figure 1 shows a typical example of the PSDs for the decomposed luminosity curve.

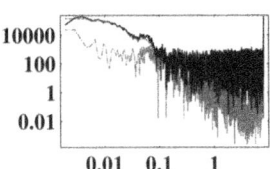

Fig. 1. BATSE07295 Channel=3. The solid line represents the PSD of the original luminosity curve. The gray dot-dashed line represents the subtracted component up to 5SD. The gray dashed line represents the skeleton. The horizontal axis is log (Hz).

The subtracted component up to 5SD is smaller than the skeleton by10-100 times. The power index in PSD is determined by the skeleton. ithe fit is not always perfect for some GRB data.j Therefore we find that a special form of a dominant shot determines the PSD and eventually the power law behavior of it.

Furthermore we calculate the peak distribution function in individual data. In this calculation, we could not find any power law distribution. The figure 2 shows a typical example. Therefore we find that there is no Levy-type noise in the luminosity curve of GRBs.

4   What Determines the Averaged Power Index?

Now we would like to average the individual PSD data. Before that we have to determine the normalization method for each PSD data. We use the normalization so that

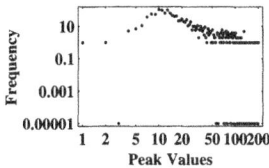

Fig. 2. Peak distribution for BATSE01626.

the peak value of the luminosity curve is identical for each data (peak normalization). Then this averaged PSD shows much clear power law than individual power law and the index turns out to be -1.67. Normalization methods seem not to change the power index of the averaged PSD so much. Actually we have checked this for the normalization by fluence. The origin of this much clear power law in the averaged PSD is not yet clarified.

However we can exclude the self similarity as the origin of this much clear power law. If many bursts were distributed self similarly, then the envelope of these PSDs would have a possibility to form a power law. In this case, we would expect the individual power index is manifestly smaller (the slope is steeper) than the averaged power index. However actually they are the same well within the dispersion of the power-index distribution. Therefore we may conclude that the power law behavior of the averaged PSD is not produced by the self similarity of the data.

5   Conclusions

We have analyzed 297 power spectrum density (PSD) of GRBs. The results are (a) Individual PSD of GRB data already shows power law. The power index obeys the Gaussian distribution with the mean -1.76, and the dispersion 0.65. (b) The individual power law behavior of PSD is due to the special form of a dominant shot. The origin will not be the Levy-type noise. (c) Averaging individual PSD yields much clear power law. The power index is almost the same as the mean of the individual power law distribution of (a).

MM thanks Alexei Pozanenko and Shin Mineshige for interesting discussions at the lighthouse conference.

References

1. Beloborodov, A. M., Stern, B. E., & Sevensson, R. 1998, ApJ, 508, L25@@———. 2000, ApJ, 535, 158.
2. Mineshige, S. & Yonehara, A. 2001, in Probing the Physics of Active Galactic Nuclei by Multiwavelength Monitoring, ed. B.M. Peterson, R.S. Polidan, and R.W. Pogge (PASP Conf. Ser.), in press.
3. ftp://cossc.gsfc.nasa.gov/pub/data/batse/ascii_data/64ms/.
4. astro-ph/0104232, Motoko Suzuki, Masahiro Morikawa, Izumi Joichi.

Part IV

Gravitational Lensing and Gravitational Waves

Lighthouses of Gravitational Wave Astronomy

Bernard F. Schutz

[1] Max Planck Institute for Gravitational Physics (Albert Einstein Institute)
14476 Golm, Germany
[2] Department of Physics and Astronomy, University of Cardiff, Wales

Abstract. Gravitational wave detectors capable of making astronomical observations could begin to operate within the next year, and over the next 10 years they will extend their reach out to cosmological distances, culminating in the space mission LISA. A prime target of these observatories will be binary systems, especially those whose orbits shrink measurably during an observation period. These systems are standard candles, and they offer independent ways of measuring cosmological parameters. LISA in particular could identify the epoch at which star formation began and, working with telescopes making electromagnetic observations, measure the Hubble flow at redshifts out to 4 or more with unprecedented accuracy.

1   Introduction

Gravitational wave interferometers now under construction at several locations around the world will soon begin making observations. Although their initial sensitivities will be marginal, a planned program of upgrades and technology development will make them powerful instruments of cosmology over the next decade. In 2011 the launch of the joint ESA-NASA gravitational wave observatory LISA will extend the reach of this form of astronomy to the entire observable universe.

Astronomy has consistently proved itself to be full of surprises for observations in new wavebands, and this may well also be true for gravitational wave astronomy. Therefore it is a little dangerous to try to predic what these new detectors will observe, but it is useful to look ahead at this point. In particular it is not too early to consider and even to begin to plan the ways in which gravitational wave detectors and other astronomical telescopes could work together.

For cosmology, one of the most interesting features of gravitational wave observations is that certain systems are standard candles: their distance can be inferred from their gravitational waveforms. These systems are chirping binaries, that is binary systems whose orbits shrink during the observations time because of the energy they lose to gravitational waves. The change of the orbit raises the frequency of the gravitational wave, producing a "chirp" waveform.

In this review I will point out a number of ways in which gravitational wave observations of these chirping waveforms, usually coupled with coordinated observations in electromagnetic wavebands, can be used to provide cosmological information. For example, chirps from neutron-star binaries in the last few minutes before coalescence should be observed frequently by advanced ground-based detectors. These detectors can give astronomers advance notice of and rough positions for such inspiral events, and optical identifications of any afterglows produced by the mergers of the neutron stars will sharpen the distance estimate made by the detectors. Redshifts to the after-glows can be used with these distance estimates to provide independent measurements of the Hubble constant to accuracies of a few percent, and of the acceleration of the universe out to redshifts of order 0.2. Chirps from mergers of stellar-mass black holes could, even in the absence of electromagnetic counterparts, provide estimates of the cosmological acceleration out to redshifts of order 1.

Chirps from very massive ($1000\ M_\odot$) black hole binaries that might have formed in the first epoch of star formation, observed by LISA, could be used to determine when star formation began. Chirps from the coalescences of massive black holes in galactic centres, again observed by LISA, could measure the cosmological deceleration out to redshifts of 4 or more, provided that electromagnetic observations can pin down the cluster of galaxies in which a coalescence occurred. This will be a real challenge to astronomy but it could have an immense payoff.

Before discussing these possibilities, I begin this article with two background sections. The first reviews gravitational wave astronomy, particularly emphasising the ways in which gravitational wave observations differ in concept and information content from electromagnetic observations, and outlining the development of detectors and the timetable on which sensitivity improvements can be expected. The second iss an intro-duction to chirp waveforms and the kind of information they carry. These two sections prepare for the subsequent discussions of how cosmological information can be ex-tracted from gravitational wave observations.

2   Gravitational Wave Observing

The principles of interferometric detectors and their current development are reviewed in a number of places in the literature [1,2]. In particular, [2] contains references to the recent literature. Several accessible textbooks [3,4] review the principles of gravitational radiation, and two recent encyclopedia articles also address these issues [5,6]. What follows here is a brief introduction to these subjects.

2.1   Action of Waves on a Detector

Figure 1 shows how gravitational waves act on a ring of free particles. The action is by tidal forces carried by the waves, which distort the ring in directions transverse to the direction of propagation of the wave. Because of the equivalence principle, the overall acceleration of the ring produces no local effects; the only measurable effects are in the relative distortions. Therefore the linear displacements of these distortions are proportional to the size of the ring: the larger the ring, the larger the displacement.

The figure can be viewed as an elementary detector. By sensing the relative dis-placements of particles on the ring, one can measure the wave. This is exactly how interferometric detectors work. They use laser interferometry to measure changes in the relative distances between the central particle of a ring and two particles in orthogonal directions along the circumference of the ring. The particles are mirrors in the interfer-ometer that are free to move along the direction of the displacement.

Older, solid-mass detectors, called bar detectors, use the stretching along one diameter of the ring. The restoring forces of the solid material means that the response to the wave is more complicated than in an interferometer, but the principle is the same.

All proposed gravitational wave detectors are linearly polarized. To measure the polarization of a wave requires either that several detectors make measurements or that the wave lasts long enough so that the motion of a detector carried by the Earth or in a space orbit changes the projected polarization of the detector, allowing it to measure two independent polarizations.

The waveforms in Fig. 1 have a simple relationship to the mass motions in the source of the gravitational wave [7]. They mimic any oscillating motions in the source, as projected on the plane of the sky as seen by the ring. If the motions are all along one line, then the polarization ellipse will have its alternating major/minor axes along that line. If the motions are circular, then the wave will have circular polarization, which is a linear combination of the two polarizations with a phase shift of 90 degrees. Thus, measuring the polarization of a gravitational wave allows one to make direct inferences

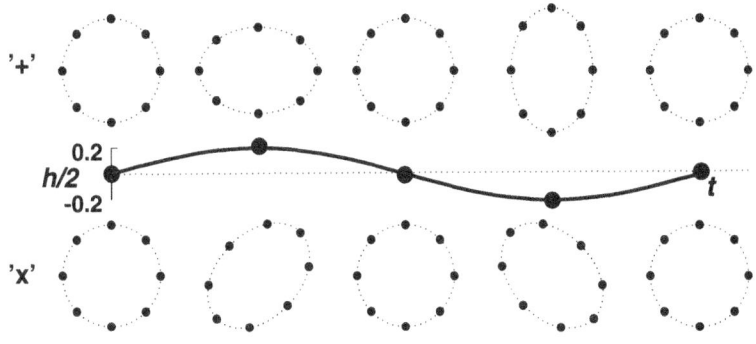

Fig. 1. Two independent polarisations of a plane gravitational wave are illustrated by their actions on a ring of free particles in empty space. The waves act transversely, so in this figure the waves approach perpendicular to the paper. The waves distort the ring into ellipses with alternating major and minor axes. The two polarizations are orthogonal because the ellipses are rotated by 45° with respect to one another. The action of any plane wave is a superposition of these two polarizations. The waves act through tidal forces, so the stretching is proportional to the size of the ring: a given gravitational wave will produce twice the relative displacement in a ring twice the size. The waveform in the centre shows the size of the strain illustrated here, which is defined as half of the amplitude h of the wave. (This is much larger than we expect, of course.) The distortions produced by a given wave mimic the motions in the source of the wave as projected onto the plane of the sky as seen from the ring of particles. For example, if the source contains stars moving back and forth along the x-axis, then the wave will produce a similar motion in the ring (top line).

about the source, such as measuring the angle of inclination of the orbital plane of a binary system.

2.2 Planned Ground-Based Detectors

I will focus here on the planned interferometers, which have the biggest potential for astronomical and particularly for cosmological observing. There are four instruments now being built [2] that should reach the target sensitivity of *first-generation* detectors: to measure $h \sim 10^{-21}$. This is a threshold that has been the goal of detector development for decades: it is the largest amplitude that could reasonably be expected from sources that might be observable in one year. Observing at this level gives no guarantees of detections: Nature has to cooperate by providing strong sources. These first-generation detectors are the first step along a planned sequence of sensitivity improvements that will produced essentially guaranteed detections by the end of this decade.

Interferometers use light to compare the lengths of the two arms. The fundamental limit on sensitivity is the amount of light, since quantum uncertainties in the arrival times of photons produce a stochastic noise called shot noise, which is less important when there are more photons. However, the main technical challenge to these detectors is to eliminate low-frequency noise from external vibrations and from internal thermal vibrations of the components [2]. These noise sources will set a lower-frequency limit of about 40 Hz on LIGO and GEO. The VIRGO detector is investing more effort in controlling vibration noise, and will have some sensitivity even at 20 Hz. All detectors go up to a few kHz before shot noise limits their sensitivity.

The largest and most ambitious project is LIGO, an American project building two 4-km detectors, one at Hanford (WA) and the other at Livingston (LA). The Hanford detector also contains a 2-km instrument for local coincidence and anti-coincidence observing. LIGO is now successfully doing interferometry, and is improving its sensitivity and reliability. LIGO may begin taking data at its planned sensitivity within the next year.

The next-largest instrument under construction is VIRGO, near Pisa. It is a cooperation between France and Italy. The timescale for operation of this 3-km instrument is about a year behind LIGO.

In Germany the GEO project is building a smaller detector, GEO600, with 600-m arms, that will nevertheless have a similar sensitivity to the LIGO instruments, and which is on the same timescale as LIGO. It achieves this sensitivity by using more advanced optical and mechanical technology. This technology will be transferred to LIGO and VIRGO when they are ready for upgrades to higher sensitivity.

GEO600 and LIGO are planning a joint test data run in December 2001, and the two projects have in fact signed a strong data-sharing and data-analysis MOU, providing for joint publication of all results. Other projects have been invited to join in this agreement.

The second-generation instruments will be upgrades of LIGO and VIRGO, which could be in place by 2007, plus a proposal in Japan that is not yet funded. These will improve the first-generation sensitivity by a factor of 10 in amplitude, and they will push the observing frequency limit down to perhaps 10 Hz. Scientists are beginning to design radical new technology for the third generation, envisioning yet a further step by a factor of 10, and a further broadening of the observing frequency window. It is possible that

VIRGO and GEO will cooperate on a joint proposal for a new third-generation detector in Europe.

2.3   LISA, the First Space-Based Detector

Ground-based detectors will never have sufficient sensitivity to do useful work below about 1 Hz, because gravity noise generated by moving masses on the Earth will be larger in amplitude than expected gravitational waves. Since gravity cannot be screened, the only solution is to put the detector into space. This is the justification for LISA, which is planned for launch in 2011. Unlike the ground-based detectors, LISA will observe many of its sources with extremely high signal-to-noise ratio. LISA is likely to be the first of a sequence of space-based detectors over the next few decades. LISA could have a mission lifetime of up to 10 years. The state of development was reviewed recently in the proceedings of the Third International LISA Symposium [8].

LISA began in 1993, when an American group led by P Bender of JILA, which had been studying space-based detectors for some time, encouraged a group of Europeans, largely in the GEO and VIRGO projects, to propose a detector for the ESA M3 mission opportunity. The mission was not selected because it was too expensive for this medium-mission limit, but the scientific potential was regarded so highly that the group was encouraged to propose for the Horizon 2000+ Cornerstone selection in 1995. The present design, based on a triangular three-armed interferometer, with a detailed plan for the optics and sensing needed, matured for that proposal.

LISA was indeed selected as a Cornerstone, but still the costs were troubling. A redesign by the European LISA team, cooperating with JPL and Bender's group at JILA, produced the current baseline design using three spacecraft, and seemed to be affordable. The project meanwhile gained considerable interest among astronomers, who were coming to the conclusion that the giant black holes that LISA could observe were ubiquitous in the centres of galaxies. This led to efforts to bring NASA into the project to share costs.

Earlier this year (2001), ESA and NASA exchanged letters of agreement to share the project equally, and ESA invited NASA to contribute to a technology demonstration mission called SMART 2, due for launch in 2006. The technology of LISA is a fascinating subject in itself, which there is no room for here. The two agencies have formed a joint LISA International Science Team (LIST), that will organise the community. It has two chairs, T Prince (Caltech and JPL) and K Danzmann (of the new branch of the Albert Einstein Institute in Hannover). Theorists and astronomers who want to contribute to the science required before LISA's launch are welcomed to join in the projects being encouraged by the LIST's Sources and Sensitivities Working Group, jointly chaired by S Phinney (Caltech) and the present author.

As mentioned above, LISA is a three-armed detector. The roughly equilateral triangle maintains its shape as the three spacecraft follow their independent orbits around the Sun. The triangle lies in a plane tilted 60° to the ecliptic, and is situated about 20° behind the Earth in its orbit. The arm-length of 5 million kilometres permits good sensitivity below 0.1 Hz. The lower limit on LISA's frequency window is about 0.1 mHz, where perturbations due to fluctuations in the solar radiation pressure dominate. Figure 2 shows the way the location and orientation of LISA change during a year.

2.4   Principles of Observation with Gravitational Waves

Gravitational wave observing is rather different from observing in the electromagnetic spectrum. This is partly because detectors cannot be pointed: they are simple quadrupoles with broad response patterns on the sky. And it is partly because detectors register the waves coherently, following the oscillations of phase. By contrast, most electromagnetic detection is bolometric, registering the energy and not the phase. Even radio interferometry, which uses phase at an early stage, eventually rectifies the signal and records the energy in the fringes. Detectable electromagnetic waves simply oscillate too fast to be recorded, and their phase in any case does not necessarily contain important information.

The gravitational wave phase oscillated at kHz frequencies or lower, and directly reflects the mass motions in the source. Almost all the useful information in a signal is in the phase $\phi(t)$. This has several implications that are not immediately obvious to astronomers used to electromagnetic observing.

First, spectroscopy and polarimetry are automatic. The detectors are linearly polarised, and spectroscopy is nothing more than taking the Fourier transform of the detected signal.

Second, detecting a gravitational wave usually means being able to measure several parameters, such as the masses of the systems, that are encoded in the phase and frequency information. Moreover, as explained earlier, the polarisation information can be used to infer source orientations.

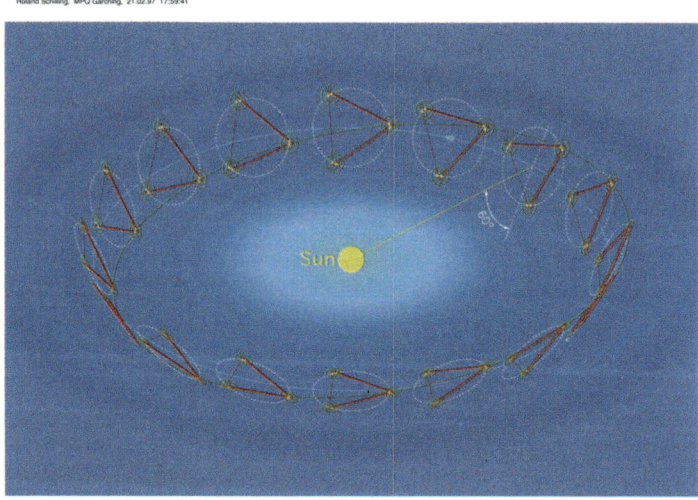

Fig. 2. LISA's configuration remains roughly equilateral and turns about an axis perpendicular to the triangle's plane, while the triangle remains in a plane tilted 60° to the ecliptic as the spacecraft orbit the Sun in one year.

Third, observing by multiple detectors brings great benefits, particularly in angular resolution, as well as in confidence when signals are weak. The angular resolution improvements are analogous to what happens in radio interferometry. Even single detectors can achieve this if they observe a continuous source long enough to take advantage of the changing detector position; in this case such a detector effectively does aperture synthesis by itself.

Fourth, data analysis on computers plays a crucial role in detection. The optimum detection strategy in gravitational wave observing is to employ matched filtering [9]. This means comparing the observed phase $\phi_{\mathrm{true}}(t)$ with the expected phase of a template signal $\phi_{\mathrm{template}}(t)$. If the two match well enough, which usually means that $|\phi_{\mathrm{true}}(t) - \phi_{\mathrm{true}}(t)| < 1$ radian over the signal duration, then one is close to optimal. Since the templates depend on parameters, which describe physical properties like sky location, source masses, orbit spindown, and other effects, a search for signals typically involves many repeated comparisons with slightly different templates. Then the availability of computer power (or the lack of it) can limit the sensitivity of an observation. This sometimes happens for other kinds of observing, for example the search for binary radio pulsars, where the parameter space that must be searched for signals is non-trivial in size.

Fifth, gravitational wave astronomers always speak about detecting *amplitudes*, not energy. This means their signal-to-noise ratios are the square-roots of energy or flux-based signal-to-noise measures. So if a gravitational wave observation with LISA can reach a signal-to-noise ratio of 10^4, then this should be compared with an optical observation with a ratio of 10^8: one photon of background for each 10^8 photons from the source! This is suggestive of how much detailed information is potentially extractable from LISA observations.

There are many analogies between long-duration gravitational wave observations and radio observations of pulsars, in that radio observations are coherent as regards the pulse period itself. This is similar to the gravitational wave period of waves from the same pulsar, so many issues are the same. For example, gravitational wave positions will be at the same accuracy level as radio positions, around the arcsecond mark.

2.5   Angular Positions

Once a source of gravitational waves has been detected, the most important information that the observation can produce is, of course, the location of the source on the sky. The accuracy of angular positions will be the crucial step in identifying sources and opening them for study by electromagnetic observation.

Since, as we remarked above, the pointing accuracy of an individual detector is poor over a short observation time, the position of the source must use more information than the instantaneous response of a single detector. The accuracy of locating a short burst comes entirely from the simultaneous observation of the event by several detectors. The accuracy of a long-duration observation is achieved, as mentioned above, by aperture synthesis.

A short burst may be defined as one in which the acceleration of the detector during the observation does not produce an overall phase-shift of the wave-form by more than one radian. The overall motion of the detector produces a constant (and usually unobservable) Doppler effect, but the acceleration of the detector distorts the waveform, and this can tell us where the wave came from. If the detector acceleration is a and the wave-vector of the radiation from the source is k, then during an observation lasting a time T, short enough to regard the acceleration as constant, the phase-shift induced by the acceleration is

$$\Delta\phi_{\text{accel}} = \frac{1}{2}(a \cdot k)T^2.$$

The condition that this should be less than 1 amounts, for a typical value of the gravitational wave frequency f_{gw} and for the acceleration produced by the rotation of the Earth, this sets a limit on the time of observation of

$$T_{\text{burst}} < 56 \left(\frac{f_{\text{gw}}}{1\,\text{kHz}}\right)^{-1/2} \text{min.} \tag{1}$$

For such bursts, the position must be triangulated by using the arrival times of the waves at several detectors. This uses the detectors as an interferometer array, and the pointing accuracy is the diffraction limit, roughly the wavelength of the waves divided by the detector spacing. Within this, a source with strong signal-to-noise ratio can be located more accurately. If SNR is the amplitude signal-to-noise ratio for a particular observation, then for detectors with a baseline between Europe and the USA, say 10^4 km, the accuracy is [10]

$$\Delta\theta_{\text{burst}} \sim \frac{2°}{\text{SNR}} \left(\frac{f_{\text{gw}}}{1\,\text{kHz}}\right)^{-1}. \tag{2}$$

A confidence limit for detection will be something like SNR > 5, so that any detected source might be triangulated to better than half a degree. A strong source with SNR $= 20$ could be located to within 5 to 10 arcminutes.

These are overly optimistic numbers, however, because there is covariance with other observational errors. If there is an error in determining the polarisation, then this could masquerade as a delay or advance in the signal by up to half a cycle. What is more, the diffraction limit applies only if there are enough detectors to determine the polarisation. With three detectors there are two possible solutions to the location on the sky. The ambiguity is resolved only with four or more detectors. Moreover, if the detectors are unusually well aligned, then they do not determine polarisation as well. Unfortunately this is the case: the LIGO detectors, in the interests of ensuring that they should see nearly identical responses, are very well aligned, so they do not contribute much to a position determination. The result is that real position determinations for bursts might be five to ten times worse than the numbers quoted above.

This situation could be significantly improved if a detector is built in Japan, with its long baseline to the others. Such a detector would improve both the detection rate and the position determinations by a factor between two and four.

A long-duration source permits a single detector to determine the position by using the phase-modulation and time-dependent polarisation projection to measure both the polarisation and position jointly. The best case is when the source lasts for a year, so that the detector synthesises a telescope with an aperture of 2 AU. The diffraction-limited position accuracy improves on the above to

$$\Delta\theta_{continuous} \sim \frac{0.5\,\text{arcsec}}{\text{SNR}} \left(\frac{f_{gw}}{1\,\text{kHz}}\right)^{-1}. \tag{3}$$

Again, this is a little optimistic because polarisation errors can add up to a cycle to the waveform. But the pointing accuracy for ground-based detectors observing pulsars, for example, is very good. However, LISA will observe in the mHz region, which degrades its position accuracy. This is compensated somewhat by the large SNR, so that the result is a resolution accuracy between 10 arcminutes and 10 degrees, depending on how strong the source is. We will come back to the importance of these errors in the next section.

2.6 Amplitude Estimates

The use of chirping binaries as standard candles depends on being able to measure the amplitude of their radiation accurately. In principle, this is just what the SNR measures, so the amplitude error would be of order $1/\text{SNR}$. But there is a strong covariance with the position error, since the antenna pattern of the detectors is broad. Roughly speaking, a position error of $\Delta\theta$ measured in radians produces a relative change in the sensitivity of the detector with respect to the source by a comparable amount. This will result in a wrong determination of the amplitude. So a good rule of thumb for amplitude errors is:

$$\Delta h = \max(1/\text{SNR}, \Delta\theta). \tag{4}$$

If the only observations of the event are from gravitational wave detectors, then $\Delta\theta$ must be inferred from the equations above. But if the event can be identified by electromagnetic observations, then the position accuracy can be much improved, and with it the amplitude accuracy. This is particularly the case for LISA, where the SNR could be as high as 10^4, but the intrinsic position accuracy could be as bad as 0.2 radians [11]. *The astronomical return from LISA observations can be greatly improved by coordinated electromagnetic observations.*

3 Chirping Binaries

3.1 Distance Determination: the Standard Candle

When LISA or a network of ground-based detectors observes a binary system, then they can determine the angular position, amplitude, and angle of inclination of the binary orbit, as described above. For simplicity let us now assume that the orbit is circular, although what we describe can be extended to elliptical orbits.

There is a remarkable coincidence in the radiation from binary systems, in that both the amplitude h of the radiation and the rate of change of the frequency of the radiation df_{gw}/dt depend on the masses of the two stars only through exactly the same combination, which is called the *chirp mass* M of the system. If the two stars have masses m_1 and m_2, with associated reduced mass μ and total mass M_T, then the chirp mass is defined by

$$\mathcal{M} := \mu^{3/5} M_T^{2/5} = (m_1 m_2)^{3/5} (m_1 + m_2)^{-1/5}. \qquad (5)$$

This was first pointed out by the present author [12], who suggested how this could be used to measure the luminosity distance d_L to any binary system that chirped, that is whose $d f_{gw}/dt$ could be measured.

One way to see how this can be done is to consider the formula for the SNR of an observation using a filter that has been perfectly matched to the incoming signal, in polarisation and chirp mass. We consider only the radiation from the orbit, not from the later coalescence event. This underestimates the SNR, but it has the advantage that the orbit is fully understood and its SNR can be characterised, while the radiation to be expected from coalescence is not yet known. Then the SNR can be written in the following way for a burst chirp [13], that is a chirp that lasts less than the time given in (1):

$$\mathrm{SNR} = 8\Theta \frac{r_0}{d_L} \left(\frac{\mathcal{M}}{1.2 M_\odot} \right)^{5/6} \zeta(f_{\mathrm{max}}). \qquad (6)$$

The following terms enter this equation:

- Θ is a factor that depends on the projection of the polarisation of the wave on the antenna pattern, so it is a function of the orientation of the binary relative to the detector. This is measurable from the polarization and direction information.
- r_0 is the *range* of the detector for this kind of observation, that is a distance that depends on the sensitivity of the detector. It is a function only of the detector.
- d_L is the luminosity distance to the source, and is what we want to determine from the observation.
- $\mathcal{M}$ is the chirp mass, and is determined from the observed rate of change of the frequency of the chirp.
- ζ is a number that depends weakly on $\mathcal{M}$, taking account of the fact that massive chirping systems reach coalescence and hence their maximum frequency at a lower frequency than less massive ones do, so the response of the detector to them is a little different. This is clearly also a function of the detector, but it is known once $\mathcal{M}$ has been determined.

From this list it is clear that all the numbers in this equation, including the value of SNR, are determined either by the detector or by the observation of the signal, except for d_L. This is therefore the unknown that can be solved for. The result is that *observations of the radiation from the orbit of a chirping binary determine its luminosity distance.*

3.2 Which Binaries Chirp?

The expression for the rate of change of the frequency of radiation from a binary alluded to above can be formulated to give a characteristic time called the *chirp time*

$\tau_{gw} = (\mathrm{d}f_{gw}/\mathrm{d}t)/f_{gw}$. Here are some useful ways to calculate this chirp time for various interesting systems:

$$\tau_{gw} = 200 \left(\frac{f}{20\,\text{Hz}}\right)^{-8/3} \left(\frac{\mathcal{M}}{1.2M_\odot}\right)^{-5/3} \text{s}, \tag{7a}$$

$$= 44 \left(\frac{f}{20\,\text{Hz}}\right)^{-8/3} \left(\frac{\mathcal{M}}{3.0M_\odot}\right)^{-5/3} \text{s}, \tag{7b}$$

$$= 44 \left(\frac{f}{10^{-4}\,\text{Hz}}\right)^{-8/3} \left(\frac{\mathcal{M}}{10^6 M_\odot}\right)^{-5/3} \text{s}, \tag{7c}$$

$$= 4 \times 10^5 \left(\frac{f}{3\,\text{mHz}}\right)^{-8/3} \left(\frac{\mathcal{M}}{0.5M_\odot}\right)^{-5/3} \text{s}. \tag{7d}$$

This list shows how long one must wait for a system to change its frequency substantially, say by a factor of about two. However, one does not have to wait that long to see a system chirp. All that one requires is that the system change its frequency by the frequency resolution of the observation, which is $1/T_{obs}$ for an observation of duration T_{obs}. Figure 3 shows the systems that chirp and those that coalesce within a one-year observing time.

From these times, it is possible also to calculate the population statistics of chirping systems if they are created at a fixed steady rate R and die away through coalescence. Then the steady-state population of systems radiating gravitational waves with a frequency larger than any given f_{gw} is

$$N(> f) = 7000 \left(\frac{f_{gw}}{0.1\,\text{mHz}}\right)^{-8/3} \left(\frac{\mathcal{M}}{1.2M_\odot}\right)^{-5/3} \left(\frac{R}{10^{-5}\text{yr}^{-1}}\right). \tag{8}$$

For example, in the Galaxy, systems like the Hulse-Taylor pulsar are expected to form once every 40^5 yr or so, so this means that there should be thousands of such systems within the LISA waveband. LISA should certainly see many if not most of them, depending on whether they are obscured by radiation from the far more common white-dwarf binary systems.

Another interesting number is to try to estimate the population of binaries of $1000M_\odot$ that might be observable by LISA. LISA should see any chirping system in its waveband anywhere in the Universe. In the first generation of stars, suppose that 10% of the present stellar mass of the universe went into such stars during a time lasting about 10^9 years. Then there would have been some 10^{18} such stars that formed. Most of them may well have evolved into black holes [14], and the gravitational collapse event might not have disrupted the binary system. Any such binaries with a local gravitational wave frequency above $6\,\text{mHz}$ would, by the above formulas, coalesce within about 1 year of our time, so LISA could follow the event most of the way to coalescence.

Suppose a fraction η of such systems formed binaries that could coalesce in the first 10^9 years. Then the rate of formation of such binaries was $10^9\eta$ per year. By (8), the number of such systems that LISA could follow to coalescence in one year of observing is about 1700η. So if the efficiency of formation of these binaries is better than one tenth of one percent, then LISA would be able to detect a few. If the efficiency is better than

1%, then LISA would have of order 200 events during a ten-year mission lifetime, and the upper limit on the luminosity distances to these events would signal the onset of this first generation of star formation.

4   Cosmology with Ground-Based Detectors

The use of chirping binaries as standard candles to discover cosmological information has been studied by a number of authors [12,15–18]. They have pointed out that there is a variety of methods to avoid the problem of identifying the galaxy in which the chirp occurred, and still extract cosmological information. I begin here, however, with the expectation that chirps may produce gamma-ray bursts, and in any case should certainly produce optical/radio/X-ray displays of some kind, which I will call "afterglows". These

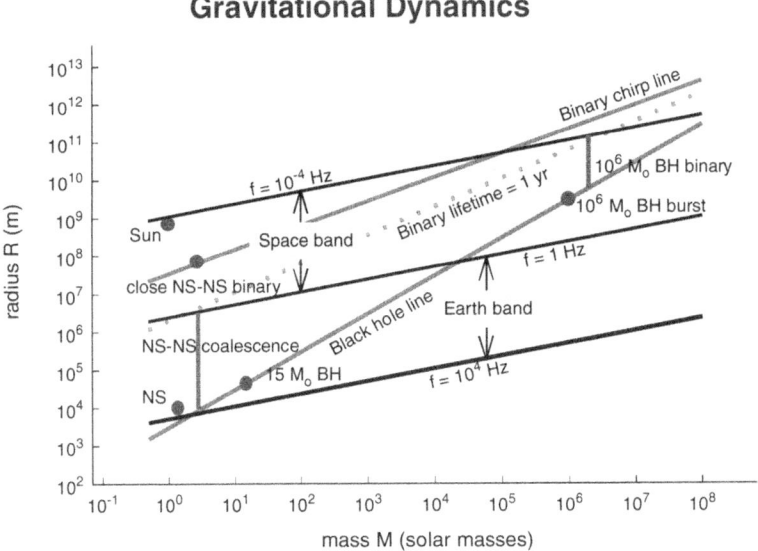

Fig. 3. A rough classification of binary systems according to their total mass (horizontal axis) and size (vertical axis). The three solid lines with slope 1/3 in this log-log chart show systems whose natural frequencies $(\pi G \varrho)^{1/2}$ are 10^{-4} Hz, 1 Hz, and 10^4 Hz, respectively. These divide the chart into systems in the frequency region that one can observe from space and the ground. The line with slope 1 is the black-hole line: systems below this line do not exist. The dotted line is the line below which systems will coalesce within one year. The solid chirp line, on the other hand, delimits systems that can be observed to change their frequency in one year. Notice that all binaries observable from the ground will coalesce within one year, while most binaries observable by LISA with masses above about 1000 solar masses will chirp.

will help the identification of the event. After this discussion, I return to the subject of identifying black hole coalescences, which should have no optical counterpart.

4.1 Afterglow Cosmology

As we have noted above, the second-generation ground-based detectors should have ranges so that confident detections of coalescences of binary neutron stars can be made out to 400 Mpc or so. Tens of events per year are to be expected [19]. However, the errors in position determination are rather large, so that if no other information is available then it will be difficult to determine the galaxies in which the event occurred.

The situation will be dramatically different if such events lead to gamma-ray bursts, or indeed if they lead to any other kind of transient event that leaves behind an afterglow. This seems very likely. From the identification of the event by electromagnetic detection of the afterglow, the position can be determined with very small errors and the redshift of the galaxy can be measured. Then the luminosity distance can be determined within errors given just by the SNR, which could be of order 10. Thus, each event leads to a value for the Hubble constant accurate to 10%. With 50 events over a few years of observing, the statistical errors could go down to a few percent. Although the Hubble constant should be known to this accuracy by other means by the time second-generation detectors operate, this method will be an important check on the systematics of other determinations.

Being able to give advance notice of a burst event by gravitational waves will also be a valuable contribution of these detectors. The gravitational wave signal will precede even the gamma-ray burst, and the detector scientists plan to build early-warning alert systems so that notice of potential events could be available to cooperating astronomers within seconds of the gravitational wave observation. In advanced detectors, the inspiral signal for two neutron stars could last several minutes, and there could be enough signal after the first minute to predict the coalescence event. In this case, astronomers would have a minute or so notice to begin observing before the actual coalescence event even occurred.

The association of gravitational wave events with afterglows will of course immensely help modelling of gamma-ray bursts, and it will also allow estimates of the beaming fraction. Gravitational wave emission by these systems is much more isotropic than the gamma radiation, and therefore detectors should give a fair sample of all coalescing systems within their range.

It is possible that gamma bursts are associated more strongly with coalescences between neutron stars and black holes than between double neutron stars. If this is the case, the second-generation detectors will have a longer range, out to redshifts of order 0.3. This will make their ability to do cosmology much more interesting, and measurements of the local acceleration of the universe to $\pm 10\%$ would be possible over 5 years.

All of these numbers improve by factors of 2 or more if a further detector is added to the network, say in Japan.

4.2   Cosmology with Binary Black Hole Observations from the Ground

It is possible that the event rate for coalescence of binary black holes of stellar mass will be comparable to that for binary neutron stars. Binaries are less likely to be disrupted by black-hole formation than by neutron-star formation because less mass is lost. And globular clusters seem to be efficient factories for black-hole binaries [20]. It may happen, then, that the first events detected by ground-based instruments will be black-hole coalescences. And if that is the case, then second-generation detectors may see many tens of such events out to redshifts of order 1. Over this distance, it is no longer appropriate to speak of the Hubble constant or the deceleration parameter, since these are just terms in the Taylor expansion of the recession velocity. The observed acceleration of the universe makes such a local approximation inadequate. The goal over cosmological distances is to sample the function $z(d_L)$, the Hubble flow, over as large a range of values of d_L as possible.

Unfortunately, distant as these black-hole events are, they do not produce afterglows or other electromagnetically detectable counterparts from which a redshift can be measured. To circumvent this, I have proposed a statistical method [12] that can still measure the parameters describing the Hubble flow over this interesting distance range. The method is interesting not only because it can determine parameters, but also because it is an example of a nonlinear statistical method whose errors improve much more rapidly with the number N of samples than the usual $N^{1/2}$ associated with linear averaging, at least at first when N is small.

The idea is best illustrated for low-redshift measurements, where the goal is simply to determine one number, the Hubble constant. After understanding how this is done, we will see how it could be generalised to larger redshifts. For each event, the detectors will produce an error box on the sky with a number of candidate clusters of galaxies in which the event may have occurred. The angular position of each candidate leads to a corresponding luminosity distance; measuring the mean redshift of each cluster then leads to a "candidate" value of the Hubble constant for that cluster. Each candidate cluster produces a candidate value. Most are wrong, but one of them should be the correct one. Now, if one has, say, ten such events, then one value of the Hubble parameter should appear in each set of candidates. As long as the number of candidate values is not so large that the observational errors create a lot of overlap between false and true values, it should be possible to zero in on the correct value of the Hubble parameter and retrospectively identify the clusters in which the events occurred. With dozens of events, this method should be very efficient.

This method is actually a one-dimensional version of the Hough transform method that was devised to analyse bubble-chamber photographs in high-energy physics (see [21]). The tracks expected of particles were parametrised, and the number of bubbles on each possible track was counted. Real tracks would have a much larger number of bubbles than random ones. The Hough transform is now being developed to search for unexpected pulsar signals in gravitational wave data [22].

For high-redshift objects, one could use the Hough transform to search for the best set of parameters for a cosmology. Appropriate parameters might be the present Hubble constant, the value of Ω_{matter}, and the value of Λ. With many tens of events, there should be enough statistics to find a set of values that all the observations are consistent with.

Of course, if by then the Hubble constant is well-enough known to determine the correct candidate from among the candidate clusters of galaxies, then this statistical method is not needed.

Once the clusters containing the events are established, then the redshift and the luminosity distance can be used to calibrate the expansion of the universe. At this point the statistical errors of the measurements of the large number of events will indeed reduce as $1/\sqrt{N}$. With data out to $z = 1$, the parametrisation of the Hubble flow will be sensitive to the turnover, where the universe changed from deceleration to acceleration. We shall see that LISA can extend this method to redshifts of 4 or more.

5   Cosmology with LISA

LISA could measure a few mergers of massive black holes in the centres of galaxies each year. It will be sensitive to masses in the range 10^3–10^7 $M_\odot$. These observations will give important insight into the processes that formed the black holes, into their population statistics, and into the role they played in galaxy formation. But here I wish to focus on the use of LISA to measure the Hubble flow itself.

For a typical system of two 10^6 $M_\odot$ black holes at $z = 1$, LISA on its own will be able to determine the position to an accuracy of about half a degree. The physical size of the error box will be of the order of 40 Mpc in all three dimensions, because LISA's distance determination is also limited by the angular errors, as explained earlier. Thus, the error box is about 1% of the distance to the source.

This error box may contain a number of rich clusters, each with one or more candidate galaxies that show evidence of a past galaxy merger that could have led to the black hole merger. We would like to identify the cluster in which the merger took place. There are at least three ways to do this.

1. If by the time LISA flies, the Hubble flow is known to an accuracy of better than 1% out to redshifts of 4 or so, then this may assist identifying the cluster. One measures the redshifts of each of the candidates, and uses the angular positions of the candidates to determine from the LISA chirp signal what the luminosity distance is to that candidate. If the Hubble flow is known accurately as a function of luminosity distance, then the expected redshift can be compared with the measured one, and if these do not coincide then the candidate can be rejected. If the expansion is known well enough, then the candidates may be narrowed down to just 1.

2. If the expansion is not known to this accuracy by the time LISA flies, then the statistical method described in the previous subsection could be brought to bear. This could work if there are of order ten events at high redshift over the mission lifetime of 10 years. The goal at this stage would be to determine the Hubble flow accurately enough to identify the galaxies in which the events have taken place.

3. Failing both of these circumstances, it will be a challenge to observers and astrophysicists to determine the galaxy in which the merger occurred by other means. Perhaps the morphology of the galaxy is special in some way. The gradual in-spiral of the two massive black holes transfers considerable energy to a number of stars in the core of the galaxy, and so it may be that the central bulge of this galaxy has a larger number of stars on nearly radial orbits than is normal. Or perhaps the two

black holes maintained accretion disks, or even jets, until they came close enough to one another for tidal forces to disrupt them. The fossil jets may still be visible in the outer regions of the galaxy, and the gas of the accretion disks may have been expelled or shocked in a way that is observable for some time after the disruption.

The identification of the merger galaxies has a large potential payoff. The accurate angular position for each galaxy will provide a very accurate value of the luminosity distance, perhaps with errors smaller than 0.1%, depending only on the SNR of the detection. With the measured redshifts, then the measurements of the Hubble flow are limited only by the proper velocities of the galaxies (inducing single-measurement uncertainties in the Hubble flow of 0.1%). With a handful of merger events spread over redshifts out to, say, 4 or more, it should be possible to go well beyond Λ-cosmology models and test quintessence and other models in which the pressure is not strictly equal to the negative of the energy density, and in which the density of dark energy/negative pressure is variable in time.

6   Stochastic Gravitational Waves from the Big Bang

Probably the most fundamental cosmological observation that gravitational wave detectors can make is of gravitational waves coming from the Big Bang. This is the gravitational analogue of the cosmic microwave background radiation, but with a key difference. Because gravitational waves couple so weakly to matter, they never thermalised. The non-thermal spectrum comes to us unchanged from whatever event(s) produced it. Using gravitational waves we can see directly to the first fraction of a second after the Big Bang.

Unfortunately, what firm predictions exist for the amount of radiation that is produced by the Big Bang are discouraging. Gravitational waves should be created at some level by inflation in the same processes that produced the fluctuations in energy density that led to galaxy formation, but the present energy density must be less than 10^{-13} of the closure density. The only observational limits are from the millisecond pulsar at frequencies of 1 cycle per 10 years, and from the requirement that the radiation not disturb nucleosynthesis. In both cases the limits require Ω_{gw} to be smaller than about 10^{-6}.

Between the prediction of inflation and the observational limits there is lots of room for other creative mechanisms, and many exist. Toy models of superstring cosmology can produce tailor-made spectra with large amounts of radiation. Cosmic defects, phase transitions, and other unknown but not implausible physics can lead to radiation confined to certain wave-bands. Even brane-world cosmologies have the potential to produce radiation up to the nucleosynthesis limit at any frequency [23].

Ground-based detectors can see this radiation best by cross-correlating the outputs of two nearby detectors. The best suited are the two LIGO instruments. In the second generation they may reach as low as , perhaps a little lower. LISA cannot cross-correlate its two independent interferometers because they share a common arm and hence common noise. It can, however, internally calibrate its instrumental noise and thereby identify any stochastic gravitational wave signal whose power is comparable to or larger than the instrumental noise. This is unlikely to take it lower than $\Omega_{gw} \sim 10^{-10}$. I have proposed a variant of the LISA mission that could go down as low as 10^{-13}, but this still does not reach the inflation prediction. A future LISA follow-on mission would be required to reach that level.

One of the problems with detecting a background from the Big Bang is that there are astrophysical backgrounds of a more recent origin. This includes radiation from white-dwarf binary systems, ordinary binaries, close neutron-star binaries, and even small objects falling into massive black holes. There could be a window around 1 Hz where the astrophysical backgrounds are weak enough to allow the cosmological background to dominate, but there are believed to be few other accessible windows [24].

Only observations will tell us what is out there. LISA will certainly measure the compact white-dwarf binary background, which is expected to stand out well above the noise below 1 mHz. LISA might also measure backgrounds at higher frequencies. Whether LISA or the ground-based detectors manages to see a cosmological background from fundamental physics near the Big Bang is one of the most unpredictable outcomes of gravitational wave astronomy.

7 Conclusions

The astronomical community has waited a considerable time for gravitational wave detectors to realise their promise. The progress has been steady but largely invisible until now. From next year, detections of some systems will be possible. But cosmological returns are likely to require another decade of development.

The second-generation ground-based detectors should make the first impact on cosmology, providing values for the Hubble constant and the acceleration of the universe that with an accuracy competitive with that of other methods. This will be a useful check on all methods. The opening up of the low-frequency window by LISA after 2011 will bring much larger potential payoffs for cosmology. With some luck (or cleverness!), LISA could measure the deceleration/acceleration history of the universe with outstanding accuracy out to redshifts of 4 or earlier. To realise this promise, coordinated observations with telescopes in the optical/IR, X-ray, radio, and other bands will be essential.

References

1. P. R. Saulson: *Fundamentals of Interferometric Gravitational Wave Detectors* (World Scientific, Singapore, 1994)
2. J Hough, S Rowan: Living Rev. Relativity **3**, 3 (2000). [Online article]: cited on 20 November 2001, http://www.livingreviews.org/Articles/Volume3/2000-3hough/
3. C. W. Misner, K. S. Thorne, J. L. Wheeler: *Gravitation* (Freeman & Co., San Francisco, 1973)
4. B. F. Schutz: *A First Course in General Relativity* (Cambridge University Press, Cambridge, 1985)
5. B. F. Schutz, C.M. Will: 'Gravitation and General Relativity'. In: *Encyclopedia of Applied Physics* **7**, 303-340 (1993)

6. B F Schutz: 'Gravitational Radiation'. In: *Encyclopedia of Astronomy and Astrophysics* (Institute of Physics Publishing, Bristol, and Macmillan Publishers Ltd., London, 2000). Electronic version: gr-qc/0003069.

7. B. F. Schutz, F. Ricci: 'Gravitational Waves, Sources and Detectors', in: *Gravitational Waves*, ed. by I. Ciufolini, V. Gorini, U. Moschella, P. Fré (Institute of Physics Publishing, Londo, 2001).

8. B. F. Schutz (ed): *Classical and Quantum Gravity*, **18**, Number 19 (7 October 2001).

9. K. S. Thorne: "Gravitational Radiation". In: *300 Years of Gravitation*, ed. by S. W. Hawking, W. Israel (Cambridge University Press, Cambridge, 1987), pp. 330–458

10. C. Cutler, E. E. Flanagan: *Phys. Rev.* **D49**, 2658 (1994)

11. A. Vecchio, C. Cutler, In: *Recent developments in theoretical and experimental general relativity, gravitation, and relativistic field theories, pt. B*, ed. by T. Piran (World Scientific, Singapore, 1999), pp. 1121–1123

12. B.F. Schutz: *Nature* **323**, 310 (1986)

13. L. S. Finn, D. F. Chernoff: *Phys. Rev.* **D 47**, 2198 (1993)

14. I. Baraffe, A. Heger, S. E. Woosley: *Astrophys. J.* **550**, 890 (2001)

15. D. Markovic: *Phys. Rev.* **D 48**, 4738 (1993)

16. D F Chernoff, L. S. Finn: *Astrophys. J.* **411**, L5 (1993)

17. L. S. Finn: *Phys. Rev.* **D 53**, 2878 (1996)

18. Y. Wang, E. L. Turner: *Phys.Rev.* **D 56**, 724 (1997)

19. D. R. Lorimer, E. P. J. van den Heuvel: *Mon. Not. Roy. astr. Soc.* **283**, L37 (1996)

20. S. F. Portegies Zwart, S. L. W. McMillan: *Astrophys. J.* **528**, L17 (2000)

21. V. E. Leavers: *CVGIP: Image Understanding*, **58**, 2, 250 (1993)

22. B. F. Schutz and M.-A. Papa In: *Proceedings of Jan 1999 Moriond meeting "Gravitational Waves and Experimental Gravity"*, ed. by J. Trân Thanh Vân, J. Dumarchez, S. Reynaud, C.Salomon, S. Thorsett, J.Y. Vinet (Hanoi, 2000). Electronic version: gr-qc/9905018.

23. C. J. Hogan: *Class. Quant. Grav.* **18**, 4039 (2001).

24. C. Ungarelli, A. Vecchio: *Phys. Rev.* **D 6306** 4030 (2001), art. no. 064030

Mimicking the Most Luminous Objects with Gravitational Lensing

Joachim Wambsganss[1]

Universität Potsdam, Institut für Physik, Am Neuen Palais 10, 14469 Potsdam, Germany
jkw@astro.physik.uni-potsdam.de

Abstract. Gravitational lensing can change the apparent brightness of distant objects. Though the probability for a high magnification is small, due to our standard flux-limited search techniques, we may occasionally find objects that are magnified by a large amount, in particular at the bright end of the luminosity function. This effect is commonly called magnification bias. After introducing some basics and general aspects of gravitational lensing, the connection between lensing and some of the most luminous objects in the universe is presented. For various classes of very luminous astronomical objects, one prototypical example is selected and discussed with respect to lensing effects, i.e. ultraluminous galaxies: FSC 10214+4724, quasars: APM 08279+5255, gamma-ray bursts: GRB000301C, supernovae: SN 1997ff and galaxy clusters: RXJ 1347.5-1145.

1 (Relevant) Basics of Lensing

The basic setup for a gravitational lens scenario is displayed in Figure 1. The three ingredients in such a lensing situation are the source S, the lens L, and the observer O. Light rays emitted from the source are deflected by the lens. For a point-like lens, there will always be (at least) two images S_1 and S_2 of the source. With external shear – due to the tidal field of objects outside but near the light bundles – there can be more images. The observer sees the images in directions corresponding to the tangents of the incoming light paths.

In Figure 1 the corresponding angles and angular diameter distances D_L, D_S, D_{LS} are indicated. In the thin-lens approximation, the hyperbolic paths are approximated by their asymptotes. In the circular-symmetric case, the deflection angle is given as

$$\tilde{\alpha}(\xi) = \frac{4GM(\xi)}{c^2}\frac{1}{\xi}. \tag{1}$$

where $M(\xi)$ is the mass inside a radius ξ. In this depiction the origin is chosen at the observer. From the diagram it can be seen that the following relation holds:

$$\theta D_S = \beta D_S + \tilde{\alpha} D_{LS} \tag{2}$$

(for $\theta, \beta, \tilde{\alpha} \ll 1$; this condition is fulfilled in practically all astrophysically relevant situations). With the definition of the reduced deflection angle as $\alpha(\theta) = (D_{LS}/D_S)\tilde{\alpha}(\theta)$, this can be expressed as:

$$\beta = \theta - \alpha(\theta). \tag{3}$$

This relation between the positions of images and source can easily be derived for a non-symmetric mass distribution as well. In that case all angles are vector-valued.

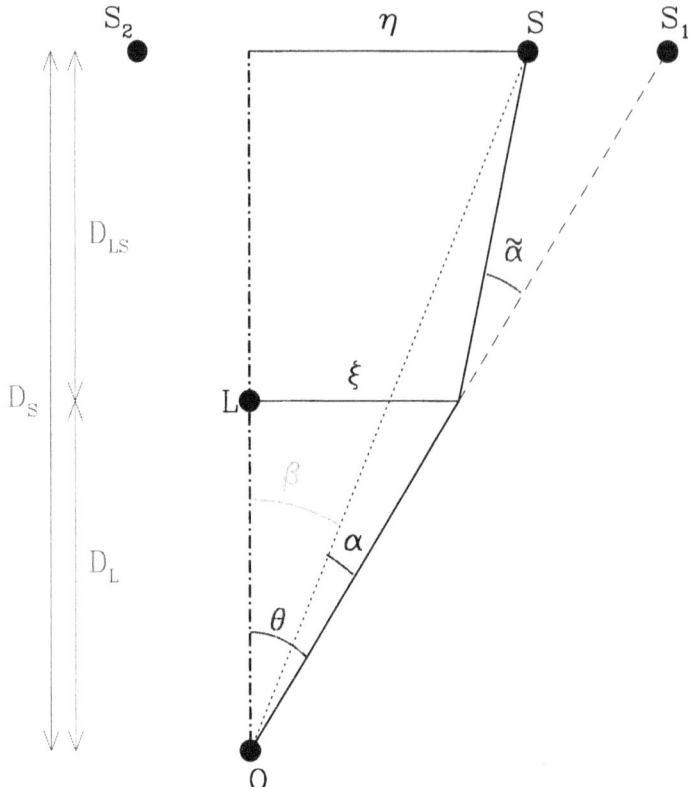

Fig. 1. a) Setup of a gravitational lens situation: The lens L located between source S and observer O produces two images S_1 and S_2 of the background source. Relations between the various angles and distances involved in the lensing setup can be derived for the case $\tilde{\alpha} \ll 1$, as formulated in the lens equation (3).

For a point lens of mass M, the deflection angle is given by equation (1). Plugging this into equation (3) and using the relation $\xi = D_L\theta$ (cf. Figure 1) one obtains:

$$\beta(\theta) = \theta - \frac{D_{LS}}{D_L D_S} \frac{4GM}{c^2\theta}. \tag{4}$$

For the special case in which the source lies exactly behind the lens ($\beta = 0$), due to the symmetry a ring-like image occurs whose angular radius is called **Einstein radius** θ_E:

$$\theta_E = \sqrt{\frac{4GM}{c^2} \frac{D_{LS}}{D_L D_S}}. \tag{5}$$

The Einstein radius defines the angular scale for a lens situation. For a massive galaxy with a mass of $M = 10^{12} M_{\odot}$ at a redshift of $z_L = 0.5$ and a source at redshift $z_S = 2.0$ (we used here a Hubble constant of $H = 50$km sec^{-1} Mpc^{-1} and an Einstein-deSitter universe), the Einstein radius is

$$\theta_{E,\text{ galaxy}} \approx 1.8 \sqrt{\frac{M}{10^{12} M_{\odot}}} \text{ arcsec} \tag{6}$$

(note that for cosmological distances in general $D_{LS} \neq D_S - D_L$!). For a galactic microlensing scenario in which stars in the disk of the Milky Way act as lenses for bulge stars close to the center of the Milky Way, the scale defined by the Einstein radius is

$$\theta_{E,\text{ galactic star}} \approx 0.5 \sqrt{\frac{M}{M_{\odot}}} \text{ milliarcsec.} \tag{7}$$

Time scales for galactic microlensing events – i.e. the duration for crossing the Einstein radius – typically range from weeks to months. For cosmological/quasar microlensing, this time scale extends to years; however, caustic crossing events can be as short as a few weeks.

The **magnification** of an image is defined by the ratio between the solid angles of the image and the source, since the surface brightness is conserved. Hence the magnification μ is given as

$$\mu = \frac{\theta}{\beta} \frac{d\theta}{d\beta}. \tag{8}$$

In the symmetric case discussed above, the image magnification can be written as (by using the lens equation):

$$\mu_{1,2} = \left(1 - \left[\frac{\theta_E}{\theta_{1,2}} \right]^4 \right)^{-1} = \frac{u^2 + 2}{2u\sqrt{u^2 + 4}} \pm \frac{1}{2} \tag{9}$$

Here u is defined as the "impact parameter", the angular separation between lens and source in units of the Einstein radius: $u = \beta/\theta_E$. The magnification of one image (the one inside the Einstein radius) is negative. This means it has negative parity: it is mirror-inverted. For $\beta \to 0$ the magnification diverges: in the limit of geometrical optics the Einstein ring of a point source has infinite magnification[1]! The sum of the absolute values of the two image magnifications is the measurable total magnification μ:

$$\mu = |\mu_1| + |\mu_2| = \frac{u^2 + 2}{u\sqrt{u^2 + 4}}. \tag{10}$$

Note that this value is (always) larger than one! (This does not violate energy conservation, since this is the magnification relative to an "empty" universe and not relative to a "smoothed out" universe. This issue is treated in detail in, e.g., [39] or in Chapter 4.5 of [43].)

---

[1] Due to the fact that physical objects have a finite size, and also because at some limit wave optics has to be applied, in reality the magnification stays finite.

Another useful concept is the **critical surface mass density**, defined as:

$$\Sigma_{\text{crit}} = \frac{c^2}{4\pi G} \frac{D_S}{D_L D_{LS}}. \tag{11}$$

This specifies the value of the (two-dimensional) surface mass density along the line of sight which is sufficient to produce multiple images. It can be visualized as the lens mass M "smeared out" over the area of the Einstein ring: $\Sigma_{crit} = M/(R_E^2 \pi)$, where $R_E = \theta_E D_L$. A typical value of the critical surface mass density is $\Sigma_{\text{crit}} \approx 0.8 \text{ g cm}^{-2}$ for lens and source redshifts of $z_L = 0.5$ and $z_S = 2.0$, respectively.

More detailed introductions to gravitational lensing including some historic aspects can be found in [51], or in the textbook [43] by Schneider et al. (1992) and in the more mathematically oriented monograph [36] by Petters et al. (2001).

2   Lensing Effects/Phenomena

Light deflection/gravitational lensing has various effects on background sources. Depending on the mass of the lensing object (from comet-like objects to clusters of galaxies), on the nature of the lensed source (point-like/unresolved or extended), and on the detection (imaging and photometry in the optical and radio regime, timing for gamma rays), the actual observations can cover quite a variety of techniques. Here we consider only "strong" lensing, where the effect can be seen for each case individually.

2.1   How Does Matter Affect the Light of Background Sources?

The consequences of strong lensing are:

- **Change of position:** This is normally not observable, since we do not have any information on the "unlensed" position of a source; only in "dynamical" situations, in which the image/lens configuration changes with time, this can be observed (e.g., the sun passing in front of stars). Exactly this was, after all, the first detection of light deflection [7] during the famous solar eclipse in 1919.
- **Distortion:** The shape of resolved sources is changed by lensing. The best visualization of this effect are the giant luminous arcs.
- **(De)Magnification:** A few sources are (highly) magnified, most sources are slightly demagnified. This means that the luminosity function of a hypothetical population of cosmological "standard candles" will unavoidably be broadened (see, e.g., [50]).
- **Multiple images:** The most dramatic lensing phenomenon: multiple quasars and multiple galaxy images are observed directly, and via microlensing we have evidence of unresolved multiple images as well.

These effects often occur in combination. In Figure 2 they are shown for a very simple lens system, the point lens plus external shear (so-called Chang-Refsdal lens, see [4]). This lensing situation produces a diamond-shaped caustic with four cusps which are connected by four folds. Depending on the exact source position relative to this caustic, the image configuration can be quite different. In particular, the image number changes from two to four between a source position outside and inside the caustic. For four different source positions, indicated by the differently colored circles, the corresponding image configurations can be seen on the right hand side of Figure 2. For a more complicated lens configuration, the images can be even more distorted, as is visualized in Figure 3: displaced, distorted, (de-)magnified and multiple images of a source shape with a particular brightness profile.

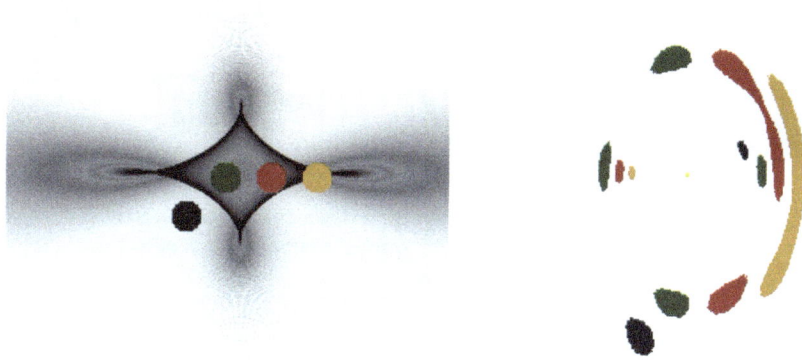

Fig. 2. Chang-Refsdal lens (see [4]): A point lens plus external shear produces a diamond-shaped caustic with four cusps, connected by four folds. In the left panel, four source positions are shown relative to the caustics, with the corresponding image configurations in the right panel. Note the difference between the blue (leftmost) source position, resulting in two images, the green (completely inside) source position, producing four images, and the intermediate positions (red, yellow), with merging images as giant luminous arcs.

2.2 Lensing Phenomena

Quite a variety of spectacular lensing phenomena have been observed in recent years. In Figure 4, four of the most spectacular recent lensing examples are presented: Multiply-imaged quasars, Giant luminous arcs, Einstein rings and quasar microlensing. These spectacular applications are now discussed in some detail.

Multiple quasars Multiply-imaged quasars were the first category of lensed systems to be discovered [49], cf. Figure 4a. By now, more than 60 multiply-imaged quasar systems have been found, most of them doubles or quadruplets, recently even a six-image configuration was discovered [44]. The angular separations range from a few tenth of an arcsecond to about 8 arcseconds. The quasars are typically at redshifts between $1.0 \leq z_Q \leq 4.5$. In almost all cases, the lens is identified to be an intermediate galaxy, in some cases "assisted" by a nearby group of galaxies. Up-to-date tables of

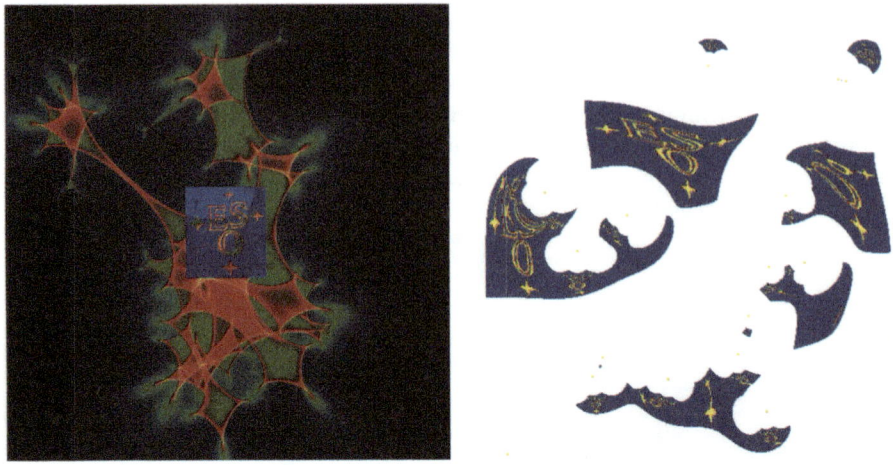

Fig. 3. Left panel: Magnification and caustic distribution in source plane due to a number of point lenses (light grey means high magnification), with specific source profile superimposed; Right panel: Corresponding distorted image configuration.

multiply-imaged quasars and gravitational lens candidates are provided, e.g., by the CASTLE group [13]).

Einstein rings A particular class of lens systems are the Einstein rings, circular images of extended background sources (cf. Figure 4b). This setup happens when there is a perfect alignment between source, lens and observer. Since the radius of the Einstein ring is proportional to the square root of the mass of the lens (see eq. 5), these systems are very good laboratories for weighing galaxies. The most remarkable example so far is the Einstein ring B1938+666 [22]. An infrared HST image shows an almost perfectly circular ring with two bright parts plus the bright central galaxy. By now about a half dozen cases have been found that qualify as Einstein rings [13]. Their diameters vary between 0.33 and about 2 arcseconds. All of them are found in the radio regime, some have optical or infrared counterparts as well.

Giant luminous arcs and arclets Rich clusters of galaxies at redshifts beyond $z \approx 0.2$ with masses of order $10^{14} M_\odot$ are very effective lenses if they are centrally concentrated. Since most clusters are not really spherical mass distributions and since the alignment between lens and source is usually not perfect, no complete Einstein rings have been found around clusters. But there are many examples known of spectacular giant luminous arcs which are curved around the cluster center, with lengths up to about 20 arcseconds. Their Einstein radii are of the order of 20 arcseconds, but cases with radii up to 35 arcseconds are known [47]. One of the best known cases is the galaxy cluster CL0024+1654 (redshift $z = 0.39$), with red cluster galaxies and nicely elongated bluish arcs [5] (see Figure 4c). Images further out are less distorted, but still clearly visibly tangentially elongated: the arclets. General results from the analysis of giant arcs

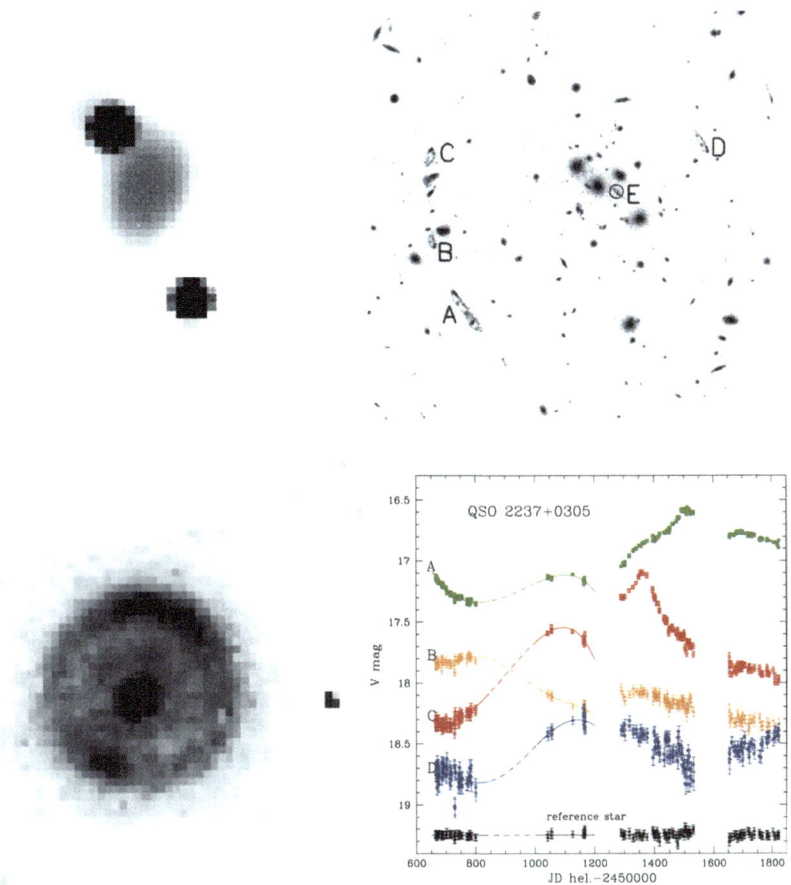

Fig. 4. Four examples of strong lensing: a) Double quasar HE1104-1805 (top left, [6]): deconvolved infrared (J-band) image of the two quasar images ($z_Q = 2.316$, $\Delta\theta = 3.2$ arcsec) and the lensing galaxy (at $z_G = 1.66$); b) Giant luminous arcs in cluster CL0024 (top right, [5]): five spectacular images of a high redshift galaxy seen lensed by a galaxy cluster (redshift $z_L = 0.39$) with radius of curvature of about 20 arcseconds. c) Einstein ring B1938+666 (bottom left, see [22]): circular image with diameter 0.95 arcseconds; d) Microlensing in Q2237+0305 (bottom right, see [52,53]): the lightcurves of the four images vary independently of each other, intrinsic variability can be excluded.

and arclets in galaxy clusters are: clusters of galaxies are dominated by dark matter, and typical "mass-to-light ratios" for clusters obtained from strong (and weak, see below) lensing analyses are $M/L \geq 100 \, M_\odot/L_\odot$.

Stellar and quasar microlensing The lensing action of stellar mass objects is usually called "microlensing". It comes in two varieties: star-star lensing, where stars in the

Galactic disk or halo deflect the light of background stars in the Galactic bulge or in nearby galaxies (LMC, SMC, M31). The second variant is star-quasar lensing, where stars in a distant (lensing) galaxy act as microlenses on a quasar at cosmological distances (cf. Figure 4d). In both cases, the action is measured as a characteristic light curve.

3   Astrophysical Objects
Affected by a Large Lensing Magnification

Lensing affects all high redshift sources to some extend. Most sources are in fact slightly demagnified, and only few sources that are highly magnified. This means that even the luminosity function of a population of intrinsically perfect standard candles (at given distance) will get broadened due to lensing, the more so the higher the redshift, cf. [50]). Most of the time, however, we are interested in the high magnification cases.

3.1   Why Are (Apparently) Luminous, High-Redshift Sources Good Lensing Candidates?

There are various factors that favour high-redshift, apparently luminous sources in a flux-limited sample to be affected by gravitational lens-induced magnification: The more distant a source, the more intervening "space" and hence the lower the value of the critical surface mass density (cf. equation 11). This means in turn, that for a given lens mass it is easier to produce multiple images and/or high magnification, if the source is a higher redshift. In addition, for a more distant source, chances are higher that there is some intervening matter along the line of sight. And although the probability for lensing is obviously independent of the absolute brightness of the source, due to the selection of most samples in astronomical research to be flux-limited (rather than volume-limited), there is always a chance that sources with apparent magnitude below the flux threshhold make it into the sample because of the gravitational lens magnification, particularly in large-area surveys that sample on the bright end of the luminosity function. This "magnification bias" depends in detail on the slope of the luminosity function, the lens and source redshifts, the finite source size etc. For very high magnification and in the approximation of point lenses and a point source, it can be shown ([41]) that the probability is $p(\mu) \propto \mu^{-3}$. Many aspects of the magnification bias have been studied in the past, see, e.g., [40,1,42,18]. In the following, one representative of high-redshift ultra-luminous galaxies, extremely luminous quasars, very luminous galaxy cluster, distant supernovae and gamma-ray bursts will be discussed with respect to their possible connection to gravitational lensing.

3.2   Galaxies: Hyperluminous IRAS Source FSC 10214+4724

The IRAS source FSC 10214+4724 was identified with an object at redshift $z = 2.286$ by Rowan-Robinson et al. (1991) and classified as an ultra-luminous infrared galaxy (ULIRG). At the time, it was the "most luminous object presently known" [38]. Its intense CO emission [3] and high polarization [25] was soon recognized. High resolution imaging in the near-infrared [29,17] and in the optical [2,10] revealed some structure: a curved arc south of a round, resolved object was detected, with length varying with wavelength: from 0.7" (optical) to 2.0" (IR). It was immediately asked whether this was related to gravitational lensing, and whether the arc could be part of an Einstein ring.

Simple lens models could easily explain the geometrical configuration (cf. Figs. 2, 5). Trendham (1995) asked "How much of the extreme luminosity of FSC 10214+4724 can be attributed to gravitational lensing?" [48], and he considered a magnification in the range $2 < \mu < 10$ probable. Subsequently, polarimetric evidence for a hidden QSO was found [16], and MgII absorption at $z = 1.316$ and $z = 0.893$, which raised the question whether either of these absorption systems could be related to the lens. High-resolution NIR spectroscopy [24] was used to explain the unusual properties of FSC 10214+4724 by assuming that lensing is preferentially magnifying one side of the inner narrow-line region "the inner 100pc". Re-mapping the length of the now resolved triple arc into the source plane, assuming a magnification of order 100 means that it corresponds to a linear length of 7 milliarcsec or 50 pc. The interpretation was a star formation rate of 4000 - 8000 $M_\odot$/yr!

With NICMOS observations in 1999, the change of arc length with wavelength could be quantitatively explored, and the counter image of the lensed quasar just beyond the "round" object could be detected [12], see Fig. 5. It was also found that the structure of the arc changes with increasing wavelength, from two maxima to one. This was interpreted as indication for variations in the source morphology with wavelength, the red stellar emission being more extended at 2.05μm. Detailed models predict mag-

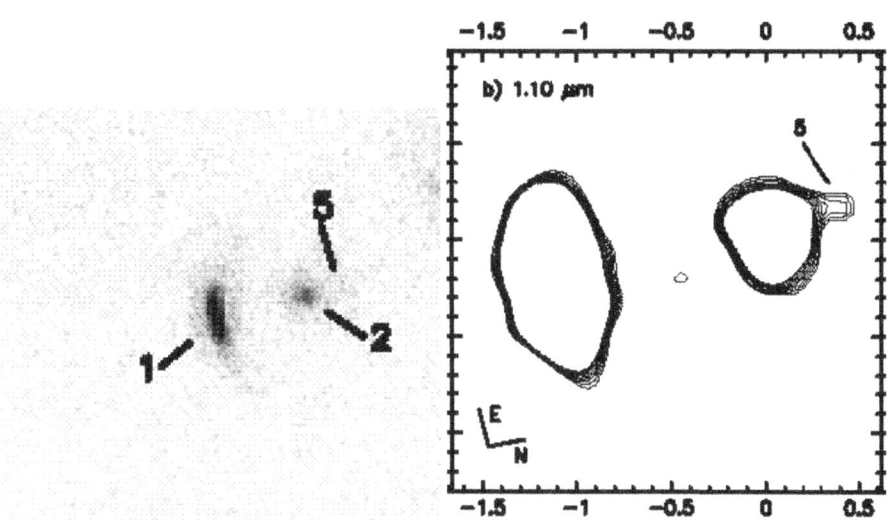

Fig. 5. Left panel: NICMOS image (at 2.05 μm) of FSC 10214+4724: the "1" indicates the arc, the "2" the round second part of the image, and the "5" the faint counter image (from [12]). Right panel: contours of FSC 10214+4724 at 1.10 μm; the faint image "5" is clearly visible as an extension in the contours (from [12]).

nifications of $\mu = 50$ and $\mu = 25$ for wavelengths of 1.10 μm and 2.05 μm, respectively. The potential lens would be consistent with an (unevolving) early-type galaxy at $z = 0.9$.

3.3 Quasars: Ultraluminous BAL Quasar APM 08279+5255

Discovered in a search for cool carbon stars (!) in the galactic halo by Irwin et al. (1998), the R = 15.2 mag BAL quasar APM 08279+5255 turned out to be positionally coincident with the IRAS source FSC 08279+5255 [21]. The redshift of $z = 3.87$ (later corrected to 3.91) makes this quasar enormously bright with an absolute brightness of $M_R = -33.2$ (for $h = 0.5$, $\Omega = 1$). With an IRAS 60μm flux of 0.51 Jy, it turned out to "easily" be the most luminous object known with $L_{bol} = 5 \times 10^{15}$ $L_\odot$! The extended structure that was visible could be fit with two discrete point sources 0.4 arcsec apart and with a flux ratio of 1.1. Three absorption systems could be identified in the spectrum: $z_{abs} = 1.18$ and 1.81 (Mg II), and $z_{abs} = 3.07$ (damped Lyα). A dust mass of $1 - 2 \times 10^9 M_\odot$ was inferred. Even correcting for the lens magnification of about a factor 40 made this quasar still a hyperluminous object.

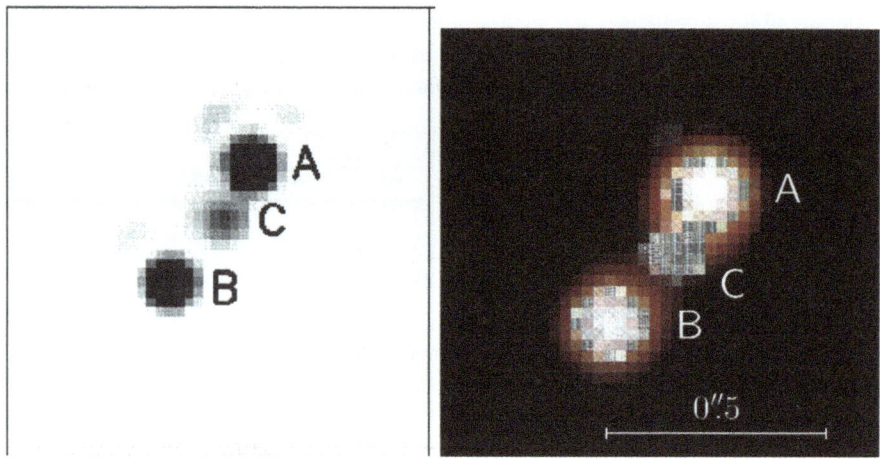

Fig. 6. NICMOS images of APM 08279+5255. Left: A 1.125"×1.125" region in filter F110W with the three images labelled. Right: color composite including all three filters F110W, F160W and F205W (from [20]).

Subsequent AO imaging and integral field spectroscopy [26] revealed clearly a double source with separation 0.35 arcsec and a flux ratio of 1.25±0.25 in the H-band (H = 12.6). The spectra are very similar, except for an Mg II absorber (λ2803) at $z_{abs} = 1.18$. No third image could be found down to H = 21.3.

Submillimeter observations with SCUBA [27] showed that the spectral energy distribution has a slope of 3.1±0.2 (in νF_ν), which is consistent with a Rayleigh-Jeans black body emission, and was interpreted as massive, optically thick distribution of dust, heated by the quasar. Simple models of ULIRGs, however, fail to explain the infrared emission.

NICMOS observations finally resolved the two point sources A and B with $\Delta\theta = 0.378$" separation and an intensity ratio of 0.782 ± 0.01 [20], see Figure 6. A faint image C was detected as well between (but not along the connecting line of) the two images (also by Keck NIR/MIR high resolution imaging, [8]). It is still not clear whether this image C is the lens or the third lensed image of the source: it is point-like (as are images A and B), colors are consistent with a brightness ratio of C/A = 0.175. If it is the lens, it would have an $M_J = -32$, (for an assumed redshift of object C of z(C) = 1.18) which is inconsistent with a velocity dispersion of $\sigma = 130$ km/sec, as derived from the image separation. The lens galaxy must have $V > 22$, and $V - K > 8(!)$. Lens models [20,8] can reproduce both a two image configuration (with a total magnification of $\mu_{tot} = 7.5 - 10$) and a three-image configuration ($\mu_{tot} = 90 - 100$), corresponding to intrinsic (unmagnified) bolometric/IR luminosities of about $L_{bol} = 5 \times 10^{13}$ $L_\odot$ and $L_{IR} = 1 \times 10^{13}$ $L_\odot$, respectively.

Keck HIRES spectroscopy [9] produced high resolution spectra ($\Delta v = 6$km/sec or $\Delta\lambda = 0.04$Å!) with a total exposure time of 31500 seconds. The wavelength coverage extends from 4400 - 9250 Ångstrom, with a variable signal-to-noise ratio of S/N = 30 ... 150. The results show a rich Lyman-α forest, the column density distribution can be fit with a power law index of $\beta = 1.27$ (for $3.11 < z_{abs} < 3.70$), and five (out of 23) Lyman-α clouds are associated with C IV absorption (spectra are available at `ftp://ftp.ast.cam.ac.uk/pub/papers/APM08279`).

3.4   Galaxy Clusters: Most Luminous X-Ray Cluster RXJ 1347.5-1145

The ROSAT source RXJ 1347.5-1145 is the most luminous galaxy cluster known so far [46,47,45]. At moderate redshift of $z_{cluster} = 0.451$, it has an X-ray luminosity of $L_{X,bol} = 2 \times 10^{46} erg/sec$. The temperature is still a bit controversial, originally determined to $T = 9keV$, more recently it was claimed to be higher than $T > 13.1keV$ [11]. The derived total mass is about $M = 10^{15}$ M$_\odot$.

In contrast to all the other cases discussed in this section, the brightness of this galaxy cluster cluster is *not* affected by lensing. Rather, due to its enormously high mass, it *acts* as a lens on background objects: Five (!) luminous arcs were discovered, each at a distance of about 35 arcsec from the center [46,47]! STIS imaging and ground-based spectroscopy (ESO 3.6m) revealed one arc redshift to be $z_{arc} = 0.81$.

3.5   Supernovae: Is SN 1997ff Magnified?

No high-redshift supernova has been proven yet to be affected by gravitational lensing, in particular, there are no multiple images yet of high-z SNe. But there is no doubt that such multiply imaged supernovae will be detected, and they will provide a wealth of new information (see, e.g., [19]).

However, even for the time being the effects of lensing on SNe is worth exploring. The cosmological significance of SNe as standard candles is increasing with respect to finding the proper cosmological model [35,37]. Hence it is worthwhile pointing out that SNe are statistically affected by lensing magnification. So far, the standard deviation of this magnification is beyond the (current) photometric uncertainty, but with increasing observational accuracy this lens-induced broadening of the luminosity function has to be considered ([50,14], see also Metcalfe, these proceedings). For individual cases, like the particularly high redshift supernova SN1997ff, it may be worthwhile to model the lensing behaviour very accurately, since this supernova carries a high weight in the determination of cosmological parameters due to its record redshift of $z \approx 1.7$ [28].

3.6 Gamma-Ray Bursts: Microlensing in GRB000301C?

For gamma-ray bursts the situation is similar to the one for SNe: being at high redshift, it is unavoidable that some will be lensed [30,31], but so far no clear signature of multiple imaging has been found [33,34]. However, with the recent discovery of optical afterglows, a new domain of "microlensing" opens up. In GRB000301C, there is a signature in the power-law decline of the afterglow, which is consistent with microlensing by stellar mass objects [15,32], though the a posteriori probability for this event is only a few percent [23]. But regardless of this particular case, due to their presumably large expansion velocity, GRB afterglows are interesting candidates for microlensing. If microlensing is detected, it can be used as a tool to explore the size and structure of the "fireball" (see also Loeb, these proceedings).

4   Summary and Outlook

Gravitational lensing and very luminous objects provide an exciting and promising combination for further study. As shown here, at least some of the known "most luminous objects" in the universe are (heavily) affected by the lensing magnification. The first consequence of this is that the real absolute magnitude is fainter than the one derived directly. This is a slightly disappointing consequence for the "record seekers" among us (though in most cases even the corrected luminosities are at the very high end). On the other hand, this is a reassuring conclusion for the "conservatives" among us, who always thought that the standard models have a hard time explaining these enormous luminosities. And it makes them particularly interesting objects for the "lensers" among us, because they provide extremely interesting laboratories for the detailed study of lens, source and intervening space, which would otherwise be difficult to explore. Summarized, we can state:

The working gravitational telescope: Fritz Zwicky has predicted in the 1930s that gravitational lensing magnification can be used effectively as an increase in the aperture of our telescope (though with partially unpleasant optical properties ...). Magnifications of order 10-100 help us study high-z quasars and galaxies in (much) more detail than would be possible without this lensing boost: the OWL-(**O**f course **W**ith **L**ensing)-Telescope already exists!

The backside-illuminated universe: highly magnified sources (5 mag!) at the far end of the universe help us study the intervening matter distribution by means of absorption line studies.

Very luminous = very massive (occassionally): bright objects acting as lenses (galaxy clusters) help us study the background objects AND the lensing mass distribution.

Lensed GRBs and SNe: the study of lensed gamma-ray bursts and supernovae just about starts, with very promising prospects for the future.

References

1. Borgeest, U., von Linde, J., Refsdal, S.: 1991 A&A 251, L35
2. Broadhurst, T., Lehár, J.: 1995, ApJ 450, L41
3. Brown, R.L., Vanden Bout, P.A.: 1991 AJ 102, 1956
4. Chang K., Refsdal S.: 1979, Nature, 282, 561
5. Colley, W.N. Tyson, J. A., Turner, E. L.: 1996, ApJ 461, L83
6. Courbin, F., Lidman, C., Magain, P.: 1998, A&A 330, 57
7. Dyson, F.W., Eddington, A.S., Davidson, C.R.: 1920, Mem. Roy. Astron. Soc. 62, 291
8. Egami, E., et al.: 2000, ApJ 535, 561
9. Ellison, S.L., et al.: 1999 PASP 111, 946.
10. Eisenhardt, P.R. et al.: 1996, ApJ 461, 72
11. Ettori, S., Allen, S.W., Fabian, A.C.: 2001, MNRAS 322, 187
12. Evans, A.S., et al.: 1999 ApJ 518, 145
13. Falco, E.E., Impey, C., Kochanek, C.S., Lehar, J., McLeod, B., Rix, H.-W.: WWW: http://cfa-www.harvard.edu/glensdata/
14. Frieman, J. A.: 1997, Comm. Astrophys., 18, 323
15. Garnavich, P.M., Loeb, A., Stanek, K.Z.: 2000, ApJ 544, L11
16. Goodrich, R.W., et al.: 1996, ApJ 456, L9
17. Graham, J.R., Liu, M.C.: 1995, ApJ 449, L29
18. Hamana, T., et al.: 1997 MNRAS 287, 341
19. Holz, D.E.: 2001 ApJ 556, L71
20. Ibata, R. A., et al.: 1999, AJ 118, 1922
21. Irwin, M.J., et al.: 1998 ApJ 505, 529
22. King, L.J., et al.: 1997, MNRAS 289, 450
23. Koopmans, L.V.E., Wambsganss, J.: 2001, MNRAS 325, 1317
24. Lacy, M., Rawlings, S., Serjeant, S.: 1998 MNRAS 299, 1220
25. Lawrence, A. et al.: 1993, MNRAS 260, 28
26. Ledoux, C., et al.: 1998 A&A 339, L77
27. Lewis, G.F., et al.: ApJ 505, L1
28. Lewis, G.F., Ibata, R.A.: 2001, MNRAS 324, L25
29. Matthews, K. et al.: 1994, ApJ 420, L13
30. Mao, S.: 1992, ApJ 389, L41
31. Mao, S.: 1993, ApJ 402, 382
32. Mao, S., Loeb, A.: 2001, ApJ 547, L97
33. Nemiroff, R.J. et al.: 1994, ApJ 432, 478
34. Nemiroff, R.J. et al.: 1998, ApJ 494, L173
35. Perlmutter, S. et al.: 1998, Nature 391, 51
36. Petters, A.O., Levine, H, Wambsganss, J.: 2001 "Singularity Theory and Gravitational Lensing" (Birkhäuser, Basel)
37. Riess, A. G.: 2000, PASP 112, 1284
38. Rowan-Robinson, M. et al.: 1991, Nature 351, 719
39. Schneider, P.: 1984, A&A 140, 119
40. Schneider, P.: 1987a, ApJ 316, L7
41. Schneider, P.: 1987b, ApJ 319, 9
42. Schneider, P.: 1992, A&A 254, 14

43. Schneider, P., Ehlers, J., Falco E.E.: 1992, "Gravitational Lenses" (Springer, Berlin)
44. Rusin, D., et al.: 2001, ApJ 557, 594
45. Sahu, K.C., et al. 1998 ApJ 492, L125
46. Schindler, S., et al.: 1995, A&A 299, L9
47. Schindler, S., et al.: 1997, A&A 317, 646
48. Trendham, N.: 1995 MNRAS 277, 616
49. Walsh, D., Carswell, R.F., Weymann, R.J.: 1979, Nature 279, 381
50. Wambsganss, J., et al.: 1997, ApJ 475, L81
51. Wambsganss J.: 1998, Living Reviews in Relativity 1/1998-12, pp. 1-80; online:
    http://www.livingreviews.org/Articles/Volume1/1998-12wamb/
52. Woźniak, P.R., et al.: 2000, ApJ 529, 88
53. Woźniak, P.R., et al.: 2000, ApJ 540, L65

Observing $z > 4$ Galaxies Through a Cosmic Lens

Narciso Benítez[1], Tom Broadhurst[2], Brenda Frye[3], Chris Lidman[4], Lindsay King[5], Georges Meylan[6], and Peter Schneider[5]

[1] Johns Hopkins University, Baltimore, USA
[2] Hebrew University, Jerusalem, Israel
[3] Princeton University, Princeton, USA
[4] European Southern Observatory, Garching, Germany
[5] University of Bonn, Bonn, Germany
[6] Space Telescope Institute, Baltimore, USA

Abstract. Practical considerations dictate two basic strategies to find very high redshift galaxies: going very deep in small fields (e.g. the HDFs) or doing shallower, wide area surveys which target the bright end of the luminosity function. A third approach which combines some of the advantages of both is to look for strongly magnified objects in the fields of lensing clusters. We present results from such an ongoing program, which has already yielded seven spectroscopically confirmed $z > 4$ galaxies, and several candidates to $z > 6$ galaxies. We also discuss the prospects for finding $z \sim 6.5$ galaxies with HST and the Advanced Camera for Surveys.

1 Introduction

High-z galaxies are quite a common sight in very deep exposures of the blank sky. For instance, the HDFN [1] has a surface density of $3 - 6$ galaxies arcmin^{-2} with $z > 4$ at $I < 27$ [2], or about $2 - 4\%$ of all the galaxies detected up to that magnitude limit. Their prominent Lyman-limit break makes them relatively easy to spot using color cuts or photometric redshifts, provided that enough filters are used for the task. However, due to their faintness and/or lack of emission lines, in many cases it is not possible to secure a spectroscopic confirmation of the redshift of the candidates. The HDFN serves as a good example again: after 5 years, only 10% of the photo-z selected $z > 4$ galaxies have been confirmed spectroscopically.

At brighter magnitudes, the number of $z > 4$ objects drops dramatically, with a density of 1 arcmin^{-2} with $25 < I < 26$ and none with $I < 25$ in the HDFN. However, by surveying large areas at moderately depth, it is possible to detect and study the bright end of the luminosity function, as has been done successfully by e.g [3].

A different strategy is to look for high-z galaxies magnified by a gravitational lens [4,5]. In this way, one can study and obtain high S/N spectra of galaxies which have moderate or even faint intrinsical luminosities. Here we present some results of such a systematic search for high-z galaxies in the fields of massive galaxy clusters.

2 Keck Observations of $z > 4$ Galaxies

Four intermediate redshift ($0.18 < z < 0.31$), X-ray luminous clusters were observed: A1689, A2390, A2219 and AC114. These massive clusters display giant arcs and ex-amples of multiply lensed images. The target selection for spectroscopy was based on

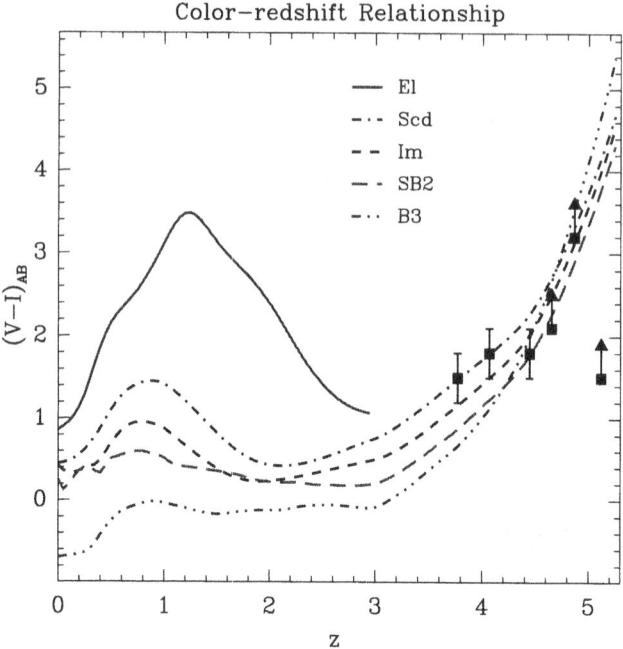

Fig. 1. Color vs. Redshift for the E1, Scd and Im galaxy types of [11], the SB2 and SB3 starbursts from [9], and a B3 stellar model [10]. The six galaxies in our sample with measured $(V - I)_{AB}$ are plotted as square-shaped points with error bars, or as lower limits for the galaxies at z=4.65, 4.87, and 5.12. Our high-z sample follows the color evolution well. Note the potential source for confusion with elliptical galaxies at z~ 1

the $V - I$ color, which was chosen to be redder than the cluster early type envelope. (Fig. 1). Using only two filters is quite inefficient to look for high-z galaxies, since most of the selected candidates are early types at intermediate redshifts. Except for some obviously interesting targets close to the critical curve, no deliberate effort to look for galaxies with elongated morphologies was done, since for intrinsically small sources, the lens distortion may leave the object unresolved even with HST.

The spectroscopic observations of our $z > 4$ objects were carried out with LRIS on the Keck telescopes between 1997 and 1999, typically with a $1''$ slit width and a resolution of 12Å at 6000Å, and with exposure times ranging from ~6hrs to ~ 2hrs. Due to the cluster magnification, the objects are quite bright, $23 < I_{AB} < 25$, producing good quality spectra. which are compared with stellar continuum fits, taking into account the intergalactic Lyman-series [12] and HI opacity at the source. Hot B-stars are shown to typically give good fits to the observations. It is found that the metal absorption lines are blueshifted with respect to the Lyman-α centroid, and that the magnitude of this blueshift correlates with the equivalent-width of the absorption. Details of these observations and their interpretation are presented in [6].

3   A1689

Our imaging data for A1689 cover the main broadband filters from the UV to the near-IR, as Table 1 shows. These data were obtained by different observers at different epochs, and the quality of the photometric calibration is quite heterogeneous, with the LRIS data having the best understood zero points. We therefore had to resort to the cluster red sequence in order to ensure a consistent calibration across the filters. With that purpose in mind, the empirical spectra of [11] were corrected using the HDFN and SDSS spectroscopic samples. Then we chose the SED which matched best the LRIS photometry for the bright cluster early types and determined the zero points for the rest of the bands. Basically this ties our relative zero points to the HDFN photometric system. Details about the reduction and analysis of these $UBVRIZJHK$ data will be presented in an upcoming paper [7]. With such a photometric information, it is possible to obtain very high quality photometric redshifts, and we used the BPZ software for this task.

We compare the galaxy colors with those of a forest opacity corrected redshifted template set, formed by the "corrected" main spectral types of [11], plus two starbursts from [9]. We take into account the intrinsic color variation or 'cosmic variance' in the galaxy colors by convolving the resulting color likelihoods with a gaussian of width $0.06(1 + z)$, empirically measured from the HDFN [2]. This makes the range of final redshift probabilities considerably wider than they would be if we included only the photometric error (see e.g. Fig. 2) but it yields a more accurate representation of the true uncertainties involved in the estimation. Since we are working on a cluster field which strongly magnifies the observed flux from background objects, we use a redshift magnitude prior $p(z|T, m)$ [2,8] for field galaxy distributions, but with the corresponding *unlensed* magnitude. A comparison between photometric redshifts and a preliminary list of spectroscopic redshifts shows good agreement (Fig. 3). Since the cluster early types

Table 1. A1689 Imaging data. The detection limits correspond to the *total* magnitude of point source

Filter	Telescope	Seeing	2σ AB
U	2.5DuPont	0.95	27.27
B	NOT	1.15	26.66
V	Keck	0.86	26.45
R	Keck	0.76	26.29
I	Keck	0.80	26.30
Z	Keck	0.93	25.00
J	NTT	0.96	24.64
H	NTT	1.16	23.92
K	NTT	0.94	23.61

Photo−z Probability Distribution

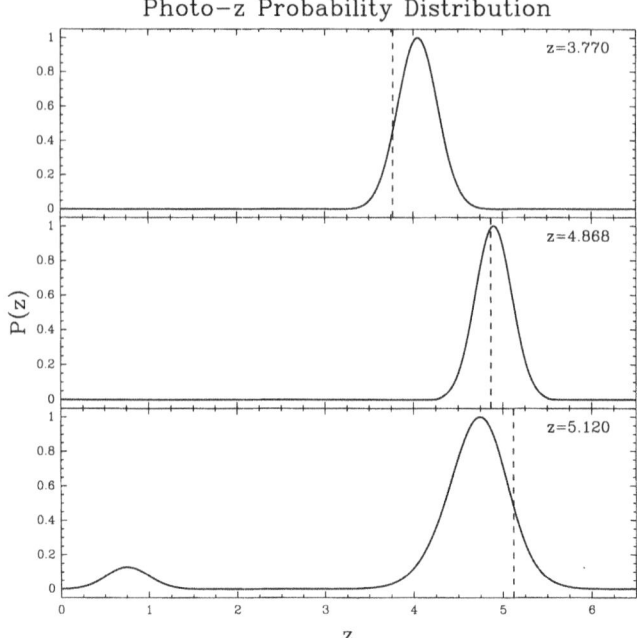

Fig. 2. Bayesian redshift probability distributions as defined in [2], based on information from 9 optical and infrared bands in A1689. The photometric redshifts agree with the spectroscopically determined ones to the 1σ level. There is a double-peaked probability for the galaxy at z=5.12, arising from its faintness and therefore relatively large photometric errors. Note, however, that the sum of the redshift probabilities lies closest to the spectroscopic redshift.

were used to calibrate the photometry for some filters, we have excluded these galaxies from the comparison. It is also noteworthy that the photometric redshift information has improved considerably the reliability of the redshifts obtained spectroscopically, at least for those objects presenting single or no emission lines.

The excellent quality of our data makes possible the selection of several candidates to $z > 6$ objects. Fig. 4 shows some of them. We recently obtained spectroscopy for these objects with VLT and are in the process of reducing and analyzing these observations.

4   What's Next?

The Advanced Camera for Surveys will be launched in February 2002. The combination of area and sensitivity of this instrument will represent a factor 10 improvement with respect to WFPC2 [14]. This fact, combined with the presence of the z-band filter which extends redward of F814W, will revolutionize the study of faint, high redshift objects. Based on the number of high-z objects with high-quality photometric redshifts in the HDFN, it is possible to estimate that the number of $4.5 < z < 6$ galaxies within a WFC field with $m_z < 27.5$ would be ~ 25. We can go one step further, by extrapolating

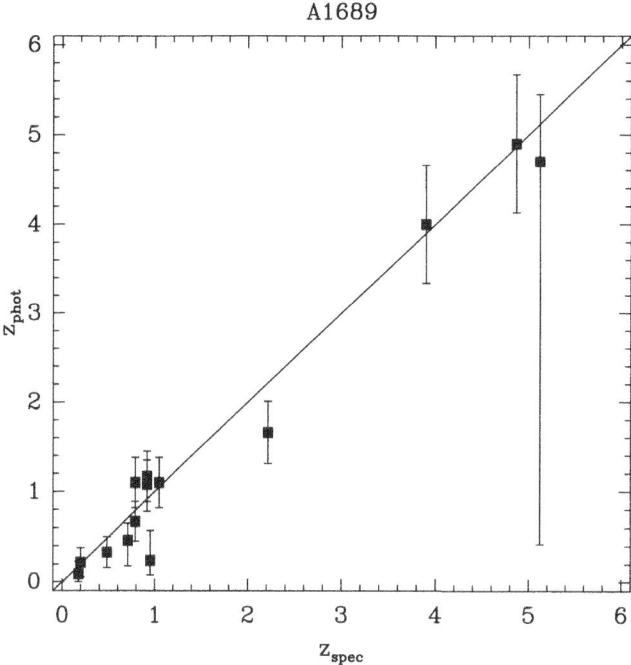

Fig. 3. Comparison between a preliminary list of spectroscopic redshifts in A1689 and photometric redshifts obtained based on $UBVRIZJHK$ photometry

these results to $z > 6$ assuming no evolution. This yields ≈ 10 objects with $6 < z < 7$ and $m_z < 27.5$ which can be expected within a WFC pointing. Of course at very faint magnitude limits it will be very difficult to obtain ground based spectroscopic confirmation of the redshift of the sources. Since the numbers counts are expected to be steep, observing a field behind a massive galaxy cluster will result in a significant increase in the surface density of sources because of the magnification bias effect[13]. Part of the orbits of the ACS GTO program [14] have been allocated to such an strategy, observing A1689 and other clusters with deep multicolor exposures combined with a grism pointing. This strategy will provide photometric redshifts for most of the sources and spectroscopic redshifts for a significant fraction of them.

References

1. Williams, R. E. et al. 1996, AJ, 112, 1335
2. Benítez, N. 2000, ApJ, 536, 571
3. Steidel, C. C., Adelberger, K. L., Giavalisco, M., Dickinson, M., & Pettini, M. 1999, ApJ, 519, 1
4. Franx, M., Illingworth, G. D., Kelson, D. D., van Dokkum, P. G., & Tran, K. 1997, ApJL, 486, L75
5. Frye, B. & Broadhurst, T. 1998, ApJL, 499, L115
6. Frye, B. & Broadhurst, T. & Benítez 2001, ApJ accepted.

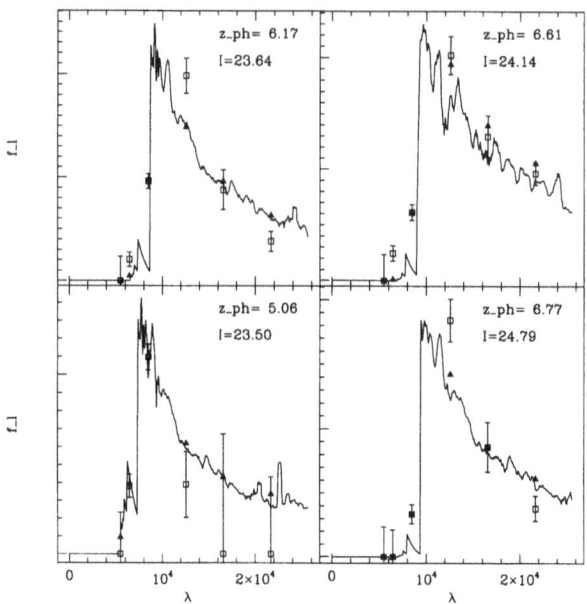

Fig. 4. Comparison between the observed fluxes and the model fluxes expected based on photometric redshifts for some of our typical $z > 6$ candidates

7. Benítez et al. 2001, in preparation
8. Benítez, N., Broadhurst, T., Bouwens, R., Silk, J., & Rosati, P. 1999, ApJL, 515, L65
9. Kinney, A. L., Calzetti, D., Bohlin, R. C., McQuade, K., Storchi-Bergmann, T., & Schmitt, H. R. 1996, ApJ, 467, 38
10. Kurucz, R. L. 1993, ASP Conf. Ser. 44: IAU Colloq. 138: Peculiar versus Normal Phenomena in A-type and Related Stars, 87
11. Coleman, G. D., Wu, C.-C., & Weedman, D. W. 1980, ApJS, 43, 393
12. Haardt, F. & Madau,P. 1996, ApJ, 461, 20
13. Broadhurst, T. J., Taylor, A. N., & Peacock, J. A. 1995, ApJ, 438, 49
14. Ford, H. C. et al. 1996, SPIE, 2807, 184

Numerical Study of the Lensing Deflection Angle Due to the Large-Scale Structures

Takashi Hamana

National Astronomical Observatory, Mitaka, Tokyo 181-8588, Japan

Abstract. We present ray tracing simulations combined with sets of large N-body simulations. Experiments were performed to explore statistical properties of fluctuations in angular separations of nearby light ray pairs (the so-called lensing excursion angle) induced by weak lensing by large-scale structures. We found that the probability distribution function (PDF) of the lensing excursion angles consists of two contributions, a Gaussian core and an exponential tail. It is, however, found that the tail, or more generally non-Gaussian nature in the PDF has no significant impact on the weak lensing of the CMB. Moreover, we found that the variance in the lensing excursion angles predicted by the power spectrum approach is in good agreement with our numerical results. These results demonstrate a validity of using the power spectrum approach to compute lensing effects on the CMB.

1 Introduction

Weak lensing effects on the temperature anisotropy in the cosmic microwave background (CMB) have been recognized as a powerful probe of cosmology (see Mellier 1999; Bartelmann & Schneider 2001 for reviews). Dark matter distribution along the line of sight between the last scattering surface (LSS) and us deflects the light ray trajectories and induces distortions in the pattern of the CMB anisotropies. Since the gravitational lensing is directly sensitive to the matter distribution up to the LSS, lensing signatures imprinted on the CMB may provide important information about the matter distribution on large scales and at high redshifts. In this point of view, various methods have been proposed (see Bartelmann & Schneider 2001 for review). Expected lensing signatures are generally small but are measurable with two planned satellite missions, MAP and Planck, and can help to break some of the parameter degeneracies in the CMB. The change in a separation angle of two nearby light rays caused by weak lensing (lensing excursion angle) plays a key role in studying weak lensing of the CMB, especially weak lensing effects on two-point statistics of the CMB.

The analytical prediction of the weak lensing effect on the CMB power spectrum in modern cosmological models was first developed by Seljak (1994) based on linear perturbation theory (the so-called power spectrum approach). While, in his subsequent paper, Seljak (1996) examined effects of the nonlinearity in the density through analytic fitting formulae (Peacock & Dodds, 1996), and pointed out that the nonlinearity is very important on sub-degree scales. Furthermore statistics of the lensing excursion angles due to weak lensing are frequently assumed to be Gaussian without a rigorous basis (e.g., Seljak 1996). Numerical simulations are, therefore, needed for testing the validity and limitation of the semi-analytic approach.

We performed ray tracing simulations combined with sets of large N-body simulations to study the cosmic shear statistics (van Waerbeke et al. 2001; Hamana et al. 2001) and statistical properties of fluctuations in angular separations of nearby light ray pairs (the so-called lensing excursion angle) induced by weak lensing by large-scale structures (Hamana & Mellier 2001, hereafter HM01). In this contribution, we summarize selected results from the latter study. We refer the reader to the above references for a detail description of our numerical technique and more detail discussions.

2   Results

The lensing excursion angle plays a crucial role on studies of the weak lensing effects on the CMB and is defined by the difference in deflection angles between nearby two light rays. We compute, from our numerical experiments, the variance in the lensing excursion angles of the two nearby light rays which would be observed with a separation θ if there was no lensing. The deflection angle of a light ray is simply the difference between its angular position in the first lens plane and that in the source plane. The lensing excursion angle between nearby rays A and B is simply the difference in deflection angles between them. In Figure 3 of HM01, the results are plotted together with the semi-analytic predictions by the power spectrum approach (Seljak 1994; 1996). The measurements agrees very well with the nonlinear semi-analytic prediction at scales larger than 10 arcmin. Below that scale, the measurements are depressed reflecting the effective resolution limit.

Figure 5 of HM01 shows the probability distribution function (PDF) of the lensing excursion angles normalized by its intrinsic separation. It is clearly shown in Figure 5 of HM01 that the PDFs consist of two contributions: A Gaussian distribution at inner part, and the exponential tail at the outer part. The inner part of the PDFs are fitted reasonably well by a Gaussian distribution with the σ computed from the PDFs (root-mean-square computed from the PDFs was used for σ). We should here noticed that the measured standard deviation agrees very well with the semi-analytical prediction as shown in Figure 2 of HM01.

The fact that the PDFs consist of two contributions suggests that there are two different processes that make the lensing excursion angle. One, which makes Gaussian distribution, might be secular small (random) deflections due to linear density fluctuations along each light ray path. This is explained by the fact that the lensing deflection angles, which are due to Gaussian random fluctuations (such as the linearly evolved density fluctuation field), are Gaussian random field. As the light ray travels longer distance, the ray can undergo more fluctuations. Therefore the width of Gaussian distribution increases as the source redshift becomes higher. The other, which makes the exponential tail, might be a single (or possibly multiple) coherent scatter by a nonlinear structure such as a galaxy or a cluster of galaxies. As the separation of a light ray pair decreases, smaller scale, strongly nonlinear structures can contribute to a coherent scatter. Therefore the exponential tail becomes more prominent for a small separation case than a larger one as shown in Figure 5 of HM01.

Does the exponential tail make a significant influence on studies of weak lensing effects on the CMB ? As far as the weak lensing effects on the CMB power spectrum concerned, it has no effect, because the crucial assumption that the lensing excursion angle is (in a statistical sense) much smaller than the intrinsic separation is true even in the presence of the exponential tail. This can not be directly demonstrated by our simulations as the light rays are not followed up to LSS, but can be proved by the semi-analytic prediction which tells that the root-mean-square of the lensing excursion angle is much smaller than the intrinsic separation angle. The validity of the semi-analytic prediction was supported by our numerical simulation (Figure 2 of HM01).

Figure 6 of HM01 shows the Kurtosis of the PDF of the lensing excursion angle shown in Figure 2 of HM01. The Kurtosis decreases as the source redshift becomes higher. On scales larger than 10 arcmin (below that, the lensing effects on the CMB will be hardly detected) the Kurtosis less than unity at $z_s = 3$, and thus, at the LSS, it must be smaller than that. It can be, therefore, said that the exponential tail or, more generally, the non-Gaussian nature in the PDF is very unlikely to have a significant effect on the weak lensing of the CMB.

Acknowledgements

The author would like to thank Y. Mellier and S. Colombi for many useful discussions. He also thank M. Bartelmann for his valuable comments. He acknowledges support from Research Fellowships of the Japan Society for the Promotion of Science. This research was supported in part by the Direction de la Recherche du Ministère Français de la Recherche.

References

1. M. Bartelmann, P. Schneider, 2001, Physics Report, **340**, 291
2. T. Hamana, S. Colombi S, A. Thion, J. Devriendt, Y. Mellier, F. Bernardeau, 2001, MNRAS submitted (astro-ph/0012200)
3. T. Hamana, Y. Mellier Y. 2001, MNRAS in press
4. Y. Mellier, 1999, ARA&A, **37**, 127
5. J. A. Peacock, S. J. Dodds, 1996, MNRAS, **280**, L19
6. U. Seljak, 1994, ApJ, **436**, 509
7. U. Seljak, 1996, ApJ, **463**, 1
8. L. van Waerbeke, T. Hamana, R. Scoccimarro, S. Colombi, & F. Bernardeau, 2001, MNRAS, **322**, 918

Detection of Lighthouses by Gravitational Waves – on Earth and in Space

Albrecht Rüdiger

Albert-Einstein-Institut, MPQ, Hans-Kopfermann-Str. 1, 85748 – Garching, Germany

Abstract. 'Lighthouses in the Universe' would be the strongest cosmic events imaginable. They are bound to emit very strong gravitational waves. Laser interferometry is the most promising scheme to detect these waves, and international efforts will provide detectors both on Earth and in space.

Ground-based detectors are nearing completion at five sites worldwide. These ground-based detectors are most sensitive between about 10 Hz and several kHz.

The high-mass events that constitute the 'Lighthouses' will emit their gravitational waves mainly at much lower frequencies.

LISA (the Laser Interferometer Space Antenna) will, with its long arms of 5 million km, be sensitive in a much lower frequency regime: 10^{-5} to 1 Hz.

The poster gave an overview of the detectors and the sensitivities that will be reached, and a notion of the international collaborations for data analysis of the ground-based detectors, as well as the collaboration of ESA and NASA in the (common) LISA project.

1 Gravitational Waves

Einstein's General Relativity predicts the emission of gravitational waves of measurable strength when large masses undergo strong accelerations. Gravitational waves are transverse quadrupole waves, and manifest themselves by a 'strain' h in space, of opposite sign in two orthogonal transverse directions.

The 'strength' of such a wave emitted by, say, a binary system (Figure 1) can be expressed as a strain h proportional to the product of the two masses, m_1, m_2, and inversely proportional to their relative distance d and the distance D from the binary to Earth.

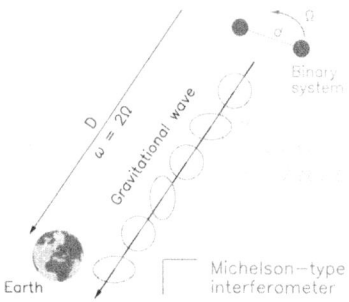

Fig. 1. Generation and propagation of a gravitational wave emitted by a binary system.

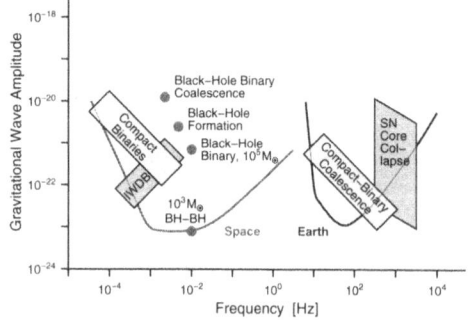

Fig. 2. Some sources of gravitational waves, with sensitivities of *Earth* and *Space* detectors.

The 'Lighthouses in the Universe' dealt with in this Conference would be the strongest cosmic events imaginable. Due to the large masses involved, and the huge accelerations of these masses, they are bound to emit very strong gravitational waves. Much information can be gained from the detection and a reliable measurement of these gravitational waves.

2   Detectors

Laser interferometry is the most promising scheme to detect these waves, and international efforts will provide detectors both on Earth and in space. In both these realizations, the principle is that of a Michelson interferometer: changes in the distance between 'free' masses at the end of two long (non-parallel) arms are monitored, the changes in the difference providing the signal of the gravitational wave. In practice, the sensitivity with which waves can be detected depends on how well noise effects can be suppressed. These noise effects are partly of identical origin on ground and in space, partly however distinctly different.

2.1   Ground-Based Detectors

Ground-based detectors to measure such gravitational waves are nearing completion at five sites, two in the USA, two in Europe, one in Japan. These detectors are most sensitive in frequency regions between about 10 Hz and several kHz.

The sensitivity of the current generation of detectors on Earth is expected to be of the order of $h = 10^{-22}$ for short bursts (supernovae) and for the 'chirps' of inspiralling binaries.

The curve shown in the right-hand part of Figure 2 is an extrapolation to future detectors such as the advanced version of the US project LIGO. That 'Advanced LIGO' will utilize techniques in interferometry and suspension technology that are being developed in the (much smaller) German-British detector GEO 600.

From a precise measurement of the time series of the strain $h(t)$, the characteristic data of binaries can be very accurately determined, and also the distance D. If such an event can be securely assigned to a given galaxy with known redshift, a new method for determining the Hubble constant H_0 is provided.

The high-mass events that constitute the 'Lighthouses' will emit their gravitational waves at low frequencies, and thus much effort goes into decreasing the lower frequency limit set by seismic and gravity-gradient effects. In an ambitious effort the French-Italian project VIRGO may push that limit down to 3 Hz.

Longer stretches of continuous operation of such a ground-based interferometer were performed with the smallest detector, the Japanese TAMA 300.

The groups in the field have agreed on an international exchange of data to be able to locate (triangulate) the sources in the sky.

2.2   Space-Borne Detector LISA

LISA (the Laser Interferometer Space Antenna) will, with its long arms of 5 million km (see Figure 3), be sensitive in a much lower frequency regime: 10^{-5} to 1 Hz. It is in that band that the strongest and most violent events are to be found. The sensitivity of LISA promises huge signal-to-noise ratios for the events involving (super)massive black holes (see Figure 2), such that failure to see them would be a serious blow to our present understanding.

LISA is a joint project of ESA and NASA, with launch anticipated for 2010 or 2011. It will be preceded by a technology demonstration, LTP, which is firmly scheduled for August 2006.

LISA consists of three spacecraft in heliocentric Earth-like orbits, designed such that they remain in a constellation of an equilateral triangle, tilted by 60° out of the ecliptic, trailing the Earth by about 20°. The laser links between the spacecraft allow the definition of up to three (but not independent) interferometers.

The annual motion of this triangle will make LISA change its sensitivity and phase relationship to a given source in a very distinct 'signature', in such a manner that the location of a (continuous wave) source can be firmly determined with only one single interferometer.

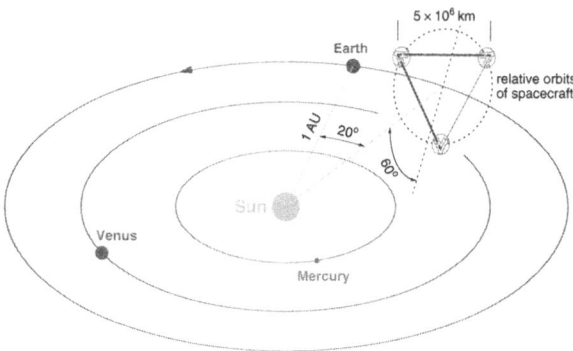

Fig. 3. Orbits of the LISA configuration

References

1. B.F. Schutz: 'Lighthouses of gravitational wave astronomy, and the chances of LIGO and LISA to detect them'. In: *Lighthouses of the Universe, Conference at Garching, Germany, August 6–10, 2001*, (Springer, Heidelberg 2001)

Part V

QSO, AGN, Blazars – Observational Data

XMM-Newton AGN Science Highlights from PV and GT Observations

Thomas Boller

Max-Planck Institut für extraterrestrische Physik, Postfach 1603, 85741 Garching, Germany

Abstract. Recent X-ray observations of active galaxies and quasars have significantly improved our ability to probe the environments of supermassive black holes. The high throughput and excellent spectral capabilities of the new generation of X-ray observatories have already revealed many new and unexpected observational results, resulting in a more precise understanding of the physics operating within a few Schwarzschild radii around the supermassive black hole, the molecular torus zone as well as the nuclear starburst activity. The talk will review the new observational results on the iron K line as an important diagnostic tool for probing matter within a few Schwarzschild radii. It appears that most of the sources may now be better described by pure continuum absorption and narrow high equivalent width iron K line emission seen in reflection from the molecular torus. Some sources show evidence for relativistically broadened red wings. We are also beginning to discover sources with sudden drops in flux at rest-frame energies above 7 keV, without any detectable narrow Fe K line emission. The energy of these features suggests a connection with the neutral or ionised Fe K photoelectric edge and the lack of any obvious Fe K reemission points to the presence of nearly neutral, high density cold gas that accompanies the active regions above an accretion disk. It is possible that the model we have developed has a wider relevance for Seyfert galaxies.

1 XMM-Newton Fe K Line Diagnostics

1.1 Historical Review

The presence of an excess between about 6 and 7 keV above an underlying power-law continuum was first established by observations with the GINGA satellite [1], [2]. This excess was interpreted as emission from the Fe K line and in combination with the Compton scattering hump above 10 keV, these observational facts were considered as strong hints for hard X-rays reflected by Compton thick matter close to the central black hole, i.e. evidence for the putative accretion disk. The higher throughput and higher spectral resolution of the ASCA satellite allowed to resolve the Fe K line complex. A strong asymmetry was found in MCG-6-30-15 [3] with orbital velocities of the X-ray emitting material of about one third of the velocity of light. In addition, an inverse correlation between the equivalent widths of the Fe K line and the X-ray luminosity (the X-ray Baldwin effect) was found [4].

1.2 XMM-Newton and Chandra Results

Fig. 1 shows the Fe K line profiles obtained with XMM-Newton (upper panel) and Chandra (lower panel) for the Narrow-Line Seyfert 1 Galaxies Ton S180, Mrk 766 and

XMM-Newton Fe K line profiles

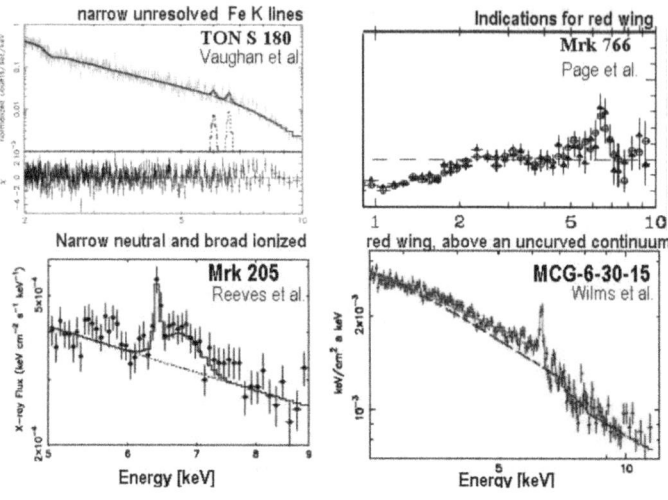

Chandra HETG Fe K line profiles

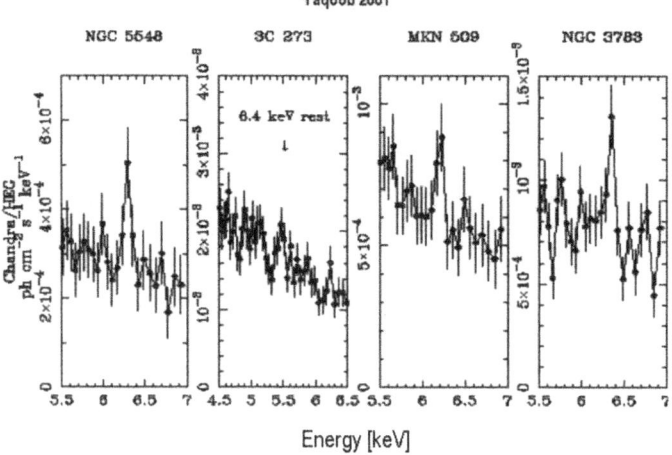

Fig. 1. XMM-Newton and Chandra HETG Fe K line profiles.

for the broad-line AGN Mrk 205, MCG-6-30-15, NGC 5548, 3C 273, Mrk 509 and NGC 3783.

Common to all objects seems to be the presence of a narrow, unresolved and neutral Fe K line at 6.4 keV in the rest frame of the objects. In addition, indications for a red wing above an uncurved power-law have been detected in MCG-6-30-15. Ionised Fe K lines are detected in TON S 180 and Mrk 205. The Chandra HETG Fe K line profiles are described in detail by [5]. The X-ray spectra are also characterized by the presence of narrow unresolved Fe K lines. However, it must be noted that the lower photon statistics of the Chandra HETG observations might be responsible of the non-detection of relativistically broadened red wings.

The origin of the *narrow* line at 6.4 keV
Reflection of primary X-rays on molecular torus

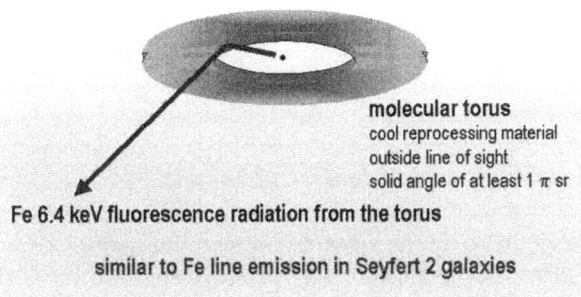

molecular torus
cool reprocessing material
outside line of sight
solid angle of at least 1 π sr

Fe 6.4 keV fluorescence radiation from the torus

similar to Fe line emission in Seyfert 2 galaxies

Fig. 2. Schematic illustration of the origin of the narrow Fe K line, which is probably due to reflection of the primary X-rays from the molecular torus.

The potential of XMM-Newton for the Fe K line diagnostics has been convincingly demonstrated by [6] for the case of Mrk 205. Mrk 205 is a nearby (z=0.07), low-luminosity, radio-quiet quasar viewed through the outer disk of NGC 4319 (z=0.005). The Fe K line complex of Mrk 205 can be modeled by Gaussian profiles for a narrow and a broad component as well as by a relativistic disk line profile. The line energy for the Gaussian modeling is found at (6.4 ± 0.03) keV for the narrow unresolved Fe K line. The broad component is centered at (6.74 ± 0.12) keV, corresponding to ionised helium and/or hydrogen like Fe. [6] argue that the narrow line at 6.4 keV most likely originates from cool reprocessing material outsitde the line of sight covering a substantial solid angle of at least 1 π sr. Hard X-rays reflected at the surface of the molecular torus into the line-of-sight seems to provide a plausible explanation for the neutral, narrow and high equivalent width Fe K line seen in Mrk 205 (and the other objects shown in Fig. 1). This type of neutral Fe K line profiles is similar to those observed in Seyfert 2 galaxies. A sketch of the origin of the neutral Fe K line in shown in Fig. 2. Based on the broad-band X-ray spectral energy distribution [6] argue that the origin of the broad Fe K line at about 6.7 keV is most likely the inner accretion disk. The spectral modeling shows that the disk is probably in a high state of ionisation with an ionising parameter of $\xi = 300 \pm 80$ erg cm s^{-1}.

XMM-Newton observations of the ultrasoft narrow-line Seyfert 1 Galaxy Ton S180 have also revealed the presence of a narrow unresolved neutral Fe K line (cf. Fig 3). The centroid line energy is found at (6.4 ± 0.3) keV with an equivalent width of (50 ± 20) eV. An ionised hydrogen Fe line is also found at (7.0 ± 0.3) keV with an equivalent width of (80 ± 40) eV. The strength of the ionised Fe line as found with XMM-Newton is smaller compared to a previous Beppo SAX observation [7]. A detailed spectral and timing analysis will be given in [8].

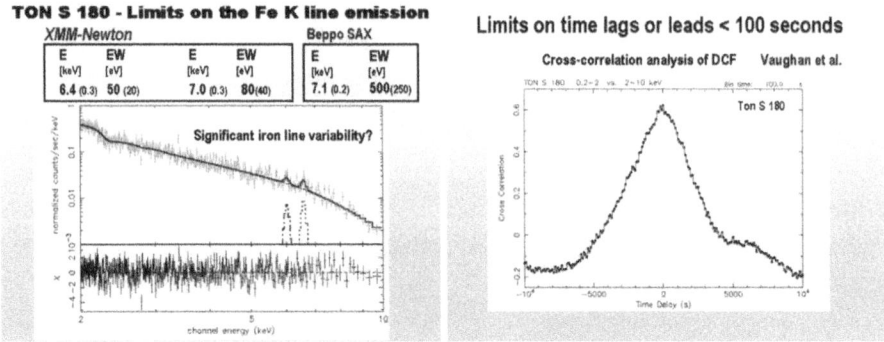

Fig. 3. XMM-Newton EPIC pn spectrum of the Narrow-Line Seyfert 1 Galaxy Ton S180 (left panel). A neutral and an hydrogen-like Fe K line have been detected. Both lines are unresolved and narrow. The source is strongly variable over the course of the observations and shows a notable lack of spectral variability. The constraints on any time leads are below 100 seconds (cf. the cross-correlation function, right panel).

XMM-Newton observations of MCG-6-30-15 are interpreted as huge emission from a redshifted disk line [9] assuming an uncurved underlying power-law. It is suggested that the extraction and dissipation of rotational energy from a spinning black hole by magnetic fields connecting the black hole or plunging region to the disk.

Mrk 766, one of the X-ray brightest AGN in the sky, has been observed with XMM-Newton during the PV phase. The spectral fitting results in the Fe K line complex are not unique as both a relativistic disk line fit (Reeves et al., private communication) and a simple Gaussian fit (this paper) provide statistically acceptable fits to the data (cf. Fig 4) . The longer-look observation of about 350 ks by K. Mason might help to progress in determining more precisely the spectral profile of the Fe K line complex emission in Mrk 766.

The high throughput of XMM-Newton allows to make detection of Fe K line in the 'high-redshift' universe. The first XMM-Newton spectrum of a high-z quasar (PKS 0537-286, z=3.1) was published by [10]. Fig. 5 shows the EPIC pn spectrum of PKS 0537-286 in the observers frame. The Fe K line is detected at about 1.5 keV, consistent with the neutral Fe K line taking into account the low photon statistics. As the Fe K line is close to the Al Kα line at 1.487 keV, a careful analysis has been performed to exclude that the line is due to instrumental effects. [10] state that the Al Kα line contributes to about 4 per cent of the quasar flux before background subtraction. Therefore, the background substracted spectrum of PKS 0537-286 is not affected by any instrumental Al Kα emission and is therefore instrinsic to the distant quasar.

In summary, the present XMM-Newton and Chandra data show (i) narrow high equivalent width Fe K lines in reflection, (ii) sometimes ionised, broad helium and/or

Spectral fits to the neutral Fe K line

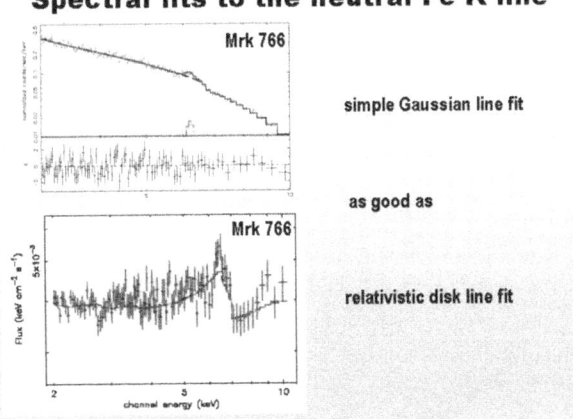

simple Gaussian line fit

as good as

relativistic disk line fit

Fig. 4. Spectral fits to the neutral Fe K line observed during the PV observation with XMM-Newton. Both, a simple Gaussian fit and a relativistic disk line fit provide statistically acceptable fits to the data.

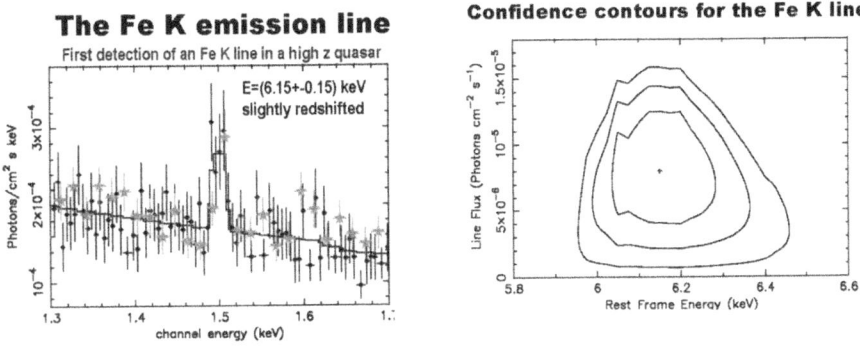

Fig. 5. XMM-Newton EPIC pn spectrum of the high-redshift quasar PKS 0537-286 (left panel). The neutral Fe K line is detected at 1.5 keV in the observers frame. The right panel shows the confidence contours for the Fe K line in the rest frame of the distant quasar.

hydrogen like Fe K line emission and (iii) a few sources with relativistically broadened red wings.

2 Discovery of a Sharp Spectral Feature at 7.1 keV in the Narrow-Line Seyfert 1 Galaxy 1H 0707-495

2.1 The Spectral Parameters of the Sharp Spectral Feature

The Narrow-Line Seyfert 1 Galaxy 1H 0707-495 was observed with XMM-Newton for about 40 ksec during the PV phase [11]. The most striking feature is the sudden drop by a factor of more than 2 at around 7 keV, visible in both pn and MOS spectra (c.f. Fig. 6). The width of the feature is not significantly larger than the energy resolution. In the 7-10 keV range the source is detected with a significance level of about 20 s, therefore the spectral feature is intrinsic to the source. The right panel of Fig. 6 shows a simple power-law fit to the data in the 2-6 keV range. The sharp spectral feature starts at an energy of (7.04 ± 0.07) keV. In addition, the EPIC pn count rate distribution at the energy of the sharp spectral cutoff is shown, demonstrating that the drop occurs within less than 200 eV.

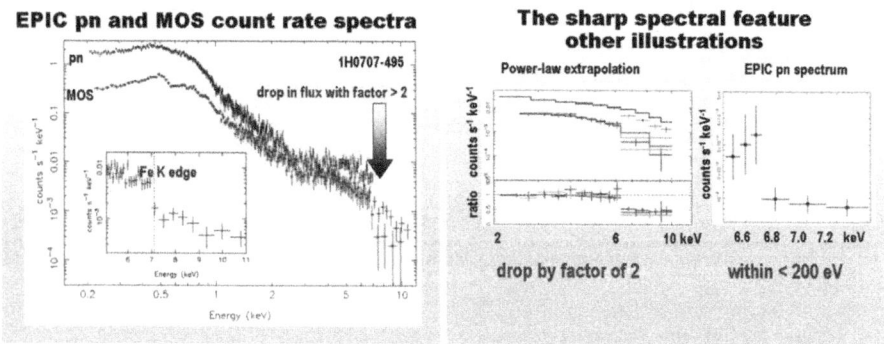

Fig. 6. Left panel: Raw EPIC pn (blue) and MOS (red) count spectra of 1H 0707-495 shifted to the rest-frame of the source. (The MOS1 and MOS2 spectra were combined for display purposes only.) These spectra clearly show a significant drop at ~ 7 keV. The inset panel shows the pn spectrum, again in the rest-frame of the source, with the expected position of the neutral iron K edge marked (dotted line). **Right panel:** The left panel shows a simple power-law fit in the 2-6 keV energy band, extrapolated up to 10 keV. The sharp spectral feature is clearly visible. The right panel shows the count rate distribution in the EPIC pn detector. The sharp spectral feature occurs within less than 200 eV

2.2   Physical Models for the Sharp Spectral Feature

We explore two alternative explanations for this unusual spectral feature: (i) partial covering absorption by clouds of neutral material and (ii) ionised disc reflection with lines and edges from different ionisation stages of iron blurred together by relativistic effects. Fig. 7 summarizes the spectral fitting results discussed below.

Absorption by neutral iron? The most natural interpretation for a drop in the spectrum at just above 7 keV is as an absorption edge from a large column of neutral iron (7.112 keV, [12]; 7.111 keV, [13]). The 2–10 keV spectrum can be adequately parameterized ($\chi^2 = 192/182 \, dof$) using a power-law and a deep absorption edge ($\tau = 1.8 \pm 0.3$). While providing an acceptable fit to the data, this model is unsatisfactory for a number of reasons. Firstly, the 2–10 keV power-law slope is unusually flat ($\Gamma = 1.07 \pm 0.09$). Secondly, the steep spectrum below 1 keV suggests there is little 'cold' absorption above that expected from the Galactic column

**Explanations for the
sharp spectral feature at ~7 keV**

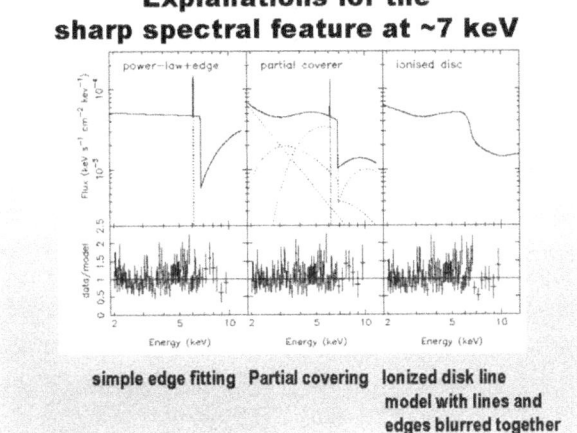

simple edge fitting   Partial covering   Ionized disk line
model with lines and
edges blurred together

Fig. 7. Spectral fits to the 2–10 keV EPIC data. The upper panels demonstrate the models (in F_ν units) and the lower panels show the data/model residuals. The left panels show the simple power-law and edge model, the middle panels show the partial covering model, and the right panels show the 3× solar Fe ionised reflection model.

($N_\mathrm{H} = 5.8 \times 10^{20}$ cm^{-2}; [14]). Indeed, the column of cold iron expected from the depth of the putative edge is $N_\mathrm{Fe} \approx 6 \times 10^{19}$ cm^{-2} which corresponds to $N_\mathrm{H} \approx 2 \times 10^{24}$ cm^{-2} for solar abundances. At this column density the spectrum should be *dominated* by absorption at low energies (L shell iron and K shell absorption from lower Z elements), whereas it actually steepens considerably in the soft band. Absorption by ionised material can produce a strong iron K edge without significant soft X-ray absorption, but the energy of the putative edge suggests the absorber is not strongly ionised. The edge energy at 7.04 ± 0.07 keV is consistent with the neutral iron K edge energies given in the literature.

The fluorescent yield for iron is close to 0.34 [15]. However, there is no strong, narrow 6.4 keV emission line in the EPIC spectrum; the 90 per cent upper limit on the photon flux in a narrow 6.4 keV line is 8×10^{-7} photons cm^{-2} s^{-1} (corresponding to an equivalent width of ~ 90 eV). The estimated photon flux absorbed by the putative edge is 1.2×10^{-5} photons cm^{-2} s^{-1}. For a spherically symmetric distribution of absorbing material one expects a ratio of fluorescence line to Fe-K absorbed flux approximately equal to the fluorescent yield, modified by the absorption of the 6.4 keV line (which is about a factor of 2.5 here). The measured ratio (of line/edge flux) is < 0.07, a factor > 2 smaller than expected for a spherical distribution. This suggests that the column density along the line-of-sight is higher than the average surrounding the X-ray source. The solid angle subtended by the putative absorber, as seen from the X-ray source, is likely to be $< \pi$ sr.

A plausible explanation for the observational effects reported above is partial covering of the central source by Compton thick clouds. Fig. 8 shows a schematic illustration of the partial covering scenario. One possibility is that the absorbers are close to, or even within, the emission region (right panel of Fig. 8). The absorbing clouds must then be very dense to remain cool and therefore held by magnetic fields. The possible presence of such clouds in AGN in general has been discussed by [16], [17], [20], [21], and [22]. [23] first suggested that such clouds might be particularly relevant for NLS1s. Cool gas trapped by the magnetic field is compressed to extreme densities by the high radiation pressure. Clouds and filaments of such cold gas may accompany the active regions above an accretion disc.

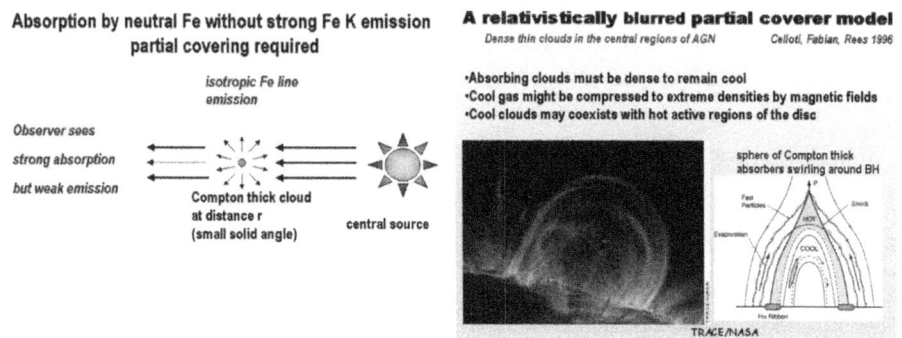

Fig. 8. Schematic illustration of the partial covering scenario. The central energy source is partially covered by Compton thick clouds while part of the flux is seen in direct emission by the observer (left panel). The right panel gives a schematic illustration for a relativistically blurred partial covering model. The absorber is part of the emitting region and it must be dense to keep neutral for iron.

Ionised disc reflection ?   Another explanation for the sharp spectral feature is ionised disc reflection with lines and edges from different ionisation stages of iron blurred together by relativistic effects. A strong Fe edge may be found in the reflected emission from an irradiated accretion disc which is not too highly ionised (right panel of Fig. 7). To investigate if such models can describe these data, the constant-density reflection models of [18] were fitted to the 2-10 keV spectrum. Since the reflected emission may originate close to the central black hole, relativistic blurring was applied to the model during fitting (using the LAOR code; [19]). Reflection from ionised material was strongly preferred over neutral reflection, an iron abundance approximately 3 times solar was needed and a Kerr space-time geometry was preferred over a Schwarzschild one. However, a model with a reflection fraction of unity can only give an adequate fit to the data between 2 and 10 keV, and it cannot fully account for the sharp spectral drop at about 7 keV. Allowing the reflection fraction R to increase to 9 resulted in a steeper continuum and an improved fit. The high value of R can be decreased by increasing the iron abundance of the reflector. The ionisation parameter is $\xi = 750$ erg cm s^{-1}. At this level of ionisation the reflection-dominated model predicts a strong He-like Fe Ka line with an equivalent width of 1.8 keV.

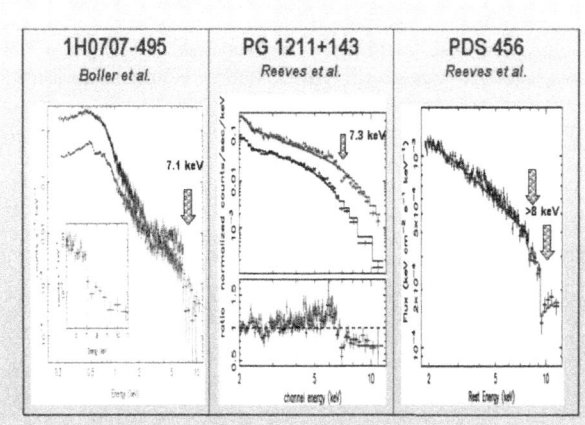

Fig. 9. Objects with detected sharp spectral features: 1H 0707-495 (this paper), PG 1211+143 [24] and PDS 654 [25]

2.3 Detection of Sharp Spectral Features in Other Objects

Similar features have been detected in other objects. The XMM-Newton spectrum of PG 1211+143 [24] appears to show a similar drop at about 7.3 keV. In the luminous quasar PDS 654 [25] sharp spectral features have been detected above 8 keV. If such features are common in other objects too, this will further constrain the range of suitable models.

References

1. Pounds K. A., Nandra K, Stewart G. C., George I. M., Fabian A. C., Nature 344, 132, 1990
2. Nandra K., Pounds K.A., MNRAS 268, 405, 1994
3. Tanaka Y, Nandra K, Fabian A. C, Inoue H, Otani C, Dotani T, Hayashida K.; Iwasawa, K, Kii T, Kunieda H, Makino, F, Matsuoka M., Nature 375, 659, 1995
4. Nandra K., George I. M., Mushotzky R. F., Turner T. J., Yaqoob T., ApJ 488, 91, 1997
5. Yaqoob T., High Energy Universe at Sharp Focus: Chandra Science, proceedings of a conference held in St. Paul, MN, 16-18 July 2001. Edited by Eric M. Schlegel and Saequ Vrtilek. Publisher: ASP Conference Series, 2001
6. Reeves J. N., Turner M. J. L., Pounds K. A., O'Brien P. T., Boller Th., Ferrando P., Kendziorra E., Vercellone S., Astronomy & Astrophysics 365, 134, 2001
7. Comastri, A., Fiore F., Guainazzi M., Matt, G., Stirpe G. M., Zamorani G., Brandt W. N., Leighly K. M., Piro L., Molendi S., Parmar A. N., Siemiginowska, A., Puchnarewicz E. M., Astronomy & Astrophysics 333, 31, 1998
8. Vaughan S., Fabian A.C., Ballantyne D.R., Boller Th., Brandt N., MNRAS, submitted
9. Wilms, J., Reynolds C., Begelman M. C., Reeves J., Molendi S., Staubert, R., Kendziorra E., MNRAS 328, 27, 2001
10. Reeves, J. N., Turner M. J. L., Bennie P. J., Pounds K. A., Short, A., O'Brien P. T., Boller Th., Kuster M., Tiengo A., Astronomy & Astrophysics 365, 116, 2001

11. Boller Th., Fabian A.C., Sunyaev R., Trümper J., Vaughan S., Ballantyne D.R., Brandt N., Keil R., Iwasawa K., MNRAS 329, L1, 2002
12. Henke B. L., Gullikson E. M., Davis J.C., ADNDT, 54, 181, 1993
13. Bearden, J. A., X-ray wavelength US Atomic Energy Commission Tennessee USA, 1964
14. Dickey J. M., Lockman F. J., ARA&A 28, 215, 1990
15. Bambynek, W. et al. *Rev. Mod, Phys.* 44, 716, 1972
16. Rees M.J., MNRAS 228, 47, 1987
17. Celotti A., Fabian A.C., Rees M.J., MNRAS 255, 419, 1992
18. Ross R. R., Fabian A. C., MNRAS 261, 74, 1993
19. Laor A., ApJ 376, 90, 1991
20. Kuncic Z., Blackman E.G., Rees M.J., MNRAS 283, 1322, 1996
21. Kuncic Z., Celotti A., Rees M.J., MNRAS 284, 717, 1997
22. Malzac J., MNRAS 325, 1625, 2001
23. Brandt W. N., Gallagher S. C., *New Astronomy Reviews* 44, 461, 2000
24. Reeves J., et al., in preparation
25. O'Brien P.T., et al., in preparation

Where Are the Lighthouses?

Halton Arp

Max-Planck-Institut für Astrophysik,Postfach 1317, 85741 Garching;
arp@mpa-garching.mpg.de

The following 24 pictures show areas on the sky surrounding what are generally regarded to be the most luminous objects in the universe. Starting from 1968, the first six maps show examples of high redshift quasars paired across active and disturbed galaxies. Such configurations have been calculated to have extremely low probabilities of being accidental background associations. (Arp 1968, 1987; Radecke 1997; Arp 1998, Burbidge 1999). The next series of pictures of ultra luminous infrared galaxies (ULIRG's) show similar ejection patterns of higher redshift quasars from very active central objects. But here the associations cover such a large area on the sky as to imply the ULIRG's to be much closer than their redshift distances. Empirically they seem to be intermediate between quasars and disturbed Seyferts and at the same, relatively close distances (Arp 2001a; Arp et al. 2001).

Next a series of examples of clusters of galaxies show pairing across, active, low redshift galaxies. Evidence for ejection of quasars and radio and X-ray sources along these same lines implies the fragmentation of ejected quasars and their evolution into clusters of galaxies. Particularly X-ray clusters constitute, together with quasars, ULIRG's and Seyfert galaxies, almost the entire class of strong extragalactic X-ray sources. Evidence of motion of very elongated X-ray galaxy clusters along the predicted ejection line is now available from Chandra observed bow shocks, or cold fronts (Arp 2001b).

Finally the recent, most accurate HST measures of galaxy distances from Cepheids is shown as a redshift - distance Hubble diagram. At higher redshifts galaxies are now encountered which have excess redshifts which cannot be accounted for by peculiar motion (too large and systematically positive). An accompanying diagram shows that companion galaxies in the best known, nearby groups have have systematically higher redshifts than the central galaxies. This appears to support the conclusion that later generation, younger objects have intrinsic redshifts which diminish as they evolve toward normal galaxies.

References

1. Arp, H. 1968, Astrofizika, 4, 59
2. Arp, H. 1987, Quasars, Redshifts and Controversies (Berkeley, Interstellar Media)
3. Arp, H. 1998, Seeing Red: Redshifts, Cosmology and Academic Science (Montreal, Apeiron)
4. Arp, H. 2001a, ApJ 549, 780
5. Arp, H. 2001b, ApJ 549, 802
6. Arp, H., Burbidge, E.M., Chu, Y., Zhu, X. 2001, ApJ 553, L11
7. Burbidge, E.M. 1999, ApJ 511, L9
8. Radecke, H.-D. 1997, A&A 319, 18

QUASARS

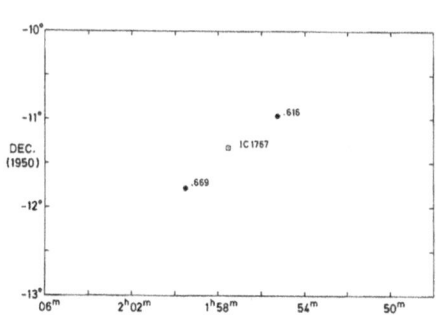

1968 - RADIO QUASARS ACROSS DISTURBED GALAXIES. THIS VERY STRONG PAIR FELL ACROSS A GALAXY WITH z = .018

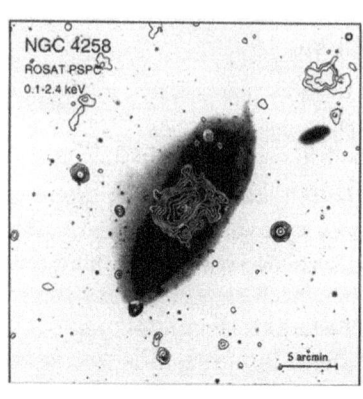

1994 - X-RAY QUASARS NEAR ACTIVE GALAXIES. THIS PAIR AT Z = .40 AND .65 FALL ACROSS A BRIGHT SEYFERT 2 GALAXY KNOWN TO BE EJECTING RADIO AND X-RAY MATERIAL

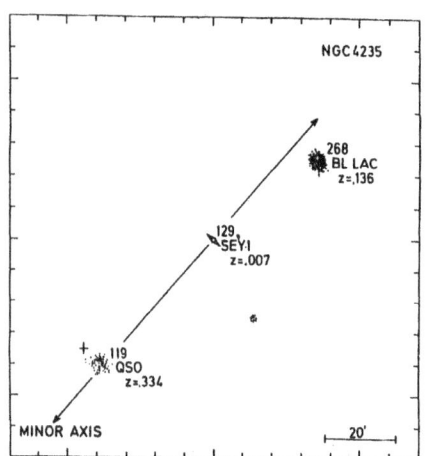

1997 - X-RAY SOURCES ASSOCIATED WITH SEYFERTS AT 7.5σ LEVEL. BRIGHTEST TEND TO BE QUASARS ALONG THE MINOR AXIS. THESE TWO SOURCES ARE 268 AND 119 COUNTS PER KILOSEC!

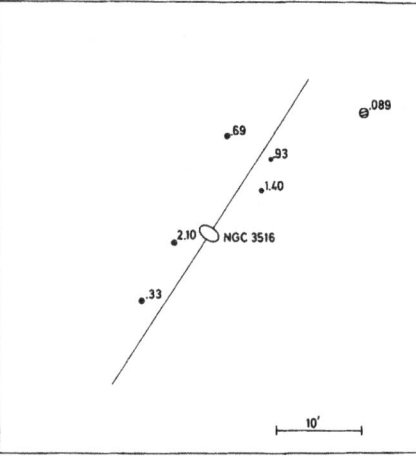

1998 - THE ROSETTA STONE. SIX BRIGHTEST X-RAY SOURCES ARE QUASARS ALIGNED ALONG MINOR AXIS IN DESCENDING ORDER OF QUANTIZED REDSHIFT. VERY ACTIVE SEYFERT HAS z = .009.

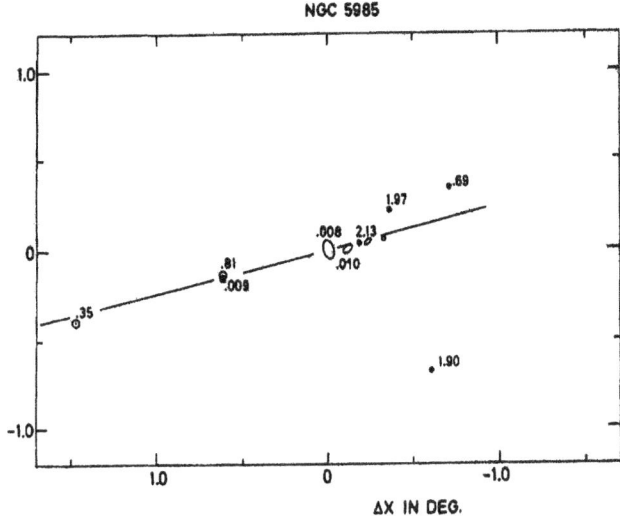

1999 - CONFIRMATION FROM FIVE QUASARS FALLING ALONG THE MI-
NOR AXIS OF THE BRIGHT SEYFERT NGC5985. SAME DESCENDING
ORDER OF QUANTIZED REDSHIFTS. NOTE FOUR LOW REDSHIFT COM-
PANIONS APPEARING ALONG EJECTION LINE.

ULIRG's

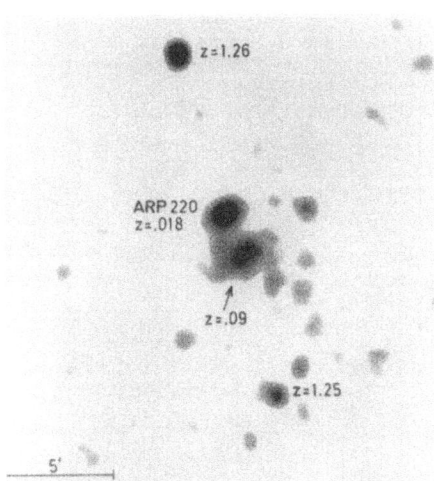

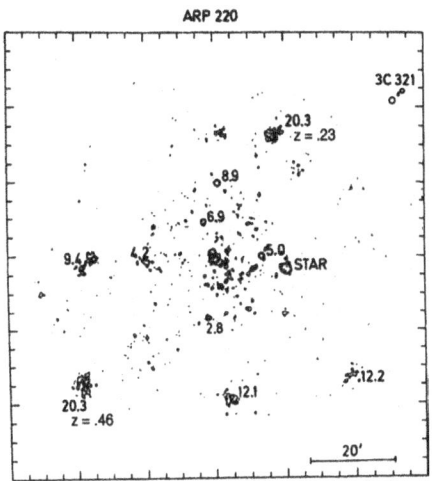

2000 - INNER REGION OF ULIRG ARP 220. X-RAY QUASARS EX-
ACTLY ALIGNED WITH ALMOST IDENTICAL SPECTRA. TRAIL OF
X-RAY SPOTS LEADING DOWN TO z=1.25 QSO. GROUP OF Z=.09
GALAXIES CONNECTED BY X-RAYS AND LOW z=.018 HYDROGEN.

BRIGHT (20.3 cts/ksec) X-RAY QUASARS FORM OUTER PAIR ACROSS
ARP 220. SMALLER SOURCES EXTEND ALONG THIS LINE. THE AS-
SOCIATION IS ALMOST 2 DEG. ON THE SKY IMPLYING CLOSER-
THAN-REDSHIFT DISTANCE AND MUCH LOWER LUMINOSITY.

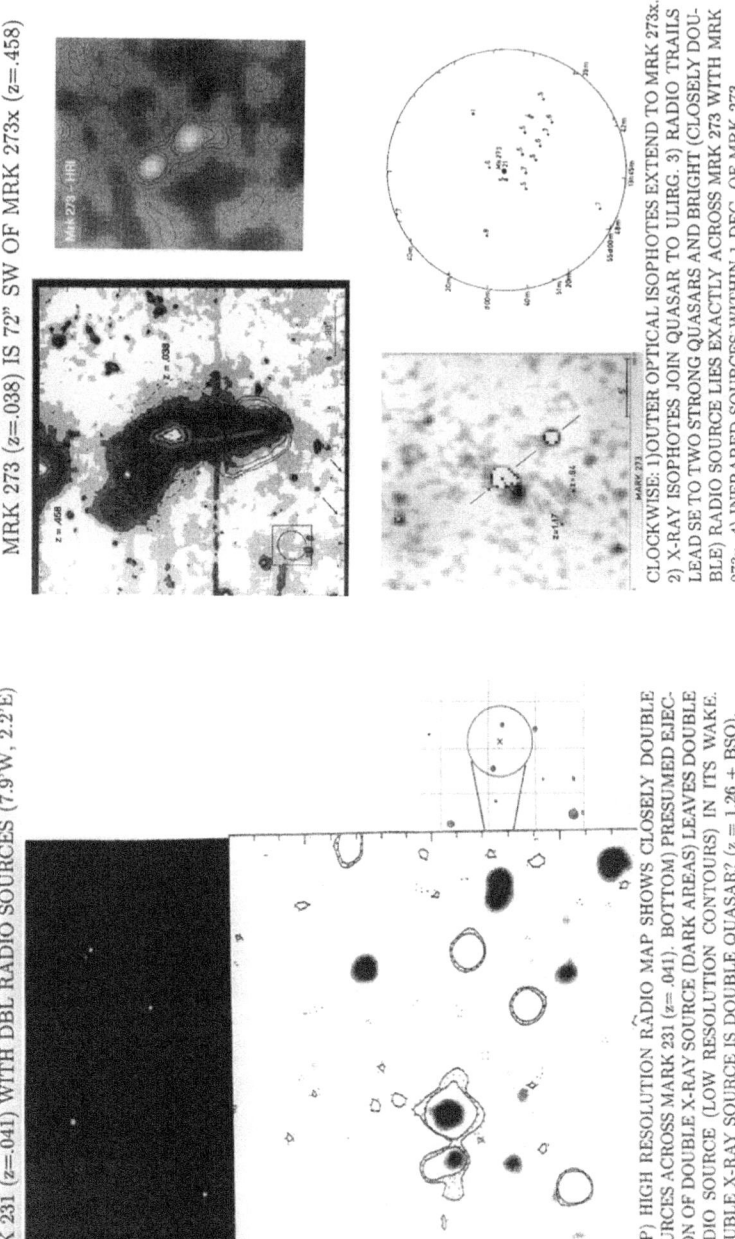

MRK 273 (z=.038) IS 72" SW OF MRK 273x (z=.458)

CLOCKWISE: 1)OUTER OPTICAL ISOPHOTES EXTEND TO MRK 273x. 2) X-RAY ISOPHOTES JOIN QUASAR TO ULIRG. 3) RADIO TRAILS LEAD SE TO TWO STRONG QUASARS AND BRIGHT (CLOSELY DOUBLE) RADIO SOURCE LIES EXACTLY ACROSS MRK 273 WITH MRK 273x. 4) INFRARED SOURCES WITHIN 1 DEG. OF MRK 273.

MRK 231 (z=.041) WITH DBL RADIO SOURCES (7.9"W, 2.2'E)

TOP) HIGH RESOLUTION RADIO MAP SHOWS CLOSELY DOUBLE SOURCES ACROSS MARK 231 (z=.041). BOTTOM) PRESUMED EJECTION OF DOUBLE X-RAY SOURCE (DARK AREAS) LEAVES DOUBLE RADIO SOURCE (LOW RESOLUTION CONTOURS) IN ITS WAKE. DOUBLE X-RAY SOURCE IS DOUBLE QUASAR? (z = 1.26 + BSO).

WHERE ARE MRK 273 AND MRK 231?

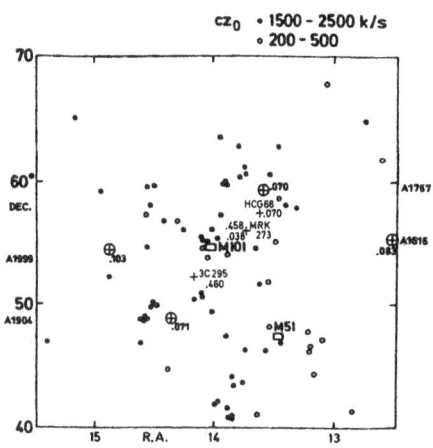

cz₀ • 1500 - 2500 k/s
     ○ 200 - 500

LINES OF HIGHER REDSHIFT GALAXIES EMERGING FROM THE VERY BRIGHT, NEARBY ScI GALAXY, M101. ABELL CLUSTERS 1767 AND 1904 AT z=.070 and .071 PAIRED ACROSS IT. COMPACT GROUP HCG 66 AT z=.070, GALAXY CLUSTER 3C295 AT z =.460 MARKED.

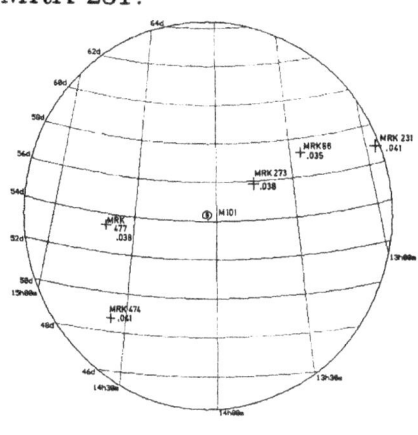

ALL MARKARIAN GALAXIES ≤ 15.3 mag., .035 ≤ z ≤ .045 PLOTTED. THESE BRIGHTEST MRK GALAXIES HAVE THE MOST LITERATURE REFERENCES, e.g. 509 AND 274 FOR MRK 231 AND 273.

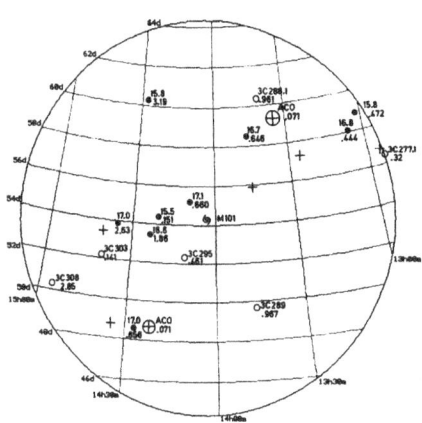

ALL QSO'S V ≤ 17.1 mag. HAVE BEEN ADDED AS FILLED CIRCLES. KNOWN 3C (BRIGHT) RADIO OBJECTS ARE OPEN CIRCLES. BRIGHTEST, ACTIVE OBJECTS TEND TO FOLLOW MRK PLUS SIGNS.

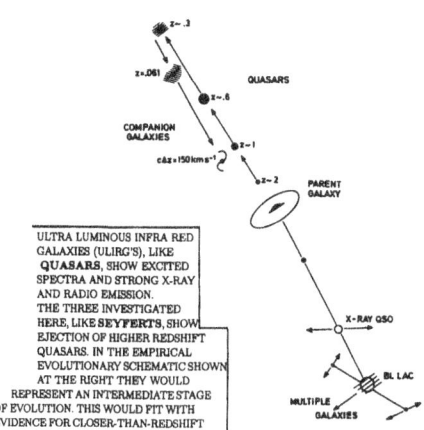

ULTRA LUMINOUS INFRA RED GALAXIES (ULIRG'S), LIKE QUASARS, SHOW EXCITED SPECTRA AND STRONG X-RAY AND RADIO EMISSION. THE THREE INVESTIGATED HERE, LIKE SEYFERTS, SHOW EJECTION OF HIGHER REDSHIFT QUASARS. IN THE EMPIRICAL EVOLUTIONARY SCHEMATIC SHOWN AT THE RIGHT THEY WOULD REPRESENT AN INTERMEDIATE STAGE OF EVOLUTION. THIS WOULD FIT WITH EVIDENCE FOR CLOSER-THAN-REDSHIFT DISTANCES AND LOWER LUMINOSITIES.

GALAXY CLUSTERS

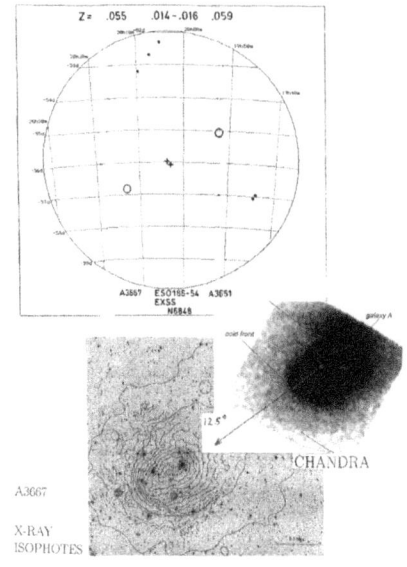

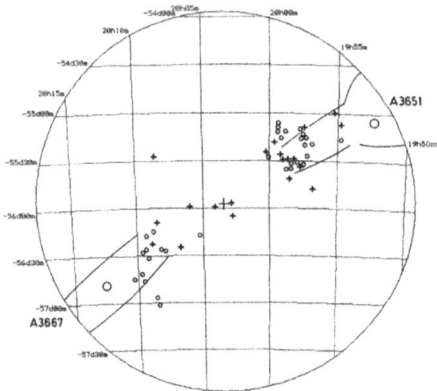

A BRIGHT PAIR OF ABELL CLUSTERS OF z=.06 CENTERED ACROSS
A BRIGHT EMISSION LINE GALAXY OF z=.015. THE CENTRAL
GALAXY IS ALSO AN X-RAY SOURCE WITH POINT SOURCES ALIGNED
TOWARD THE STRONG X-RAY CLUSTERS (A 3667 HAS 2240 cts/ks).
GALAXIES WITH cz=6,000 km/sec EXTEND IN BOTH DIRECTIONS
FROM THE CENTER TO MINGLE WITH THE 16,000 TO 18,000 km/sec
GALAXIES OF THE TWO OUTER CLUSTERS (PLUS SIGNS AND OPEN
CIRCLES). CHANDRA OBSERVATIONS SHOW A COLD FRONT MOV-
ING DOWN THE LENGTH OF A3667 AND EXACTLY AWAY FROM
THE CENTRAL, LOW REDSHIFT GALAXIES.

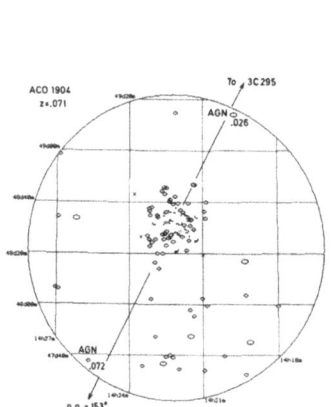

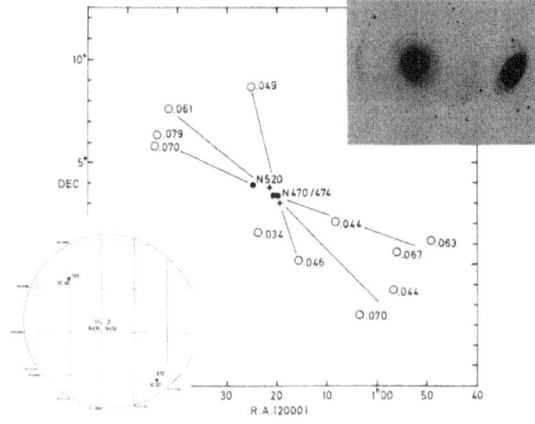

THE SE CLUSTER OF THE PAIR, A1904, IS ELONGATED BACK TO-
WARD 3C295 AND M101. SOME OF THE GALAXIES TO THE SW
ALSO HAVE z's NEAR .07, OTHERS SHOULD BE MEASURED.

NGC 470/474 (ARP 227), TWO DISTURBED GALAXIES WITH A PAIR
OF STRONG RADIO QUASARS (3C 37 AND 3C 39) PAIRED ACROSS
THEM. ACCIDENTAL PROBABILITY OF THIS ALIGNMENT IS $2x10^{-9}$.
BRIGHT ABELL CLUSTERS BETWEEN $15.0 \leq m_{10} \leq 16.4$ mag. FALL
ALONG EJECTION LINE DEFINED BY QUASARS.

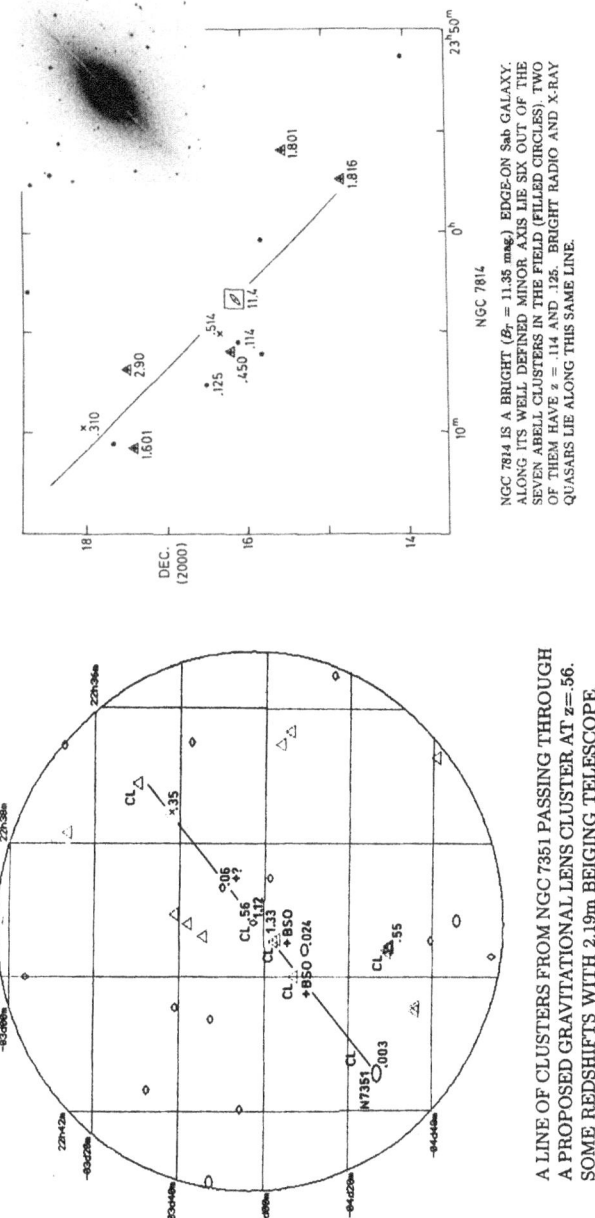

NGC 7814 IS A BRIGHT (B_T = 11.35 mag.) EDGE-ON Sab GALAXY. ALONG ITS WELL DEFINED MINOR AXIS LIE SIX OUT OF THE SEVEN ABELL CLUSTERS IN THE FIELD (FILLED CIRCLES) TWO OF THEM HAVE z = .114 AND .125. BRIGHT RADIO AND X-RAY QUASARS LIE ALONG THIS SAME LINE.

A LINE OF CLUSTERS FROM NGC 7351 PASSING THROUGH A PROPOSED GRAVITATIONAL LENS CLUSTER AT z=.56. SOME REDSHIFTS WITH 2.19m BEIJING TELESCOPE

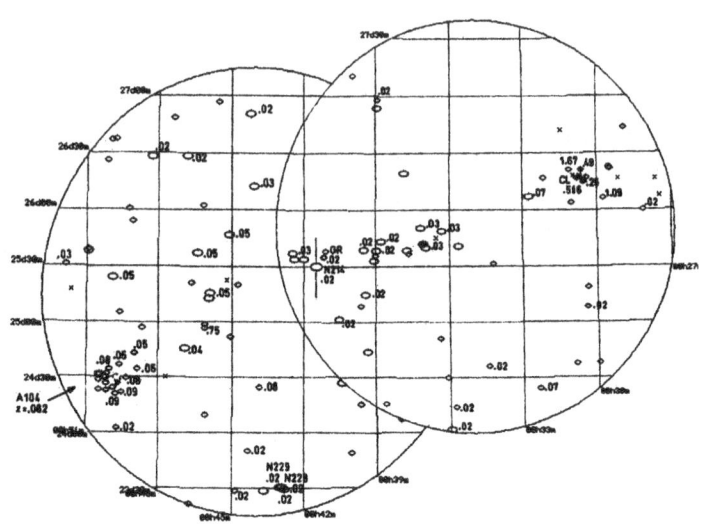

ALL GALAXIES FROM SIMBAD AROUND THE ScI, NGC 214. THE
REDSHIFTS INCREASE TOWARD THE CLUSTER AT $z = .516$ AND ALSO
OPPOSITELY TOWARD THE ELONGATED CLUSTER ABELL 104.

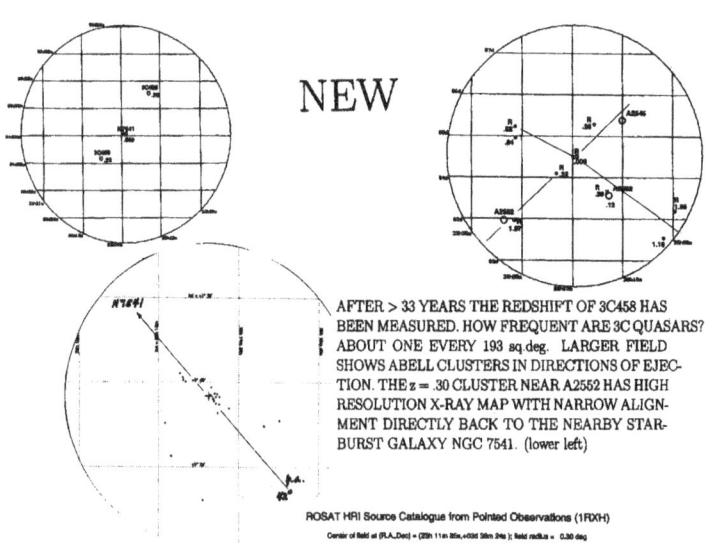

NEW

AFTER > 33 YEARS THE REDSHIFT OF 3C458 HAS
BEEN MEASURED. HOW FREQUENT ARE 3C QUASARS?
ABOUT ONE EVERY 193 sq.deg. LARGER FIELD
SHOWS ABELL CLUSTERS IN DIRECTIONS OF EJEC-
TION. THE $z = .30$ CLUSTER NEAR A2552 HAS HIGH
RESOLUTION X-RAY MAP WITH NARROW ALIGN-
MENT DIRECTLY BACK TO THE NEARBY STAR-
BURST GALAXY NGC 7541. (lower left)

ROSAT HRI Source Catalogue from Pointed Observations (1RXH)

Center of field at (R.A.,Dec) = (23h 11m 35s,+03d 38m 24s); field radius = 0.30 deg

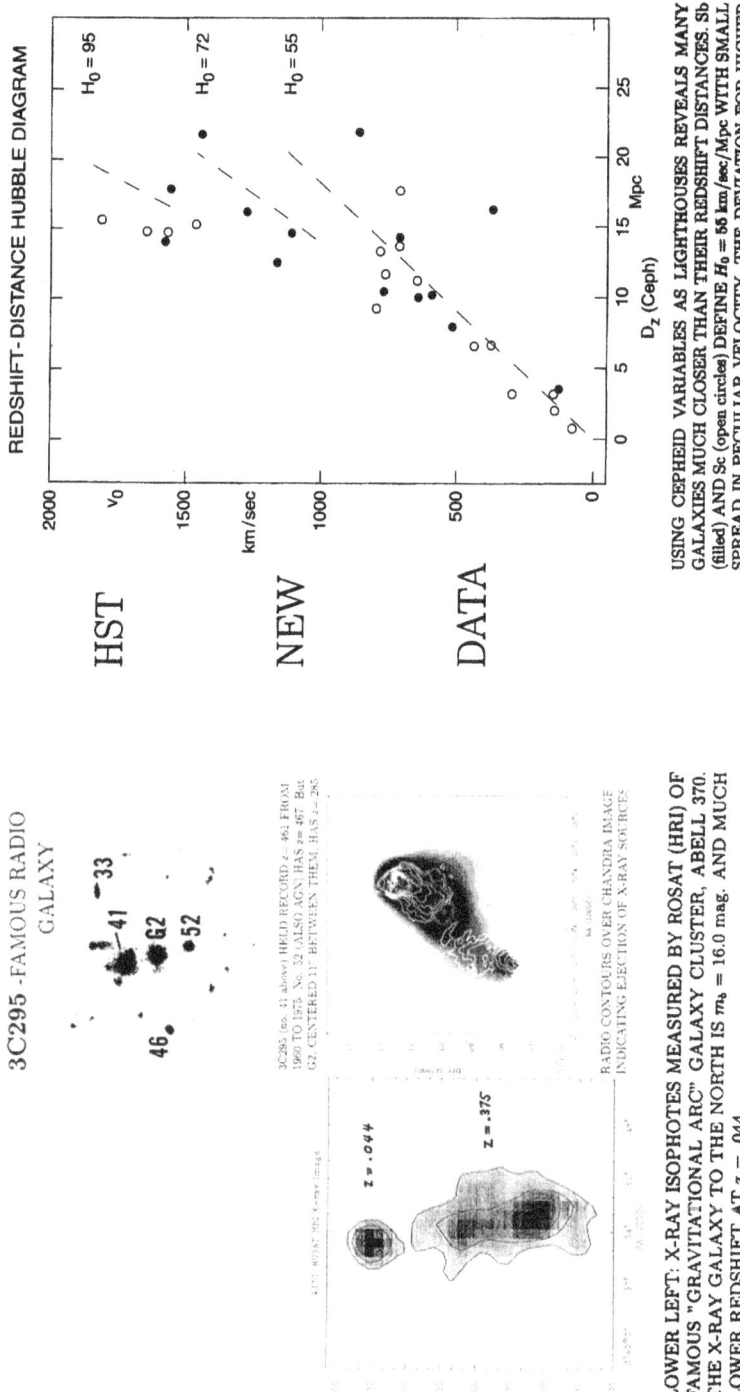

REDSHIFT- DISTANCE HUBBLE DIAGRAM

$H_0 = 95$

$H_0 = 72$

$H_0 = 55$

v_0

km/sec

D_z (Ceph)

Mpc

HST

NEW

DATA

USING CEPHEID VARIABLES AS LIGHTHOUSES REVEALS MANY GALAXIES MUCH CLOSER THAN THEIR REDSHIFT DISTANCES. Sb (filled) AND Sc (open circles) DEFINE $H_0 = 55$ km/sec/Mpc WITH SMALL SPREAD IN PECULIAR VELOCITY. THE DEVIATION FOR HIGHER REDSHIFTS CANNOT BE DUE TO PECULIAR VELOCITY.

3C295 - FAMOUS RADIO GALAXY

33

41

62

52

46

3C295 (no. 41 above) HELD RECORD z = 461 FROM 1960 TO 1975. No. 52 (ALSO AGN) HAS z= 467. But 62. CENTERED 11" BETWEEN THEM. HAS z= 285.

RADIO CONTOURS OVER CHANDRA IMAGE INDICATING EJECTION OF X-RAY SOURCES.

z = .044

z = .375

LOWER LEFT: X-RAY ISOPHOTES MEASURED BY ROSAT (HRI) OF FAMOUS "GRAVITATIONAL ARC" GALAXY CLUSTER, ABELL 370. THE X-RAY GALAXY TO THE NORTH IS $m_b = 16.0$ mag. AND MUCH LOWER REDSHIFT AT z = .044.

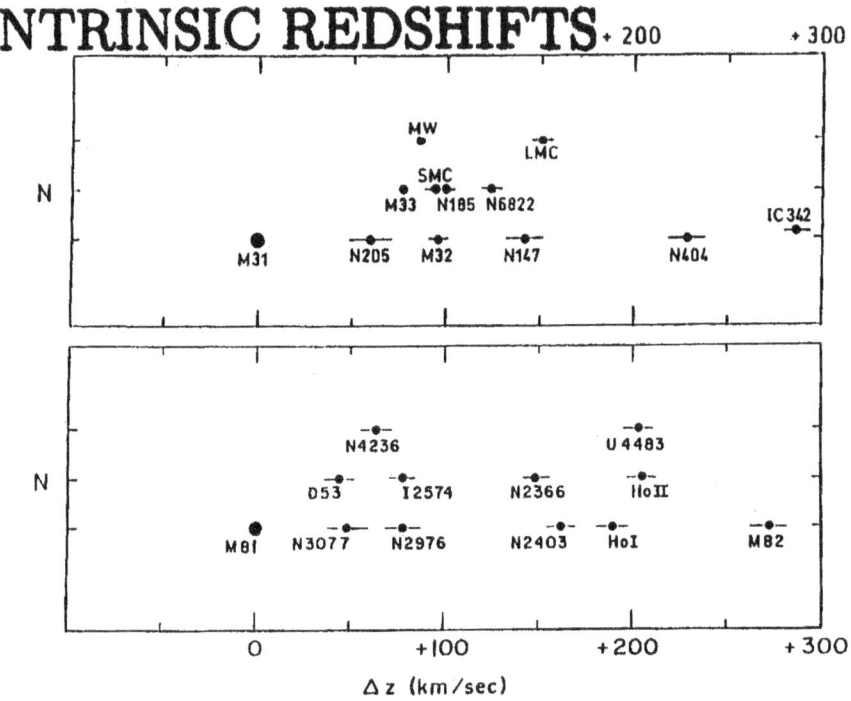

DIFFERENTIAL REDSHIFTS FROM THE DOMINANT GALAXY FOR
MAJOR COMPANIONS OF BEST KNOWN LOCAL GROUP (M31) AND
M81 GROUP. Sc AND LATER COMPANIONS ARE AMONG 22 OUT
OF 22 WHICH ARE POSITIVELY REDSHIFTED. INTERPRETATION:
AS IN ALL PRECEDING EXAMPLES, YOUNGER GALAXIES ARE IN-
TRINSICALLY REDSHIFTED RELATIVE TO OLDER GALAXIES.

X-Rays from the Jet in 3C 273: Clues from the Radio-Optical Spectra

Sebastian Jester[1], Hermann-Josef Röser[1], Klaus Meisenheimer[1], and Rick Perley[2]

[1] Max-Planck-Institut für Astronomie, Königstuhl 17, 69117 Heidelberg, Germany
[2] NRAO, P. O. Box 0, Socorro, NM 87801, USA

Abstract. Using new deep VLA and HST observations of the large-scale jet in 3C 273 matched to $0.''3$ resolution, we have detected excess near-ultraviolet emission ($\lambda\,300$ nm) above a synchrotron cutoff spectrum accounting for the jet's emission from radio through optical ($\lambda\lambda 3.6$ cm–620 nm). This necessitates a two-component model for the jet emission. The radio–optical–X-ray spectral energy distributions suggest a common origin for the UV excess and the X-rays from the jet. We discuss how to constrain the emission mechanism for this high-energy component.

Abstract of a paper submitted to A&A Letters.

The Size of Blazar Radio Cores from Intraday Variability

Thomas Beckert, Lars Fuhrmann, Thomas P. Krichbaum, Arno Witzel, and Anton Zensus

Max-Planck-Institut für Radioastronomie, 53121 Bonn, Germany

1 Intraday Variability

Flux-density variations on time scales of about 1 day (Wagner & Witzel 1995) and amplitudes up to 20% in radio and/or optical frequency bands are known as Intraday Variability (IDV). Since its discovery in 1985 (Witzel et al. 1986) it has been shown that IDV is a common phenomenon among compact flat-spectrum extragalactic radio sources ($\sim 25\%$). The light-travel time argument $\Delta l = c\Delta t$ for intrinsic variations has been used to infer an extremely small physical size for the radio-emitting regions. The apparent brightness temperatures of $T_{\mathrm{app}} = 10^{17}$–$10^{19}$ K are much higher[1] than expected from relativistic jets in AGNs emitting Doppler-boosted emission with Doppler-factors $D = 5$–20.

2 No intrinsic Variations?

Several models for intrinsic variations on IDV timescales have been suggested and provoked new questions. Rapid particle acceleration and radiation cooling on timescales of 1 day at GHz-frequencies require intrinsic brightness temperatures of $T_{\mathrm{in}} \geq 10^{15}$ K. This is above the Inverse Compton (IC) Limit $T_{\mathrm{IC}} = 10^{12}$K and catastrophic cooling predicts large X-ray and γ-ray fluxes, which are not observed. Furthermore p-e$^-$-Jets with $T_{\mathrm{in}} > 10^{13}$K are intrinsically depolarized or fine tuning of $\gamma_{\mathrm{min}} > 0.4\gamma_{\mathrm{rad}}$ for the power-law population is required. The observed fast polarization variability is therefore crucial for our knowledge of the low-energy electrons in jets. Another suggestion invokes large Doppler factors $D > 100$ to explain the short timescale by time dilation. But until now the observed superluminal motions do not support this idea. Finally the propagation of thin relativistic shocks ($\Delta z/R < 0.1$) in the jet can relax the light travel time argument. The model predicts time lags between frequencies and requires fine tuning of the direction of shock normal to the line of sight (Beckert et al. 2002).

3 Interstellar Scintillation

The most successful explanation of IDV at radio frequencies at present is Interstellar Scattering (ISS) known from scattering of pulsar and some extragalactic sources at low

---

[1] Even more extreme sources with $T_{app} \approx 10^{21}$ K – PKS 0537-441 (Romero et al. 1994), PKS 0405-385 (Kedziora-Chudczer et al. 1997) and J1819+384 (Dennett-Thorpe & de Bruyn 2000) – have been found in the last years.

frequencies (e.g. Bondi et al. 1994). The ionized gas in the galactic disk is known to be turbulent with a scale height of 0.5 kpc.

3.1 Results of Structure Functions Analysis

The appropriate tool to analyse the variability in IDV lightcurves in terms of ISS are auto- and cross-correlation function, when polarization variability is seen (Rickett et al. 2001), and structure functions

$$\mathrm{SF}(\tau) = \left\langle [I(t) - I(t+\tau)]^2 \right\rangle_t \qquad \tau = sv \quad , \tag{1}$$

where τ is the time lag, s is the spatial scale of ISS variability pattern and v the projected perpendicular velocity of the observer. A characteristic timescale τ_c of the relevant process is measured by the first peak of SF, while the height of the peak gives the modulation index $m = \sqrt{\mathrm{SF}/2}/\langle I \rangle$ of the variability. Analysis of multi-frequency data of 0917+62 (Beckert et al. 2001) indicate that the components of the ISM responsible for IDV are a quasi-continuous medium with an enhanced level of turbulence out to 100 pc. The turbulent spectrum $\Phi(k) = C_N^2 k^{-\beta}$ has to be steep $\beta > 4$, much steeper than a Kolmogorov-like spectrum ($\beta = 11/3$).

3.2 Annual Modulation of IDV Time Scale

Due to the Earth's orbit and the motion of the solar system relative to the local standard of rest (LSR) the variability time scale is expected to show seasonal changes. With the LSR velocity from Hipparcos data (Dehnen & Binney 1998) a prediction of the annual modulation of the variability time scale (Qian & Zhang 2001) is derived (Fig. 1). For velocities of sub-components in the ISM of 10 km/s in various directions the modulation can be enhanced or reduced. The largest modulations are due to gas moving in the direction ($l = 45°, b = 35°$) in galactic coordinates. The broad peak at day 300 for 0917+624 is different from the work of Rickett et al. (2001) due to the LSR assumptions.

3.3 Lessons from 0917+62

Interstellar Scintillation of sources smaller than the Fresnel scale in the ISM would show larger amplitude variations than observed for 0917+62. For a larger source size the modulation index is reduced $m \propto \theta$, where θ is the angular size of the source. We can therefore derive a model dependent source size of the radio core of 0917+62. At $\lambda 6$ cm the size is $\theta \approx 40\,\mu$as, while at $\lambda 20$ cm we derive a size $\theta \approx 160$–$400\,\mu$as. This implies brightness temperatures in the radio core of $T_{\mathrm{app}} = 7.5 \cdot 10^{13}$ K at 6 cm and $T_{\mathrm{app}} = 0.1$–$5.2 \cdot 10^{13}$ K at 20 cm still above the Inverse Compton Limit. A strong Doppler boost is required to reduce the intrinsic brightness temperature to the IC Limit. If so the optical thick-thin transition–seen at ~ 90 GHz in 0917+62–implies a size of the emitting region at the 'base of the jet' of $L = 4 \cdot 10^{17}$ cm $= 1.3 \cdot 10^3 R_S$ when extrapolated within a conical jet model from a 10^9 M$_\odot$ Black Hole.

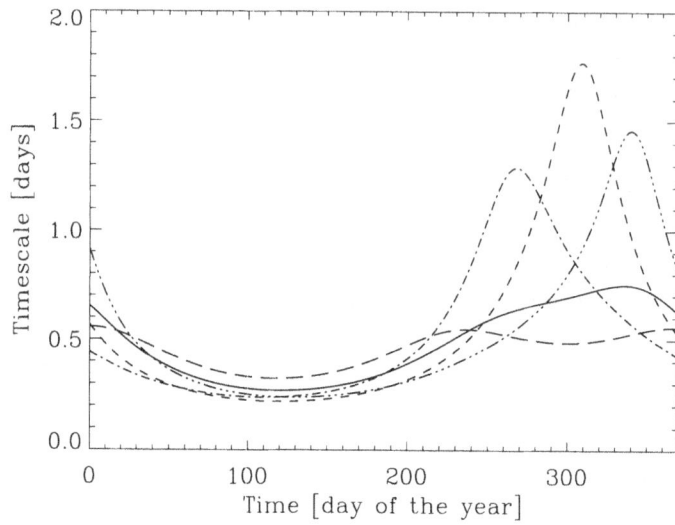

Fig. 1. Annual modulation of the variability timescale in the direction of 0917+62. We use the LSR based on Hipparcos data (Dehnen & Binney, 1998), which results in smaller amplitudes (solid line) than derived by Rickett et al. (2001). The other modulation curves arise for ISM velocities of 10 km/s in different directions.

4   Summary

While intrinsic explanations for IDV face unsolved problems, the hypothesis of interstellar scattering does not invoke extreme physical conditions in the scattering medium, but it does require an anomalous steep turbulent spectrum in the ISM. ISS reduces the observed brightness temperatures, but the implied steady sources are much brighter than allowed by the IC Limit and strong Doppler boosting is needed.

References

1. Beckert, T., Krichbaum, T. P., Cimò, G., et al., 2002, PASA Vol. 19, in press
2. Bondi M., Padrielli L., Gregorini L. et al., 1994, A&A 287, 390
3. Dehnen, W., Binney, J. J., 1998, MNRAS 298, 387
4. Dennett-Thorpe, J., de Bruyn, A. G., 2000, ApJ 529, L65
5. Kedziora-Chudczer, L., Jauncey, D. L., Wieringa, M. H. et al., 1997, ApJ490, L9
6. Qian, S. J., Zhang, X. Z., 2001, ChJA&A 1, 133
7. Rickett, B. J., Witzel, A., Kraus, A., Krichbaum, T. P., Qian, S. J., 2001, ApJ 550, L11
8. Rickett, B. J., Kedziora-Chudczer, L., Jauncey, D. L., 2001, PASA, Vol. 19, No. 1
9. Romero, G. E., Combi, J. A., Colomb, F. R., 1994, A&A 288, 731
10. Witzel A., Heeschen D. S., Schalinski C. et al., 1986, Mitt Astron. Ges. 65, 239
11. Wagner S.J., Witzel A., 1995, ARA&A 33, 163

The Luminosity Function of the Host Galaxies of QSOs and BL Lac Objects

Nicoletta Carangelo[1], Renato Falomo[2], and Aldo Treves[1]

[1] Università dell'Insubria, via Valleggio 11, Como, Italy
[2] Osservatorio Astronomico di Padova, Vicolo dell'Osservatorio 5, Padova,Italy

1 Introduction

A clear insight of the galaxies hosting active galactic nuclei is of fundamental importance for understanding the processes of galaxies and nuclei formation and their cosmic evolution. A good characterization of the host galaxies properties requires images of excellent quality in order to disentangle the light of the galaxy from that of the bright nucleus. To this aim HST has provided a major improvement of data on QSOs (Disney et al. 1995; Bahcall et al. 1996, 1997; Boyce et al. 1998; McLure et al. 1999; Hamilton et al. 2000; Kukula et al. 2001) and BL Lacs (Scarpa et al. 2000, Urry et al. 2000).

We present a comparative study of low redshift QSO and BL Lac host galaxy luminosity function (HGLF). To this aim we have considered samples of BL Lacs (Urry et al. 2000) and QSOs (Bahcall et al. 1997; Boyce et al. 1998; McLure et al. 1999) that have been well resolved by images obtained with WFPC2 on board of HST.

2 The Datasets

We have collected data for BL Lacs and QSOs at $z<0.5$ observed with WFPC2 of HST. All magnitudes are converted into R (Cousins) band; absolute magnitudes are calculated assuming for $H_0=50$ Km s^{-1} Mpc^{-1} and $\Omega_0=0$ and applying uniform k and galactic reddening corrections.

The HST snapshot imaging survey of BL Lacs (Urry et al. 2000, Scarpa et al. 2000) has provided a homogeneous set of 110 short exposure high resolution images through the F702W filter. From this dataset we have considered all objects at $z<0.5$ that are resolved in the HST images. This yields 57 sources with $0.027< z <0.495$ ($< z >=0.2\pm0.1$). For these objects the associated host galaxy morphology is always well described by de Vaucouleurs modelling.

There is not a comparable large set of HST observation for QSOs, therefore we have considered two representative datasets (Hamilton et al. 2000 and Treves et al. 2001) constructed from a collection of various sources reporting QSO images secured by HST. Hamilton et al. 2000 have investigated HST archival images of 71 QSOs (26 RLQs and 45 RQQs) with $M_V \leq$-23 mag and redshift $0.06\leq z \leq0.46$. Treves et al. 2001 have reported on 15 RLQs (in the redshift in the range $0.158< z <0.389$)

collected from the samples of Bahcall et al. 1997, McLure et al. 1999 and Boyce et al. 1998, and homogenized to the sample of BL Lacs (see Treves et al. 2001). The average luminosities of the above samples are summarized in Table 1.

Table 1. Properties of the datasets

Dataset	Reference	N_{obj}	$<z>$	$<M_B(nuc)>$	$<M_R(host)>$
BL Lacs	Urry et al. 2000	57	0.20	-22.3	-23.7
RLQs	Treves et al. 2001	15	0.26	-25.1	-24.3
RQQs	Hamilton et al. 2000	45	0.22	-23.9	-23.8
RLQs	Hamilton et al. 2000	26	0.29	-25.3	-24.8

3   The Host Galaxy Luminosity Function (HGLF)

Assuming the host galaxy luminosity is independent of nuclear luminosity we consider that the present datasets are representative of the general population of host galaxies of the respective classes and therefore apt to produce a rough luminosity function of the host galaxies. To set the normalization factor of HGLF for QSOs we took the value of the QSO luminosity function (Boyle et al. 2000) corresponding to the average value of nuclear magnitude in B band (taking B-R~0.56) and assumed that RLQs are 10% of QSO population (Kellermann et al. 1989) at $M_B(nuc)$=-25.3. For BL Lacs we refer to the FRI luminosity function given by Padovani et al. 1991 and normalized the HGLF at $M_R(host)$=-22.8. We fit the luminosity function of the host galaxies with a modified Schechter function $\Phi = K \times \Phi_S \times (L/L^*)^\beta$, where Φ_S is the Schechter function for elliptical galaxies (Metcalfe et al. 1998): $\Phi_S = \Phi^* \times (L/L^*)^\alpha \times \exp(-L/L^*)$, with $\Phi^* = 8.5 \times 10^{-2}$ Mpc^{-3}, α=-1.2 and L^*=2.25 $\times 10^{44}$ erg s^{-1}. The best fit has been estimated minimizing χ^2 for the function Φ. We find $\beta \sim 3$ for BL Lac and $\beta \sim 5$ for RLQ hosts. The derived HGLFs are given in Fig. 1.

4   Main Conclusions

- The HGLFs of QSOs and BL Lacs are remarkably different in shape from the one of inactive ellipticals and indicates that these AGNs are preferentially drawn from the bright tail of elliptical galaxy luminosity function.
- The HGLFs of RLQs and BL Lacs have similar shape but with a trend for brighter galaxies in RLQs. This is quantified by comparison of the β parameter: $\beta_{RLQ} \sim 5 > \beta_{BLL} \sim 3$.
- There is some indication of a different shape for the HGLFs of RLQs and RQQs with a larger fraction of low luminosity galaxies in RQQs.

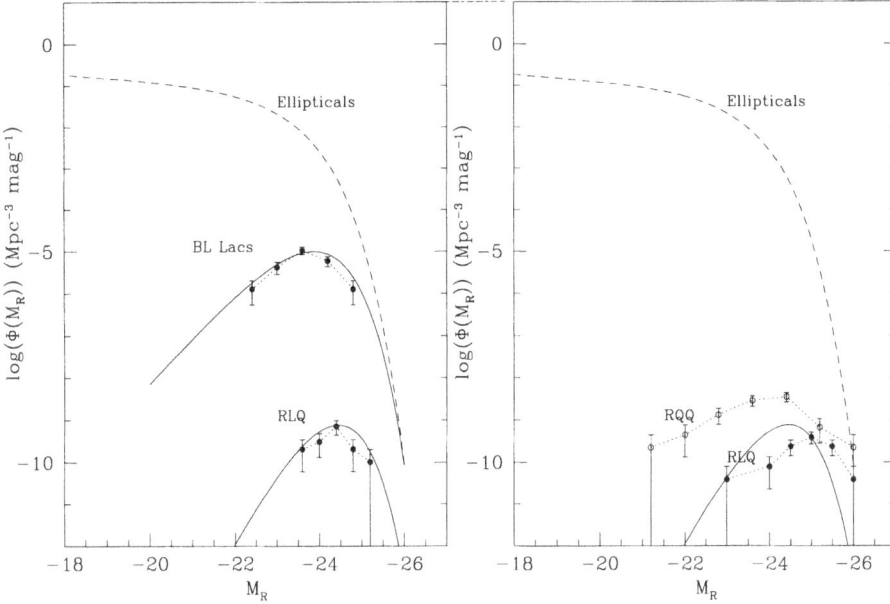

Fig. 1. (**Left**) The HGLF of RLQs (Treves et al. 2001) and BL Lacs (Urry et al. 2000). The *dashed curve* is the elliptical galaxy luminosity function of Metcalfe et al. 1998 and the *solid lines* are the fits of the HGLF of RLQs and BL Lacs. (**Right**) The HGLF of RQQs (empty dot) and RLQs (filled dot) of Hamilton et al. 2000 compared to the luminosity function of Metcalfe et al. (*dashed curve*). The *solid line* is the fit with a modified Schechter function Φ for the HGLF of the Treves et al. (2001) sample

References

1. J.N. Bahcall, S. Kirhakos, D.H. Saxe & D. P. Schneider 1997 ApJ 479, 642
2. P.J. Boyce et al. 1998, MNRAS 298, 121
3. B.J. Boyle, T. Shanks, S.M. Croom, R.J. Smith et al. 2000 MNRAS 317, 1014
4. M.J. Disney, P.J. Boyce et al. 1995 Nature 376, 150
5. T.S. Hamilton, D.A. Turnshek, S. Casertano 2000 astro-ph 0011255
6. K.I. Kellermann, R. Sramek, M. Scamidt, D.B. Shaffer, R. Green 1989 AJ 98, 1195
7. J. M: Kukula, J.S: Dunlop, R.J. McLure et al. 2001 astro-ph 0010007
8. R.J. McLure et al. 1999 MNRAS 308, 377
9. N. Metcalfe, A. Ratcliffe, T: Shanks and R. Fong 1998 MNRAS 294, 147
10. P. Padovani & C. M. Urry 1991 ApJ 368, 373
11. R. Scarpa, C.M. Urry, R. Falomo, J. Pesce & A. Treves 2000 ApJ 532, 740
12. A. Treves, N. Carangelo, R. Falomo 2001 astro-ph 0107129
13. C. M. Urry, R. Scarpa, M. O'Dowd, R. Falomo et al. 2000 ApJ 532, 816
14. C.M. Urry, P. Padovani 1995 PASP 107, 803

AGN at MeV Gamma-Ray Energies

Werner Collmar

Max-Planck-Institut für Extraterrestrische Physik, Postfach 1312, D-85741 Garching, Germany

Abstract. The COMPTEL experiment had explored the, previously unknown, MeV sky for about 9 years. It has detected 11 AGN: 10 blazars and the radio galaxy Centaurus A. No Seyfert galaxy was detected. With these results COMPTEL opened the field of extragalactic γ-ray astronomy at MeV energies.

1 Introduction

Prior to the Compton Gamma-Ray Observatory (CGRO) mission (April 1991 to June 2000) four Active Galactic Nuclei (AGN) had been reported to emit detectable γ-rays at energies above 1 MeV: two Seyfert galaxies and the radio galaxy Centaurus A up to ~20 MeV, and the blazar-type quasar 3C 273 at energies above 50 MeV. Soon after the launch of CGRO in 1991, the EGRET experiment, mainly observing at γ-ray energies above 100 MeV, reported the detection of several radio-loud AGN. With the exception of Cen A, all belong to the blazar subclass. EGRET has detected about 90 of these sources in total ([1]). Several of them are also detected by the COMPTEL experiment aboard CGRO, measuring γ-rays at MeV energies (0.75 - 30 MeV). These source detections and the source properties at MeV energies will be briefly summarized in the following.

2 Summary of COMPTEL AGN Results

COMPTEL has found evidence for 11 AGN emitting at MeV energies: 10 blazars and the radio galaxy Centaurus A. No Seyfert galaxy was 'seen' by COMPTEL. So, only strong jet sources are detected at these energies indicating the dominance of non-thermal processes in these sources at γ-ray energies. The COMPTEL-detected AGN are listed in Table 1. Apart from Markarian (Mkn) 421, a BL Lac object, all blazars belong to the class of flat-spectrum radio quasars (FSRQs). They are often detected during γ-ray-flaring events reported by EGRET at energies above 100 MeV. The AGN MeV luminosities (isotropy assumed; see Table 1) range from ~10^{42} erg/s for the nearby radio galaxy Cen A, ~7×10^{44} erg/s for the BL Lac object Mkn 421, ~10^{47} erg/s for 3C 273, the lowest luminosity FSRQ of the detected sources, up to several 10^{49} erg/s for PKS 0528+134 during γ-ray flaring periods.

No blazar is always detected when it is located favourably (i.e. near the pointing direction) within the COMPTEL field-of-view which suggests time variability for all sources. The shortest variability with a flux change of a factor of ~4 within 10 days was observed from 3C 279 ([2]) during a large γ-ray flaring period in 1996. Because

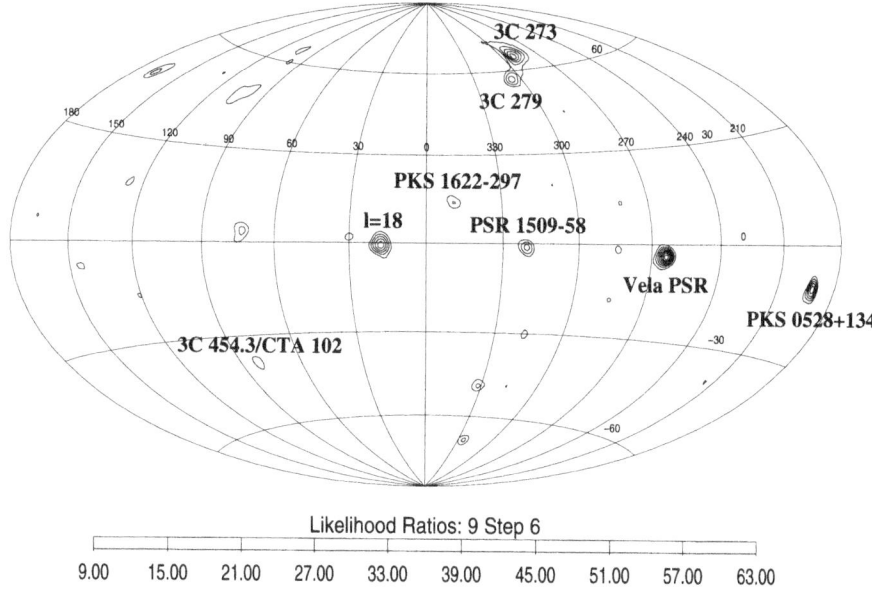

Fig. 1. COMPTEL time-averaged (April 1991 to November 1997) maximum-likelihood point source all-sky map in the 10-30 MeV band. The galactic and extragalactic diffuse emission as well as the emission from the Crab, which is by far the strongest MeV point source, are subtracted off. About half of the significant γ-ray sources are AGN.

Table 1. Summary of the COMPTEL blazar detections at energies between 0.75 and 30 MeV. The list is ordered according to source distance (z). Several parameters are listed. Their meanings are explained in the footnotes.

Source ID	z	Det.[1]	Energy[2] Bands	Spectral Shape	O/E[3] Det.	Type[4] FS,AT,BT	Lum.[5] [10^{47} erg/s]	
Cen A	0.0007	m	yyyy	2.3±0.1	yy	nR.	8.7×10^{-6}	
Mkn 421	0.031	s	nnny	—	ny	yBT	7×10^{-3}	
3C 273	0.158	m	yyyy	2.0±0.1	yy	yQE	0.9	
PKS 1222+216	0.435	s	nnyn	—	yy	yQE	1.4	
3C 279	0.538	m	yyyy	1.8±0.2	yy	yQE	4.5	[1]Detections: s -
PKS 1622-297	0.815	s	nnyy	hard	yy	yQE	16	
3C 454.3	0.859	s	nnny	—	yy	yQE	17	
PKS 0208-512	1.003	m	yyyy	—	ny	yQM	45	
CTA 102	1.037	s	nnny	hard	yy	yQE	22	
GRO J0516-609	1.09	s	nyyn	soft	yy	yQM	48	
PKS 0528+134	2.06	m	yyyy	1.9±0.4	yy	yQE	151	

single, m - multiple; [3]OSSE/EGRET detection

[2]Detection in the 4 standard COMPTEL energy bands: y - detected, n - not detected

[4]Source type: FS: flat radio spectrum; AT (AGN type): Q - quasar, R - radio galaxy, B - BL Lac Object; BT (blazar type): E - EGRET type, M/T MeV/TeV blazar

[5]time-averaged (if possible), isotropic MeV-luminosity for the energy band of detection

the sources are often detected near threshold in only 1 or 2 of the COMPTEL bands, the knowledge on their MeV spectra is limited. Nevertheless, some trends are apparent. The time-averaged spectra are well described by power-law shapes ($E^{-\alpha}$) with a photon index of ~2. However, during flaring periods observed by EGRET above 100 MeV the MeV spectra are usually harder ($\alpha < 2$) indicating that mainly the high-energy (>3 MeV) part of the COMPTEL band is following the flux increase. Combining COMPTEL blazar spectra with the ones from neighbouring energy bands (OSSE: 50 keV to 1 MeV, EGRET: 30 MeV to 20 GeV) shows that for COMPTEL-detected sources often a spectral turnover occurs within or near the MeV band.

One of the main EGRET blazar results is that these sources are strongly variable at γ-ray energies; some sources even show strong flaring activity with flux variations up to a factor of ~100 (e.g. 3C 279, [3]). The cause of such flares is still unknown. To investigate this flaring behaviour in more detail, simultaneous COMPTEL and EGRET spectra for selected periods of low and high fluxes have been combined for different blazars. Spectral variability at MeV energies is observed in PKS 0528+134 and 3C 279, which is correlated to the flux above 100 MeV as observed by EGRET: a hard MeV spectrum ($\alpha_{ph} \sim1.4$) during flaring periods and a soft one ($\alpha_{ph} >2.0$) otherwise (e.g., [4]). A γ-ray flare in the EGRET range of 3C 273 had no counterpart at MeV energies showing that only the high energies were flaring ([5]). In these three cases, PKS 0528+134, 3C 279 and 3C 273, mainly the high-energy part of the γ-ray spectrum increased, suggesting that such flares are predominantely a high-energy γ-ray phenomenon. This might indicate a multicomponent emission scenario for γ-rays in blazars, in the sense that a parameter change in the emitting region preferentially causes flux changes at the higher (e.g. >100 MeV) γ-ray energies.

3   Conclusion

In general, COMPTEL pioneered the MeV band (0.75 - 30 MeV) of the electromagnetic spectrum, showing that interesting phenomena can be studied in this band. A follow-up mission with improved sensitivity would be scientifically very valuable.

References

1. Hartman, R.C., et al. *ApJS* **123**, 79 (1999).
2. Collmar, W., et al., *AIP Conference Proceedings* **410**, 1341 (1997).
3. Wehrle, A.E., et al., *ApJ* **497**, 178 (1998).
4. Collmar, W., et al., *A&A* **328**, 33 (1997).
5. Collmar, W., et al., *A&A* **354**, 513 (2000).

A New Analysis of the BeppoSAX Observation of NGC 5548

Pascal Favre[1,2] and Thierry Courvoisier[1,2]

[1] INTEGRAL Science Data Center, 16 Ch. d'Ecogia, CH-1290 Versoix, Switzerland
[2] Observatoire de Genève, Switzerland

Abstract. We fit the BeppoSAX data of the bright Seyfert 1 galaxy NGC 5548 using an advanced minimization algorithm to obtain a correlation matrix between the model parameters. The matrix allows us to assess whether parameters are univocally measured.
We discuss whether the cut-off and the reflection hump are needed in the data.

1 Introduction

Up to the launch of INTEGRAL, BeppoSAX will remain a unique instrument to investigate the high energy components of AGN spectra, i.e. the reflection component, the iron line and the high energy cut-off.
We concentrate on the MECS-PDS, high energy spectra (3–220 keV) of the bright Seyfert 1 galaxy NGC 5548 (z = 0.017). We used a mean spectrum of the 4 observations made in 1997 (during approximately a week). The 1997 observation (including LECS data that show the presence of warm absorber, edges, etc) has been discussed in Nicastro et al. (2000).

2 Advanced Minimization and the Correlation Matrix

As our models are usually non-linear in the parameters (we will use the notation $\underline{\alpha}$ for the vector of free parameters), we have to use numerical methods to search the parameter space for minimizing the χ^2 function (hereafter $F = \chi^2$). A Taylor expansion

$$F(\underline{\alpha}) = F(\underline{\alpha_0}) + G^{\mathrm{T}}(\underline{\alpha} - \underline{\alpha_0}) + \frac{1}{2}(\underline{\alpha} - \underline{\alpha_0})^{\mathrm{T}} H (\underline{\alpha} - \underline{\alpha_0}) + \dots \qquad (1)$$

introduces the gradients vector G (that tells in which direction to go to minimize F) and the Hessian matrix H (matrix of the second derivatives; it tells if the extremum is a minimum or a maximum). This gives $V = H^{-1}$, the error matrix which has the variance of the parameters $\sigma_{\alpha_i}^2$ on the diagonal and $\varrho_{\alpha_i \alpha_j} \sigma_{\alpha_i} \sigma_{\alpha_j}$ elsewhere, where $\varrho_{\alpha_i \alpha_j}$ is the linear correlation coefficient between α_i and α_j. It is then straightforward to extract the correlation matrix.
We used a variable metric algorithm in which F can be seen as the intrinsic property of a new parameter space with metric $\mathrm{d}s^2 = \mathrm{d}\underline{\alpha}^{\mathrm{T}} H(\underline{\alpha}) \mathrm{d}\underline{\alpha}$. It is the first method to give a good convergence criterion, not based on the gradient. This algorithm is implemented in a program called MINUIT (James, 1998), part of the CERN tools, the variable metric method being available through the function MIGRAD. It has been made available in XSPEC by Keith Arnaud.

3   The Matrix

We used the method described above to analyse the average 3–220 keV spectrum of NGC 5548 observed in 1997. We used a simple, phenomenological model

$$e^{-N_H\sigma}(C_2E^{-\Gamma}e^{-E/E_c} + C_3\frac{1}{\sigma\sqrt{2\pi}}e^{-(E-E_l)^2/2\sigma^2} + C_4\frac{8.0525E^2\mathrm{d}E}{(kT)^4(e^{E/kT}-1)}), \quad (2)$$

where a black-body modelizes the Compton hump (N_H was frozen to the galactic value). The matrix shows 5 significant correlations between the parameters of the model and in particular between the cut-off (E_c) and Γ, between the cut-off and C_2 and between the cut-off and the intercalibration variable C_5 (intercalibration MECS-PDS). The first two are due to the analytic form of the model. The fit led to $C_5 = 1.20$, out of the allowed range 0.77–0.95.

We then used the Compton reflection model of Magdziarz & Zdziarski (1995) to modelize the hump and found the following matrix for the linear correlation coefficients $\varrho_{\alpha_i\alpha_j}$

$$
\begin{array}{l}
\Gamma \\
E_c \\
R \\
C_2 \\
E_l \\
\sigma \\
C_3 \\
C_5
\end{array}
\left(
\begin{array}{cccccccc}
1 & & & & & & & \\
0.431 & 1 & & & & & & \\
0.253 & 0.366 & 1 & & & & & \\
-0.520 & -0.844 & -0.385 & 1 & & & & \\
0.078 & 0.172 & 0.071 & -0.159 & 1 & & & \\
0.139 & 0.084 & 0.097 & -0.359 & 0.067 & 1 & & \\
-0.267 & -0.321 & -0.118 & 0.329 & -0.061 & -0.172 & 1 & \\
0.036 & 0.005 & 0.166 & -0.007 & -0.029 & 0.052 & 0.122 & 1
\end{array}
\right) \quad (3)
$$

where $R = \Omega/2\pi$ expresses the amount of reflection.

Here the cut-off (second column) is significantly correlated with 4 other parameters. Therefore, the fitting process does not allow us to determine the cut-off or any of the 4 other parameters univocally, but a linear combination of the parameters : E_c, Γ, R, C_2, C_3

Here $C_5 = 0.99$, closer to the recommended value.

4   Discussion

We investigate if the cut-off is needed when the curvature induced by a reflection component is included in the model. This is an important discussion in the context of reprocessing since this parameter gives the temperature kT_e of the scattering electrons of the corona.

A simple model composed of a power-law plus gaussian line (absorption fixed at the galactic value in what follows) is not statistically acceptable (χ^2_ν=1.53(73)) when applied to the data. When an exponential cut-off is added to the power-law, the fit is improved but remains a poor fit (χ^2_ν=1.40(72) with F-test probability of 7.47 10^{-3}). In the context of a reflection model, when curvature is added to the spectrum we tested if this was still the case. We used the Compton reflection model of Magdziarz & Zdziarski (1995), without cut-off in the incident spectrum (χ^2_ν=1.23(72)), and then added one (χ^2_ν=0.84(71), F-test : $1.16\,10^{-7}$). In conclusion, the reflection hump and the cut-off are needed by the data. In the previous fits, we checked that the intercalibration variable was in the range 0.77–0.95 as recommended. If it was above or below, we froze it to the corresponding limit.

The results are quite different if we let that variable free to vary : the cut-off is really needed in the data ; the F-test gives $4.49\,10^{-9}$ when adding a cut-off to the power-law and $1.5\,10^{-6}$ when adding it to the Compton reflection model, but the existence of the Compton hump is doubtful. We show below the residuals of the fit without a reflection hump (just a power-law with cut-off plus a gaussian and absorption, as above) followed on the right panel by the residuals of a fit with the Compton reflection model with cut-off (Magdziarz & Zdziarski, 1995) plus a gaussian and absorption. Comparison of the structure of the residuals on both graphs does not show evidence of the existence of a component between 15 and 40 keV.

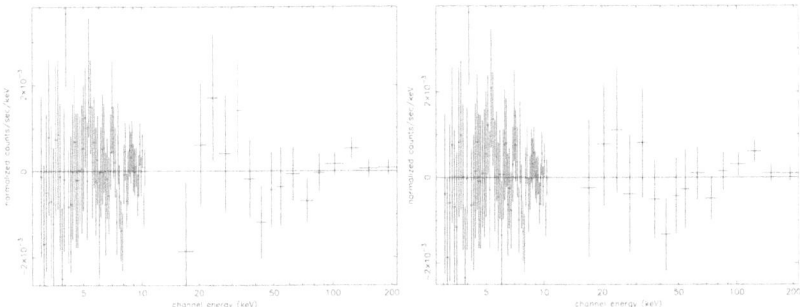

We conclude that the cut-off is really needed in the data when the intercalibration variable is let free to vary as well as when it's not, but in addition to the reflection hump in the last case. It's not measured univocally, but a linear combination of parameters is, for two different hump models. The reflection hump is however not needed by the data if the intercalibration variable is considered a free parameter.

References

1 Arnaud K.A., 1996, ASP Conf. Series Vol. 101, 17
2 James J., 1998, CERN Program Library, MINUIT Reference Manual
3 Magdziarz P., Zdziarski A.A., 1995, MNRAS, 273, 837
4 Nicastro F.,Piro L.,De Rosa A. et al., 2000, ApJ, 536, 718

Study of the BL Lac Object Markarian 501

Marcello Giroletti[1,2], Gabriele Giovannini[1,3], and Philip G. Edwards[4]

[1] Istituto di Radioastronomia del CNR, via Gobetti 101, I-40129 Bologna
[2] Dipartimento di Astronomia dell'Università di Bologna, via Ranzani 1, I-40127
[3] Dipartimento di Fisica dell'Università di Bologna, via Berti Pichat 6/2, I-40127
[4] Institute of Space and Astronautical Science, 3-1-1 Yoshinodai, Sagamihara, Kanagawa 229-8510, Japan

Abstract. Due to its low redshift, Markarian 501 is an ideal target to study parsec scale properties of BL Lac objects. Based on multi epoch, multi frequency VLBI and VSOP observations, we present milliarcsecond resolution images of this TeV BL Lac. Inner jet morphology, proper motion and spectral index are discussed in the light of unified models.

1 Introduction

The BL Lac object Markarian 501 is one of the most remarkable sources in the sky at all wavelenghts. Detected even at TeV energies, it's the second closest BL Lac and the closest between those with radio flux density greater than 1 Jy. Its redshift is $z = 0.03366$; at this distance, assuming $H_0 = 65\,\mathrm{km s^{-1}\,Mpc^{-1}}$ and $q_0 = 0.5$, 1 mas$= 0.71\,\mathrm{pc}$. To perform a parsec scale study, we have observed this source with Space VLBI at closely separated epochs to investigate the inner jet region and possibly detect proper motion of the components.

2 Observations: VSOP and RRFID

Our data were obtained at epochs 1997.8, 1998.4 and 2001.3 from space VLBI observations. The HALCA orbiting 8 m antenna was used in conjunction with a ground array consisting of the VLBA at 5 GHz and 1.6 GHz. In some observations the Robledo and Goldstone DSN telescopes were also used.

In order to densely cover the time interval between 1996 and 1998, we also collected data from the Radio Reference Frame Image Database at 8.4 GHz and 15 GHz. Table 1 lists the six observations spanning this time interval that we have considered to date.

Table 1. List of observations considered for proper motion

epoch	observation	frequency	epoch	observation	frequency
1995.4	RRFID	8.4 GHz	1997.8	VSOP	1.6 GHz
1996.4	RRFID	15 GHz	1998.4	VSOP	5 GHz
1996.6	VLBA	8 GHz	1998.6	RRFID	8.4 GHz

Moreover we studied the source spectral index with the new VSOP observations, taken in 2001 March 5 at 1.6 GHz and March 6 at 5 GHz. Because of the large time gap elapsed since the previous observations, we did not use here these last images to study the proper motion.

3 Radio Images

Figure 1 shows the extended morphology of the source as seen from the 1.6 GHz VSOP observation of March 2001.

While the dramatic bend in the jet direction at about 20 mas from the core is in agreement with the presence of projection effects, the study of the six 1995-1998 images reveals also a proper motion in the inner jet region that allows to estimate the angle θ and the velocity v.

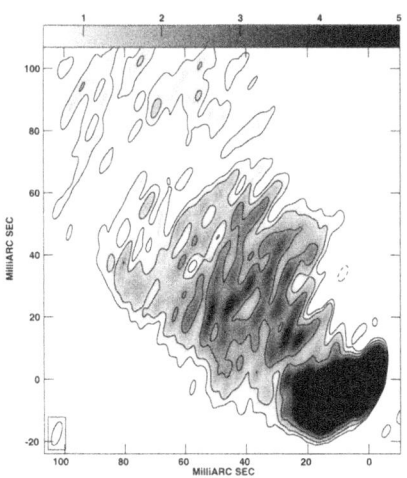

Fig. 1. 1.6 GHz VSOP image: lowest contour is 0.5 mJy/beam, peak is 465 mJy/beam

By means of using different values for the ROBUST parameter in the AIPS task IMAGR, we could reach an homogeneous resolution of 1.2 mas for five of the six observations, the only exception being the 1997 1.6 GHz VSOP observation, which has a beam of 2.9 mas x 1.5 mas at PA $-9°$. We nevertheless included this image along with the others as the motion should be approximately in the direction of the beam minor axis.

The closest component in the first image can be seen in all of the following and seems to move with apparent speed $v_{app} = 2c$. Since the jet/counterjet ratio sets a constrain on $\beta \cos\theta > 0.87$, we therefore find $\beta > 0.92$ and $\theta < 16°$, in agreement with values expected for the unification of BL Lac objects and FR I radio galaxies. Given this jet velocity and orientation, we estimate a Doppler factor $\delta > 4$. If we correct the core and jet radio power for the Doppler boosting and estimate the intrinsic

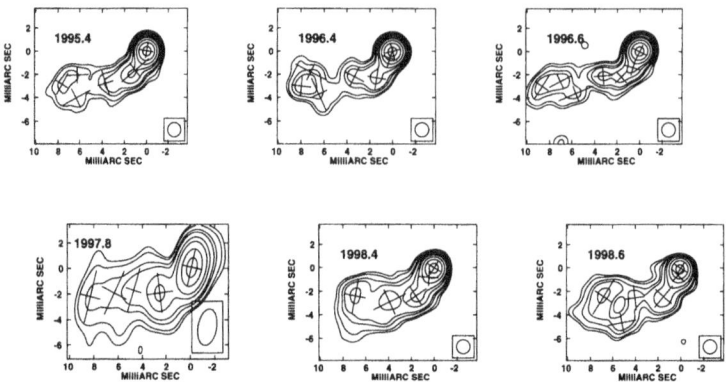

Fig. 2. Inner jet of Mkn 501: lowest contour level is 2.5 mJy/beam in the highest resolution images and 10 mJy/beam in the 1.6 GHz image (bottom left). Crosses indicate position and extension of Gaussian components as fitted by AIPS task JMFIT

values, we find typical values of FR I sources. That is, if we saw Mkn 501 lying in the plane of the sky, its total, jet and core radio power would be the typical one of FR I radio galaxies.

4   Spectral Index Map

Using VSOP observations at 1.6 GHz and 5 GHz taken at the same epoch (only one day apart from each other in March 2001), it was possible to study the spectral index α (defined such that $S(\nu) \propto \nu^{\alpha}$). Positions of a strong and not self absorbed component in the jet were aligned before combining the two images.

While the milliarcsecond core seems to have an inverted spectrum and not much can be said about the first few mas of the jet, we then find a narrow flat spectrum region lying between two steeper limbs, in agreement with a change in the jet morphology.

A small spot with inverted spectrum is found on the south-east edge and could either be due to errors (larger in regions farther from the core) or be related to the bending of the jet. From that point on, where the jet broadens and gets fainter, the spectrum becomes quite steep, with $-1.5 < \alpha < -0.5$.

Acknowledgements

We gratefully acknowledge the VSOP Project, which is led by ISAS in cooperation with many organizations and radio telescopes around the world. Also, this research has made use of the USNO Radio Reference Frame Image Database.

M.G. thanks Denise Gabuzda for much helpful advice in the reduction of the 2001 VSOP data and JIVE for support and warm hospitality.

Nonlinear Analysis of AGN Light Curves

Mario Gliozzi[1], Wolfgang Brinkmann[2], and Christoph Räth[2]

[1] MPE (Garching),
[2] Centre for Interdisciplinary Plasma Science, MPE (Garching)

Abstract. A timing analysis can provide useful information to characterize the central engines of AGN. We use the data from a recent long ASCA observation of the Narrow Line Seyfert 1 galaxy Ark 564 to investigate in detail its timing properties. The main result is that the the process causing the variability is very likely to be linear. Using a nonlinear statistic based on the scaling index as a tool to discriminate time series, we demonstrate that the high and low count rate states are intrinsically different, with the high state characterized by higher complexity.

1 Nonlinearity

Active Galactic Nuclei (AGN) are variable in every observable wave band. The X-ray flux exhibits variability on time scales shorter than any other energy band, indicating that the emission occurs in the innermost regions of the central engine. Therefore, a study of the X-ray variability is a powerful tool to probe the extreme physical processes operating in the accretion flow close to the accreting black hole. Narrow-Line Seyfert 1 galaxies (NLS1) often display rapid, large amplitude X-ray variability as well as extreme long-term changes ([1], [2], [3]), and therefore represent ideal objects for an X-ray temporal analysis. Although X-ray variability has been observed in AGN for more than two decades, its origin and nature is still poorly understood.

One of the most critical open questions related to this variability concerns the nature of process causing the variability: is it linear or nonlinear? A positive detection of nonlinearity in the light curve would immediately rule out all models explaining the variability in terms of a superposition of independent emission regions. Various mathematical approaches have been taken to search for nonlinearity in time series; one of the best ways is to apply the method of surrogate data [7]: one creates a number of surrogate data sets with the same linear properties as the measured data and compare them using nonlinear statistics. If the statistical properties of the real time series are significantly different from those of the surrogate data, nonlinearity is inferred. However, the standard surrogate method is not suited for unevenly sampled satellite data, because it utilizes Fourier transformations and their inverses. We have applied a more general surrogate method (where the desired linear properties of surrogates are imposed by constraints; [6]) to a long light curve of the NLS1 Ark 564, observed by ASCA from 2000 June 1 to 2000 July 5, taking into account the uneven sampling and the possible nonstationarity. We find no clear evidence for the presence of nonlinearity in the time series.

2 Nonlinear Analysis

Linear methods assume intrinsically that the dynamics of the system is governed by the linear paradigm: small causes lead to small effects. Since linear equations can only lead to exponentially growing (or decaying) solutions or to periodic oscillations, an irregular behavior of the system has to be attributed to some random external input. However, nonlinear, chaotic systems can produce very irregular data from deterministic equations of motion. In other words, the apparently irregular and unpredictable behavior typical of X-ray light curves of AGN is not necessarily due to stochastic dynamics, i.e the action of a large number of excited degrees of freedom. It might be generated by chaotic dynamics of a limited number of collective modes: dissipation, coupling between different degrees of freedom, and the action of an external field may lead to a collective behavior. When methods from "nonlinear dynamics" are applied to time series analysis, the concept of phase space reconstruction represents the basis for most of the analysis tools. In fact, for studying chaotic deterministic systems it is important to establish a vector space such that specifying a point in this space specifies the state of the system, and vice versa.

We have applied the phase space reconstruction method to the Ark 564 light curve, converting the time series (i.e. a scalar sequence of measurements) into a set of state vectors. In particular, we have selected two intervals of the light curve, which are locally stationary, contain a sufficiently large number of data points (10000, using a time bin size of 16 s) for a meaningful statistical analysis. The two data segments have significantly different mean count rates in order to characterize the intrinsic temporal differences from the low and the high state of the source.

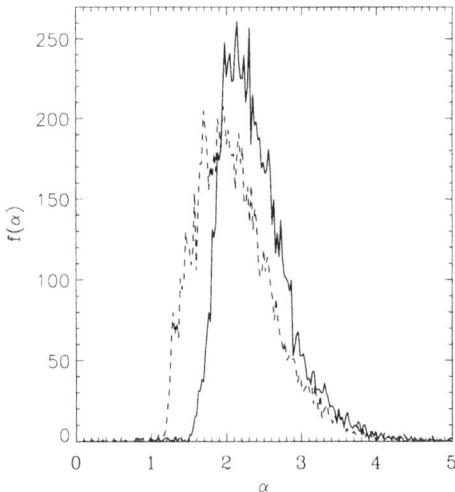

Fig. 1. Spectrum of the scaling index distribution of Ark 564 during the high (solid line) and low (dashed line) states.

For quantifying the difference between the phase space distributions in the high and low count rate states, a suitable quantity is the correlation integral $C(r)$, which basically counts the number of pairs of points with distances smaller than r. This algorithm was introduced by Grassberger & Procaccia [4] and was first used in astronomy to analyze light curves of variable stars and binary systems, in order to put constraints on the origin of the irregular variability. An exhaustive description of the correlation integral method and its application to X–ray light curves of AGN is given by Lehto and collaborators [5]. An important property is that local estimates of $C(r)$ at small r behave as a power law and the exponent, the scaling index α, is closely related to the correlation dimension $D^{(2)}$, which provides an estimate of the number of the degree of freedom excited for deterministic dynamical systems. More generally, it gives an indication of the "complexity" of the system.

We have calculated the scaling index for all the points of the two selected parts in the light curve using an embedding space of dimension four and radii corresponding to a region in the $\log C(r) - \log r$ plane, where the slope of the correlation integral stays constant. The resulting histograms are plotted in Fig. 1. The low and high states of Ark 564, which share similar properties (scaling and time correlation properties, as well as the probability density function), show significant differences on the basis of a nonlinear analysis, with the high state characterized by a higher complexity (possibly due to the action of a larger number of excited degrees of freedom) with respect to the low state.

Acknowledgements

We would like to thank Herbert Scheingraber for his advices and help to get accustomed with the nonlinear analysis of time series.

References

1. Th. Boller, W.N. Brandt, A.C. Fabian, H.H. Fink, 1997, MNRAS 289, 393
2. W.N. Brandt, Th. Boller,A.C. Fabian, M. Ruszkowski, 1999, MNRAS 303, L53
3. K. Forster , J.P. Halpern, 1996, ApJ 468, 565
4. P. Grassberger, I. Procaccia, 1983, Phys Rev. Letters 51, 346
5. H.J. Lehto, B. Czerny, I.M. McHardy, 1993, MNRAS 261, 125
6. A. Schmitz, T. Schreiber, 1999, Phys. Rev. E 59, 4, 4044
7. J. Theiler, S. Eubank , A. Longtin, B. Galdrikian, J. Farmer, 1992, Physica D 58, 77

X-Ray Transient Galaxies and AGN

Dirk Grupe

MPI für extraterrestrische Physik, PO Box 1312, D-85741 Garching, Germany

Abstract. X-ray transience is the most extreme form of variability observed in AGN or normal in-active galaxies. While factors of 2-3 on timescales of days to years are quite commen among AGN, X-ray transients appear only once and vanish from the X-ray sky years later. The ROSAT All-Sky Survey was the tool to discover these sources. X-ray transience in AGN or galaxies can be caused by dramatic changes in the accretion rate of the central black hole or by changes of the properties of the accretion disk.

1 Introduction

Our sample of soft X-ray AGN contains 113 sources selected from the ROSAT All-Sky Survey (RASS, [12]) by the PSPC count rates CR > 0.5 cts s^{-1} and the hardness ratio HR <0.0. Pointed PSPC and HRI observations are available for 60 and 50 sources, respectively. All in all, for more than 80 sources at least one pointed observation is available [7]. In this way we have a tool to search for long-term large amplitude variations. Fig. 1 displays the RASS vs. pointed observation count rates. HRI count rates have been converted into PSPC count rates assuming no spectral change between both observations. The solid line marks no change, the short-dashed line a change by a factor of 10 and the long-dashed line by a factor of 100 between the RASS and the pointed observation. Four sources turned out to vary by factors of almost 100 or even more: **WPVS 007, IC 3599, RX J1624.9+7554**, and **RX J2217.9–5941**. The first three are X-ray transients while RX J2217.9–5941 is a possible X-ray transient candidate.

2 X-ray Transient Sources

2.1 WPVS 007

The Narrow-Line Seyfert 1 galaxy (NLS1) WPVS 007 was *the* softest AGN observed during the RASS [5] and had a mean count rate of about 1 cts s^{-1}. In later pointings the source shows only the flux expected from a normal inactive galaxy. A possible scenario to explain this dramatic turn-off is a temperature change in the Comptonization layer of the accretion disk that shifts the soft X-ray photons out of the ROSAT energy window.

2.2 IC 3599

The Seyfert 2 galaxy IC 3599 has shown an X-ray outburst during the RASS followed by a response in its optical emission lines ([4], [2]). A possible explanation of this X-ray outburst is an accretion event either caused by an instability of the accretion disk or by a tidal disruption of a star orbiting around the central black hole.

2.3   RX J1624.9+7554

RX J1624.9+7554 has shown a dramatic derease of its X-ray flux by at least a factor of more than 200 between the RASS and a pointed observation 1.5 years later (Grupe et al. 1999). Optical spectroscopy identified this source as a normal in-active galaxy. The most plausible explanation for this X-ray event is the tidal disruption of a star by the central black hole. Other in-active galaxies in which an X-ray outburst have occured are NGC 5905, RX J1242.6–1119 ([10], [9]; see also S. Komossa's contribution in these proceedings), and RX J1420.4+5334 [3].

2.4   RX J2217.9–5941

The NLS1 RX J2217.9–5941 is a possible X-ray transient candidate. It is highly variable on time scales of days as well as years [8]. During its two-day RASS observation the count rate decrease by a factor of 15. Observed several times in pointed observations by ROSAT and ASCA the source has become fainter over the years. It is not clear yet if this source will become a transient. However, due to the black hole mass of $\approx 10^8 M_\odot$ the timescales are larger than in e.g. IC 3599 ($M_{BH} \approx 10^6 M_\odot$).

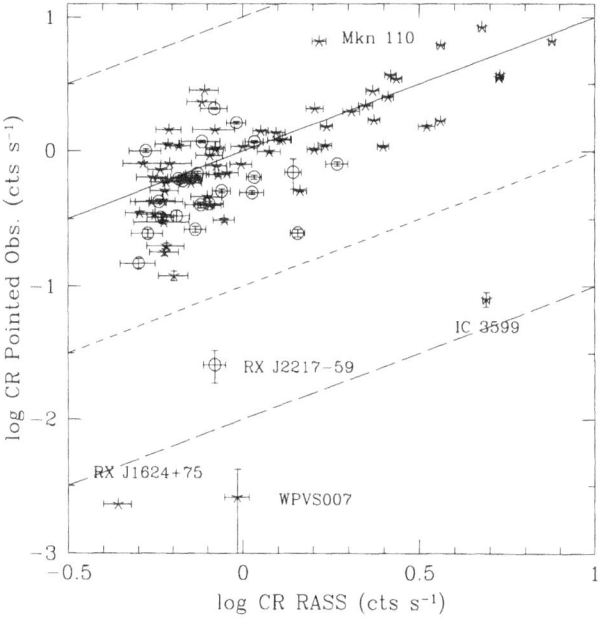

Fig. 1. RASS vs. pointed observation count rates

3 Discussion

The nature of X-ray transient sources is of different origin. While in WPVS007 the question is why its X-ray source is off right now, the question for IC 3599 and RX J1624.9+7554 is what was the reason for the outburst. The X-ray flux seen in WPVS007 during the RASS is in good agreement what is expected from the mean optical to X-ray flux ratio (see [1]). A possible explanation is a temperature change in the Comptonization layer above the disk that shifts the UV photons into the soft X-ray range [5] and caused that the observed spectrum is shifted out of the ROSAT PSPC energy window. The outburst in IC3599 and RX J1624.9+7554 can be explained by accretion events, e.g. the tidal disruption of a star that comes to close to the central black hole, a scenario that has been suggested by e.g. [11]. Such X-ray outburst events are rare. However, performing new soft X-ray surveys in order to find more of these x-ray transient sources. Quick follow-up observations in the optical and in X-rays would provide as with a powerful tool to map the inner region of an AGN or galaxy while the light front is passing through the inner region of the AGN.

References

 1. Beuermann,K., Thomas, H.-C., Reinsch, K., et al., 1999, A&A 347, 47
 2. Brandt, W.N., Pounds, K., Fink, H.H., 1995, MNRAS 273, L47
 3. Greiner, J., Schwarz, R., Zharikov, S., Orio, M., 2000, A&A 362, L25
 4. Grupe, D., Beuermann, K., Mannheim, K., et al., 1995a, A&A 299, L5
 5. Grupe, D., Beuermann, K., Mannheim, K., et al., 1995b, A&A 300, L21
 6. Grupe, D., Thomas, H.-C., Leighly, K.M., 1999, A&A 350, L31
 7. Grupe, D., Thomas, H.-C., Beuermann, K., 2001a, A&A 367, 470
 8. Grupe, D., Thomas, H.-C., Leighly, K.M., 2001b, A&A 369, 450
 9. Komossa & Greiner, 1999, A&A 349, L45
10. Komossa, S., Bade, N., 1999, A&A 343, 775
11. Rees, 1990, Science 247, 817
12. Voges, W., Aschenbach, B., Boller, Th., et al., 1999, A&A 349, 389

Do Quasars Cluster on Small Scales ($\lesssim 15h^{-1}$ Mpc)?

Leopoldo Infante[1], Mariano Moles[2], and Jesus Varela[2]

[1] Pontificia Unversidad Católica de Chile, Santiago, Chile
[2] Consejo Superior de Investigaciones Científicas, España

Abstract. We carry out a photometric search for bright ($V < 23.5$) Lyα emitters on scales $\lesssim 15h^{-1}$ Mpc, of a sample of 48 $z \approx 3$ known QSOs. Wide and narrow band filter color-magnitude diagrams were generated for each of the $6'.6 \times 6'.6$ fields. Non resolved objects with a significant color excess were considered as QSO candidates at a redshift similar to that of the target. Candidates within 2 magnitudes of the central one were detected in 17 fields, or 35% of the surveyed fields. In most cases more than one candidate was detected, indicating that high redshift QSOs could frequently form small groups. A number of resolved objects, candidates to be Lyα emitters at the redshift of the QSOs have been also detected.

1 Introduction

Whatever the mechanisms to form and feed QSOs is, they must be associated with matter density enhancements and strong dynamical activity. It is well known that $z \leq 0.6$ QSOs are not frequently found in clusters, and very rarely in dense groups (Kombeg and Lukash 1994, MNRAS 269, 277). However, at high-z, at the time of cluster formation, the situation could be different. Quasars could be associated either with massive primordial galaxies, or could be the result of violent galaxy interactions, more frequent in the central parts of massive forming clusters. In either case, high redshift QSOs would tend to appear in pairs or small groups, particularly in the vicinity of over-dense regions. In this paper, the hypothesis whether high redshift QSOs would preferentially appear in small groups or pairs, associated with massive, young cluster is tested

2 Observations and Reductions

In order to detect Lyα emitters in the vicinity of known Quasars we use a combination of a narrow filter, n (HeII/3000, HeII/6500, [OIII], [OIII]/3000 & [OIII]/6000), to cover the wavelength of the redshifted ($z \approx 3$) Lyα line, and a wide, w (g_{Gunn}&F_{485}) filter, to measure the continuum. If we set the zero point of the ($w - n$) color index as that corresponding to pure continuum objects, any significant, positive color excess will be used to identify the Lyα Emission Candidates (LEC).

We selected all known QSOs in the southern hemisphere with redshift within the ranges $2.863 \leq z \leq 2.959$, and $3.090 \leq z \leq 3.222$. There are 27 and 36 objects respectively, which also have RA between 21h and 5h. Observations were performed with FORS1 at ANTU VLT on 22 and 23 October, 2000. Seeing ranged between 0.4 and 1 *arcsecs*. For each QSO we acquired images with a narrow band filter (3×200 seconds) covering the wavelength of the redshifted Lyα line, as well as a broader band filter (3×20 sec.) to get the $(w - n)$ color index. Image detection, photometry and classification were performed using the Source Extractor (SExtractor) software program. Magnitude zero points were determined for each field from the known QSO V magnitude. Typically the 1σ error amounts to $0.18\ mag$ at $V = 23.5$. In Fig.1 we explain how the detection threshold was determined.

3   QSO Candidates

In turn, results are listed according to QSO candidate rank. Rank 3 candidates are the strongest cases.

(1) Within the magnitude limit, $V \leq 23.5$, the total number of candidates with a detected color excess at the 3.5σ level is 68, in 31 fields; 18 fields have more than 1 candidate. (2)To pick stronger cases we considered only the candidates within 2 magnitudes of the central QSO. We find 28 candidates in 16 different fields, representing 33% of all the observed cases.(3)In 7 fields, candidates are as bright or brighter than the central QSO. In three fields there are 3 candidates, and 2 candidates in other 3 cases. Only in one of those 7 fields there is a unique candidate.

4   Discussion and Conclusions

Adopting a cosmological model with $H_0 = 65$ km s^{-1} Mpc^{-1}, $\Omega_m = 0.2$, $\Omega_\Lambda = 0.8$, and $\Omega_k = 0$, the size of the field covered by our frames is 14.6×14.6 Mpc^2 (co-moving) $i.e.$, about 1.5 times the scale length of the clustering of QSOs found by Croom $et\ al.$(2001, astro-ph/0105399). In the radial direction, the redshift range covered by our narrow filters is $\sim$50 Mpc.

In most cases, we detect more than 1 candidate; indicating that QSO groups are present in a significant fraction of the cases. The number of detected pairs in the 2dF (Croom et al. 2000), in volumes similar to that surveyed here, is significantly lower, particularly the number of triplets or quartets. It is clear that spectroscopic follow-up is needed, since, if confirmed, the clustering tendency of QSOs at high-z indicated by our data could be most relevant to understand galaxy formation.

We conclude that QSOs at z$\approx$3 tend to have companions in volumes similar to that of a (present day) supercluster. Triplets and Qaurtets are far more frequent than pairs. And, the distribution of distances between the targeted QSO and the companions is not random, but present a higher density at relatively short distances (Fig. 2).

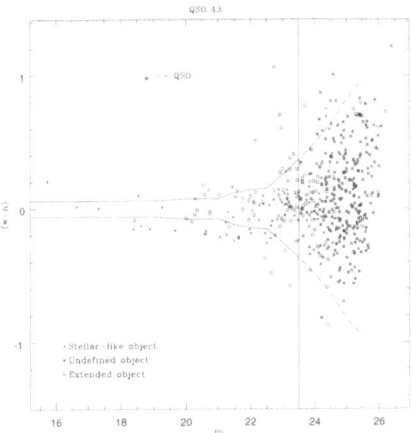

Fig. 1. A typical Color magnitude diagrams. The solid lines are the average detection threshold. The threshold to accept a non-zero color index was established as follows: First, we consider the with of the zero color sequence that would be produced by any slope of the continuum. From simulations this was fixed to 0.02 mag. Then, we added to the above, in quadrature, the 3.5 σ uncertainty associated with the measured color index; $(w - n)$, was plotted against the wide filter magnitude, w. Any object with a color index above that threshold, and having $V \le 23.5$ mag, was kept as a LEC. Objects with negative color indices were also selected as Lyα absorbers (LAC). Among them, the stellar-like (non-resolved) would be QSO candidates at the same redshift.

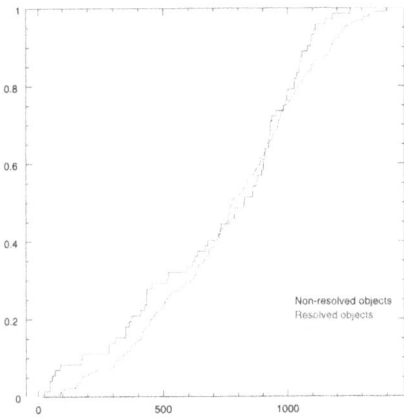

Fig. 2. Cumulative histogram on radial distances for resolved (red), non-resolved objects (blue) and for a simulation of random distribution (green). It is clear the excess of none-resolved objects at small radial distances from the central QSO as compared to a random distribution of points.

XMM-Newton Study of the ULIRG NGC 6240

Ralf Keil[1], Thomas Boller[1], and Ryuichi Fujimoto[2]

[1] Max-Planck-Institut für extraterrestrische Physik,
   Postfach 1312, 85741 Garching, Germany
[2] Institute of Space and Astronautical Science,
   3-1-1 Yoshinodai, Sagamihara, Kanagawa 229-8510, Japan

Abstract. A recently performed *XMM-Newton* observation of the ULIRG NGC 6240 clearly indicates the presence of an AGN contribution to its X-ray spectrum. In the $5.0 - 7.0$ keV energy range there is a clear signature of the fluorescent Fe K α lines at 6.4, 6.7 and 6.9 keV, respectively. The line strength of the 6.4 keV line cannot be produced by a thermal component.

The $0.3 - 10.0$ keV spectral energy distribution is characterized by the following components: (I) two hot thermal components (the starburst), (II) one direct component (heavily absorbed; AGN is hidden), (III) one reflection component (the AGN), (IV) three narrow Fe lines. The model parameters for the broad-band spectral energy distribution are consistent with the results of previously works.

1 Introduction

Many if not all high-luminous infrared galaxies (ULIRGs, $L_{FIR} > 10^{12} L_{\odot}$) possess regions hidden by huge amounts of dust. This makes it difficult to ascertain whether this enormous energy output is due to a starburst activity or an accretion process onto a supermassive black hole. One of the best known objects to study this relationship is the nearby ULIRG NGC 6240 (assuming $H_0 \leq 65$ km s^{-1} Mpc^{-1}). Infrared observations favour an energy source dominated by starburst processes, whereas observations in the X-ray range point to an AGN as the central engine ($L_X \sim 10^{11} L_{\odot}$).

2 Spectral Analysis

We have analyzed the data of NGC 6240 taken from an 24 ksec observation with *XMM-Newton* using the EPIC-PN and EPIC-MOS instruments. In order to investigate the Fe line complex around 6.4 keV and the 0.3 - 10.0 keV spectrum as a whole the high sensitivity and therefore the good photon statistics - especially in the 6.4 keV range - in combination with a higher energy resolution enables us to examine this feature in unprecedented detail.

2.1 The Fe K α Line Complex

Table 1 summarizes some basic parameters (powerlaw - Γ, line energies) of different models (first column) after fitting to the data. The first of the leading three models includes line profiles with no line width ($\sigma = 0$), whereas eachone of the last two models uses a second powerlaw, but with a different number of line profiles. Each model contains a 6.4 keV line as an indication of an AGN contribution. A prove of an Compton-thick AGN has been reported by [3] using BeppoSax and by [1] using RXTE. However, the last model seems to have the best statistical acceptance (see Fig. 1, left).

Table 1. Spectral fitting results to the Fe line complex

Emission lines	powerlaw Γ	gaussian lines [in keV]			$\chi^2/$ d.o.f.
		Energy-line 1	Energy-line 2	Energy-line 3	
lines : $\sigma = 0$	-0.18	6.40 (0.01)	6.67 (0.02)	6.98 (0.04)	38.5/53
lines : $\sigma \neq 0$	-0.16	6.40 (0.02)	6.67 (0.03)	6.97 (0.04)	38.4/51
lines : 2nd broad	-0.27	6.40 (0.03)	6.60 (0.11)	6.66 (0.06)	43.1/53
Emission lines + absorp. edge:	powerlaw Γ	gaussian lines [in keV]			$\chi^2/$ d.o.f.
		Energy-line 1	Energy-line 2	Energy-line 3	
po + 2 lines	0.47	6.39 (0.02)	6.65 (0.04)	-	39.7/54
po + 3 lines	0.47	6.40 (0.02)	6.65 (0.04)	7.05 (0.02)	39.1/54

2.2 Models to the Overall X-ray Spectrum

The analysis of the spectral data $(0.3-10.0\,\mathrm{keV})$ indicates at least two models providing an statistically acceptable fit: Each of them contains two thin thermal plasmas (kT $\approx$ 0.6 keV and kT $\approx$ 1.1 keV), a direct component (absorbed powerlaw with $\Gamma = 1.8$ and $N_H = 2.18 \cdot 10^{24}\,\mathrm{cm}^{-2}$, both fixed) as well as a reflection component (absorbed powerlaw, either reflected from neutral matter or not). Finally, three gaussian lines have been added to the models (neutral + ionized K α and K β). The right plot of Fig. 1 shows the components of the second model (incl. reflection) and their deviations from the data points.

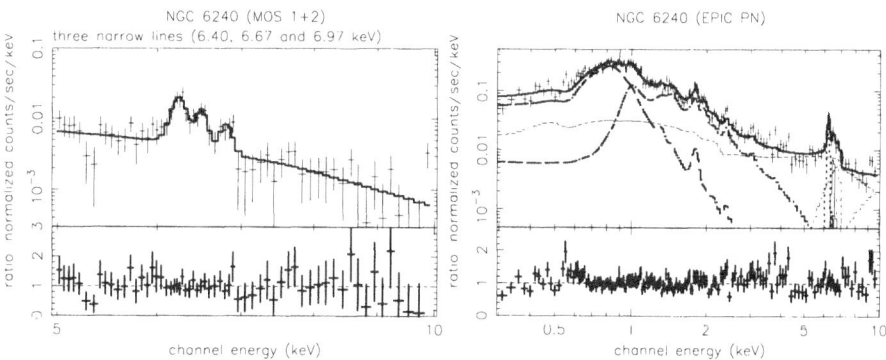

Fig. 1. The Fe K line complex and the overall spectrum

The model parameters for the broad-band spectral energy distribution is consistent with the results previously reported by [2], except for the Fe line complex.

References

1. Y. Ikebe, K. Leighly, Y. Tanaka, T. Nakagawa, Y. Terashima, S. Komossa: MNRAS **316**, 433 (2000)
2. K. Iwasawa, A. Comastri: MNRAS **297**, 1219 (1998)
3. P. Vignati, S. Molendi, G. Matt, M. Guainazzi, L. A. Antonelli, L. Bassani, W. N. Brandt, A. C. Fabian, K. Iwasawa, R. Maiolino, G. Malaguti, A. Marconi, G. C. Perola: A&A **349**, L 57 (1999)

High Energy Properties of the γ-Ray Blazars PKS 1622-297, 3C 454.3 and CTA 102

Shu Zhang[1,2], Werner Collmar[1], and Volker Schönfelder[1]

[1] Max-Planck-Institut für Extraterrestrische Physik Postfach 1603, D-85740 Garching, Germany

[2] High Energy Astrophysics Lab Institute of High Energy Physics P.O.Box 918-3, Beijing 100039, China

Abstract. The MeV-properties of the blazars PKS 1622-297, 3C 454.3 and CTA 102 have been analyzed using COMPTEL data (0.75 - 30 MeV). Our results show significant detections of PKS 1622-297 (5.9σ) in the 10-30 MeV band during its γ-ray flare in 1995, and of 3C 454.3 (5.6σ) in the sum of 8 years observations in the 3-10 MeV band. By comparing the MeV results to the ones obtained in neighboring bands by the Compton Gamma-Ray Observatory (CGRO) experiments EGRET and OSSE, we reveal some properties of these 3 blazars at γ-ray energies.

1 Introduction

The blazars PKS 1622-297, 3C 454.3 and CTA 102 were firstly detected as γ-ray sources by EGRET/CGRO [1–3]. We have consistently analyzed the COMPTEL data (0.75-30 MeV) on these 3 blazars. In our analyses we selected 28 CGRO viewing periods (VPs) for PKS 1622-297 between '91 and '97, and 22 VPs for 3C 454.3 and CTA 102 between '91 and '99. The analyses were carried out using the standard COMPTEL maximum-likelihood method including a filtering technique for background generation and including models to subtract the galactic and extra-galactic diffuse γ-ray emission.

2 Results

2.1 PKS 1622-297

We defined two states for PKS 1622-297 according to the EGRET detections above 100 MeV [3]: a flare state (June 6 - July 10, 1995) and a quiescent state (the rest). In the flare state is PKS 1622-297 detected with $\sim 6\sigma$ in the 10-30 MeV band. Below 10 MeV only a hint (3-10 MeV) and non-detections are found. In the quiescent state only upper limits are obtained.

The COMPTEL spectrum (Fig. 1) of the flare state indicates a hard MeV-shape ($\alpha_{ph} < 2$, $\sim E^{-\alpha}$). If combined with the simultaneous EGRET (30 MeV - 10 GeV) spectrum, a spectral break ($\Delta\alpha$: ~ 1.2) around 20 MeV becomes obvious. The summed COMPTEL spectrum of the γ-ray quiescent state is also shown in Fig. 1. One notices that the flux limit in the 10-30 MeV band is about a factor of 4 below the flux measured during the flare state. The broad-band spectrum of the flare state is given in Fig. 2. It shows that the peak of the radiated power per logarithmic frequency interval is located at energies around 10 MeV.

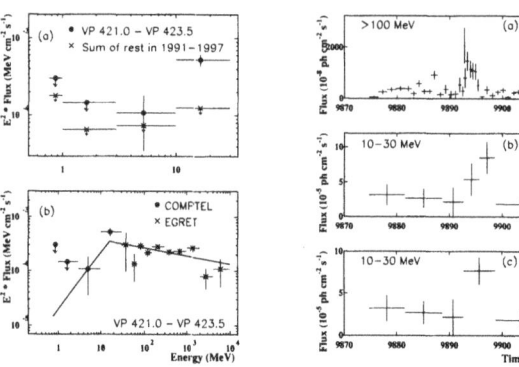

Fig. 1. Left: The COMPTEL spectra of the flare and quiescent state (upper) as well as the combined COMPTEL/EGRET spectrum of the flare state (lower). Right: Light curve of PKS 1622-297 during its flare at energies > 100 MeV (a) and the 10-30 MeV range (b). The two high fluxes of (b) are combined in (c).

During the flare state, no significant flux variability is seen in both, COMPTEL and EGRET, γ-ray light curves with each bin averaged over $\sim$ 1 week. EGRET detected a short huge flare of $\sim$ 3 days during the flare state in 1995 [3]. To investigate it at MeV energies, we subdivide the VP in which the short flare occurred in 3 parts according to the EGRET lightcurve: pre-flare (TJD 9888.8-9892.7), on-flare (TJD 9892.7-9895.6) and post-flare (TJD 9895.6-9898.6). The 10-30 MeV flux peaks in the post-flare part, indicating the MeV photons lag those above 100 MeV. By combining the on-flare and post-flare parts, we derive a $\sim 5\sigma$ detection of this short flare (Fig. 1).

2.2   3C 454.3 and CTA 102

Combining all the COMPTEL data from 1991 to 1999 provides evidence for 3C 454.3 at energies above 1 MeV and for CTA 102 at energies above 10 MeV. A 4σ-excess is detected between 3C 454.3 and CTA 102 in the 10-30 MeV band, which is consistent with emission from both. No signal can be derived from CTA 102 in the 3-10 MeV band, while an excess is detected around 3C 454.3 at a significance level of $\sim 5.6\sigma$.

Searching for flux variability shows that all light curves are consistent with a constant flux for both sources over a period of 8 years. While only marginal hints or upper limits are found along the CGRO mission for CTA 102, 3C 454.3 seem to be emitting at a low but rather stable level in the 3-10 MeV band.

In non-simultaneous broad-band spectra in frequencies of 10^8 - 10^{24} Hz (Fig. 2), two maxima are obvious for both sources. The one at the MeV energies dominates the bolometric output. The other one is at frequencies between 10^{11} Hz and 10^{13} Hz. Both spectra show a turnover region at MeV energies with a change in spectral index of >0.5 from high energy γ-rays to hard X-rays.

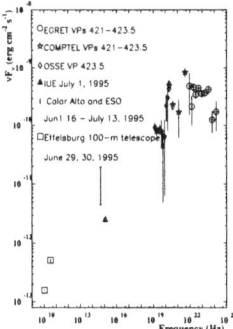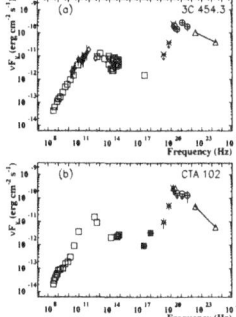

Fig. 2. Left: Broad-band spectrum of PKS 1622-297 during its flare state in 1995. Right: Non-simultaneours broad-band spectra of 3C 454.3 (a) and CTA 102 (b).

3 Discussion and Summary

PKS 1622-297 shows a 'hard' MeV spectrum and a spectral break in the combined COMPTEL/EGRET (0.75 MeV - 10 GeV) spectrum. This indicates that the flare is more efficient at higher γ-ray energies. Such a behavior can generally be explained in a multi-component γ-ray emission scenario, consisting of synchrotron-self Compton and external Comptonization components. They have different dependencies on the bulk Lorentz factor; a change of it could result in increased emission at mainly the higher-energy γ-rays (e.g., [4]).

3C 454.3 and CTA 102 show the typical behaviour of γ-loud flat-spectrum radio quasars (FSRQs): the broad-band spectra display two spectral humps with peaks at infrared and MeV γ-ray energies. They are usually interpreted as synchrotron and inverse-Compton (IC) emission from a relativistic jet. For both sources the IC hump dominates the bolometric output and they show a strong spectral transition from a hard to a soft spectrum with a change in spectral index of larger than 0.5.

The different MeV properties observed in the 3 γ-ray blazars provide important constraints on the emission mechanisms at work in the γ-ray band.

References

1. P.L. Nolan, D.L. Bertsch, C.E. Fichtel, et al.: ApJ **414**, 82 (1993)
2. R.C. Hartmann, D.L. Bertsch, B.L. Dingus, et al.: ApJ **407**, L4 (1993)
3. J.R. Mattox, S.J. Wagner, M. Malkan, et al.: ApJ **476**, 692 (1997)
4. M. Böttcher, W. Collmar, A&A **329**, 57 (1998)

X-Ray Variability of Blazar PKS 2155–304: Probing the Dynamics of the Jet

You Hong Zhang[1], Aldo Treves[1], Annalisa Celotti[2], Lucio Chiappetti[3], Giovanni Fossati[4], Gabriele Ghisellini[5], Laura Maraschi[6], Elena Pian[7], Gianpiero Tagliaferri[5], and Fabrizio Tavecchio[6]

[1] Università dell'Insubria, via Valleggio 11, I-22100 Como, Italy
[2] SISSA/ISAS, via Beirut 2-4, I-34014 Trieste, Italy
[3] IFCTR/CNR, via Bassini 15, I-20133 Milano, Italy
[4] CASS, UCSD, 9500 Gilman Drive, La Jolla, CA 92093-0424, USA
[5] OAB, via Bianchi 46, I-22055 Merate, Italy
[6] OAB, via Brera 28, I-20121 Milano, Italy
[7] OAT, via G.B.Tiepolo 11, I-34131 Trieste, Italy

Abstract. We present temporal and spectral analysis of the X-ray bright blazar PKS 2155–304 observed with *Beppo*SAX. Power density spectra show strong red noise feature with steep slope of $\sim$ 2–3. Structure functions suggest typical timescale of $\sim$ 0.5 days. Inter-band time lags differ from flare to flare. Peak energies of synchrotron component increase with increasing flux, and complexities of spectral evolution are detected. The implications of our results are discussed in the context of synchrotron cooling model of relativistic electrons accelerated through internal shocks taking place in the jet.

1 X-Ray Variability

PKS 2155-304 was monitored with *Beppo*SAX in November of 1996, 1997 and 1999, with $\sim$ 2 days pointing each time. Light curves for 1996 and 1997 have been presented in [1]. We show in Figure 1 (left panel) the light curve for 1999. The source underwent different brightness changes and showed significant variations with recurrent flares detected.

We calculate the normalized power density spectrum (NPDS) utilizing the standard discrete Fourier transform. Each NPDS generally shows quick decrease of power with increasing frequency and can be fitted with a power-law model ($P(f) \propto f^{-\alpha}$). The behavior of strong *red noise* variability is indicated by the steep slope ranging from $\sim$ 2 to 3. We also calculate the first order Structure Function (SF). One example is shown in Figure 1 (right panel) derived from the 2–10 keV light curve of 1999. The most powerful ability of SF is to estimate the characteristic τ_c, identified as the timescale where the first "turn-over" of the SF occurs. To do so, we fit the SF with a broken power-law model. One example of this fitting is also shown in the same figure as SF.

Cross-correlation analysis using two techniques suited to unevenly sampled time series, the Discrete Correlation Function (DCF) and the Modified Mean Deviation (MMD), are performed to evaluate the degree of correlation and inter-band time lags of variations. The results show that soft lag (lower energy photons lagging higher energy ones) of the source differs from flare to flare, ranging from a few hundred seconds to one hour or so. We further quantify the energy-dependence of the soft lags. One example is shown in Figure 2 (left panel) for the #2 flare of 1999.

Soft lag is thought to be due to the energy-dependence of synchrotron cooling timescale, t_{cool}, of relativistic electrons, and in turn to the emitted photon energies. In the observer's frame, the dependence of t_{cool} on photon energies is $t_{cool}(E) = 3.04 \times 10^3 (1 + z)^{3/2} B^{-3/2} \delta^{-1/2} E^{-1/2}$ s, where B and δ are the magnetic field and Doppler factor of the emitting region, and E_{keV} is the observed photon energy in keV. The soft lag is then defined as $\tau_{lag} = t_{cool}(E_s) - t_{cool}(E_h)$, where E_s and E_h refer to soft and hard energy, respectively. Soft lag can give a constraint to B and δ by fitting the observed $\tau_{lag}(E)$ with $t_{cool}(E)$. The best fit suggests that $B\delta^{1/3}$ is 0.37 and 0.40 Gauss for the 1999 #1 and #2 flare, respectively. The fit for the second flare is also shown in Figure 2 (left panel).

2 Spectral Evolution

To reveal in detail the spectral evolution, we perform time-resolved spectral analysis with the curved model to account for the spectral curvature in PKS 2155–304. The special features of such curved model are to find the energy at which the νF_ν spectrum peaks (E_{peak} of synchrotron component) and the local spectral index at desired energies. The correlation between E_{peak} and 2–10 keV flux is shown in Figure 2(right panel). The best-fit power law gives $E_{peak} \propto F^{-0.75\pm0.04}$, indicating that E_{peak} shifts to higher energy with increasing flux. We further probe spectral evolution on the plane of $\alpha - F$, and complex behaviors are found. Apart from the normal clockwise direction showing soft lag feature in both soft and hard energy band, the main flare of 1996 shows evidence of hard lag in the soft energy band and soft lag in the hard energy band. Moreover, one flare of 1997 shows the opposite behavior, i.e., soft and hard lag in the soft and hard energy band, respectively.

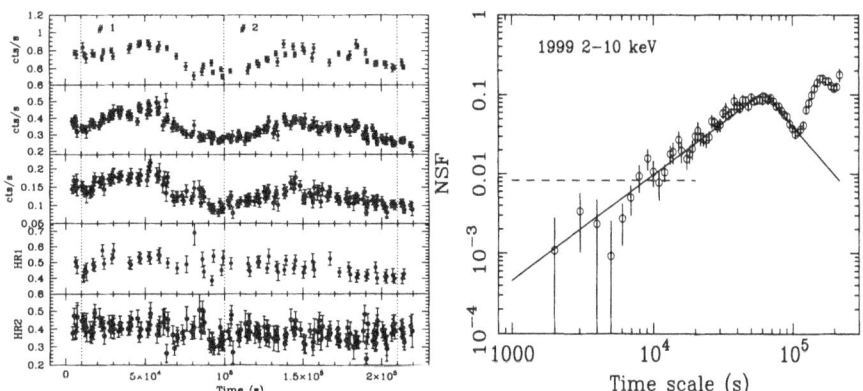

Fig. 1. (Left) Light curves and hardness ratios of 1999 November 4–6 observation. From top to bottom: light curve in 0.1–1.5, 1.5–3.5 and 3.5–10 keV, and hardness ratios of 1.5–3.5/0.1–1.5 and 3.5–10/1.5–3.5 keV, respectively. **(Right)** Structure function of the 2–10 keV light curve. Solid lines are the best fit with a broken power law model, and the dashed line indicates the level of measurement noise.

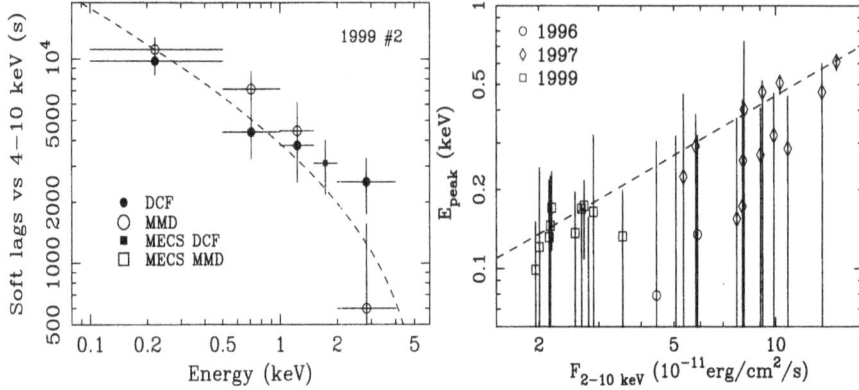

Fig. 2. (Left) Soft lags as function of photon energies. The dashed line is the best fit with synchrotron cooling time, $t_{\rm cool}(E)$. **(Right)** Peak energies of synchrotron component versus the 2–10 keV fluxes. The dashed line represents the best fit with a power law model, with a slope of 0.75 ± 0.04.

3   Discussion

We discuss our main findings, i.e., $\tau_{\rm c}$, soft lag and $E_{\rm peak}$, in line with the internal shock scenario. The key idea of this scenario is to assume that the central engine of a blazar is working in a intermittent rather than in a continuous way to expel discrete shells of plasma with slightly varying velocities. The shock will be formed due to collision when a later faster shell catches up a earlier slower one. The dissipation of bulk kinetic energy carried by the shells during shock period are used to accelerate particles and generate magnetic field, from which the radiation is produced by synchrotron and Compton processes. The main conclusions are (1) $\tau_{\rm c}$ (in the observer's frame) is a measurement of the initial time interval of the two shells ejected from central engine, $\tau_{\rm c} \simeq (2a^2/(a^2-1))t_0$, where $a > 1$ is the ratio of bulk Lorentz factor of the two shells, and t_0 is the initial time interval expelled from the central source; (2) soft lags could increase steeply with $\tau_{\rm c}$, $\tau_{\rm lag} \propto \delta^{5/2}\tau_{\rm c}^{9/4}$; (3) $E_{\rm peak}$ decreases with collision distance, and in turn $\tau_{\rm c}$. (4) the energy where soft/hard lag occurs depends on the value of $t_{\rm acc}/t_{\rm cool}$, where $t_{\rm acc}$ is the accelerating time. (5) more than one emitting region may also play a role if observed simultaneously.

References

1. Y.H. Zhang et al.: ApJ, **527**, 719 (1999)

Part VI

First Bright Objects and Their Role in the End of Dark Ages

$z \sim 6$ Quasars from the SDSS: Probing the End of the Dark Ages

Xiaohui Fan[1], Vijay K. Narayanan[2], Michael A. Strauss[2], Robert H. Lupton[2], Robert H. Becker[3], Richard L. White[4], Laura Pentericci[5], and Hans-Walter Rix[5]

[1] Institute for Advanced Study, Princeton, NJ 08540, USA
[2] Princeton University Observatory, Princeton, NJ 08544, USA
[3] Physics Department, University of California, Davis, CA 95616, USA
[4] Space Telescope Science Institute, Baltimore, MD 21218, USA
[5] Max-Planck-Institut für Astronomie, Königstuhl 17, D-69171 Heidelberg, Germany

Abstract. We present the recent results on a survey for quasars at $z \sim 6$, selected from ~ 1500 deg^2 of multicolor imaging data from the Sloan Digital Sky Survey (SDSS). Four luminous ($M \sim -27$) quasars at $5.7 < z < 6.3$ are discovered. The existence of strong metal lines in the quasar spectra suggests early metal enrichment in the quasar environment. We find that at $z = 6$, the comoving density of luminous quasars declines by a factor of ~ 20 from $z \sim 3$. The strength of Lyα absorption increases strongly with redshift. In particular, Keck and VLT spectroscopy of the highest redshift quasar at $z = 6.28$ reveal the first detection of a complete Gunn-Peterson trough. By comparing with cosmological simulations, we find that the ionizing background declines by a factor of > 20 from $z \sim 3$ to 6, while the neutral faction increases by > 100. The ionization state of the IGM at $z \sim 6$ is typical of those in at the end of overlap stage of the reionization epoch. Thus, $z \sim 6$ markes the end of the cosmic dark ages.

1 Introduction

High-redshift quasars provide direct probes of the epoch when the first generation of galaxies and quasars formed. The existence of very luminous quasars at high redshift, likely residing in massive halos, constrains models of the formation of massive black holes and quasar evolution. The absorption spectra of high-redshift quasars reveal the state of the intergalactic medium and the evolution of the ionization background at early epochs. As shown in this proceeding (see also Fan et al. 2001b,c, Becker et al. 2001, Pentericci et al. 2001), the observations of the highest redshift quasars known today may finally constrain when the reionization happened, thus ended the cosmic dark ages, and how the universe was transformed from mostly neutral to mostly ionized by the light of first generation galaxies and quasars.

Sloan Digital Sky Survey (SDSS; York et al. 2000) uses a dedicated 2.5m telescope and a large format CCD camera (Gunn et al. 1998) at the Apache Point Observatory in New Mexico to obtain images in five broad bands (u, g, r, i and z) over 10,000 deg^2 of high Galactic latitude sky. The goal of the SDSS quasar survey is to establish a color-selected, flux-limited sample of $\sim 10^5$ quasars over the entire survey area. Quasar candidates are selected based on their broad-band colors from the SDSS five color photometry (Fan et al. 1999, Richards et al. 2001). Over 200 quasars at $z > 3.5$ have been discovered to date from the SDSS multicolor imaging data (Anderson et al.

2001 and reference therein). The inclusion of the reddest band, z, in principle enables the discovery of quasars up to $z \sim 6.5$ from the SDSS data (Fan et al. 2000).

2   Discovery of $z \sim 6$ Quasars

We have carried out a systematic survey of i-dropout quasars ($z > 5.7$, whose Lyα forest is entirely in the i band) based on the SDSS imaging data (Fan et al. 2001b). The observed colors of quasars evolve strongly with redshift, as the intrinsic emission and the intervening absorption spectral features move through the SDSS filter system. At $z > 5.7$, the Lyα emission line begins to move out of the SDSS i filter. With an average $i^* - z^* > 2$, these high-redshift quasars have undetectable flux in the bluer u, g and r bands, and little flux (sometimes also undetectable) in i band. They become i-dropout objects with only one measurable color in SDSS photometry.

Quasars at $z > 5.8$ are extremely rare on the sky. We found four i-dropout quasars in 1550 deg^2. For comparison, over the same area, there are about 15 million objects detected in the z band above 6-σ, and about 6.5 million cosmic ray hits in the z band. The quasars are faint, with low signal-to-noise ratio (S/N) photometry from the SDSS imaging data. Therefore, the key to a successful selection process is an efficient elimination of potential contaminants. We face three technical challenges:

(1) Elimination of false z-band only detections, composed primarily of cosmic rays, but also including satellite trails, electronic ghost images and bleed trails from bright stars. We have developed sensitive cosmic ray classification routine to reject these false detections.

(2) Reliability of z^* photometry. Objects in the tail of the error distribution could scatter into the selected region of color space, and pose a serious contamination when a large number of objects is searched. We carry out an additional round of z band photometry on the i-dropout objects using separated telescope to improve the photometric accuracy.

(3) Separating quasars and cool dwarfs. The surface density of cool dwarfs, with spectral types ranging from mid-L to T, which have $i^* - z^* > 2.2$ (Geballe et al. 2001, Leggett et al. 2001), is ~ 15 times higher than that of $z > 5.8$ quasars. We use J-band photometry to separate these two classes of objects.

We discovered four quasars at $z > 5.7$ in the 1500 deg^2 of the survey area: SDSS 1044-0103 at $z = 5.8$, SDSS 0836+0054 at $z = 5.82$, SDSS 1306+0356 at $z = 5.99$, and SDSS 1030+0524 at $z = 6.28$. Their spectra are presented in Figure 1. As a by-product of the survey, a large number of L dwarfs and a dozen T dwarfs are discovered. They form a complete sample of L and T dwarfs, enable study of the luminosity function and mass function of substellar objects (Knapp et al. 2002).

3   Quasar Evolution at $z \sim 6$

Emission line ratios can be used to measure the metallicity of the gas in the Broad Emission Line Region (BELR). There is growing evidence from those measurements that BELR have roughly solar or higher metallicities even out to $z > 4$ (Hamman &

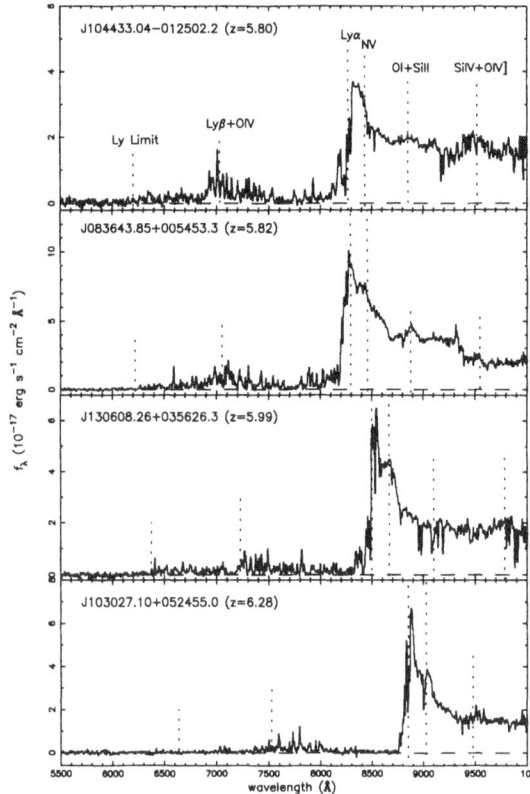

Fig. 1. Keck/ESI spectra of four SDSS quasars at $z > 5.7$.

Ferland 1999). We use the spectra of our quasars to estimate the quasar metallicity at $z \sim 6$. Of various line ratios, the NV $\lambda 1240$/CIV $\lambda 1549$ ratio and NV $\lambda 1240$/HeII $\lambda 1640$ ratio are particularly useful abundance diagnostics (Hamman & Ferland 1993, 1999). The sensitivity of NV/CIV to metallicity is due to the fact that N is a secondary element, with abundance proportional to Z^2, while C is a primary element with abundance proportional to Z. Using the optical/IR spectrum of SDSS 1030+0524 (Fan et al. 2001b, Becker et al. 2001), we find the flux ratios NV/CIV ~ 0.7 and NV/HeII > 1.5. Both values imply super-solar metallicity of the BELR region, suggesting rapid chemical enrichment and multiple generation of star formation in the quasar environment at these very early epochs.

The four quasars at $z > 5.7$ form a complete color-selected, flux-limited sample at $z \sim 6$. We calculate the color selection completeness of this sample, and derive the spatial density of luminous quasars at $z \sim 6$, $\rho(M_{1450} < -26.8) = 1.1 \times 10^{-9}$ Mpc^{-3}, for $\Omega = 1$ and $H_0 = 50$ km s^{-1} Mpc^{-1}. As shown in Figure 2, this density is about a factor of two lower than that at $z \sim 5$, and a factor of ~ 20 lower than that at $z \sim 3$. It is consistent with an extrapolation of the observed redshift evolution of quasars at

$3 < z < 5$. The low quasar density at $z > 5$ further implies that UV photons from high-redshift quasars are unlikely to keep the universe ionized at early epochs.

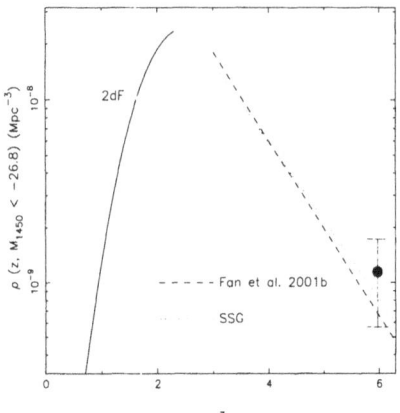

Fig. 2. The evolution of quasar comoving spatial density at $M_{1450} < -26.8$ in the $\Omega = 1$ model (h=0.5). The large dot represents the result from this survey. The dashed and dotted lines are the best-fit models from Fan et al. (2001a) and Schmidt et al. (1995), respectively. The solid line is the best-fit model from the 2dF survey (Boyle et al. 2000) at $z < 2.5$.

The black hole masses of these quasars are probably several times $10^9 M_{\odot}$. The quasars are likely to reside in very massive systems, with the minimum mass of host dark halos $\sim 10^{13} M_{\odot}$. These massive dark halos represent rare peaks in the density field at high redshift, and are in the steep tail of the mass function.

4   Evolution of Lyman Absorption and Detection of a Complete Gunn-Peterson Trough

We have obtained high S/N, moderate resolution spectroscopy of these four $z > 5.7$ quasars using Keck (Fan et al. 2000, Becker et al. 2001) and VLT (Pentericci et al. 2001) telescopes to study the evolution of HI absorption at high-redshift. Figure 3 illustrate the evolution of transmitted flux and effective optical depth in the Lyα forest region along the four lines of sight. We find that the Lyα absorption in the spectra of these quasars evolves strongly with redshift. To $z \sim 5.7$, the Lyα absorption evolves as expected from an extrapolation from lower redshifts. However, in the highest redshift object, SDSS 1030+0524 ($z = 6.28$), the average transmitted flux is 0.0038 ± 0.0026 times that of the continuum level over $8450\,\text{Å} < \lambda < 8710\text{Å}$ ($5.95 < z_{abs} < 6.16$), consistent with zero flux. Thus the flux level drops by a factor of > 150, and is consistent with zero flux in the Lyα forest region immediately blueward of the Lyα emission line, compared with a drop by a factor of ~ 10 at $z_{abs} \sim 5.3$. A similar break is seen at Lyβ; because of the decreased oscillator strength of this transition, this allows us to put a considerably stronger limit, $\tau_{eff} > 20$, on the optical depth to Lyα absorption at $z = 6$.

This is a clear detection of a complete Gunn-Peterson (1965) trough, caused by neutral hydrogen in the intergalactic medium. Even a small neutral hydrogen fraction

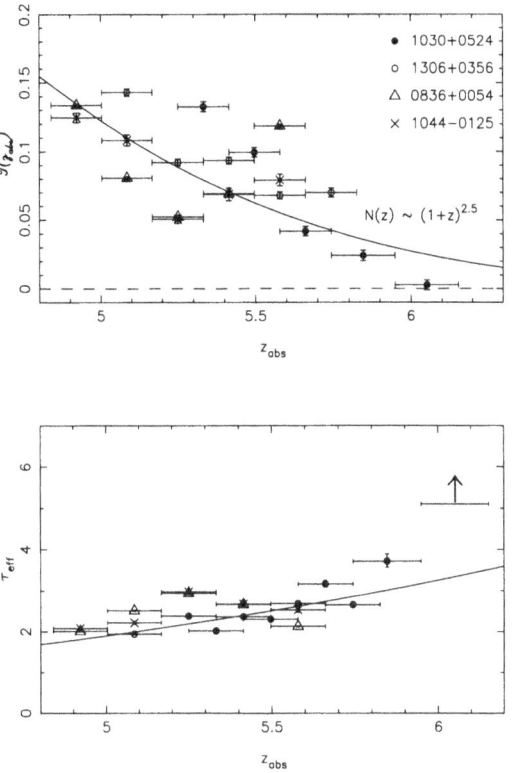

Fig. 3. Evolution of transmitted flux ratio and effective Gunn-Peterson optical depth as functions of redshift. The solid line is the expected evolution if the number density of Lyα clouds increases as $N(z) \propto (1+z)^{2.5}$. No flux is detected in the spectrum of SDSS 1030+0524 at $z_{abs} \sim 6$, indicating $\tau_{eff} > 5.0$.

in the intergalactic medium would result in an undetectable flux in the Lyα forest region. Therefore, the existence of the Gunn-Peterson trough by itself does not indicate that the quasar is observed prior to the reionization epoch. However, the fast evolution of the mean absorption in these high-redshift quasars suggests that the mean ionizing background along the line of sight to this quasar has declined significantly from $z \sim 5$ to 6, and the universe is approaching the reionization epoch at $z \sim 6$.

5   Constraining the Epoch of Reionization

We study the process of cosmic reionization and estimate the ionizing background in the intergalactic medium (IGM) using the Lyman series absorption in the spectra of the four quasars at $5.7 < z < 6.3$. Figure 4 illustrates the evolution of the ionization state of the IGM at high redshift inferred from observations. First, we derive the redshift evolution of the ionizing background at high redshifts, using both semi-analytic techniques and cosmological simulations to model the density fluctuations in the IGM. The existence of the complete Lyα Gunn-Peterson trough in the spectrum of the $z = 6.28$ quasar SDSS 1030+0524 indicates a photoionization rate (Γ_{-12} in units of 10^{-12}s^{-1}) at $z \sim 6$ lower than 0.08, at least a factor of 6 smaller than the value at $z \sim 3$. The Lyβ and Lyγ Gunn-Peterson troughs give an even stronger limit $\Gamma_{-12} < 0.02$ due to their smaller oscillator strengths, indicating that the ionizing background in the IGM at $z \sim 6$ is more than 20 times lower than that at $z \sim 3$.

We also find that the volume-averaged neutral hydrogen fraction increases from 10^{-5} at $z \sim 3$ to $> 10^{-3}$ at $z \sim 6$. At this redshift, the mass-averaged neutral hydrogen fraction is larger than 1%; the mildly overdense regions ($\delta > 3$) are still mostly neutral and the comoving mean free path of ionizing photons is shorter than 8 Mpc. In Figure 4, we compare these values with the simulation of cosmological reionization by Gnedin (2000), which includes 3-D radiative transfer, star formation process and detailed atomic and molecular physics. We find that that the observed properties of the IGM at $z \sim 6$ are typical of those in the era at the end of the overlap stage of reionization when the individual HII regions merge. Thus, $z \sim 6$ marks the end of the reionization epoch. Even though the current observations certainly cannot probe deeper into the reionization epoch, the near phase-transition behavior of the ionization state of the IGM at $z \sim 6$, and the narrow redshift range over which this process occurs in cosmological reionization simulations both imply that the reionization redshift *cannot be at a redshift much higher than six.*

However, we emphasize that the complete Gunn-Peterson trough is observed in only one (the highest-redshift) object, and if the reionization is non-uniform, one may not see exactly the same dependence of the Gunn-Peterson trough with redshift along other lines of sight. In order to gain a more complete understanding of the IGM at $z > 6$, more lines of sight are clearly needed. The SDSS is expected to find ~ 20 quasars at $z > 6$ over the course of the survey (Paper I), so we can look forward to opportunities to study this question in detail over the next few years.

References

1. Anderson, S. F. et al. 2001, AJ, 122, 503
2. Becker, R. H., et al. 2001, AJ, in press (astro-ph/0108097)
3. Boyle, B. J., et al. 2000, MNRAS, 317, 1014
4. Fan, X. 1999, AJ, 117, 2528
5. Fan, X., et al. 2000, AJ, 120, 1167
6. Fan, X., et al. 2001a, AJ, 121, 54
7. Fan, X., et al. 2001b, AJ, in press (astro-ph/0108063)
8. Fan, X., et al. 2001c, AJ, submitted (astro-ph/0111184)
9. Geballe, T., et al. 2001, ApJ, in press (astro-ph/0108443)
10. Gnedin, N. 2000, ApJ, 535, 530
11. Gunn, J. E., et al. 1998, 116, 3040
12. Gunn, J. E., & Peterson, B. A. 1965, ApJ, 142, 1633
13. Hamman, F., & Ferland, G. 1993, ApJ, 418, 11
14. Hamman, F., & Ferland, G. 1999, ARA&A, 37, 487

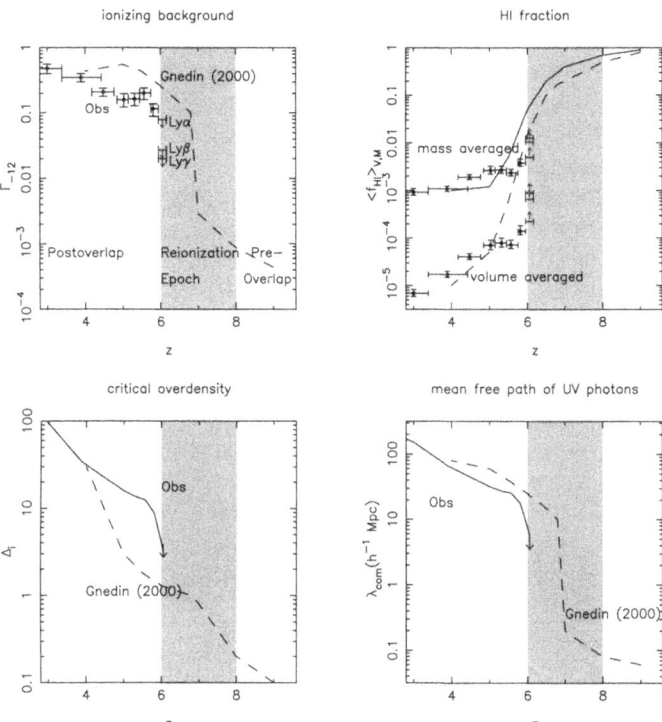

Fig. 4. Evolution of the ionization state of the IGM inferred from the observations. It is compared with simulation of Gnedin (2000).

15. Leggett, S. K., et al. 2001, ApJ, in press (astro-ph/0108435)
16. Knapp, G. R., et al. 2001, in preparation
17. Pentericci, L., et al. 2001, AJ, submitted
18. Richards, G. T., et al. 2001, AJ, 121, 2308
19. Schmidt, M., Schneider, D. P., & Gunn, J. E. 1995, ApJ, 110, 68
20. York, D. G., et al. 2000, AJ, 120, 1579

From Darkness to Light: The First Stars in the Universe

Volker Bromm[1], Paolo S. Coppi[2], and Richard B. Larson[2]

[1] Harvard-Smithsonian Center for Astrophysics
   Cambridge, MA 02138, USA
[2] Astronomy Department, Yale University
   New Haven, CT 06520-8101, USA

Abstract. Paramount among the processes that ended the cosmic 'dark ages' must have been the formation of the first generation of stars. In order to constrain its nature, we investigate the collapse and fragmentation of metal-free gas clouds. We explore the physics of primordial star formation by means of three-dimensional simulations of the dark matter and gas components, using smoothed particle hydrodynamics. We find characteristic values for the temperature, $T \sim$ a few 100 K, and the density, $n \sim 10^3 - 10^4$ cm^{-3}, characterising the gas at the end of the initial free-fall phase. The corresponding Jeans mass is $M_J \sim 10^3 M_\odot$. The existence of these characteristic values has a robust explanation in the microphysics of H_2 cooling, and is not very sensitive to the cosmological initial conditions. These results suggest that the first stars might have been quite massive, possibly even very massive with $M_* > 100 M_\odot$.

1 Introduction

The history of the universe proceeded from an extremely uniform initial state to the highly structured present-day one. When did the crucial transition from simplicity to complexity first occur? Recently, this question has become the focus of an intense theoretical effort (e.g., [1–6,10,12]). There must have been a time, between the last scattering of the CMB photons at $z \sim 1000$ and the formation of the first luminous objects at $z > 6$, when the universe contained no visible light. This era has been called the cosmic 'dark ages', and the key question is how and when it ended (e.g., [15]).

In the context of hierarchical scenarios of structure formation, as specified by a variant of the cold dark matter (CDM) model, the collapse of the first baryonic objects is expected at redshifts $z \simeq 50 - 10$, involving dark matter halos of mass $\sim 10^6 M_\odot$ [16]. The question arises how one can make any progress in understanding primordial star formation, given the lack of direct observational constraints. The physics of the first stars, however, is characterized by some important simplifications, as compared to the extreme complexity of present-day star formation [9]. The absence of metals, and consequently of dust, leaves atomic and molecular hydrogen as the main agent of radiative cooling and the source of opacity. Magnetic fields were likely to be dynamically insignificant, prior to the onset of efficient (stellar) dynamo amplification. The chemistry and heating of the primordial gas was not yet complicated by the presence of a UV radiation background. The intergalactic medium (IGM) must have been a rather quiescent place, with no source to sustain turbulent motion, as long as the density perturbations remained in their linear stage. Only after the explosion of the first supernovae, and the associated input of mechanical and thermal energy, is this state of primordial tranquility bound to change [8]. Therefore, the physics of primordial star formation is mainly governed by gravity, thermal pressure, and angular momentum. This situation renders the problem theoretically more straightforward and tractable than the highly complex present-day case which continues to defy attempts to formulate a fundamental theory of star formation. Finally, the initial conditions for the collapse of a primordial star forming cloud are given by the adopted model of cosmological structure formation.

The importance of the first stars and quasars derives from the crucial feedback they exert on the IGM. A generation of stars which formed out of primordial, pure H/He gas (the so-called Population III) must have existed, since heavy elements can only be synthesized in the interior of stars. Population III stars, then, were responsible for the initial enrichment of the IGM with heavy elements. From the absence of Gunn-Peterson absorption in the spectra of high-redshift quasars, we know that the universe has undergone a reionization event at $z > 6$. UV photons from the first stars, perhaps together with an early population of quasars, are expected to have contributed to the reionization of the IGM [8,7,11].

To probe the time when star formation first started entails observing at redshifts $z > 10$. This is one of the main purposes of the *Next Generation Space Telescope* (NGST) which is designed to reach $\sim$ nJy sensitivity at near-infrared wavelengths. In preparation for this upcoming observational revolution, the study of the first stars is very timely, providing a theoretical framework for the interpretation of what NGST might discover, less than a decade from now.

2 Simulations

Our code is based on a version of TREESPH which combines the Smoothed Particle Hydrodynamics (SPH) method with a hierarchical (tree) gravity solver. To study primordial gas, we have made a number of additions. Most importantly, radiative cooling due to hydrogen molecules has been taken into account. In the absence of metals, H_2 is the main coolant below $\sim 10^4$ K, the typical temperature range in collapsing Population III objects. The efficiency of H_2 cooling is very sensitive to the H_2 abundance. Therefore, it is necessary to compute the nonequilibrium evolution of the primordial chemistry (see [4] for details).

We have devised an algorithm to merge SPH particles in high density regions to overcome the otherwise prohibitive timestep limitation, as enforced by the Courant stability criterion. To follow the simulation for a few dynamical times, we allow SPH particles to merge into more massive ones, provided they exceed a pre-determined density threshold, typically $10^8 - 10^{10}$ cm^{-3} [4].

We have carried out a comprehensive survey of the relevant parameter space, and focus in the following on one select example that is representative for the overall results.

2.1 The Fragmentation of Primordial Gas

We model the site of primordial star formation as an isolated overdensity, corresponding to a high$-\sigma$ peak in the random field of density perturbations. The numerical simulations are initialized at $z_i = 100$ by endowing a spherical region, containing dark matter and gas, with a uniform density and Hubble expansion. Small-scale density fluctuations are imprinted on the dark matter according to the CDM power spectrum. We assume that the halo is initially in rigid rotation with a given angular velocity, prescribed in accordance with the prediction for the spin parameter λ, as found in cosmological N-body simulations.

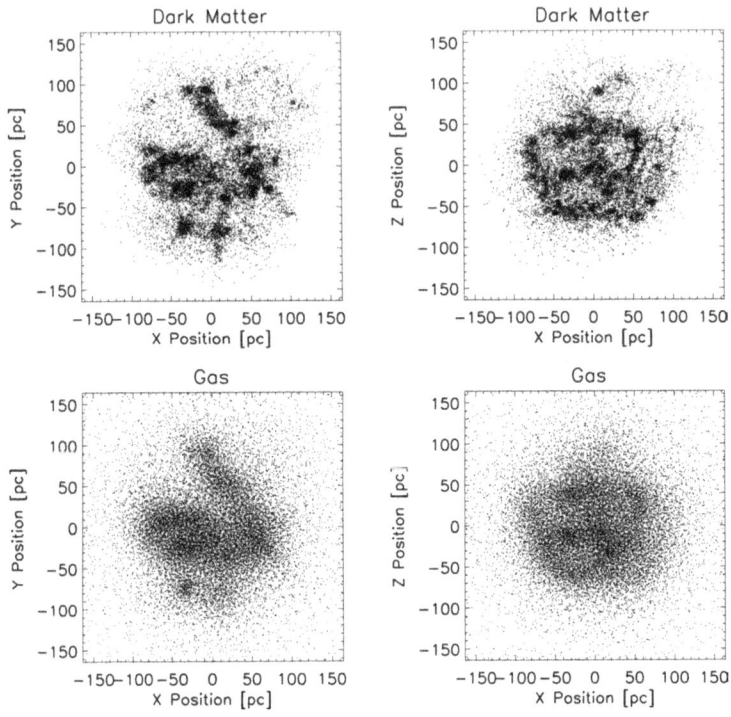

Fig. 1. Collapse of a primordial star forming cloud (from [4]). The halo has a total mass of $2 \times 10^6 M_\odot$, and will collapse at $z_{vir} \simeq 30$. Shown here is the morphology at $z = 33.5$. *Top row:* DM particles. *Bottom row:* Gas particles. *Left panels:* Face-on view. *Right panels:* Edge-on view. The DM has developed significant substructure, and the baryons are just beginning to fall into the corresponding potential wells.

Fig. 1 shows the situation at $z = 33.5$, briefly before the virialization of the dark matter. In response to the initially imprinted k^{-3}-noise, the dark matter has developed a pronounced substructure. The baryons have just begun to fall into the potential wells which are created by the DM substructure. Thus, the DM imparts a 'gravitational head-start' to certain regions of the gas, which subsequently act as the seeds for the formation of high-density clumps.

At the end of the free-fall phase, the gas has developed a very lumpy, filamentary structure in the center of the DM potential. The gas distribution is very inhomogeneous, and the densest regions are gravitationally unstable. The ensuing runaway collapse leads to the formation of high-density sink particles or clumps. These clumps are formed with initial masses close to $M \sim 10^3 M_\odot$, and subsequently gain in mass by the accretion of surrounding gas, and by merging with other clumps (see Fig. 2).

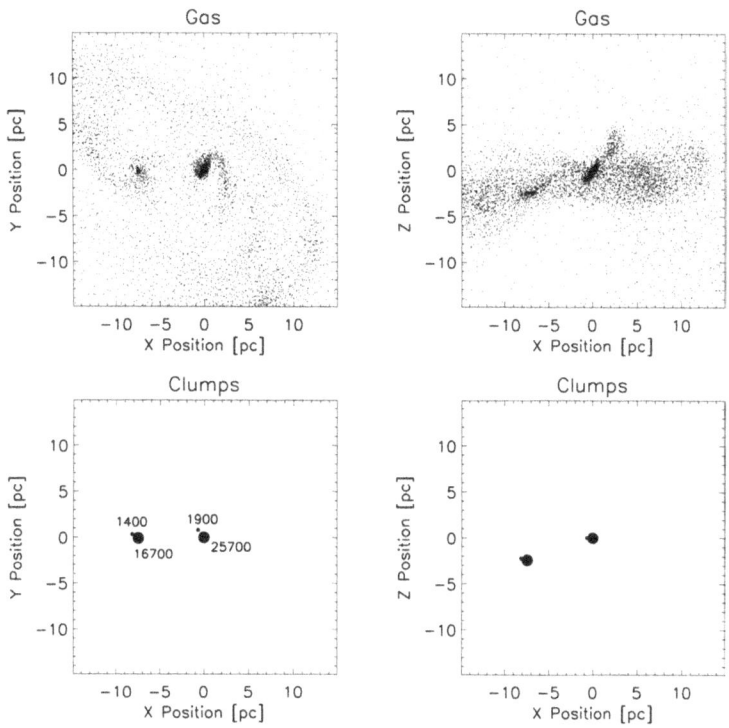

Fig. 2. Fragmentation into high density clumps (from [4]). Shown here is the situation at $z = 28.9$ in the innermost 30 pc. *Top row:* The remaining gas in the diffuse phase. *Bottom row:* Distribution of clumps. The numbers next to the dots denote clump mass in units of $M_\odot$. *Left panels:* Face-on view. *Right panels:* Edge-on view. The gas has settled into an irregular configuration with two dominant clumps. The clumps are formed with initial masses of $\sim 10^3 M_\odot$, and subsequently grow in mass by accretion and merging with other clumps, up to $\sim 20,000 M_\odot$.

There is a good physical reason for the emergence of high-density clumps with initial masses $\sim 10^3 M_\odot$. To understand this, consider the thermodynamic and chemical state of the gas, as summarized in Fig. 3. Since the abundances, temperature and density are plotted for every SPH particle, this mode of presentation has an additional dimension of information: Particles accumulate ('pile up') in those regions of the diagram where the evolutionary timescale is long. In panel (c) of Fig. 3, one can clearly discern such a preferred state at temperatures of a few 100 K, and densities of $10^3 - 10^4$ cm^{-3}. These characteristic values have a straightforward physical explanation in the microphysics of H$_2$ cooling. A temperature of $T \sim 100 - 200$ K is the minimum one attainable via H$_2$ cooling. The corresponding critical density, beyond which the H$_2$ rotational levels are populated according to LTE, is then $n_{crit} \simeq 10^3 - 10^4$ cm^{-3}. Due to the now inefficient cooling, the gas 'loiters' and passes through a phase of quasi-hydrostatic, slow contraction.

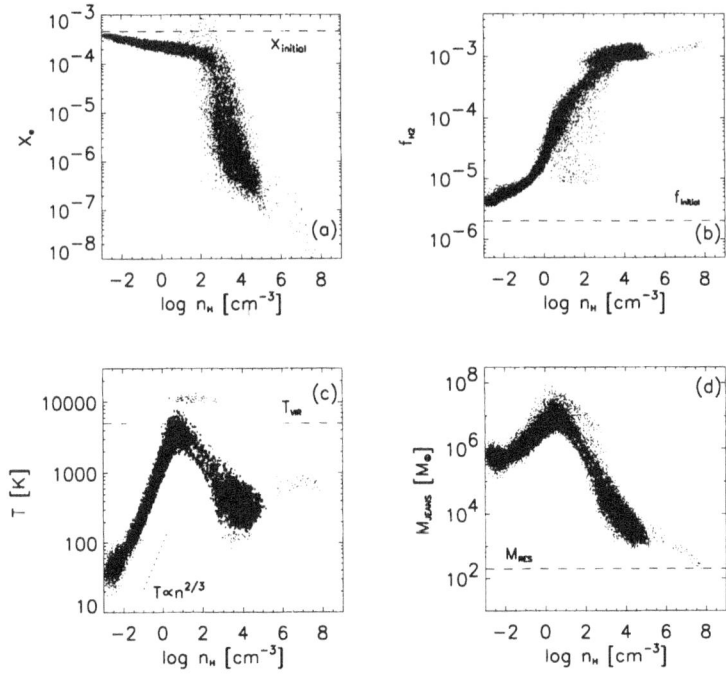

Fig. 3. Gas properties at $z = 31.2$ (from [4]). (**a**) Free electron abundance vs. hydrogen number density (in cm^{-3}). (**b**) Hydrogen molecule abundance vs. number density. (**c**) Gas temperature vs. number density. At densities below ~ 1 cm^{-3}, the gas temperature rises because of adiabatic compression until it reaches the virial value of $T_{vir} \simeq 5000$ K. At higher densities, cooling due to H$_2$ drives the temperature down again, until the gas settles into a quasi-hydrostatic state at $T \sim 300$ K and $n \sim 10^4$ cm^{-3}. Upon further compression due to the onset of the gravitational instability, the temperature experiences a modest rise again. (**d**) Jeans mass (in $M_\odot$) vs. number density. The Jeans mass reaches a value of $M_J \sim 10^3 M_\odot$ for the quasi-hydrostatic gas in the center of the DM potential well.

To move away from this loitering regime, and to attain higher densities, the gas has to become gravitationally unstable. Evaluating the Jeans mass for the characteristic values $T \sim 200$ K and $n \sim 10^3 - 10^4$ cm^{-3} results in $M_J \sim 10^3 M_\odot$. When enough gas has accumulated in a given region to satisfy $M > M_J$, runaway collapse of that fluid region ensues. We find that the gas becomes self-gravitating ($\rho_B > \rho_{DM}$) coincident with the onset of the Jeans instability.

2.2 Protostellar Collapse

To further constrain the characteristic mass scale for Population III stars, we have investigated the collapse of a clump to even higher densities. As our starting configuration, we select the first region to undergo runaway collapse in one of the lower resolution simulations and employ a technique of refining the spatial resolution in the vicinity of this region, which in the unrefined simulation would have given rise to the creation of a sink particle.

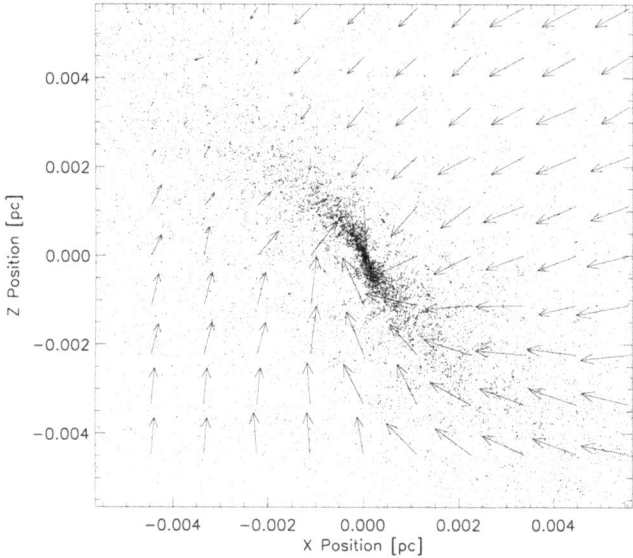

Fig. 4. Gas morphology and kinematics in the vicinity of the density maximum. Shown is the situation ~ 5000 yr after the onset of runaway collapse in a box of linear size ~ 2500 AU. The small dotted symbols give an indication of the gas density, and the overplotted arrows depict the velocity field in the x-z plane. The length of an arrow scales with speed such that the largest one corresponds to ~ 14.5 km s^{-1}. It is evident that a highly concentrated, spindle–like structure has formed in the center which comprises a few tens of solar masses at this instant. The surrounding flow field is supersonic with typical Mach numbers of $\sim 3 - 5$.

In the refined simulation, three-body reactions become important, and lead to the almost complete conversion of the gas into molecular form [14].

Only ~ 5000 yr after the onset of runaway collapse, the central, highest-density region has already evolved significantly. As can be seen in Fig. 4, an elongated, spindle-like structure has formed, which comprises a mass of $\sim 20 M_\odot$, and has a characteristic size of $L_{char} < 10^{-4}$ pc $\simeq 20$ AU. By examining the surrounding velocity field, which is characterized by Mach numbers of $\sim 3 - 5$, it is evident that matter continues to fall onto the central object. To derive an estimate for the accretion rate, we consider the average mass flux, $< \rho v_r >$, through a spherical surface around the density maximum with radius $r = 10^{-3}$ pc. Here, v_r is the radial velocity component, and velocities are measured relative to the density maximum. Assuming spherical accretion, one finds for the accretion rate

$$\dot{M}_{acc} = -4\pi r^2 < \rho v_r > \sim 1 M_\odot \, \text{yr}^{-1} \quad . \tag{1}$$

This is likely to be an overestimate, since the assumption of spherical symmetry is only a rough approximation to the complex kinematics of the flow. Nevertheless, it is clear that the central object will rapidly grow in mass, on a timescale $t_{acc} \sim M/\dot{M}_{acc} \sim 20$ yr. We find no indication for further subfragmentation in this simulation. These results suggest that the first stars might have been quite massive, possibly even very massive with $M_* > 100 M_{\odot}$.

Important caveats remain, however. The question of how massive the incipient star in the center of the collapsing clump will eventually be, cannot be answered with any certainty at present. Our attempts in doing so are foiled by our ignorance of the complex and rather unexplored physics of accretion from a dust free envelope. This, then, is the frontier of our current knowledge [13].

References

1. T. Abel, G.L. Bryan, M.L. Norman: Ap. J., **540**, 39 (2000)
2. R. Barkana, A. Loeb: Physics Reports, **349**, 125 (2001)
3. V. Bromm, P.S. Coppi, R.B. Larson: Ap. J., **527**, L5 (1999)
4. V. Bromm, P.S. Coppi, R.B. Larson: Ap. J., in press; astro-ph/0102503 (2001)
5. V. Bromm, A. Ferrara, P.S. Coppi, R.B. Larson: M.N.R.A.S., in press; astro-ph/0104271 (2001)
6. V. Bromm, R.P. Kudritzki, A. Loeb: Ap. J., **552**, 464 (2001)
7. A. Ferrara: Ap. J., **499**, L17 (1998)
8. Z. Haiman, A. Loeb: Ap. J., **483**, 21 (1997)
9. R.B. Larson: M.N.R.A.S., **301**, 569 (1998)
10. A. Loeb, R. Barkana: A.R.A.&A., **39**, 19 (2001)
11. J. Miralda-Escudé, M. Haehnelt, M.J. Rees: Ap. J., **530**, 1 (2000)
12. F. Nakamura, M. Umemura: Ap. J., **548**, 19 (2001)
13. K. Omukai, F. Palla: Ap. J. Lett., in press; astro-ph/0109381 (2001)
14. F. Palla, E.E. Salpeter, S. W. Stahler: Ap. J., **271**, 632 (1983)
15. M. J. Rees: *New Perspectives in Astrophysical Cosmology* (Cambridge University Press, Cambridge 2000)
16. M. Tegmark, J. Silk, M.J. Rees, A. Blanchard, T. Abel, F. Palla: Ap. J., **474**, 1 (1997)

Search for the First Quasars and Supernovae

Wei Zheng[1], Holland Ford[1], Zlatan Tsvetanov[1], Arthur Davidsen[1], Alexander Szalay[1], Jeffrey Kruk[1], George Hartig[2], Peter Stockman[2], Marc Postman[2], Hans-Walter Rix[3], Rainer Lenzen[3], and Peter Shu[4]

[1] Department of Physics and Astronomy, Johns Hopkins University, Baltimore MD 21218, USA
[2] Space Telescope Science Institute, Baltimore MD 21218, USA
[3] Max-Planck-Institut fr Astronomie, Heidelberg, Germany
[4] Goddard Space Flight Center, Greenbelt MD 20771, USA

Abstract. Quasars at $z > 7$ and supernovae at $z > 1$ are very rare and faint, and they can be found only from large surveys in the near-infrared band. We describe the concept of *PRIME*, a satellite that will survey one quarter of the sky in four bands between 0.9 and 3.5 μm, to an AB magnitude of 24, and its deep survey will reach AB magnitude of 27. Such a survey is capable of finding targets for *NGST* as well as the largest ground-based telescopes, including quasars up to $z = 25$ and Type-Ia supernovae at $z > 3$.

1 How to Find Luminous Objects in the Early Universe

Our vision of the universe has received new boost from the discovery of quasars and supernovae at unprecedented distances. Information gathered from these objects, even handful, provided invaluable insight into the unseen universe and the cosmological structure. We are more convinced than ever than these lighthouses may be found at even higher redshifts.

1.1 Role of Surveys

Even quasars and supernovae are among the most luminous objects, they, at high redshifts, are very faint and rare. The recent discovery of $z \sim 6$ quasars [2] from the Sloan Digital Sky Survey implies that the total number of quasars at $z \geq 6$ is only handful. The surface density of these objects is so low that they cannot be found in just a few deep fields. Take the example of 3C 273: Its brightness (V=12.8) should have made its signature on many photo plates, but it was not until a radio survey that this object was discovered. Assuming a constant luminosity, the flux of quasars and supernovae decreases at least by a factor with $(1 + z)^3$. Therefore, a Type-Ia supernova at $z = 3$ should have an apparent AB magnitude of ~ 25 (depending on cosmological constants), and quasars at $z \sim 10$ magnitude of 21. These objects are new challenges to surveys in terms of a large sky coverage and very faint limiting magnitudes.

Many lighthouse objects in the early universe, such as supernovae and Gamma-ray bursts are highly transient. If a survey is aimed at these objects, its mapping strategy should include repetitive observations of certain areas.

1.2 Importance of Infrared Bands

To make the situation more complicated, the spectroscopic signature of these objects moves out of the optical band at higher redshifts. The Lyman-α absorption of the intergalactic medium effectively cuts off fluxes below $0.12(1 + z)$ μm for any objects at $z > 5$. At $z > 7$, objects virtually cannot be detected in the optical band. Supernovae do not emit much flux below the Balmer limit in the rest frame. Therefore at $z > 1$, most of their fluxes are at 1μm or longward.

Many early-type objects such as star-forming regions contain a significant amount of dust, and they are much easier to detect in infrared bands because of reduced extinction. The Next Generation Space Telescope (*NGST*) is set to study the youngest galaxies and quasars, therefore its primary waveband is between 1 and 5 μm.

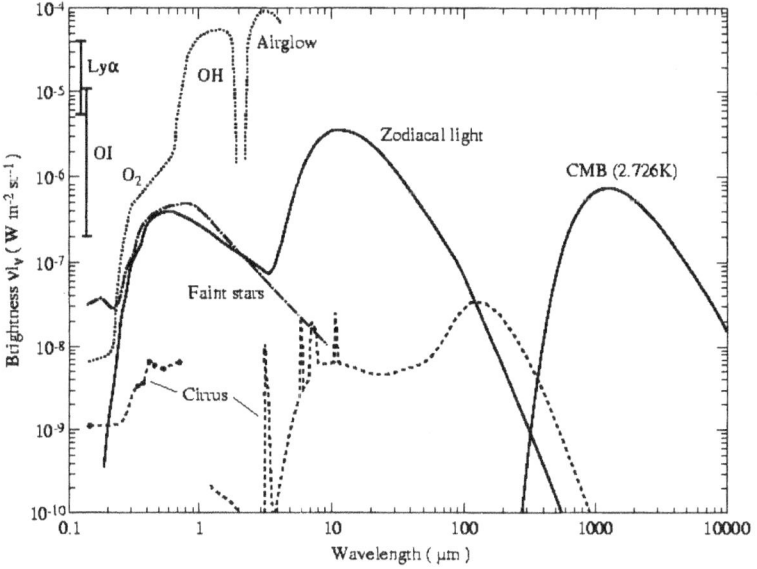

Fig. 1. Sky background in space and on ground. Adapted from [3].

1.3 Advantage from Space

While near-infrared surveys have been carried out from the ground, their capability is hampered by a very high sky background. 2MASS and DENIS can reach limiting magnitudes of AB $16 - 17$, which are not sufficient for most extragalactic objects. As shown in Figure 1, the sky background around 2.3 μm is about 1000 times fainter than that from the ground. The figure also indicates that the sky background reaches a minimum at $\sim 3.5\mu$m. At 3.5 μm, the gain from space is $\sim 3 \times 10^5$. Therefore, a space-based survey is advantageous for infrared, including near-infrared.

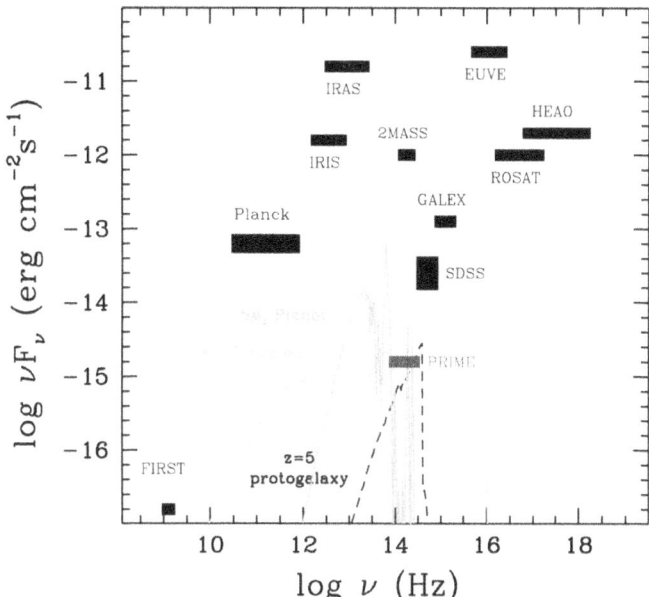

Fig. 2. Sensitivity of large sky surveys. Only surveys that cover at least 1/4 of the sky are included. Overplotted are the simulated spectra of a quasar at $z = 25$, with its flux scaled with respect to the observed $z \sim 6$ quasars, a $z = 5$ protogalaxy, scaled to those at $z \sim 4$, and a $5 M_{Jupiter}$ planet of 1 Gyr age at 10 parsec. Because of intergalactic and Galactic absorption, there is a significant blank region between $\log \nu \sim 14.5 and 16$.

Table 1. *PRIME* Performance

Mode	Area (deg^2)	AB Magnitude
Large	10000	24.0
Medium	200	26.5
Deep	12	27.0

2   *PRIME* – A Deep Near-Infrared Survey

We have developed a mission concept *PRIME*, the Primordial Explorer. *PRIME* will carry out a deep survey in the near infrared by obtaining images in four bands between 0.9 and 3.5 μm. As shown in Figure 2, *PRIME* is in the right band and right sensitivity to catch the signature of some most interesting objects. Table 1 shows the general performance of the *PRIME* survey, assuming 5σ detection for point sources. This third generation will be approximately 600 times more sensitive than 2MASS [5] and make the first survey around 3.5 μm.

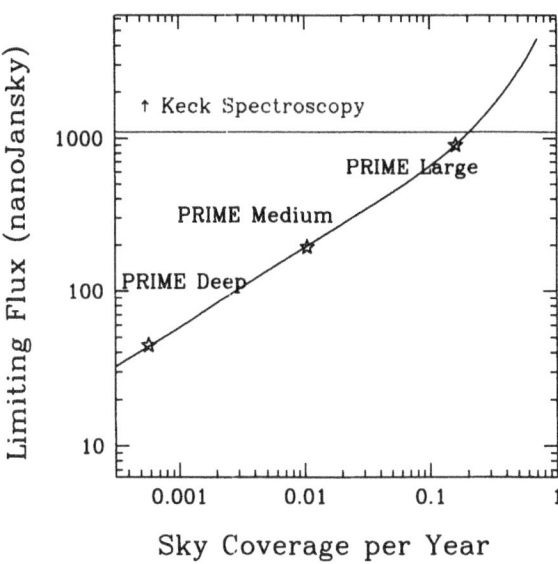

Fig. 3. Survey rate for *PRIME*. The two horizontal lines represent the capability of the Keck spectroscopic observations at low resolution ($R = 100$). The stars mark the anticipated limiting fluxes for *PRIME*.

The participating institutions in the *PRIME* project are Johns Hopkins University, Space Telescope Science Institute, Max-Planck-Institut fr Astronomie, and Goddard Space Flight Center. The basic design is a 75-cm telescope flying in low-altitude, sun-synchronous orbit. Large-format HgCdTe detectors will be used. These detectors require a working temperature of about 80 K, and the 3.5 μm band also requires that the optical system is cooled to below 150K. These thermal requirements may be achieved by passive cooling, which is reliable and inexpensive. A SMEX-Lite type spacecraft will be used to carry out altitude control, and data and command control.

As demonstrated in Figure 3, the sensitivity range of *PRIME* is between Keck and the *NGST* capability of low-resolution spectroscopy. In other words, the targets found by *PRIME* can feed into the world largest telescopes for follow-up observations. *PRIME* therefore profoundly enhances the capability of these telescopes.

Figure 4 and 6 plot the sensitivity range of the *PRIME* surveys. They demonstrate that the surveys are capable of finding quasars up to redshift of 25 and supernovae of redshift greater than 3. High-z supernovae may be found in the *PRIME* medium and

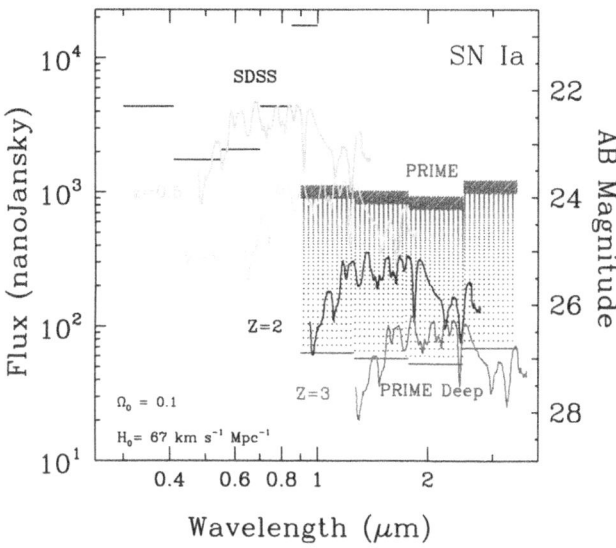

Fig. 4. Prime sensitivity and its application in the search for supernovae.

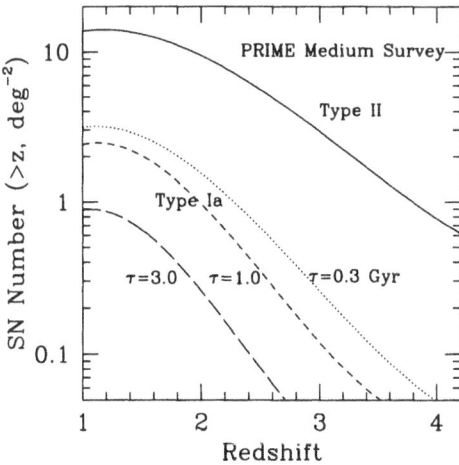

Fig. 5. Predicted cumulative number of supernovae in the *PRIME* medium survey. SNe II: solid line; SNe Ia with time delay from stellar birth $\tau = 0.3$ Gyr: dashed-dotted line, $\tau = 1.0$ Gyr: dotted line, and $\tau = 3.0$ Gyr: dashed line. Adapted from [1] and [4].

deep surveys (Figure 5). The surface density of these objects is highly speculative at higher redshifts, but according to the current models, our surveys should be able to find them or set clear cut to their luminosity functions.

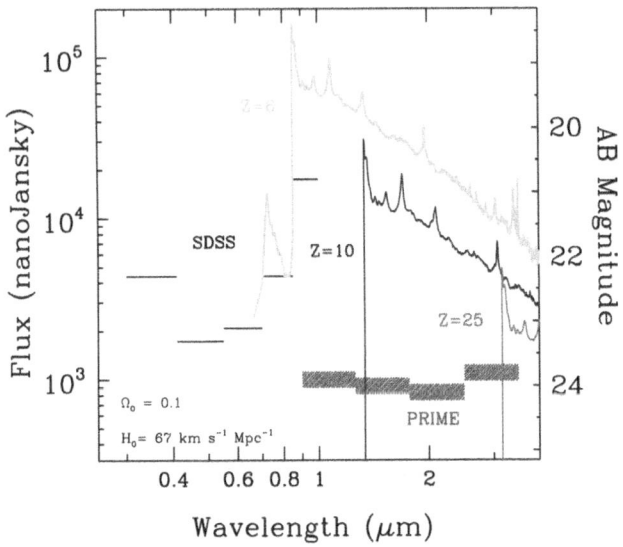

Fig. 6. *PRIME* sensitivity and the brightness of quasars at very high redshift, assuming a constant luminosity between $z = 5$ and 25.

The nominal mission will last approximately three years. The total amount of data will exceed 20 terabytes, and includes catalogues of several billion objects. The *PRIME* database will become one of the largest in the astronomical community. *PRIME* will open a new window of the early universe, and it will lead us to the discoveries of new classes of objects that have never been known to us.

References

1. T. Dahlén, & C. Fransson, A&A, **350**, 349, (1999)
2. X. Fan, et al. AJ, in press (2001)
3. Ch. Leinert, et al. A&ApS, **127**, 1 (1998)
4. P. Madau, M. Della Valle, & N. Panagia, MNRAS, **297**, L17, (1998)
5. M. F. Skrutskie, BAAS, **33**, 827, (2001)

Quasar Evolution and Star Formation History

Matthias Dietrich[1], Fred Hamann[1], Immo Appenzeller[2], Marianne Vestergaard[3], and Stefan Wagner[2]

[1] Department of Astronomy, University of Florida, Gainesville, FL 32611, USA
[2] Landessternwarte Heidelberg-Königstuhl, D-69117 Heidelberg, Germany
[3] Department of Astronomy, The Ohio State University, Columbus, OH 43210, USA

Abstract. We observed a sample of high redshift quasars ($z \simeq 3.4$, $z \simeq 4.5$) in the near-infrared to measure the relative iron and Mg II emission strengths and thereby date the first star formation epoch. A detailed comparison with quasars at low-redshift show essentially the same Fe II/Mg II emission ratios and very similar continuum and line spectral properties, suggesting a lack of evolution of the relative iron to magnesium abundance of the gas since $z \simeq 4.5$ in bright quasars. These results are consistent with major star formation activity in the host galaxies beginning at an epoch corresponding to $z_f \simeq 10$, when the age of the universe was less than 0.5 Gyrs ($H_o = 72$ km s^{-1} Mpc^{-1}, $\Omega_M = 0.3$, $\Omega_\Lambda = 0.7$).

1 Introduction

Assuming that the gas near quasars is related to the interstellar medium of the host galaxy, the elemental abundances can be used to trace star formation history out to redshifts of $z \gtrsim 6$. The prominent emission line spectrum of quasars provides several diagnostic line ratios of the chemical composition and enrichment history of quasar host galaxies at all cosmological epochs. Earlier studies of the emission line spectra demonstrated that quasars at redshift $z \gtrsim 3$ have metallicities of typically solar or higher [10–12,8,16,2,3,13]. These results show that before the epoch corresponding to $z = 3$ significant star formation must have taken place in the galactic or proto-galactic cores where these quasars reside.

Measurements of specific abundance ratios, such as Fe/α (where α= O, Mg, Si, etc.), can constrain the ages of the stellar population near quasars and thereby define the epoch of the first (local) star formation [12]. α-elements are produced predominantly by massive stars on short time scales via supernovae of types II,Ib,Ic. The dominant source of iron is ascribed to intermediate mass stars in binary systems, ending in supernova type Ia explosions [22,17,24,26]. The significantly different time scales of the release of α-elements and iron to the interstellar medium results in a time delay of the order of ~ 1 Gyr. Strong Fe II emission from quasars might therefore be an indication that the star formation began $\gtrsim 1$ Gyr earlier.

2 Results and Discussion

The best indicator of α-elements vs. iron in quasars is the strength of Mg II 2798 emission compared to broad blends of Fe II multiplets spanning several hundred Ångstroem (rest-frame) on either side of the Mg II line [25,28,1]. Very few quasars at redshifts larger than $z = 3$ have been observed across the Mg II and Fe II features with wavelength coverage that is wide enough to define the underlying continuum [14,7,15,20,27,18,21]. We observed a sample of 11 quasars in the near infrared ($\lambda \simeq 1$ - 2.5μm) with SoFi (3.5 m NTT, ESO), NIRSPEC (9.8 m Keck), and OSIRIS (4 m CTIO). Fig. 1 shows an example of our multi-component analysis of these spectra [4,6]. The quasar spectra were analysed by simultaneously fitting a powerlaw continuum, a Balmer continuum emission spectrum, and an empirical Fe-emission template [23]. We measured the Fe-emission for a rest wavelength range of $\lambda\lambda$ 2200 - 3090Å. For the six quasars at z$\simeq$3.4 we obtain an average ratio of I(Fe II UV)/I(Mg II) = 3.72$\pm$0.20 and for five the quasars at z$\simeq$4.5 I(Fe II UV)/I(Mg II) = 4.44$\pm$0.59. To compare the spectral properties of the high-z quasars with the quasar population at lower redshift, we calculated a mean quasar spectrum using 101 quasar spectra taken from a large quasar sample [5]. The lower redshift quasars were selected to have $z \leq 2$ and to cover a luminosity range comparable to the quasars under study. The criterion of comparable luminosity minimizes luminosity effects in the emission line strengths like the Baldwin effect [19]. Within the uncertainties there is no significant difference between the high redshift results and the ratio I(Fe II UV)/I(Mg II) = 3.82$\pm$0.40 we derive from the lower redshift mean quasar spectrum.

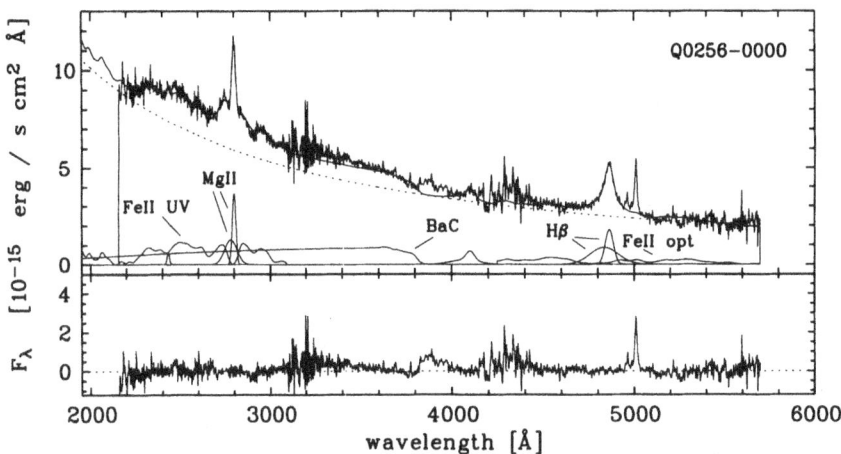

Fig. 1. In the top panel the spectrum of Q0256-0000 is shown together with the power law continuum fit (dotted line), the scaled and broadened Fe-emission template, the scaled Balmer continuum emission, and the Gaussian components to fit the Mg II2798 and Hβ4861 emission line profiles. The resulting fit is overplotted as a solid line. In the bottom panel the quasar spectrum is shown after subtraction of these components.

The lack of evolution in the relative iron emission strength for quasars out to $z \simeq 3.4$ and $z \simeq 4.5$ suggests that Fe is already significantly abundant at these redshifts and therefore an episode of major star formation began $\gtrsim 1$ Gyr before the observed quasar epochs. The age of the universe at the time when the light was emitted by our $z \simeq 3.4$ and $z \simeq 4.5$ quasars was about 1.8 Gyrs and 1.3 Gyrs, respectively. Given parameters and the stellar population age of $\gtrsim 1$ Gyr then implies an epoch for the beginning of star formation of $\lesssim 0.5$ Gyrs, corresponding to $z_f \gtrsim 10$. This epoch is consistent with model predictions [9], which also explain high metal abundances ($\gtrsim Z_\odot$) in bound massive objects (proto-galaxies) at $z \simeq 6$.

References

1. Boroson, T.A. & Green, R.F. 1992, ApJS , 80, 109
2. Dietrich, M., Appenzeller, I., Wagner, S.J., et al. 1999, A&A , 352, L1
3. Dietrich, M. & Wilhelm-Erkens U., 2000, A&A, 354, 17
4. Dietrich, M., et al., 2001, ApJ in press, astro-ph/0109208
5. Dietrich, M. & Hamann, F. 2001a, in prep.
6. Dietrich, M. & Hamann, F. 2001b, in prep.
7. Elston, R., Thompson, K.L., & Hill. G.J., 1994, Nature, 367, 250
8. Ferland et al.,1996, ApJ, 461, 683
9. Gnedin, N.Y. & Ostriker, J.P. 1997, ApJ, 486, 581
10. Hamann, F. & Ferland, G.F., 1992, ApJ, 381, L53
11. Hamann, F. & Ferland, G.F., 1993, ApJ, 418, 11
12. Hamann, F. & Ferland, G.F., 1999, ARAA, 37, 487
13. Hamann, F. et al., 2001, ApJ, in press, astro-ph/0109006
14. Hill, G.J., Thompson, K.L., & Elston, R., 1993, ApJ, 414, L1
15. Kawara, K., Murayama, T., Taniguchi, Y., & Arimoto, N., 1996, ApJ, 470, L85
16. Korista, K.T., Hamann, F., Ferguson, J., & Ferland, G.J., 1996, ApJ, 461, 641
17. Matteucci, F. & Greggio, L. 1986, A&A , 154, 279
18. Murayama, T., Taniguchi, Y., Evans, A.S., et al., 1998, AJ, 115, 2237
19. Osmer, P.S. & Shields, J.C. 1999, in 'Quasars and Cosmology', ASP Conf. Ser. 162, eds. G. Ferland & J.A. Baldwin, p.235
20. Taniguchi, Y., Murayama, T., Kawara, K., & Arimoto, N., 1997, PASJ, 49, 419
21. Thompson, K.L., Hill, G.J., Elston, R., 1999, ApJ, 515, 487
22. Tinsley, B.M. 1979, ApJ , 229, 1046
23. Vestergaard, M. & Wilkes, B.J. 2001, ApJS , 134, 1
24. Wheeler, J.C., Sneden, C., & Truran, J.W., 1989, ARAA, 27, 279
25. Wills, B.J., Netzer, H., & Wills, D. 1985, ApJ , 288, 94
26. Yoshii, Y., Tsujimoto, T., & Nomoto, K., 1996, ApJ, 462, 266
27. Yoshii, Y., Tsujimoto, T., & Kawara, K., 1998, ApJ, 507, L113
28. Zheng, W. & O'Brien, P.T. 1990, ApJ, 353, 433

This work was supported by NASA through their Long Term Space Astrophysics program (NAG5-3234), NSF AST99-84040, and an archival research grant from the Space Telescope Science Institute (AR-07988.01-96A).

Results from the Mount Stromlo Abell Cluster Supernova Search

Lisa M. Germany[1], David J. Reiss[2], Brian P. Schmidt[3], and Christopher W. Stubbs[2]

[1] European Southern Observatory, Vitacura, Santiago, Chile
[2] University of Washington, Seattle, USA
[3] Research School of Astronomy and Astrophysics, Australian National University, Canberra, Australia

1 The Goal

The Mount Stromlo Abell Cluster Supernova Search was a 3 year project to discover and follow type Ia supernovae (SNe) in rich, southern Abell Clusters. One of the primary goals of this search, was to measure accurate distances to these type Ia SNe to determine their peculiar velocities and derive the bulk motion for the sample.

2 The Results

Despite uncooperative weather, the Mount Stromlo Abell Cluster Supernova Search uncovered 48 SNe during its 3 year campaign. Unfortunately, due to the weather, telescope scheduling constraints and the faintness of many of the SNe, we were unable to obtain spectra for every object. For this reason, the redshift of each supernova was taken as the redshift of its host galaxy, and for those SNe without spectra, the supernova class was determined by the goodness of fit of the template light curves to the data. Of the 48 SNe discovered, 23 were deemed to be type Ia by these methods. In addition, we augmented our relatively small sample of type Ia SNe with all the other well-observed type Ia data available, bringing the total number of SNe used to determine the Hubble constant and the bulk flow to 61. We used the Δm_{15} template fitting technique [1][5] to determine the redshift-independent distance to each supernova in this extended sample.

2.1 The Hubble Constant

Figure 1 shows the Hubble diagram for the extended sample of type Ia SNe. We measured a Hubble constant of 70 ± 4(internal)± 7(external) km s^{-1} Mpc^{-1} which is a little larger but still within the error budgets of the values derived by several other researchers [3][6][8].

Figure 1 was also used to check our supernova classification based on the goodness of fit of the template light curves. If a supernova of some other type was wrongly classified as a type Ia, it would be fainter than expected and would stand out in the Hubble diagram. The two outliers seen in Fig. 1 are actually spectroscopically confirmed type Ia SNe which do not contain enough photometric information to tie down the time of maximum in the template fit. Although there is no way of knowing whether some true type Ia SNe have been eliminated by our methods, we are confident that those used in the determination of the bulk flow were indeed type Ia.

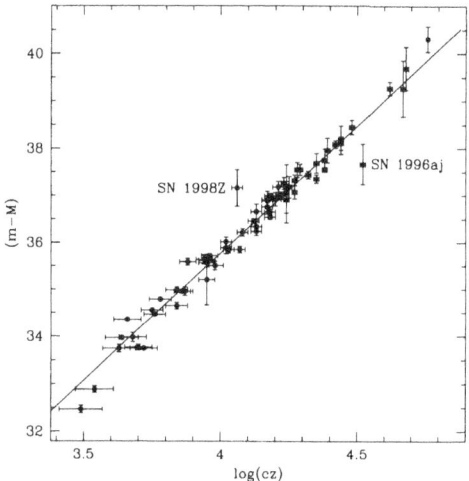

Fig. 1. The Hubble diagram constructed with the type Ia SNe from the Mount Stromlo Abell Cluster Supernova Search, and those from [6] with $z > 0.01$. The derived Hubble constant is 70 ± 4 km s^{-1} Mpc^{-1}

2.2 The Bulk Flow

Figure 2 shows the positions of the type Ia SNe, the derived bulk motion and the directional uncertainty of the motion based on 2000 monte carlo simulations. The maximum likelihood bulk flow, taking into account survey geometry and small-scale flows, was found to be 763 ± 455 km s^{-1} in the direction $l = 202 \pm 36°$, $b = -35 \pm 27°$. This result was robust against the deletion of any one supernova, as well as the division of the sample into redshift shells. Also shown are the derived bulk flow directions from other surveys [2][4][10]. In order to properly compare the different surveys, one must take into account sparse sampling and suvey geometry of each sample. This detailed analysis between surveys has not yet been done, but will be carried out in the future.

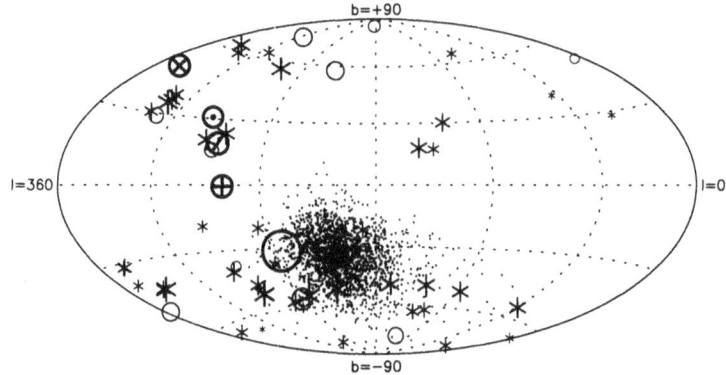

Fig. 2. The peculiar velocities of the type Ia SNe in Galactic coordinates, and the bulk motion of the extended sample (the black square). SNe with positive(negative) peculiar velocities are indicated by asterisks(circles) whose sizes are related to the amplitude of the peculiar velocity. The cloud of small dots shows the directional error determined from Monte-Carlo simulations. Also shown with respect to the CMB frame are the Local Group dipole (⊙), the Lauer & Postman dipole (⊗)[4], the Hudson dipole (⊕)[2], and the Willick dipole (⊘)[10]

3 Conclusion

Based on the large errors in the bulk flow measurement and the shape of the window function derived for this sample, many more type Ia SNe are needed to properly tie down the bulk flow. In accordance with the findings of Watkins & Feldman [9], we agree that a sample of 200 SNe within a volume of $z < 0.03$ could accurately determine the bulk flow, provided one could reduce the effect of cosmic covariance. A sample of this many SNe could be constructed within the next few years given the current rate of discovery of these objects.

References

1. M. Hamuy et al.: AJ **109**, 1 (1995)
2. M. J. Hudson et al.: 1999, ApJ **512**, L79 (1999)
3. S. Jha et al.: ApJS **125**, 74 (1999)
4. T. R. Lauer, M. Postman: ApJ **425**, 418 (1994)
5. M. M. Phillips: ApJ **413**, L105 (1993)
6. M. M. Phillips et al.: AJ **118**, 1766 (1999)
7. D. J. Reiss, L. M. Germany, B. P. Schmidt, C. W. Stubbs: AJ **115**, 26 (1998)
8. N. B. Suntzeff, et al.: AJ **117**, 1175 (1999)
9. R. Watkins, H. A. Feldman: ApJ **453**, L73 (1995)
10. J. A. Willick: ApJ **522**, 647 (1999)

The Stellar Population of High Redshift Galaxies

Dörte Mehlert[1], Stefan Noll[1], Immo Appenzeller[1], and the FDF team[1,2,3,4]

[1] Landessternwarte Heidelberg, [2] Universitäts-Sternwarte München,
[3] Universitäts-Sternwarte Göttingen, [4] ESO Garching, [5] MPIA Heidelberg

1   The Project

The consortium of the institutes listed above, which combined forces to built the FORS instruments at the ESO VLT used a significant fraction of their guaranteed observing time to observe a "FORS Deep Field". One of the scientific objectives of the FDF is to study the dependence of the physical properties of galaxies on the cosmic age. Hence we obtained deep multi-band images of the FDF as well as spectra of a subsample of galaxies in the FDF. With these spectroscopic data we will investigate the stellar population (age of the starburst, IMF, metallicity, dust reddening) of distant galaxies with the aim of deriving new information on the evolution of the young universe.

2   The Observations

During 3 nights of MOS observations with FORS2 at the VLT we obtained 389 object spectra. Using the grism 150I and a slitwidth of 1" we covered the spectral range from $\lambda = 3000...9200$ Å with a spectral scale of 5 Å/pixel. Depending on the object magnitude the integration times ranged between 2 and 10 hours with average seeing of 0.69". Standard reduction (bias subtraction, flatfielding, cosmic ray elimination, sky subtraction, rebinning, etc.), was performed using MIDAS routines.

3   First Results

For ≈ 203 objects we have spectra with sufficient S/N to determine the type and redshift. Among these we found 169 galaxies, 71 with $z > 1$. Fig. 1 shows a typical spectrum of a high redshift galaxy ($z = 2.437$) in our sample. The most prominent feature is the Ly_α absorption line. Additionally several metal absorption lines can be identified. Furthermore the intense (rest frame) UV continua indicates intensive starburst activity. In Fig. 2 we present a comparison of three of our galaxies with synthetic spectra from Leitherer et al. (2001), showing the spectral region of the CIV resonance line. For galaxy 1747 ($z = 2.37$) and 2403 ($z = 2.77$) the spectral features of CIV are well represented by the solar and LMC model, respectively. On the other hand, for galaxy 2160 ($z = 3.26$) neither the solar nor the LMC models fit the observed feature. Hence we conclude that the metallicity of object 2160 is much lower than the LMC value.
Note that with increasing redshift the galaxies' metallicity seems to decrease. To investigate this trend for the whole sample we measured the equivalent width (EW) of the two

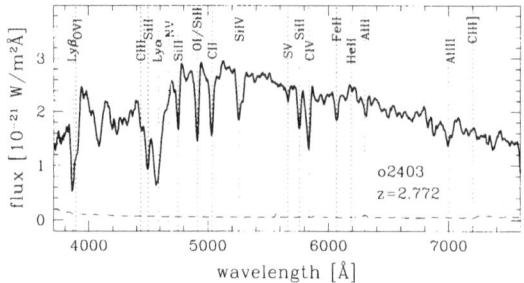

Fig. 1. Spectrum of a distant (z = 2.772) galaxy in the FDF with a S/N $\approx$ 35 per resolution element (= 5 pixel). The dashed line indicates the noise level, which varies with wavelength due to the night sky spectrum and the wavelength dependent instrumental efficiency. The position of some prominent spectral lines are indicated by the vertical dotted lines.

prominent stellar absorption lines CIV and SiIV. To increase the sample we included measurements for some galaxies in the field of the lensing cluster 1E0657, the HDF-S and the AXAF Deep Field, which we had observed during FORS commissioning runs with the same spectroscopic setup as described above.

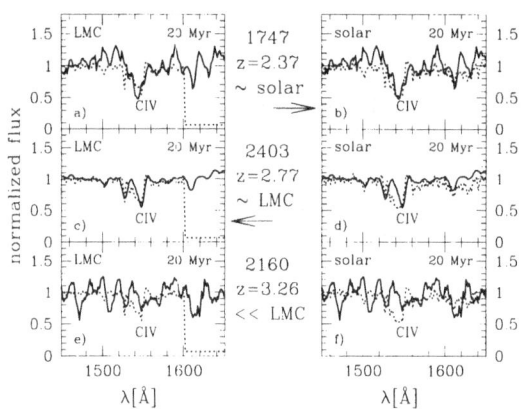

Fig. 2. Comparison between the observed spectra of galaxy 1747, 2403, 2160 (solid line) and synthetic spectra from Leitherer et al. (2001; dotted line). The synthetic spectra are based on 20 Myr old starbursts with continuous star formation (1 M$_\odot$/yr) and the parameter α_{IMF} = 2.35 and M$_{up}$ = 100 $_\odot$. The left and right panels indicate models with LMC and solar metallicity, respectively.

Fig. 3a shows that for $z > 1$, EW(CIV) increases with decreasing redshift. A weighted χ^2 fit gives a slope of $\alpha = -1.44 \pm 0.16$ (dashed line). Including the measured EW of 5 nearby ($z = 0$) starburst galaxies, arbitrarily chosen from the IUE archive, extends the trend to the local universe. A weighted χ^2 fit gives a slope of $\alpha = -1.14 \pm 0.12$ (solid line).

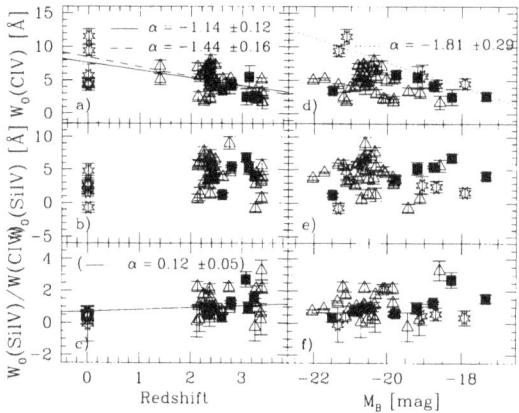

Fig. 3. Measured CIV 1550 and SiIV 1400 rest frame equivalent widths as well as their ratio SiIV/CIV versus redshift (a, b, c) and absolute B-magnitude (d, e, f). The absolute magnitudes are determined with $H_0 = 50\,\mathrm{km/s/Mpc}$ and $q_0 = 0.5$. For the k-correction we used values provided by Möller et al. (2001). Evolutionary corrections are not applied. Open triangles: FDF galaxies; filled squares: Galaxies in the field of the cluster 1E0657, in the HDF-S and the AXAF Deep Field, respectively (see text); stars: Nearby starburst galaxies. Dashed line, solid line, dotted line: weighted χ^2 for objects with $z > 1$ only, for all shown galaxies and for the local ones ($z > 1$), respectively.

Fig 4a and b show that both EW(CIV) and EW(SiIV) mainly depend on the metallicity of the stellar population but only little on the age of a starburst. Hence the existing anticorrelation of EW(CIV) with z indicates an increase of metallicity with decreasing redshift (i.e., increasing age of the universe). A similar behavior has been found for damped Lyα systems (Savaglio et al. 2001).

The fact that for SiIV no correlation with z is present may be due to the fact that EW(CIV) is universally present for O supergiants, main sequence stars and dwarfs, while EW(SiIV) is luminosity dependent and decreases rapidly from supergiants to dwarfs (Walborn & Panek 1984; Pauldrach et al. 1990).

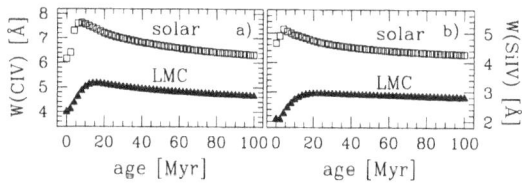

Fig. 4. Measured CIV 1550 (a) and SiIV 1400 (b) equivalent widths within the synthetic spectra of Leitherer et al. (2001) versus the age of the starburst. Open squares and filled triangles correspond to solar and LMC metallicity, respectively.

Therefore the SiIV is more strongly affected by population differences (i.e., stellar age differences) than the CIV line. EW(SiIV) has its maximum in B0/B1 stars, while EW(CIV) is mainly universally present in O and bright B stars (e.g., Leitherer, et al. 1995). Hence the ratio of EW(SiIV)/EW(CIV) contains information about the star formation history (instantaneous or continuous) and the stellar population itself (e.g., IMF and cutoff masses). Unfortunately the different parameters that determine the value of EW(SiIV)/EW(CIV) cannot be disentangled easily. As seen in Fig. 3c the ratios of EW(SiIV)/EW(CIV) show a weak trend corresponding to an increasing ratio with z seems (a weighted χ^2 fit gives $\alpha = 0.12 \pm 0.05$). This could be understood in terms of increase of the relative importance of continuous star formation (decrease of instantaneous starbursts) at low z. Finally, Figs. 3d to f show that there is no overall dependence of the measured equivalent width on the galaxies' luminosity. If, in fact, EW(CIV) is mainly determined by the metallicity, the nearby starburst galaxies seem to follow the well known local metallicity-luminosity relation (e.g. Kobulnicky & Zaritsky 1998), while the high-z galaxies seem not to conform this relation. As also found for Lyman break galaxies by Pettini et al. (2001), the high-z galaxies investigated in this paper seem to be overluminous for their metallicity (EW(CIV)), which may indicate that their mass-to-light ratios are low compared to present-day galaxies.

4   Conclusions

• Our observed high-z galaxy spectra agree with synthetic ones.
• EW(CIV) is a good indicator for the galaxies metallicity.
• Our high-z starburst galaxies show increasing metal content with decreasing redshift and are overluminous for their metallicity compared with local starburst galaxies.

References

1. Kobulnicky, H.A., Zaritsky, D. 1998, ApJ, 511, 188
2. Leitherer, C., Leao, J.R.S., Heckman, T.M., et al. 2001, ApJ, 550, 724
3. Leitherer, C., Robert, C., Heckman, T.M. 1995, ApJS, 99, 173
4. Möller, C.S., Fritze-v.Alvensleben, U., Fricke, K.J. 2001, A&A, submitted
5. Pauldrach, A. W. A., Puls, J., Kudritzki, R.-P., Butler, K. 1990, A&A 228, 125
6. Pettini, M., Shapley, A.E., Steidel, C.C., et al. 2001, ApJ, in press
7. Savaglio, S., Panagia, N., Stiavelli, M. 2001, in ASP Conference Series, Cosmic evolution and galaxy formation: Structure, Interactions, and Feedback, eds. J. Franco, E. Terlevich, O. Lopez-Cruz, I. Aretxaga, in press
8. Walborn, N. R., Panek, R. J. 1984, ApJL 280, L27

Formation of First Stars and First Luminous Objects

Ryoichi Nishi

Department of Physics, University of Kyoto, Sakyo-ku, Kyoto 606-8502, Japan

Abstract. We investigate the formation site of the primordial stars, the formation process of them and the feedback on the host clouds. We also investigate the formation condition of first luminous objects. Considering 3 σ perturbations, the mass of the first luminous objects is estimated to be several times $10^7 M_\odot$, of which virialised redshift is about 30. However, they can be enough luminous only after polluted by heavy elements enough and the expected redshift redshift when they become luminous is about 10.

1 Formation Site of Primordial Stars

First, we consider the cooling diagram of collapsed clouds. including H_2 cooling (Nishi and Susa 1999). With this diagram, we can predict whether a cloud cools and becomes a star formation site.

In order to estimate the cooling rate at $T < 10^4$K, we need the fraction of H_2. The number fraction of H_2 is not generally in equilibrium for $T < 10^4$ K in the epoch of galaxy formation. Using recent reaction rates and the cooling rate of H_2 (Galli and Palla 1998), we estimate the fraction of H_2. There are four important time scales, dissociation and formation time of H_2, cooling time, and recombination time, respectively. Comparing these time scales we can estimate the fraction of H_2 and cooling efficiency for cosmological low mass objects.

There are three different star formation sites and the star formation process is probably different. (1) The cloud in the region $t_{ff} < t_{cool} < H^{-1}$ cools faster than the Hubble expansion, but cannot collapse dynamically, initially. Therefore, star formation of this site may not be so efficient. (2) The cloud in the region $t_{cool} < t_{ff}$ cools faster than the free fall time and can collapse dynamically. (3) The first collapse of the cloud with the virial temperature higher than about 10^4 K is highly dynamical and the ionization at the bounce epoch is important (e.g., Shapiro and Kang 1987, Yamada and Nishi 1998, Susa et al. 1998).

2 Dynamical Collapse of Primordial Protostellar Clouds

Here, we show the results of the hydrodynamical calculation of primordial protostellar clouds and the formation of stellar cores (Omukai and Nishi 1998).

The collapse proceeds almost self-similarly and is analogous to the Larson-Penston similarity solution (Larson 1969, Penston 1969) until the central region reaches the stellar density. The density profile outside the core is very resemble to $\rho \propto r^{-2.2}$, which correspond to the effective adiabatic index is 1.1.

After the central part of the cloud contracts almost adiabatically to some extent, a hydrostatic core with very small mass ($\sim 5 \times 10^{-3} M_\odot$) forms at the center. At that time, the central number density and temperature are $\sim 10^{22}$ cm^{-3} and $\sim 3 \times 10^4$ K. These values are almost the same as those of present-day stellar cores at the formation epoch. But before the formation of the stellar core, no transit core is formed, in contrast to the case of present-day star formation.

The mass accretion rate is estimated as

$$\dot{M}_* = 8.3 \times 10^{-2} M_\odot \, \mathrm{yr}^{-1} \left(\frac{t}{1 \, \mathrm{yr}}\right)^{-0.27} = 3.7 \times 10^{-2} M_\odot \, \mathrm{yr}^{-1} \left(\frac{M_*}{M_\odot}\right)^{-0.37} . \quad (1)$$

The estimated mass accretion rate is very large and diminishes with time. The main reason for this large mass accretion rate is that the temperature of primordial gas clouds ($\sim 10^3$ K) is higher (which is the result of weak cooling) than that of the present-day molecular clouds (~ 10 K). The final stellar mass is expected to be large because of this large mass accretion rate (e.g., Omukai and Palla 2001).

3   Feedback by the First Generation Massive Stars

Now, we investigate feedback by the first generation massive stars. The feedback effects considered here are UV radiation from the stars and energy input by SNe.

3.1   Star Formation Regulation by UV Radiation

Around an OB star, hydrogen is photoionized, and an HII region is formed. Ionizing photons hardly escape from the HII region, but photons whose radiation energy are below the Lyman limit can get away. Such FUV photons photodissociate molecular hydrogen, and a photodissociation region (PDR) is formed just outside the HII region. It is shown that H_2 is dissociated in such a large region that the whole of an ordinary low mass cloud is influenced by one O5 type star (Omukai and Nishi 1999). This indicates that FUV radiation from one or a few OB stars prohibits the whole small pregalactic cloud from H_2 cooling and quenches subsequent star formation in it.

For low metallicity clouds, if metallicity is lower than about 10^{-2} of the solar metallicity, the host cloud depletes main coolant (H_2) by one O5 type star (Nishi and Tashiro 1999). This indicates that stars do not form efficiently before the metallicity becomes larger than about 10^{-2} of the solar metallicity. When the metallicity is lower than this value, star formation can occur only after previous massive star died. If these massive stars died with SN explosion, we can consider that SN-induced star formation is the only star formation process, since stars can form only after SN explosion.

3.2   Disruption of the Host Cloud by SNe

After massive first generation stars form, evolution of the host clouds become slower because of strong regulation by UV radiation as the previous section (Nishi and Susa 1999). Thus, next generation stars are hardly formed before the first generation stars die. Subsequent SNe might disrupt the gas binding before significant amount of total gas transferred into stars. We derived the cloud disruption condition by SNe, with the assumption that the cloud is spherical and the density is constant, for the simplicity. Clouds with the mass below about $10^{7.5} M_\odot$ are almost disrupted.

4 Formation of First Luminous Objects

Considering the previous investigations, the mass of the clouds, which can evolve into luminous objects, are fairly massive (virial temperature is higher than about 10^4 K) (Nishi et al. 1998). Considering 3 σ perturbations, the mass of the first luminous objects is estimated to be several times $10^7 M_\odot$, of which virialised redshift is about 30. However, the star formation rate is very low when the metallicity is lower than about 1% of solar metallicity and they can be enough luminous only after polluted by heavy elements enough.

Before the metallicity is higher than the above critical value, about 50 cycles of Star Formation and SN phase are necessary (Nishi and Tashiro 2000). The interval time scale between the SN and the next generation star formation is about the free-fall time scale of the surrounding medium (Nishi and Kamaya 2000). Thus, the necessary time for enough pollution is $\sim 50 t_{ff}(\rho_B) \sim 6 t_H(\rho/\rho_B)^{1/2}$. Here $t_{ff}(\rho_B) (= (32 G \rho_B/3\pi)^{-1/2})$ is the free-fall time scale for the mean baryon density of a cloud and $t_H (= (8\pi G \rho_b/3)^{-1/2})$ is the Hubble time scale and ρ_b is the background mean density of the universe. And the mean density of a cloud is estimated as $\rho = 18\pi^2 \rho_b$. Before star formation, baryons are cooled and the density becomes higher and we estimate as $\rho_B \sim \rho$. Thus, the redshift when first luminous objects becomes luminous is $z_L \sim z_{vir}/6^{2/3} - 1 \sim 10$, since $1 + z \propto t^{-2/3}$ and $z_{vir} \sim 30$. This redshift may correspond to the reionization redshift of the universe.

References

1. D. Galli, F. Palla: AA **335**, 403 (1998)
2. R.B. Larson R.: MNRAS **145**, 271 (1969)
3. R. Nishi, Kamaya, H.: ApJ **532**, 1172 (2000)
4. R. Nishi, H. Susa: ApJL **523**, L64 (1999)
5. R. Nishi, H. Susa, H. Uehara, M. Yamada, K. Omukai Prog. Theor. Phys. **100**, 881 (1998)
6. R. Nishi, M. Tashiro: ApJ **537**, 50 (2000)
7. K. Omukai, R. Nishi: ApJ **508**, 141 (1998)
8. K. Omukai, R. Nishi: ApJ **518**, 64 (1999)
9. K. Omukai K., F. Palla: ApJ in press (2001)
10. M.V. Penston: MNRAS **144**, 425 (1969)
11. P.R. Shapiro, H. Kang: ApJ **318**, 32 (1987)
12. H. Susa, H. Uehara, R. Nishi, M. Yamada: Prog. Theor. Phys. **100**, 63 (1998)
13. M. Yamada, R. Nishi: ApJ **505**, 148 (1998)

Bike tour of Munich

Part VII

Formation and Growth of Supermassive BH

Formation and Growth of Supermassive Black Holes

Martin J. Rees

Institute of Astronomy, Madingley Road, Cambridge, CB3 OHA

Abstract. The supermassive collapsed objects in the cores of most galaxies pose still-unanswered questions. How did they form, what is the reason for the close correlation between their masses and the properties of the host galaxy, and what will their implications be for cosmology and the study of relativistic gravity?

1   Introduction

Compact dark objects, with deep gravitational potential wells, lurk in most galactic centres; they are implicated in the power output from active galactic nuclei (AGNs), and in the production of relativistic jets that energise strong radio sources. The demography of these massive holes has been clarified by studies of relatively nearby galaxies: the centres of most of these galaxies display either no activity or a rather low level, but most seem to harbour dark central masses.

There are two spectacularly convincing cases of massive collapsed objects in nearby galaxies. The first, in the peculiar spiral NGC 4258, has been revealed by amazingly precise mapping of gas motions via the 1.3 cm maser emission line of H_2O. [1,2]. The spectral resolution of this microwave line is high enough to pin down the velocities with accuracy of 1 km/sec. The Very Long Baseline Array achieves an angular resolution better than 0.5 milliarc seconds (100 times sharper than the HST, as well as far finer spectral resolution of velocities!). These observations have revealed, right in NGC 4258's core, a disc with rotational speeds following an exact Keplerian law around a compact dark mass. The inner edge of the observed disc is orbiting at 1080 km/sec. It would be impossible to circumscribe, within its radius, a stable and long-lived star cluster with the inferred mass of $3.6.10^7$ $M_\odot$.

The second utterly convincing candidate lies in our own Galactic Centre. As discussed elsewhere in these proceedings, there is direct evidence from stellar proper motions, observed in the near infrared band, where obscuration by intervening material is less of an obstacle. The speeds scale as $r^{-1/2}$ with distance from the centre, consistent with a hole of mass $2.6.10^6$ $M_\odot$. Corroboration that comes from the compact radio source that has long been known to exist right at the dynamical centre of our Galaxy, which can be interpreted in terms of accretion onto a massive hole (see [3-5] and Narayan in these proceedings). Direct evidence used to be ambiguous because intervening gas and dust in the plane of the Milky Way prevents us from getting a clear optical view of the central stars, as we can in, for instance, M31. A great deal was known about gas motions, from radio and infrared measurements, but these were hard to interpret because gas does not move ballistically like stars, but can be influenced by pressure gradients, stellar winds, and other non-gravitational influences.

As other speakers discuss (see also [6] and references cited therein), there is a remarkably close proportionality between the hole's mass and the velocity dispersion of the central bulge or spheroid in the stellar distribution (which is of course the dominant part of an elliptical galaxy, but only a subsidiary component of a disc system like M31 or our own Galaxy). How did the holes form?

2   Scenarios for Black Hole Formation

We have got used to the idea that black holes exist within most galaxies, and it now seems unsurprising that material should accumulate, and perhaps undergo runaway collapse. but it is rather depressing that we still cannot decide which formation route for supermassive black holes is most likely. Let us briefly consider the main options

2.1   Monolithic Formation of Supermassive Objects

One possibility is that the gas in a 'proto-spheroid' does not all break up into stellar-mass condensations, but that some of it (including, perhaps, material that has already been reprocessed through high-mass stars, and consequently already enriched in heavy elements) undergoes monolithic collapse into a supermassive star, which then collapses to a black hole. As the gas evolves (through loss of energy and angular momentum) to a state of higher densities and more violent internal dissipation, radiation pressure would prevent fragmentation, and puff it up into a single superstar. Ordinary star formation may be suppressed even at less extreme densities – i.e. before the gas has become a single superstar – by other effects. For example, a magnetic field, even if not dynamically important overall, could inhibit fragmentation, especially because the free-electron concentration is unlikely to fall low enough to permit ambipolar diffusion, whereby the magnetic flux can escape from protostars in present-day molecular clouds.

Once a large mass of gas started to behave like a single superstar, it would continue to contract and deflate. Some mass would inevitably be shed, carrying away angular momentum, but the remainder would undergo complete gravitational collapse. This could be a substantial fraction – for example, if 10 percent of the mass had to be shed in order to allow contraction by a factor of 2, about 20 percent could form a black hole.

The mass of the hole would depend on that of its host galaxy, though not necessarily via an exact proportionality: the angular momentum of the protogalaxy and the depth of its central potential well are relevant factors too. Firmer and more quantitative conclusions will have to await elaborate numerical simulations. But it certainly seems in no way implausible that massive black holes form directly from gas (some, albeit, already processed through stars), perhaps after a transient phase as a supermassive object.

2.2   Mergers of Smaller Holes (Stellar Mass or 'Intermediate Mass')

The first-generation stars may leave massive remnants. As discussed in detail by, for instance Heger and Woosley ([7] and these proceedings), there are two distinct mass ranges that do this:

Ordinary massive stars, with helium core masses up to $64\,M_{\odot}$; and
'Very Massive Objects' (VMOs) with helium cores above $130\,M_{\odot}$.
Stars in between these two ranges leave no compact remnant at all, instead ending their lives by a disruptive explosion induced by the onset of electron-positron pair production.

It is important to pin down the mass range of these first stars, by pursuing the lines of work discussed by Abel and Bromm in these proceedings. If most of the UV background comes from such objects, we might expect $\sim 10^{-4}$ of the mass of a galaxy to be in the form of their remnants, and they could have interesting implications in the present universe [8]. The effects of biasing would concentrate these holes in the bulges of early-forming galaxies (by a factor ~ 4). Dynamical friction would induce extra concentration within the central 100 pc; such objects, captured by supermassive holes, would yield gravitational radiation signals detectable by LISA out to redshifts of order unity, possibly dominating the event rate (see further discussion in section 3).

Could there be a link between these 'intermediate mass' holes and supermassive holes? There are two possibilities. The most obvious, at first sight, is that a cluster of such objects might merge into one. But it is not easy for a cluster of black holes to merge into a single one. To see this, we can note that one binary with orbital speed 10^4 km/sec, (which would have a separation of 1000 Schwarzschild radii) would have just as much binding energy as a cluster of 10000 holes with velocity dispersion 100 km/sec. Thus, if a cluster accumulated in the centre of a galaxy, the likely outcome would be the expulsion of most objects, as the consequence of straightforward N-body dynamics, leaving only a few.

The prospects of build-up by this route are not quite as bad as this simple argument suggests, because the binding energy of the compact cluster could be enhanced by dynamical friction on lower-mass stars, by gas drag, or by gravitational radiation. This nonetheless seems an inefficient route towards supermassive holes.

Ordinary stars, with large geometrical cross-sections, have a larger chance of sticking together than pairs of black holes; indeed, in clusters with velocity dispersion of a few hundred km/sec, a star is likely to undergo a coalescence event on a timescale of order the two-body relaxation time. We therefore cannot exclude a 'scenario' where a supermassive star builds up within a dense central cluster of ordinary stars. The most detailed calculations were done by Quinlan and Shapiro ([9] and other references cited therein). These authors showed that stellar coalescence, followed by the segregation of the resultant high-mass stars towards the centre, could trigger runaway evolution without (as earlier and cruder work had suggested) requiring clusters whose initial parameters were unrealistic (i.e. already extremely dense, or with implausibly high velocity dispersions). It would be well worthwhile extending these simulations to a wider range of initial conditions, and also to follow the build-up from stellar masses to supermassive object.

2.3 Runaway Growth of a Favoured Stellar-Mass Hole to Supermassive Status

Even if a large population of low-mass holes is unlikely to merge together, is it, alternatively, possible for one of them, in a specially favoured high-density environment, to undergo runaway growth via accretion? An often-cited constraint on the growth rate is based on the argument that, however high the external density was, growth could not happen on a timescale less than the classic 'Salpeter time'

$$t_{\text{Sal}} = 4 \times 10^7 (\varepsilon/0.1) \, \text{yrs} \tag{1}$$

For an efficiency ε of 0.1 this would yield an e-folding timescale of 4×10^7 years. If these holes started off with stellar masses, or as the remnants of Population III stars, there would seem to be barely enough time for them to have grown fast enough to energise quasars at $z = 6$.

This is not, however, a generic constraint; even twenty years ago there were several suggestions in the literature for evading it. For instance, a super-Eddington luminosity could be driven off, along the spin axis, via a magnetically-dominated wind. Moreover, there are two distinct ways whereby, even without magnetic fields, the efficiency can adjust itself to a low value so as to allow a hole to accept a high inflow rate, in accordance with (1), without emitting a super-Eddington luminosity. The first of these was Begelman's suggestion [10] that radiation would be trapped and advected inwards. The value of r_{trap} scales in proportion to $\dot{M}$ (for high inflow rates) so that the efficiency with which radiation escapes automatically remains lower than the Eddington limit. This is clearly relevant for spherical inflow; less so when angular momentum is important. However Abramowicz and his colleagues noted (eg [11]) that for a thick torus, the inner cusp (on which the angular velocity is Keplerian) can adjust its position so that it corresponds to an orbit of small binding energy: i.e., for a torus around a Schwarzschild hole, it moves in from $6r_g$ (the innermost stable orbit, and the orbit of highest binding energy) towards $4r_g$, the circular orbit of zero binding energy (where $r_g = GM/c^2$).

Although these very simple models illustrate that (1) does not offer a simple constraint, the actual situation is still confusing. Recent work suggests that there are circulation patterns and it is unclear that a black hole can accept a high $\dot{M}$ without 'choking'. On the other hand, Begelman [12] and Shaviv [13] have discussed ingenious models where the gas develops a small-scale blobby or filamentary structure, such that a 'super-Eddington' luminosity can escape along low-density channels, whereas the coupling between high-density and low-density gas (perhaps via magnetic fields) is strong enough that the gravitational energy of both components can be extracted.

3   The Galactic Context

3.1   The Key Issues

Physical conditions in the central potential wells of young and gas-rich galaxies should be propitious for black hole formation: such processes, occurring in the early-forming galaxies that develop from high-amplitude peaks in the initial density distribution, are presumably connected with high-z quasars. It now seems clear that most galaxies that existed at $z = 3$ would have participated subsequently in a series of mergers; giant present-day elliptical galaxies are the outcome of such mergers. Any black holes already present would tend to spiral inwards, and coalesce (unless a third body fell in before the merger was complete, in which case a Newtonian slingshot could eject all three: a binary in one direction; the third, via recoil, in the opposite direction).

The issues – discussed in Kauffmann's contribution, which presents detailed semi-analytic models – are then:
(a) How much does a black hole grow by gaseous accretion (and how much electromagnetic energy does it radiate) at each stage? Models based on semi-analytic schemes for galaxy evolution have achieved a good fit with the luminosity function and z-dependence of quasars. Less gas is available at later epochs, and this accounts for the scarcity of high-luminosity AGNs at low z.
and
(b) How far back along the 'merger tree' did this process start? A single big galaxy can be traced back to the stage when it was in hundreds of smaller components with individual internal velocity dispersions as low as 20 km/sec. Did central black holes form even in these small and weakly bound systems?

Perhaps black holes form with the same efficiency in small galaxies (with shallow potential wells), or maybe their formation had to await the buildup of substantial galaxies with deeper potential wells (i.e. with velocity dispersion σ_v above some threshold). This issue is important because it determines whether there is a population of high-z miniquasars; it also determines whether the ionizing UV background at high redshifts has a strong enough nonthermal component to doubly-ionize He as well as to ionize H.

3.2 Physics of Mergers of Supermassive Holes

When two massive holes spiral together, energy is carried away by dynamical friction (leading, when the binary is 'hard', to expulsion of stars from the galaxy) and also by interaction with gas. Eventually, after a time which is uncertain, the members of the binary get close enough for gravitational radiation to take over and drive them towards coalescence. There would be a recoil due to the non-zero net linear momentum carried away by gravitational waves in the coalescence. If the holes have unequal masses, a preferred longitude in the orbital plane is determined by the orbital phase at which the final plunge occurs. For spinning holes there may be a rocket effect perpendicular to the orbital plane, since the spins break the mirror symmetry with respect to the orbital plane. The dynamics (and gravitational radiation) when two holes merge has so far been computed only for cases of special symmetry. The more general problem – coalescence of two Kerr holes with general orientations of their spin axes relative to the orbital angular momentum – is one of the 'grand challenge' computational projects.

This recoil could displace the hole from the centre of the merged galaxy – it might therefore be relevant to the low-z quasars that seem to be asymmetrically located in their hosts (and which may have been activated by a recent merger). The recoil might even be so violent that the merged hole breaks loose from its galaxy and goes hurtling through intergalactic space. Even galaxies that do not harbour a central hole may, therefore, once have done so in the past. The core of a galaxy that has experienced such an ejection event may retain some trace of it (perhaps, for instance, an unusual profile), because of the energy transferred to stars via dynamical friction during the merger process (cf [14, 15]).

3.3   Role of Stars in the Surrounding 'Cusp'

Black holes dominate the gravitational potential out to

$$r_h = \frac{GM_h}{\sigma_v^2} \qquad (2)$$

Within this radius, the stellar density profile has an approximately power-law form. Within this 'cusp' the number of stars is $\sim (M_h/m_*)$ where m_* is the typical stellar mass. If these stars have an isotropic velocity distribution, and can be treated as point masses, then a fraction $\sim (r_g/r_h)$ would be on bound orbits that would fall into the hole within a dynamical time. This approximate argument yields a capture rate, which, if maintained, would cause the hole to grow to a mass

$$M \simeq 10^8 \left(\frac{\sigma_v}{200\,\text{km/sec}}\right)^5 \left(\frac{t}{10^{10}}\,\text{yrs}\right) M_\odot \qquad (3)$$

(Note that growth would be even faster if the capture radius were $\gg r_g$, e.g. because of tidal capture, or gravitational radiation).

In practice, the rate of capture is reduced below (3) by loss cone depletion effects. However, the capture and tidal disruption of stars is a potentially-observable phenomena, and a good probe of massive black holes that are otherwise quiescent because of slow gaseous accretion. (See the discussion by Komossa in these proceedings).

At first sight equation (3) seems to offer a seductively attractive argument for the relation between M and σ_v claimed by Merritt and Ferrarese [6]. However, detailed scrutiny reveals that there are problems.

The first is that the inner parts of galaxies have density profiles flatter than isothermal. This means that σ_v, the velocity dispersion at r_h, is lower than in most of the bulge. The second reason is that the loss cone would actually be depleted, because orbits cannot alter fast enough to replenish it. The theory of the loss cone is straightforward for a spherically-symmetric system where replenishment occurs via star-star interactions (cf [16] and references cited therein); in the case of an axisymmetric system there is instead a 'loss wedge' [17], But even then the capture rate is well below the full-cone rate.

The process encapsulated in (3) could, however, lead to the observed correlation, if two conditions were fulfilled:
If (i) bulges of early-generation galaxies had isothermal profiles (with density varying with radius as r^{-2});
and if (ii) the stellar orbits are kept sufficiently churned up that the loss cone is always full.

As I have speculated in a recent paper with Haehnelt and Zhao [18], the latter condition may be fulfilled by the repeated capture of smaller galaxies, whose inner parts would avoid tidal disruption until they reached a radius r.

The profiles of bulges are now flatter than r^{-2}. This could be reconciled with hypothesis (i) if the inner part of the bulge had been scoured out by the dynamical friction of a coalescing supermassive black hole during the final merger, the gas by then being sufficiently depleted that the inner bulge cannot be re-formed by further star formation.

So is not impossible that stellar capture could play a role in establishing this surprisingly close correlation between hole mass and stellar velocity dispersion, by preferentially boosting the mass of those holes that had been 'disadvantaged' in their earlier growth. If this hypothesis were valid, it would imply a much higher rate of stellar captures in high-z galaxies than is predicted in nearby galaxies. These captures are observable if the star is destroyed before falling in (as happens, for solar-type stars, when the hole mass is below $10^8 M_\odot$)

3.4 Tidally-Disrupted Stars

When the central hole mass is below $10^8 M_\odot$, solar-type stars are disrupted before they get close enough to fall within the hole's horizon. A tidally disrupted star, as it moves away form the hole, develops into an elongated banana-shaped structure, the most tightly bound debris (the first to return to the hole) being at one end. There would not be a conspicuous 'prompt' flare signalling the disruption event, because the energy liberated is trapped within the debris. Much more radiation emerges when the bound debris (by then more diffuse and transparent) falls back onto the hole a few months later, after completing an eccentric orbit. The dynamics and radiative transfer are then even more complex and uncertain than in the disruption event itself, being affected by relativistic precession, as well as by the effects of viscosity and shocks.

The radiation from the inward-swirling debris would be predominantly thermal, with a temperature of order 10^5 K; however the energy dissipated by the shocks that occur during the circularisation would provide an extension into the X-ray band. High luminosities would be attained – the total photon energy radiated (up to 10^{53} ergs) could be several thousand times more than the *photon* output of a supernova. The flares would, however, not be standardised – what is observed would depend on the hole's mass and spin, the type of star, the impact parameter, and the orbital orientation relative to the hole's spin axis and the line of sight; perhaps also on absorption in the galaxy. To compute what happens involves relativistic gas dynamics and radiative transfer, in an unsteady flow with large dynamic range, which possesses no special symmetry and therefore requires full 3-D calculations – still a daunting computational challenge.

Supernova-type searches should either detect flares due to this phenomenon, or else place limits on its nature or the number of massive black holes in nearby galaxies. This possible bonus should be an added incentive for such searches. It is not clear whether the best strategy involves monitoring nearby galaxies over a large area of sky or larger numbers of more remote galaxies. Large numbers of distant galaxies are, for instance, being routinely monitored by programmes aimed at discovering supernovae at redshifts of order 0.5. It would be surprising if such programmes 'did not detect such flares – a negative result will itself be interesting. However, if a 'flare' (with the expected duration of months) happened in a distant galaxy, one would not be able to check just how quiescent the galaxy had previously been. It would be easier to be sure that a detected flare was actually due to a disrupted star (and not just an upward fluctuation in the gaseous accretion rate) if it were observed in a closer galaxy that was known to have previously been inactive.

The detection of possible flares of this kind is discussed by Komossa in her contribution to these proceedings. These flares offer a probe of massive holes in quiescent galaxies. Note that, for the reasons conjectured in 3.3 above, the disruption rate could be far higher in galaxies at redshifts $z > 1$.

3.5 'Fossil Events' in Our Galactic Centre?

The rate of tidal disruptions in our Galactic Centre would be no more than once per 10^5 years. But each such event could generate a luminosity several times 10^{44} erg/s for about a year. Were this in the UV, the photon output, spread over 10^5 years, could exceed the current ionization rate: the mean output might exceed the median output.

The resultant fossil ionization would set a lower limit to the electron density. The radiation emitted from the event might reach us after a delay if it were reflected off surrounding material. Sunyaev and his collaborators have already used this argument to set a non-trivial constraint on the history of the Galactic Centre's X-ray output over the last few thousand years.

Half the debris from a disrupted star would be ejected on hyperbolic orbits in a fan (which may intersect an orbiting disc in a line). The structure in the central 2 pc could be a single spiral feature. One speculative possibility [19] is that this feature may be a 'vapour trail' created by such an event.

3.6 Gravitational Waves from Newly-Forming Massive Holes?

At first sight, the formation of a massive hole from a monolithic collapse might seem an obvious source of strong gravitational wave pulses. The wave emission would be maximally intense and efficient if a hole formed on a timescale as short as (rg/c)– something that might happen if it formed via a merger of two holes of comparable mass.

If, on the other hand, supermassive black holes formed directly from gas (some, albeit, already processed through stars), perhaps after a transient phase as a supermassive object – then the process may be too gradual to yield efficient gravitational radiation. That is because post-Newtonian instability is triggered at a radius $r_i \gg r_g$. Supermassive stars are fragile because of the dominance of radiation pressure: this renders the adiabatic index γ only slightly above 4/3 (by an amount of order $(M/M_\odot)^{-1/2}$). Since $\gamma = 4/3$ yields neutral stability in Newtonian theory, even the small post-Newtonian corrections then destabilise such 'superstars'. The characteristic collapse timescale when instability ensues is longer than r_g/c by the 3/2 power of the collapse factor.

The foregoing estimate, based on the assumption of spherical symmetry, may be pessimistic. Since the post-Newtonian instability is suppressed by rotation, a differentially rotating supermassive star could in principle support itself against collapse until it became very tightly bound. It could then perhaps develop a bar-mode instability and collapse within a few dynamical times. To achieve this tightly-bound state, the object would need to have deflated over a long timescale, losing energy at no more than the Eddington rate.

The formation of a hole 'in one go' from a supermassive star thus seems an unpromising source of gravitational waves. On the other hand, strong signals are expected when already-formed holes coalesce, as the aftermath of mergers of their host galaxies. The gravitational waves associated with supermassive holes would be concentrated in a frequency range around a millihertz, accessible to the Laser Interferometric Spacecraft (LISA) discussed by Schultz at this meeting. Many galaxies have experienced a merger since the epoch $z > 2$. The holes in the two merging galaxies would spiral together (see (3.3) above) emitting, in their final coalescence, up to 10 per cent of their rest mass as a burst of gravitational radiation in a timescale of only a few times r_g/c. Such events yield a 1000:1 ratio of signal-to-noise even from large redshifts. Whether such events happen often enough to be interesting can to some extent be inferred from observations (we see many galaxies in the process of coalescing), and from simulations of the hierarchical clustering process whereby galaxies and other cosmic structures form. Haehnelt ([20] and later references) has calculated the merger rate of the large galaxies believed to harbour supermassive holes: it is only about one event per century, even out to redshifts $z = 4$. However, big galaxies are probably the outcome of many successive mergers. As mentioned in (3.1), we still have no direct evidence on whether these small galaxies harbour black holes (nor, if they do, of what the hole masses typically are). However it is certainly possible that enough holes of (say) $10^5\,M_\odot$ lurk in small early-forming galaxies to yield, via subsequent mergers, more than one event per year detectable by LISA.

3.7   What is the Expected Spin?

The spin of a hole affects the efficiency of 'classical' accretion processes; the value of a/m in a Kerr hole also determines how much energy is in principle extractable by the Blandford/Znajek effect. Moreover, the orientation of the spin axis may be important in relation to jet production, etc.

Spin-up is a natural consequence of prolonged disc-mode accretion: any hole that has (for instance) doubled its mass by capturing material that is all spinning the same way would end up with a/m being at least 0.5. A hole that is the outcome of a merger between two of comparable mass would also, generically, have a substantial spin. On the other hand, if it had gained its mass from capturing many low-mass objects (holes, or even stars) in randomly-oriented orbits, a/m would be small.

4   Do Supermassive Black Holes Obey the Kerr Metric?

4.1   Probing Near the Hole

The observed molecular disc in NGC 4258 lies a long way out [1,2]: at around 10^5 gravitational radii. Its dynamics provide compelling evidence for a central dark mass, in that we can exclude all conventional alternatives (dense star clusters, etc); however, the measurements tell us nothing about the innermost region where gravity is strong – certainly not whether the putative holes actually have properties consistent with the Kerr metric. The stars closest to our Galactic Centre likewise lie so far out from the putative hole (their speeds are less than 1 percent that of light) that their orbits are essentially Newtonian.

We can infer from the high luminosity of AGNs that 'gravitational pits' exist, which must be deep enough to allow several percent of the rest mass of infalling material to be radiated from a region compact enough to vary on timescales as short as an hour. But we still lack quantitative probes of the relativistic region. We believe in general relativity primarily because it has been resoundingly vindicated in the weak field limit (by high-precision observations in the Solar System, and of the binary pulsar) – not because we have evidence for black holes with the precise Kerr metric.

The emission from most accretion flows is concentrated towards the centre, where the potential well is deepest and the motions fastest. Such basic features of the phenomenon as the overall efficiency, the minimum variability timescale, and the possible extraction of energy from the hole itself all depend on inherently relativistic features of the metric – on whether the hole is spinning or not, how it is aligned, etc. Relativists would seize eagerly on any relatively 'clean' probe of the strong-field domain, and the possibilities, including X-ray periodicities, and quasi-periodic oscillations, are discussed by other speakers. We would occasionally expect to observe, even in quiescent nuclei, the tidal disruption of a star (see (3.3)). Exactly how this happens would depend on distinctive precession effects around a Kerr metric, but the gas dynamics are so complex that even when a flare is detected it will not serve as a useful diagnostic of the metric in the strong-field domain.

4.2   Compact Stars in Relativistic Orbits?

Gas-dynamical phenomena are complicated because of viscosity, magnetic fields etc. A 'clean' and quantitative probe would be a small star orbiting close to a supermassive hole. Such a star would behave like a test particle, and its precession would probe the metric in the 'strong field' domain [21,22]. An ordinary star certainly cannot reach such an orbit by the kind of 'tidal capture' process that can create close binary star systems. This is because the binding energy of the final orbit (a circular orbit with the same angular momentum as an initially near-parabolic orbit with pericentre at the tidal-disruption radius) would have to be dissipated within the star, and that cannot happen without destroying it. An orbit can be 'ground down' by successive impacts on a disc (or any other resisting medium) without being destroyed [23]: the orbital energy then goes almost entirely into the material knocked out of the disc, rather than into the star itself.

These stars would not be directly observable, except maybe in our own Galactic Centre. But they might have indirect effects: such a rapidly-orbiting star in an active galactic nucleus could signal its presence by quasiperiodically modulating the AGN emission.

For ordinary stars, the 'point mass' approximation breaks down for encounter speeds $\gtrsim$ 1000 km/s – physical collisions are then more probable than large-angle deflections. But there is no reason why a 'cusp' of tightly bound compact stars should not extend much closer to the hole. Neutron stars or white dwarfs could exchange orbital energy by close encounters with each other until some got close enough that they either fell directly into the hole, or until gravitational radiation became the dominant energy loss. When stars get very close, gravitational radiation losses become significant, and tend to circularise an elliptical orbit with small pericentre. Most will be swallowed by the hole before circularisation, because the angular momentum of a highly eccentric orbit 'diffuses' faster than the energy does due to encounters with other stars, but some would get into close circular orbits. LISA is potentially so sensitive that it could detect the nearly-periodic waves waves from stellar-mass objects orbiting a $10^5 - 10^6 \, M_\odot$ hole, even at a range of 100 Mpc. The orbits would precess in a manner that (if it could be inferred from the gravitational radiation signal or (better still) by modulation of the electromagnetic emission) would probe the form of the metric. The long quasi-periodic wave trains from such objects, modulated by orbital precession (cf refs [21,22]) in principle carries detailed information about the metric.

The gravitational radiation has an amplitude that involves a factor $(m * / M_h)$; if, therefore, there were a population of intermediate-mass holes weighing hundreds of solar masses, surviving as a remnant of an early generation of massive stars, then they could lead to capture events detectable by LISA out to redshifts of order unity [8].

4.3 Precession and Alignment

Most of the literature on flows around Kerr holes assumes that the flow is axisymmetric. This assumption is motivated not just by simplicity, but by the expectation that Lense-Thirring precession would impose axisymmetry close in, even if the flow further out were oblique and/or on eccentric orbits. Plausible- seeming arguments, dating back to the pioneering 1975 paper by Bardeen and Petterson [24] suggested that the alignment would occur, and would extend out to a larger radius if the viscosity were low because there would be more time for Lense-Thirring precession to act on inward-spiralling gas. However, later studies, especially by Pringle, Ogilvie, and their associates, have shown that naive intuitions can go badly awry. The behaviour of the 'tilt' is much more subtle; the effective viscosity perpendicular to the disc plane can be much larger than in the plane. In a thin disc, the alignment effect is actually weaker when viscosity is low. What happens in a thick torus is a still unclear, and will have to await 3-D gas-dynamical simulations.

The orientation of a hole's spin and the innermost flow patterns could have implications for jet alignment. An important paper by Natarajan and Pringle [25] shows that 'forced precession' effects due to torques on a disc can lead to swings in the rotation axis that are surprisingly fast (i.e. on timescales very much shorter than the timescale for changes in the hole's mass).

5 Acknowledgements

I am grateful to several colleagues, especially Mitch Begelman, Roger Blandford, Andy Fabian, and Martin Haehnelt for discussions and collaboration on topics mentioned here. I thank Rashid Sunyaev and his co-organisers for the opportunity to participate in this splendid and fascinating conference.

References

1. W.D. Watson, B.K. Wallin: Astrophys. J. (Lett) **432**, L35 (1994)
2. K. Miyoshi et al.: Nature **373**, 127 (1995)
3. M.J. Rees: 'The Compact Source at the Galactic Center'. In *The Galactic Center*, ed. by G. Riegler, R.D. Blandford (AIP) pp. 166–176 (1982)
4. F. Melia: Astrophys. J. **426**, 577 (1994)
5. R. Narayan, I. Yi, R. Mahadevan: Nature **374**, 623 (1995)
6. D. Merritt, L. Ferrarese. In *The Central Kpc of Starbursts and AGN, 2001*, ed. by J.H. Knapen et al. in press.
7. A. Heger, S. Woosley: Astrophys. J. in press
8. P. Madau, M.J. Rees: Astrophys. J. **551**, L27 (2001)
9. G.D. Quinlan, S.L. Shapiro: Astrophys. J. **356**, 483 (1990)
10. M.C. Begelman: MNRAS **187**, 237 (1979)
11. M. Abramowicz, M. Jaroszynski, M. Sikora: Astron. Astrophys. **63**, 221 (1980)
12. M.C. Begelman: Astrophys. J. in press
13. N. Shaviv: Astrophys. J. **494**, L193 (1998)
14. T. Ebisuzaki, J. Makino, S.K. Okumura: Nature **354**, 212 (1991)
15. M. Milosavljevic, D. Merritt: Astrophys. J. in press
16. D. Syer, A. Ulmer: MNRAS **306**, 35 (1999)
17. J. Magorrian, S. Tremaine: MNRAS **309**, 447 (1999)
18. H. Zhao, M.G. Haehnelt, M.J. Rees: MNRAS submitted
19. M.J. Rees: 'The Central Object: Some Comments and Speculations'. In *The Galactic Center*, ed. by D.C. Backer, R. Genzel (AIP, 1987)
20. M.G. Haehnelt: MNRAS **269**, 199 (1994)
21. V. Karas, D. Vokrouhlicky: MNRAS **265**, 365 (1993)
22. V. Karas, D. Vokrouhlicky: Astrophys. J. **422**, 208 (1994)
23. D. Syer, C.J. Clarke, M.J. Rees: MNRAS **250**, 505 (1991)
24. J. Bardeen, J.A. Petterson: Astrophys. J. (Lett.) **195**, L65 (1975)
25. P. Natarajan, J.E. Pringle: Astrophys. J. **506**, 97 (1998)

Supermassive Stars: Fact or Fiction?

Hans-Thomas Janka

Max-Planck-Institut für Astrophysik, Karl-Schwarzschild-Str. 1, D-85741 Garching, Germany

Abstract. Supermassive black holes are now realized to exist in the centers of most galaxies. The recent discoveries of luminous quasars at redshifts higher than 6 require that these black holes were assembled already when the Universe was less than a billion years old. They might originate from the collapse of supermassive stars, a scenario which could ensure a sufficiently rapid formation. Supermassive stars are dominated by photon pressure and radiate at their Eddington limit, which drives their quasi-static evolution to a final relativistic instability. Above some critical value of the metallicity, their collapse can lead to a gigantic explosion, powered by the energy release due to hydrogen burning, but below this critical metallicity their collapse inevitably ends in the formation of a black hole, accompanied by the emission of huge amounts of energy in the form of neutrinos. Although collapsing supermassive stars are the most powerful known burst sources of neutrinos, the associated conditions do not appear favorable for producing highly relativistic outflows that can explain cosmic gamma-ray bursts.

1 Introduction

The existence of supermassive black holes (SMBHs) in most galaxies is now becoming a generally accepted fact [38]. Increasingly better resolved observations reveal large Doppler shifts of spectral lines from hot gas swirling around the galactic centers with huge velocities, indicating extraordinarily high mass concentrations in remarkably small volumes (Fig. 1). The rapid orbital motions of the stars in the cluster surrounding Sgr A* in the Milky Way require the stabilizing gravitational attraction of a dark object with a mass of about $3 \times 10^6 \, M_\odot$ [13]. This, though, is near the lower end of the empirical mass distribution of supermassive galactic black holes. The black hole masses, the largest of which exceed $10^9 \, M_\odot$, correlate with the luminosity of the elliptical-galaxy-like bulge part of the host galaxy and with the average line-of-sight random velocity ("velocity dispersion") of the stars in the host galaxy. This indictes a correlation between black hole mass and galaxy mass and is interpreted as a hint to a direct connection between galaxy formation and black hole fueling [28,34].

SMBHs with masses of a million to a few billion solar masses are believed to be the engines that power active galactic nuclei ranging from faint, compact radio sources to quasars that are brighter than the whole galaxy in which they live. The detection of quasars with redshifts larger than 6 requires that these objects were formed only several hundred million years after the Big Bang. The growth of black holes by accretion is exponential, $M_{\mathrm{BH}} = M_{\mathrm{BH},0} \exp(t/\tau)$, with a timescale $\tau \sim 4 \times 10^7 \, (\epsilon/0.1)/\eta$ years, where $\epsilon = L/(\dot{M}_{\mathrm{BH}} c^2)$ is the radiation efficiency of the accreting black hole and $\eta =$

Fig. 1. A 3,700 light-year-diameter dust disk around the 300 million solar-mass black hole in the center of the elliptical galaxy NGC 7052 as observed by the Hubble Space Telescope. (Credits: Roeland P. van der Marel (STScI), Frank C. van den Bosch (University of Washington), and NASA)

$L/L_{\rm Edd}$ is the ratio of the luminosity L to the Eddington luminosity. Therefore SMBHs with 10^7–10^9 solar masses need at least 10–20 e-foldings to assemble by accretion on seed black holes of 10~$1000\,M_\odot$, which might be the compact remnants of an early generation of massive or very massive stars [23]. Provided enough gas were available in the surroundings to be swallowed by such a stellar mass black hole, and the efficiency of the black hole to absorb the gas flow were as high as assumed above, the time for its growth by accretion seems marginally short enough to explain the existence of quasars in the early Universe.

A consistent and satisfactory picture of the formation process of quasar black holes has not been developed yet [39]. A variety of different routes were suggested (for a review, see Ref. [37]), including scenarios based on gas hydrodynamics, stellar dynamics, or combinations of both. Primordial gas clouds could collapse directly to SMBHs when their fragmentation is inhibited by radiation pressure or magnetic fields [31,22]. Alternatively, when angular momentum plays a role and the cooling timescale is much smaller than the viscous timescale, they might contract to form a supermassive disk [47,14]. If instead fragmentation of a primordial gas cloud occurred and a dense cluster of stars were born, stellar collisions and mergers might lead to a runaway growth of intermediate-mass black holes (e.g., Refs. [6,36,12,35]). A dense cluster composed of neutron stars or black holes as the compact remnants of massive stars might be driven by the secular "gravothermal catastrophe" to the point of a final relativistic instability that happens on a dynamical timescale [50,44] (see, however, Ref. [6] for arguments against this picture).

Some of the proposed scenarios are envisioned to lead to the build-up of a supermassive star (SMS) as an intermediate stage of the evolution before the final gravitational instability sets in and the collapse to a SMBH takes place. Further growth of a seed black hole, possibly a SMBH, by accretion, could be linked to processes during the formation or evolution of the bulge of a galaxy (e.g., Refs. [42,46]). Galaxy formation or interaction could directly result in the black hole feeding that makes quasars shine, and bigger galaxies might be able to provide more fuel. This might explain why more massive galaxies contain more massive black holes [28].

2   Supermassive Stars: Some Basic Facts

Supermassive stars (SMSs) are equilibrium configurations that are dominated by radiation pressure. Their temperature is low enough that electron-positron pairs do not play a role. Baryons yield only a minor contribution to the equation of state. At some point of their evolution SMSs collapse due to a general relativistic gravitational instability [25,26,11,15,16,49,7,43].

SMSs can have masses between $\sim 10^4\, M_\odot$ and about $10^8\, M_\odot$. Since they are expected to be fully convective [43] (a formal argument can be found in Ref. [31]), they are isentropic and their structure can be well described by a Newtonian polytrope with $n = 3$ or $\gamma = 1 + 1/n = 4/3$. With their specific entropy being nearly constant, the adiabatic polytropic constant γ is roughly equal to the local adiabatic index $\Gamma = [\mathrm{d}(\ln P)/\mathrm{d}(\ln \rho)]_s$. To good accuracy the entropy per nucleon in a SMS, s_{SMS}, is given by the radiation entropy (in units of Boltzmann's constant k)

$$\frac{s_{\mathrm{SMS}}}{k} \approx \frac{s_{\mathrm{r}}}{k} = (1.22 \times 10^{-22}) \frac{T^3}{\rho} = 0.94 \left(\frac{M}{M_\odot} \right)^{1/2}, \tag{1}$$

where M is the mass of the star, ρ the density in $\mathrm{g\,cm^{-3}}$, and T the temperature in K (see, e.g., Refs. [43,21]). SMSs with large masses have entropies much higher than typical astrophysical plasmas. This suggests that dissipative processes (shocks, cloud-cloud collisions, turbulence) must be invoked for generating appropriate conditions for the formation of such supermassive equilibrium configurations.

The evolution of SMSs proceeds on a Kelvin-Helmholtz timescale and is driven by the loss of energy and entropy through radiation and — in case of rotation being important — the loss of angular momentum, e.g. via mass shedding. Because the pressure is dominated by radiation, the luminosity of SMSs is close to the Eddington limit,

$$L_{\mathrm{SMS}} = L_{\mathrm{Edd}} = \frac{4\pi G c M m_p}{\sigma_{\mathrm{T}}} = 1.3 \times 10^{38} \frac{M}{M_\odot}\ \mathrm{ergs\,s^{-1}}, \tag{2}$$

with σ_T being the Thompson cross section and m_p the proton mass.

Although plasma corrections (due to nuclei and electrons[1]) and general relativistic effects are small, neither of both can be neglected for the evolution of a configuration which is so close to the pathological, i.e., dynamically marginally stable, case of a $\gamma = 4/3$ polytrope (for a discussion, see Ref. [43]). Pressure contributions by plasma components raise the adiabatic index of the equation of state, Γ, to a value above 4/3 [10]:

$$\Gamma_{\mathrm{SMS}} = 1 + \frac{P}{\varepsilon} \approx \frac{4}{3} + \frac{\beta}{6}. \tag{3}$$

Here ε is the total internal energy density without rest-mass energy, and the second expression is correct to first order in the ratio of the gas pressure to the radiation pressure, $\beta = P_{\mathrm{g}}/P_{\mathrm{r}} = 8/(s_r/k) \ll 1$.

General relativity leads to the existence of a maximum for the equilibrium mass as a function of the central density for SMSs with constant entropy. This implies that general relativistic stars of mass M, which are supported both by radiation and gas pressure, have a maximum, i.e., "critical", central density:

$$\rho_{\mathrm{crit}} = 2 \times 10^{18} \left(\frac{M}{M_\odot} \right)^{-7/2} \mathrm{g\,cm}^{-3}. \tag{4}$$

For higher central densities the nonlinear effects of gravity have a destabilizing influence to radial perturbations. The central temperature at the onset of the gravitational instability is

$$T_{\mathrm{crit}} = 2.5 \times 10^{13} \left(\frac{M}{M_\odot} \right)^{-1} \mathrm{K}, \tag{5}$$

and the corresponding equilibrium energy can be found to be

$$E_{\mathrm{crit}} = -3.6 \times 10^{54} \mathrm{\ ergs}, \tag{6}$$

which is independent of M. The redshift factor at the surface of a supermassive star (with radius $R = R_{\mathrm{crit}}$) at this point of the evolution is

$$\left(\frac{GM}{Rc^2} \right)_{\mathrm{crit}} = \frac{1}{2} \frac{R_{\mathrm{s}}}{R_{\mathrm{crit}}} = 0.63 \left(\frac{M}{M_\odot} \right)^{-1/2}, \tag{7}$$

which is in the range $10^{-2}....10^{-4}$ for $M = 10^4\,M_\odot....10^6\,M_\odot$ and thus indeed very small so that the configuration is much larger than its Schwarzschild radius, $R_{\mathrm{crit}}/R_{\mathrm{s}} = 0.794\sqrt{M/M_\odot}$. Nevertheless, general relativity plays a crucial role for the evolution.

The gravitational instability sets in when the effective adiabatic index of the configuration (Eq. 3) drops below a critical value, which is approximately given by

$$\Gamma < \Gamma_{\mathrm{crit}} \approx \frac{2}{3} \frac{2 - 5\eta}{1 - 2\eta} + 1.12 \frac{R_{\mathrm{s}}}{R}. \tag{8}$$

---

[1] For reasons of simplicity, a pure hydrogen plasma is assumed here. The expressions for the general case can be found in Refs. [21,45]).

$R_s = 2GM/c^2$ is the Schwarzschild radius of the star and $\eta = T/|W|$ is the ratio of the rotational energy to the gravitational potential energy. For $\eta = 0$, Eq. (8) becomes $\Gamma_{\mathrm{crit}} = 4/3 + 1.12 R_s/R$ (see Ref. [32] and references therein). Therefore the gravitational collapse begins when the plasma contribution to the equation of state does not raise the adiabatic index sufficiently above $4/3$ to compensate for the destabilizing influence of general relativity. The latter grows as the radius of the star shrinks during its evolution. Rotation has a stabilizing effect and can hold up the collapse if

$$\eta > \eta_{\mathrm{crit}} \approx \frac{1}{2} \frac{4 - 3(\Gamma - 1.12 R_s/R)}{5 - 3(\Gamma - 1.12 R_s/R)} . \tag{9}$$

If $\eta > 0$ remains a constant during the equilibrium evolution, the final instability is again reached when the relativistic terms become too large.

Since the central density and temperature are higher for stars with smaller masses, and both increase during the quasistatic evolution, the creation of electron-positron pairs can be neglected for $M \gtrsim 10^4 \, M_\odot$ (i.e., $T \leq T_{\mathrm{crit}} \lesssim 2.5 \times 10^9$ K; Eq. 5) [9]. A comparison of the nuclear energy generation rate with the photon luminosity (Eq. 2) shows that nuclear burning is also irrelevant prior to the gravitational instability for SMSs with masses in excess of a few $10^5 \, M_\odot$ [16,49].

While radiating energy at its Eddington limit, the SMS evolves on a timescale

$$t_{\mathrm{KH}} = \frac{|E_{\mathrm{crit}}|}{L_{\mathrm{SMS}}} = 2.8 \times 10^{16} \left(\frac{M}{M_\odot} \right)^{-1} \mathrm{s} \tag{10}$$

with essentially constant mass but decreasing entropy and energy until the critical configuration for the general relativistic instability is reached and the catastrophic collapse sets in. Only when the thermal (Kelvin-Helmholtz) timescale is longer than the hydrodynamical timescale,

$$t_{\mathrm{hydro}} \sim (G\rho)^{-1/2} = 2.7 \times 10^{-6} \left(\frac{M}{M_\odot} \right)^{7/4} \mathrm{s} , \tag{11}$$

there is an equilibrium phase of the evolution of the supermassive object. The two timescales become equal (about 10 years) for $M \sim 10^8 \, M_\odot$. Above this mass no hydrostatic phase is possible. More typical SMSs with $M \sim 10^6 \, M_\odot$ have a lifetime of about 1000 years.

Since SMSs are very close to the edge of instability, rotation can appreciably stretch their equilibrium evolution. Considering the secular evolution of a uniformly rotating configuration along the mass-shedding sequence, Baumgarte and Shapiro [5] found a lifetime independent of the stellar mass, $t \approx 9 \times 10^{11}$ s. Moreover, they showed that the key nondimensional ratios R/R_s, $T/|W|$, and $Jc/(GM^2)$ (J is the angular momentum) for a maximally and rigidly rotating $n = 3$ polytrope at the onset of radial instability are universal numbers, independent of the mass, spin, radius, or history of the star:

$$\left(\frac{T}{|W|} \right)_{\mathrm{crit}} \approx 0.009 , \quad \left(\frac{Jc}{GM^2} \right)_{\mathrm{crit}} \approx 0.97 , \quad \left(\frac{R_p}{R_s} \right)_{\mathrm{crit}} \approx 214 , \tag{12}$$

with the polar radius R_p being roughly $2/3$ of the equatorial radius R_e. This deformation reduces the luminosity by about 36% below the usual Eddington luminosity of a corresponding nonrotating star [4]. The effects of differential rotation, where mass loss during the quasi-stationary evolution could be ignored, were discussed by New and Shapiro [33].

3   The Death of Supermassive Stars

Driven by energy loss through radiation and angular momentum loss due to mass shedding, SMSs contract slowly in a quasi-static manner and approach the point of dynamical instability as a consequence of the increasing effects of general relativity. The subsequent catastrophic collapse can lead to the formation of a SMBH [1] or, for sufficiently large initial metallicity, to a violent explosion that is powered by the release of nuclear energy through hydrogen burning in the CNO cycle [8,2,17,18,21].

These events are associated with the emission of gigantic amounts of energy in neutrinos [48], a fact that has nourished speculations that SMSs collapsing to black holes might be the sources of cosmic gamma-ray bursts [20]. In the case of rotationally deformed configurations, which might encounter a dynamical bar mode instability that triggers the growth of nonaxisymmetric bars, the generation of long-wavelength gravitational waves can be expected. Such gravitational waves could be detectable by future space-based laser interferometers like LISA [5,33,40]. Moreover, SMSs that were disrupted by explosions might have contributed to the enrichment of the gas in the young Universe with elements heavier than helium, in particular of comparatively rare isotopes like ^{13}C, ^{15}N, ^{17}O, and ^{22}Ne [3], and in case of SMSs with high metallicities of ^{26}Al [24].

The last investigation of nuclear burning during the collapse and explosion of SMSs was undertaken by Fuller et al. [21]. They performed hydrodynamical calculations with a post-Newtonian approximation to general relativistic gravity, used a detailed equation of state including electron-position pairs, and took into account photon and neutrino losses as well as all nuclear reactions for describing hydrogen burning by the CNO cycle (limited by β^+ decays of ^{14}O and ^{15}O) and by the rapid proton capture (rp-) process that characterizes hydrogen burning at very high temperatures ($T \gtrsim 10^9$ K; at such temperatures, however, neutrino losses become dominant). Considering nonrotating configurations, Fuller et al. [21] found that stars with a mass $M \gtrsim 10^5 M_\odot$ and initial metallicities $Z \lesssim 0.005$ do not explode, whereas objects with $Z \gtrsim 0.005$ (a value near the solar mass fraction of heavy elements!) do explode. The explosion energies range from 2×10^{56} ergs for stars of mass $M \approx 10^5 M_\odot$ to 2.5×10^{57} ergs for $M \approx 10^6 M_\odot$, and the photon luminosities can exceed 10^{45} ergs s^{-1} for a period of more than ten years. The nucleosynthesis in exploding SMSs is characterized by the production of a large amount of ^4He and trace amounts of ^{15}N and ^7Li. Deuterium production turned out to be negligible because this nucleus is too fragile to survive the high temperatures during the explosion of supermassive objects (see, however, Ref. [19] for neutrino-induced deuterium generation in the ejected envelope of exploding SMSs). Since zero metallicity (nonrotating) SMSs do not blow up, they, however, cannot be considered as a source of pre-galactic helium.

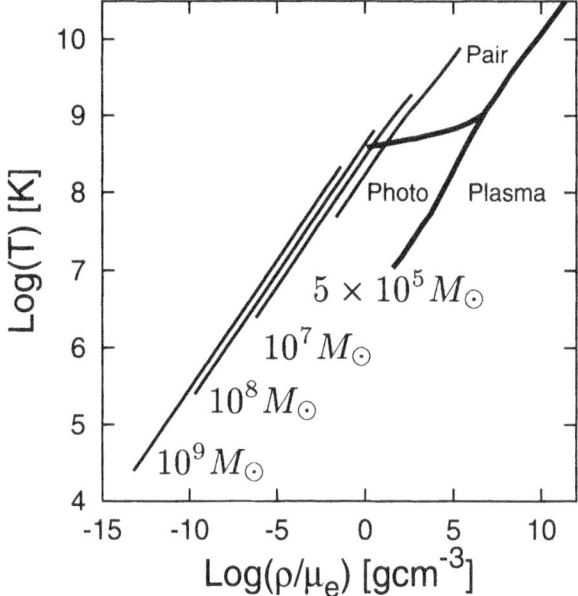

Fig. 2. Evolutionary tracks of the center of collapsing SMSs with different masses in the ρ-T-plane (μ_e is the mean molecular weight). Initially the photo-neutrino production yields the major contribution to the neutrino loss, during the later phases of the collapse it is the pair annihilation process. Plasmon neutrinos are negligible

Recently Linke et al. [29,30] have performed simulations in full general relativity of the collapse of nonrotating SMSs to black holes with the aim to calculate in detail the neutrino (and antineutrino) emission of such events until the point when the emission is quickly terminated by the formation of the event horizon. Their models also included electron-positron pairs and plasma contributions besides photons in the equation of state and were focused on cases where the energy release by nuclear burning is unimportant because it is dwarfed by neutrino losses. In Fig. 2 the evolutionary tracks of the central density and temperature of SMSs in the mass range between $5 \times 10^5\,M_\odot$ and $10^9\,M_\odot$ are plotted on top of the regions of dominant energy loss by the neutrino-antineutrino pair production through the photo-neutrino process ($\gamma + e^\pm \rightarrow e^\pm + \nu + \bar{\nu}$), electron-positron pair annihilation ($e^- + e^+ \rightarrow \nu + \bar{\nu}$), and plasmon decay ($\tilde{\gamma} \rightarrow \nu + \bar{\nu}$).

Energy losses by neutrino production become important only during the later stages of the collapse. Initially the photo-neutrino process plays the dominant role, whereas shortly before the black hole forms, when most of the neutrino emission occurs, the temperature is so high that the pair-neutrino process takes over. Plasmon neutrinos yield a negligible contribution in all cases. The neutrino emission rates are extremely temperature dependent. The energy production rate by electron-positron annihilation, Q_ν, rises like T^9 above the threshold temperature for pair formation and even more steeply ($Q_\nu \propto T^{20}$) for temperatures just below 10^9 K when e^+e^- pair creation sets in [27].

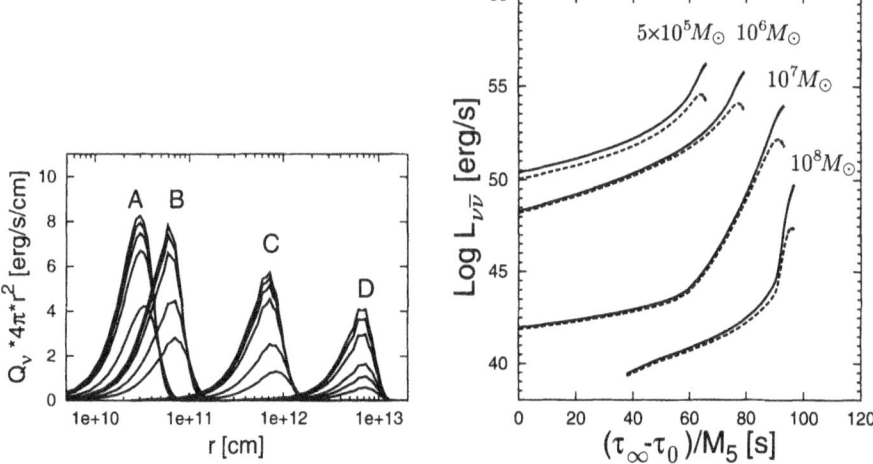

Fig. 3. Radial profiles of the neutrino plus antineutrino emission rate (times $4\pi r^2$; left) and luminosities of neutrinos plus antineutrinos as a function of time (right) for collapsing SMSs with different masses. The left plot gives the quantity $dL_{\nu\bar\nu}(r)/dr$ for $M = 5 \times 10^5\ M_\odot$ (A), $10^6\ M_\odot$ (B), $10^7\ M_\odot$ (C), and $10^8\ M_\odot$ (D) for different epochs of the model evolution with the end of the simulations being represented by the uppermost lines. The corresponding scaling factors are 5×10^{44} (A), 10^{44} (B), $2 \times 10^{41}\ M_\odot$ (C), and 2×10^{36} (D), respectively. In the right figure the dashed lines include Doppler shifts and general relativistic redshift effects, the solid lines do not. The time is measured in seconds with τ_∞ being the proper time for an observer at infinity, shifted by the overall collapse timescale ($\tau_0 = 8 \times 10^5$ s, 1.7×10^6 s, 8.0×10^7 s, and 3.2×10^9 s, respectively) and scaled by $M_5 = M/(10^5\ M_\odot)$

Although enormous amounts of energy are radiated away in neutrinos, these energy losses are small compared to the internal energy or the gravitational potential energy of the star. The neutrino losses are therefore essentially irrelevant in the energy budget and the collapse can well be considered as adiabatic. It proceeds nearly homologously so that the density profile evolves in a self-similar manner. Deviations from this ideal behaviour of a Newtonian $n = 3$ polytrope at a later stage of the collapse are caused by the increasing influence of general relativity and to a minor degree also by the possible formation of electron-positron pairs and the corresponding reduction of the adiabatic index. Except for such effects that are associated with the equation of state or with the neutrino emission — both of which are very sensitive to the maximum value of the temperature that is reached during the implosion — the collapse of SMSs of different masses is found to be very similar. The stellar mass therefore simply acts as a scaling parameter, a fact which will be made use of in the following discussion.

The collapse timescale t_{coll} is roughly proportional to $R_s/c \propto M$ and lasts between about 9 days for a SMS with $5 \times 10^5\ M_\odot$ and several years for $10^7\ M_\odot$ stars. For more massive stars a meaningful calculation of the duration of the phases of contraction and implosion requires the inclusion of photon emission (cf. Eq. 10), which in fact was ignored in our models. The collapse of the innermost $\sim 25\%$ of the mass proceeds in

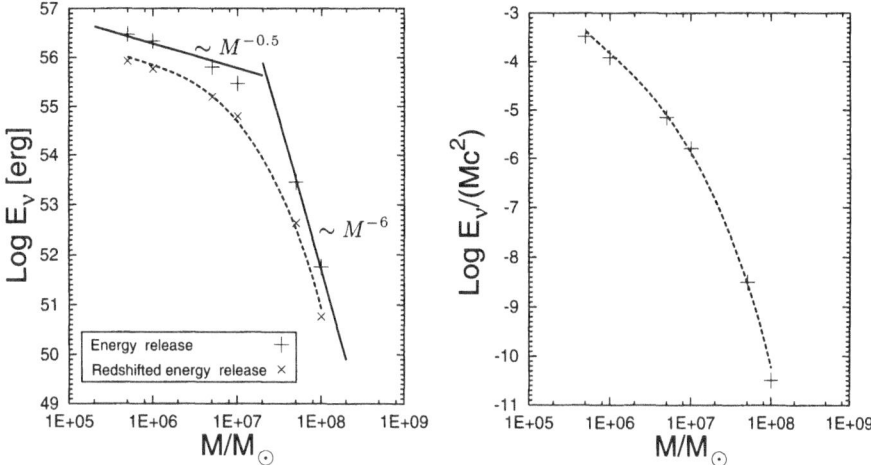

Fig. 4. Total energy release in form of neutrinos during the collapse of SMSs to SMBHs (left). The mass of the star is given on the abscissa. The dashed curve interpolates the computed values (symbols) which include the effects of Doppler shift and general relativistic redshift, while the solid lines interpolate the results for the integral energies as measured by local observers. The right plot shows the efficiency of the conversion of rest-mass energy to neutrinos, $E_\nu/(Mc^2)$. In case of less massive stars the higher core temperatures before black hole formation imply much larger values for the total energy emitted in neutrinos and for the conversion efficiency

a nearly coherent way with an approximately homologous velocity profile. This part of the star therefore forms a black hole first and determines the radius of the innermost apparent horizon, which is proportional to M.

Because the highest temperatures are reached at the end of the collapse, the time interval of strongest neutrino emission is also given by $t_{em} \sim R_s/c \propto M$. The peak of the neutrino production is located in a shell just outside the innermost apparent horizon (Fig. 3). The corresponding "neutrino emission radius" is therefore also proportional to the event horizon of the SMS, $R_\nu \propto R_s \propto M$ (Fig. 3) and thus much smaller than the stellar radius (Eq. 7). Since the collapse proceeds very nearly adiabatically ($T^3/\rho \approx$ const; Eq. 1), one can also easily estimate how the maximum core temperature, which determines the neutrino emission, depends on the mass of the star. Mass conservation implies for the final density $\rho_f \sim (R_{crit}/R_s)^3 \rho_{crit}$ with $R_{crit}/R_s \propto M^{1/2}$ (Eq. 7) and $\rho_{crit} \propto M^{-7/2}$ (Eq. 4). Using this in Eq. (1) one finds for the final temperature $T_f \propto M^{-1/2}$ [45]. Neutrinos are emitted with a mean energy $\langle \epsilon_\nu \rangle$ that scales with the stellar temperature in the main production region. A detailed discussion yields $\langle \epsilon_\nu \rangle \sim 6kT_f$ [45].

The maximum neutrino plus antineutrino luminosity decreases steeply with higher stellar masses:

$$L_{\nu\bar\nu} \sim Q_\nu \frac{4\pi}{3} R_\nu^3 \propto \begin{cases} M^{-3/2} & \text{for } 10^5\, M_\odot \lesssim M \lesssim 5 \times 10^6\, M_\odot, \\ M^{-7} & \text{for } 5 \times 10^7\, M_\odot \lesssim M \lesssim 10^8\, M_\odot. \end{cases} \tag{13}$$

The right plot in Fig. 3 shows this trend. The change in the slope of the luminosities as a function of time that is visible for the cases $M = 10^7 M_\odot$ and $M = 10^8 M_\odot$ near $L_{\nu\bar{\nu}} \sim 10^{43}\,\mathrm{ergs\,s^{-1}}$ is associated with the transition from the photo-neutrino dominated to the pair-neutrino dominated regime (compare Fig. 2). The total energy release in neutrinos and antineutrinos exceeds 10^{56} ergs for SMS with $M \lesssim 10^6 M_\odot$. Stars with smaller masses are the clearly stronger neutrino sources:

$$E_\nu \sim L_{\nu\bar{\nu}}\, t_{\mathrm{em}} \propto \begin{cases} M^{-1/2} & \text{for } 10^5 M_\odot \lesssim M \lesssim 5 \times 10^6 M_\odot\,, \\ M^{-6} & \text{for } 5 \times 10^7 M_\odot \lesssim M \lesssim 10^8 M_\odot\,. \end{cases} \quad (14)$$

As displayed in Fig. 4, the observable energies are somewhat lowered by effects due to Doppler shift and gravitational redshift. SMSs near the lower end of the investigated mass range convert a fraction of more than 10^{-4} of their rest-mass energy to neutrinos, whereas it is less than 10^{-10} in case of stars with $M = 10^8 M_\odot$ (right plot in Fig. 4).

4 Conclusions

Although the energy emitted in neutrinos is huge in case of SMSs that form SMBHs with masses $M \lesssim 10^7 M_\odot$, it is very unlikely that these neutrinos can produce a highly relativistic outflow to power cosmic gamma-ray bursts. On the one hand, the efficiency of neutrino-antineutrino annihilation to electron-positron pairs is extremely low ($\lesssim 10^{-5}$ of the neutrino energy are converted to $e^\pm$ pairs [29,30]). On the other hand, 99.8% of this energy are deposited in the close vicinity of the neutrino emitting shell, i.e. in matter which is swept inward in the collapse with velocities up to 60% of the speed of light. The deposited energy is much too small to invert this rapid infall and to create a highly relativistic outflow. A black hole-disk configuration might provide a more suitable environment, but is very unlikely to form even in the case of the collapse of SMSs with rotation [40]. Moreover, SMSs as gamma-ray burst sources are unable to account for the short-time variability of the observed emission, which is directly linked to the activity of the source [41]. This requires a very compact energy source with a size which is typical of a neutron star or stellar mass black hole.

Investigations of the evolution and collapse of nonrotating SMSs are a somewhat academic exercise. Since the configurations are so close to dynamical instability, a small amount of rotation may have a significant impact. This has indeed been found for the quasi-static evolution of uniformly rotating [5] and differentially rotating stars [33]. The collapse of stars which rotate uniformly at the onset of gravitational instability, however, turns out to be very similar to the nonrotating case. Saijo et al. [40] found that such a collapse is likely to form a SMBH coherently, with almost all of the matter falling into the hole on a dynamical timescale and only very little matter possibly ending up in a disk. They did not discover an unstable growth of a nonaxisymmetric bar. Certainly such investigations of the death of SMSs should be extended to differentially rotating configurations and to models which include a detailed microphysical description of the equation of state (instead of using a simple Γ-law equation $P = (\Gamma - 1)\varepsilon$). Moreover, the neutrino emission and possible energy release by nuclear burning should be taken into account. The latter might be more important for rotating SMSs than for nonrotating ones. Centrifugal forces might hold up the collapse long enough for nuclear reactions to generate sufficient energy to influence the dynamics even in case of zero initial metallicity [21].

Acknowledgements. Stimulating conversations with T. Abel and A. Heger are acknowledged. The author thanks F. Linke, J.A. Font, E. Müller and P. Papadopoulos for a pleasant collaboration and is very grateful to R. Sunyaev for inspiring discussions and the opportunity to present this review at the conference "Lighthouses of the Universe". This work was supported by the Sonderforschungsbereich 375 "Astroparticle Physics" of the Deutsche Forschungsgemeinschaft.

References

1. I. Appenzeller, K. Fricke: Astron. Astrophys. **18**, 10 (1972)
2. I. Appenzeller, K. Fricke: Astron. Astrophys. **21**, 285 (1972)
3. J. Audouze, K.J. Fricke: Astrophys. J. **186**, 239 (1973)
4. T.W. Baumgarte, S.L. Shapiro: Astrophys. J. **526**, 937 (1999)
5. T.W. Baumgarte, S.L. Shapiro: Astrophys. J. **526**, 941 (1999)
6. M.C. Begelman, M.J. Rees: Mon. Not. R. Astr. Soc. **185**, 847 (1978)
7. G.S. Bisnovatyi-Kogan, Ya.B. Zel'dovich, I.D. Novikov: Soviet Astronomy–AJ **11**, 419 (1967)
8. G.S. Bisnovatyi-Kogan: Soviet Astronomy–AJ **12**, 58 (1968)
9. J.R. Bond, A.D. Arnett, B.J. Carr: Astrophys. J. **280**, 825 (1984)
10. S. Chandrasekhar: *An Introduction to the Study of Stellar Structure* (University of Chicago Press, Chicago 1939)
11. S. Chandrasekhar: Astrophys. J. **140**, 417 (1964)
12. T. Ebisuzaki, J. Makino, T.G. Tsuru et al: Astrophys. J. **562**, L19 (2001)
13. A. Eckart, R. Genzel: Mon. Not. R. Astr. Soc. **284**, 576 (1997) R. Genzel, A. Eckart, T. Ott, F. Eisenhauer: Mon. Not. R. Astr. Soc. **291**, 219 (1997)
14. D.J. Eisenstein, A. Loeb: Astrophys. J. **443**, 11 (1995)
15. W.A. Fowler: Rev. Mod. Phys. **36**, 545, 1104err (1964)
16. W.A. Fowler: Astrophys. J. **144**, 180 (1966)
17. K.J. Fricke: Astrophys. J. **183**, 941 (1973)
18. K.J. Fricke: Astrophys. J. **189**, 535 (1974)
19. G.M. Fuller, X. Shi: Astrophys. J. **487**, L25 (1997)
20. G.M. Fuller, X. Shi: Astrophys. J. **502**, L5 (1998)
21. G.M. Fuller, S.E. Woosley, T.A. Weaver: Astrophys. J. **307**, 675 (1986)
22. O.Y. Gnedin: Class. Quantum Grav. **18**, 3983 (2001)
23. A. Heger, S.E. Woosley: Astrophys. J., in press (2002) (astro-ph/0107037)
24. W. Hillebrandt, F.-K. Thielemann, N. Langer: Astrophys. J. **321**, 761 (1987)
25. F. Hoyle, W.A. Fowler: Mon. Not. R. Astr. Soc. **125**, 169 (1963)
26. I. Iben: Astrophys. J. **138**, 1090 (1963)
27. N. Itoh, H. Hayashi, A. Nishikawa, Y. Kohyama: Astrophys. J. Suppl. **102**, 411 (1996)
28. J. Kormendy, D. Richstone: Ann. Rev. Astron. Astrophys. **33**, 581 (1995) J. Kormendy, K. Gebhardt: 'Supermassive Black Holes in Galactic Nuclei'. In: *Proceedings of The 20th Texas Symposium on Relativistic Astrophysics, Austin, Texas, Dec. 10–15, 2000*, ed. by H. Martel, J.C. Wheeler, in press (2002) (astro-ph/0105230)
29. F. Linke: General Relativistic Simulation of Collapsing Supermassive Stars. Diploma Thesis, Technical University, Munich (2000)

30. F. Linke, J.A. Font, H.-T. Janka, E. Müller, P. Papadopoulos: Astron. Astrophys. **376**, 568 (2001)
31. A. Loeb, F.A. Rasio: Astrophys. J. **432**, 52 (1994)
32. C.W. Misner, K.S. Thorne, J.A. Wheeler: *Gravitation* (W.H. Freeman and Company, San Francisco 1973)
33. K.C.B. New, S.L. Shapiro: Astrophys. J. **548**, 439 (2001)
34. M.J. Page, J.A. Stevens, J.P.D. Mittaz, F.J. Carrera: Science **294**, 2516 (2001)
35. S.F. Portegies Zwart, S.L.W. McMillan: Astrophys. J., submitted (astro-ph/0201055)
36. G.D. Quinlan, S.L. Shapiro: Astrophys. J. **356**, 483 (1990)
37. M.J. Rees: Ann. Rev. Astron. Astrophys. **22**, 471 (1984)
38. M.J. Rees: 'Astrophysical Evidence for Black Holes'. In: *Black Holes and Relativistic Stars, Proceedings of Chandrasekhar Memorial Conference, Chicago, Dec. 1996*, ed. by R.M. Wald (University of Chicago Press, Chicago 1998) pp. 79–100
39. M.J. Rees: 'Supermassive Black Holes: Their Formation, and Their Prospects as Probes of Relativistic Gravity'. In: *Black Holes in Binaries and Galactic Nuclei, Proceedings of ESO Workshop, Garching, Germany, Sept. 6–8, 1999*, ed. by L. Kaper, E.P.J. van den Heuvel, P.A. Woudt (Springer, New York 2001) pp. 351–364
40. M. Saijo, T.W. Baumgarte, S.L. Shapiro, M. Shibata: Astrophys. J., in press (2002)
41. R. Sari, T. Piran: Astrophys. J. **485**, 270 (1997)
42. H. Schulz, S. Komossa: 'Formation of Massive Black Holes in Early Mergers?'. In: *Trends in Astrophysics and Cosmology, Proceedings of Workshop, Bad Honnef, Germany, Aug. 24–28, 1998*, ed. by W. Kundt, C.v.d.Bruck (Lecture Notes in Physics, Springer, Heidelberg) in press (astro-ph/9905118)
43. S.L. Shapiro, S.A. Teukolsky: *Black Holes, White Dwarfs, and Neutron Stars* (John Wiley & Sons, New York 1983)
44. S.L. Shapiro, S.A. Teukolsky: Astrophys. J. **292**, L41 (1985)
45. X. Shi, G.M. Fuller: Astrophys. J. **503**, 307 (1998)
46. M. Umemura: Astrophys. J. **560**, L29 (2001)
47. R.V. Wagoner: Ann. Rev. Astron. Astrophys. **7**, 553 (1969)
48. S.E. Woosley, J.R. Wilson, R. Mayle: Astrophys. J. **302**, 19 (1986)
49. Ya.B. Zel'dovich, I.D. Novikov: *Stars and Relativity* (University of Chicago Press, Chicago 1971)
50. Ya.B. Zel'dovich, M.A. Podurets: Soviet Astronomy **9**, 742 (1965)

Evolution and Explosion
of Very Massive Primordial Stars

Alexander Heger[1], Stan Woosley[1], Isabelle Baraffe[2,3], and Tom Abel[4]

[1] Department of Astronomy and Astrophysics, University of California, 1156 High Street, Santa Cruz, CA 95064, U.S.A.
[2] Ecole Normale Supérieure, C.R.A.L (UMR 5574 CNRS), 69364 Lyon Cedex 07, France
[3] Max-Planck-Institut für Astrophysik, Karl-Schwarzschild-Str. 1, 85741 Garching, Germany
[4] Institute of Astronomy, Madingley Road, Cambridge, CB3 0HA, England

Abstract. While the modern stellar IMF shows a rapid decline with increasing mass, theoretical investigations suggest that very massive stars ($\gtrsim 100\,M_\odot$) may have been abundant in the early universe. Other calculations also indicate that, lacking metals, these same stars reach their late evolutionary stages without appreciable mass loss. After central helium burning, they encounter the electron-positron pair instability, collapse, and burn oxygen and silicon explosively. If sufficient energy is released by the burning, these stars explode as brilliant supernovae with energies up to 100 times that of an ordinary core collapse supernova. They also eject up to $50\,M_\odot$ of radioactive ^{56}Ni. Stars less massive than $140\,M_\odot$ or more massive than $260\,M_\odot$ should collapse into black holes instead of exploding, thus bounding the pair-creation supernovae with regions of stellar mass that are nucleosynthetically sterile. Pair-instability supernovae might be detectable in the near infrared out to redshifts of 20 or more and their ashes should leave a distinctive nucleosynthetic pattern.

1 Introduction

Owing to the lack of any metals, the cooling processes that govern star formation are greatly reduced for first generation of stars (Pop III). Magnetic fields and turbulence may also be less important at these early times [1]. Consequently, theoretical studies [11] indicate that the Jeans mass for primordial stars in their special environment may have been as great as $\sim 1000\,M_\odot$. Numerical simulation of primordial star formation predict the occurrence of such stars at red shifts ~ 20 and an initial mass function (IMF) that either peaks at $\sim 100\,M_\odot$ [1,4] or is bimodal [12], i.e., also contains stars of a few $M_\odot$.

Once formed, at solar metallicity, massive stars ordinarily experience significant mass loss [5] and may end as relatively small objects, but for low metallicity mass loss is suppressed. In § 2 we discuss the peculiarities of mass loss and evolution of very massive primordial stars. Figure 1 gives an overview of expected final fates of metal-free stars as a function of initial mass. In § 3 we examine the expected light curve of a pair-creation supernova from a $250\,M_\odot$ star at a redshift of $z = 20$ and in § 4 we review nucleosynthetic yields from Pop III. Some conclusions are given in § 5.

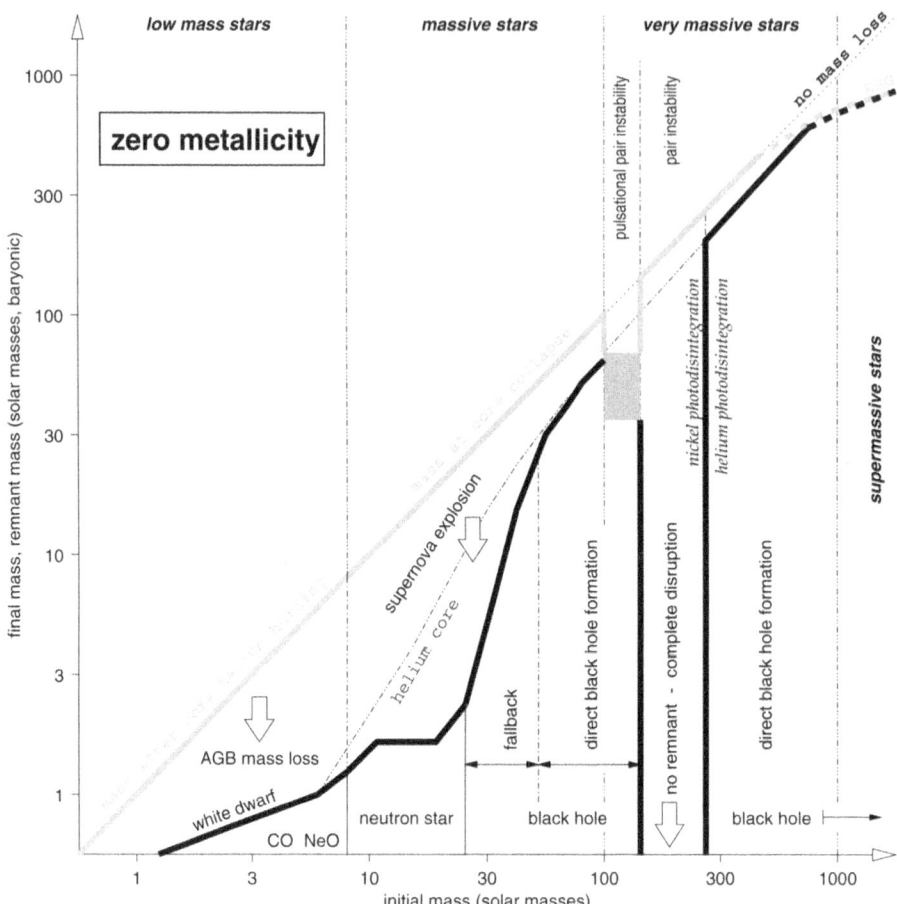

Fig. 1. Initial-final mass function of non-rotating Pop III stars. The *x-axis* gives the initial mass. The *y-axis* gives both the final mass of the collapsed remnant (*thick black curve*) and the mass of the star when the event begins that produces that remnant (e.g., mass loss in AGB stars, supernova explosion for those stars that make a neutron star, etc.; *thick gray curve*). We distinguish four regimes of initial mass: *low mass stars* below $\sim 10\,M_\odot$ that end as white dwarfs; *massive stars* between $\sim 10\,M_\odot$ and $\sim 100\,M_\odot$ that form an iron core that eventually collapses; *very massive stars* between $\sim 100\,M_\odot$ and $\sim 1000\,M_\odot$ that encounter the pair instability; and *supermassive stars* (arbitrarily) above $\sim 1000\,M_\odot$. Since no mass loss is expected for $Z = 0$ stars before the final stage, the grey curve is approximately the same as the line of no mass loss (*dotted*). Exceptions are $\sim 100 - 140\,M_\odot$ where the pulsational pair instability ejects the outer layers of the star before it collapses, and above $\sim 500\,M_\odot$ where pulsational instabilities in red supergiants may lead to significant mass loss [2]. Since the magnitude of the latter is uncertain, lines are drawn *dashed*. For a more detailed description, please refer to [7]

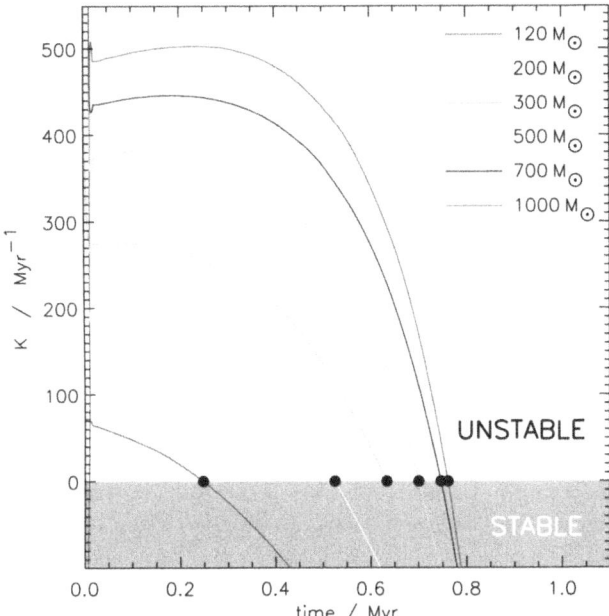

Fig. 2. Growth rate, K, as a function of the age of the star. The amplitude of the pulsation grows with time as $A(t) = A_0 \exp\{\frac{2\pi i}{P}t\} \exp\{Kt\}$ [2].

2   Mass Loss

Current studies of winds driven from hot stars by interactions with atomic lines indicate that their mass loss rate decreases with metallicity as $Z^{1/2}$ [8,9] or even more [16]. This scaling seems to hold down to 0.1 % solar metallicity. Extrapolating to zero, it seems reasonable that mass loss from this wind driving mechanism becomes negligible. Winds driven by continuum opacity in low metal stars are not very well understood (Kudritzki, priv. com.), and will be neglected here.

Stars above $\sim 60\,M_\odot$ may also be unstable to the epsilon mechanism - pulsations driven by the high temperature sensitivity of nuclear burning [10,13]. However, historical studies of stellar stability focused on stars of solar composition and did not take into account the peculiarities of the first generation of stars. We have recently studied the pulsational instability due to epsilon mechanism in massive metal-free stars between 120 and $1000\,M_\odot$ [2]. Their structure is uniquely different from later stellar generations. Since hydrogen burning through the "pp-chain" is not sufficient, the star contracts to a high enough temperature to produce carbon by the triple-alpha reaction, more than 1×10^8 K in the stars investigated here. A mass fraction of $\sim 1\times10^{-9}$ is sufficient to stop the contraction and supply energy by the CNO cycle for hydrogen burning. The stars, however, stay very compact and hot in their centers throughout their hydrogen burning lifetime. The higher temperature causes a lower temperature sensitivity of the nuclear energy generation. From pulsational analysis we find that the epsilon mechanism is weak in these compact stars and operates only for a fraction of the hydrogen-burning life-time (Fig. 2) [2]. Radial pulsations driven by opacity or re-combination are not found. We estimate that the resulting mass loss due to the epsilon mechanism should be less than 5 % for a 500 $M_\odot$ star, and perhaps 10 % for a 1000 $M_\odot$ star, but quite negligible for stars of lower mass. Stars of $\sim 500\,M_\odot$ or more may en-counter a red supergiant phase towards the end of central helium burning. This could result in an additional, maybe significant, mass loss, but its strength is not yet known. Therefore stars of $\lesssim 500\,M_\odot$ can be safely assumed to reach carbon burning without significant mass loss.

3   Supernovae at the Edge of the Universe

Stars that reach carbon burning with a helium core mass of $\sim 64\ldots133\,M_\odot$ (corre-sponding to ZAMS masses of $\sim 140\ldots260\,M_\odot$ for stars without mass loss) are unsta-ble to pair-creation (see Fig. 1), collapse, then explode as a supernova (SN). These are the most powerful thermonuclear explosions in the universe. Pair-creation SNe release energies ranging from 3×10^{51} erg for a 64 $M_\odot$ He core up to almost 100×10^{51} erg for a 133 $M_\odot$ He core [7] – enough energy to disrupt a small proto-galaxy. In Fig. 3 we show, in the observer frame, the early light curve of an exploding 250 $M_\odot$ star, neglecting in-tergalactic and interstellar absorption. This star has a He core of 120 $M_\odot$ and a total explosion energy of 65×10^{51} erg, producing 21 $M_\odot$ of ^{56}Ni. The visible brightness is not quite as impressive since most of the energy is kinetic, but nevertheless should be a few times brighter than a typical Type Ia SN. The bolometric luminosity of an event at $z = 20$ is not much dimmer than at a red shift of a few, where Type Ia SNe have al-ready been observed. The main difference, however, is the significantly larger red shift and time dilation. Given the bigger intrinsic time scale associated with the large ejected mass, they also last much longer.

For a current standard cosmology ($\Omega_\Lambda = 0.7$, $\Omega_\mathrm{m} = 0.3$, $H_0 = 65\,\mathrm{km/s/Mpc}$, $\Omega_\mathrm{b} = 0.02/h^2 = 0.047$) and assuming that $f_\mathrm{1st} = 10^{-6}$ of all baryons goes into stars of $M_\mathrm{1st} = 250\,M_\odot$, at a red shift of $z = 20$, we estimate the pair-creation supernova rate by

$$r_\mathrm{2SN} = 4\pi \left(\frac{d_\mathrm{lum}}{1+z}\right)^2 \frac{c}{1+z}\rho_\mathrm{c}\Omega_\mathrm{b}(1+z)^3 \frac{f_\mathrm{1st}}{M_\mathrm{1st}} = \frac{4\pi d_\mathrm{lum}^2 c\rho_\mathrm{c}\Omega_\mathrm{b}f_\mathrm{1st}}{M_\mathrm{1st}}$$

where $\rho_\mathrm{c} = 3H_0{}^2/8\pi G$, and the luminosity distance $d_\mathrm{lum} = 243\,\mathrm{Gpc}$. This gives ~ 0.16 events per second per universe, i.e., $\sim 3.9\times10^{-6}$ events per second per square degree. The first peak of the light curve (Fig. 3) lasts for about a month, the second peak ("plateau"; not depicted here), probably brighter in the far infrared, would peak for about 10 yr (based on preliminary calculations by Phil Pinto, priv. com.). Statistically, at any time per square degree about a dozen of these supernovae should be at the peak of the light curve, and more than ~ 1000 in the plateau phase of the light curve. Note that $d_\mathrm{lum} \propto z$ for high z and thus this rate depends critically on the red shift adopted as well as on the baryon fraction assumed to go into these stars.

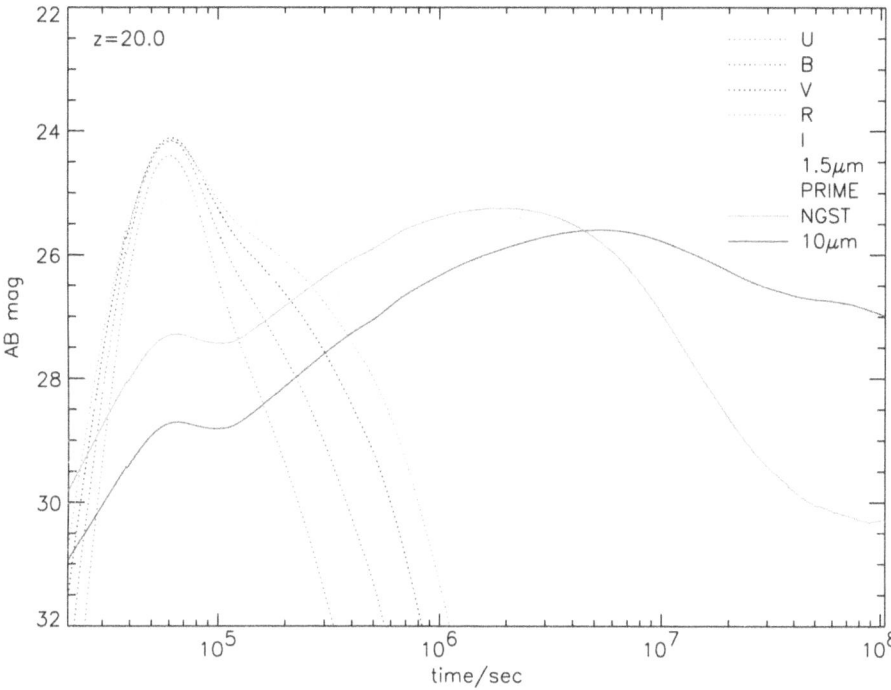

Fig. 3. Preliminary light curve of pair-creation supernova from a $250\,M_\odot$ star at $z = 20$ as computed by the KEPLER code [17]. Time, wave lengths and magnitudes (without internal or intergalactic extinction) are given in observer rest frame. Wave lengths that are beyond the IGM Ly–α absorption (2.55μm) are displayed as *dotted lines*. "PRIME" and "NGST" corresponds to 3.5 and 5.0μm. The "spherically symmetric" emission has been folded to account for the extent of the "photosphere". The first "bump" at $\sim 10^3$ s is from the shock breakout, the right "peak" is the peak of the SN light curve.

4 Nucleosynthesis

In massive stars most of neutrons responsible for the "weak component" of the s-process are made from initial CNO seeds during helium burning. Since Pop III stars are devoid of these seeds, both the s-process and the production of a "neutron excess" for the advanced burning stages is strongly suppressed [14]. In massive stars the odd-Z elements are therefore underproduced with respect to the α elements [7]. Above $\sim 25\,M_\odot$ significant fallback onto the remnant already occurs after the SN [6] (Fig. 1), and few or no heavy elements can be ejected. From ~ 40 to $\sim 100\,M_\odot$ a black hole is formed directly [6] and essentially the whole star will be swallowed — no nucleosynthesis products are ejected. Between ~ 100 and $\sim 140\,M_\odot$ the pulses of pair instability [3] will eject the outer layers of the star, possibly including parts of the CO core, especially at the high-mass end of the regime, but nothing heavier than magnesium leaves the star. After the pulses these objects probably encounter the same fate as their lighter cousins, i.e., the remaining part of the star falls into a black hole. For initial masses of $\sim 140\ldots 260\,M_\odot$, the pair instability completely disrupts the star when explosive

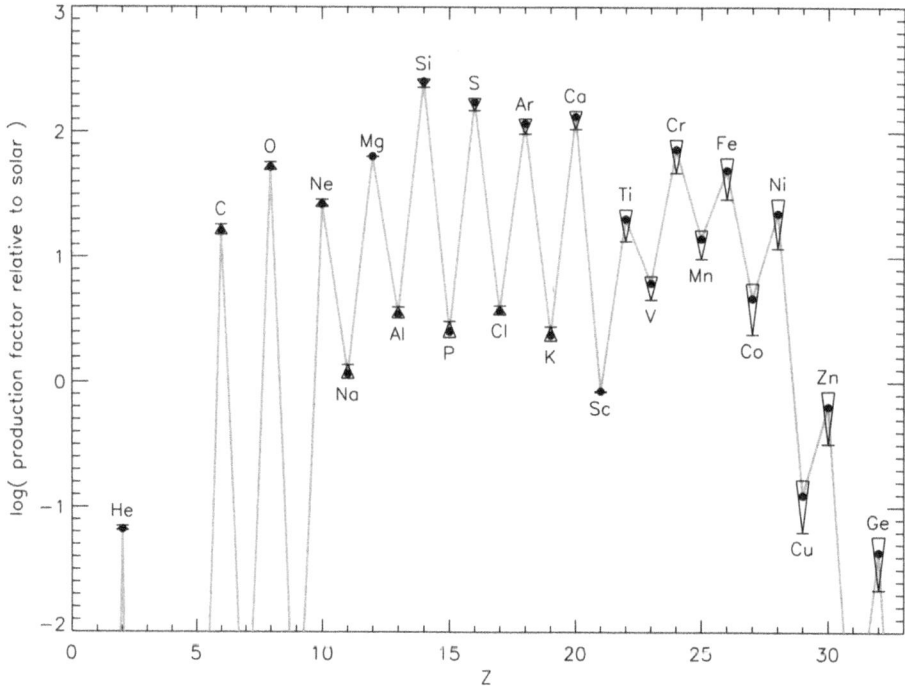

Fig. 4. Elemental production factors relative to solar in Pop III pair-creation supernovae (helium core mass 65–130 $M_\odot$). The thin and thick ends of the triangle and the dots correspond to IMF slopes of $d \log N / d \log M = -1.5$, -4.5, and -2.5, respectively.

burning of oxygen and silicon release enough energy to reverse the collapse into an explosion. However, this explosion is too rapid and the star not dense enough at the point when the implosion turns around to lead to significant neutronization. A very marked odd-even pattern results (Fig. 4) with silicon being the biggest overproduction and elements above germanium are essentially not produced. In stars more massive than $\sim 260\,M_\odot$, photo-disintegration of nuclei becomes important and the collapse is not reversed. The whole star falls into a black hole (Fig. 1).

5   Conclusions

Due to the absence of metals the first generation of star will likely not experience significant mass loss by radiation-driven stellar winds or opacity-driven pulsations. Their unique structure also prevents significant mass loss by the epsilon mechanism. Therefore very massive Pop III single stars reach carbon burning with enough mass to encounter the pair instability. Since current theoretical studies indicate that such star may constitute a significant, if not dominant, fraction of Pop III, we predict that their nucleosynthetic yields may have a unique imprint on the chemical evolution of the early universe. Their production of odd-Z elements is by ~ 2 orders of magnitude lower than that of even-Z. Elements heavier than zinc are not produced.

Since stars of lower mass ($\lesssim 140\,M_\odot$) or higher mass ($\gtrsim 260\,M_\odot$) collapse into black holes without significant heavy element creation. Pair-creation SNe are thus a "clean" source of nucleosynthesis in the sense that neighboring mass ranges do not "pollute" the sample. In case of a bimodal Pop III IMF, massive stars in the range $\sim 8\ldots40\,M_\odot$ will also contribute to the resulting abundance pattern, though on a slightly slower time-scale (factor ~ 2). Also in them, the lack of initial CNO "seeds" will lead to an elemental odd-even pattern, though much less expressed [7], and they possibly contribute r-process isotopes. It is even conceivable that Pop III AGB stars contribute some s-process [15]. The interaction of the ejecta with the surrounding matter, a possible enrichment of the intergalactic medium and mixing of contributions from different-mass sources before the formation of the first Pop II stars will have to be studied in more detail in the future.

We predict that Pop III pair-creation SNe might be detectable by future near infrared space experiments — all the way out to the edge of universe — to redshifts of 20 or more. Combined with the challenge to find old Pop II stars that show the predicted abundance pattern from the ashes of these explosions, this should allow deeper insight into the happenings at the times when the first sparks of stellar light terminated the "dark ages".

Acknowledgments This research has been supported by the NSF (AST 97-316569), the DOE ASCI Program (B347885), the DOE SciDAC Program, and the Alexander von Humboldt-Stiftung (FLF-1065004).

References

1. T. Abel, G. L. Bryan, M. L. Norman: Science (2001)
2. I. Baraffe, A. Heger, S. E. Woosley: ApJ **550**, 890 (2001)
3. J. R. Bond, W. D. Arnett, B. J. Carr: ApJ **280**, 825 (1984)
4. V. Bromm, P. S. Coppi, R. B. Larson: ApJL **527**, L5 (1999)
5. D. F. Figer, F. Najarro, M. Morris, I. S. McLean, T. R. Geballe, A. M. Ghez, N. Langer: ApJ **506**, 384 (1998)
6. C. L. Fryer: ApJ **522**, 413 (1999).
7. A. Heger, S. E. Woosley: ApJ, accepted; astro-ph/0107037 (2001)
8. R.-P. Kudritzki: 'Wind Models and Ionizing Fluxes of Massive Stars'. In: *The First Stars*, ed. by A. Weiss, T. Abel, V. Hill (Springer, Berlin 2000).
9. R.-P. Kudritzki, J. Puls: ARA&A **38**, 613 (2000)
10. P. Ledoux: ApJ, **94**, 537 (1941)
11. R. B. Larson: MNRAS **301**, 569 (1998)
12. F. Nakamura, M. Umemura: ApJ **548**, 19 (2000)
13. M. Schwarzschild, R. Härm: ApJ **129**, 637 (1959)
14. Truran, J. W., & Arnett, W. D. 1971, Ap&SS, **11**, 430
15. P. Ventura, F. D'Antona, I. Mazzitelli, R. Gratton: ApJ, **550**, L65 (2001)
16. J. S. Vink, A. de Koter, H. J. G. L. M. Lamers: A&A **369**, 574 (2001)
17. T. A. Weaver, G. B. Zimmerman, S. E. Woosley: ApJ **225**, 1021 (1978)
18. S. E. Woosley, A. Heger, T. A. Weaver: Rev. Mod. Phys., accepted (2002)

Super-Massive Stars: Dense Star-Gas Systems

Pau Amaro-Seoane, Rainer Spurzem, and Andreas Just

Astronomisches Rechen-Institut, Mönchhofstraße 12-14
D-69120 Heidelberg, Germany
pau, spurzem, just@ari.uni-heidelberg.de

Abstract. We use a gaseous model and a semi-analytical approach to study the evolution of a super-massive central gaseous object (a super-massive star, SMS from now on) in an AGN and its evolution by interactions with the surrounding stellar system. Our future work in this field is outlined, which aims at a more detailed study of energy flows in the interstellar medium, stellar evolution and the relation between QSOs and galaxy formation.

Super-Massive Stars, Galaxies and Cosmology

Several theoretical models have been proposed in order to explain the properties of quasars and other types of active galactic nuclei (hereafter AGNs). In the 60s and 70s super-massive central objects (from now onwards SMOs) were thought to be the main source of their characteristics. SMSs and super-massive black holes (SMBHs) are two possibilities to explain the nature of these SMOs being harboured in the AGNs. These super-massive gaseous objects may be the precursors of the SMBHs and those may be an intermediate step towards the formation of these. Large amounts of gas lost by stars during their evolution will stock in galactic centres. In previous work (Amaro-Seoane & Spurzem, 2001) we analyze such a physical system in spherical symmetry. We aim to carry out in the next time the study of the evolution of the SMS and the stellar system by performing a more detailed semi-analytical evaluation of the interaction between stars and gas (cf. Deiss, Just, & Kegel 1990), examining the global stability of the central object, and its dependence on the loss-cone stars that heat it and determine its contraction rate. A detailed numerical model is the via to do it.

Nowadays there is strong evidence that the formation of central black holes in galaxies can be qualitatively understood in the framework of cosmological hierarchical galaxy formation (Haehnelt & Kauffmann 2000, Kauffmann & Haehnelt 2000, see also earlier work of e.g. Eisenstein & Loeb 1995). We stress that, although the embedding into cosmological scenarios had advanced very much during the past two or three decades since the times of Spitzer, Colgate and others, the detailed study of star-gas interactions in dense nuclei, is still incomplete and very important for an understanding of the physical processes at work.

Core Structure and Evolution of SMSs

Hoyle & Fowler (1963) dropped already the hint that maybe at the centres of galaxies star-like objects exist with masses of up to $10^8 M_\odot$. As regards the stability and/or existence of such an object, they "turned a blind eye". They developed their suggestion with the argument that wholly convective stars could "do the job". When radiation pressure is dominant, the convective condition is expressed by a polytrope of index three. The polytropic index is very near to the stability limit and post-Newtonian corrections lead to an instability before the central temperature allows nuclear burning for masses

$$\mathcal{M} > 3 \cdot 10^5 M_\odot; \quad \mathcal{R}_{\mathrm{crit}} = 0.086\mathrm{pc} \cdot \mathcal{M}_8^{3/2} \tag{1}$$

where $\mathcal{M}$ is the mass of the $\mathcal{SMS}$ and $\mathcal{R}_{\mathrm{crit}}$ is the corresponding critical radius, where the instability sets in ($\mathcal{M}_8$ is the mass in units of $10^8 M_\odot$). Therefore, an isolated $\mathcal{SMS}$ eventually run into a catastrophic collapse. The outcome (super-massive black hole, re-bounce, disruption of the $\mathcal{SMS}$, ...) depends very much on the details of its history. Thus, the study of not just the time evolution of the system, but of its stability and, thence, of the general dimensions of the central object seems to be of crucial importance to get a rigorous understanding of the whole problem. Energy and mass transport processes, among them so-called "energy diffusion" and trapping of loss-cone stars (Frank & Rees, 1976, Hara 1978) control the global polytropic structure of the $\mathcal{SMS}$ and its evolution. They strongly depend on a correct physical description of the interactions between the gaseous and stellar components by, for instance, dynamical friction.

Following Hara (1978) and Langbein et al. (1990) we select two competing star-gas interaction processes in our dense gas-star system. The first one is the contraction (or collapse with re-bounce) of the mixed core with stars that are trapped by friction within the gas and slow down additionally to build a new highly condensed stellar core. This core may re-heat the gas or decouple to build a new core. Second, a core-halo interaction is considered, where loss-cone stars of the surrounding stellar system heat the core and feed it by means of new trapped stars. This core collapse can be conked for a while until the loss-cone is "empty" (not rigorously speaking, but for the practice we can assume it is empty) or the core becomes too massive. The result may be a sequence of $\mathcal{SMS}$-stars cores contained one inside the other until the relatively low mass innermost $\mathcal{SMS}$ initiates hydrogen burning or collapses to a BH "seed". In Fig. 1 we give an intuitive scheme for this scenario.

The Gaseous Model

To go from stationary to dynamical models we use a gaseous model of star clusters in its anisotropic version (see e.g. Louis & Spurzem 1991, Spurzem 1994, Giersz & Spurzem 1994, and also http://www.gaseous-model.de for the more recent developments). The basic idea is that of a kinetic equation of the Boltzmann type with the inclusion of a self-consistent collisional term of the Fokker-Planck type. By taking velocity moments of such Fokker-Planck equation of up to order two and closing the system with a heat flux closure we get a set of moment equations which is very similar to gas-dynamical equations and it is then coupled to Poisson's equation for the total gravitational potential. Such a model is very well suitable to treat collisional and collisionless evolution of coupled stellar and star-gas systems, because the stellar dynamical equations resemble gas dynamical ones.

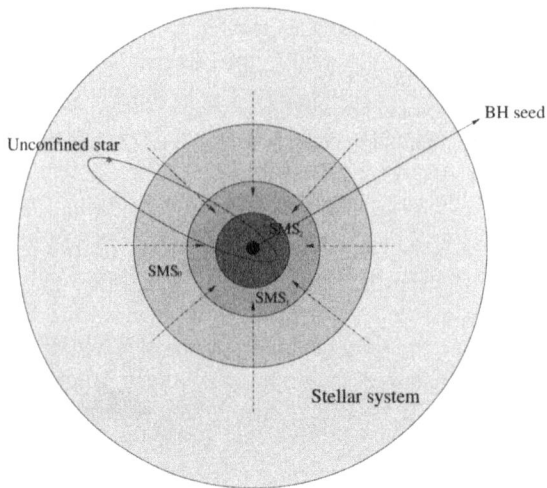

Fig. 1. Sequence of $\mathcal{SMS}$-stars cores: $\mathcal{SMS}_2 \subset \mathcal{SMS}_1 \subset \mathcal{SMS}_0$. The non-confined stars which belong to the loss-cone sink onto the central object after a number of crossings, which we call the "trap" number. The innermost $\mathcal{SMS}$ may collapse to a BH seed.

Acknowledgements

This work has been performed in SFB439 "Galaxies in the Young Universe" at the Univ. of Heidelberg. We are grateful to S. Deiters for providing the gaseous model web page.

References

1. Amaro-Seoane, P., Spurzem, R.: MNRAS in press (2001), astro-ph/0105251
2. Deiss, B.M.; Just, A.; Kegel, W.H.: A&A, **240**, 123 (1990)
3. Eisenstein, D.J.; Loeb, A.: ApJ **448**, 17L (1995)
4. Frank, J.; Rees, M. J.: MNRAS, **176**, 633 (1976)
5. Giersz, M.; Spurzem, R.: MNRAS, **269**, 241 (1994)
6. Hara, T.: Prog. Theor. Phys. **60**, 711 (1978)
7. Haehnelt, M.; Kauffmann, G.: MNRAS, **318**, 35L (2000)
8. Hoyle, F.; Fowler, W. A.: MNRAS, **125**, 169 (1963)
9. Kauffmann, G.; Haehnelt, M.: MNRAS, **311**, 576 (2000)
10. Langbein, T.; Spurzem, R.; Fricke, K.J.; Yorke, H.W.: A&A, **227**, 333 (1990)
11. Louis, P.D.; Spurzem, R.: MNRAS, **251**, 408 (1991)
12. Spurzem, R.: in "Ergodic Concepts in Stellar Dynamics", D.Pfenniger, V.G. Gurzadyan (eds.), Springer-Vlg., Berlin, Heidelberg, p. 170 (1994).

Part VIII

Activity Connected with the Presence of Supermassive Black Holes

To the Lighthouse

Roger D. Blandford

Caltech, Pasadena, CA 91125

Abstract. The extreme hypothesis that essentially all types of ultrarelativistic outflow – specifically AGN jets, pulsar wind nebulae and GRB –are electromagnetic, rather than gas dynamical, phenomena is considered. Electromagnetic flows are naturally anisotropic and self-collimating so as to produce jet-like features. The relativistic force-free description of these flows, which is simpler than the relativistic MHD description, is explained. It is suggested that the magnetic field associated with AGN jets and GRB is quite extensively distributed in latitude, without necessarily increasing by much the total power. It is also proposed that the observed emission from these sources traces out regions of high current density where global instabilities drive a turbulence spectrum that is ultimately responsible for the particle acceleration and the synchrotron, inverse Compton and synchro-Compton emission. The direct extraction of spin energy from a black hole is re-examined and an electromagnetic model of GRB explosions is developed. It is also suggested that some GRB "lighthouses" be identified with accretion-induced collapse of a neutron star to form a black hole in a binary system.

1   Introduction

For an international audience, I should explain that my title is taken from a novel written by Virginia Woolf. It is one of the premier examples of the "stream of consciousness" school of narrative fiction, a style that lends itself well to scientific conference proceedings. If I take literally the metaphor that inspired this meeting, lighthouses draw attention to themselves (and warn of incipient danger) by shining brightly through obscuring material and varying, just like their cosmic counterparts. Now, nearly all of the variable, cosmic beacons that we have discussed here have, implicitly, involved the formation or activation of a black hole. However, few of the other talks have remarked upon this fact, let alone analyzed its physical implications. To most observers and phenomenologists, a black hole might just as easily be a black box - a flexible source of power whose properties are limited only by its total mass and the referees of theoretical papers. It is my task to "deconstruct" the lighthouse.

I must first emphasize an important point. Provided that general relativity is the correct theory of strong gravity, (and there is a great need to test this proposition directly), we have a very good understanding of the theoretical properties of black holes. The Kerr spacetime, the stage upon which epic cosmic dramas are enacted, introduces distinctive but quite calculable modifications to fundamental physics which we may even be observing directly. Our interpretation of these observations remains quite sketchy, on account of our poor understanding of the nineteenth century subjects of gas dynamics, electromagnetism and statistical mechanics, not the general relativity and quantum mechanics of the twentieth!

In this article, I will focus on two linked issues. The first involves revisiting the extraction of black hole spin energy through the application of electromagnetic torques. Here general relativity is of crucial importance. The second is to consider the time-dependent, special relativistic evolution of a magnetic shell containing toroidal magnetic field that has been wound up by a central, spinning, compact object, specifically a black hole plus accretion disk or a neutron star. I will report on an analysis being carried out with Max Lyutikov and argue that this model provides an alternative explanation for gamma ray bursts and is clearly contrasted with the standard, fireball picture. These considerations are also relevant to pulsar wind nebulae (PWN) and possibly to relativistic jets in AGN.

2 Ultrarelativistic Outflows

2.1 AGN

Jets We have known, since the earliest days of VLBI, that quasars create ultrarelativistic outflows[76] [26] with Lorentz factors $\Gamma \sim 10$ and that this flow is typically collimated into two antiparallel jets with opening angles $\sim 5°$. We know now that the radio emission is only the smoke [24]. The γ-ray fire [39] is far more powerful and originates from smaller radii than the radio waves. The γ-rays are believed to be produced by inverse Compton scattering (of scattered, disk photons in the case of high power sources and jet synchrotron photons in the case of low power sources [72]). γ-rays with energies as high as ~ 1 TeV have been recorded, with variability as rapid as 15 min [33].

Although jets are commonly modeled as stationary outflows, they are quite variable and this variation can be observed at radio wavelengths using VLBI. The phenomenon of superluminal expansion is usually attributed to outwardly propagating, relativistic, internal shocks, that are caused by outbursts close to the central black hole [18]. The kinematics of these shocks can be quite complex and many intricate models of their emission properties have been constructed. The energies associated with individual outbursts, lasting of order a week, can be as high as $\sim 10^{54} f$ erg, where $f \sim 10^{-3}$ is the beaming fraction.

Disk and Holes The lighthouses are the central black holes with masses up to $\sim 10^9$ M$_\odot$ and their attendant accretion disks [36], [30]. In most scenarios, the radiant power derives from the binding energy of the accreted gas. Formally this is limited by the existence of a smallest, stable, circular, Keplerian orbit and ranges from $\sim 0.5 - 4 \times 10^{20}$ erg g^{-1} as the hole angular velocity changes from 0 to $0.5/m$. This energy channel is unavoidable while black holes are feeding voraciously and growing rapidly and it is probably responsible for most of the AGN luminosity density in the universe (although the details of how the photons are produced remain unclear [3]). However, it is far from obvious that it can account for ultrarelativistic outflows. The main objection is that gas dynamical winds from gravitationally bound reservoirs, like the solar wind and bipolar outflows from young stellar objects, usually have terminal speeds that are not much more than the gravitational escape velocity at their footpoints [49], whereas AGN jets are clearly ultrarelativistic.

There are several ways out. One hypothesis is that, instead of being radiated efficiently, the binding energy is dissipated in a corona as a high entropy – radiation and pair-dominated fluid that forms the base of a thermally-driven, hot wind. As this fluid expands, the radiation decompresses and the pairs annihilate, just like in the expanding universe, leaving the flow baryon-dominated. The difficulty with this hypothesis, in the case of AGN, is that most of the prominent radio jets that we observe seem to be relatively low entropy sources. For example, in the radio galaxy M87 (which forms jets within $\sim 100m$ [44]), the hole mass is $M \sim 3 \times 10^9$ M$_\odot$ while the bolometric luminosity appears to be $L \sim 10^{43}$ erg s^{-1} $\sim 3 \times 10^{-5} L_{\mathrm{Edd}}$, and is probably less than the jet power. (This objection does not apply to the "Galactic Superluminal Sources", which do appear to be formed in radiation-dominated environments and only have mildly relativistic speeds.)

Another hypothesis is that the working substance is a pure, electron-positron pair plasma. However, this is problematic because radiative drag in the nucleus precludes acceleration to more than mildly relativistic speed. Yet another possibility is that the jet momentum is carried by ultrarelativistic protons, which may have been accelerated by shock fronts formed close to the black hole. There are two difficulties here. The first is that VLBI polarization observations are best interpreted in terms of a pair plasma [95]. The second is in collimating the outflow. This is effectively impossible using gas pressure without violating observational constraints on the X-ray luminosity. (Electro)magnetic field with an energy density comparable with the protons therefore have to be invoked to effect the collimation under the conditions thought to obtain in AGN.

For these and other reasons, many astrophysicists have suggested that AGN jets are powered and collimated hydromagnetically, as also appears to be the case in protostellar jets. A variety of detailed mechanisms have been proposed [15]. In most of these mechanisms, the magnetic field derives from the gas in the accretion disk. The field lines may be primarily poloidal, they may be toroidal or they may involve poloidal field becoming toroidal with increasing distance from the disk. The magnetic field may have large scale order over many octaves of radius [49] or be quite tangled [89]. When the disk orbits with a near-Keplerian velocity, it is possible to launch gas centrifugally [51] and, although it may be possible to create an ultrarelativistic terminal velocity, the natural presumption is that the terminal velocity of an outflow emanating from an accretion disk is no more than mildly relativistic.

These considerations have, in turn, motivated the investigation of an alternative possibility, that the power and speed of the jets follow from their direct magnetic connection to the spinning black hole. (The collimation can still be caused by a mildly relativistic, disk outflow, though.) Black holes possess rotational energy   the differ ence between the mass $m = m_0(1 - \beta^2)^{-1/2}$ and the irreducible mass m_0, where $\beta = 2m_0 \Omega_H < 2^{-1/2}$, with Ω_H being the hole angular velocity [87]. Up to 29 percent of the mass is, in principle, available. For a billion solar mass hole this is $\sim 6 \times 10^{62}$ erg, ample for the most profligate of extragalactic radio sources. Furthermore, connecting the jet to the event horizon makes it very difficult for plasma from the disk to attach itself to these field lines and it is natural to produce an ultrarelativistic, baryon-starved outflow under these conditions. We explore this idea further below.

2.2   Pulsar Wind Nebulae

Pulsars Pulsars were first discovered through their pulsed radio emission and quickly identified as spinning, magnetized neutron stars. More recently, some of them have been identified as far more powerful γ-ray sources. However most of the energy that they release is in the form of a relativistic outflow which inflates a pulsar wind nebula, like the famous Crab Nebula. The outflow Lorentz factors in pulsar winds have sometimes been estimated to exceed $\sim 10^6$ and it has been argued that they become strongly particle-dominated by the time they reach the nebula [45]. On this view, which is like a high speed version of the fluid model of AGN jets, there is a transformation of the power from Poynting flux to bulk kinetic energy of a fluid at fairly small radius, so that a strong, fluid shock can be produced when the momentum flux in the outflow balances the ambient pressure in the nebula. However, recent X-ray observations of the Crab Nebula exhibit strong synchrotron emission concentrated along the axes and the equatorial plane [97] and there is little evidence for a strong shock. These observations encourage us to re-examine magnetic models of PWN.

Electromagnetic Model of the Crab Nebula The Crab pulsar is a spinning, magnetized, neutron star with an inclined dipole moment. It presumably has a force-free magnetosphere through which currents flow. The complete electrodynamical description of this magnetosphere remains an unsolved problem. However, it seems plausible that, somewhere beyond the light cylinder, the electromagnetic field becomes essentially axisymmetric and that variation on the scale of a wavelength dies away. There are at least three ways by which this can occur. There can be steady reconnection in the outflowing, "striped" wind [93], [60]. Alternatively, the waves can decay successively through parametric instability into higher frequency waves [65]. These two processes are essentially dissipative and the magnetic energy will ultimately be converted into heat. However, a third option is essentially non-dissipative. Consider the minority of the open magnetic field lines that emanate from the neutron star's southern magnetic pole, and which can be followed into the northern hemisphere. These may be pulled back across the equatorial plane by magnetic tension into the southern hemisphere (and *vice versa*). Any one of these three mechanisms will destroy the AC component of the electromagnetic field leaving behind only the DC component in which the magnetic field will become increasingly toroidal and axisymmetric with radius. If the magnetic field develops this structure, there must be an associated current flow out along the poles and eventually returning in the equatorial plane, or *vice versa*.

Now consider what must happen to this relativistic, magnetic wind. As the PWN expands with a speed of only $\dot{R} \sim 1000 \text{kms}^{-1} << c$, where $R(t)$ is the radius of the bubble, it is easy to see that the pulsar must be producing magnetic flux at a rate that is roughly $(c/\dot{R})^{1/2}$ times too large to account for the strength of the magnetic field in the nebula. Therefore, most (95 percent in the case of the Crab Nebula) of the flux must be destroyed [77]. On topological grounds, the natural places for this destruction to occur are on the axis and the equatorial plane [6] and these regions are, in any case, formally unstable to pinch and tearing mode instabilities. (The contact discontinuity, although formally stable to Kruskal-Schwarzschild modes is, in practice likely to be unstable to other hydromagnetic instabilities.) There must be a fairly rapid migration of magnetic flux towards the poles and the equator, where it will become tangled on progressively smaller lengthscales in a turbulence spectrum until it can reconnect at a resistive inner scale. Looking at the same process from a current point of view, in the context of the Crab Nebula, we can imagine that the X-ray synchrotron emission delineates regions where a quadrupolar current flow becomes dissipative as a consequence of these instabilities and accelerates relativistic electrons and positrons which diffuse outwards into the body of the nebula. (The current flow observed within the heliosphere, as infered from observations by the Ulysses spacecraft, is also mostly confined to the axis and the equator.)

What would happen if we were to prevent the dissipation required in the Crab Nebula through having the current flow along rigid, perfect conductors? The answer turns out to be that the Poynting flux will be reflected by the outer boundary of the PWN and channeled back onto its pulsar source where it would react back upon the power supply.

Magnetars Magnetars are most convincingly interpreted as slowly rotating ($P \sim 5 - 10$ s, high field $B \gtrsim 10^{14}$ G pulsars) [50]. Most of the energy that they lose is probably derived from the internal magnetic field and released explosively [80], [86]. The timescales for energy release are almost surely much longer than the light crossing times of the magnetosphere and they will form a complex, anisotropic, electromagnetic pulse that expands essentially at the speed of light. As the pulsar spin period is small, the pulse carries essentially no angular momentum, just like a model of a pulsar wind in which we ignore the poloidal magnetic field.

Implications for AGN Jets In the above model of the Crab Nebula, the dynamics is dominated by electromagnetic field everywhere within the contact discontinuity that separates the PWN from the shocked interstellar medium; the inertia and the pressure of the plasma is of minor importance. (We actually know that some thermal plasma does evaporate off the expanding debris from the original supernova explosion so it is not totally ignorable in practice [99].) Can we imagine that the same is true of AGN jets? Can it be that the "jet" features that we image at radio, optical and X-ray wavelengths are really just delineating the regions of high current density? In other words, are these jets surrounded by extensive, evacuated sheaths of toroidal field that flow radially inward to compensate the flux that is destroyed as a result of electromagnetic instabilities [13], [59], [8]?

One of the merits of this viewpoint is that it provides a dynamical rationalisation of the doctrine of equipartition. Pinched currents, on all scales, may continue to become unstable until their stresses are balanced by pressure. Another merit is that it provides a natural explanation for the helical structures that are often seen in VLBI maps. A third advantage is that currents can account for distributed particle acceleration in well-resolved jets, as recent spectral studies suggest may be required [100].

2.3 GRB

The most interesting and the most pressing challenge, though, is to understand γ-ray bursts (GRB) [69]. Most contemporary models invoke a high entropy fireball which eventually creates a baryon-dominated, fluid outflow as discussed above. Internal shocks in this outflow, derived from source variability, are responsible for the γ-ray burst itself; the relativistically expanding shell of shocked interstellar medium being the site of the afterglow. The source itself is most popularly associated with "hypernovae" or "collapsars" [64] – evolved, rapidly rotating massive stars that form a "twin exhaust" [22] through which a pair of ultrarelativistic, fluid jets escape. The best evidence in support of this model is the observation of achromatic breaks in some afterglows. These are expected to arise when the expansion Lorentz factor falls to a value equal to the reciprocal of the jet opening angle. Put another way, the emission declines more rapidly after a sound wave can cross the jet on the expansion timescale.

However, it is possible to consider electromagnetic models [93] in the case of GRB as well - cold, steady state models rather than hot big bangs! In other words, the electromagnetic view is that GRB are much more like the other ultrarelativistic flows that we observe than the early universe. (It is possible that the reason that the fireball interpretation took hold in the context of GRB and not in AGN jets, is that the former were first discovered and studied as γ-ray sources whereas the latter were initially observed as radio sources.)

There are several reasons for considering electromagnetic models. Perhaps the strongest argument is that the outflow has to have a very large Lorentz factor, $\Gamma \sim 300$ [58], in order to reduce the pair production opacity for the highest energy γ-rays to a value below unity at a radius where internal shocks can still operate. This implies that the ratio of bulk kinetic energy to internal energy in the jet is $\gtrsim 10^5$. This is unprecedented in gas dynamics. It is extremely hard to create hypersonic, high Mach number flow in the laboratory using a carefully machined nozzle. There are always transient rarefactions and compression waves associated with the walls that cause large fluctuations in the velocity field and subsequent heating.

It should be noted that the relatively well-studied long duration bursts last for $\sim 10^6$ source light crossing times, if they really are associated with the formation of a ~ 10 M$_\odot$ black hole. (By contrast we have only observed quasars for $\sim 10^5$ light crossing times!) The sources are quasi-steady on the small scale just like AGN. It is therefore possible that the transport of energy from the source to the γ-ray emission region and the expanding blast wave is entirely electromagnetic in this case as well. There are further advantages of the electromagnetic model, which we discuss further below. As the dissipation is initiated by electromagnetic instabilities, the γ-ray emission region does not have to be tied to the central source through internal shocks and can, consequently, be located much further away, where the Lorentz factor, beaming and pair production opacity can be less extreme, as long as the variations do not average out to produce a smooth burst Another advantage is that the electromagnetic field is naturally self-collimating and creates an anisotropic explosion. It is not necessary to invoke collimating channels as in fluid models. A third benefit is that the internal sound speed is effectively c rather than $3^{-1/2}c$ as in the fluid model. This implies that there is more causal contact across the blast wave when it is pushed electromagnetically.

However, it also implies that shocks are not likely to be responsible for the emission of γ-rays.

2.4 Electromagnetic Lighthouses

We have already identified massive black holes as the lighthouses for AGN jets and neutron stars as being responsible for PWN. For GRB, we cannot see the sources directly (though this could change with the advent of neutrino or gravitational radiation observations). Consequently there are many possibilities. In the context of the hypernova model, it can be envisaged that the fireball is created by large scale electromagnetic interactions involving the newly-formed, spinning black hole and its surrounding torus. It is commonly supposed that the energy is quickly transformed into a fireball and a baryonic jet [69]. The major concern with this hypernova explanation is whether the baryons can be excluded efficiently from the flow, especially where it becomes trans-sonic, so that the jet can achieve as large a Lorentz factor as is inferred.

Now, the model that I am discussing here is initially the same as at least some versions of the hypernova model that I have just described [55]. However it departs from it by assuming that the energy remains in an electromagnetic form all the way out to the region where the γ-rays are emitted. There is no fireball or hadronic intermediate state. It also deviates from other electromagnetic models by assuming that it is only the DC field component that is important well beyond the light surface. Now it is possible that all this could happen inside a star and that the entrainment of gas from the star into the electromagnetic outflow could be negligible. However, as this does seem rather improbable, at least to this reviewer, it is worth considering some alternative choices of prime mover for GRB.

One possibility is that a GRB is a newly formed magnetar [29] – a strongly magnetised, rapidly spinning, neutron star that is able to blow away the surrounding stellar envelope before it slows down. To be specific, for a long duration GRB, if the star is to have an electromagnetic power of $\sim 10^{49}$ erg s^{-1} for a time $t_{\text{source}} \sim 100$ s, then a period of ~ 4 ms and a field $\gtrsim 10^{15}$ G is necessary at breakout. Alternatively, a GRB may be formed as a result of the operation of an r-mode instability in a spun-up neutron star [80].

Another possibility is accretion-induced neutron star collapse [63]. Suppose that, a neutron star in a binary system accretes gas from its companion over the course of time and is able to accept the additional $\sim 0.4 - 1$ M$_\odot$ required to exceed the Oppenheimer-Volkov limit, in much the same way that white dwarfs are now thought to grow to become SNIa. Suppose that not all of the mass in the star crosses the event horizon and ~ 0.1 M$_\odot$ of gas is left behind in a relativistic torus, as a result of neutrino or centrifugal stresses, and that this torus traps a flux $\sim 10^{26}$ G cm^2. The spin energy of the black hole may be extracted to release the requisite electromagnetic energy over the observed timescale. One advantage of this model is that there will be no stellar envelope to impede the escape of an ultrarelativistic outflow.

A further idea, that is only credible for the long duration GRB, is that the source be a massive ($\sim 10^5$ M$_\odot$) black hole that captures a white dwarf which is tidally disrupted and generates a large magnetic field $\sim 10^{10}$ G when its orbit becomes relativistic [17]. In this case, the electromagnetic energy must be released on an orbital timescale. (It is in principle possible to release an energy $>> M_\odot c^2$ by tapping the spin energy of the hole in this type of model.)

For the remainder of this article I shall explore the extreme hypothesis that all ultrarelativistic outflows are essentially electromagnetic rather than fluid in character and that the dissipating/accelerating regions coincide with the most intense current densities rather than shock fronts [56]. I shall emphasize the complementarity between field and the current formulations. As most of this will be described in greater detail elsewhere, I shall mostly confine our attention to two key aspects – the direct extraction of energy from a spinning hole and the dynamics of GRB.

3   Some Formal Preliminaries

3.1   Force-Free Formalism

The simplest way to describe electromagnetic field both in the vicinity of a black hole event horizon and in an electromagnetic outflow is using the force-free approximation. In a flat space, this amounts to solving the two Maxwell equations:

$$\frac{\partial E}{\partial t} = \nabla \times B - \mu_0 j \tag{1}$$

$$\frac{\partial B}{\partial t} = -\nabla \times E \tag{2}$$

where the current density perpendicular to the local magnetic field is determined by the force-free condition, which drops the inertial terms

$$\rho E + j \times B = 0 \tag{3}$$

(Note that this implies that the invariant $E \cdot B = 0$ and that electromagnetic energy is conserved, $E \cdot j = 0$.)

Under stationary and axisymmetric conditions, these equations guarantee that the angular velocity Ω is conserved along field lines. They also require a space charge density $\rho = \mu_0^{-1} \nabla \cdot E$ of magnitude $\sim \Omega B$ to develop. (Formally this, like the equation $\nabla \cdot B = 0$, is just an initial condition.)

The force-free condition can be re-expressed by setting the divergence of the electromagnetic stress tensor to zero [53]. This form has the merit that it brings out the analogy with fluid mechanics. Electromagnetic stress pushes and pulls electromagnetic energy which moves with an electromagnetic velocity $E \times B/B^2$, perpendicular to the electric and magnetic fields. This is the velocity of the frames (only defined up to an arbitrary Lorentz boost along the magnetic field direction) in which the electric field vanishes, (provided that the other invariant $B^2 - E^2 > 0$).

If we combine Eq. (3) with the equation

$$\frac{\partial (E \cdot B)}{\partial t} = 0 \tag{4}$$

we obtain the (linear) constitutive relation

$$j = \frac{(E \times B)\nabla \cdot E + (B \cdot \nabla \times B - E \cdot \nabla \times E)B}{\mu_0 B^2} \tag{5}$$

which can be substituted into Eq. (1), (3) to obtain a five dimensional (using $\nabla \cdot \boldsymbol{B} = 0$) hyperbolic set of equations for the electromagnetic field [91] which can be solved numerically [48]. Generalizing to a curved spacetime, specifically that associated with a Kerr hole, presents no difficulties of principle. Boundary conditions have to be specified on conducting surfaces and at the horizon. The latter is tantamount to requiring that the electromagnetic field be non-singular when measured in an infalling frame [102].

3.2 Relativistic MHD

It is instructive to contrast this electromagnetic approach with the relativistic MHD formalism that is currently used by most investigators [74], [25], [83], [82], [73]. In the MHD formulation, which is almost certainly required for field lines that connect to the accretion disk and the plasma surrounding the outflow, the force-free equation must be modified to include inertial terms and, perhaps, pressure gradients. This means that a fluid velocity field must be tracked, along with an enthalpy density and a pressure. In addition, it is generally supposed that the current density satisfies an Ohm's law in the rest frame of the fluid.

$$\frac{\boldsymbol{j}}{\sigma} = \boldsymbol{E} + \boldsymbol{v} \times \boldsymbol{B} \to 0 \tag{6}$$

Under relativistic MHD, the constitutive relation, Eq. (5) must be augmented with inertial terms. The equations are still evolutionary, though more complex.

The introduction of these extra complications, when the inertia of the plasma is relatively small, can be questioned on several grounds. Firstly, it is assumed that the electric field vanishes in the center of momentum frame of the plasma. This is not guaranteed by the plasma physics. Secondly, it is usually assumed that the plasma slides without friction along the magnetic field. In other words, there is no dissipation. This is unlikely to be the case in the face of instabilities and radiative drag [10]. Thirdly, it is generally supposed that plasma is conserved. This is untrue of a black hole ergosphere, where pair creation is going on all the time. Finally, there is the assumption that the particle pressure tensor is isotropic. In practice this is rarely the case in observed plasmas, including the solar wind. This is relevant to the discussion of the sound speed and is of crucial importance to the asymptotic behaviour of characteristics.

3.3 Gaps

There is a potential problem with both the force-free and the relativistic MHD approaches. This was uncovered in the original, Goldreich-Julian[37] model of an axisymmetric pulsar. The sign of the space charge in a force-free magnetosphere is given by the sign of $\boldsymbol{\Omega} \cdot \boldsymbol{B}$ and can change along a magnetic field line. This means that positive current must stream through outward regions of negative charge density (or *vice versa*). This, in turn, implies that positive and negative charges have a substantial relative velocity (as required in some models of pulsar emission). These conditions may lead to the the formation of "gaps", within which enough fresh pairs are created to allow these conditions to be satisfied.

Actually, there is no kinematic requirement that any pairs be created to satisfy the current and charge density change in any stationary outflow as long as the density is large enough. The currents of the positrons (or protons) and electrons can be separately conserved and it is always possible to adjust the mean velocities of these two species so as to produce any required charge density [13]. However it is not clear that the plasma will cooperate electrodynamically when we include inertia and instabilities. If, for example, all the particles travel ultrarelativistically with Lorentz factors $\gamma \gg 1$, then either the particle density must exceed the Goldreich-Julian density by a factor $= O(\gamma^2)$ or there will be pair production in a gap. Under these circumstances there may well be time-dependence, dissipation and emission of radiation, as has been proposed in the case of pulsars.

Despite this concern, the potential differences required to create pairs through a vacuum breakdown are typically ~ 1 GV and never more than ~ 1 TV. These are orders of magnitude smaller than the EMFs required to account for ultrarelativistic outflows (~ 30 PV for the Crab pulsar, ~ 100 EV for a powerful AGN jet and ~ 10 ZV for a long GRB). Therefore, I shall assume that the plasma solves these problems without expending much of its motional EMF along the magnetic field lines or invalidating the force-free approach inside magnetospheres.

3.4 Currents and Fields

Electromagnetic models exhibit some simple generalities that derive from treating them as electromagnetic circuits. The source region can be thought of as a generator capable of sustaining an EMF $\mathcal{E}$. Under most conditions the maximum energy to which a particle can be accelerated will be limited by $\sim Ze\mathcal{E}$ and the load will be electromagnetic implying that its effective impedance Z_{load} is typically that of free space $Z_{\text{load}} \sim 100\Omega$. This implies that the power lost to the load is $\sim \mathcal{E}^2/Z_{\text{load}}$. For example if an electromagnetic source is invoked to account for the ultra high energy cosmic ray protons, an EMF $\mathcal{E} \sim 0.3$ ZV must be invoked and a quasar-like power of $\sim 10^{46}$ erg s^{-1} will usually be present. This is not an invariable law, but a necessary feature of most particle acceleration models. (Of course as we discuss below, this power is not necessarily radiated, it might appear as bulk kinetic energy or heat.)

3.5 Characteristics

In order to analyze the causal behaviour of electromagnetic models of ultrarelativistic outflows, it is necessary to consider the characteristics. These are simplest to consider in the zero inertia, force-free case. There are then just two characteristics, a fast mode and an intermediate mode.

The fast mode is easiest to discuss in the (primed) frame in which the background electric field vanishes. In this frame, it is simply an electromagnetic wave polarised such that its electric vector is parallel to $\boldsymbol{k}' \times \boldsymbol{B}'$. Transforming into a general, unprimed frame we find that the common phase and group velocities are both $\boldsymbol{V}_f = \hat{\boldsymbol{k}}$ and that the electric and magnetic perturbations are given by

$$\delta \boldsymbol{B} = A\boldsymbol{V}_f \times [\boldsymbol{E} + \boldsymbol{V}_f \times \boldsymbol{B}] \tag{7}$$

$$\delta \boldsymbol{E} = A\boldsymbol{V}_f \times [\boldsymbol{B} - \boldsymbol{V}_f \times \boldsymbol{E}] \tag{8}$$

where A is a dimensionless amplitude [92]. Note that the electric and magnetic perturbations are equal in magnitude and form an othogonal triad with the wave vector k. The quantity $|\delta B|/\omega$ is Lorentz invariant.

The intermediate mode is also best understood in the primed frame. The electric and magnetic perturbations are also equal in magnitude and form an orthogonal triad with $\hat{B}'$ so that the group velocity is $\hat{B}'$, (assuming that $k \cdot B > 0$, without loss of generality). When we transform to the unprimed frame, the group velocity becomes

$$V_i = \frac{E \times B + \left(B^2 - E^2\right)^{1/2} B}{B^2}. \tag{9}$$

This is what is important for the propagation of information. The electric and magnetic perturbation are still equal and form an orthogonal triad with V_i in this frame, so that

$$\delta B = \frac{AB}{\omega} k \times V_i \tag{10}$$

$$\delta E = \delta B \times V_i \tag{11}$$

This time, the quantity $(1 - E^2/B^2)^{1/2}|\delta B|$ is Lorentz invariant.

Next, consider a force-free electromagnetic field rotating with angular velocity Ω in flat space so that $E = -(\Omega \times r) \times B$. There are two intermediate group velocities with components parallel and antiparallel to B. A calculation shows that the inward-directed group velocity has a radial component that changes from negative to positive at the *intermediate critical surface*, (also called the light cylinder), where $|\Omega \times r| = 1$. By contrast, in a stationary axisymmetric, black hole magnetosphere, where spacetime is curved, there are two intermediate critical surfaces, an inner one, within which outward-directed intermediate modes cannot propagate out, and an outer one beyond which inward-directed modes cannot propagate in. There are also fast critical surfaces. The innner one coincides with the horizon; the outer one is located at infinity. The addition of a small amount of inertia, under the rubric of relativistic MHD, changes this, in ways that have proven to be controversial [71]. It leads to the formation of fast mode critical surfaces at finite locations beyond which the electromagnetic field is isolated. However, providing the inertial loading is small, this should not lead to a large change in the magnetospheric structure.

Now, consider the information carried by these waves in the far field under the force-free approximation. The intermediate modes can only communicate information outward, beyond their critical surface. However, the group velocities of fast modes can point in all directions, including inward. If the background magnetic field is nearly toroidal and k is poloidal, (as it will be for axisymmetric disturbances), then the magnetic perturbation is also almost toroidal. As such it carries information about the poloidal current and allows inward communication of this information beyond the light cylinder. (It is not necessary for the wave to include an electrical current perturbation for this to happen *cf.*[75].) Similar considerations apply in a force-free black hole magnetosphere within the inner, intermediate critical surface.

4 Electromagnetic Extraction of Energy from Spinning Black Holes

4.1 Direct Extraction

There are several ways through which the rotational energy associated with the spinning spacetime can be tapped electromagnetically [23], [74], [62], [81], [87], [101], [55], [90], [16], Fig. 1. The particular choice that I have emphasised, because I believe that it represents the dominant energy channel, is that the horizon is threaded by a large flux of open magnetic field, supported by external electrical currents flowing in the inner accretion disk. The magnetic flux may have accumulated as a result of being carried inward by the inflowing gas or as result of dynamo action in the disk [52]. (These two processes are unlikely to be cleanly separable.)

However, if flux does thread the hole, (in some appropriate infalling coordinate system [48]), a continuous, electromagnetic Penrose process operates in the ergosphere of the black hole which results in Poynting flux flowing inward across the horizon and, simultaneously, propagating away from the hole to infinity. How can this be? The

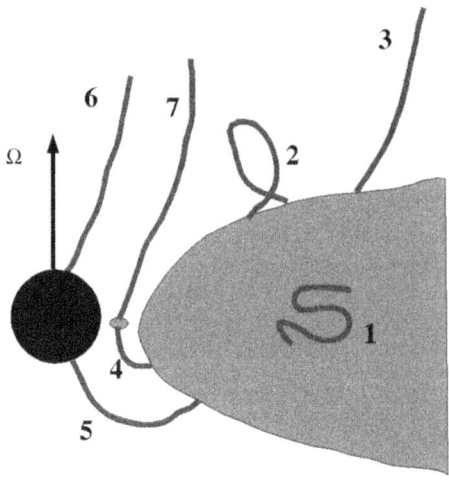

Fig. 1. Different types of magnetic field line discussed in the text. 1. The interior torque is contributed by magnetic field amplified through the magnetorotational instability. 2. Short loops of toroidal field will energise the disk corona, through having their footpoints being twisted in opposite senses and creating small scale flares. 3. Open field lines that connect the disk to the outflow may drive a hydromagnetic wind. Loops of field from the inner disk that connect to plunging gas (4.) or the event horizon (5.) of the hole can effectively remove energy from the hole. 6. Open field lines that cross the event horizon can power a relativistic jet, which may be collimated by and possibly decelerated by the outflow. 7. Open field lines that connect to plunging gas in the ergosphere may likewise contribute to the jet power.

answer is that that energy flux is conserved in a frame at rest with respect to Boyer-Lindquist coordinates, which becomes singular at the horizon. Therefore power appears to emerge from the horizon in the Boyer-Lindquist frame. However, physical observers must orbit with respect to this coordinate system. Doppler boosting the energy flux into a frame moving with a physical observer produces an inwardly directed energy flux, as we expect. The source of the power is ultimately the spacetime, against which the electromagnetic field in the ergosphere is doing work.

Another concern is that the current-carrying charged particles, presumably pairs, should flow inwards at the horizon and presumably outwards, far from the hole. This requires pair creation in the region between the ingoing and outgoing light (intermediate) surfaces. As discussed above, this will happen quite readily either through cross-field diffusion, pair production by ambient γ-rays or by opening up a gap [42]. The ratio of the minimum pair density required to carry the current to the maximum that could be present without inertia being important is of order the ratio of Ω_H to the electron gyro frequency $\sim 10^{-15}$ in the context of a powerful AGN.

The net efficiency of energy extraction can be considered using a circuit analogy. The Poynting flux that flows into the black hole (in a frame rotating with angular velocity Ω_H can be considered as a form of dissipation or internal resistance. Likewise, the Poynting flux far from the hole represents a second (load) resistance. The efficiency of energy extraction – the fraction of the reducible mass that is actually dissipated in the load – depends upon the angular velocity of the field lines. In a stationary, axisymmetric electromagnetic model, this is a resistance-weighted average of the hole and load angular velocities. It depends upon the shape of the magnetic surfaces far from the hole. For example, for a slowly rotating hole, with a radial field, the efficiency is 0.5; when the magnetic surfaces are paraboloidal, the average efficiency is 0.38. Collimating outflows are generally less efficient.

However, not all the field lines that thread the horizon need connect with the outflow. Some low latitude field lines may connect directly to the accretion disk and provide a supplementary power source for the disk as well as a possible driving torque for exciting quasi-periodic oscillations. [14], [57]. This energy channel could be important, especially if the disk is thick. However, it is unlikely to lead to an ultrarelativistic outflow.

The process that I have just described – the direct electromagnetic extraction of energy from the black hole – is distinct from (though can operate simultaneously with) an alternative process, the extraction of binding energy from the accreting gas both in the disk [13], [21], [54] and in the plunging region between the inner edge of the disk and the horizon [47], [67]. The extra power that this process produces can be charged to the spin energy of the hole, which increases at a slower rate than it would do so in the absence of magnetic stress. However the intermediate working substance that effects this transformation is the inertia of the plasma not the electromagnetic field. One way of accounting for the greater energy release of power from an accreting, Kerr hole is that it is "borrowed" from the extractable spin energy of the hole. After the gas eventually crosses the horizon the hole mass increases by a smaller amount than it would have done in the absence of the magnetic stress. (The magnetic connection of the disk to the plunging gas seems to be a less promising source of power because the magnetic field lines which are likely to have a far greater inertial load than open field lines threading the hole, either quickly reconnect or become super-fast, leaving the inflowing gas effectively disconnected from the disk [43], [2], [35], [4].)

There are three reasons for emphasising direct extraction of energy from the hole over extraction from the infalling gas at least for a rapidly spinning hole. The first is that the event horizon has a larger effective area than the annular ring between the hole and the disk. The second is that any gas-driven outflow is likely to be contaminated with baryons and consequently is unlikely to achieve the ultrarelativistic outflow velocity required. The third is that holes probably rotate much faster than orbiting gas, except quite close to the horizon, from where the extraction of energy will be quite inefficient.

4.2 Criticisms and Variations

The general idea that the spin energy of a black hole can be tapped electromagnetically as just discussed, has received an observational boost from the discovery that black holes are commonplace on both the stellar and the massive scale (as well as, perhaps, on the intermediate scale [27]) and that the second parameter, the spin, appears to be large so as to allow gas to orbit close to the horizon and to form strongly redshifted emission lines [84], [98]. However, the more specific notion that ultrarelativistic outflows are powered by direct, electromagnetic extraction of energy from a spinning hole has been criticized on several grounds and some alternative models have been developed. These criticisms and alternatives include, in addition to those discussed above:

- Black hole magnetospheres develop gravitationally significant space charge particularly just after the formation of the event horizon. The foregoing considerations imply that it is necessary for large, field-parallel electric fields to develop[78], [90], [75]. In the limiting case the potential differences along the magnetic field lines approach the impressively large, fundamental "gravitational" value, $(G\mu_0)^{1/2}c^3 \sim$ 1 XV. (1 xenna eV $\equiv 10^3$ yotta eV $\equiv 10^{27}$ eV!)
- The electromagnetic field has a vacuum configuration with $E \cdot B \neq 0$. The potentials here are significantly smaller $\sim GMB/c$ though still comparable with the full EMF developed in a force-free model, as discussed above and, on the face of it, ample to ensure discharge of the vacuum. If an electromagnetic configuration like this can be sustained then interesting, new physical effects can be contemplated [88], [40].
- The stationary, force-free, electromagnetic configuration is unstable because the region within the inner light surface is effectively out of causal contact with the horizon and the magnetosphere cannot respond to changes in the spin of the hole [75]. Probably the most convincing way to decide if this really happens is to perform time-dependent numerical calculations of a black hole magnetosphere and determine if it settles down to a stationary state with electromagnetic energy being steadily extracted. Preliminary computations show no sign of instability [48], [88].
- A rather different type of instability is to non-axisymmetric screw modes. These are likely to be present especially when the magnetic field becomes tightly wound as we discuss further below. What is important is what happens in the nonlinear regime.

- The force-free equations are fundamentally inadequate and only relativistic MHD can provide a complete description of a magnetosphere and its outflow particularly when considering the role of critical points [74], [71].
- The angular velocity of the magnetic field lines is determined by the details of the pair creation not the boundary conditions at the horizon and large radius. If so, the efficiency of energy extraction is expected to be seriously reduced [9], [74], [75].
- Electromagnetic extraction of energy from the hole is irrelevant because too little flux threads the horizon for the power extracted to be significant relative to the power extracted from the disk [70]. This is undoubtedly the case for slowly rotating holes with thin disks that only extend down to $r \sim 6m$ [23]. It will not be clear what is the disk - hole power ratio for rapidly spinning holes until we have a better understanding of disk structure and solve the cross-field stress equation in the magnetosphere. Note that it is still possible for the jet power to exceed the bolometric luminosity and yet be less than the total power lost by the accreting gas if the disk does not radiate efficiently, which is commonly assumed to be the case for radio galaxies and quasars [66].

Many of these issues are unlikely to be satisfactorily resolved until there are an extensive series of time-dependent numerical simulations using either the force-free or the relativistic MHD formalism, or preferably both [47], [48], [66].

5   Electromagnetic Shells

5.1   Electromagnetic Solution

Suppose that a magnetic rotator - black hole plus torus or a magnetised neutron star spins off magnetic flux for a time t_{source} into the far field, and that this creates a relativistically expanding shell of electromagnetic field that drives a blast wave into the surrounding medium. The blast wave is supposed to be bounded on its outside by a strong shock front that moves with Lorentz factor Γ and, on its inside, by a contact discontinuity, separating it from the electromagnetic shell, that moves with Lorentz factor Γ_c. Suppose further, for simplicity, that the conditions near the light cylinder are such that the current well beyond the light cylinder flows along the axes, then flows along the contact discontinuity and finally returns to it source through the equatorial plane (Fig. 2), just as we dicsussed above in the context of the Crab Nebula and the solar wind.

If we ignore the poloidal component of the magnetic field, (and consequently the flux of angular momentum), then the relevant solution of the force-free equations associated with this current flow has the form

$$B_\phi = \frac{f_+(t-r) + f_-(t+r)}{r \sin \theta} \tag{12}$$

$$E_\theta = \frac{f_+(t-r) - f_-(t+r)}{r \sin \theta} \tag{13}$$

The two terms in each expression are fast modes propagating outward and inward. The charge and current density vanish in the interior of the shell. We can determine the functions f_+, f_- by specifying the electromagnetic field at some small radius beyond the light surface and by matching to an ultrarelativistic blast wave expanding into the

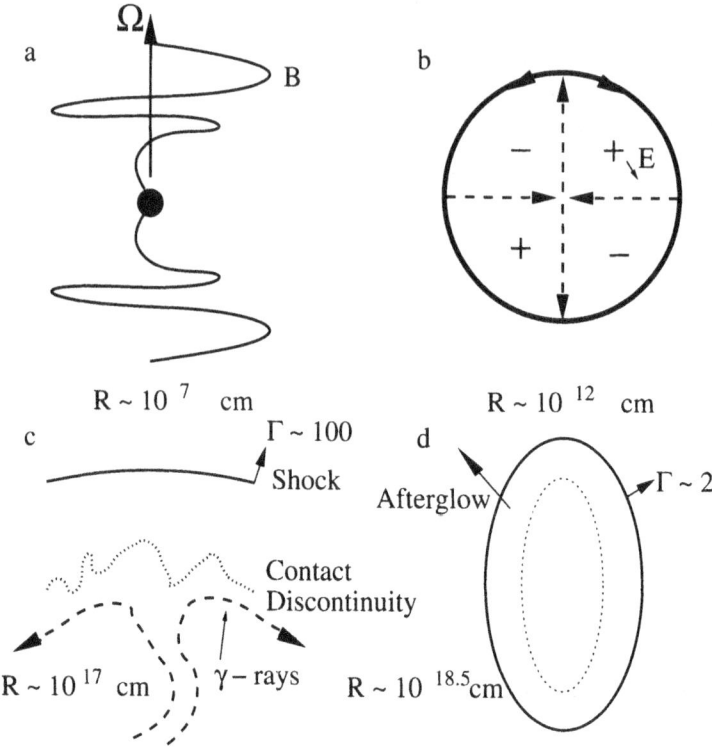

Fig. 2. Four stages in the expansion of a magnetic shell with scales appropriate to a long duration GRB. a) The magnetic field B changes from poloidal to toroidal close to the outgoing light surface of the magnetic rotator at a radius $R \sim 10^6$ cm. The alternating component of the electromagnetic field decays relative to the DC toroidal field. b) The source is active for ~ 100 s. By this time, it will have inflated a magnetic bubble with radius $R \sim 3 \times 10^{12}$ cm, expanding with Lorentz factor $\Gamma \sim 3 \times 10^4$. The magnetic field is mostly toroidal, with the signs shown, while the electric field E is poloidal. The quadrupolar current flow is shown dashed. The shocked circumstellar medium is compressed into a thin shell of thickness $\sim 10^3$ cm. c) By the time the shell has expanded to $R \sim 10^{17}$ cm, $\Gamma \sim 100$ and most of the electromagnetic pulse has caught up with the blast wave. This phase is observed a time ~ 100 s after the initial explosion. The current flow is still largely quadrupolar, though it is unstable along the axis and the equator and this drives an electromagnetic turbulent cascade. which ultimately creates electrical resistance and dissipation in the form of pair production, particle acceleration and intermittent, γ-ray emission. These instablities also promote corrugation of the contact discontinuity and incorporation of the magnetic field into the shocked interstellar medium where it can mix with relativistic electrons accelerated at the bounding shock front. d) When the blast wave has expanded by a further factor ~ 30, its speed is only mildly relativistic. Its shape will be quite prolate as the expansion is fastest along the axis. Most of the energy released by the central, spinning, magnetic rotator is now carried by the shocked interstellar medium.

surrounding medium. This last requires that the outer surface of the shell move at the same speed as the inner surface of the blast wave and that the magnetic stress normal to this surface match the pressure in the blast wave.

The simplest assumption to make is that the strength of the magnetic rotator is constant, (f_+ = const), for $t \lesssim t_{\text{source}}$ and that the external density is constant in radius. These assumptions imply that the Lorentz factor of the blast wave's outer shock front varies with radius R according to $\Gamma \propto \csc \theta R^{-1/2}$ [19]. (The Lorentz factor of the contact discontinuity, Γ_c, exhibits a similar variation.) The electromagnetic velocity (the velocity of the frame in which the electric field vanishes) in the body of the shell $\boldsymbol{E} \times \boldsymbol{B}/B^2$ is radial and is equal in magnitude to $(f_+ - f_-)/(f_+ + f_-)$ and the magnetic stress in a frame moving with this velocity is $\propto f_+ f_- \csc^2 \theta$. The expansion of the blast wave is anisotropic, being faster along the poles, giving an electromagnetic power per steradian $L_\Omega \propto \csc^2 \theta$.

It is instructive to relate this simple solution to our earlier discussion of characteristics. The backward-propagating, reflected wave is a fast mode but has a small amplitude when $\Gamma_c \gg 1$. (In this approximation, the intermediate mode's group velocity is purely toroidal and carries no radial information.) However the fast mode is able to propagate radially inward and, in effect, create an electromagnetic pressure wave which decelerates the electromagnetic velocity of the outflow and reduces it to match that of the contact discontinuity. The magnetic field is toroidal and carries information about the current flow. Put another way, if we were to change the properties of the load, e.g. by encountering a sudden increase in the ambient gas density, which would cause sudden jumps in Γ, Γ_c, then the boundary conditions on the interior flow would change, along with changes in the amplitude of the reflected wave. This allows the interior solution to adjust.

However, it takes a very long time for a wave to be reflected by the blast wave and return to the origin. This is generally true of ultrarelativistic flows and stationary solutions, which take a long time to be established, can be quite misleading. This has a second consequence. There is no necessity to destroy magnetic flux through ohmic dissipation, until the wave can actually propagate back to the source. (This is in contrast to what happens with PWN, where it is necessary to destroy most of the magnetic flux.) Stated another way, there *need* be little resistance in the electrical circuit. The effective load can consist of the performance of work on the expanding blast wave. This is where most of the power that is generated by the central magnetic rotator ends up. Until the blast wave becomes nonrelativistic, the distinction between the inertial load of this solution and a different, dissipative load is quite unimportant for the behavior of the magnetic rotator. As long as the expansion remains ultrarelativistic, it is a very good approximation to impose a Sommerfield radiation condition on the solution of the magnetosphere and to ignore the reflected wave. Analogous remarks apply to the horizon boundary condition, in the specific case of a black hole magnetosphere.

This simple, electromagnetic solution will only remain valid until the end of the outward-propagating, electromagnetic pulse catches up with the contact discontinuity. This occurs at some radius $R_{\text{free}} \sim \Gamma(R_{\text{free}})^2 ct_{\text{source}}$. Thereafter, the surrounding blast wave which, by now, has acquired almost all of the energy in the explosion, will expand freely with $\Gamma \propto R^{-3/2}$ until it becomes non-relativistic, after which point, the blast wave will follow a Sedov solution.

5.2 Gamma Ray Bursts

As well as bring out some formal points, the electromagnetic solution just described provides a possible model for GRB [69], [93], [61]. Suppose that a magnetic rotator spins off toroidal magnetic field as it slows down and that this magnetic field fills an anisotropic, relativistically expanding shell in a uniform medium. Suppose, further, that the flux distribution near the light cylinder is consistent with the current being concentrated along the axis and in the equatorial plane, as described above. The current density is most intense on the axis and, although there is no requirement that flux be destroyed as long as the expansion is relativistic, in practice the magnetic pinch will become hydromagnetically unstable to sausage and kink modes (in the comoving frame) after expansion beyond a radius where the stabilising, poloidal field becomes insignificant [5]. These global instabilities, which should have a longitudinal wavelength comparable in size to the width of the current distribution, may sustain an electromagnetic turbulence spectrum which should ultimately be responsible for particle acceleration and the excitation of transverse gyrational motion. This turbulence may have already been seen in the measured fluctuation power spectrum [7]. One reason why particles can be accelerated is that, when the power cascades down to short enough wavelengths, there may be too few charged particles to carry the electrical current and field-parallel electric fields will develop [85]. In addition, nonlinear wave-wave interactions will lead to an efficient, stochastic acceleration [12]. This particle acceleration is the microscopic source of the electrical resistance that is invoked in the global electrical circuit.

Of course the acclerated particles will also radiate in this electromagnetic maelstrom. There will be a mean magnetic field in the comoving frame which will be responsible for regular synchrotron radiation if transverse gyrational motion is excited. However, in the absence of a strong, high frequency, resonant turbulence spectrum, the two particle acceleration mechanisms identified above are more likely to pump energy into the longitudinal motion. In this case, the emitted spectrum is more likely to be determined by the synchro-Compton process [11] where the amplitudes of lower frequency wave modes are so strong that they cause the radiating electrons to oscillate with angular amplitudes in excess of $\sim \gamma^{-1}$. In addition, inverse Compton scattering is likely to produce a strong γ-ray spectrum above the threshold for pair production. All of this will take place in a frame moving outward with the electromagnetic velocity and so the photon energies will suffer one final Doppler boost and the emission will be strongly beamed outward in the frame of the explosion. (By contrast, no such boost will be necessary in the case of PWN jets which explains why they can be seen from large inclination angles.)

Global pinch instabilities also provide a plausible explanation for the large $\sim 1 - 10$ ms fluctuations in the observed γ-ray flux that are observed. The reason why it is possible to circumvent the usual argument that the emission profile will be smooth [79], is that the emitting elements (essentially fast or intermediate wavepackets) move with relativistic speed in the electromagnetic frame. A distant observer may therefore only see the few small intense beams at a given time that are beamed towards him.

The number and emissivity of these patches can vary through the total burst while the radiative efficiency can remain high.

Actually, although the current density is strongest along the pole, it should also be quite strong along the contact discontinuity at the outer boundary of the pulsar wind nebula and in the equatorial plane. γ-ray emission could also arise as a result of local instabiliites from these regions as well and this will change the predicted beaming properties.

5.3 Afterglows

The afterglow is formed after the blast wave becomes free of its electromagnetic driver. Now, in most afterglow models, including those involving jets, it is supposed that the expansion velocity does not vary with angle. However, an electromagnetically-driven blast wave necessarily creates an anisotropic explosion and this has important consequences for observations of the afterglow, especially in the ultrarelativistic phase of expansion. If we continue to use our simple model, we find that the afterglow expansion varies most rapidly and remains relativistic for longest closest to the symmetry axis. As $L_\Omega \propto \csc^2 \theta$ the energy contained in each octave of θ is roughly constant. This means that the most intense bursts and afterglows in a flux-limited sample will be seen pole-on and can exhibit achromatic breaks when $\Gamma \sim \theta^{-1}$, which might be mistaken for jets. The inferred explosion energy with our simple model will be roughly independent of θ and characteristic of the total energy [32]. When the expansion becomes non-relativistic, the remnant will have a prolate shape which might be measurable. (It is tempting to associate some of the barrel-shaped supernova remnants observed in our Galaxy with the remnants of electromagnetic explosions [46].)

This electromagnetic model provides a solution to the puzzle of how to launch a blast wave that extends over an angular scale $>> \Gamma^{-1}$ and where the individual parts are out of causal contact. (Something similar happens in the early universe. Indeed the computation of the temporal fluctuations in GRB has some features in common with the computation of angular fluctuations in the microwave background.) In the electromagnetic model, the energy is transferred to the blast wave by a magnetic shell that pushes (unevenly) on the surrounding gas all the way out to R_{free}. It also supplies an origin for the magnetic flux in the blast wave, for which the alternative origin in the bounding shock front seems very hard to explain [20], [38]. In the present model, magnetic field can simply be mixed into the blast wave (and the shock-accelerated relativistic electrons) at the contact discontinuity through instabilities, much like what seems to happen in regular supernova remnants.

5.4 Some Numbers

Let us give some illustrative, orders of magnitude for one model of a long duration GRB. The electromagnetic energy flux near the pole is $L_\Omega \sim 10^{50}$ erg s^{-1} sterad^{-1} and lasts for a time $t_{\text{source}} \sim 100$ s. The associated EMF in the electrical circuit $\sim 10\,Z$ V and the current is ~ 100 EA. (A potential difference this large, made available along the contact discontinuity, provides one of the few astrophysical options for accounting for UHE cosmic rays [96].) If the external density is uniform with $n \sim 1$ cm^{-3}, then the blast wave is driven by the electromagnetic pulse with Lorentz factor $\Gamma \propto R^{-1/2}$ until $R \sim R_{\text{free}} \sim 10^{17}$ cm, $\Gamma \sim \Gamma_{\text{free}} \sim 100$. Thereafter there is a freely expanding blast wave with $\Gamma \propto R^{-3/2}$ until the expansion becomes non-relativistic when $R \sim R_{\text{NR}} \sim 3 \times 10^{18}$ cm.

If most of the GRB emission (around ~ 1 MeV) is produced when $R \lesssim R_{\text{free}}$ by the synchrotron emission then this requires ~ 100 GeV electrons and a comoving magnetic field of strength $B \gtrsim 30$ G. The comoving cooling time of these electrons is ~ 3 s, a fraction $\lesssim 10^{-4}$ of the expansion timescale and so if the ~ 100 GeV pair energy density is maintained at a significant fraction of the equipartition energy density, then the magnetic energy can be efficiently transformed into γ-rays. The opacity to pair production for a γ-ray of energy E is $\sim 0.1(E/1\text{GeV})$. The Thomson optical depth depends upon the details of the particle acceleration but is plausibly much smaller than unity so that the observed γ-rays can freely escape without erasing the variability.

5.5 Some Possible Generalizations

This simple model of GRB was predicated upon a very simple current flow. In practice, it is the detailed electrodynamics in the vicinity of the outgoing light surface that fixes the poloidal magnetic field and electrical current distributions. We can therefore change these (still, of course, maintaining the force-free condition) and solve for a new evolution of the magnetic shell and blast wave. A broader distribution of currents will generally produce a less pronounced expansion along the axis and change somewhat the statistics of the observed afterglows. In the solution above, we ignored the poloidal magnetic field and, consequently, the angular momentum. These can be reinstated perturbatively into the solution. Their influence wanes with increasing radius.

Other ways to obtain different solutions include changing the temporal variation of the source from the simple step function considered above and allowing the external density to vary with radius and latitude. As the individual parts of the blast wave expand essentially independently, when ultrarelativistic, there are no new issues of principle to address in solving these problems. However, all this changes when the blast wave becomes non-relativistic. At this point the interior gas will be roughly isobaric and the shell will become more spherical with time.

6 Discussion

In this brief overview, I have explored the strong version of the "electromagnetic hypothesis" for ultrarelativistic outflows, namely that they are essentially electromagnetic phenomena which are driven by energy released by spinning black holes or neutron stars and that this electromagnetic behavior continues into the source region even when the flows become non-relativistic. (If this hypothesis is falsified in the resolved, emission regions, then there is the fallback position- that the outflows are only initially electromagnetic and quickly convert to a baryonic – jet [94].) The most striking implications of the electromagnetic hypothesis are that particle acceleration in the sources is due to electromagnetic turbulence rather than shocks and that the outflows are cold, electromagnetically dominated flows, with very few baryons at least until they become strongly dissipative.

There are many possible tests of this hypothesis. GRB, should not be associated with strong, high energy neutrino sources. (A gravitational wave signal is expected in some, though not all, source models and would be strongly diagnostic if ever detected.) It is possible, though not very likely, that electromagnetic GRB will be associated with core-collapse supernovae. In this case there may be a detectable MeV neutrino pulse. More immediate, though less specific, diagnostics include relating the duration and character of the GRB with the inferred observation angle of the burst – higher inclination should be associated with longer and less intense bursts.

The spectrum and polarisation of the afterglow emission might also contain some clues, though the lack of a usable theory of particle acceleration at relativistic shocks is a handicap. (Recent, promising progress on this problem produces a power law distribution function with a logarithmic slope of 2.2 [1], provided that the scattering is essentially normal to the shock front. What is not yet clear is whether these scattering conditions are present or whether rather different principles [34] might be at work.) On general grounds, it would be of strong interest to determine if the particle acceleration properties of ultrarelativistic shocks propagating into interstellar gas are just a function of the shock Lorentz factor Γ, while scaling with the external density, as should be the case. It is also important to try to understand kinematically if the magnetic field is introduced into the blast wave at the shock, as conventionally assume, or at the contact discontinuity as proposed here. A more detailed discussion of the GRB emission, than presented here should also account for the MeV breaks observed in γ-ray spectra.

From a more theoretical perspective, there is much to be learned about the properties of force-free electromagnetic fields, especially their stability. The possible relationship of the GRB fluctuation power spectrum to an underlying turbulence spectrum is especially tantalising. Undoubtedly, numerical simulations will be crucial as the problem is essentially three-dimensional. Force-free electromagnetism is easier to study than relativistic MHD and may well be a very good approximation in many of these sources.

Turning to PWN, the most direct, observational challenge is to see if there really is a strong, dissipating shock as expected with a fluid wind or a flow of electromagnetic energy towards the axis and the equatorial plane as predicted by the electromagnetic model and as appears to be exhibited by the Crab Nebula. It is worth trying to model the images from radio to X-ray wavelengths phenomenologically adopting the electromagnetic hypothesis. For the pulsar, the oblique rotator remains an unsolved, theoretical problem and which can now be tackled numerically, at least in the time-dependent, electromagnetic limit [28]. If a satisfactory, global, electromagnetic solution can be found, then this provides a framework for carrying out microphysical investigations to revisit such question as to how currents manage to change their space charge density as they flow along magnetic field lines.

Finally, for AGN, it is predicted that the jets are not hadronic and should also not be powerful high energy neutrino sources. This is refutable. In addition, we need to see if the well-resolved jets and their lobes can be re-interpreted in terms of an unstable Z-pinch. A good place to start is through mapping the magnetic field using polarisation measurements and the internal mass density through internal depolarisation data. In addition, detailed imaging spectra from radio to X-ray energies can be used to determine where the particles are being accelerated – at shock surfaces or in volumes containing strong, unstable currents – and how they diffuse away from these acceleration sites at different energies.

The prospects are pretty good for determining if cosmic lighthouses are powered by electricity or fire.

Acknowledgments

I thank the organisers of this workshop for their hospitality and patience and Mitch Begelman, Paolo Coppi, Arieh Königl, Amir Levinson, Max Lyutikov, Chris McKee, David Payne, Martin Rees and Roman Znajek for their collaboration. Brian Punsly and Maurice van Putten are also thanked for stimulating debates on some of the issues raised here. Support under NASA grant 5-2837 is gratefully acknowledged.

References

1. Achterberg, A. Gallant, Y. A., Kirk, J. G. & Guthmann, A. W. 2001 MNRAS 328 393
2. Agol, E. & Krolik, J. H. ApJ 528 161
3. Agol, E., Krolik, J. H., Turner, N. J. & Stone, J. M. 2001 ApJ 558 543
4. Armitage, P. J., Reynolds, C. S. & Chiang, J. 2001 ApJ 548 868
5. Begelman, M. C. & Li, Z.-H. 1992 ApJ 397 187
6. Begelman, M. C. 1998 ApJ 493 291
7. Beloborodov, A., Stern, B.& Svennson, R. 2000 ApJ 535 158
8. Benford, G. 1978 MNRAS 183 29
9. Beskin, V. S. & Kuznetsova, I. V. 2001 Nuovo Cim. in press
10. Beskin, V. S. & Parév, V. I. 1993 Physics Uspekhi 36 529
11. Blandford, R. D. 1972 Astron. Astrophys. 20 135
12. Blandford, R. D. 1973 Astron. Astrophys. 26 161
13. Blandford, R. D. 1976 MNRAS 176 465
14. Blandford, R. D. 1999 Astrophysical Disks ed. J. Sellwood & J. Goodman New York:ASP p265
15. Blandford, R. D. 2000 Phil. Trans. Roy. Soc. A 358 1
16. Blandford, R. D. 2001 Prog. Theor. Phys. Suppl 143 182
17. Blandford, R. D., Ostriker, J. & Mészáros, P. 2002 in preparation
18. Blandford, R. D. & Königl, A. 1979 ApJ 232 34
19. Blandford, R. D. & McKee, C. F. 1976 Phys. Fluids 19 1130
20. Blandford, R. D. & McKee, C. F. 1977 MNRAS 180 343
21. Blandford, R. D. & Payne, D. G. 1982 MNRAS 199 883
22. Blandford, R. D. & Rees, M. J. 1974 MNRAS 169 395
23. Blandford, R. D. & Znajek, R. L. 1977 MNRAS 179 433
24. Bridle, A. H. & Perley, R. A. 1984 ARAA 22 319
25. Camenzind, M. 1986 Astron. Astrophys. 162 32
26. Cohen, M. H. et al.1971 ApJ 170 207
27. Colbert, E. J. M. & Mushotzky, R. F. 1999 ApJ 519 89
28. Contopoulos, J., Kazanas, D. & Fendt, L. 1999 ApJ 511 351
29. Duncan, R. C. & Thompson, A. C. 1992 ApJ 392 L9

30. Ferrarese, L. *et al.*2001 ApJ 555 L79
31. Ferrari, A. 1998 ARAA 36 539
32. Frail, D. *et al.*2001 ApJ 562 L55
33. Gaidos, J. A. *et al.*1996 Nature 383 319
34. Gallant, Y. A. & Arons, J. 1994 ApJ 435 430
35. Gammie, C. F. 1999 ApJ 516 177
36. Gebhardt, K. *et al.*2000 ApJ 539 L13
37. Goldreich, P. M. &Julian, W. H. 1969 ApJ 157 869
38. Gruzinov, A. 2001 ApJ 563 L15
39. Hartman, R. C. *et al.*1992 ApJ 385 L1
40. Heyl, J. S. 2001 Phys. Rev. D 63 4028
41. Hirotani, K. ApJ 549 495
42. Hirotani, K. & Shibata, S. 2001 ApJ 558 216
43. Hirotani, K., Takahashi, M., Nitta S.-Y. & Tomimatsu, A. 1992 ApJ 386 455
44. Junor, W., Biretta, J. A. & Livio, M. 1999 Nature 401 891
45. Kennel, C. F. & Coroniti, F. V. 1984 ApJ 283 694
46. Kesteven, M. J. & Caswell, J. L. 1987 Astron. Astrophys. 183 118
47. Koide, S., Meier, D. L., Shibata, K. & Kudoh, T. 2000 ApJ 536 638
48. Komissarov, S. S. 2001 MNRAS 326 L41
49. Königl, A. & Pudritz, R. E. 1999 Protostars and Planets IV ed. V. Mannings, A. P. Boss & S. S. Russell Tucson:University of Arizona Press
50. Kouveliotou, C. 1999 PNAS 96 5351
51. Krasnopolsky, R., Li, Z.-Y. & Blandford, R. D. 1999 ApJ 526 631
52. Kudoh, T. & Kaburaki, O. 1996 ApJ 460 199
53. Landau, L. D. & Lifshitz, E. M. 1975 Electrodynamics of Continuous Media Oxford:Pergamon
54. Lee, H. K. 2001 Phys. Rev. D in press
55. Lee, H. K., Wijers, R. A. M. J. & Brown, G. E. 2000 Phys. Rep. 325 83
56. Levinson, A. & van Putten, M. H. P. 1997 ApJ 488 69
57. Li, L.-X. & Paczynski, B. 2000 ApJ 534 L197
58. Lithwick, Y. & Sari, R. 2001 ApJ 555 540
59. Lovelace, R. V. E. 1976 Nature 262 649
60. Lyutikov, M. & Blackman, E. G. 2001 MNRAS 321 177
61. Lyutikov, M. 2002 Phys. Fluids in press
62. MacDonald, D. A. 1984 MNRAS 211 313
63. MacFadyen, A. & Woosley, S. 1999 ApJ 524 262
64. MacFadyen, A., Woosley, S. & Heger, A. ApJ 550 410
65. Max, C. E. & Perkin, F. W. 1972 Phys. Rev. Lett. 29 1731
66. Meier, D. L. 2001 ApJ 548 L9
67. Meier, D. L., Koide, S. & Uchida, Y. 2001 Science 291 84
68. Melia, F. V. & Falcke, H. 2001 ARAA 39 309
69. Mészáros, P. 2001 ARAA in press
70. Ogilvie, G. I., Livio, M. & Pringle, J. E. 1999 ApJ 512 100
71. Okamoto, I. 2001 MNRAS 327 550
72. Padovani, P. & Urry, C. M. 2000 PASP 112 1516
73. Park, S. J. 2001 Current High Energy Emission around Black Holes ed. C. H. Lee Singapore:World Scientific in press
74. Phinney, E. S. 1982 Astrophysical Jets ed. A. Ferrari & A. Pacholczyk Dordrecht:Reidel p201
75. Punsley, B. 2001 Black Hole Gravitohydromagnetics Berlin:Springer

76. Rees, M. J. 1966 Nature 211 468
77. Rees, M. J. & Gunn, J. E. 1973 MNRAS
78. Ruffini, R., Bianco, C. L., Fraschetti, F., Xue, S-S, Chardonnet, P. 2001 ApJ 555 L113
79. Sari, R. & Piran, T. 1998 ApJ 485 270
80. Spruit, H. C. 1999 Astron. Astrophys. 341 L1
81. Suen, W.-M. & MacDonald, D. A. 1985 Phys. Rev. D 32 848
82. Takahashi, M. 2000 Il Nuovo Cimento 115 843
83. Takahashi, M., Niita, S., Tatematsu, Y. & Tomimatsu, A. 1990 ApJ 363 206
84. Tanaka, Y. *et al* 1995 Nature 375 659
85. Thompson, A. C. & Blaes, O. M. 1998 Phys. Rev. D 57 3219
86. Thompson, A. C., Lyutikov, M. & Kulkarni, S. R. 2002 ApJ in press
87. Thorne, K. S., Price, R. H. & MacDonald, D. A. 1986 Black Holes: The Membrane Paradigm New Haven:Yale University Press
88. Tomimatsu, A. & Takahashi, M. 2001 ApJ 552 710
89. Tout, C. A. & Pringle, J. E. 1996 MNRAS 281 219
90. van Putten, M. H. P. 2001 Phys. Rep. 345 1
91. Uchida, T. 1997 Phys. Rev. E 56 2181
92. Uchida, T. 1997 MNRAS 291 125
93. Usov, V. V. 1994 MNRAS 267 1035
94. Vlahakis, N. & Königl, A. 2001 ApJ 563 L129
95. Wardle, J. F. C. *et al.*1998 Nature 395 457
96. Waxman,E. 1995 ApJ 452 L1
97. Weisskopf, M. *et al.*2000 ApJ 536 L81
98. Wilms, J. *et al.*2001 MNRAS 328 L27
99. Wilson, A. S. 1974 MNRAS 166 617
100. Wilson, A. S. & Yang, Y. 2002 ApJ in press
101. Zhang, X.-H. 1989 Phys. Rev. D 39 3933
102. Znajek, R. L. 1978 MNRAS 185 833

Why Do AGN Lighthouses Switch Off?

Ramesh Narayan

Harvard-Smithsonian Center for Astrophysics, 60 Garden Street, Cambridge, MA 02138, U.S.A.

Abstract. Nearby galactic nuclei are observed to be very much dimmer than active galactic nuclei in distant galaxies. The Chandra X-ray Observatory has provided a definitive explanation for why this is so. With its excellent angular resolution, Chandra has imaged hot X-ray-emitting gas close to the gravitational capture radius of a handful of supermassive black holes, including Sgr A^* in the nucleus of our own Galaxy. These observations provide direct and reliable estimates of the Bondi mass accretion rate $\dot{M}_{Bondi}$ in these nuclei. It is found that $\dot{M}_{Bondi}$ is significantly below the Eddington mass accretion rate, but this alone does not explain the dimness of the accretion flows. In all the systems observed so far, the accretion luminosity $L_{acc} \ll 0.1 \dot{M}_{Bondi} c^2$, which means that the accretion must occur via a radiatively inefficient mode. This conclusion, which was strongly suspected for many years, is now inescapable. Furthermore, if the accretion in these nuclei occurs via either a Bondi flow or an advection-dominated accretion flow, the accreting plasma must be two-temperature at small radii, and the central mass must have an event horizon. Convection, winds and jets may play a role, but observations do not yet permit definite conclusions.

1 Introduction

At high redshift, many galactic nuclei are extremely bright, with luminosities in excess of 10^{45} erg s^{-1}. These active galactic nuclei (AGN) are believed to be powered by accretion of gas onto supermassive black holes (SMBHs) at nearly the Eddington rate [1]. The accretion very likely occurs via a Shakura-Sunyaev thin disk [2],[3],[4],[5]. The Big Blue Bump, which is present in the spectra of all bright AGN, is identified with blackbody emission from the optically thick disk [6], while the X-ray emission (roughly 10% of the luminosity) is thought to be produced by an optically thin corona above the disk [7],[8]. In addition, most AGN have substantial infrared emission, usually the result of dust reprocessing at a relatively large distance from the SMBH. Some AGN also have significant radio emission from relativistic jets.

Most galactic nuclei at low redshift are very different. These nearby nuclei are much less active — sometimes not active at all. The nuclear source in our own Galaxy, Sagittarius A^*, is a particularly good example of a dim galactic nucleus. Studies of AGN demographics suggest that a significant fraction of (perhaps all?) SMBHs must have gone through an AGN "lighthouse" phase at some early stage in their lives. Why and how did these early lighthouses switch off to become the dormant nuclei we see today?

The reduced activity in nearby galactic nuclei is certainly not because of a lack of SMBHs. Recent observations have provided ample evidence that virtually every galaxy in our local neighborhood has a SMBH [9]. The lack of activity must therefore be the result of reduced gas supply. While this is certainly part of the answer, we shall see that it is not the whole story. A second, equally important, reason is that the very mode of accretion is different in dim nuclei: the accretion occurs via a radiatively inefficient mode, such that even what little gas reaches the SMBH produces much less radiated energy per unit accreted mass than in bright AGN. We review below the evidence for this conclusion. Our primary emphasis is on the nearest and best-studied dim SMBH, Sgr A*, but we also discuss briefly other dim nuclei. The focus is on the observations and on what we can or cannot tell with confidence from the presently available data.

2   Observations of Sgr A*

2.1   Spectrum

Figure 1 summarizes the available data on the spectrum of Sgr A* [10],[11],[12],[13],[14]. Until recently, Sgr A* had been firmly detected only in the radio and millimeter bands, where the source has a hard rising spectrum. There are useful upper limits in the submillimeter and infrared bands which indicate that the emission peaks in the submillimeter/far infrared region of the spectrum, with a luminosity of about 10^{36} erg s^{-1}. Sgr A* may have been detected during one observation at 2.2 μm [15], with a luminosity of 10^{35} erg s^{-1}, but the detection was not confirmed in other observations. If the source is variable, the lone detection may represent its brightest state. The data point is shown as an upper limit in Fig. 1. In the optical, UV and soft X-ray bands, the source is strongly extincted (visual extinction $A_V \approx 30$ magnitudes [20]) and there are no direct constraints on the spectrum. Nevertheless, it

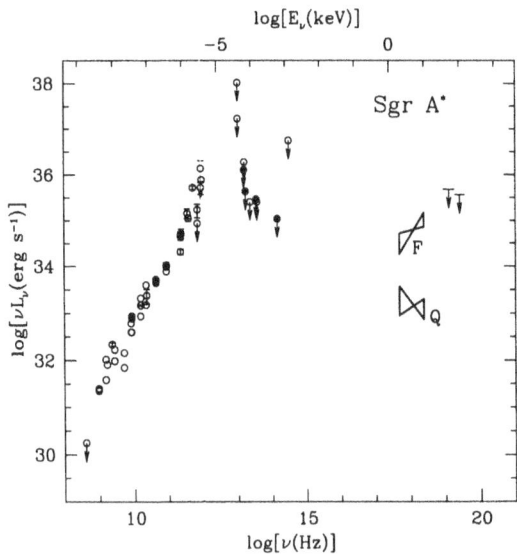

Fig. 1. Spectral data on Sgr A*. Shown is νL_ν, the luminosity per logarithmic interval of frequency ν, versus ν (lower scale) or photon energy $E_\nu = h\nu$ (upper scale).

is generally agreed that the emission in these hidden bands is unlikely to exceed that in the submillimeter peak.

Over the years, Sgr A* has been observed many times in X-rays [16],[18],[17],[19]. However, because the source is very weak and the field is highly crowded, there was no unambiguous detection until the recent observations of Baganoff et al. [13],[14] with the Chandra X-ray Observatory. The exquisite angular resolution of Chandra allowed Sgr A* to be isolated within the crowded field and its flux and spectrum to be measured. Out of a total of 76 ks of observations, extending over two epochs, the source was in a quiescent state (marked Q in Fig. 1) for nearly 70 ks, with a flux of 2.2×10^{33} erg s^{-1} and a relatively soft spectrum (photon index $\Gamma \sim 1.6 - 2.8$ [14]). For a brief period of several ks during the second epoch of observation, the source went into a flare state (F in Fig. 1), during which the flux went up by a factor of a few tens and the spectrum became quite hard ($\Gamma \sim 0.7 - 1.8$). Integrated over time, the emission in the flare was small compared to the quiescent emission. We therefore take the quiescent data as representative of the average properties of the source. We discuss the flare briefly in §5.3.

The mass of the SMBH at the center of our Galaxy has been measured to be $2.6 \times 10^6 M_\odot$ [21],[22],[23],[24], which means that the Eddington luminosity of Sgr A* is $L_{Edd} = 3 \times 10^{44}$ erg s^{-1}. On the scale of L_{Edd}, the peak emission in the submillimeter is $L_{submm} \sim 10^{-8.5} L_{Edd}$. The luminosity in the infrared is even less, $L_{IR} < 10^{-9.5} L_{Edd}$, and the quiescent X-ray luminosity is a pitiful $L_X \sim 10^{-11} L_{Edd}$. Thus, Sgr A* is an extremely dim galactic nucleus. Indeed, it is the dimmest nucleus for which we have useful data, which explains why this source plays such a central role in all discussions of dim galactic nuclei.

2.2 Bondi Accretion Rate

In a famous paper, Bondi [25] discussed the problem of spherical hydrodynamical accretion onto a black hole (BH) of mass M immersed in a uniform medium of density ρ_0 and sound speed $c_{s,0}$. He showed that the sphere of influence of the BH extends out to the gravitational capture radius $R_c = GM/c_{s,0}^2$. Gas external to R_c is only mildly perturbed by the BH, whereas any gas that falls within R_c is gravitationally captured by the BH and free-falls down to the center. The radial velocity in the free-fall zone is

$$v_R \sim (2GM/R)^{1/2} \sim c_{s,0}(R/R_c)^{-1/2}, \tag{1}$$

and so the mass accretion rate onto the BH is approximately (see Bondi's original article or [26] for exact results)

$$\dot{M}_{Bondi} \sim 4\pi R_c^2 \rho_0 c_{s,0}. \tag{2}$$

Bondi's model applies strictly to a uniform infinitely extended medium. However, even if the external medium is not truly uniform, the above formula for $\dot{M}_{Bondi}$ is still valid, provided we use for ρ_0 the density near the capture radius R_c.

Until recently, the Bondi accretion rate in Sgr A* could be estimated only indirectly, by estimating what fraction of the winds ejected by surrounding stars is captured by the SMBH (see [27],[28]). The estimate had large uncertainties. Chandra has changed the situation dramatically by directly imaging hot X-ray-emitting thermal gas in the vicinity of the capture radius of the SMBH [13]. The observations show that there is extended ~ 1 keV gas with a number density $n_0 \sim 30$ cm^{-3} over an area in the sky of about 10 arcsec around Sgr A*. In addition, there is ~ 2 keV gas with $n_0 \sim 100$ cm^{-3} spread over about an arcsec around the source (and clearly resolved by Chandra [13]). The capture radius of Sgr A* for keV gas is about 1 arcsec (for a BH mass of $2.6 \times 10^6 M_\odot$ and a distance of 8.5 kpc), so the 2 keV emission presumably comes from gas that has just been captured by the BH and been mildly heated by compression. Thus, Chandra provides a direct measurement of the properties of the gas (n_0 and $c_{s,0}$) right at the capture radius. Using this, one can obtain a reliable estimate of the Bondi accretion rate [13]:

$$\dot{M}_{Bondi} \sim (0.3 - 1) \times 10^{-5} M_\odot yr^{-1} \sim 3 \times 10^{20} \text{ g s}^{-1} \sim 10^{-4} \dot{M}_{Edd}, \qquad (3)$$

where we have formally assumed a radiative efficiency of 10% in the definition of the Eddington accretion rate, i.e. $\dot{M}_{Edd} \equiv L_{Edd}/0.1c^2$.

3   There Is No Thin Disk in Sgr A*

Could Sgr A* have a Shakura-Sunyaev thin disk? In the most straightforward version of this model, mass would flow onto the disk at the Bondi rate $\dot{M}_{Bondi}$ estimated in (3) and would flow steadily through the disk onto the SMBH. The model would predict a luminosity $L_{disk} \sim 0.1 \dot{M}_{Bondi} c^2 \sim 10^{40.5}$ erg s^{-1}, with the bulk of the emission appearing in the near infrared and optical bands. Figure 2 shows model spectra corresponding to a thin disk with three choices of $\dot{M}$: $(10^{-6}, 10^{-7}, 10^{-8}) \times \dot{M}_{Edd}$. Even

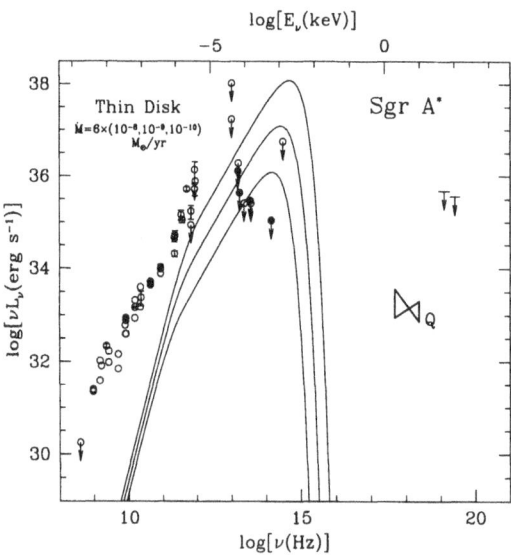

Fig. 2. Spectra corresponding to thin disk models of Sgr A* with three choices of $\dot{M}$. All three models shown have $\dot{M} \ll \dot{M}_{Bondi}$. Yet, all three models are too bright to fit the data.

models with such low accretion rates, which are far lower than $\dot{M}_{Bondi}$, are ruled out by the infrared limits.

One way of trying to save the thin disk model is to assume that gas flows in at the Bondi rate at R_c, but then condenses onto a cold "dead" disk where it sits without accreting onto the BH. This would require the disk to be in a very quiet state with an extremely low viscosity. This is not unreasonable — for instance, cataclysmic variables and soft X-ray transient binaries have quiescent states in which the gas accretes onto the central mass at a much lower rate than the rate at which gas is fed on the outside by mass transfer from the companion star. In the case of Sgr A*, however, such a model runs into difficulties because the inflowing gas would produce a fair amount of luminosity in the infrared, and possibly also X-rays, as it crashes onto the thin disk and loses its thermal and kinetic energy [29]. For example, cataclysmic variables and X-ray transients in quiescence have optical and UV emission from the "hot spot" where the incoming gas stream hits the disk, even though the disk itself may be hardly accreting. The lack of any evidence for thermalization radiation from the inflowing gas rules out the thin disk model quite strongly in Sgr A*. In the opinion of this author, any variant of the model that succeeds in getting round this constraint is likely to be very contrived. We will therefore take it as given that there is no thin disk in Sgr A*.

As an aside, we note that the above argument does not apply if the accretion occurs via an advection-dominated accretion flow (see §§5.1,5.2) instead of a thin disk (despite claims to the contrary [12]). The reason is that in the case of an advective flow, there is no free-falling Bondi-like zone. At the capture radius the gas directly makes a transition to a subsonic rotating accretion flow, and there is no supersonic zone or shock. (In addition, even if there were a shock it is not clear that the hot gas would radiate very much, as it is advection-dominated.)

Returning to the discussion on thin disk models, the essence of the argument against the presence of a thin disk in Sgr A* is the fact that the accretion luminosity L_{acc} is very low,

$$L_{acc} \ll 0.1 \dot{M}_{Bondi} c^2, \tag{4}$$

whereas a thin disk, being radiatively efficient, will normally have $L_{acc} \sim 0.1 \dot{M}_{Bondi} c^2$. If Sgr A* does not have a thin disk, then what kind of a flow does it have? Whatever the flow is, the data tell us that it has to be radiatively highly inefficient. Radiatively inefficient gas will be hot, because the energy has nowhere to go except into thermal energy. The gas will also be quasi-spherical rather than thin. We now turn to a consideration of models with these properties.

4   Bondi Model

4.1   Spherical Accretion

The most famous example of a radiatively inefficient accretion flow is Bondi spherical accretion [25]. The radial velocity of the accreting gas and the mass accretion rate in this model are given by (1) and (2) above. From these, the density profile is easily obtained: $\rho \sim \rho_0 (R/R_c)^{-3/2}$, where ρ_0 is the density of the external medium.

Bondi originally assumed that the accreting gas is neither heated nor cooled. This assumption is not valid if the gas is magnetized. As Shvartsman [30] first argued, any magnetic field frozen in the gas is amplified as $B \propto R^{-2}$, and so the magnetic energy density grows as R^{-4}. The gas energy density, however, varies only as $R^{-5/2}$, which means that even if the gas starts off with a sub-equipartition strength B, the field quickly grows to equipartition strength at some radius. Inside this radius, the field will reconnect so as to maintain rough equipartition. The reconnection will heat the gas [31], and as a result, the thermal energy will come roughly into equipartition with the magnetic energy. Since there is negligible cooling, the sum of the thermal, magnetic and kinetic energies should equal the potential energy of the gas. It is then easy to show that the temperature of the gas will vary roughly as

$$T \sim 10^{12} (R/R_S)^{-1} \text{ K}, \qquad c_s \sim c \, (R/R_S)^{-1/2}, \qquad (5)$$

where $R_S = 7.7 \times 10^{11} (M/2.6 \times 10^6 M_\odot)$ cm is the Schwarzschild radius of the accreting BH.

The heating of the gas leads to convection; we discuss this topic in §7.1.

4.2 Bondi Models of Sgr A*

Given the above profiles of density and temperature, it is straightforward to calculate the spectrum of a Bondi accretion flow. The primary emission mechanisms are synchrotron and bremsstrahlung, plus Comptonization of each. The solid line in Fig. 3 shows the predicted spectrum of Sgr A* if accretion occurs via a Bondi flow. We have normalized

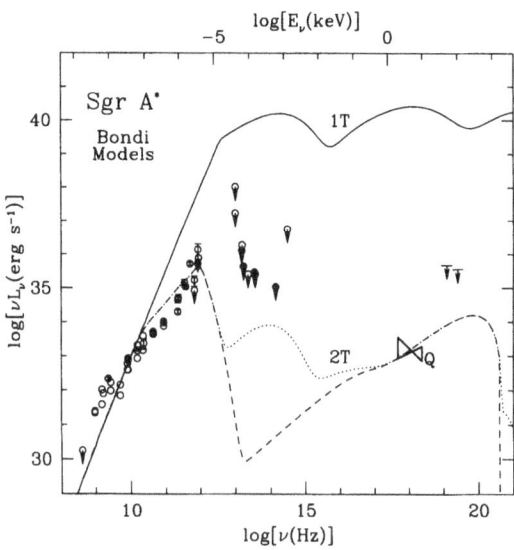

Fig. 3. Spectra of Sgr A* corresponding to Bondi spherical accretion models (see the text for details)

the density of the model such that, at $R = R_c$, the density is equal to ρ_0 as measured by Chandra. The model clearly predicts far too much flux all across the spectrum. The reason is the very high temperature of the electrons, especially close to the BH (see eq 5). The hot electrons radiate synchro-Compton emission very efficiently.

A Bondi model for Sgr A* was proposed by Melia [32],[33] many years ago and pursued in detail by his group. This work was the first to propose that a hot accretion flow emitting synchrotron and bremsstrahlung radiation could explain the strange properties of Sgr A*.

The published models of Melia fit the observed spectrum quite well, which is rather surprising considering that we find the Bondi model to overpredict the luminosity by many orders of magnitude (solid line in Fig. 3). More surprising still, the Melia models used an accretion rate 100 times larger than the one used here in Fig. 3, and yet agreed with the observations. One reason for the large discrepancy is an error (at least in the early work) in the computation of the synchrotron spectrum (see [34]). Another reason could be that Compton-scattering was not included in some of the models. It is also possible that the models did not correctly include heating due to magnetic dissipation. Strong heating is generic to any spherical flow that falls over many decades of radius (§4.1). The heating can be avoided if the magnetic field is extraordinarily weak in the external medium so that the field does not grow to equipartition strength even at the BH horizon, but this is very unlikely. Alternatively, heating would be weak if the field reconnects while it is still well below equipartition, as proposed in some models [35]. But this requires the reconnection to proceed at much faster than the Alfven speed, which seems unlikely (e.g., [36],[37]).

The electrons in a Bondi model of Sgr A* are too luminous because they are too hot. This suggests a possible fix: make the accreting plasma two-temperature, with the electrons much cooler than the protons. Two-temperature models of spherical flows were developed in detail by Shapiro and co-workers for accretion flows onto neutron stars [38] and BHs [39]. If one assumes that ions and electrons couple only via Coulomb collisions — a not unreasonable assumption — then at the low density of the accretion flow in Sgr A*, the plasma would automatically develop a two-temperature structure.

To illustrate how strong an effect the electron temperature has on the spectrum, we consider an artificial problem in which we modify the electron temperature profile in the Bondi flow such that it is capped at 10^{10} K. The model predicts the spectrum shown by the dotted line in Fig. 3 (the dashed line is the contribution due to bremsstrahlung and synchrotron alone, without Comptonization). This model is clearly in much better agreement with the observations than the one-temperature model (solid line). We thus conclude that any successful model of Sgr A* is likely to involve a two-temperature plasma.

Before closing this section, we note that the spherical accretion model, while good for developing insight into the properties of hot flows, is unlikely to describe real accretion flows since it ignores the angular momentum of the accreting gas. In almost any accretion flow, the gas is likely to possess sufficient angular momentum that it would hit the centrifugal barrier at some radius R_{cb} before it can fall into the BH. Inside R_{cb}, accretion is possible only if there is some agency to transfer angular momentum outward; we will call this agency "viscosity" though it is probably magnetic stresses [40]. Some authors have proposed models in which they invoke a Bondi flow from the capture radius R_c down to R_{cb}, and they then introduce a thin disk or some other kind of rotating solution from R_{cb} down to the marginally stable orbit (e.g., [33],[41]). The problem with any such model is that there is nowhere for the angular momentum to go. A viscous rotating flow can accrete only if it can get rid of angular momentum to the outside, but angular momentum cannot be transported across supersonic zones such as a Bondi flow [42],[43]. Therefore, if the supersonically infalling gas in a Bondi-like flow is arrested by centrifugal forces, then whatever rotating structure forms at the center must extend out at least to the capture radius in order to be able to transfer angular momentum to the outside. Many models of Sgr A* in the literature fail to satisfy this simple consistency condition. (The ADAF model discussed next does satisfy the condition.)

5   Advection-Dominated Accretion Flow (ADAF) Model

5.1   The ADAF Model

The thin accretion disk is a radiatively efficient model in which all the energy released through viscous dissipation is radiated away. One could describe it as a cooling-dominated accretion flow. An "advection-dominated accretion flow" (ADAF) is one in which most of the heat energy released by viscosity and compression is retained in the gas and advected to the center, and only a very small fraction of the energy is radiated.

Technically, any radiatively inefficient accretion flow, including the Bondi flow, is advection-dominated. However, the term ADAF is conventionally applied only to a particular class of quasi-spherical models that include rotation, viscosity and a two-temperature plasma [44],[45],[46],[47],[48], [49][50]. We will follow this convention here.

The reason ADAFs are advection-dominated is that the accreting gas has a low density (because of the low mass accretion rate) and the thermal structure of the plasma is two-temperature. Because of the low density, very little of the heat energy in the ions gets transferred to the electrons through Coulomb collisions [39],[48], [51]. Since the ions hardly radiate at all, they retain their thermal energy and advect essentially all of it to the center. The ion temperature thus scales essentially as in (5) and becomes of order 10^{12} K near the BH. The electrons do radiate some of their energy. They also have a different equation of state once they become relativistic (at small radii). For both reasons, they are cooler than the ions, and do not get much hotter than few $\times 10^{10}$ K near the BH. The lower temperature makes the electrons less efficient radiators than in the one-temperature Bondi model shown in Fig. 3. Indeed, at sufficiently low $\dot{M}$ (as in Sgr A*), the electrons become radiatively quite inefficient, though not as inefficient as the ions, and advect a large part of their energy. In this regime, the overall radiative efficiency of the accretion flow can be very low.

Most ADAF models in the literature assume that the ions and electrons have thermal energy distributions, which is reasonable under many conditions [52]. A few papers have considered the effects of nonthermal particles, e.g., [53],[54],[55].

Because the gas in an ADAF has angular momentum, accretion is driven by viscosity (not just gravity as in the Bondi flow). The radial velocity of the gas then scales roughly as $v_R \sim \alpha(GM/R)^{1/2}$ [46], where α is the usual dimensionless viscosity parameter [2],[5]. The velocity is thus lower than in a Bondi flow by a factor $\sim \alpha$, whose typical value is ~ 0.1. The lower radial velocity means that, for a given ambient density in the external medium, $\dot{M}_{ADAF}$ in an ADAF is lower than $\dot{M}_{Bondi}$ in a Bondi flow by a factor $\sim \alpha \sim 0.1$.

The ADAF solution is allowed only for relatively low values of $\dot{M}$, less than a few per cent of the Eddington rate for $\alpha \sim 0.1$ [45],[48],[56] (but see [57]). This makes the solution a natural choice for modeling low-luminosity accretion flows. Unlike other hot accretion flow solutions [39],[58], the ADAF does not suffer from any serious thermal or viscous instability [59],[60],[61],[62].

The reader is referred to the following reviews for more details of the ADAF model and its application to dim accretion sources: [63],[64],[65],[66].

5.2 ADAF Models of Sgr A*

ADAF models of Sgr A* have been described in a number of papers in the literature [67],[68],[11],[69],[54],[55],[70]. These studies have shown that it is possible to explain both the very low luminosity of Sgr A* and its spectrum without invoking an unreasonably low mass accretion rate. The above studies were done before the recent Chandra observations, when there was only a rough estimate of the mass accretion rate. Here we present some new results using the Chandra measurement of the density and temperature of the external medium.

Figure 4 shows spectra corresponding to three ADAF models of Sgr A*. The models assume $\alpha = 0.1$ (the results are not sensitive to the precise value) and take the magnetic

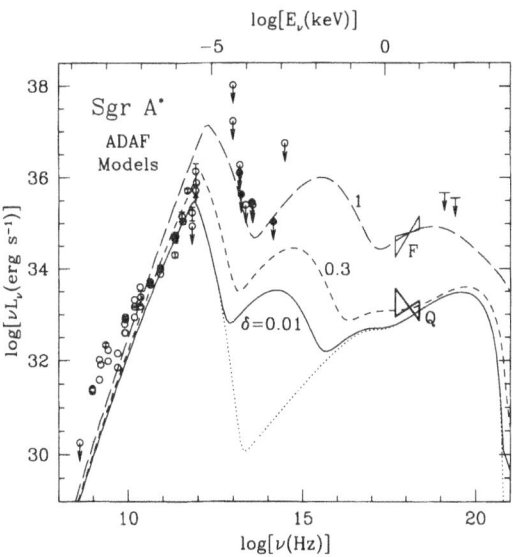

Fig. 4. ADAF models of Sgr A* (see the text for details)

pressure in the gas to be a tenth of the total pressure. The density profile is matched to the ambient density of the external medium at the capture radius; more precisely, the models have been adjusted so that the predicted bremsstrahlung emission agrees with the spatially resolved X-ray flux measured by Chandra. Because we have used a boundary condition on the density, these ADAF models have a lower mass accretion rate, $\dot{M}_{ADAF} \sim 10^{-6} M_\odot \mathrm{yr}^{-1}$, than the Bondi models described in §4.2. This is to be expected since the radial velocity in an ADAF is lower by a factor $\sim \alpha$ than in a Bondi flow. Note that, because we allow the ADAF to extend all the way out to the capture radius, there is no Bondi-like segment in the accretion flow. The external gas directly makes a transition to the ADAF. The flow is subsonic all the way down to a sonic radius close to the BH, and there is viscous coupling between the accretion flow and the external medium. This allows angular momentum to be expelled from the system to the outside.

The three models shown in Fig. 4 differ in the value of a parameter δ, which measures the fraction of the viscous energy that goes into the electrons. Early ADAF models assumed that nearly all the viscous heat goes into the ions and that δ is very small, say ~ 0.01. The solid curve in Fig. 4 shows the spectrum for this choice of δ. The spectrum was calculated by the methods described in [11]. We see that the calculated spectrum is generally consistent with the data, though the X-ray spectrum may be a little too hard [13].

Is it reasonable to assume that electrons receive such a small fraction of the viscous heat energy [71]? Several authors have investigated this question [72],[73],[74],[75],[76]. In brief, while it appears that negligible electron heating is possible under some conditions, under other conditions electrons are expected to receive a good fraction (few tenths) of the energy. The short-dashed curve in Fig. 4 shows the predicted spectrum for a model with $\delta = 0.3$. It is reassuring that this model, which uses a "reasonable" value of δ, is again in pretty good agreement with the data. In fact, the model predicts a somewhat softer X-ray spectrum than the $\delta = 0.01$ model (because part of the emission is now due to Comptonization), which agrees better with the Chandra data.

The dotted line in Fig. 4 corresponds to the contributions from synchrotron and bremsstrahlung emission alone for the $\delta = 0.01$ model. The synchrotron peak on the left nicely fits the radio/mm data, and cuts off where it should in order to satisfy the infrared upper limits. This emission comes from fairly small radii in the flow (few to few tens of R_S). The bremsstrahlung peak on the right explains the quiescent X-ray emission in Sgr A* and is almost entirely from large radii $\sim R_c \sim 10^5 R_S$, which corresponds to an angular scale of order an arcsec. The rest of the emission in the spectrum is from Compton scattering, which fills in the region between the synchrotron and bremsstrahlung peaks. The Compton emission comes primarily from radii inside $100 R_S$. It is very reassuring that the model does not produce too much Compton emission in X-rays, since the Compton emission would be spatially unresolved, whereas Chandra has clearly shown that at least 50% of the observed X-ray emission in quiescence is resolved. Overall, we conclude that an ADAF model with $\delta < 0.3$ is consistent with all the presently available data on Sgr A*. Note that, since Chandra has fixed the accretion rate, and since the results are insensitive to α and the details of the magnetic field strength, the model is almost parameter-free. The only parameter that has been adjusted is δ, and even this parameter has a reasonably wide range of allowed values.

As an aside, we note that the argument we made in §3 (second paragraph) against the presence of a "dead" thin disk in Sgr A* does not apply to the ADAF model. Since the ADAF extends all the way out to the capture radius, there is no supersonically infalling region, and therefore there is no shock where the infalling gas may thermalize and radiate the incoming kinetic energy. Another point is that the ADAF model does not require the incoming gas to have a high specific angular momentum. As discussed in §4.2, so long as there is sufficient angular momentum to prevent the gas from falling into the BH directly, a rotating viscous ADAF will grow to the size of the capture radius. Only then can angular momentum be transported out into the ambient medium. For the same reason, the CDAF solution discussed in §7 again has to extend out to the capture radius. In that case, it is not only angular momentum that needs to be transported to the outside, but also a great deal of energy.

Figure 5 shows the profiles of the ion and electron temperatures in the ADAF models discussed above. These profiles have been calculated self-consistently (see [11]), with separate energy equations for the ions and the electrons [77], and assuming that the two species couple only via Coulomb collisions. Notice how a two-temperature structure develops naturally in this model, so that the electrons do not get any hotter than few$\times 10^{10}$ K. As already discussed in §4, this is the kind of electron temperature needed to fit the spectral data. The brightness temperature of Sgr A* at mm wavelengths is also measured to be of this order [78].

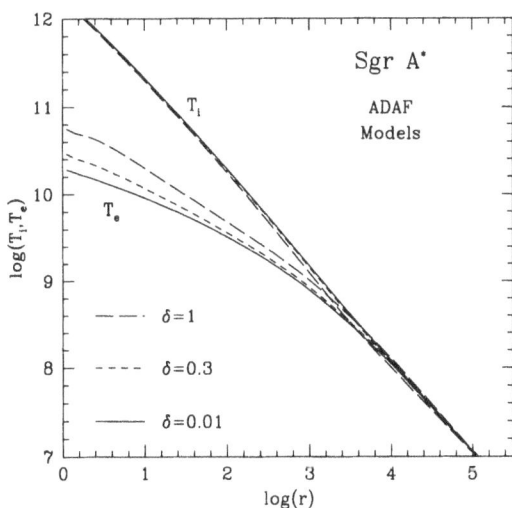

Fig. 5. Variation of the ion and electron temperatures with radius in the three ADAF models shown in Fig. 4

5.3   The Flare in Sgr A*

The long-dashed curve in Fig. 4 shows the spectrum of a model with $\delta = 1$, with all the viscous heat assumed to go into the electrons. The electrons are hotter in this model than in the other two models. Correspondingly, both the synchrotron peak and the Compton-scattering component are significantly stronger. The bremsstrahlung emission is not affected since it is emitted from large radii, where δ has very little effect on the temperature (Fig. 5).

It is interesting that the model with $\delta = 1$ fits both the X-ray flux and the X-ray spectral slope of the flare state of Sgr A*. Moreover, the spectrum passes through the 2.2 μm upper limit. As discussed in §2.1, this limit corresponds to a claimed detection of Sgr A*. Could it be that Sgr A* underwent a flare during that particular observation?

It is certainly a coincidence that the $\delta = 1$ model fits the flare data so well, all the more so since the models are not very reliable in this limit. However, the result does show that any minor perturbation that heats up the electrons at small radii by as little as a factor of 2 will produce a spectrum similar to that seen in the flare. Such a temperature change could be caused by many effects. Apart from an increase in δ, other possibilities are (i) a sudden enhancement in the coupling between ions and electrons, (ii) a major reconnection event, or (iii) a sudden injection of a population of nonthermal electrons.

5.4   Does Sgr A* Have an Event Horizon?

Even allowing for the fact that the accretion rate in an ADAF model is smaller than in the Bondi model by a factor $\sim \alpha$, Sgr A* is still highly underluminous: $L_{acc} \ll 0.1\dot{M}_{ADAF}c^2$. Energy advection is the reason why the radiative efficiency of the accretion flow is so low. But advection alone is not enough, because we still have to decide what happens to the advected energy when it finally reaches the center. In the models shown in Fig. 4, it was assumed that the mass at the center is a BH which swallows the advected energy.

What would happen if the object were not a BH? If the object had a hard surface, then the advected energy would be radiated when the hot gas hit the surface (because the density would go up and the radiative efficiency would increase suddenly), and the predicted spectrum would disagree violently with the data [11],[79]. Figure 6 shows some results. The various curves correspond to ADAF models in which the object in the center is postulated to have a hard surface with radius equal to $10^{0.5}, 10^1, 10^{1.5}, 10^2, 10^{2.5}, 10^3$ Schwarzschild radii. Notice that all the models are ruled out by the infrared limits.

We thus conclude that, if the ADAF model is the correct description of the accretion flow in Sgr A*, then the central object must be a black hole with an event horizon. The same is true even if the accretion proceeds via a Bondi flow.

6   Other Dim Galactic Nuclei

6.1   Nuclei of Giant Ellipticals

It has long been recognized that the nuclei of nearby giant elliptical galaxies are unusually dim [80], with $L_{acc} \ll 0.1\dot{M}_{Bondi}c^2$. Soon after the successful application of the two-temperature ADAF model to Sgr A* [67], it was proposed that the same model would also solve the riddle of dim elliptical nuclei [81],[82],[83], [84].

Until recently, $\dot{M}_{Bondi}$ in the giant ellipticals was estimated indirectly by modeling the X-ray cooling flow in the central regions of the galaxy and extrapolating the model to the nucleus. However, the Chandra X-ray Observatory has improved matters significantly by providing in a few ellipticals direct images of hot X-ray emitting gas close to the capture radius of the SMBH. Good quality data are presently available for NGC 1399, NGC 4472, NGC 4636 [85] and NGC 6166 [86]. From these observations, reliable estimates of $\dot{M}_{Bondi}$ have been obtained, and we are now in a position to state with great confidence that L_{acc} is indeed very much less than $0.1\dot{M}_{Bondi}c^2$ in all these galactic nuclei.

In fact, L_{acc} is so low and $\dot{M}_{Bondi}$ is so large that even a two-temperature ADAF model with $\dot{M}_{ADAF} = \dot{M}_{Bondi}$ cannot satisfy the X-ray constraints [85]. This is shown for NGC 1399 by the dashed line in Fig. 7. The predicted spectrum is far too bright in X-rays compared to the Chandra upper limit. However, as explained earlier, it is not correct to set $\dot{M}_{ADAF} = \dot{M}_{Bondi}$. Rather, one must match the density of the ADAF model to the ambient density of the external medium at $R = R_c$ and solve consistently for $\dot{M}_{ADAF}$. When this is done, one obtains a lower value of $\dot{M}_{ADAF}$, and correspondingly a significantly dimmer source. The solid line in Fig. 7 shows the spectrum thus obtained for NGC 1399 (using $\alpha = 0.1$, but the results are not very sensitive to this choice). The model is consistent with the X-ray data. Similar results are obtained for other nuclei for which there is Chandra data.

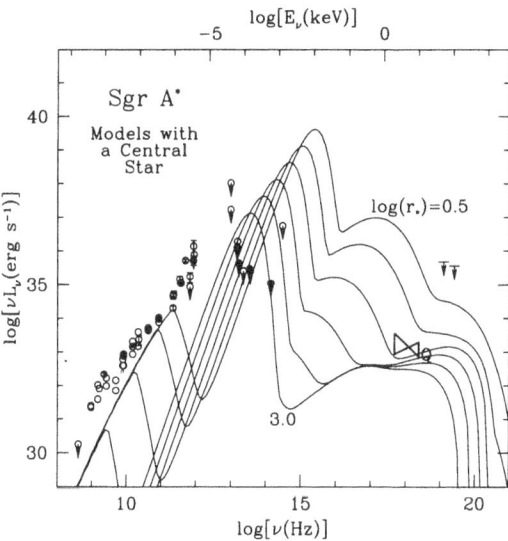

Fig. 6. Spectra of models in which accretion occurs via an ADAF and the central accreting mass is assumed to have a surface that radiates the advected energy with a blackbody spectrum. The curves are labeled by the radius r_* of the central mass in units of the Schwarzschild radius.

The matching of the ADAF solution to the external medium was done somewhat approximately here. Ideally, one should obtain a consistent solution of the viscous accretion equations, extending from the external medium/cooling flow through the capture radius down to the BH. The analogous problem for spherical accretion has been solved [87], but the viscous problem is yet to be analysed.

6.2   LINERs

Low-Ionization Nuclear Emission Region (LINER) galaxies, and more generally low luminosity AGN (LLAGN), have unusual spectral properties [88]. It has been proposed that the observations could be understood if accretion proceeds via an ADAF in these galactic nuclei [89],[91]. A feature of the proposed models is that the accretion flow consists of two zones: an outer thin disk that extends from some large radius down to a transition radius R_{tr} and an inner ADAF that extends from R_{tr} to the BH horizon.

There is some direct evidence for such a geometry of the accretion flow in NGC 4258 [90], M81 and NGC 4579 [91]. M81 and NGC 4579, in particular, lack the Big Blue Bump that is seen in AGN. Instead, they have a red bump, which can be fitted only with a truncated disk [91]. Furthermore, the absence of short time scale variability [92] and unusually strong radio emission [93] in LLAGN appear to confirm the presence of radiatively inefficient accretion in these galactic nuclei.

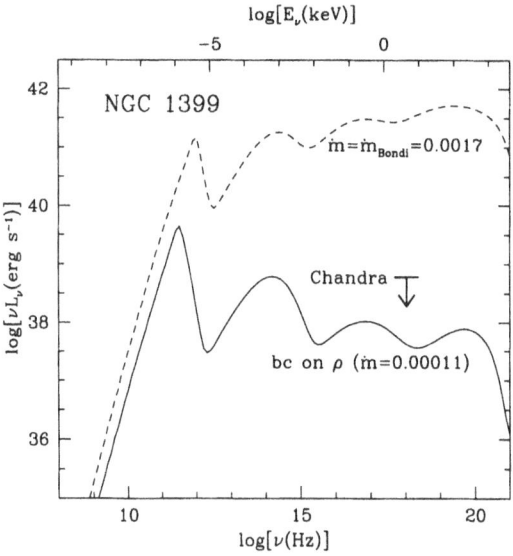

Fig. 7. ADAF models of NGC 1399. For the dashed curve, $\dot{M}$ has been set equal to the estimated Bondi accretion rate (0.0017 in Eddington units). For the solid line, the ADAF model has been adjusted so as to match the external density as measured by Chandra ($\dot{m} = 0.00011$).

6.3   Transition From Thin Disk to ADAF

So far in this article, we have discussed three kinds of accretion flows. First, we briefly mentioned bright AGN, with accretion luminosities of say $L_{acc} \sim 0.1 - 1 L_{Edd}$. We said that these objects have Shakura-Sunyaev thin accretion disks. Next, at the other extreme, we discussed Sgr A*, with $L_{acc} \sim 10^{-8.5} L_{Edd}$. We said that this source might have an ADAF extending from the capture radius at $R \sim 10^5 R_S$ down to the BH. Finally, in the previous subsection, we discussed intermediate cases like M81, with $L_{acc} \sim 10^{-4} - 10^{-5} L_{Edd}$. We argued that the accretion occurs via a thin disk at large radii ($R > R_{tr} \sim 10^2 R_S$ [91]) and vai an ADAF at smaller radii ($R < R_{tr}$).

The pattern suggested by the above facts is very reminiscent of a paradigm that was developed for BH X-ray binaries [94],[56],[95]. The key idea is that the geometry of the accretion flow is determined primarily by the Eddington-scaled mass accretion rate $\dot{m} = \dot{M}/\dot{M}_{Edd}$. For $\dot{m}$ greater than a critical value $\dot{m}_{crit} \sim 0.01 - 0.1$, the accretion occurs via a radiatively efficient thin disk, with perhaps a hot corona. At these accretion rates, the ADAF solution is not allowed [48] because the gas density is too high to permit a radiatively inefficient flow. Once $\dot{m}$ falls below $\dot{m}_{crit}$, an ADAF is allowed at small radii and the thin disk develops a small hole in the inner regions where the cold gas switches to a hot ADAF. The range of radius over which an ADAF is allowed increases as $\dot{m}$ decreases [48], and correspondingy the hole in the thin disk also becomes larger. For an extremely low accretion rate, e.g., $\dot{m} \sim 10^{-5}$ in Sgr A*, the hole is so large that the outer thin disk disappears altogether and the accretion occurs via a pure ADAF over a wide range of radius. Such flows are radiatively extremely inefficient. This paradigm is in qualitative agreement with a variety of observations, e.g., [96], though undoubtedly there are parameters other than $\dot{m}$ that also affect the geometry of the flow.

An important question is: how does the cold gas in a thin disk switch to a hot quasi-spherical ADAF? A number of papers have been written on this "evaporation" process [97],[98],[99],[100],[101], [102],[103],[104], and there is some qualitative understanding of the relevant physics. However, there is no reliable quantitative model yet that can calculate the dependence of the transition radius R_{tr} on the accretion rate $\dot{m}$.

7   Other Possibilities

7.1   Bondi Accretion With Magnetic Fields

Recently, the first three-dimensional numerical simulations of spherical accretion with magnetic fields were reported [105]. The results turned out to be unexpected. The density profile of a magnetized spherical flow was found to differ significantly from the $R^{-3/2}$ behavior predicted by the Bondi model (§4), and the mass accretion rate was found to be much less than the value given in (2). The reason for the discrepancy is as follows [105].

As already noted in §4, when magnetic fields are frozen into the inflowing gas, the fields are amplified, causing them to reconnect, and the energy released thereby heats up the gas. Since the gas radiates very little of its energy, the entropy of the gas increases inward. The negative radial entropy gradient causes the gas to be convectively unstable, so violent convection is set up. Figure 8 shows velocity streamlines in the meridional

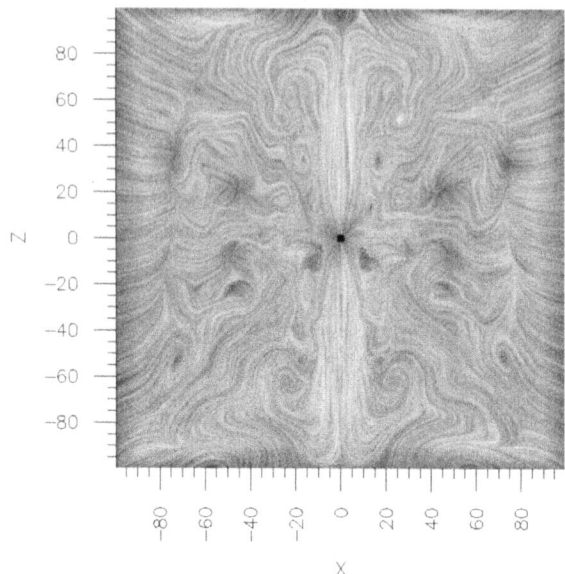

Fig. 8. Velocity streamlines in the meridional plane in a simulation of a magnetized spherical accretion flow [105]. Note the obvious turbulent eddies which are due to convection.

plane from one of the numerical simulations. Notice how turbulent the flow is, and how different it is from the purely radial streamlines of Bondi's hydrodynamic model. In this convective flow, the outward energy flux due to convection dominates the physics at large radii. As a result, the global structure of the flow is strongly affected and the mass accretion rate onto the BH is much reduced from the Bondi rate.

An analytical model of spherical accretion with convection has been developed [105]. The model predicts that the density should vary as $\rho \propto R^{-1/2}$ rather than the Bondi scaling of $\rho \propto R^{-3/2}$. The scaling drops out naturally just from the assumption that the convective energy flux dominates over other fluxes. Because of the modified scaling, the accretion rate onto the BH is much reduced from the Bondi rate (2):

$$\dot{M}_{MHD} \sim (R_S/R_c)\dot{M}_{Bondi}. \qquad (6)$$

The numerical simulations do not have sufficient dynamic range, nor have they been run long enough, to unequivocally confirm this result. What is clear from the simulations, however, is that magnetic fields can modify even such a long-cherished paradigm as the Bondi model.

As discussed in §4.2, the Bondi model is unlikely to be relevant for real accretion flows because it ignores the angular momentum of the gas. It might therefore appear that the above results are not of practical importance. However, as we discuss next, convection causes similarly large effects even in rotating viscous flows. (Historically, the rotating flows were studied first.)

7.2   Convection-Dominated Accretion Flow (CDAF)

An ADAF has two unusual properties that strongly influence the global structure of the flow.

First, the gas has a positive Bernoulli parameter (sum of potential energy, kinetic energy and enthalpy), which means that the gas is technically not bound to the BH and is liable to flow out of the system [46],[47],[106]. If the outflow is strong enough, then the amount of mass reaching the BH could be a great deal less than the mass supplied on the outside [107],[108] (but see [109],[110]). In such an "advection-dominated inflow outflow solution" (ADIOS), the density profile can be strongly modified: instead of $\rho \propto R^{-3/2}$ as in a standard ADAF, one could have $\rho \propto R^{-3/2+p}$, with $0 < p < 1$ [107]. The limit $p = 0$ corresponds to the standard ADAF model described in §5.

Second, because viscous dissipation in an ADAF adds heat energy to the accreting gas, and since there is negligible cooling, the entropy of the gas increases inward (see the analogous discussion in §7.1). As a result, the gas in an ADAF is convectively unstable [111],[46],[47]. The effect of this convection has been studied via numerical hydrodynamic simulations [113],[114],[115],[116]. As in the case of the magnetic Bondi flow, convection alters the density profile to $\rho \propto R^{-1/2}$ and causes the mass accretion rate onto the BH to be reduced drastically compared to the standard ADAF model [115]. The physics of these "convection-dominated accretion flows" (CDAFs) is fairly well understood [117],[118],[119]. Convection introduces a strong flow of energy outward, just as in the magnetized Bondi problem. In addition, it also introduces a flow of angular momentum *inward* (i.e. opposite to the direction of viscous transport), as anticipated in some previous work [120],[121],[122]. All of this leads to novel and quite interesting properties [117],[118]. Analysis of the global structure of CDAFs indicates that the gas switches from a convection-dominated state to a more traditional ADAF-like state for radii less than about $50R_S$ [123].

Preliminary work has been done on the effect of magnetic fields on rotating advection-dominated flows [124],[125],[126],[127],[128],[129]. Depending on the initial configuration of the magnetic field, the dynamical properties of the flow appear to be quite distinct. For a predominantly "vertical" initial field, there are strong jets [124], whereas for a toroidal initial field there are only weak outflows [126]. Regardless, convection seems to be present in both cases, and some studies [128] find quite good agreement between numerical MHD simulations and analytical work on hydrodynamical CDAFs. Others, however, claim that there are large differences (see especially [130] where it is argued that convection in a differentially rotating magnetized medium behaves very differently from the unmagnetized case).

Work in this area is very important and needs to be pursued vigorously. The reason is that pure hydrodynamical simulations are ultimately limited by the fact that they invoke an artificial viscosity (parameterised via α) to transport angular momentum. If the shear stress is generated by the magneto-rotational instability [40], as universally believed, then it is obviously much better to carry out full MHD simulations so that the magnetic shear stress and the associated "viscosity" are computed self-consistently.

The variability properties of MHD flows have been investigated [131],[132]. It is clear that magnetic flares associated with reconnection events occurring over a wide range of scales can give broad-band fluctuations as well as occasional strong outbursts (as in the Sgr A* flare).

7.3   ADIOS/CDAF Models of Sgr A*

Spectral calculations with the ADIOS model using different values of the parameter p (which measures the strength of the outflow) are described in [112]. The results are as follows. Consider a sequence of models in which p is varied while keeping the density at the outer boundary and all other model parameters fixed. With increasing p, the density and the temperature of the gas near the BH decrease. This causes the synchrotron emission and the Compton emission to drop, without affecting the bremsstrahlung emission (which comes from the outside). The good agreement between the ADAF model and the data in Fig. 4 would then be lost. However, if the parameter p as well as the electron heating parameter δ (§5.1) are both increased simultaneously, then one can recover a good fit with the observations [112]. According to Fig. 4, a standard ADAF model ($p = 0$) with $\delta \sim 0.3$ gives a good fit to the data. By increasing p and δ simultaneously, larger values of p are also likely to give acceptable results.

Although the CDAF model has very different physics than the ADIOS model, nevertheless, as far as spectral calculations are concerned, it behaves very much like an ADIOS model with $p = 1$. Spectra corresponding to CDAF models have been presented in the literature [133], but no results are available specifically for Sgr A*. It is possible that even with $\delta = 1$, i.e. with all the viscous energy going into the electrons, the spectrum may still be deficient in the radio/mm band relative to the data. The X-ray spectrum might also be a little too soft. These are not necessarily bad, but they imply that one would need to include additional components in the model, or additional physics, to explain the observations. In the former category would be jet models such as those described below, and in the latter category would be models that include non-thermal electrons.

Although the theoretical motivations behind the ADIOS and CDAF models are strong, there is as yet no unambiguous observational evidence for these models in astrophysical sources. If a claimed detection [134] of linear polarization of Sgr A* in mm waves is confirmed (but see [138]), it would strongly suggest that the density of gas close to the BH is much less than that predicted by the ADAF model [135],[136]. An ADIOS model with a largish value of p or a CDAF model (which is equivalent to $p = 1$) would then be indicated. The discovery of radio variability with a possible 106 day cycle [137] independently suggests the presence of a turbulent CDAF. However, none of these indications is particularly robust at this time.

7.4   Jets, Other Components

In the discussion so far, it has been tacitly assumed that the accretion flow is the source of all the observed radiation. In bright AGN, it is known that radio emission is usually associated with relativistic jets. Since Sgr A* is brightest in radio and mm waves, it is not unreasonable to suppose that some of the observed radiation at these wavelengths originates in a jet, or some other component that is external to the accretion flow. Models of this kind have been developed by several groups (see [12] for a review).

In the context of jet models, we note that a pure accretion model with an ADAF does a pretty good job of explaining the data on Sgr A* (Fig. 4). As explained in §5.2, the model has almost no free parameters, since the accretion rate is fixed by the Chandra observations. If Sgr A* does have an ADAF, then there is very little room for additional emission from a jet.

On the other hand, if the accretion flow corresponds to an ADIOS or a CDAF, then the accretion flow might produce very little radio or mm emission, especially if δ is small. In this case, almost all the radio and mm emission could be from something other than the accretion flow. Jet models then become attractive (but see [140]).

Note that most of the quiescent X-ray emission in Sgr A* is spatially resolved [13]. Therefore, jet models of the quiescent emission need to ensure that they do not predict too much emission in X-rays, since the jet emission is expected to be unresolved. During the X-ray flare in Sgr A*, all the observed excess emission was point-like and so this emission could in principle be entirely from a jet.

Successful disk-plus-jet models have been developed for Sgr A* in which synchro-Compton emission from a compact nozzle region of a jet produces the observed radiation [141],[142],[143],[144]. The "disk" in these models is unlikely to be a standard thin disk for the reasons discussed in §3. Indeed, recent work suggests that a jet-plus-ADAF model is able to combine many of the attractive features of both models [140]. A jet also provides a convincing explanation for the flare in Sgr A* [145].

One interesting result is that the electron energy distribution in Sgr A* has to be nearly mono-energetic in order to fit the sharp cutoff of the spectrum in the infrared [146],[147]. This is a somewhat curious result, since jets in bright AGN almost always have power-law energy distributions. If the relativistic electrons in jets are accelerated via shocks, there must be something very different about the shock in very low-luminosity systems like Sgr A*.

8   Summary and Conclusions

The Chandra X-ray Observatory has eliminated a major uncertainty that has hampered our understanding of dim galactic nuclei. Thanks to Chandra's excellent angular resolution, we now have direct measurements of the density and temperature of ambient gas close to the gravitational capture radius of the SMBHs in Sgr A* and a few nearby galactic nuclei. This information allows us to estimate for these nuclei the Bondi mass accretion rate $\dot{M}_{Bondi}$, as well as the accretion rate in an advection-dominated accretion flow $\dot{M}_{ADAF}$. With the uncertainty in $\dot{M}$ removed, we are now in a position to answer the question posed in the title of this article. The answer consists of three parts:

- In all the dim galactic nuclei for which Chandra has provided a direct estimate of $\dot{M}_{Bondi}$, the accretion rate is found to be well below the Eddington rate. This is in contrast to bright AGN which are believed to accrete at close to the Eddington rate.

Thus, the first, and obvious, reason why AGN switch off is that the gas supply to the SMBH is reduced, presumably because most of the gas has been converted into stars.

- Even after allowing for the reduced $\dot{M}$, the objects studied are still anomalously dim: $L_{acc} \ll 0.1 \dot{M}_{Bondi} c^2$. Therefore, we can state with great confidence that the accretion *must* proceed via a radiatively inefficient mode. A two-temperature ADAF model fits the available data quite well (Fig. 4), with almost no adjustable parameters (only δ needs to be adjusted, and even it is loosely constrained: §5.2). In this model, there is not much room for additional emission from a jet. A Bondi model can also be made to fit the data, provided the accreting gas is taken to be two-temperature (§4.2). However, the neglect of angular momentum of the gas is a serious weakness of the model. Both the ADAF and Bondi models work only if the central object has an event horizon (§5.4). Independently of these results, one can state with high confidence that there is no Shakura-Sunyaev thin disk in Sgr A* (§3).
- Outflows and convection may be important, in which case the mass accretion rate onto the BH may be significantly less than the mass supply on the outside, i.e. $\dot{M}_{BH} \ll \dot{M}_{Bondi}, \dot{M}_{ADAF}$ (see the discussion of ADIOS/CDAF models in §§7.1– 7.3). There is as yet no compelling observational evidence for these models, but there are strong theoretical reasons for favoring them. If the accretion flows in Sgr A* and other dim nuclei are of the ADIOS or CDAF type, then the accretion flow may be very dim in radio/mm, and the observed emission in these bands may come from a relativistic jet or some other component external to the accretion flow.

Thus, it appears that three different effects all conspire to make nearby galactic nuclei extraodinarily dim: there is less gas available, the gas accretes via a radiatively inefficient mode, and (perhaps) less gas reaches the BH than is available for accretion.

Acknowledgements: The author thanks Shin Mineshige and Eliot Quataert for useful comments on the manuscript and the W.M. Keck Foundation for support as a Keck Visiting Professor at the Institute for Advanced Study, Princeton. This research was supported by NSF grant AST-9820686.

References

1. J.H. Krolik: *Active Galactic Nuclei* (Princeton University Press, Princeton 1999)
2. N.I. Shakura, S.A. Sunyaev: A&A, 24, 337 (1973)
3. I.D. Novikov, K.S. Thorne: in *Blackholes*, ed. by C. DeWitt, B. DeWitt (Gordon & Breach, 1973) p343
4. J.E. Pringle: ARAA, 19, 137 (1981)
5. J. Frank, A. King, D. Raine: *Accretion Power in Astrophysics* (Cambridge Univ. Press, Cambridge 1992)
6. A. Koratkar, O. Blaes: PASP, 111, 1 (1999)
7. F. Haardt, L. Maraschi: ApJ, 380, L51 (1991)
8. F. Haardt, L. Maraschi: ApJ, 413, 507 (1993)
9. D. Richstone et al.: Nature, 395A, 14 (1998)

10. P.G. Mezger, W.J. Duschl, R. Zylka: A&A Rev., 7, 289 (1996)
11. R. Narayan, R. Mahadevan, J.E. Grindlay, R.G. Popham, C. Gammie: ApJ, 492, 554 (1998)
12. F. Melia, H. Falcke: ARAA, 39, 309 (2001)
13. F.K. Baganoff et al.: ApJ, in press (2001) (astro-ph/0102151)
14. F.K. Baganoff et al.: Nature, 413, 45 (2001)
15. R. Genzel, A. Eckart, T. Ott, F. Eisenhauer: MNRAS, 291, 219 (1997)
16. M.G. Watson, R. Willingale, J.E. Grindlay, P. Hertz: ApJ, 250, 142 (1981)
17. A. Goldwurm et al.: Nature, 371, 589 (1994)
18. P. Predehl, J. Trumper: A&A, 290, L29 (1994)
19. K. Koyama et al.: PASJ, 48, 249 (1996)
20. M. Morris, E. Serabyn: ARAA, 34, 645 (1996)
21. A. Eckart, R. Genzel: MNRAS, 284, 576 (1997)
22. A.M. Ghez, B.L. Klein, M. Morris, E.E. Becklin: ApJ, 509, 678 (1998)
23. R. Genzel, C. Pichon, A. Eckart, O.E. Gerhard, T. Ott: MNRAS, 317, 348 (2000)
24. A.M. Ghez, M. Morris, E.E. Becklin, A. Tanner, T. Kremenek: Nature, 407, 349 (2000)
25. H. Bondi: MNRAS, 112, 195 (1952)
26. S.L. Shapiro, S.A. Teukolsky: *Black Holes, White Dwarfs, and Neutron Stars* (Wiley Inter-science, New York, 1983)
27. R.F. Coker, F. Melia: ApJ, 488, L149 (1997)
28. E. Quataert, R. Narayan, M.J. Reid: ApJ, 517, 101 (1999)
29. H. Falcke, F. Melia: ApJ, 479, 740 (1997)
30. V.F. Shvartsman: Sov. Astr., 15, 37 (1971)
31. P. Meszaros: A&A, 44, 59 (1975)
32. F. Melia: ApJ, 387, L25 (1992)
33. F. Melia: ApJ, 426, 577 (1994)
34. R. Mahadevan, R. Narayan, I. Yi: ApJ, 465, 327 (1996)
35. V. Kowalenko, F. Melia: MNRAS, 310, 1053
36. A. Lazarian, E.T. Vishniac: ApJ, 511, 193 (1999)
37. A. Lazarian, E.T. Vishniac: ApJ, 517, 700 (1999)
38. S.L. Shapiro, E.E. Salpeter: ApJ, 198, 671 (1975)
39. S.L. Shapiro, A.P. Lightman, D.M. Eardley: ApJ, 204, 187 (1976)
40. S.A. Balbus, J.F. Hawley: Rev. Mod. Phys., 70, 1 (1998)
41. F. Melia, S. Liu, R. Coker: ApJ, 545, L117 (2000)
42. J.E. Pringle: MNRAS, 178, 195 (1977)
43. R. Popham, R. Narayan: ApJ, 394, 255 (1992)
44. S. Ichimaru: ApJ, 214, 840 (1977)
45. M.J. Rees, M.C. Begelman, R.D. Blandford, E.S. Phinney: Nature, 295, 17 (1982)
46. R. Narayan, I. Yi: ApJ, 428, L13 (1994)
47. R. Narayan, I. Yi: ApJ, 444, 231 (1995)
48. R. Narayan, I. Yi: ApJ, 452, 710 (1995)
49. M. Abramowicz, X. Chen, S. Kato, J.P. Lasota, O. Regev: ApJ, 438, L37 (1995)
50. X. Chen, M.A. Abramowicz, J.P. Lasota, R. Narayan, I. Yi: ApJ, 443, L61 (1995)
51. D. Tsiklauri: New Astron., 6, 487 (2001)
52. R. Mahadevan, F. Quataert: ApJ, 490, 605 (1997)
53. R. Mahadevan, R. Narayan, J. Krolik: ApJ, 486, 268 (1997)
54. R. Mahadevan: MNRAS, 304, 501 (1999)
55. F. Ozel, D. Psaltis, R. Narayan: ApJ, 541, 234 (2000)
56. A.A. Esin, J.E. McClintock, R. Narayan: ApJ, 489, 865; 500, 523 (1997)
57. F. Yuan, Q. Peng, J. Lu, J. Wang: ApJ, 537, 236 (2000) ???
58. T. Piran: ApJ, 221, 652 (1978)

59. S. Kato, M.A. Abramowicz, X. Chen: PASJ, 48, 67 (1996)
60. S. Kato, T. Yamasaki, M.A. Abramowicz, X. Chen: PASJ, 49, 221 (1997)
61. X. Wu, Q. Li: ApJ, 469, 776 (1996)
62. X. Wu: MNRAS, 292, 113 (1997)
63. R. Narayan, R. Mahadevan, E. Quataert: in *The Theory of Black Hole Accretion Discs*, ed. by M.A. Abramowicz, G. Bjornsson, J.E. Pringle (Cambridge Univ. Press, Cambridge, 1998) p148
64. S. Kato, J. Fukue, S. Mineshige: *Black Hole Accretion Disks* (Kyoto Univ. Press, Kyoto, 1998)
65. E. Quataert: in *Probing the Physics of Active Galactic Nuclei by Multiwavelength Monitoring*, ed. by B.M. Peterson, R.S. Polidan, R.W. Pogge (Astr. Soc. Pacific, San Francisco, 2001) p71
66. R. Narayan, M.R. Garcia, J.E. McClintock: in *Proc. IX Marcel Grossmann Conference*, in press (2001) (astro-ph/0107387)
67. R. Narayan, I. Yi, R. Mahadevan: Nature, 374, 623 (1995)
68. T. Manmoto, S. Mineshige, M. Kusunose: ApJ, 489, 791 (1997)
69. R. Mahadevan: Nature, 394, 651 (1998)
70. T. Manmoto: ApJ, 534, 734 (2000)
71. E.S. Phinney: in *Plasma Astrophysics*, ed. by T. Guyenne (ESA SP-161, 1981) p337
72. G.S. Bisnovatyi-Kogan, R.V.E Lovelace: ApJ, 486, L43 (1997)
73. E. Quataert: ApJ, 500, 978 (1998)
74. A. Gruzinov: ApJ, 501, 787 (1998)
75. E. Quataert, A. Gruzinov: ApJ, 520, 248 (1999)
76. E. Blackman: MNRAS, 302, 723 (1999)
77. K.E. Nakamura, M. Kusunose, R. Matsumoto, S. Kato: PASJ, 49, 503 (1997)
78. S.S. Doeleman et al.: AJ, 121, 2610 (2001)
79. K. Menou, E. Quataert, R. Narayan: in *Black Holes, Gravitational Radiation and the Universe*, ed. by B.R. Iyer, B. Bhawal (Kluwer, Dordrecht, 1999) p265
80. A.C. Fabian, C.R. Canizares: Nature, 333, 829 (1988)
81. A.C. Fabian, M.J. Rees: MNRAS, 277. L5 (1995)
82. C.S. Reynolds, T. di Matteo, A.C. Fabian, U. Hwag, C.R. Canizares: MNRAS, 283, L111 (1996)
83. R. Mahadevan: ApJ, 477, 585 (1997)
84. T. di Matteo, A.C. Fabian: MNRAS, 286, 50 (1997)
85. M. Loewenstein, R.F. Mushotzky, L. Agnelini, K.A. Arnaud, E. Quataert: ApJ, 555, L21 (2001)
86. T. di Matteo, R.M. Johnstone, A.C. Fabian, S.W. Allen: ApJ, 550, L19 (2001)
87. E. Quataert, R. Narayan: ApJ, 528, 236 (2000)
88. L.C. Ho: ApJ, 516, 672 (1999)
89. J.P. Lasota, M.A. Abramowicz, X. Chen, J. Krolik, R. Narayan, I. Yi: ApJ, 462, 142 (1996)
90. C.F. Gammie, R. Narayan, R. Blandford: ApJ, 516, 177 (1999)
91. E. Quataert, T. di Matteo, R. Narayan, L.C. Ho: ApJ, 525, L89 (1999)
92. A. Ptak, T. Yaqoob, R. Mushotzky, P. Serlemitsos, R. Griffiths: ApJ, 501, L37 (1998)
93. J.S. Ulvestad, L.C. Ho: ApJ, 562, L133 (2001)
94. R. Narayan: ApJ, 461, 136 (1996)
95. A.A. Esin, R. Narayan, W. Cui, J.E. Grove, S.N. Zhang: ApJ, 505, 854 (1998)
96. A.A. Esin, J.E. McClintock, J.J. Drake, M.R. Garcia, C.A. Haswell, R.I. Hynes, M.P. Muno: ApJ, 555, 483 (2001)
97. F. Meyer, E. Meyer-Hofmeister: A&A, 361, 175 (1994)
98. F. Honma: PASJ, 48, 77 (1996)

99. C.P. Dullemond, R. Turolla: ApJ, 503, 361 (1998)
100. B.F. Liu, W. Yuan, F. Meyer, E. Meyer-Hofmeister, G.Z. Xie: ApJ, 527, L17 (1999)
101. A. Rozanska, B. Czerny: A&A, 360, 1170 (2000)
102. B. Czerny, A. Rozanska, P.Y. Zycki: New Astron. Rev., 44, 439 (2000)
103. F. Meyer, B.F. Liu, E. Meyer-Hofmeister: A&A, 361, 175 (2000)
104. H.C. Spruit, B. Deufel: A&A, submitted (2001) (astro-ph/0108497)
105. I.V. Igumenshchev, R. Narayan: ApJ, in press (2001) (astro-ph/0105365)
106. R. Narayan, S. Kato, F. Honma: ApJ, 476, 49 (1997)
107. R.D. Blandford, M.C. Begelman: MNRAS, 303, L1 (1999)
108. T. Beckert: ApJ, 539, 223 (2000)
109. M.A. Abramowicz, J.P. Lasota, I.V. Igumenshchev: MNRAS, 314, 775 (2000)
110. R. Turolla, C.P. Dullemond: ApJ, 531, L49 (2000)
111. M.C. Begelman, D.L. Meier: ApJ, 253, 873 (1982)
112. E. Quataert, R. Narayan: ApJ, 520, 298 (1999)
113. I.V. Igumenshchev, X. Chen, M.A. Abramowicz: MNRAS, 278, 236 (1996)
114. I.V. Igumenshchev, M.A. Abramowicz: MNRAS, 303, 309 (1999)
115. J.M. Stone, J.E. Pringle, M.C. Begelman: MNRAS, 310, 1002 (1999)
116. I.V. Igumenshchev, M.A. Abramowicz: ApJ, 537, L27 (2000)
117. R. Narayan, I.V. Igumenshchev, M.A. Abramowicz: ApJ, 539, 798 (2000)
118. E. Quataert, A. Gruzinov: ApJ, 539, 809 (2000)
119. I.V. Igumenshchev, M.A. Abramowicz, R. Narayan: ApJ, 537, L27 (2000)
120. D. Ryu, J. Goodman: ApJ, 388, 438 (1992)
121. P. Kumar, R. Narayan, A. Loeb: ApJ, 453, 480 (1995)
122. J.M. Stone, S.A. Balbus: ApJ, 464, 364 (1996)
123. M.A. Abramowicz, I.V. Igumenshchev, E. Quataert, R. Narayan: ApJ, in press (2001) (astro-ph/0110371)
124. R. Matsumoto: in *Numerical Astrophysics*, ed. by S.M. Miyama, K. Tomisaka, T. Hanawa (Kluwer, Dordrecht, 1999), p195
125. J.F. Hawley: ApJ, 528, 462 (2000)
126. M. Machida, M.R. Hayashi, R. Matsumoto: ApJ, 532, L67 (2000)
127. J.M. Stone, J.E. Pringle: MNRAS, 322, 461 (2001)
128. M. Machida, R. Matsumoto, S. Mineshige: PASJ, 53, L1 (2001)
129. J.F. Hawley, S.A. Balbus, J.M. Stone: ApJ, 554, L49 (2001)
130. S.A. Balbus: ApJ, 562, 909 (2001)
131. T. Kawaguchi, S. Mineshige, M. Machida, R. Matsumoto, K. Shibata: PASJ, 52, L1 (2000)
132. S. Mineshige, H. Negoro, R. Matsumoto, M. Machida: this volume (2001)
133. G.H. Ball, R. Narayan, E. Quataert: ApJ, 552, 221 (2001)
134. D.K. Aitken et al.: ApJ, 534, L173 (2000)
135. E. Quataert, A. Gruzinov: ApJ, 545, 842 (2000)
136. E. Agol: ApJ, 538, L121 (2000)
137. J.H. Zhao, G.C. Bower, W.M. Goss: ApJ, 547, L29 (2001)
138. G.C. Bower, M.C.H. Wright, H. Falcke, D.C. Backer: ApJ, 555, L103 (2001)
139. H. Falcke: in *Reviews in Modern Astronomy 14: Dynamical Stability and Instabilities in the Universe*, ed. by R.E. Schielicke (Astronomische Gesellschaft, Hamburg, 2001) p15
140. F. Yuan, S. Markoff, H. Falcke: A&A, in press (2001) (astro-ph/0112464)
141. H. Falcke, K. Mannheim, P.L. Biermann: A&A, 278, 71 (1993)
142. H. Falcke: in *IAU Symp. 169: Unsolved Problems in the Milky Way*, ed. by L. Blitz, P.J. Teuben (Kluwer, Dordrecht, 1996) p163
143. H. Falcke, P.L. Biermann: A&A, 342, 49 (1999)
144. H. Falcke, S. Markoff: A&A, 362, 113 (2000)
145. S. Markoff, H. Falcke, F. Yuan, P.L. Biermann: A&A, 379, L13 (2001)
146. W.J. Duschl, H. Lesch: A&A, 286, 431 (1994)
147. T. Beckert, W.J. Duschl: A&A, 328, 95 (1997)

The Power of Jets: New Clues from Radio Circular Polarization and X-Rays

Heino Falcke[1], Thomas Beckert[1], Sera Markoff[1], Elmar Körding[1], Geoffrey C. Bower[2], and Rob Fender[3]

[1] Max-Planck-Institut für Radioastronomie, Auf dem Hügel 69, D-53121 Bonn, Germany
[2] UC Berkeley, RAL, 601 Campbell Hall, Berkeley CA 94720, USA
[3] Astronomical Institute "Anton Pannekoek"and Center for High Energy Astrophysics, University of Amsterdam, Kruislaan 403, 1098 SJ Amsterdam, The Netherlands

Abstract. Jets are ubiquitous in accreting black holes. Often ignored, they may be a major contributor to the emitted spectral energy distribution for sub-Eddington black holes. For example, recent observations of radio-to-X-ray correlations and broad band spectra of X-ray binaries in the low/hard state can be explained by a significant synchrotron contribution from jets also to their IR-to-X-ray spectrum as proposed by [14]. This model can also explain state-transitions from low/hard to high/soft states. Relativistic beaming of the jet X-ray emission could lead to the appearance of seemingly Super-Eddington X-rays sources in other galaxies. We show that a simple population synthesis model of X-ray binaries with relativistic beaming can well explain the currently found distribution of off-nucleus X-ray point sources in nearby galaxies. Specifically we suggest that the so-called ultra-luminous X-ray sources (ULXs, also IXOs) could well be relativistically beamed microblazars. The same model that can be used to explain X-ray binaries also fits Low-Luminosity AGN (LLAGN) and especially Sgr A* in the Galactic Center. The recent detection of significant circular polarization in AGN radio cores, ranging from bright quasars down to low-luminosity AGN like M81*, Sgr A* and even X-ray binaries, now places additional new constraints on the matter contents of such jets. The emerging picture are powerful jets with a mix of hot and cold matter, a net magnetic flux, and a stable magnetic north pole.

1 Introduction

Quasars are probably the most important lighthouses of the universe, because they are extremely luminous, compact, and emit at all observable wavelengths. Most of the radio and high energy emission in quasars can be attributed to relativistic jets produced in the vicinity of the central black hole. In blazars also infrared, optical and X-ray emission is produced by these jets. Despite their prominence at all these energies, some basic properties of jets have not been clarified despite intense observing campaigns over the last three decades. Some of these questions are: Why and how are jets produced? What matter are they made of? How much energy is carried in the jet? Why are some jets more radio-loud than others? What is the exact magnetic field configuration?

A possible explanation for the slow progress in answering some of these questions could be that, in contrast to many other astrophysical phenomena, jets were first discovered at rather large distances, i.e. in distant quasars and only in a few cases closer to home. The reason for this is most likely that the radio luminosity of jets scales non-linearly with jet power and is in fact relatively weaker in low-luminosity jets than in quasars (see for example [6]). Only in recent years have we become to appreciate that relativistic jets exist over a wide range of distances and over a wide range of black hole masses and accretion rates. This now opens up the possibility to study the physics of jets in a much larger parameter range and to revisit some of the early questions in this context. Here we will concentrate on two relatively new ways of approaching jet physics, namely through their X-ray emission and their circular polarization.

2 X-Rays from Microquasar Jets

One important finding of recent years was that X-ray binaries (XRBs) possess relativistic jets as well [17], leading to the term "microquasar". So far these jets have mainly been seen in radio emission, but we will argue here that emission in other wavebands, specifically NIR and X-rays, is almost unavoidable.

A characteristic feature of jets, also in X-ray binaries, is their flat-spectrum radio core which is best seen during phases of relative quiescence (e.g. as in GRS 1915+105, [5]). Fender (2001) found that the low/hard-state of the X-ray spectrum is correlated with the presence of a persistent flat-spectrum radio core. He also argued that this flat spectrum of the synchrotron emitting radio core extends up into the near-infrared and perhaps optical regime. This is, in fact, not surprising. The standard model for flat radio spectra [1,11,7] suggests that the emission arises from self-absorbed sections in a conical jet, where the smallest scales contribute at the highest frequencies (size $\propto \nu^{-1}$). This is schematically shown in Fig. 1. The interesting point here is that the smallest scale in a system will be set by the size of the black hole. Hence stellar mass black holes will be able to produce flat spectra up to a maximum frequency $\nu_{\mathrm{ssa,max}}$ that is much higher than the supermassive black holes in quasars – by a factor given by the ratio of the black hole masses which is of order 10^{5-8}. Therefore XRBs should be much more likely to exhibit direct synchrotron emission in NIR, UV and even X-rays than normal AGN.

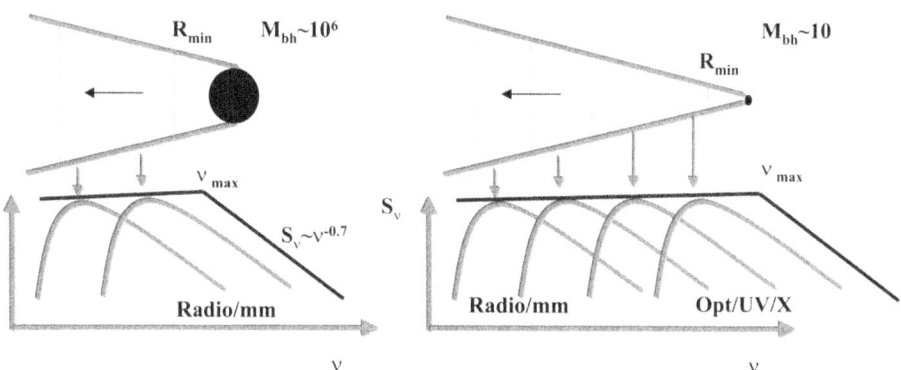

Fig. 1. In a self-similar conical jet model the frequency is inversely proportional to the size. The turn-over frequency of the flat, optically thick part of the jet spectrum is determined by the smallest scale in the system, which should scale with the gravitational radius and the mass of the black hole. Hence, the turnover which in Sgr A* ($M \sim 10^6 M_{\odot}$) or a quasar occurs somewhere in the submm-range should shift into the optical/X-ray regime for a stellar mass black hole.

This is particularly interesting for the hard X-ray power-laws observed in some XRBs. In a jet spectrum, beyond the turnover point of the flat radio spectrum at $\nu >$ $\nu_{\mathrm{ssa,max}}$, the synchrotron emission is optically thin. The shape of the spectrum depends on the electron distribution on the smallest scales in the jet. A thermal distribution leads to an exponential cutoff, but a power-law distribution which is typically observed in luminous AGN jets leads to a hard spectrum with spectral indices ranging from $\alpha =$ -0.5 to -1 (energy flux density index: $S\nu \propto \nu^{\alpha}$). The maximum frequency of the optically thin spectrum can be found by balancing acceleration and radiation loss times and, as Markoff, Falcke and Fender, 2001 (MFF01) showed, can in principle easily reach several 100 keV fairly independent of the jet power or shock location. MFF01 also showed that such a model can well explain the broad-band spectrum of the X-ray binary XTE J1118+480 (see Fig. 2, top left). Therefore X-ray emission from jets in XRBs is something that has to be dealt with in understanding the spectra of stellar mass black holes.

Of course, the situation may not always be so simple as described in MFF01 for XTE J1118+480, a source which shows almost no sign of reflection components from an accretion disk in the observed spectra [16]. Under different circumstances, the disk itself should have a more direct or indirect influence on the X-ray spectrum. One such effect is radiation cooling of highly relativistic electrons in the jet due to photons from the accretion disk. This is important when the accretion rate increases, the disk luminosity increases and, in the truncated disk model, the inner edge of the optically thick disk approaches the inner region of the jet. The drastic increase in photon density near the black hole can then cool most hot electrons and leave the jet very radio quiet as well as suppress the hard X-ray power-law. This situation is shown in Fig. 2, where we start with the published jet spectrum in XTE J1118+480 and move the transition radius arbitrarily closer to the black hole, taking radiation cooling into account. As expected, radio and hard X-rays disappear, while the black-body emission of the accretion disk appears in soft X-rays. There is only some jet contribution to the EUV and soft X-ray spectrum from the jet nozzle left. Therefore, a change in transition radius of the accretion disk could at least qualitatively explain the transition from low-hard to high-soft state and explain why radio and hard X-rays seemingly disappear while the accretion rate goes up.

Such a spectral transition would be difficult to observe in AGN due to the much longer timescales. However, one cannot help to wonder whether such a behavior could also explain the differences between BL Lacs and luminous blazars (or between FR I and FR II radio galaxies/radio-loud quasars as their respective host populations). After all, BL Lacs are intrinsically less luminous than beamed radio-loud quasars but produce much harder and more energetic spectra, extending up to TeV energies with little evidence for disk emission.

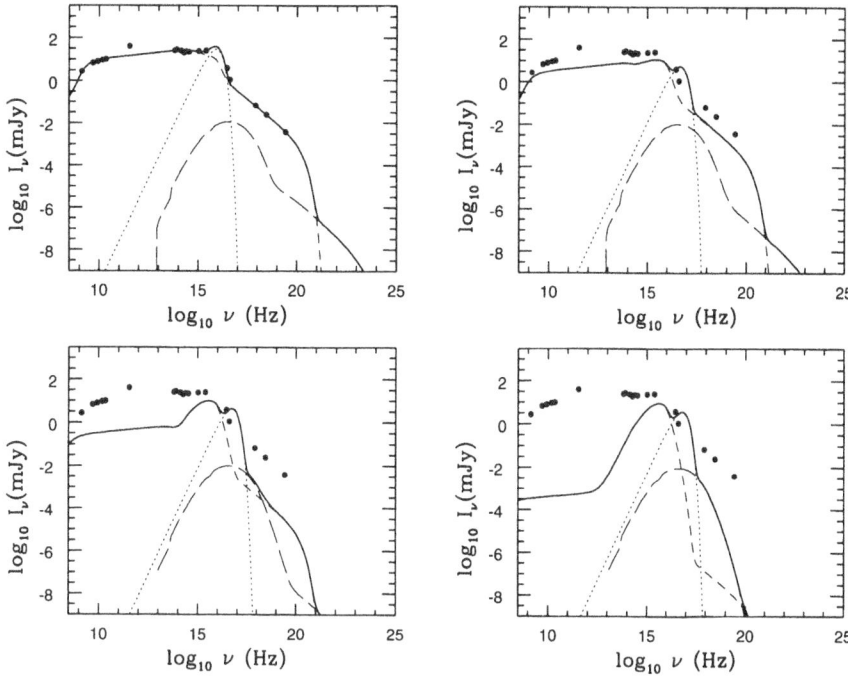

Fig. 2. The self-similar jet model for XRBs: The top left panel is the fit of a jet model to the broad-band data of XTE J1118+480 as shown in [14] together with black body emission from a standard optically thick disk at a large transition radius. The subsequent panels (left to right and top to bottom) show how the spectrum theoretically would evolve from a low/hard-tate to a high/soft-state if one moves the transition radius closer to the black hole. Inverse Compton cooling by disk photons strongly quenches the hard synchrotron emission in the X-rays and suppresses the flat radio spectrum. (From Markoff et al., in prep.)

3   Microblazars as Lighthouses of the Nearby Universe

The idea that jets in XRBs contribute to the X-ray spectrum[1] has some other interesting consequences. If jet-emission is significant for edge-on sources like XTE J1118+480 it will be even more important when the source points towards the observer. In analogy to AGN, where jets pointing toward the observer are believed to cause the strong blazar emission from radio to gamma-rays, a microquasar pointing towards the observer should appear as a 'microblazar' [18].

This immediately raises the question whether (some of) the ultra-luminous off-nuclear X-ray sources (ULXs), also called Super-Eddington sources or intermediate X-ray objects (IXOs; [4]), that have been discussed at this conference (Makishima),

---

[1] In this context it is interesting to note that the possibility of an angle-dependent jet contribution in X-rays was already raised by Shakura & Sunyaev (1973).

are in fact such microblazars. As commonly done in unified schemes for AGN [20] the population of relativistically beamed sources is rather well defined, once one specifies a host population and appropriate parameters for the jets – particularly the bulk Lorentz factor γ_j. This allows one to check the validity of such ideas.

Along these lines [13] have investigated a simple population synthesis model for XRBs and compared it with X-ray point source populations in nearby galaxies detected in recent Chandra observations.

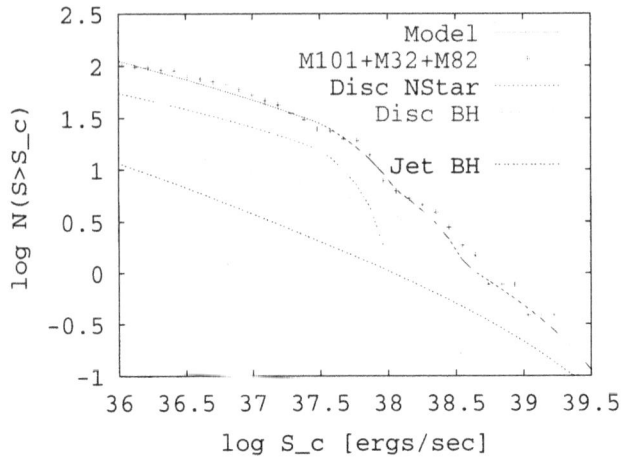

Fig. 3. Log N/Log S distribution of X-ray sources (dots) in three nearby galaxies together with a simple population synthesis model (lines) for isotropic disk and anisotropic, relativistically beamed jet emission for XRBs in low/hard and high/soft states. (From Körding et al. 2001)

The model consists of basically two populations of XRBs, neutron stars and black holes, which emit X-ray emission isotropically from an accretion disk and anisotropically from a jet due to relativistic beaming. Below a critical accretion rate $\dot{M}_c$ (10% of Eddington) the disk luminosity is assumed to scale as $L_{\text{disk}} \propto \dot{M}^2$ (according to the ADAF paradigm) and above $\dot{M}_c$ as $L_{\text{disk}} \propto \dot{M}$. The jet emits with $L_{\text{jet}} \propto \dot{M}^{1.4}$ (according to Falcke & Biermann (1995)) and above $\dot{M}_c$ as $L_{\text{jet}} \propto \dot{M}$ in the radiation-cooling dominated regime mentioned above. Accretion rates ($\dot{M}$) are assumed to be distributed in a power-law. Figure 3 shows the result of this beaming model together with the data for a $\gamma_j = 5$ jet. This shows that the high-luminosity end can indeed be explained by microblazars without violating the low-luminosity end of the distribution or making extreme assumptions about the properties of jets. Hence, as blazars are the most luminous lighthouses of the distant universe, microblazars could be the lighthouses of the local universe.

4 Circular Polarization and the Nature of Jets

So far we have merely used the fact that jets exist and radiate to discuss their observable impact. Their emission and kinetic power depends significantly on their mass content and the electron distribution – specifically the distribution of hot electrons. Recent observations of radio circular polarization in AGN, X-ray binaries, and the Galactic Center black hole [21,2,9] suggest that this may only be the tip of the iceberg.

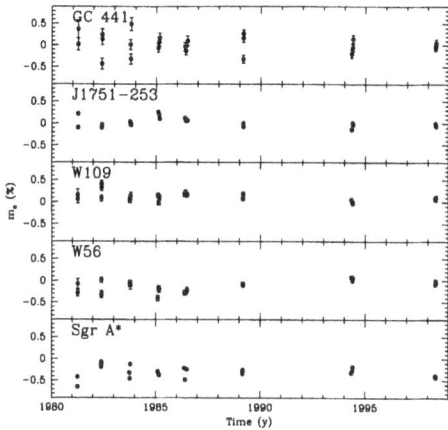

Fig. 4. 20 Years light curve of circular polarization measurements in Sgr A* and nearby calibrator sources with the VLA at 5 GHz. Evidently the level and sign of circular polarization remains constant for Sgr A* over the entire period, while short-term variability may be present. (From Bower et al., in prep.)

One example of such measurements is shown in Fig. 4, where we plot a 20 year light curve of circular polarization in Sgr A* [3] – the radio source coincident with the supermassive black hole at the Galactic Center [15]. The source shows about 0.5% circular polarization with a stable sign despite strong flux variations during this time. There is also no detectable linear polarization. The overall spectrum of the source can be understood in terms of a jet model [8].

Circular polarization can be produced through conversion by a bi-refringent medium (such as a magnetized plasma; see [12]) where the magnetic field has a component transverse to the line-of-sight and the radio waves are Faraday rotated. Depolarization of linear polarization can be obtained by random Faraday rotation in a turbulent plasma where field components are along the line-of-sight. Both processes are sensitive to the presence of low-energy electrons. The ratio of linear to circular polarization in a jet can be calculated with an appropriate radiation transfer code for various parameters (Beckert & Falcke, in prep.). Fig. 5 shows the result of such a calculation for the case of Sgr A* and two magnetic field configurations: a helical and a purely poloidal magnetic field on top of a turbulent field. Only the former can produce the observed level of circular polarization. Since field-reversals cancel any effect of Faraday rotation and conversion, one needs a field configuration with a dominating

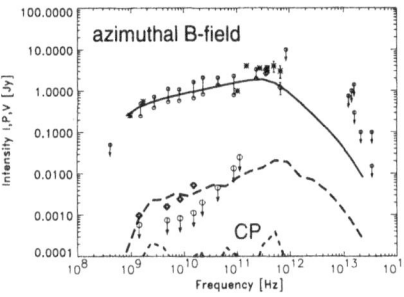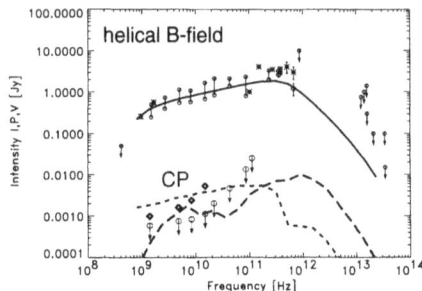

Fig. 5. Data and jet model for polarization in Sgr A*: small circles – average measured total intensity, big circles – measured upper limits for linear polarization, small diamonds – measured circular polarization, solid line – jet model total intensity, long-dashed line – jet model linear polarization, short-dashed line – jet model circular polarization. The left panel is for a turbulent plus a purely azimuthally ordered magnetic field, the right panel is for a turbulent plus a helically ordered magnetic field. A pure azimuthal (plus turbulent) magnetic field is not able to produce significant circular polarization. (From Beckert & Falcke, in prep.)

component of *one polarity* along the line-of-sight (or along the jet axis for moderately inclined jets).

Such a configuration is most naturally achieved by a helical magnetic field as is presumed to exist in jets. In addition, the number of low-energy electrons producing the conversion and de-polarization needs to be significantly (by 2-3 orders of magnitude) higher than the number of radiating hot electrons. For Sgr A* this increase in particle numbers means that the mass outflow rate and total jet power can be orders of magnitude higher than inferred so far. The stability of circular polarization also indicates that the polarity of the magnetic field (the magnetic north pole) has remained constant over the last two decades. Given the rather short accretion time scale in this source one could speculate that this polarity is related to the accretion of a stable large-scale magnetic field which is accreted and expelled by a jet close to the black hole.

5 Conclusions

Jets are a major source of luminous emission in AGN. Our modeling of jet spectra suggests that this is also the case for X-ray binaries. Jets can contribute to the hard X-rays, and in microblazars X-rays could be beamed to apparent super-Eddington luminosities. The power of these jets could be relatively large, as indicated by the recent detection of circular polarization in various sources which implies a large number of low-energy electrons. The long-term stability of the sign of circular polarization indicates that these jets have a non-vanishing magnetic flux and stable north pole.

References

1. Blandford R. D., Königl A., 1979, ApJ, 232, 34
2. Bower G. C., Falcke H., Backer D. C., 1999, ApJ Lett., 523, L29

3. Bower G. C., Falcke H., Sault R. J., Backer D. C., 2002, ApJ, to be submitted
4. Colbert E. J. M., Mushotzky R. F., 1999, ApJ, 519, 89
5. Dhawan V., Mirabel I. F., Rodríguez L. F., 2000, ApJ, 543, 373
6. Falcke H., Biermann P. L., 1995, A&A, 293, 665
7. Falcke H., Biermann P. L., 1999, A&A, 342, 49
8. Falcke H., Markoff S., 2000, A&A, 362, 113
9. Fender R., Rayner D., Norris R., Sault R. J., Pooley G., 2000, ApJ Lett., 530, L29
10. Fender R. P., 2001, MNRAS, 322, 31
11. Hjellming R. M., Johnston K. J., 1988, ApJ, 328, 600
12. Jones T. W., O'Dell S. L., 1977, ApJ, 214, 522
13. Körding E., Falcke H., Markoff S., 2001, A&A Letters, submitted
14. Markoff S., Falcke H., Fender R., 2001, A&A, 372, L25
15. Melia F., Falcke H., 2001, ARAA, 39, 309
16. Miller J. M., Ballantyne D. R., Fabian A. C., Lewin W. H. G., 2001, MNRAS, submitted
17. Mirabel I. F., Rodriguez L. F., 1994, Nat, 371, 46
18. Mirabel I. F., Rodríguez L. F., 1999, ARAA, 37, 409
19. Shakura N. I., Sunyaev R. A., 1973, A&A, 24, 337
20. Urry C. M., Padovani P., 1995, PASP, 107, 803
21. Wardle J. F. C., Homan D. C., Ojha R., Roberts D. H., 1998, Nat, 395, 457

X-Ray Evidence for Supermassive Black Holes in Non-Active Galaxies:
Detection of X-Ray Flare Events, Interpreted as Tidal Disruptions of Stars by SMBHs

Stefanie Komossa

Max-Planck-Institut für extraterrestrische Physik, Postfach 1312, 85741 Garching; skomossa@mpe.mpg.de

Abstract. It has long been suggested that supermassive black holes in non-active galaxies might be tracked down by occasional tidal disruptions of stars on nearly radial orbits. A tidal disruption event would reveal itself by a luminous flare of electromagnetic radiation. Theorists argued that the convincing detection of such a tidal disruption event would be the observation of an event which fulfills the following three criteria: (1) the event should be of finite duration (a 'flare'), (2) it should be very luminous (up to $L_{max} \approx 10^{45}$ erg/s in maximum), and (3) it should reside in a galaxy that is otherwise perfectly *non*-active (to be sure to exclude an upward fluctuation in gaseous accretion rate of an *active* galaxy). During the last few years, several X-ray flare events were detected which match exactly the above criteria. We therefore consider these events to be excellent candidates for the occurrences of the theoretically predicted tidal disruption flares. In this contribution, we review the previous observations of giant X-ray flares from normal galaxies, present new results on these objects, critically discuss alternatives to the favored outburst scenario, and report results from our ongoing search for further tidal disruption flares based on the *ROSAT* all-sky survey database.

1 Flares from Tidally Disrupted Stars as Probes for the Presence of SMBHs in *Non-Active* Galaxies

There is strong evidence for the presence of massive dark objects at the centers of many galaxies. Does this hold for *all* galaxies ? Questions of particular interest in the context of AGN evolution are: what fraction of galaxies have passed through an active phase, and how many now have non-accreting and hence unseen supermassive black holes (SMBHs) at their centers (e.g., Lynden-Bell 1969, Rees 1988)? Several approaches were followed to study these questions. A lot of effort has concentrated on the determination of central object masses from studies of the *dynamics of stars and gas* in the nuclei of nearby galaxies. Earlier (ground-based) evidence for central quiescent dark masses in non-active galaxies has been strengthened by recent HST results (see Kormendy & Richstone 1995 for a review).

Whereas the dynamics of stars and gas probe rather large volumes, i.e., distances from the SMBH, high-energy *X-ray emission* originates from the very vicinity of the SMBH (see Komossa 2001 for a review). In *active* galaxies, excellent evidence for the presence of SMBHs is provided by the detection of luminous hard power-law like X-ray emission, rapid variability, and the detection of relativistically broadened FeKα lines (e.g., Tanaka et al. 1995). How can we find *dormant* SMBHs in *non-active* galaxies ?

Lidskii & Ozernoi (1979) and Rees (1988) suggested to use the flare of electromagnetic radiation produced when a star is tidally disrupted and accreted as a means to detect SMBHs in nearby non-active galaxies.

A star on a radial 'loss-cone' orbit gets tidally disrupted after passing a certain distance to the black hole (e.g., Hills 1975, Lidskii & Ozernoi 1979, Diener et al. 1997), the tidal radius, given by

$$r_t \simeq 7 \, 10^{12} \, (\frac{M_{BH}}{10^6 M_\odot})^{\frac{1}{3}} (\frac{M_*}{M_\odot})^{-\frac{1}{3}} \frac{r_*}{r_\odot} \, \text{cm} \, . \tag{1}$$

The star is first heavily distorted, then disrupted. About 50%–90% of the gaseous debris becomes unbound and is lost from the system (e.g., Young et al. 1977, Ayal et al. 2000). The rest will eventually be accreted by the black hole (e.g., Cannizzo et al. 1990, Loeb & Ulmer 1997). The stellar material, first spread over a number of orbits, quickly circularizes (e.g., Rees 1988, Cannizzo et al. 1990) due to the action of strong shocks when the most tightly bound matter interacts with other parts of the stream (e.g., Kim et al. 1999). Most orbital periods will then be within a few times the period of the most tightly bound matter (e.g., Evans & Kochanek 1989). A star will only be disrupted as long as its tidal radius lies outside the Schwarzschild radius of the BH, else it is swallowed as a whole (this happens for BH masses larger than $\sim 10^8$ M$_\odot$).[1] More massive BHs may still disrupt or strip the atmospheres of giant stars.

2 Tidal Disruption Flares from Non-Active Galaxies: Observational Evidence

2.1 Summary of X-Ray and Optical Observations

With the X-ray satellite *ROSAT*, some rather unusual observations have been made in the last few years: the detections of giant-amplitude, non-recurrent X-ray outbursts from a handful of *optically non-active* galaxies, starting with the case of NGC 5905 (Bade et al. 1996, Komossa & Bade 1999). Based on the huge observed outburst luminosity, the observations were interpreted in terms of tidal disruption events. Below, we first give a brief review of the properties of all published X-ray flaring non-active galaxies, and then present results from a search for radio emission from these galaxies. There are now four X-ray flaring 'normal' galaxies

---

[1] Numerical simulations of the disruption process, the stream-stream collision, the accretion phase, and the depletion of loss-cone orbits and disruption rates have been studied in the literature (e.g., Nolthenius & Katz 1983, Carter & Luminet 1985, Evans & Kochanek 1989, Laguna et al. 1993, Diener et al. 1997, Ayal et al. 2000, Kim et al. 1999, Hills et al. 1975, Kato & Hoshi 1978, Gurzadyan & Ozernoi 1980, Cannizzo et al. 1990, Loeb & Ulmer 1997, DiStefano et al. 2001, Frank & Rees 1976, Magorrian & Tremaine 1999; see Komossa & Dahlem 2001 for more references, incl. a few observations of *active* galaxies that might be related to tidal disruption events). Renzini et al. (1995; R95) reported the detection of a UV flare from the (only mildly active) galaxy NGC 4552. The luminosity was several orders of magnitude weaker than what could have been expected from a tidal disruption event. Tidal stripping of a star's atmosphere is one possible explanation (R95).

Table 1. Summary of the X-ray properties of the flaring non-active galaxies during outburst (NGC 5905: Bade et al. 1996, Komossa & Bade 1999, RXJ1242−1119: Komossa & Greiner 1999, RXJ1624+7554: Grupe et al. 1999, RXJ1420+5334: Greiner et al. 2000; for first results on another candidate see Reiprich & Greiner 2001. Based on the position they report, we refer to this source as RXJ1331−3243). T_{bb} denotes the black body temperature derived from a black body fit to the data (cold absorption was fixed to the Galactic value in the direction of the individual galaxies). $L_{x,bb}$ gives the intrinsic luminosity in the (0.1–2.4) keV band, based on a black body fit. We note that this is a lower limit to the actual peak luminosity, since we most likely have not caught the sources exactly at maximum light, since the spectrum may extend into the EUV, and since we have conservatively assumed no X-ray absorption intrinsic to the galaxies.

galaxy name	redshift z	kT_{bb} [keV]	$L_{x,bb}$ [erg/s]
NGC 5905	0.011	0.06	$3 \ 10^{42}$*
RXJ1242−1119	0.050	0.06	$9 \ 10^{43}$
RXJ1624+7554	0.064	0.097	$\sim 10^{44}$
RXJ1420+5334	0.147	0.04	$8 \ 10^{43}$
RXJ1331−3243	0.051		

* Mean luminosity during the outburst; since the flux varied by a factor ~ 3 during the observation, the peak luminosity is higher.

(NGC 5905, RXJ1242−1119, RXJ1624+7554, RXJ1420+ 5334[2]), and a possible fifth candidate (RXJ1331−3243), all of which show similar properties (see Table 1 for a summary):

- huge X-ray peak luminosity (up to $\sim 10^{44}$ erg/s),
- giant amplitude of variability (up to a factor ~ 200),
- ultra-soft X-ray spectrum ($kT_{bb} \simeq 0.04$-0.1 keV),
- absence of optical signs of Seyfert activity (the spectrum of NGC 5905 is of HII-type; the other galaxies do not show any emission lines).

2.2 Radio Observations

Radio observations are important for two reasons: Firstly, they allow the search for a peculiar, optically hidden AGN at the center of each flaring galaxy. Secondly, radio emission could possibly be produced in relation to the X-ray flare itself. We have performed a search for radio emission from the X-ray flaring galaxies, based on the NRAO VLA Sky Survey (NVSS) catalogue (Condon et al. 1998) at 1.4 GHz, and the FIRST VLA sky survey at 1.5GHz (e.g., Becker et al. 1995). With the exception of NGC 5905, no flaring galaxy is radio-detected. For NGC 5905, several radio observations from the literature are available, summarized in Fig. 1. The bulk of the radio emission is extended, and

---

[2] The X-ray position error circle of RXJ1420+53 contains a second galaxy for which a spectrum is not yet available. Based on the galaxy's morphology, Greiner et al (2000) argue that it is likely non-active

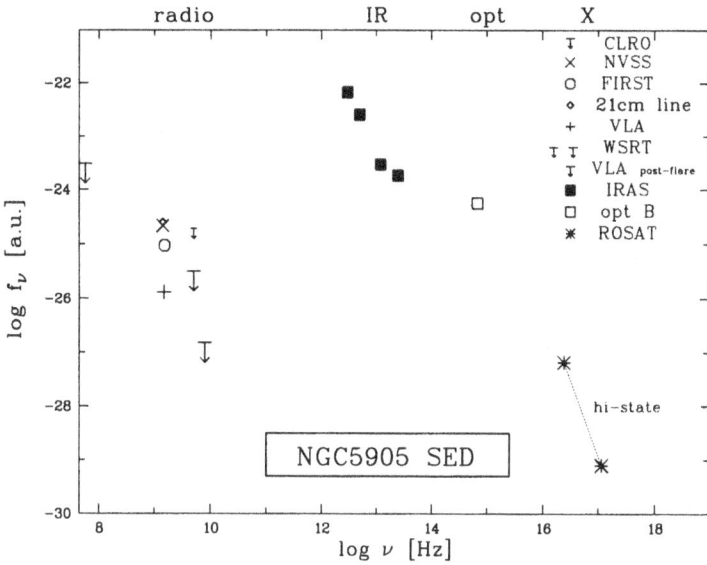

Fig. 1. Multiwavelength continuum spectrum of NGC 5905 (symbols represent data from Israel & Mahoney 1990, van Moorsel 1982, Condon et al. 1998, Becker et al. 1995, Hummel et al. 1987, Brosch & Krumm 1984, Komossa & Dahlem 2001, NED, Bade et al. 1996, Komossa & Bade 1999; the solid/dotted line corresponds to the continuum of the galaxy NGC 4151, shown for comparison). Note: data were taken with different aperture sizes and resolution, and at different times.

NGC 5905 does not show any peculiar radio properties as compared to other similar spiral galaxies. Dedicated VLA radio observations of the nucleus of NGC 5905, performed at a frequency of 8.46 GHz several years after the X-ray outburst, provided an upper limit of 0.15 mJy for the presence of a central point source (Komossa & Dahlem 2001).

3   Outburst Scenarios

Most outburst scenarios do not survive scrutiny (Komossa & Bade 1999), because they cannot account for the huge maximum luminosity (e.g., X-ray binaries within the galaxies, or a supernova in a dense environment), are inconsistent with the optical observations (gravitational lensing), or predict a different temporal behavior (X-ray afterglow of a Gamma-ray burst; see, e.g., Fig. 2 of Bradt et al. 2001). A critical discussion of AGN-related scenarios (presence of a dusty warm absorber, or other absorption-related variablity), and why they are highly unlikely, is given by Komossa & Dahlem (2001).

3.1   Tidal Disruption Model

Except for some types of GRB-related emission mechanisms, the huge peak outburst luminosity nearly inevitably calls for the presence of a SMBH. This, in combination with the complete absence of any signs of AGN activity, makes tidal disruption of a star by a SMBH the most plausible outburst mechanism.

Intense electromagnetic radiation will be emitted in three phases of the disruption and accretion process: First, during the stream-stream collision when different parts of the bound stellar debris interact with themselves (e.g., Rees 1988, Kim et al. 1999). Secondly, radiation is emitted during the accretion of the stellar material. Finally, the unbound stellar gas leaving the system may shock the surrounding interstellar matter and cause intense emission, like in a supernova remnant (Khokhlov & Melia 1996).

Although many details of the actual tidal disruption process are still unclear, some basic predictions have been repeatedly made in the literature how a tidal disruption event should reveal itself observationally: (1) the event should be of finite duration (a 'flare'), (2) it should be very luminous (up to $L_{\max} \approx 10^{45}$ erg/s in maximum), and (3) it should reside in a galaxy that is otherwise perfectly *non*-active (to be sure to exclude an upward fluctuation in gaseous accretion rate of an *active* galaxy). All three predictions are fulfilled by the X-ray flaring galaxies; particularly by NGC 5905 and RXJ1242−1119, which are the two best-studied cases so far.

In addition, we can do some further order of magnitude estimates and consistency checks. The luminosity emitted if the black hole is accreting at its Eddington luminosity can be estimated by

$$L_{\text{edd}} \simeq 1.3 \times 10^{38} M/M_\odot \text{ erg/s}. \tag{2}$$

In case of NGC 5905, a BH mass of at least a few $\sim 10^4$ M$_\odot$ would be required to produce the observed luminosity, and a higher mass if L_x was not observed at its peak value. For comparison, a BH mass of NGC 5905 of $M_{\text{BH}} \approx 10^7 M_\odot$ would be inferred, based on the correlation between bulge blue luminosity and BH mass for spiral galaxies (Salucci et al. 2000), or even up to a few $10^8 M_\odot$ if we use the correlation reported mostly for ellipticals by Ferrarese & Merritt (2000; their 'sample A', their Fig. 1). For the other galaxies, using again L_{edd}, we infer BH masses reaching up to a few 10^6 M$_\odot$. Alternative to a complete disruption event, the atmosphere of a giant star could have been stripped. It is also interesting to note that NGC 5905 possesses a complex bar structure (Friedli et al. 1996) which might aid in the fueling process by disturbing the stellar velocity fields.

In a simple black body approximation, the temperature of the accretion disk scales with black hole mass as

$$T \simeq 8\,10^4 \left(\frac{M_{\text{BH}}}{M_\odot}\right)^{\frac{1}{12}} \text{ K (at } r_t), \quad T \simeq 2\,10^7 \left(\frac{M_{\text{BH}}}{M_\odot}\right)^{-\frac{1}{4}} \text{ K (at } 3\,r_S). \tag{3}$$

This gives $T_{r_{\text{tidal}}} \simeq 3\,10^5$ K, $T_{3r_S} \simeq 7\,10^5$ K for M=10^6 M$_\odot$, where r_S is the Schwarzschild radius. Using black body fits of the X-ray flare spectra we find temperatures in a similar range; $T_{\text{obs}} \simeq (4\text{-}10)\,10^5$ K. Like in AGN, X-ray powerlaw tails are possible. They might have escaped detection during the observations, since weak, or they may develop only after a certain time after the start of the accretion phase. We soon expect first results from a *Chandra* and *XMM* observation of RXJ1242−1119, which will give valuable constraints on the post-flare evolution.

The Eddington time scale for the accretion of the stellar material is given by

$$t_{\text{edd}} \simeq 4\,\eta_{0.1}(M_{\text{BH}}/10^6 M_\odot)(M_*/0.1 M_\odot) \text{ yrs}. \tag{4}$$

Uncertainties in estimating the total duration of the tidal disruption event arise from questions like: how much material is actually accreted or expelled, does a strong wind develop, etc. The events are expected to last for months to years (e.g., Rees 1988). Observationally, the duration of the events was at least several days, followed by gaps in the observations. The source fluxes were then significantly down several years later (e.g., Fig. 9 of Komossa & Bade 1999).

Finally, we note that the redshift distribution of the few sources observed so far is consistent with the predicted tidal disruption rate, in the sence that the events are sufficiently distant to define a large volume of space, in which the detection of a few events would be expected.

3.2 Search for Further X-Ray Flares

We performed a search for further X-ray flaring activity using the sample of nearby galaxies of Ho et al. (1995) and *ROSAT* all-sky survey (Voges et al. 1999) and archived pointed observations. 136 out of the 486 galaxies in the catalogue were observed at least twice with *ROSAT*. We do not find another flaring normal galaxy in this sample, entirely consistent with the expected tidal disruption rate of one event in at least $\sim 10^4$ years per galaxy (e.g., Magorrian & Tremaine 1999).

4 Outlook

X-ray outbursts from non-active galaxies provide important information on the presence of SMBHs in these galaxies, and the link between active and normal galaxies. Future X-ray surveys, like those planned with the *LOBSTER* ISS X-ray all-sky monitor, *ROSITA* and *MAXI*, will be valuable in finding more of these outstanding sources. Rapid follow-up observations at all wavelengths will then be important. In particular, X-ray observations with high spectral and temporal resolution might open up a chance to probe the realm of strong gravity, since the temporal evolution of the stellar debris will depend on relativistic precession effects around the Kerr metric.

References

1. Ayal S., Livio M., Piran T., 2000, ApJ **545**, 772
2. Bade N., Komossa S., Dahlem M., 1996, A&A **309**, L35
3. Becker R.H., White R.L., Helfand D.J., 1995, ApJ **450**, 559
4. Bradt H., Levine A.M., Marshall F.E., et al. 2001, astro-ph/0108004
5. Brosch N., Krumm N., 1984, A&A **132**, 80
6. Cannizzo J.K., Lee H.M., Goodman J., 1990, ApJ **351**, 38
7. Carter B., Luminet J.P., 1985, MNRAS **212**, 23
8. Condon J.J., Cotton W.D., Greissen E.W., et al., 1998, AJ **115**, 1693
9. Diener P., Frolov V.P., Khokhlov A.M., et al., 1997, ApJ **479**, 164
10. Di Stefano R., Greiner R., Murray S., Garcia M., 2001, ApJL, in press
11. Evans C.R., Kochanek C.S., 1989, ApJ **346**, L13
12. Ferrarese L., Merritt D., 2000, ApJ, 539, L9
13. Frank J., Rees M.J., 1976, MNRAS **176**, 633

14. Friedli D., Wozniak H., Rieke M., Martinet L., Bratschi P., 1996, A&AS **118**, 461
15. Greiner J., Schwarz R., Zharikov S., Orio M., 2000, A&A **362**, L25
16. Grupe D., Leighly K., Thomas H., 1999, A&A **351**, L30
17. Gurzadyan V.G., Ozernoi L.M., 1980, A&A **86**, 315
18. Hills J.G., 1975, Nature **254**, 295
19. Ho L.C., Filippenko A.V., Sargent W.L.W., 1995, ApJS **98**, 477
20. Hummel E., van der Hulst J.M., Keel W.C., et al., 1987, A&AS **70**, 517
21. Israel F.P., Mahoney M.J., 1990, ApJ **352**, 30
22. Kato M., Hoshi R., 1978, Prog. Theor. Phys. **60/6**, 1692
23. Khokhlov A., Melia F., 1996, ApJ **457**, L61
24. Kim S.S., Park M.-G., Lee H.M., 1999, ApJ **519**, 647
25. Komossa S., 2001, in *IX. Marcel Grossmann Meeting on General Relativity, Gravitation and Relativistic Field Theories*, V. Gurzadyan et al. (eds), World Scientific, in press [astro-ph/0101289]
26. Komossa S., Bade N., 1999, A&A **343**, 775
27. Komossa S., Greiner J., 1999, A&A **349**, L45
28. Komossa S., Dahlem M., 2001, in *MAXI workshop on AGN variability*, in press [astro-ph/0106422]
29. Kormendy J., Richstone D.O., 1995, ARA&A **33**, 581
30. Laguna P., Miller W.A., Zurek W.H., Davies M.B., 1993, ApJ **410**, L83
31. Lidskii V.V., Ozernoi L.M., 1979, Sov. Astron. Lett. **5(1)**, 16
32. Loeb A., Ulmer A., 1997, ApJ **489**, 573
33. Lynden-Bell D., 1969, Nature **223**, 690
34. Magorrian J., Tremaine S., 1999, MNRAS **309**, 447
35. Nolthenius R.A., Katz J.I, 1983, ApJ **269**, 297
36. Rees M.J., 1988, Nature **333**, 523
37. Reiprich T., Greiner J., 2001, in *ESO workshop on black holes*, 168
38. Renzini A., Greggio L., Di Serego Alighieri S., et al., 1995, Nature **378**, 39
39. Salucci P., Ratnam C., Monaco P., Danese L., 2000, MNRAS **317**, 488
40. Tanaka Y., Nandra K., Fabian A.C., et al., 1995, Nature **375**, 659
41. van Moorsel G.A., 1982, A&A **107**, 66
42. Voges W., et al., 1999, A&A **349**, 389
43. Young P., Shields G., Wheeler J.C., 1977, ApJ **212**, 367

Chandra and the Black Hole in M87

Tiziana Di Matteo[1,2], Steven W. Allen[3], Andrew C. Fabian[3], Andrew S.Wilson[4], and
Andrew J. Young[4]

[1] Harvard-Smithsonian Center for Astrophysics,
   60 Garden St., Cambridge, MA 02138;
[2] Dept. of Physics, Carnegie Mellon University,
   5000 Forbes Avenue, Pittsburgh, PA 15232
[3] Institute of Astronomy,
   Madingley Road, Cambridge, CB3 OHA, UK
[4] Astronomy Department, University of Maryland,
   College Park, MD 20742

Abstract. *Chandra* observations of the nucleus of M87 allow us to resolve the thermal state of its hot interstellar medium well within the accretion radius of its central $3 \times 10^9 M_\odot$ black hole. We measure gas temperature and density in the accretion region and calculate a Bondi accretion rate of $\dot{M}_{\mathrm{Bondi}} \sim 0.2 M_\odot \mathrm{yr}^{-1}$. With *Chandra* we measure the X-ray luminosity of the nucleus of M87 to be $L_x \sim 3 \times 10^{40}$ erg s^{-1}. Accretion onto the black hole at the Bondi rate predicts a nuclear luminosity of $L_B \sim 7 \times 10^{44}$ erg s^{-1} for a canonical accretion radiative efficiency of 10%. These new *Chandra* and longer wavelength constraints, provide the most direct evidence that the black hole in M87 is likely to accrete at a much lower radiative efficiency than the canonical value (unless not much of the mass fed into the outer region of its accretion flow reaches the black hole). We show the Bondi accretion rate is consistent with low-radiative efficiency accretion flow models.

1 Introduction

The nucleus of M87 (NGC 4486) contains a black hole of mass $M \sim 3 \times 10^9 M_\odot$ directly determined from *HST* observations (Ford et al. 1995; Harms et al. 1995; Macchetto et al. 1997). It is the most nearby active nucleus with a one sided jet and giant radio lobes. However, the activity displayed by its nucleus is far less than what is predicted if the central black hole was accreting mass from its hot interstellar medium (with a standard radiative efficiency of ~ 10 per cent; see e.g. Fabian & Rees 1995; Reynolds et al. 1996; Di Matteo et al. 2000). M87 is probably the most illustrative case of a low luminosity system otherwise common in nearby galaxies known to contain supermassive black holes (e.g, Magorrian et al. 1998; Ferrarese & Merrit 2000).

There are only two possible explanations for the low luminosities of nearby black holes: (a) the accretion occurs at extremely low rates or (b) the accretion occurs at low radiative efficiencies as predicted, for example, by advection dominated accretion flow models (ADAFs; e.g. Rees et al. 1982; Narayan & Yi 1994,1995; Abramowicz et al. 1995). In order to discriminate between these two possibility, It is necessary to measure reliably both accretion rates and nuclear luminosity. Direct measurement of either of these quantities has never been possible with previous X-ray satellites. Here we show this can be achieved with *Chandra* for the nucleus of M87.

The massive black holes at the center of elliptical galaxies are likely to accrete primarily from their hot, quasi-spherical interstellar medium (ISM). Accretion rates can therefore be simply estimated using Bondi accretion theory. This requires accurate measurement of both the density and sound speed (i.e.; temperature for the ISM gas) of the hot, X-ray emitting ISM at the 'Bondi accretion radius', the radius where the gravitational potential of the central black hole begins to dominate the dynamics of the hot gas and gas itself starts falling onto the black hole.

Thanks to its high spatial resolution and sensitivity, the *Chandra* X-ray observatory provides the most stringent constraints to the problem of low-luminosity black holes. In particular, *Chandra* resolution at the distance of M87 corresponds to a radius of a few tens of parsec or equivalently to about 10^5 Schwartzchild radii from the black hole. For the case of M87 this allows us to measure for the first time the density and temperature of the ISM directly *at the accretion radius* of the black hole and hence estimate the mass supply into its accretion flow. *Chandra* also provides the best measure of its nuclear point source luminosity: corresponding to $L_x \sim 3 \times 10^{40}$ erg s^{-1} at 1 keV. Here we show that if M87 is accreting at the Bondi rate with a standard mass to energy conversion efficiency of about 10 per cent, its black hole should display a luminosity of $L_x \sim 6 \times 10^{44}$ erg s^{-1}. This is ~ 20000 times more luminosity than what observed. Using the calculated Bondi accretion rate we show that ADAF models are consistent with the required low-radiative efficiency.

2   Gas Properties at the Black Hole Accretion Radius

2.1   Central Temperature and Density

M87 was observed twice with the Advanced CCD Imaging Spectrometer on *Chandra*, on 2000 July 29 and 2000 July 30. The first observation was made in the standard full-frame mode with a net exposure, after cleaning, of 33.7ks. The second, shorter (12.8ks) observation was made in a reduced 1/8 subarray mode to alleviate the effects of pileup on the central source. The back-illuminated CCD detectors were used with a focal plane temperature of -120C. The details of the data analysis and properties of the ISM will be presented in future work. Wilson & Yang (2001) and Yound et al. (2001) discuss the jet emission.

The spectra were extracted from 5 annuli. The central region has a radius of 2 arcsec (0 - 0.2 kpc) and is dominated by the central AGN (see Section 3). The ISM gas properties are obtained from the 4 successive annuli (0.2 - 1; 1-4; 4-8; 8-12 kpc). The spectra were modelled using the 'vmekal' model in XSPEC (with the abundances of O, Mg, Si, S and Fe included as independent free fit parameters) with an additional power-law component for the central region, and following the spectral deprojection procedure discussed by Allen et al. (2001). The resulting ISM gas temperature profile is shown in Figure 1. The solid points show the deprojected temperature profile. The temperature is higher in the outer regions, where the cluster potential dominates and decreases in the inner regions where the galaxy potential takes over. In the innermost annulus the temperature $kT = 0.77 \pm 0.04$ keV.

The density profile, shown in Figure 1, was derived by deprojecting the surface brightness profile (e.g. Fabian et al. 1981) and matching the observed deprojected tem-

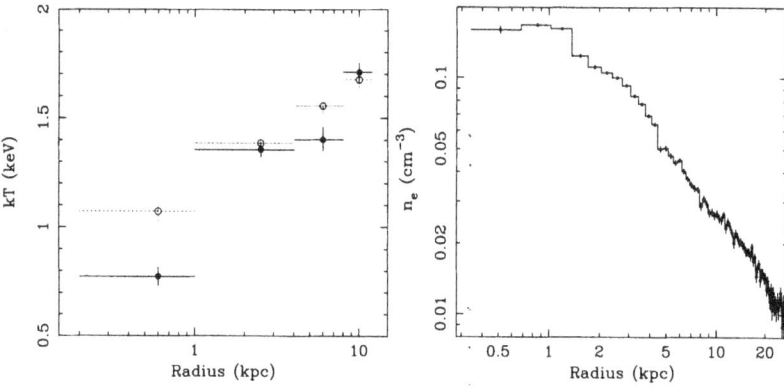

Fig. 1. Left: The temperature profile of the ISM in M87 measured by *Chandra*. The solid points show the deprojected temperature measurements and the dotted ones the observed (projected) values. Right: The density profile of the ISM in M87 obtained from the observed surface brightness deprojected according to the temperature profile.

perature profile. The density profile (Fig. 1) is steeper at radii larger than a few kpc and follows the standard profile of $\rho \propto r^{-1}$. Inside these radii the density profile flattens within a kpc or so of the nucleus. Fitting a constant value to the central two annuli we get $n = 0.167 \pm 0.003\,\mathrm{cm}^{-3}$ (or $n = 0.160 \pm 0.006\,\mathrm{cm}^{-3}$ for the inner most annulus).

2.2 $\dot{M}$

We estimate the accretion rate from the ISM gas onto the central black hole using Bondi accretion theory. Matter passing within the accretion radius given by:

$$R_A \sim \frac{GM}{c_s^2} \sim (c/c_s)^2 R_S \sim 0.2 T_{0.77}^{-1} M_9 \quad \mathrm{kpc} \sim 10^6 R_S, \tag{1}$$

is assumed to be accreted onto the black hole. Here $M_9 = 10^9 M_\odot$ is the mass of the black hole ($M_9 = 3$), c is the speed of light, $c_s \sim 10^4 T^{1/2}$ cm s^{-1} the sound speed (where T is the ISM gas temperature), and $R_S = 2GM/c^2$ the Schwarzschild radius of the black hole. Equation 1 shows that the accretion radius of M87 which corresponds to ~ 2 arcsec, lies within *Chandra* resolution. This implies that for M87 we are able to accurately measure the properties of the gas within the black hole accretion radius.

The accretion rate is related to the density and temperature at the accretion radius by the continuity equation, where, at R_A the velocity $v = c_s$;

$$\dot{M} = 4\pi R_A^2 \rho_A c_s(R_A); , \tag{2}$$

where ρ_A and $c_s(R_A)$ are the density and sound speed at the accretion radius. With density and temperature measured directly at the accretion radius (Section 2.1) the Bondi accretion rate to M87 is given by

$$\dot{M}_{\text{Bondi}} = 7.2 \times 10^{24} \, M_9^2 \, T_{0.77}^{-3/2} \, n_{0.17} \quad \text{gs}^{-1}$$
$$\sim 0.16 \quad \text{M}_\odot \text{yr}^{-1}. \tag{3}$$

Eqn. (3) provides the accretion rate at the outer edge of the accretion flow. As discussed in the next session, the mass accretion rate onto the black hole may be smaller if e.g. it decreases with radius because of an outflow.

The Bondi accretion rate (Eq. 3) implies a luminosity

$$L_{\text{Bondi}} = \eta \dot{M}_{\text{Bondi}} c^2 \sim 6 \times 10^{44} \text{erg s}^{-1} \tag{4}$$

if $\eta = 0.1$, as in a standard radiatively efficient thin disk.

3 L_x

The spectrum of the central AGN was obtained taking a 2 arcsec (4 pixels) radius circle and using only the data from the shorter non-piled-up, windowed observations. With the high resolution and sensitivity of *Chandra* we measure the flux density of the core separate from the bright jet knots. The central point source is best fit by a power law. The power-law model gives a photon index $\Gamma = 2.33^{+0.03}_{-0.04}$ assuming Galactic absorption, and a flux density at 1 keV of $(7.8 \pm 0.2) \times 10^{-13}$ erg cm^{-2} s^{-1} keV^{-1}, where all parameters are quoted at a 1σ confidence. Note that we have assumed that the intrinsic ISM gas properties in the central 2 arcsec region are the same as in the next annulus out as obtained from the deprojection analysis. This does not affect the results as the power law flux totally dominates the emission in the 0-2 arcsec region. Our results on the central core are consistent with those of of Marshall et al. (2001).

The observed luminosity is over four orders of magnitude smaller than the predicted Bondi luminosity, implying (unless the Bondi estimate is inappropriate) that the radiative efficiency has to be $\eta \sim 10^{-5}$. In the next section we examine the prediction from hot accretion flow models with low radiative efficiencies.

4 Accretion Models

Hot accretion flow around a supermassive black hole radiate in the radio to X–ray band. In the radio band the emission results from synchrotron radiation. At higher energies and up to the X-ray band the emission is produced to bremsstrahlung processes and inverse Compton scattering of the soft synchrotron photons (e.g. Narayan, Barret & McClintock 1998)

We measure radii in the flow in Schwarzschild units: $R = rR_S$. We measure black hole masses in solar units and accretion rates in Eddington units: $M_{\text{BH}} = mM_\odot$ and $\dot{M} = \dot{m}\dot{M}_{\text{Edd}}$. We take $\dot{M}_{\text{Edd}} = 10L_{\text{Edd}}/c^2 = 2.2 \times 10^{-8} mM_\odot$ yr^{-1}, i.e., with a canonical 10% efficiency. We take $r = 10^4$ to be the outer radius of the flow. The Bondi accretion rate for M87 in these units is $\dot{m}_{\text{Bondi}} = 1.7 \times 10^{-3}$.

The predicted spectrum from an ADAF depends (weakly) on the ratio of gas to magnetic pressure β, the viscosity parameter α, and the fraction of the turbulent energy in the plasma which heats the electrons, δ. Here, we fix $\alpha = 0.1$, $\beta = 10$, and take $\delta = 0.3$ or 0.01. The two major parameters, though, are the accretion rate $\dot{M}$ and the black hole mass M, both of which are constrained. With M given for M87, we normalise the models to the observed *Chandra* flux. This gives us the $\dot{m}$ required by the models to explain the X-ray emission. The model is is ruled out if it requires $\dot{m} \ll \dot{m}_{\mathrm{Bondi}}$ to account for the observed luminosity.

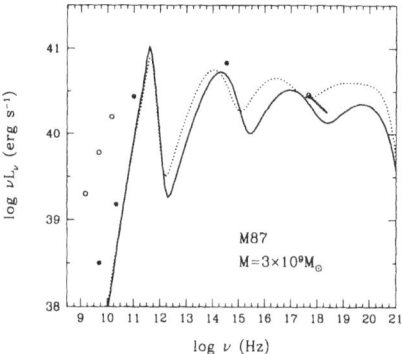

Fig. 2. Spectral models calculated for hot accretion flows fitted to the *Chandra* X-ray flux. The solid line is an ADAF model with $\delta = 0.3$. The required accretion rate, $\dot{m} = 7 \times 10^{-4}$, is consistent with the Bondi value. The dotted line shows a model with $\delta = 0.1$ with $\dot{m} = 10^{-3}$. The filled dots are VLBI nuclear flux density measurements by Reid et al. (1989); Paulini-Toth et al. (1981); Spencer & Junor (1986) and Bääth et al. (1992) in increasing frequency order respectively. The optical nuclear continuum measurements is from HST observations by Harms et al. (1994). The radio and optical data are tabulated by Reynolds et al. 1996. The thick solid line shown the the Chandra flux (in the $0.2 - 10$ keV band).

The solid line in Figure 2 shows the predicted spectrum for a pure inflow ADAF model fitted to the 1 keV flux. In this model $\dot{m} = 6 \times 10^{-4}$, (with $\delta = 0.3$) consistent with the Bondi estimate (within the model uncertainties which should be taken to be $\sim 50\%$). A model with $\delta = 0.01$ and corresponding $\dot{m} = 10^{-3}$ is shown by the dotted line. Because of the relatively large accretion rate, Comptonization of the synchrotron emission dominates the X-ray emission in these models (see e.g., Harayan et al. 1998). The exact positions of the Comptonization bumps present in the optical and X-ray bands are a function of temperature, hence δ. Models with higher values of electron heating are better consistent with the measured 2–10 keV spectral slope of $\Gamma \sim 2$. Such values are also preferred from a theoretical point of view.

Although the jet is inevitably contributing to the observed emission (see e.g. Marshall et al. 2001; Di Matteo et al. 1999) the accretion flow is potentially able to explain a significant fraction of the observed spectral energy distribution. The solid dots in Figure 2 are the high resolution VLBI, HST and Chandra high resolution measurements of the nuclear flux in the radio, optical and X-ray band respectively. The open dots show the lower resolution VLA radio measurements which are likely to include a more significant contribution from the jet.

5 Discussion

Chandra observations of the nucleus of M87 resolve the state of the ISM gas all the way into the accretion radius of its central black hole. This makes M87 the only black hole system in which both gas temperature and density are measured reliably down to the accretion region and for which a Bondi accretion rate can be directly calculated. M87 also possesses a relatively bright nuclear source the flux of which can be accurately measured in both the X-ray band with *Chandra* and throughout the whole of the spectral energy distribution. At present, it is therefore the best constrained low-luminosity black hole system. With both L_x and $\dot{M}$ given, we were able to show unambiguously that the black hole in M87 is highly underluminous with respect to its mass supply: the expected power output exceeds by over 4 orders of magnitude the observed value. If the black hole in M87 is indeed accreting close to its Bondi rate, the accretion efficiency, η, must be low. We have shown that the required values of η are consistent with predictions from ADAF models.

Note that this result is independent of whether or not the observed nuclear luminosity is attributed to an ADAF or entirely dominated by jet emission. The requirement for low-radiative efficiency of the accreting gas can only be invalidated if the actual mass accreted by the black hole were to be a very small fraction of the mass supplied in the outer region of the flow (i.e. of $\dot{M}_{\mathrm{Bondi}}$). This could be the case if strong outflows or convection were important in the accretion flow such as it has been discussed by e.g.; Blandford & Begelman (1999), Stone et al. (1999) and Quataert & Gruzinov (2000); Narayan et al. (2000) respectively. In such models, the accretion flow would produce a much lower luminosity than a pure inflow ADAF and make a negligible contribution to the observed luminosity (which, in the case of M87, would be totally dominated by the jet). At present, for M87 there are no observational constraints which point to a decrease in mass supply with radius. The Bondi rate could also be decreased if the jet heated the ISM gas inside the accretion radius (i.e. in the region that with *Chandra* resolution is dominated by the nuclear point source). Again this hypothesis also cannot be tested.

Note that using radio observations Di Matteo et al. 1999; 2000 had favoured outflow models for the nuclei of ellipticals. With *Chandra* the properties of the central regions of the ISM gas have been determined more accurately for a number of elliptical galaxies (e.g., Loewenstein et al. 2001). In particular, the new observations have shown that the ISM density profiles tend to flatten off significantly in the central regions of ellipticals, implying Bondi accretion rates much smaller than previously estimated. In most cases (Loewenstein et al. 2001) *Chandra* also does not detect nuclear point sources. Given that black hole mass determinations are a lot less certain the constraints on other low luminosity systems remain fairly inconclusive. The only two *Chandra* detections of X-ray nuclear sources in low luminosity elliptical nuclei are in M87 and NGC 6166 (Di Matteo et al. 2001). For both of these nuclei the pure inflow ADAF models are consistent with the observational constraints. Further constraints on the ADAF model in these cases requires knowledge of the contribution, at all wavelengths, of emission from the base of the jet.

References

1. Allen S.W., Ettori S., Fabian A.C., 2001, MNRAS, 324, 877
2. Blandford R.D., Begelman M.C., 1999, MNRAS, 303, L1
3. Di Matteo T., Fabian A.C, Rees M.J., Carilli C., Ivison R J., 1999, MNRAS, 305, 49
4. Di Matteo, T., Quataert, E., Allen, S. W., Narayan, R., Fabian, A.C., 2000, MNRAS, 311, 507
5. Di Matteo, T., Johnstone, R.M., Allen S.W., Fabian A.C., 2001, ApJ, 550, L19
6. Ferrarese, L., Merritt, D., 2000, ApJ, 539, L9
7. Ford H. C. et al. 1995, ApJ, 1994, 435, L27
8. Harms R. J. et al., 1994, ApJ, 435, L35
9. Loewenstein, M., Mushotzky, R. F., Angelini, L., Arnaud, K. A., & Quataert, E. 2001, ApJ, 555, L21
10. Magorrian, J. et al. 1998, AJ, 115, 2285
11. Narayan R., Igumenshchev, I. V. Abramowicz, M. A., 2000, ApJ, 539, 798
12. Pauliny-Toth I. I. K., Preuss E., Witzel A., Graham D., Kellerman K. I., Ronnang B., 1981, AJ, 86, 371
13. Quataert E., Gruzinov I., 2000, ApJ, 539, 809
14. Reid M. J., Biretta J. A., Junor W., Muxlow T. W. B., Spencer R. E., 1989, ApJ, 336, 112
15. Stone, J. M., Pringle, J. E. Begelman, M. C. 1999, MNRAS, 310, 1002
16. Young A.J., Wilson A.S., Mundell C.G., ApJ, submitted
17. Wilson, A. S. & Yang, Y. 2001, ApJ, submitted.

Intergalactic Magnetic Fields from Quasar Outflows

Steven R. Furlanetto and Abraham Loeb

Department of Astronomy, Harvard University, Cambridge MA 02138, USA

Abstract. Outflows from quasars inevitably pollute the intergalactic medium (IGM) with magnetic fields. The short-lived activity of a quasar leaves behind an expanding magnetized bubble in the IGM. We model the expansion of the remnant quasar bubbles and calculate their distribution as a function magnetic field strength at different redshifts. We find that by a redshift $z \sim 3$, about 5–80% of the IGM volume is filled by magnetic fields with an energy density $> 10\%$ of the mean thermal energy density of a photo-ionized IGM (at $\sim 10^4$ K). As massive galaxies and X-ray clusters condense out of the magnetized IGM, the adiabatic compression of the magnetic field could result in the fields observed in these systems without a need for further dynamo amplification.

1 Introduction

Clusters of galaxies contain substantial magnetic fields with strengths $B \sim 0.1 - 10~\mu$G and coherence lengths $\ell \sim 10$ kpc [1]. The origin of such fields could have important implications for structure formation. Assuming flux conservation, a cluster field $B_{\rm cl} \sim 10^{-7}$ G would imply $B_{\rm IGM} \sim 10^{-9}$ G in the diffuse intergalactic medium (IGM), which would constitute $\sim 5\%$ of the thermal energy density of a photoionized IGM. The observational constraints on an intergalactic magnetic field (IGMF) are weak, requiring only that $B_{\rm IGM} < 10^{-8}(\ell/~{\rm Mpc})^{-1/2}$ G in the currently popular ΛCDM model [1].

Unfortunately, there is no convincing model for the formation of cluster fields or the IGMF. Because the rotation times of clusters exceed the Hubble time, dynamos are expected to be ineffective. Primordial generation scenarios cannot explain the large coherence lengths [2]. Models that generate magnetic fields during structure formation [3] or in starbursts [4] suffer from similar problems.

We have examined the possibility that the IGMF was originally produced near supermassive black holes and expelled into the IGM through mechanical outflows [5] from radio-loud quasars (RLQs) and broad absorption line quasars (BALQs). Supermassive black holes are one of the few classes of astrophysical objects with energy reservoirs large enough to account for the large-scale fields in clusters, and the relatively small number of powerful sources can accomodate the large observed coherence lengths, as emphasized by [6]. For more details on the model, see [7].

2 Filling Factor of Magnetized Regions & Mean Fields

While a quasar is active, its outflow is in the form of twin collimated jets (for a RLQ) or equatorial winds (for a BALQ). After the quasar becomes dormant, the outflow remnant is overpressured with respect to the IGM and continues to expand in comoving coordinates until its outward velocity matches the Hubble flow velocity. We assume that the outflow remnant isotropizes and expands adiabatically as a spherical shell during this late phase.

Simple energy conservation provides a surprisingly accurate estimate of the final comoving bubble size $\hat{R}_{max}$. Balancing the energy input of the quasar and the final kinetic energy of the shell, we find that $\hat{R}_{max} \propto [L_q \tau_q \varepsilon_K (1 + \varepsilon_B)]^{1/5}$, where L_q is the luminosity of the quasar, τ_q is its lifetime, ε_K is the ratio of the mechanical and radiative luminosities, and ε_B is the ratio of the magnetic and mechanical energy outputs. Note the weak dependence on the quasar parameters, making our results robust to uncertainties in their measurement. In particular, we find that magnetic fields do not play an important role in the expansion for realistic values of ε_B. We therefore ignore the geometric and magnetohydrodynamic effects of the magnetic field and assume simple flux conservation.

We next calculate the number of quasar sources. For $z < 4$, we use the observed optical luminosity function of quasars [8] together with an assumed incidence rate of outflows f. For $z > 4$, we assume that the incidence rate of quasars is proportional to the Press-Schechter mass function, with the proportionality constant set by matching to the observed luminosity function at $z \sim 4$ [9].

We examine two quasar models: the RLQ model, with $\varepsilon_B = 0.1$ and $f = 0.1$ [10], and the BAL model, with $\varepsilon_B = 0.01$ and $f = 1$ [11]. Note that there are no existing observations of magnetic fields in BALQs, though the fields may play an important role in the wind mechanism [12].

We then calculate the total filling factor of the quasar bubbles as a function of redshift, $F(z)$, and the global volume-averaged magnetic energy density, $\bar{u}_B(z)$, for our two models by numerically integrating the equation of motion of each remnant and summing over all quasar sources. Our results are shown in Fig. 1, along with the distribution of magnetic field strengths for a series of redshifts.

3   Results

We predict a cellular IGMF filling a substantial fraction of space by $z \sim 3$ (Fig. 1), with each cell a fossil bubble produced by a single magnetized quasar outflow. Cells generated by RLQ quasars are more highly magnetized than those from BAL quasars but fill a smaller volume. We predict that $B_{IGM} \sim 10 f \varepsilon_B$ nG when averaged over all of space; for each of our models, simple adiabatic compression of such an IGMF can account for the observed cluster magnetic fields.

Direct detection of the cells via Faraday rotation measurements will be difficult, but electron acceleration by shocks in the cells will cause synchrotron emission. The same accelerated electrons produce γ-rays through inverse-Compton scattering of the cosmic microwave background. Correlated maps of the γ-ray and radio skies may allow us to calibrate the magnetic field in the shocks [13].

Finally, we note that the non-thermal pressure of the magnetic fields may help to resolve the discrepancy between simulated and observed line widths in the Lyα forest [14].

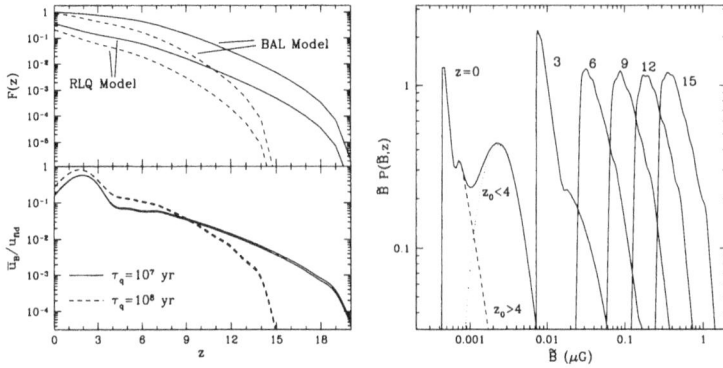

Fig. 1. *Top Left:* Volume filling fraction of magnetized bubbles $F(z)$. *Bottom Left:* Ratio of magnetic energy density, $\bar{u}_B$, to the fiducial thermal energy density $u_{fid} = 3n(z)kT_{IGM}$, where $T_{IGM} = 10^4$ K. In each of these panels, results are shown for both the RLQ and BALQ models. *Right:* Probability distributions of bubble magnetic field, $P(\tilde{B}, z)$, for the RLQ model, at various redshifts. Here $\tilde{B} = B/(\sqrt{\epsilon_B/0.1})$; in these units the curves are independent of ϵ_B. The dotted curve shows the contribution to the $z = 0$ distribution function from quasars forming at $z_0 < 4$ and the dashed curve shows the contribution from quasars at $z_0 > 4$. All curves are normalized to have unit area.

References

1. P. P. Kronberg: Rep. Prog. Phys. **57**, 325 (1994)
2. J. Quashnock et al.: Ap. J. **344**, L49 (1989)
3. R. M. Kulsrud et al.: Ap. J. **480**, 481 (1997)
4. P. P. Kronberg, H. Lesch, & U. Hopp: Ap. J. **511**, 56 (1999)
5. M. J. Rees & G. Setti: Nature **219**, 127 (1968)
6. S. Colgate & H. Li: 'The Magnetic Fields of the Universe and Their Origin'. In *Highly Energetic Physical Processes and Mechanisms for Emission from Astrophysical Plasmas*, ed. by P. C. H. Martens, S. Tsuruta, & M. Weber (ASP, San Francisco 2000), pp. 255–264
7. S. R. Furlanetto & A. Loeb: Ap. J. **556**, 619 (2001)
8. Y. Pei: Ap. J. **438**, 623 (1995)
9. Z. Haiman & A. Loeb: Ap. J. **503**, 505 (1998)
10. D. Stern et al.: A.J. **119**, 1526 (2000)
11. Weymann, R. J.: 'BAL QSOs: Properties and Problems – An Optical Spectroscopist's Perspective'. In *Mass Ejection from AGN*, ed. by N. Arav, I. Shlosman, & R.J. Weymann (ASP, San Francisco 1997), pp. 3–12
12. M. de Kool & M. C. Begelman: Ap. J. **455**, 448 (1995)
13. E. Waxman & A. Loeb: Ap. J. **545**, L11 (2000)
14. G.L. Bryan et al.: Ap. J. **517**, 13 (1999)

Simple Scaling and the Non-Linear Mass-Luminosity Relation in Radio Sources

Sebastian Heinz[1] and Rashid A. Sunyaev[1,2]

[1] Max-Planck-Institut für Astrophysik
[2] Space Research Institute (IKI), Profsouznaya 84/32, 117810 Moscow, Russia

Abstract. Since the fundamental physical parameters in the inner disk and the jet should assume a simple linear (or inverse) scaling with mass, one might assume that the same holds for the observable quantities as well. However, the radio luminosity of radio sources shows a strongly non-linear dependence on the mass of the central black hole, roughly scaling like $L_{5\,\mathrm{GHz}} \sim M^{1.4-1.9}$, observed to hold over more than 8 orders of magnitude in M. Starting from first principles, we show that such a non-linear dependence arises naturally in the inner jet region, even in the simplest possible models for a wide array of input parameters. The non-linearity is mainly a consequence of the non-linear mass-dependence of the location of the jet photosphere.

1 Simple Scaling Model

Most fundamental parameters of relativistic jets close the black hole that launches them should scale in a very simple fashion with black hole mass M, most notably their kinetic luminosity, which should be linearly proportional to M. One might therefore naively think that the observable characteristics of jets, namely its radio luminosity, should also follow a simple linear scaling. However, it is observatoinally well established that radio loudness in relativistic jets is a strongly non-linear function of black hole mass M.

In the case of selected classes of radio galaxies, this relation has been shown to take on the form $F_{5\,\mathrm{GHz}} \sim M^{1.8}$ (Lacy et al. 2001), but the most impressive display of this non-linearity can be made comparing the relativistic jets in Galactic microquasars (where $L_{5\,\mathrm{GHz}}/L_{\mathrm{bol}} < 10^{-6}$) with their AGN counterparts (where $L_{5\,\mathrm{GHz}}/L_{\mathrm{bol}} < 10^{-2}$), spanning over 8 orders of magnitude in M. Below (see Heinz & Sunyaev 2001 for a detailed discussion), we shall present arguments that such a non-linear scaling has to be expected even in the simplest possible models (see Falcke & Biermann 1999 and references therein for a series of pioneering paper in this field).

From the early days of accretion disk theory (Shakura & Sunyaev 1973) it was well known that the inner, radiation dominated part of an accretion disk is decoupled from the outer regions of the disk. Conditions in the inner disk should thus depend only on a few fundamental parameters: black hole mass M and spin a, and accretion rate $\dot{M}$. The fundamental parameters in the inner accretion disk take on a very simple scaling with the mass of the central black hole: time and size scales vary linearly with mass, $\tau \propto r \propto M$, while density and pressure (particle and magnetic) will scale inversely with mass, $n \propto p \propto B^2/8\pi \propto M^{-1}$. Since jets are produced in the inner accretion disk, the fundamental physical parameters in the inner jet should follow the same simple scaling.

In the following we will concentrate on the M dependence only, neglecting $\dot{m}$ (which is easy to include in the scaling but will be neglected for brevity) and a. Then, jet parameters at the injection point (where the jet is created, denoted by subscript 0) will scale as follows: Jet cross section R_0 and injection distance from the hole $r_0 \sim 10\, r_g$ follow the mass linearly, $R_0 \propto r_0 \propto r_g \propto M$, density and pressure (both particle and magnetic) at injection scale like $n_0 \propto p_0 \propto B_0^2 \propto M^{-1}$, and kinetic jet power follows $L_{kin} \propto M$. As a consequence, the terminal Lorentz factor Γ_{jet} of the jet should be roughly independent of mass, typically $\Gamma_{jet} \sim 5$.

Guided by the remarkable spatial self-similarity of observed AGN jets (see, e.g., Bridle & Perley 1984), we will assume that all jets simply follow a self-similar behavior with distance from the central engine (while not a good assumption for the outer jet, dominated by interactions with the environment, this should be reasonable for the inner jet, which emits most of the radiation). We write jet radius R, particle density n, and mean magnetic field B (which is likely not completely ordered) as

$$
\begin{aligned}
R &= R_0 (r/r_0)^\beta, & 0 < \beta < 1 \\
n &= n_0 (r/r_0)^{-\lambda}, & 2\beta \le \lambda \le 3\beta \\
B &= B_0 (r/r_0)^{-\delta}, & 1/2\lambda \le \delta \le \lambda
\end{aligned}
\tag{1}
$$

with $\beta = 1$ for a ballistic jet, i.e., with constant opening angle. For a continuous jet, we have $\lambda \sim 2\beta$, while a jet made of discrete ejections would satisfy $\lambda \sim 3\beta$. For a completely tangled field, we expect $\delta = 2/3\lambda$.

2   Jet Emission

Radio emission from jets is produced by synchrotron radiation. Based on the above self-similar picture, we can deduce the radiative properties at any point in the jet assuming only that the electrons follow a powerlaw distribution with

$$
n(\gamma) \propto \gamma^{-s}, \quad p \sim 2
\tag{2}
$$

and adiabatic evolution of this electron spectrum (i.e., at frequencies low enough to be unaffected by radiative cooling; note that we do not hold the lower cutoff γ_{min} of the spectrum fixed, which changes the behavior of the equations under expansion/contraction).

Since the inner jet is optically thick to synchrotron self-absorption while the outer jet is optically thin, a photosphere must exist, located at a distance r_ν away from the core, which, based on the above relations, follows the scaling

$$
\frac{r_\nu}{r_0} \propto \left(M^{\frac{p+2}{4}} \nu^{\frac{p+4}{2}} \right)^{-6/[(2\lambda+3\delta)(p+2)-6\beta]} \sim
\begin{cases}
M^{-0.23} \nu^{-0.69} & \text{(cont. jet)} \\
M^{-0.14} \nu^{-1.21} & \text{(discr. eject.)}
\end{cases}
\tag{3}
$$

For reasonable parameters, one can show that the bulk of the radiation emitted by the jet is coming from this region (this is true as long as $1 + 2\beta - \delta(1+p) - \lambda(p+2)/3 < 0$, which is satisfied for reasonable paramter choices). The total flux from the jet then follows the scaling

$$
F_\nu \propto
\begin{cases}
M^{\frac{9}{4}} \nu^{\frac{5}{2}} \left(\frac{r_\nu}{r_g}\right)^{1+\beta+\frac{\delta}{2}} & \text{continuous jet} \\
M^{\frac{9}{4}} \nu^{\frac{5}{2}} \left(\frac{r_\nu}{r_g}\right)^{2\beta+\frac{\delta}{2}} & \text{discrete ejections}
\end{cases}
\sim
\begin{cases}
M^{1.64} \nu^{0.65} \\
M^{1.82} \nu^{1.21}
\end{cases}
\tag{4}
$$

Note that since jet emission is not actually stationary, i.e., the emission is dominated by knots travelling down the jet. This is natural in the context of a jet made up of discrete ejections. In the case of a continuous jet this would correspond to the internal shock model (e.g., Kaiser et al. 2000), and the above scaling still holds. In both cases, the integrated flux in eq. (4) corresponds to the temporal *peak* flux. In other words: The maximum observable flux at a given frequency is following the above scaling.

One can see from eq. (4) that the peak jet flux should follow a non-linear scaling with black hole mass, and for a reasonable range in parameters, this non-linear relation falls into the range $F_{\nu,\text{peak}} \propto M^\zeta \sim M^{1.4}$ to $M^{1.8}$.

Certainly, a large number of complications will broaden such a relation. For example, the effects of Doppler boosting will increase the scatter in jet fluxes by several orders of magnitude. Also, it is impossible to observe the true peak fluxes of AGN jets, since the involved variability time scales are far too long (and it is possible that we are also missing the peak flux for microquasar jets).

However, since it is reasonable to assume that Γ_{jet} is mass independent, we can circumvent these complications by only considering the upper envelope of the mass-radio luminosity relation of as large a sample of radio jets as possible. The slope of this upper envelope should then reveal the exponent ζ. Alternatively, once can consider a very carefully selected sample of sources, as was done by Lacy et al. 2001, which already implies that viewing angle and other secondary parameters of the sources are very similar. As mentioned above, the observed values of $\zeta \sim 1.5$ (over the entire mass range from microquasars to AGN jets) to $\zeta \sim 1.9$ (Lacy et al. 2001) are entirely consistent with this picture.

As more complicated emission models will most likely not lead to a simpler mass dependence, we arrive at the consequence that the observable characteristics of radio jets must take on a non-linear scaling with black hole mass.

References

1. Bridle, A. H. & Perley, R. A. 1984, ARAA, 22, 319
2. Falcke, H. & Biermann, P. L. 1999, A&A, 342, 49
3. Heinz, S. & Sunyaev, R. A. 2001, A&A, in preparation
4. Kaiser, C. R., Sunyaev, R., & Spruit, H. C. 2000, A&A, 356, 975
5. Lacy, M., Laurent–Muehleisen, S. A., Ridgway, S. E., Becker, R. H., & White, R. L. 2001, astro-ph/0103087, L01
6. Shakura, N. I. & Sunyaev, R. A. 1973, A&A, 24, 337

Simulation of Jet Formation from Magnetized Accretion Disk Around Kerr Black Hole

Shinji Koide[1], Kazunari Shibata[2], Takahiro Kudoh[3], and David L. Meier[4]

[1] Department of Engineering, Toyama University, 3190 Gofuku, Toyama 930-8555, Japan
[2] Kwasan and Hida Observatory, Kyoto University, Yamashina, Kyoto 607-8471, Japan
[3] National Astronomical Observatory, Mitaka, Tokyo 181-8588, Japan
[4] Jet Propulsion Laboratory, 4800 Oak Grove Drive, Pasadena, CA 91109, USA

Abstract. We present a general relativistic magnetohydrodynamic simulation of jet formation from accretion disk around Kerr black hole.

1 Introduction

Relativistic jets now have been discovered in several different classes of objects, including active galactic nuclei [1,2], micro-quasars [3], and gamma-ray bursts [4]. It is believed that a rapidly spinning black hole exists at the center of each of these objects and that the violent phenomena that occur near the hole is responsible for the jets. Dynamics of magnetized plasma around Kerr black hole is one of the most promising candidates of the process in the violent phenomena. In order to understand the magnetic mechanism of jet formation around a black hole, we have performed general relativistic magnetohydrodynamic (GRMHD) simulations in Kerr space-time [5–7].

2 Numerical Result

We show a GRMHD simulation of jet formation from magnetized accretion disk around a rapidly rotating ($a = 0.95$) Kerr black hole (Koide et al. 2000). The simulations were performed for two cases in which the disk co-rotates and counter-rotates with respect to the black hole rotation. Figures 1a-c illustrate the time evolution of the counter-rotating disk case and Fig. 1d the final state of the co-rotating case. These figures show the proper mass density (gray-scale), velocity (vectors), and magnetic field (solid lines) in $0 \le R \equiv r\sin\theta \le 7r_\mathrm{S}$, $0 \le z \equiv r\cos\theta \le 7r_\mathrm{S}$. The black region at the origin shows the inside of the black hole horizon, whose radius is $r_\mathrm{H} = 0.656r_\mathrm{S}$. The initial state in the simulation consists of a hot corona and a cold accretion disk around the black hole (Fig. 1a). In the corona, plasma is assumed to be in nearly stationary infall, with the specific enthalpy $h/\rho c^2 = 1 + \Gamma p/[(\Gamma - 1)\rho c^2] = 1.3$, where specific heat ratio $\Gamma = 5/3$. The initial velocity of the accretion disk is assumed to be the Kepler velocity. Except for the disk rotation direction, we use the same initial conditions in both cases. The mass density of the disk is 100 times that of the corona at the inner edge of the disk. The mass density profile is given by that of a hydrostatic equilibrium corona with a scale height of $r_\mathrm{c} \sim 3r_\mathrm{S}$. The disk is in pressure balance with the corona, and the magnetic field lines are perpendicular to the accretion disk. We use the azimuthal component of the vector potential A_ϕ of the Wald solution [8] to set the magnetic field. Here the magnetic field strength is $B_0 = 0.3\sqrt{\rho_0 c^2}$, where ρ_0 is the initial corona density at $r = 3r_S$. The Alfvén velocity and plasma beta value at the disk ($r = 3.5r_S$) are $v_A = 0.03c$ and $\beta \equiv 2p/\hat{B}^2 \sim 3.4$, respectively.

Figure 1b shows the state at $t = 30\tau_S$, where τ_S is defined as $\tau_S \equiv r_S/c$. In the counter-rotating disk case, there is no stable circular orbit at $R \leq 4.4r_S$, the disk falls into the black hole rapidly dragging the magnetic field lines. The disk enters the ergosphere and then crosses the horizon, as shown by the crowded magnetic field lines near $r = 0.75r_S$ (Fig. 1c). The jet is ejected almost along the magnetic field lines. Its maximum total and poloidal velocities are the same, $\hat{v} = \hat{v}_p = 0.44c$ at $R = 3.2r_S$, $z = 1.6r_S$. Inside the ergosphere, the velocity of frame dragging exceeds the local light speed, causing the disk to rotate in the same direction of the black hole rotation, even though it was initially counter-rotating. The rapid, differential frame dragging enhances the azimuthal magnetic field [9]. This enhanced magnetic field pressure blows off the plasma upward and pinches it into a powerful collimated jet.

Figure 1d shows a snapshot of the co-rotating disk case at $t = 47\tau_S$. The disk stops its infall near $R = 3r_S$ due to the centrifugal barrier with a shock at $r = 3.4r_S$. The high pressure behind the shock causes a gas pressure-driven jet with total and poloidal velocities of $\hat{v} = \hat{v}_p = 0.30c$ at $R = 3.4r_S$, $z = 2.4r_S$. The centrifugal barrier makes the disk take much long time to reach the ergosphere, which causes the difference between the co-rotating and counter-rotating disk cases.

References

1. J. J. Pearson et. al Nature **290**, 365 (1981)
2. J. A. Biretta, W. B. Sparks, & F. Macchetto: ApJ **520**, 621 (1999)
3. I. F. Mirabel & L. F. Rodriguez Nature, **374**, 141 (1994)
4. S. R. Kulkarni Nature, **398**, 389 (1999)
5. S. Koide, K. Shibata, & T. Kudoh ApJ, **495**, L63 (1998)
6. S. Koide, K. Shibata, & T. Kudoh ApJ, **522**, 727 (1999)
7. S. Koide, D. L. Meier, K. Shibata, & T. Kudoh ApJ, **536**, 668 (2000)
8. R. M. Wald Phys. Rev. D, **10**, 1680 (1974)
9. D. Meier ApJ, **522**, 753 (1999)

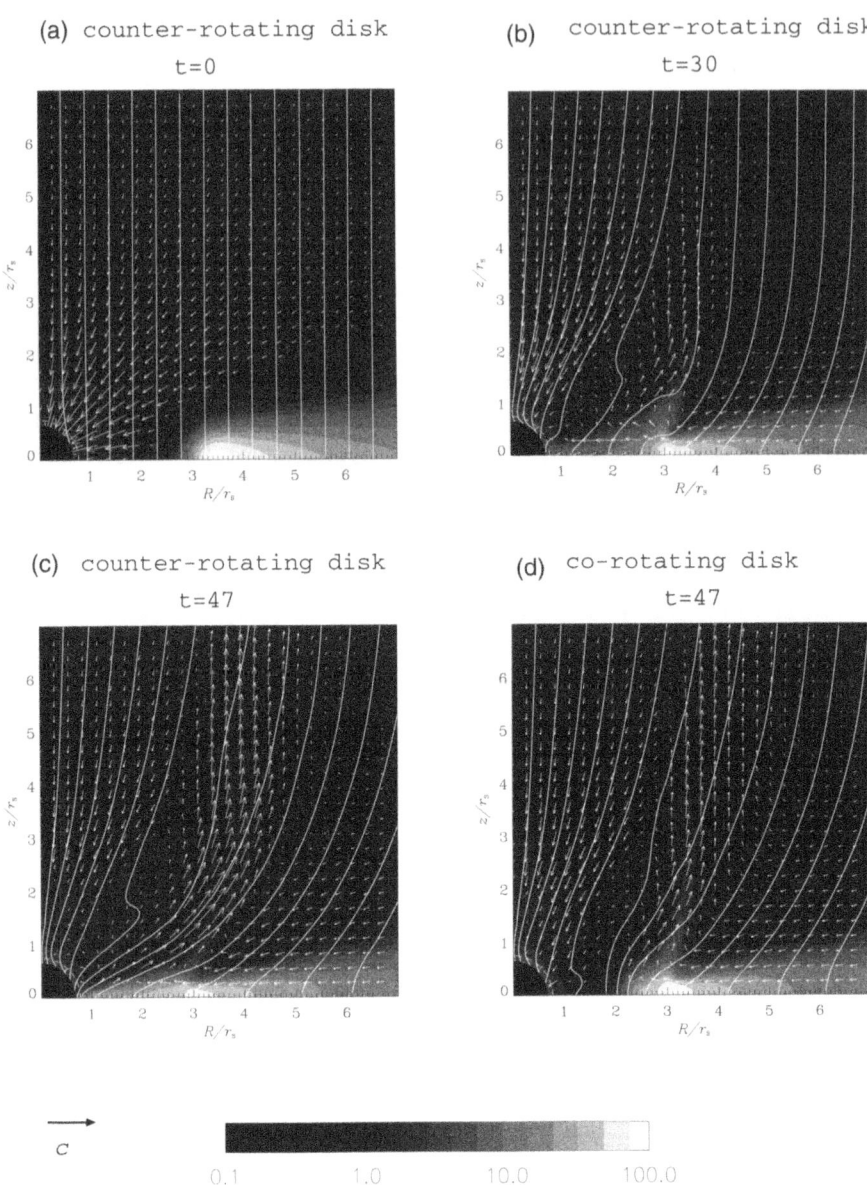

Fig. 1. Time evolution of jet formation in the counter-rotating disk case and the final state of the co-rotating disk case. Gray-scale shows the logarithm of the proper mass density; vectors indicate velocity; solid lines show the poloidal magnetic field. The black fan-shaped region at the origin shows the horizon of the Kerr black hole ($a = 0.95$). The dashed line near the horizon is the inner boundary of the calculation region. At $t = 47\tau_S$, while the infall of the disk in the co-rotating disk stops (due to a centrifugal barrier), the unstable orbits of the counter-rotating disk plasma continue to spiral rapidly toward the black hole horizon. This difference causes the magnetohydrodynamic jet formation mechanisms in the two cases to differ drastically, resulting in a powerful jet emanating from deep within the ergosphere.

Coexistence of Corona and Disk in AGN?

Bifang Liu[1], Shin Mineshige[1], Emmi Meyer-Hofmeister[2], and Friedrich Meyer[2]

[1] Yukawa Institute for Theoretical Physics, Kyoto University, Kyoto 606-8502, Japan
[2] Max-Planck-Institut für Astrophysik, Karl-Schwarzschild-Str.1, D-85741 Garching, Germany

Abstract. We studied the disk corona around a supermassive black hole by including decoupling of ions and electrons and Compton scattering. We found that the mass evaporation rate from the disk to the corona depends strongly on viscosity and total mass accretion rate, smaller viscosity and larger accretion rate result in less efficient evaporation and hence weaker corona. The model predicts weaker contributions to the hard X-rays and a steeper X-ray spectrum in narrow-line Seyfert 1 galaxies (NLS1), consistent with observations. To maintain a corona lying above a thin disk at the innermost region in active galactic nuclei (AGN), a large amount of energy deposit/supply to the corona is required.

1 Introduction

There are strong observational evidences for the existence of hot gas with $T \sim 10^9$ K in the neighborhood of an inner accretion disk in AGNs. It is postulated that the hot gas is a corona lying above the thin disk. A self-consistent model, including how the mass and energy exchange between cool disk and hot corona and how the disk and corona are maintained, is to be established.

In this investigation, we extended the disk corona model (Meyer & Meyer-Hofmeister 1994; Meyer et al. 2000) to regions near the central black hole by taking into account both decoupling of ion and electron temperature and Compton scattering. We aim at explaining the AGN spectrum.

2 Properties of the Corona

Numerical calculations of the structure of the disk corona were performed by solving the differential equations of continuity, momentum and energy for a system with black hole mass $M = 10^8 M_\odot$ and viscosity α=0.3 (For detail see Liu et al. 2001). The changes of coronal temperatures with distances are shown in Fig. 1. The ion temperature increases towards the central black hole, at values of about $(1/5)T_v$ (T_v virial temperature) in the outer region ($r \sim 10^{17}$cm$\sim 3000 r_g$, r_g Schwarzschild radius) and $(1/3)T_v$ close to the black hole, while the electron temperature almost stops to increase at $T_e \sim 10^9$ K from $r \sim 100 r_g$ inwards. This is because viscous heating of ions hardly goes to electrons by collisions but is carried as entropy of ions. The electrons, however, can cool by thermal conduction and radiation and hence keep their temperatures almost constant in the inner region. Such a feature is very similar to the ADAF (e.g. Narayan & Yi 1995). Similarity can also be seen from the energy flow in the corona shown in Fig. 2. It is shown that the viscous heat almost advects inwards with the accretion flows

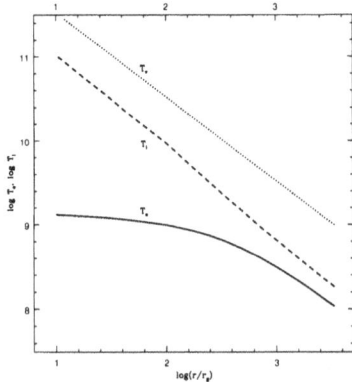

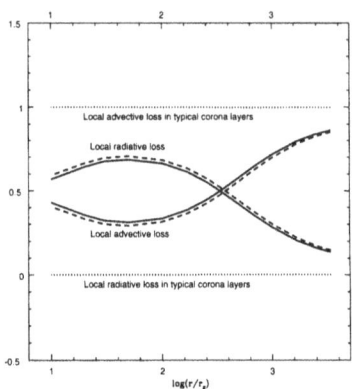

Fig. 1. Temperature distribution along distances. Solid line: electron temperature T_e; dashed line: ion temperature T_i; Dotted line: Virial temperature T_v.

Fig. 2. Radiative fraction and advective fraction of viscous heat along distances. Dotted line: corona layers; Dashed line: transition layers; Solid lines: the transition and corona layers.

with little radiation (a few thousandths at $r \sim 10^{17}$cm). However, if the transition layers between the disk and the upper corona are included, the local radiative fraction of viscous heat is larger in the inner region than that in the outer region. The corona is virtually identical to the ADAFs except there exist transition layers between the disk and corona and vertical advection.

Fig. 3 shows the rate of mass evaporating from the disk to the corona along distances, where the Compton cooling is not included. The evaporation rate, which is roughly the coronal flow rate, increases with decreasing distance r, reaches a maximum of 0.012 Eddington rate ($\dot{M}_{\text{Edd}}$) at distance $r \sim 340r_g$, and then decreases towards the black hole. The maximum is caused by the change in the dominant process that removes the frictional heat.

Fig. 3 also shows the evaporation rate for a larger viscosity, $\alpha = 0.9$. The maximal rate for $\alpha = 0.9$ is $0.14\dot{M}_{\text{Edd}}$ occurring at the distance $r = 30r_g$, about 10 times larger at a distance about 10 times smaller than that for $\alpha = 0.3$. The increase of viscosity increases the mass evaporation since larger viscosity leads to more efficient heating and faster accretion in the corona, both of which need larger mass supply by evaporating the underlying disk.

The effect of Comptonization with a total mass accretion rate of $0.05M_{\odot}$/yr in both the disk and corona is shown in Fig. 4 along with the non-Compton results. The Compton cooling largely restrains the evaporation at small distances, the evaporation rate at $r = 10^{16}$cm decreases more than 10 times and influences are even stronger with larger mass accretion rate in the disk. At very high accretion rate the coronal gas in the innermost region may fall back to the disk by efficient Compton cooling.

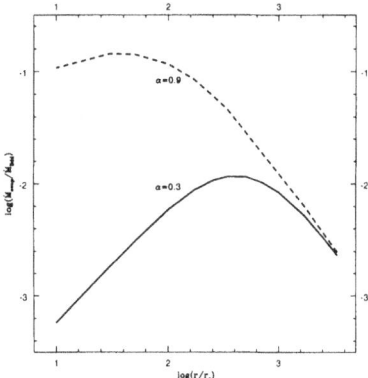

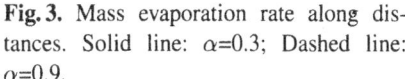

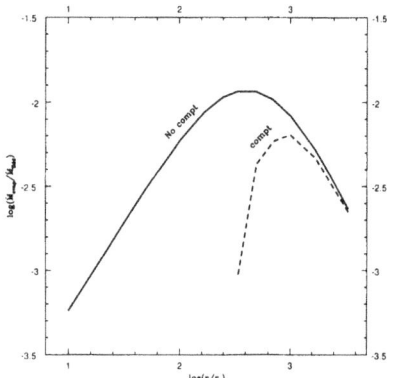

Fig. 3. Mass evaporation rate along distances. Solid line: α=0.3; Dashed line: α=0.9.

Fig. 4. Mass evaporation along distances. Solid line: without Comptonization; Dashed line: with Comptonization.

3   Discussion and Conclusion

The disk evaporation by the corona above provides a mechanism for transition from the outer disk to the inner ADAFs. At small accretion rates, a large part of the disk is depleted by efficient mass evaporation and the accretion is dominated by the ADAFs; At rates around the maximal evaporation rate, the corona coexists with an outer disk and probably an interior disk fed by the corona; At even large rates, the innermost corona cools down through strong Compton cooling and the thin disk dominates the accretion. Since the critical evaporation rate depends on viscosity and total mass flow rate, the dominant accretion mode is determined by and depends strongly on total accretion rate and viscosity.

In the AGN with high mass accretion rate, Comptonization by strong radiation from the disk removes viscous heat in the corona and restrains evaporation. Instead of evaporation, coronal gas falls back to the disk in the region close to the black hole and no corona is maintained in this region. Consequently there is little radiative contribution to hard X-rays from the innermost region. This implies a steeper hard X-ray spectrum in brighter objects as observed in NLS1.

To keep the corona coexistent with the disk down to the last stable orbit, a large fraction of accretion energy released in corona is necessary. The model shows that the corona could extend close to the black hole with very large viscosity, the required value of α increases with increasing mass incoming rate.

References

1. Liu B. F., Mineshige, S., Meyer F., et al., 2001, ApJ.
2. Meyer F. & Meyer-Hofmeister E. 1994, A&A 288,175
3. Meyer F., Liu B.F., & Meyer-Hofmeister E. 2000, A&A 361,175
4. Narayan, R. & Yi, I., 1995, ApJ, 452, 710

Radiative Effects in Active Galactic Nuclei

Ken Ohsuga[1] and Masayuki Umemura[2]

[1] Yukawa Institute for Theoretical Physics, Kyoto University, Kyoto 606-8502, Japan
[2] Center for Computational Physics, University of Tsukuba, Tsukuba, Ibaraki 305-8577, Japan

Abstract. By taking account of radiation force by a circumnuclear starburst and an active galactic nucleus (AGN), we analyze the equilibrium configuration and the stability of the obscuring wall which consists of dusty gas. As a result, it is found that the wall is supported by the radiation force and hide the nucleus from our line of sight in the case of luminous starburst. Thus, Seyfert 2 nuclei are more frequently accompanied by the circumnuclear starburst than Seyfert 1 nuclei. If the nucleus itself is much brighter than the starburst, as a quasar is, the formation of the wall is prevented by the strong radiation force of the nucleus. Therefore, there is a tendency that the quasars are identified as type 1 AGNs. However, when star-formation rate is very high, even luminous nucleus is obscured by a large amount of dusty gas and this object would be observed as ultra luminous infrared galaxy (ULIRG). This picture provides a physical explanation for the putative starburst-AGN connection, and expects the evolution of AGN type from type 2 to type 1 according to the stellar evolution in the starburst.

1   Introduction

Recent observations have been revealed that Seyfert 2 galaxies are more frequently associated with the circumnuclear starbursts than type 1 galaxies (Heckman et al. 1989; Maiolino et al. 1995, 1997, 1998; Pérez-Olea & Colina 1996; Hunt et al. 1997; Malkan, Gorjian, & Tam 1998; Storchi-Bergmann, Schmitt, & Fernandes 1999; Storchi-Bergmann et al. 2000; González Delgado, Heckman, & Leitherer 2001). On the other hand, most quasars are observed as type 1 AGNs regardless of starburst activity in their host galaxies (Barvainis, Antonucci, & Coleman 1992; Ohta et al. 1996; Omont et al. 1996; Schinnerer, Eckart, & Tacconi 1998; Brotherton et al. 1999; Canalizo & Stockton 2000a, 2000b; Dietrich & Wilhelm-Erkens 2000). These observations imply that the starburst event plays an important role in the apparent type of AGNs. In this paper, we consider a physical mechanism that may connect circumnuclear star-forming activities with AGN type, by taking account of radiation force of circumnuclear starburst regions and an AGN.

2   Starburst-AGN Connection

In the circumnuclear regions, radiation force would play an important role in the distribution of dusty gas, since the total luminosity of an AGN and starburst regions is comparable to the Eddington luminosity for the dust opacity. Hence, the stable equilibrium branches between the radiation force and the gravity show the actual configuration of dusty structure. We calculate the radiation fields by assuming the ring-shaped starburst at the size of a few hundred parsecs, and analyze the equilibrium configuration of dusty gas. Here, we also take account of stable condition, which is satisfied when the effective potential is local minimum. The force balance and stable conditions are given by Ohsuga & Umemura (2001). Resultant conditions for the formation of the obscuring wall are

$$L_{\mathrm{SB}} > \frac{4\pi c G M_{\mathrm{SB}}}{\chi_{\mathrm{V}}} A_{\mathrm{V}}, \tag{1}$$

and

$$L_{\mathrm{AGN}} < (M_{\mathrm{IB}} + M_{\mathrm{BH}}) \frac{L_{\mathrm{SB}}}{M_{\mathrm{SB}}}, \tag{2}$$

where L_{SB} and M_{SB} are the luminosity and the mass of the starburst ring, L_{AGN} is the AGN luminosity, M_{IB} is the mass of the bulge within the starburst ring, M_{BH} is the mass of the central black hole, χ_{V} is the mass extinction at the V-band, and A_{V} is the optical depth, respectively.

These conditions show that the wall can form and the nucleus is hidden when the starburst is luminous. However, the dusty gas is blown out if the luminosity of the nucleus is much larger than that of the circumnuclear starburst. Thus, type 2 nuclei tend to be accompanied by the starburst event on condition that the nuclei are relatively faint, as Seyfert nuclei. In contract, starburst activity around the type 1 nuclei would be calmer than that around the type 2 nuclei. On the other hand, the luminous nuclei, as quasars, tend to be identified as type 1 AGN. But, even quasars might be obscured by a large amount of the dusty gas in the case of vigorous star-forming activities. These objects would be recognized as ULIRGs. To conclude, this picture provides a physical explanation for the putative starburst-AGN connection, whereby Seyfert 2 galaxies are more frequently associated with circumnuclear starbursts than type 1 galaxies, whereas quasars are mostly observed as type 1 AGNs.

3 Evolution of AGN

In previous section, we have shown that apparent type of AGN depends on the starburst luminosity and AGN luminosity. Since starburst luminosity decrease with time according to the stellar evolution, AGN type is also time-dependent. Assuming the simple stellar evolution in the starburst (Ohsuga & Umemura 1999; Ohsuga & Umemura 2001), the wall is supported by the starburst radiation within several times 10^7 yr. Therefore, Seyfert 2 nuclei would evolve onto Seyfert 1 nuclei in several times 10^7 yr. In addition, it is inferred from this picture that the number density of Seyfert 1 nuclei is comparable to that of Seyfert 2 nuclei, because of the typical lifetime of AGNs, $\sim 10^8$ yr.

In the case of quasar, no wall forms and nucleus is directly observed, since the nucleus is much brighter than circumnuclear starburst. Thus, the type of the quasar does not vary. However, if the vigorous star-formation occurs in a very early evolutional stage, the nucleus is obscured by a large amount of dusty gas and strong IR radiation could be emitted. This object would be observed as an ULIRG and it evolves onto a quasar. Such a paradigm of the evolution of ULIRGs to quasars is proposed by Sanders et al. (1988) and Norman & Scoville (1988). Here, we do not consider the evolution of AGN luminosity. There might be a seed quasar at the center of an ULIRG. In this case, the obscuring wall can form easily because of faint AGN. However, black hole might grow up rapidly and the stable dusty wall would disappear after a while. The physical mechanism of feeding of black hole in ULIRGs is suggested by Umemura (2001). As a result, ULIRGs destined to evolve onto quasars because of decrease of starburst luminosity as well as increase of AGN luminosity.

References

1. Barvainis, R., Antonucci, R., Coleman, P. 1992, ApJ, 399, L19
2. Brotherton, M. S., et al. 1999, ApJ, 520, L87
3. Canalizo, G., Stockton, A. 2000a, ApJ, 528, 201
4. Canalizo, G., Stockton, A. 2000b, AJ, 120, 1750
5. Dietrich, M., Wilhelm-Erkens, U. 2000, A&A, 354, 17
6. González Delgado, R. M., Heckman, T., Leitherer, C. 2001, ApJ, 546, 845
7. Heckman, T. M., Blitz, L., Wilson, A. S., Armus, L., Miley, G. K. 1989, ApJ, 342, 735
8. Hunt, L. K., Malkan, M. A., Salvati, M., Mandolesi, N., Palazzi, E., Wade, R. 1997, ApJS, 108, 229
9. Maiolino, R., Krabbe, A., Thatte, N., Genzel, R. 1998, ApJ, 493, 650
10. Maiolino, R., Ruiz, M., Rieke, G. H., Keller, L. D. 1995, ApJ, 446, 561
11. Maiolino, R., Ruiz, M., Rieke, G. H., Papadopoulos, P. 1997, ApJ, 485, 552
12. Malkan, M. A., Gorjian, V., Tam, R. 1998, ApJS, 117, 25
13. Norman, C., Scoville, N. 1988, ApJ, 332, 124
14. Ohsuga, K., Umemura, M. 1999, ApJ, 521, L13
15. Ohsuga, K., Umemura, M. 2001, ApJ, 559, 157
16. Ohta, K., Yamada, T., Nakanishi, K., Kohno, K., Akiyama, M., Kawabe, R. 1996, Nature, 382, 426
17. Omont, A., Petitjean, P., Guilloteau, S., McMahon, R. G., Solomon, P. M., Pecontal, E. 1996, Nature, 382, 428
18. Pérez-Olea, D. E., Colina, L. 1996, ApJ, 468, 191
19. Sanders, D. B., Soifer, B. T., Elias, J. H., Madore, B. F., Matthews, K., Neugebauer, G., Scoville, N. Z. 1988, ApJ, 325, 74
20. Schinnerer, E., Eckart, A., Tacconi, L. J. 1998, ApJ, 500, 147
21. Storchi-Bergmann, T., Raimann, D., Bica, E. L. D., Fraquelli, H. A. 2000, ApJ, 544, 747
22. Storchi-Bergmann, T., Schmitt, H. R., Fernandes, R. C. 1999, in IAU Symp. 194, Activity in Galaxies and Related Phenomena, ed. Y. Terzian, D. Weedman, E. Khachikian (San Francisco: ASP), 295
23. Umemura, M. 2001, ApJL, in press

Accretion Scenarios and Evolution of AGN

Ewa Szuszkiewicz

Institute of Physics, University of Szczecin, Wielkopolska 15, 70-451 Szczecin, Poland

1 Understanding Standard Accretion History

An understanding of the nature of active galaxies and quasars requires an explanation of the source of their tremendous energies ($10^{43} - 10^{48}$ erg s^{-1}). The nuclear energy, the principal energy source in the universe at the beginning of last century, is wholly inadequate. The argument that such energies might be produced by accretion onto massive compact objects was first made by Salpeter [1] and Zeldovich [2]. Since that time the existence of supermassive black holes has been confirmed in the nuclei of nearby galaxies. The masses of black holes determined dynamically in these galaxies are now fully consistent with masses inferred in active galaxies and quasars and with the density of dark relic objects produced by accretion during the quasar epoch [3]. There is no need for non-standard accretion histories in order to reproduce a large density of black holes in the current universe. There is, however, need to understand the standard accretion history itself. We must be able to construct a model which explains all the observational data. One of the characteristic properties of the accreting black holes is their variability and this is the property which we are concentrated on in the first phase of our study.

2 Quintessence and Few Examples

The aim of our research programme is to study the origin of the incipient instabilities operating in accretion discs around black holes and to investigate their observational consequences for diverse astronomical phenomena. For doing this, we use a class of simple vertically-integrated, non-self-gravitating models of transonic accretion discs. This is well justified because, despite the ongoing development of increasingly sophisticated large-scale numerical models, simple disc modelling still remains the central link between theory and observations. We have started our programme of work by investigating the thermal instability driven by radiation pressure, which is relevant for intrinsically bright sources. Our main result is that the thermally-unstable discs undergo limit-cycle behaviour with successive evacuation and refilling of the central parts of the disc [4,5]. Moreover, we found [6] that models predicted to be stable by local analysis do indeed remain stable and stationary. The same applies for models which, according to local analysis, would have a potentially unstable region smaller than the minimum wavelength for unstable perturbations. We have calculated radiation spectra emitted from thermally-unstable discs to provide detailed theoretical predictions for the time dependent spectral changes in accreting objects [7]. Systematic studies of the radiation pressure driven thermal instability for different black hole masses, accretion rates and viscosity parameters are in progress. The physics learned from studying individual objects should be consistent with observational characteristics derived from whole populations of accreting black holes. One of the possible ways to constrain accretion scenarios is to construct the mass functions [8] derived from investigations of massive dark objects resident in local galaxies and the mass function of the black holes inferred from the past activity of AGN (active galactic nuclei). The observational consequences of different types of instabilities (secular, pulsational and dynamical) which can be responsible for the observed variability in active galactic nuclei will be studied in the context of the cosmological evolution of AGN. And now, let us briefly describe our most interesting results obtained so far.

Supermassive black holes (relevant for QSOs, BL LACs, NLS1, ULIRGs). It is widely believed that in quasars the accretion disc is the main emission component and driver of the quasar activity. We calculated a disc evolution around a supermassive black hole with mass 10^8 M$_\odot$ in the presence of the thermal instability driven by radiation pressure. As a consequence of this instability the inner part of the accretion disc heats up, its density decreases and as a result it expands forming an optically thin toroidal structure. The characteristic time scale of such behaviour has been found to be about 11 years [9]. A different type of thermal instability, driven by partial ionization of the disc material, has been found to operate in most of the accretion disc models relevant for AGN (with the exception of the brightest sources for which the discs are completely ionized). Two important consequences follow as suggested in [10]: quiescent AGN must appear as quite normal galaxies, and the average mass fuelling rate in many if not all AGN is much lower than implied by their current luminosities. This in turn limits the masses that their central black holes are expected to reach. These two types of instability and other physical processes occurring in the accretion disc around a black hole must be taken into account in calculating the quasar luminosity function and quasar evolution.

Stellar mass black holes (relevant for LMXBs, XRBs and MicroQuasars). Both observations and theory point to strong similarities in the behaviour of accretion discs around supermassive and Galactic black holes. Accretion discs have been studied in detail in many types of binary systems and they usually show complex variabilities. We try to learn as much as possible from these similarities. We have calculated the time evolution of the initially sub-Eddington model ($L = 0.06 L_{Edd}$) around a 10 solar mass black hole (viscosity parameter $\alpha = 0.1$) and found that its character is cyclic [5]. Integrating the radiated flux per unit area over the disc at successive times we have obtained the bolometric light curve. The disc luminosity exhibits a burst-like time variation with a burst duration of about 20 s and a quiescent phase lasting for the remaining 767 s of the cycle. The amplitude of the variation is around two orders of magnitude.

Intermediate mass black holes (relevant for ULX). Supermassive black holes have masses in the range $10^6 - 10^9 M_\odot$ and Galactic black holes around $10 M_\odot$. Black holes with masses in the range $10^2 - 10^3 M_\odot$ have been hypothesized to hide within off-nuclear "Ultra luminous X-ray Sources" (ULXs) [11,12]. The nature of these sources is still unknown. We have performed a calculation for a super-Eddington model with black hole mass $M = 10 M_\odot$, initial accretion rate $\dot{M} = 2$ in critical units which is equivalent to $\dot{M} = 32$ in Eddington units, and viscosity parameter $\alpha = 0.1$. We have found limit cycle solution with the following characteristics: the burst duration is roughly 300s and the quiescence lasts around 900s [13]. The sub-Eddington accretion discs around intermediate mass black holes are under investigation.

3 Final Remarks

Here we reported on the present status of our project. The results obtained till now provide some strengthened motivation for saying that the type of dics behaviour seen in our calculations might truly have some fundamental importance and its effects might be present among the range of different variability patterns observed for accreting black holes.

Acknowledgements

I gratefully acknowledge financial support from the grant 2P03D01817 of Polish State Committee for Scientific Research (KBN).

References

1. E.E. Salpeter: ApJ **140**, 796 (1964)
2. Ya.B. Zeldovich: Soviet Physics -Doklady **9**, 195 (1964)
3. D. Merritt, L. Ferrarese: astro-ph/0107134 (2001)
4. E. Szuszkiewicz, J.C. Miller: MNRAS **298**, 888 (1998)
5. E. Szuszkiewicz, J.C. Miller: MNRAS accepted (2001) (astro-ph/0107257)
6. E. Szuszkiewicz, J.C. Miller: MNRAS **287**, 165 (1997)
7. L. Zampieri, R. Turolla, E. Szuszkiewicz: MNRAS **325**, 1266 (2001)
8. P. Salucci, E. Szuszkiewicz, P. Monaco, L. Danese: MNRAS **307**, 637 (1999) (erratum: MNRAS **311**, 448 (2000))
9. E. Szuszkiewicz: Mem. Soc. Astron. Ital. **70**, 95 (1999)
10. L. Burderi, A.R. King, E. Szuszkiewicz: ApJ **509**, 85 (1998)
11. H. Matsumoto, at al.: ApJ **547**, L25 (2001)
12. G. Fabbiano, A. Zeza, S.S. Murray: astro-ph/0102256 (2001)
13. E. Szuszkiewicz, J.C. Miller: Advances in Space Research, accepted (2001)

Conference dinner

Part IX

Ultra-Luminous X-Ray Sources and Stellar Mass Black Holes

Spectral Transition for ULXs – Super-Eddington Luminosity from Rapidly Spinning Moderate Mass BHs?

Friedrich Meyer and Emmi Meyer-Hofmeister

MPI für Astrophysik, Karl-Schwarzschild-Str. 1, D85740 Garching, Germany

Abstract. The X-ray luminosity of ultraluminous compact X-ray sources (ULXs) reaches 10^{39-40} erg/s. To avoid super-Eddington luminosity high black hole masses of up to 100 M_{solar} are required. But severe problems are encountered with this interpretation. We suggest that the high luminosities are truly super-Eddington, but that the source of energy is not accretion, but magnetic extraction of energy from a rapidly rotating spinning black hole. The dissipation of this energy fills the corona above the disk with moderately relativistic particles, part of which impinge on the accretion disk underneath and provide its super-Eddington luminosity. At the same time the super-Eddington radiation pressure from below is confined by the particle impact pressure from above. We discuss requirements and derive order of magnitude estimates for magnetic field strength, particle density, relativistic γ-factor and surface density of the disk.

1 Observations of ULXs and interpretations

In the arms of nearby spiral galaxies luminous point-like X-ray sources were detected. Due to their variability these sources are thought to be single objects. The X-ray luminosity reaches 10^{39-40} erg/s. These objects are therefore called ultraluminous compact X-ray sources (ULXs). If accretion is the source of energy and the Eddington luminosity should not be exceeded relatively high black hole masses, of up to 100 M_{solar} are required. ASCA spectroscopy brought further information for several sources. The majority of the spectra can be reproduced by a multi-color accretion disk. The observed high inner disk temperatures are a severe problem. To solve this problem the inner disk geometry around a rapidly rotating black hole was discussed. Another approach uses the concept of a slim accretion disk. - For a review see Makishima, 2001, these Proceedings.

The discovery of spectral transitions from ultraluminous compact X-ray sources strongly supports the scenario of accreting black holes. We show in Figure 1 the distribution of X-ray sources in soft state (standard accretion disk reaches inward to the last stable orbit) or hard state (standard disk outside + coronal flow/ADAF in the inner region) in the $L_x - M$ plane. ULXs are drawn where their luminosity would be the Eddington luminosity.

2 ULXs moderate mass black holes, but truly super-Eddington luminosity ?

The discovery of hard/soft spectral transitions, expected for an accretion rate $\dot{M} \approx \frac{1}{10}\dot{M}_{\mathrm{Edd}}$ (compare modeling of spectral transitions by Esin et al. 1997; analysis of

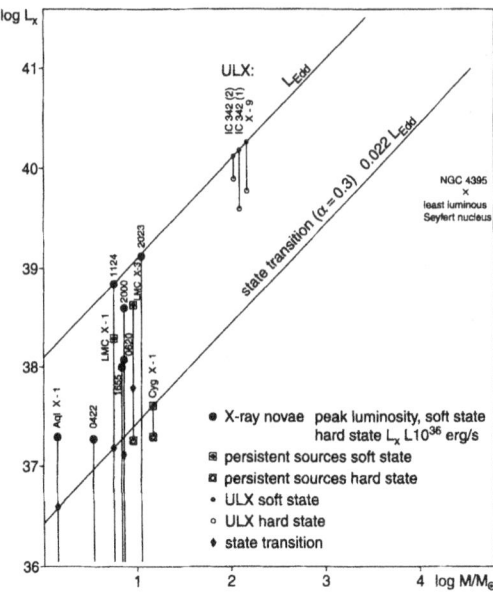

Fig. 1. X-ray luminosity as a function of black hole mass for LMXRBs, HMXRBs and ULXs. For comparison the least luminous Seyfert nucleus is added in the diagram. The upper line indicates the Eddington luminosity, the lower line the luminosity of spectral transition, soft state above, hard state below

the transition by Meyer & Meyer-Hofmeister 2000) implies an accretion luminosity of $L_x = 10^{38}$, observed are $L_x \approx 10^{40}$erg/s $\approx 10 \cdot L_{\mathrm{Edd}}$. The high luminosity is clearly not accretion luminosity. We suggest it could be extracted from a spinning black hole.

The total spin energy available is $E_{\mathrm{rot}} \leq \frac{1}{2}Mc^2 \approx 10^{54}$erg. If a fraction f goes into L_x (the larger part might go into a jet), the lifetime is t=$fE_{\mathrm{rot}}/L_x = f \cdot 10^{6.5}$yr. This points to young objects and agrees with the location in young star forming regions.

But how can the super-Eddington luminosity be possible? Our suggestion is the following: relativistic particles in a strong magnetic field impinge from above onto the innermost accretion disk. Their energy is deposited in the accretion disk and reprocessed into X rays. Then the super-Eddington radiation pressure can be balanced by the particle impact pressure.

Consistency of the model requires:

- (a) the particles in the corona reach relativistic energies
- (b) the energy extraction gives the high luminosity
- (c) the confinement of the relativistic particles
- (d) the disk surface density is high enough to stop the particle penetration
- (e) the confinement of the magnetic field by the disk below

With obvious assumptions we estimate the following values:

(a) The density ρ of the gas can be estimated from the accretion rate in the hot corona $\dot{M} \approx 4\pi r^2 \rho v_r$, $\rho = nm$ with n, m particle number density and mass, radial drift velocity $v_r \approx \alpha^* c$, $\alpha^* < 1$, c velocity of light. The luminosity from the reprocessing of the energy of the impinging particles can be written as $L_x = 2\pi r^2 \eta \gamma n m c^3$ (πr^2 disk area, $\gamma m c^2$ energy of particles) if a fraction η reaches the disk surface below. With the assumption $\dot{M} = \frac{1}{10} \dot{M}_{\mathrm{Edd}}$ (spectral transitions) n can be deduced from $\dot{M}$. The value of γ follows as $20 \alpha^* / \eta$.

(b) In our scenario the flux tubes connecting the black hole region with the inner disk are twisted due to the differential rotation between black hole and disk. The extracted energy $\dot{E}$ is proportional to the volume, magnetic stress and the twisting rate of the field. The energy extracted near the black hole horizon (Schwarzschild radius $r_S = \frac{2GM}{c^2}$) is then of order of magnitude

$$\dot{E} = \frac{4\pi}{3} r_S{}^3 \frac{B^2}{4\pi} \Omega_{\mathrm{BH}} = \frac{B^2}{3\pi} \sqrt{GM r_S{}^3}. \tag{1}$$

Equating $\dot{E}$ to the luminosity L_x with $L_{40} = \frac{L_x}{10^{40}} \mathrm{erg/s}$ the magnetic field estimate becomes

$$B = 10^{8.5} \frac{L_{40}{}^{1/2}}{M/10 M_\odot} \mathrm{gauss}. \tag{2}$$

(c) The relativistic particles require a magnetic energy density for their confinement similar or larger than their own. This leads to a similar estimate for the magnetic field as above.

(d) The strong magnetic field suppresses the (magnetic) friction ("Balbus-Hawley instability"). If the magnetic field energy density becomes comparable to the gas pressure frictional heating is quenched, the disk then is practically isothermal, the disk temperature is determined by the reprocessing of the energy of impinging particles. From the temperature corresponding to the luminosity, the gas pressure equal to the magnetic pressure ($B = 10^{8.5}$ gauss) and surface density equal to density times pressure scaleheight we get $\Sigma_d \approx 10^{4.6} (r/r_s)^{5/4} (M/10 M_\odot)^{3/4} L_{40}{}^{1/8} \mathrm{g/cm}^3$, large enough to stop the penetration of the impinging relativistic particles.

(e) Taking the values for the mass in the disk derived above we see that the gravitational pull exceeds the expansion force of the magnetic field by far, so that the magnetic field can be held by the disk.

From these estimates it appears that this might be a viable model deserving further investigation.

References

1. A. Esin, E. McClintock, R. Narayan et al. ApJ 489, 865 (1997)
2. F. Meyer, B.F. Liu, E. Meyer-Hofmeister 354, L67 (2000)

X-Ray Spectral Variability of Cygnus X–1 and Simulated MHD Flow

Shin Mineshige[1], Hitoshi Negoro[2], Ryoji Matsumoto[3], and Mami Machida[3]

[1] Yukawa Institute, Kyoto University, Sakyo-ku, Kyoto 606–8502, Japan
[2] RIKEN, Hirosawa, Wako, Saitama 351–0198, Japan
[3] Dept. of Physics, Chiba University, Inage-ku, Chiba 263–8522, Japan

Abstract. X-ray hardness ratio variations of Cyg X–1 in the hard state comprise two components with different timescales; one characterizing the light variation and the other rapid hardening around the peak. These seemingly indicate two (or more) physical processes being involved; e.g., disturbance propagation and magnetic flares. Such features may be consistent with the view of MHD accretion flow.

1 Observed X-Ray Spectral Variability

If we have a close look at the X-ray intensity variations of Cyg X–1, we notice rather spiky features (referred to as shots) present on the timescale of several sec. The upper panel of figure 1 shows an average time profile of the shots of Cyg X-1 in the 1.2–58.4 keV band obtained by superposing 872 large shots (with peak intensity of exceeding twice its average) by aligning their peaks.

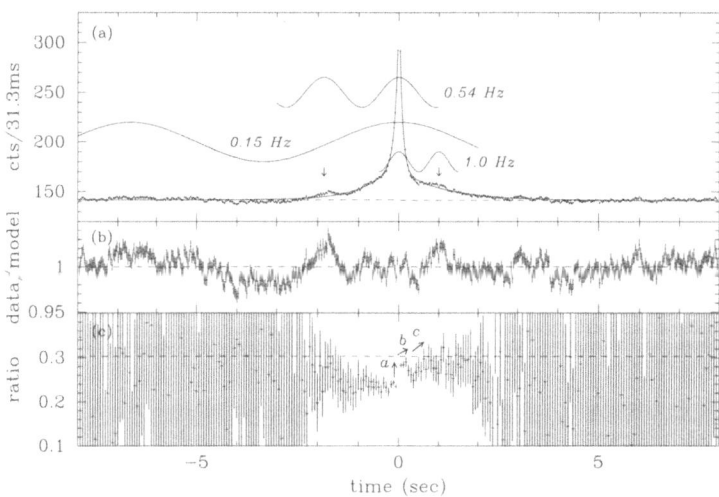

Fig. 1. (a) Average shot profile of the shots, (b) the ratio of the observed counts to that expected by the best-fit two-exponential model, and (c) the hardness ratio variations, all being based on the *Ginga* data. (From Negoro et al. 2001)

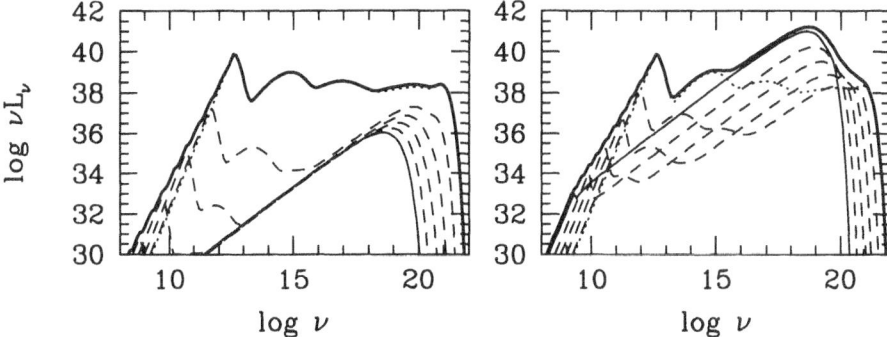

Fig. 2. Typical ADAF (left) and CDAF (right) spectra. Two models have different density profiles but other parameters are set to be the same. The thick solid line represents the total spectra, while thin solid, dashed, and dotted lines represent the contributions from the outermost, middle, and innermost rings, respectively.

Surprisingly, the profiles are nearly time-symmetric. It is also interesting to note that the hardness ratio (7.3–14.6 keV/1.2–7.3 keV) variation of the excess shot depicted in the bottom panel of figure 1 clearly indicates a soft spectrum before the peak intensity, and a rapid hardening in $\ll 0.1$ s around the peak, followed by complex changes.

Our current analysis strongly points the presence of two (or more) physical timescales involved with X-ray variability: $\tau \simeq 0.1 - 1.0$ s, on which shot profile rises and decays, and $\tau < 10$ ms, on which a spectral hardening occurs around the shot peak. We should caution that the hard X-ray peak does not significantly lag behind the soft X-ray peak even within a shorter time resolution of ~ 2 ms of the RXTE (Feng et al. 1999). It seems difficult to explain both different timescales by means of a single physical mechanism. Thus, the hot spot model and Compton cloud model are not acceptable, since they predict comparable hardening timescale to the shot timescale (few sec). Rather, it seems that the relatively long shot timescales could be somehow related to accretion timescales, while the short one could be due to magnetic reconnection.

2   MHD Flow Spectra

Next, we turn the subject to the spectra of hot flow. Recently, 2D/3D simulations have been intensively performed and new paradigms of ADAF and CDAF have emerged (see Narayan in these proceedings). The critical test of these flows are to examine their spectral properties. Figure 2 displays the typical ADAF (left) and CDAF (right) spectra (cf. Ball et al. 2001). The adopted parameters are: the inner-edge radius is $1.0 \, r_{\rm g}$ (with $r_{\rm g}$ being the Schwarzschild radius), the outer-edge radius is $10^{2.5} \, r_{\rm g}$, the mass-flow rate is $\log \dot{m} \equiv \log(\dot{M}c^2/L_{\rm E}) = -3.29$ (with $L_{\rm E}$ being the Eddington luminosity), the mass of black hole is $M_{\rm BH} = 10^8 M_\odot$, the ion temperature profile is $10^{12}(r/r_{\rm g})^{-1.0}$ K, the electron temperature profile is $10^{10}(r/r_{\rm g})^{-0.6}$ K, the density profile is $\rho \propto (r/r_{\rm g})^{-p}$ with $p = 1.5$ (ADAF) and $p = 0.5$ (CDAF), respectively, and magnetic field strength is the equipartition values; i.e., $B^2 \propto (r/r_{\rm g})^{-(1+p)}$.

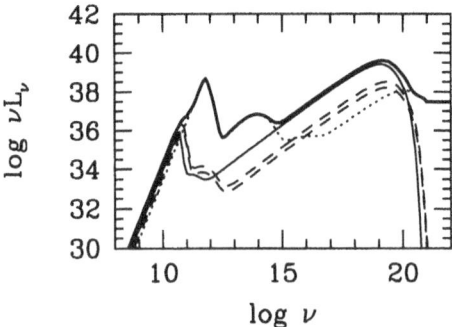

Fig. 3. Same as figure 2 but for the case of the MHD flow. The resultant spectrum is more like that of CDAF than that of ADAF (see figure 2).

In ADAF emission from the innermost ring is dominant because of its steep density profiles ($\rho \propto r^{-3/2}$, leading to the bremsstrahlung emissivity of $dE \propto \rho^2 T^{1/2} r^2 dr \propto r^{-1.5} dr$). In the case of CDAF, in contrast, X-ray emission comes from the outer parts due to less steep density profile ($\rho \propto r^{-1/2}$, leading to $dE \propto r^{0.5} dr$). This is well demonstrated in figure 2.

How about the case of MHD flow? Machida et al. (2001) already noted similar density and temperature profiles of MHD flow to those of CDAF. Therefore, the spectra are more like that of CDAF (see figure 3). However, such CDAF and MHD flow models cannot account for the observations of Cyg X–1 during its hard state in the two senses. (1) CDAF predicts rather flat ($f_\nu \propto \nu^0$) spectrum, unless we assume significant electron heating as well as ion heating (Ball et al. 2001), but the observations clearly show a power-law decline (e.g. $f_\nu \propto \nu^{-0.7}$). (2) Rapid variability seems to arise via time-dependent processes probably associated with magnetic flares in the innermost region (see §1), but X-ray emission from the inner part is negligible in CDAF (figure 2).

The key to resolving this issue is whether dissipated energy by magnetic reconnection goes directly to plasma or radiation. So far, all the MHD simulations postulate no radiative cooling, but if dissipated energy can be radiated away immediately, MHD flow structure should be more like that of ADAF, thus explaining both of the spectral shape and the presence of X-ray variability.

References

1. G.H. Ball, R. Narayan, E. Quataert: Astrophys. J. **552**, 221 (2001)
2. Y.X. Feng, T.P. Li, L. Chen: Astrophys. J. **514**, 373 (1999)
3. M. Machida, R. Matsumoto, S. Mineshige: Pub. Astr. Soc. Japan **53**, L1 (2001)
4. H. Negoro, S. Kitamoto, S., S. Mineshige: Astrophys. J. **554**, 528 (2001)

Evaporation of the Inner Disk in Black Hole Candidates

Henk C. Spruit and Bernhard Deufel

Max-Plank-Institut für Astrophysik, Postfach 1317, 85741 Garching Germany

1 The Disk-ADAF Transition

One of the promising models for the inner accretion flow in black hole accreters is a combination of an ion supported flow (ADAF) inside, and fed by, a cool standard disk extending to larger distances. This model is suggested by the observed correlations between the reflection components in the X-ray spectrum (Fe line and Compton bump) with the slope of the power law component in the same spectrum and with the frequency of quasiperiodic oscillations in the X-rays (Churazov et al. 2001, Zdziarski et al. 1999).

Conceptually, this model has the difficulty that a plausible way has to be found how a cool disk with inner edge at, say, 30 Schwarzschild radii (r_g) from the hole (as suggested by the correlations mentioned), would transform itself into a flow with a 10^5 times higher ion temperature. We summarize here the processes, described in more detail elsewhere (Deufel et al. 2001, Spruit & Deufel 2001) by which such evaporation occurs. It turns out that all that is needed are the known physics of energy exchange between charged particles by Coulomb interaction in an ionized plasma, and a standard α-prescription for viscosity in accretion disks.

2 Interaction Between ADAF and Cool Disk

At larger distances (from about $300\,r_g$) the 'coronal' evaporation mechanism is available, at temperatures up to some 10^8 K. This process (Meyer et al. 1994, 2000) operates by electron conduction from the hot corona to a very thin transition layer. The converging heat flux in this transition zone heats the plasma rapidly enough for it to evaporate, and feed the corona. For the physics at ADAF temperatures (around 10^{11}–10^{12} K), however, this process does not work. A simple argument would suggest that here, instead, the cool disk is an almost arbitrarily efficient cooling surface onto which any prospective ADAF would quickly *condense*. At the conditions in an ADAF, the energy and conductive heat flux in the ADAF electrons (around 10^9K) can be neglected. At the high temperature of the protons, their mean free path when penetrating the disk is not small, so they spread the incoming energy over a significant depth. One would then conclude that the efficient bremsstrahlung emissivity of a cool disk would be sufficient to radiate away the incident proton energy flux by low energy radiation. It turns out that in some cases (namely, at very low incident energy flux) this is indeed how the energy balance works (Deufel et al. 2001a).

For conditions corresponding to X-ray binaries in their hard X-ray state, the physics is more interesting. The first thing to notice is that protons at energies much above the temperature of the target plasma lose their energy by small-angle scattering on the target *electrons*. This contrasts with the situation at incident energies closer to the target temperature, in which case the energy exchange is mainly between the protons. This is described by Spitzer's (1962) general results for Coulomb interaction in an ionized plasma. The rate of energy loss of an incoming proton of energy E to the target electrons in this theory is given by

$$\frac{1}{E}\frac{\mathrm{d}E}{\mathrm{d}\tau} = 3\ln \Lambda \, \frac{m_\mathrm{e}}{m_\mathrm{p}} (\frac{c}{v})^4 [\psi(x) - x\psi'(x)], \tag{1}$$

where τ is the Thomson optical depth, $\ln \Lambda$ a Coulomb logarithm, v the velocity of the incoming proton, ψ the error function, ψ' its derivative, and the argument $x = v[m_\mathrm{e}/(2kT)]^{1/2}$ is the ratio of the incident proton's velocity to the thermal velocity of the target electrons. A similar expression holds for the energy loss by transfer to the target protons, and comparing the two expressions one finds that energy loss to the electrons dominates when the energy of the incoming proton is larger than about 10 times the thermal energy of the target protons. This is amply satisfied for ADAF protons entering a cool disk.

The surface layer of the disk heats up by the proton energy flux, and expands under hydrostatic equilibrium, thus lowering the electron density. This reduces the cooling by bremsstrahlung, and the layer heats up further until at $T \sim 100$ keV a stable balance is reached between heating and cooling by Comptonization of soft photons (Spruit 1997, Spruit and Haardt 2000, Deufel et al. 2001a,b). The result is a Comptonizing layer of optical depth of order unity overlying the cool disk.

This stable balance exists only if there are enough soft photons. Near its inner edge, the total optical depth of the cool disk is not sufficient to produce enough soft photons, and these regions heat up to a somewhat higher temperature (a few hundred keV), such that pair production limits further temperature increase. We call this part of the disk a 'warm disk', to indicate that its temperature is large compared with the temperature of the cool disk, but still small compared with the ion temperature in the ADAF. When this 'warm' state is thermally stable (more about this in the next paragraphs) it is only because the heating is external, provided by the proton energy flux of the ADAF. If the heating were internal, by viscous dissipation, such a state would be thermally unstable: it would lie on the SLE (Shapiro-Lightman-Eardley) branch of viscous disk models (e.g. Chen et al. 1995).

3   Thermal Stability of the Warm State, Evaporation

If it can be assumed that viscous heating in disks goes into the protons, or at least a fraction of order unity does so, it can cause instability in a warm disk. At the low surface density near the inner edge of the cool disk, viscous heating can be shown to be small compared with the heating flux from the ADAF, hence energetically unimportant. The viscously dissipated heat ends up in the protons, however, and this energy can be radiated away only after being transferred to the electrons. At the temperatures and densities of the warm state, this transfer is slow. The protons in the disk thus lie 'outside the main energy channel': they don't receive much energy, and don't loose much.

Most of the energy entering the warm disk is delivered to the electrons, which radiate it away by Comptonization of soft photons. The temperature of the protons in the warm state can become unstable, because the efficiency of energy transfer to the electrons decreases with increasing proton temperature. As the protons in the warm disk heat up, their cooling rate goes down, and the proton temperature runs away until the virial temperature. From the on, further heating is balanced by advective cooling: the warm disk has become a part of the ADAF surrounding it. From detailed calculations (Spruit and Deufel 2001), we find that this instability first affects the upper layers of the warm disk, where the densities are lowest. The upper layers thus 'evaporate' into the ADAF. At temperatures of about 300 keV, this evaporation rate becomes large enough to balance the accretion flow from the cool disk.

Because the physics of the Coulomb interactions in an ionized plasma, the evaporation process depends (apart from geometrical factors and a Coulomb logarithm) only on three dimensionless parameters: the distance r from the hole in units of the Schwarzschild radius, the accretion rate in units of the Eddington rate, and the α-viscosity of the disk.

The processes described have the potential to account for the origin of much of the X-rays in the hard state, as well as providing a natural mechanism for the transition of a cool disk to an ion supported accretion flow.

References

1. Chen X.M., Abramowicz M.A., Lasota J.-P., Narayan R., Yi I. 1995, ApJ, 443, L61
2. Churazov, E., Gilfanov, M. & Revnivtsev, M. 2001, MNRAS 321, 759
3. Meyer, F. & Meyer-Hofmeister, E. 1994, A&A, 288, 175
4. Meyer, F., Liu, B.F. & Meyer-Hofmeister, E. 2000, A&A, 361, 175
5. Spitzer, L. 1962, Physics of fully ionized gases, Wiley, New York, Ch. 5
6. Spruit, H.C. 1997 in Accretion disks: new aspects, E. Meyer-Hofmeister & H. Spruit (eds.), Lecture Notes in Physics, Springer, Berlin, 487, 67
7. Spruit, H.C. & Haardt, F. 2000, MNRAS, 351, 751
8. Deufel, B. & Spruit, H.C. 2000, A&A, 262, 1
9. Deufel, B., Dullemond, C.P. & Spruit, H.C. 2001a, A&A, 377, 955
10. Deufel, B., Dullemond, C.P. & Spruit, H.C. 2001b, A&A, submitted (astro-ph/0108496)
11. Spruit, H.C. & Deufel B. 2001, A&A, submitted (astro-ph/0108497)
12. Zdziarski, A.A., Lubinski, P., Smith, D.A. 1999, MNRAS 303, L11

Slim Disk Model for Ultraluminous X-Ray Sources

Ken-ya Watarai[1], Shin Mineshige[2], and Tsunefumi Mizuno[3]

[1] Department of Astronomy, Graduate School of Science, Kyoto University, Sakyo-ku, Kyoto 606-8502, Japan
[2] Yukawa Institute for Theoretical Physics, Kyoto University, Sakyo-ku, Kyoto 606-8502, Japan
[3] Department of Physics, Hiroshima University, Higashihiroshima, Hiroshima, 739-8526, Japan

Abstract. Ultraluminous X-ray sources (ULXs) are very mysterious objects because of their extremely large X-ray luminosity for stellar mass black holes, but too faint for low luminosity active galactic nucleus. Our super-critical accretion disk model can explain some features of ULXs, for example, relatively high temperature ($T_{in} \sim 1.0$–2.0 keV), and small emitting region (~ 1–$3\ r_g$) and so on. In particular, IC342 source 1 moves along with our mass-constant lines, and it may undergo ~ 33–$100\ \dot{M}_{crit}$ super-critical accretion.

1 Introduction

Already in the late 1980's, $Einstein$ satellite had found luminous X-ray sources with X-ray band, $L_x \sim 10^{39-40}$ erg s^{-1}, in nearby spiral galaxies (Fabbiano. 1989). At present day, these luminous, non-AGN and non-supernova-remnant sources are called "Ultra Luminous X-Ray Sources (ULXs)".

Analysis of high-quality spectra with $ASCA$ have been accumulated for numbers of ULXs. In particular, a rather extensive study has been performed by Makishima et al. (2000), Mizuno et al. (2000). According to their study, the spectra of most ULXs can be well fitted with multi-color disk (MCD) model (Mitsuda et al. 1984), reinforcing its black-hole X-ray binaries (BHXBs) interpretation. However, obtained disk temperatures, typically ranging 1.0–2.0 keV, somewhat contradict the BHXBs hypothesis.

Theoretically, they are too high to be achieved by a standard-type disk (Shakura & Sunyaev 1973) around such a high-mass, Schwarzschild (non-rotating) black hole. The observed high disk temperatures are hence the most severe problem regarding ULXs. We need both the detailed calculations of spectra from super-critical accretion disk, so called "slim disk" (Abramowicz et al. 1988; see also Watarai et al. 2000) and the careful comparison with the observations.

2 Basic Equations

The basic equations are almost same as those used in Matsumoto et al. 1984 (see also Kato et al. 1998, chapter 8). We solved the height-integrated equations in the radial direction based on the pseudo-Newtonian potential (Paczyńsky, Wiita, 1980), and neglected self-gravity (we assume α-viscosity prescription). The momentum in the radial direction, angular momentum conservation, continuity, hydrostatic balance, and energy equations are

$$v_r \frac{dv_r}{dr} + \frac{1}{\Sigma}\frac{dW}{dr} = \frac{\ell^2 - \ell_K^2}{r^3} - \frac{W}{\Sigma}\frac{d\ln\Omega_K}{dr}, \tag{1}$$

$$\dot{M}(\ell - \ell_{\rm in}) = 2\pi r^2 \alpha W, \tag{2}$$

$$\dot{M} = -2\pi r v_r \Sigma = {\rm const.}, \tag{3}$$

$$H^2 \Omega_K^2 = \frac{W}{\Sigma}, \tag{4}$$

$$-r\alpha W \frac{d\Omega}{dr} = \frac{8acT_0^4}{3\tau} + v_r \Sigma T_0 \frac{ds_0}{dr}, \tag{5}$$

where v_r is the radial velocity of gas, Σ is surface density, and W is the pressure integral, $W = R/\bar{\mu}\Sigma T_0 + aHT_0^4/3$, where R, $\bar{\mu}$, a, T_0 and H are the gas constant, the mean molecular weight, the radiation constant, the temperature on the equatorial plane, and the scale height, respectively. The specific angular momentum is defined by ℓ ($= rv_\varphi$), and ℓ_K is the Kepler angular momentum. The energy equation involves the viscous heating, radiative cooling and advective cooling. Here, s_0 is the entropy on the equatorial plane, and τ is the optical depth. We solved the above set of equations (1)–(5) from the outer edge of the disk ($r_{\rm out} = 10^4 r_{\rm g}$) to the vicinity of the central black hole through the transonic point. For evaluating the effective temperature, we assumed that all of the radiative energy is emitted with a blackbody, $Q_{\rm rad}^- = 2\sigma T_{\rm eff}^4$. Finally, we define the normalized parameter $\dot{m} = \dot{M}/\dot{M}_{\rm crit}$ with $\dot{M}_{\rm crit} = L_E/c^2$, $m = M/M_\odot$.

3 Result of Spectral Fitting

Figure 1 displays the X-ray "H-R diagram" representing the relationship between X-ray luminosities and peak temperatures. In high-$\dot{M}$ regions, our calculation results systematically shift towards the lower M direction in the frame of the standard-disk relation for given $\dot{M}$. This is due to the apparent shift of the inner boundary in transonic flows with high $\dot{M}$. The behavior of the two models on the X-ray H-R diagram is thus qualitatively distinct at high L. Here, we pay particular attention to the multiple points of IC342 source 1 in figure 1, which indicate that this source moves along the constant-M lines not of the standard disk but of the slim disk. Obviously, black-hole masses cannot change, while $\dot{M}$ can on short time-scales, say, < 1 s. That is, the slim disk is more relevant here. Other objects, such as $M81$ X-8 and $NGC1313$ source B, also move along our constant M lines.

4 Discussion & Conclusion

1. All the ULXs so far observed fall on the regions with high mass ($M \sim 30 M_\odot$) and high accretion rate (~ 33–$100\, L_E/c^2$) on the X-ray H-R diagram (figure 1).
2. It is not always clear whether a central black hole is rotating or not. What we can conclude is that high luminosity, $\sim L_E$, is realized in ULXs, independent of the nature of black holes.
3. More extensive observational studies, particularly independent estimates of inclination angles, are indispensable for fully understanding the nature of ULXs.

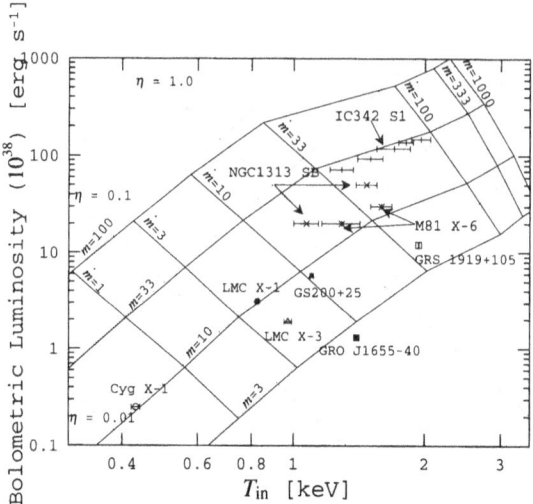

Fig. 1. X-ray H-R diagram of X-ray sources. Solid lines represent the constant m (black-hole mass) and constant $\dot{m}$ loci according to our model, while dotted lines are the same but based on the standard accretion disk (SSD) theory. When calculating both lines, the boundary (ξ) and Compton ($T_{\text{col}}/T_{\text{eff}}$) effects (Shimura & Takahara (1995)) are both included. Other symbols denote the $ASCA$ data of ULXs and BHXBs taken from Mizuno (2000) and Makishima et al. (2000). The inclination angle is assumed to be $i=0$ (face on).

References

1. M.A. Abramowicz, B. Czerny, J.P. Lasota, E. Szuszkiewicz: ApJ, **332**, 646 (1988)
2. G. Fabbiano: AR&A, **27**, 87 (1989)
3. S. Kato, J. Fukue, S. Mineshige: $Black - HoleAccretionDisks$ (Kyoto University Press, Kyoto 1998)
4. K. Makishima, et al.: ApJ **535**, 632, (2000)
5. R. Matsumoto, S. Kato, J. Fukue, A. Okazaki: PASJ, **36**, 71, (1984)
6. K. Mitsuda, et al.: PASJ, **36**, 741 (1984)
7. T. Mizuno, A. Kubota, K. Makishima : ApJ, **554**, 1282 (2001)
8. B. Paczyńsky, P.J. Wiita : A&A, **88**, 23 (1980)
9. N.I. Shakura, R.A. Šunyaev : A&A, **24**, 337 (1973)
10. T. Shimura, F. Takahara : ApJ **445**, 780 (1995)
11. K. Watarai, J. Fukue, M. Takeuchi, S. Mineshige: PASJ **52**, 133 (2000)

Luminous Hot Accretion Disks

Feng Yuan

Max-Planck-Institut für Radioastronomie, Auf dem Hügel 69, 53121 Bonn, Germany; Email:
fyuan@mpifr-bonn.mpg.de

Abstract. We find a new two-temperature hot branch of accretion disk solutions. These solutions correspond to accretion rates higher than the critical rate of ADAF. In these solutions, the energy loss rate of the ions by Coulomb collision with electrons is larger than the viscous dissipation and it is the advective heating together with the viscous dissipation that balances the Coulomb cooling of ions. Compared to ADAF, the electron temperature in this solution is moderately lower while the scattering optical depth is larger. This solution is much more luminous than ADAF.

1   Introduction: The Physics of Luminous Hot Accretion Disks

The equation of the energy of ions is

$$\rho v T_i \frac{ds}{dr} = \rho v \frac{d\epsilon_i}{dr} - q^c \equiv q_{\mathrm{adv}} = q^+ - q_{ie}. \tag{1}$$

In a typical ADAF, almost all of the viscously dissipated energy is advected into the centre black hole rather than transferred to the electrons and radiated away: $q_{\mathrm{adv}} \approx q^+ \gg q_{ie}$. Since $q^+ \propto \dot{m}$ and $q_{ie} \propto \dot{m}^2$, when the accretion rate increases to a certain critical value, the electrons-ions Coulomb coupling becomes so efficient that a large fraction of the viscously dissipated energy is transferred to the electrons and radiated away therefore the accretion flow ceases to be an ADAF. The critical accretion rate of ADAF $\dot{m}_1$ is determined by (Narayan, Mahadevan & Quataert 1998):

$$q^+ \approx q_{ie}. \tag{2}$$

Because of its low accretion rate and efficiency, ADAF is dim. It succesfully explain some sources with very low luminosity.

It is generally assumed that only the standard cold disk solution exists when the mass accretion rate $\dot{m} > \dot{m}_1$. However, from eq. (1), it is clear that the accretion flow can remain hot if q_{ie} is lower than the sum of q^+ and the compression work q^c (note that q^c is always comparable or larger than the q^+). Therefore, a new hot solution should exist in this case. The critical accretion rate of this new solution $\dot{m}_2$ is determined by,

$$q_{ie} \approx q^c + q^+. \tag{3}$$

Different with ADAF, in this solution the entropy decreases with the decreasing radii and the energy advection serves as a heating term in the Lagrangian point of view. So it is the decrease of the entropy of the plasma together with the viscous dissipation that balance the radiation of the accretion flow. Since the new solution corresponds to higher accretion rates and efficiency compared to ADAF, we call it Luminous Hot Accretion Flow (LHAF).

2 Results of Global Solution

We obtain the global slution of the Luminous Hot Accretion Flow (Yuan 2001). We use the exactly same equations with the canonical ADAF. The two-temperature assumption is adopted and the same radiative mechanisms with ADAF are included. The coupled radiation hydrodynamical equations are solved completely self-consistently as in Nakamura et al. (1997).

Figure 1 shows the dynamics of some examples of ADAF and LHAF. Figure 2 shows their energy relationship. The main results are as follows.

- A new hot solution exists when the accretion rate $\dot{m}$ satisfies $\dot{m}_1 < \dot{m} < \dot{m}_2 \sim 5\dot{m}_1$. In this case, the flow remains hot throughout ths disk from the outer boundary to the horizon;
- When $\dot{m} > \dot{m}_2$, up to Eddington rate, Coulomb interaction will cool the inner region of the disk within a certain radius $\sim$ several–tens of Schwarzschild radii, therefore the disk collaps onto the equatorial plane and form an optically thick cold disk;
- Compared with ADAF, the electrons temperature in the luminous hot solution is lower, while the scattering optical depth is larger.

3 Discussion

Previous authors didn't find the existence of this solution because they apriori set the advection factor "f" (Narayan & Yi 1995; Esin et al. 1997) or advection parameter "ξ"(Abramowicz et al. 1995) as positive.

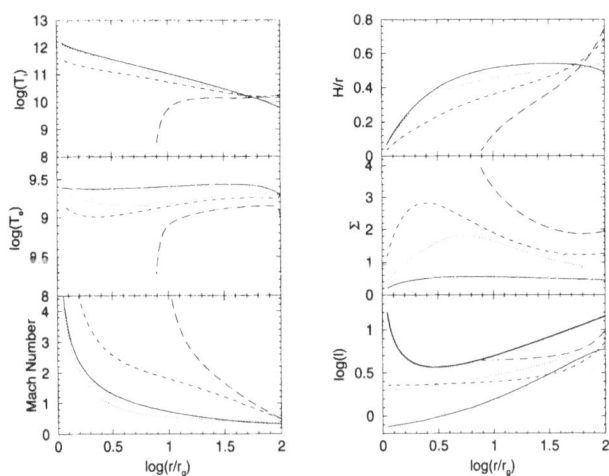

Fig. 1. The variations with the radii of some quantities for some *hot* accretion solutions. Solid (ADAF): $\dot{m} = 0.05 < \dot{m}_1$; dotted (critical ADAF): $\dot{m} = 0.1 \approx \dot{m}_1$; dashed (new hot solution): $\dot{m} = 0.3 < \dot{m}_2$; long-dashed (new hot solution): $\dot{m} = 0.5 > \dot{m}_2$; All are for $\alpha = 0.3$, $M = 10 M_\odot$ and $r_{\text{out}} = 100 r_g$. The units of Σ and T are g cm^{-2} and K while l is in c=G=M=1 units.

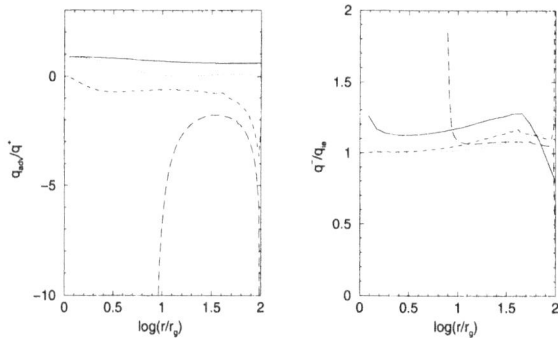

Fig. 2. Energy balance relationship for ions(left) and electrons(right) for the four solutions presented in Figure 1. The electron energy equation is $q_{\mathrm{adv,e}} = q_{\mathrm{ie}} - q^-$.

We show that LHAF is thermally unstable under local perturbation (Yuan et al., in preparation). By analogy to cooling flow in galaxtic clusters, which is dynamically similar to LHAF, we expect some cold dense clumps may form in the hot flow as a result of thermal instability. Then the Comptonization of the soft photons emitted by the clumps may dominate the cooling of LHAF.

LHAF can be applied to luminous X-ray sources such as black hole X-ray binary and AGNs. Assuming that thermal Comptonization is responcible for the X-ray emission of Seyfert 1 galaxies (Zdziarski 2000), we have shown that LHAF predicts exactly the parameters needed to fit the observed X-ray spectrum of these source while ADAF doesn't (Yuan et al., in preparation).

References

1. Abramowicz, M.A., Chen, X., Kato, S., Lasota, J.-P., & Regev, O. 1995, ApJ, 438, L37
2. Esin, A.A., McClintock, J. E., & Narayan, R. 1997, ApJ, 489, 865
3. Nakamura, K. E., Kusunose, M., Matsumoto, R., & Kato, S. 1997, PASJ, 49, 503
4. Narayan, R., Mahadevan, R., & Quataert, E. 1998, in "The Theory of Black Hole Accretion Discs", eds. M.A. Abramowicz, G. Bjornsson, and J.E. Pringle, (Cambridge University Press)
5. Narayan, R. & Yi, I. 1995, ApJ, 444, 231
6. Yuan, F., 2001, MNRAS, 324, 119
7. Zdziarski, A.A., 2000, Invited review for IAU Symp. 195, Highly Energetic Physical Processes and Mechanisms for Emission from Astrophysical Plasmas, P. C. H. Martens, S. Tsuruta, & M. A. Weber, eds., ASP, pp. 153-170

Part X

QSO, AGN, Blazars as Probes of the Universe

Exploring the Intergalactic Medium with VLT/UVES

Stefano Cristiani[1,2], Simone Bianchi[3], Sandro D'Odorico[3], and Tae-Sun Kim[3]

[1] ST European Coordinating Facility, K.-Schwarzschild-Str. 2, D-85748 Garching
[2] Osservatorio Astronomico di Trieste, via Tiepolo 11, I-34131 Trieste
[3] European Southern Observatory, K.-Schwarzschild-Str. 2, D-85748 Garching

Abstract. The remarkable efficiency of the UVES spectrograph at the VLT has made it possible to push high-resolution, high-S/N ground observations of the Lyα forest down to $z \sim 1.5$, gaining new insight into the physical conditions of the intergalactic medium and its evolution over more than 90% of the cosmic time. The universal expansion, the UV ionizing background and the gravitational condensation of structures are the driving factors shaping the number density and the column density distribution of the absorbers. A (limited) contribution of UV photons produced by galaxies is found to be important to reproduce the observed evolutionary pattern at very high and low redshift. The Lyman forest contains most of the baryons, at least at $z > 1.5$, and acts as a reservoir for galaxy formation. The typical Doppler parameter at a fixed column density is measured to slightly increase with decreasing redshift, but the inferred temperature at the mean density is increasing with redshift. The signatures of HeII reionization and feedback from the formation of galactic structures have possibly been detected in the Lyman forest.

1 The Observations

A sample of 8 QSOs with $1.7 < z_{\rm em} < 3.7$ has been observed [11,12] with VLT/UVES at a typical resolution 45000 and S/N $\sim 40 - 50$. Thanks to the two-arm design of the spectrograph [4], a high efficiency has been achieved in the whole optical range, from the atmospheric cutoff to $1\mu m$, which translates immediately in the possibility of obtaining new results on the Lyman forest, especially at $z \lesssim 2.5$. The data have been reduced with the UVES pipeline [1] - an non-negligible factor in maximizing the scientific output per unit time - and analyzed with the VPFIT package [3].

2 The Number Density and Column Density Evolution of Lyα Lines

The swift increase of the number of absorptions (and the average opacity) with increasing redshift is the most impressive property of the Lyα forest. Fig. 1 shows the number density evolution of the Lyα lines [11,12] in the column density interval [1] $N_{HI} = 10^{13.64-16}$ cm^{-2}. The maximum-likelihood fit to the data at $z > 1.5$ with the customary power-law parameterization provides $N(z) = N_0(1+z)^\gamma = (6.5 \pm 3.8)\,(1+z)^{2.4 \pm 0.2}$. The UVES observations imply that the turn-off in the evolution does occur at $z \sim 1$, not at $z \sim 2$ as previously suggested.

---

[1] This range in N_{HI} has been chosen to allow a comparison with the HST Key-Programme sample at $z < 1.5$ [22] for which a threshold in equivalent width of 0.24 Å was adopted.

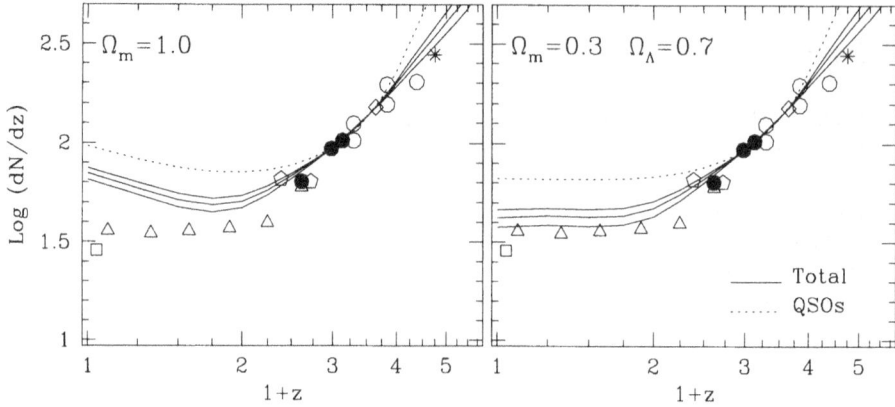

Fig. 1. Number density evolution of the Lyα forest with $N_{HI} = 10^{13.64-16}$ cm^{-2}. Dotted lines refer to the evolution compatible with an ionising UV background due only to QSOs. Solid lines show the expected evolution when both QSOs and galaxies contribute to the background, for models with f_{esc}=0.05 (upper line), 0.1 and 0.4 (lower line). Data points come from several observations in the literature, as given by [11]. The modelled evolution has been normalized to the observed evolution in the redshift range $2 < z < 3$. [2]

While the opacity is varying so fast, the column density distribution stays almost unchanged. The differential density distribution function measured by UVES [11,12], that is the number of lines per unit redshift path and per unit N_{HI} as a function of N_{HI}, basically follows a power-law $f(N_{HI}) \propto N_{HI}^{-1.5}$ extending over 10 orders of magnitude with little, but significant deviations: the slope β of the power-law in the range $14 \lesssim \log N_{HI} \lesssim 16$ goes from about -1.5 at $< z >= 3.75$ to -1.7 at $z < 2.4$. Recent HST STIS data [6] confirm that this trend continues at lower redshift, measuring a β of -2.0 at $z < 0.3$.

3   The Evolution of the Lyα Forest and the Ionizing Background

The evolution of the $N(z)$ is governed by two main factors: the Hubble expansion and the metagalactic UV background (UVB). At high z both the expansion, which decreases the density and tends to increase the ionization, and the UVB, which is increasing or non-decreasing with decreasing redshift, work in the same direction and cause a steep evolution of the number of lines. At low z, the UVB starts to decrease with decreasing redshift, due to the reduced number and intensity of the ionizing sources, counteracting the Hubble expansion. As a result the evolution of the number of lines slows down.

Up to date numerical simulations [18] have been remarkably successful in qualitatively reproducing the observed evolution, however they predict the break in the dN/dz power-law at a redshift $z \sim 1.8$ that appears too high in the light of the new UVES results. This suggests that the UVB implemented in the simulations may not be the correct one: it was thought that at low redshift QSOs are the main source of ionizing photons, and, since their space density drops below $z \sim 2$, so does the UVB. However, galaxies can produce a conspicuous ionizing flux too, possibly more significant than it was thought[17]. The galaxy contribution can keep the UVB relatively high until at $z \sim 1$ the global star formation rate in the Universe quickly decreases, determining the qualitative change in the number density of lines.

Under relatively general assumptions, it is possible to relate the observed number of lines above a given threshold in column density or equivalent width to the expansion, the UVB, the distribution in column density of the absorbers and the cosmology [5]:

$$\left(\frac{dN}{dz}\right)_{>N_{HI,\lim}} = C\left[(1+z)^5 \Gamma_{\mathrm{HI}}^{-1}(z)\right]^{\beta-1} H^{-1}(z), \tag{1}$$

where Γ_{HI} is the photoionization rate and β the power-law index of the N_{HI} distribution.

To estimate Γ_{HI} we have investigated the contribution of galaxies to the UVB[2], exploring three values for the fraction of ionizing photons that can escape the galaxy interstellar medium, $f_{esc} = 0.05, 0.1$ and 0.4 (the latter value corresponds to the Lyman-continuum flux detected by [17] in the composite spectrum of 29 Lyman-break galaxies). Measurements of the UVB based on the proximity effect at high-z and on the Hα emission in high-latitude galactic clouds at low-z provide an upper limit on $f_{esc} \lesssim 0.1$, consistent with recent results on individual galaxies both at low-z [7,10] and at $z \sim 3$ [8]. Introducing a contribution of galaxies to the UVB, the break in the Lyα dN/dz can be better reproduced than with a pure QSO contribution [2]. The agreement improves considerably also at $z \gtrsim 3$. Besides, models with $\Omega_\Lambda = 0.7$, $\Omega_M = 0.3$ describe the flat evolution of the absorbers much better than $\Omega_M = 1$. A consistency check is provided by the evolution of the lower column density lines. For $\log N_{HI} \lesssim 14$ the N_{HI} distribution follows a flatter slope β, and according to Eq. 1 this translates directly into a slower evolutionary rate, which is consistent with the UVES observations[11]: $dN/dz_{(13.1 < N_{HI} < 14)} \propto (1+z)^{1.2 \pm 0.2}$. Another diagnostic can be derived from the spectral shape of the UVB and its influence on the intensity ratios of metal lines [14,16].

4   Mapping the Column Density Distribution into the Mass Distribution of the Gas

It is instructive to transform the observed column density distribution in the mass distribution of the photoionized gas (Fig. 2) and interpret it, following Schaye[15], as a function of the matter density contrast: 1) the flattening at $\log N_{HI} \lesssim 13.5$ is partly due to line crowding and partly to the turnover of the density distribution below the mean density; 2) the steepening at $\log N_{HI} \gtrsim 14$, with a deficiency of lines that becomes more and more evident at lower z, reflects the fall-off in the density distribution due to the onset of rapid, non-linear collapse 3) the flattening at $N_{HI} \gtrsim 10^{16}$ cm^{-2} can be attributed to the flattening of the density distribution at density contrast $\gtrsim 10^2$ due to the virialization of collapsed matter. The differential mass density distribution has a sort of universal form when plotted as a function of the density contrast. A given density contrast, however, corresponds to lower and lower column densities with decreasing redshift, and this causes a translation of the mass density distribution (Fig. 2) towards

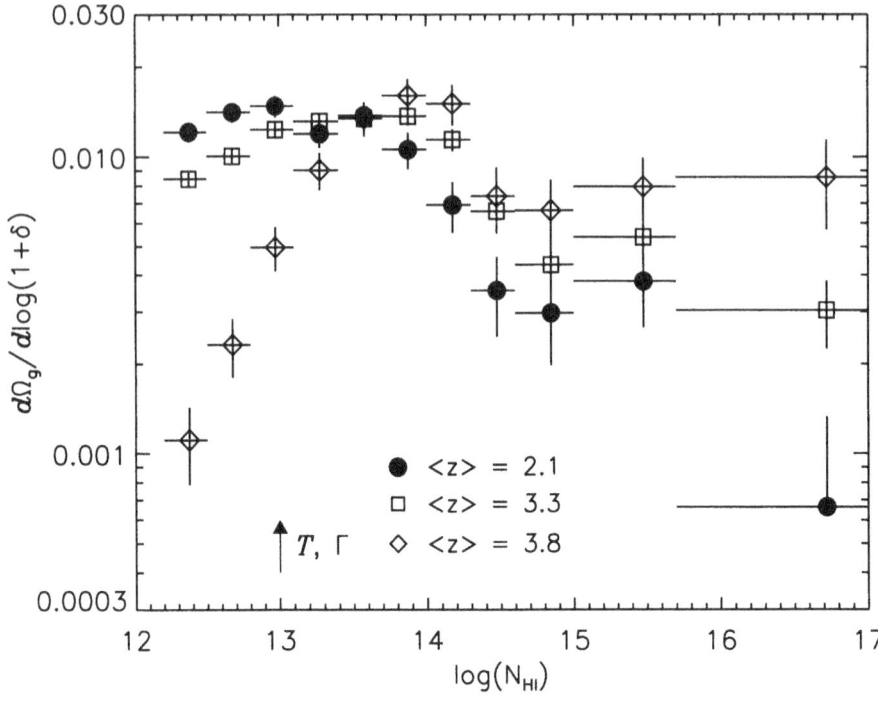

Fig. 2. The differential mass density distribution of the Lyα forest as a function of N_{HI}. The arrow indicates the direction towards which the points move if the temperature or the ionization rate increase [12],

the left with decreasing redshift, which explains the steepening of the slope β reported in Sect.2. Hydrodynamical simulations successfully reproduce this behaviour, indicating that the derived matter distribution is indeed consistent with what would be expected from gravitational instability.

5   The Cosmic Baryon Density

The amount of baryons required in a given cosmological scenario to produce the observed opacity of the Lyman forest can be computed [21] under general assumptions. A lower-bound to the cosmic baryon density can be derived from the mean Lyα flux decrement, $\overline{D}$,[13] and/or from the distribution of the Lyα optical depths. The limits derived from the effective optical depths measured in the UVES spectra at $1.5 < z < 4$ are reported in Tab. 1. They are consistent with the BBN value for a low D/H primordial abundance. Most of the baryons reside in the Lyman forest at $1.5 < z < 4$ with little change in the contribution to Ω as a function of z. Conversely, given the observed opacity, a higher UVB requires a higher Ω_b. As pointed out by [9], an escape fraction as large as 0.4, as measured by [17], would result in an $\Omega_b \sim 0.06$ in conflict either with the primordial D/H abundance or in general with the BBN or with the Lyα opacity measurements.

Table 1. Lower limits to $\Omega_b h^{1.5}$ derived from the UVES spectra at $1.5 < z < 4$ (for a universe with $\Omega_m = 0.3$, $\Omega_\Lambda = 0.7$)

UVB	$T = 2 \cdot 10^4$ K	$T = 6 \cdot 10^3$ K
QSOs	0.017	0.011
QSOs + GALs ($f_{esc} = 0.1$)	0.028	0.018

6 The Temperature of the IGM

If the Lyα forest is in thermal equilibrium with the metagalactic UV background, the line width of the absorption lines, described by the b parameter of the Voigt profile, is directly related to the gas temperature of the absorbing medium determined by the balance between adiabatic cooling and photo-heating: $b = \sqrt{2kT/m_{ion}}$. Additional sources of broadening exist, such as the differential Hubble flow across the absorbers, peculiar motions, Jeans smoothing. However, there is a lower limit to the line widths, set by the temperature of the gas, that is in principle measurable. In practice the situation is slightly more complex because for a photoionized gas there is a temperature-density relation, the so-called equation of state: $T = T_0 (1 + \delta_b)^{\gamma_T - 1}$, where T is the gas temperature, T_0 is the gas temperature at the mean gas density, δ_b is the baryon over-density, $(\rho_b - \bar{\rho}_b)/\bar{\rho}_b$ and γ_T is a constant which depends on the ionization history. The equation of state translates into a lower cutoff $b_c(N_{HI})$ in the N_{HI}-b distribution.

The observed cut-off Doppler parameter at a fixed column density of $\log_{N_{HI}} = 13.6$, $b_c(13.6)$ is measured to increase with decreasing redshift, while the slope of the cutoff does not change significantly. A typical value of $b_c \sim 18$ km s^{-1} at $1.5 < z < 4$ corresponds to a reference temperature of $2 \cdot 10^4$K. This does not mean that the temperature of the IGM increases with decreasing z: on the contrary, taking into account the equation of state and the fact that a given column density corresponds to higher and higher over-densities with decreasing redshift, it turns out that the temperature at the mean density is actually decreasing with decreasing z. Furthermore, evidence has been found [20] for a general increase of the temperature around the redshift $z = 3.3 \pm 0.15$, attributed to the reionization of HeII. Temperature and ionization fluctuations are also expected due to feedback processes from the formation of galactic structures [19] and might have been observed in correspondence of some voids in the Lyα forest [11].

References

1. Ballester P., Modigliani A., Boitquin O., Cristiani S., Hanuschik R., Kaufer A., Wolf S., 2000 ESO Mess. 101, 31

2. Bianchi S., Cristiani S., Kim. T.-S., 2001, A&A 376, 1
3. Carswell R.F., Webb J.K., Cooke A.J., Irwin M.J.,
   http://www.ast.cam.ac.uk/ rfc/vpfit.html
4. D'Odorico S., Cristiani S., Dekker H., Hill V., Kaufer A., Kim T.-S., Primas F., 2000, SPIE 4005, 121
5. Davé R., Hernquist L., Katz N., Weinberg D., 1999, ApJ 511, 521
6. Davé R., Tripp T.M., 2001, ApJ 553, 528
7. Deharveng J.-M., Buat V., Le Brun V., et al., 2001, A&A 375, 805
8. Giallongo E., Cristiani S., Fontana A., D'Odorico S., 2001, in preparation
9. Haehnelt M.G., Madau P., Kudritzki R., Haardt F., 2001, ApJ 549, L151
10. Heckman T.M., Sembach K.R., Meurer G.R. et al., 2001, ApJ 558, 81
11. Kim T.-S., Cristiani S., D'Odorico S., 2001, A&A 373, 757
12. Kim T.-S., Carswell R.F., Cristiani S., D'Odorico S., Giallongo E., 2001, MNRAS submitted
13. Oke J. B., Korcyansky D.G., 1982, ApJ 255, 11
14. Savaglio S., Cristiani S., D'Odorico S. et al., 1997, A&A 318, 347
15. Schaye J., astro-ph/0104272
16. Songaila A., 1998, AJ 115, 2184
17. Steidel C.C., Pettini M., Adelberger K.L., 2001, ApJ 546, 665
18. Theuns T., Leonard A., Efstathiou G., 1998, MNRAS 297, L49
19. Theuns T., Mo H. J., Schaye J., 2001, MNRAS, 321, 450
20. Theuns T., Zaroubi S., Kim T.-S., et al. astro-ph/0110600
21. Weinberg D.H., Miralda-Escudé J., Hernquist L., Katz N., 1997, ApJ 490, 564
22. Weymann R.J., Jannuzi B.T., Lu L. et al., 1998, ApJ 506, 1

Tracing the Remnants of Powerful Quasars to Probe the IGM

Torsten A. Enßlin[1], Rashid A. Sunyaev[1,2], and Marcus Brüggen[1,3]

[1] Max-Planck-Institut für Astrophysik, Karl-Schwarzschild-Str.1, Postfach 1317, 85741 Garching, Germany
[2] Space Research Institute (IKI), Profsoyuznaya 84/32, Moscow 117810, Russia
[3] Institute of Astronomy, Madingley Road, Cambridge CB3 0HA, United Kingdom

Abstract. Powerful quasars and radio galaxies are injecting large amounts of energy in the form of radio plasma into the inter-galactic medium (IGM). Once this nonthermal component of the IGM has radiatively cooled the remaining radio emission is difficult to detect. Two scenarios in which the fossil radio plasma can be detected and thus be used to probe the IGM are discussed: a) re-illumination of the radio emission due to the compression in large-scale shock waves, and b) inverse Compton scattered radiation of the cosmic microwave background (CMB), cosmic radio background (CRB), and the internal very low frequency synchrotron emission of the still relativistic low energy electron population. We present 3-D magneto-hydrodynamical simulations of scenario a) and compare them to existing observations. Finally, we discuss the feasibility of the detection of process b) with upcoming instruments.

1 Fossil Radio Plasma

The jets of powerful radio galaxies inflate large cavities in the IGM that are filled with relativistic particles and magnetic fields. Synchrotron emission at radio frequencies reveals the presence of electrons with GeV energies. These electrons have radiative lifetimes of the order of 100 Myr before their observable radio emission extinguishes due to radiative energy losses. The remnants of radio galaxies and quasars are called 'fossil radio plasma' or a 'radio ghosts' [1]. Their existence as a separate component of the IGM is supported by the detections of cavities in the X-ray emitting galaxy cluster gas [2–11, and others]. In many cases associated radio emission and in a few cases a lack of such emission was found, as expected for aging bubbles of radio plasma. Such bubbles should be very buoyant and therefore rise in the atmosphere of a galaxy cluster. It is not clear yet if they break into pieces during their ascent and thereby are slowed down. Another possibility is that they are able to ascend up to the accretion shock of a galaxy cluster, where their further rise will be prohibited by the infalling gas of the accretion onto the cluster.

2 Shock Wave Re-Illumination

Whenever a radio ghosts is hit by a shock wave, which may originate either from a cluster merger or from the steady accretion of gas onto the still forming large scale structure, it is strongly compressed. The compression should be adiabatic since typical IGM shock speeds of a few 1000 km/s are expected to be well below the internal sound

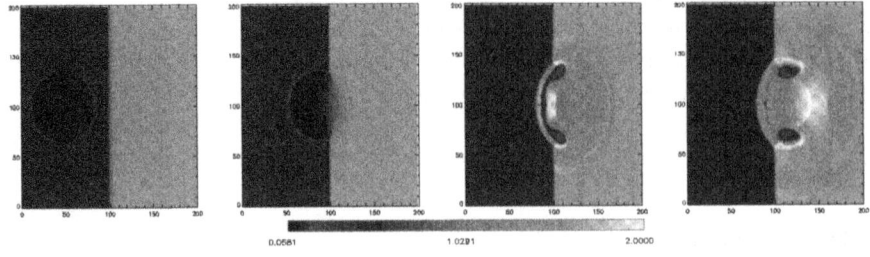

Fig. 1. Shock passage of a hot, magnetized bubble (a radio ghost) through a shock wave. The flow goes from the left to the right. The evolution of the mid plane gas density is displayed (white is dense, black is dilute gas)

speed of the fossil but still relativistic plasma. Since the radio plasma has to adapt to the new ambient pressure, the compression factor can be high and the particles and magnetic fields can gain a substantial amount of energy. The synchrotron emission can go up by a large factor, especially at frequencies which were only a little bit higher than the cutoff frequency of the uncompressed fossil plasma [12]. Thus, the radio plasma can be revived to emit at observing frequencies if it was not too old, a few 100 Myr inside and few Gyr at the boundary of galaxy clusters.

3-D magneto-hydrodynamical simulations [13] show that during the traversal of a shock wave, the radio plasma is first flattened and then breaks up into small filaments, often in form of one or several tori. The formation of a torus can be seen in Fig. 1. At places where the hot, under-dense radio plasma bubble touches the shock wave the balance of pre-shock ram-pressure and post-shock thermal pressure disappears due to the lack of substantial mass load of the advected radio plasma. The post shock gas therefore starts to break through the radio plasma and finally disrupts it into a torus or a more complicated filamentary structure.

The magnetic fields becomes mostly aligned with the filaments leading to a characteristic polarization signature which can be seen in the synthetic radio map displayed in Fig. 2.

Polarized radio emitting regions of often filamentary morphologies could be found in a (recently strongly growing) number of merging clusters of galaxies. In most cases they are near those places where shock waves are expected from either observed temperature structures or comparison of X-ray maps to simulated cluster merger. These radio sources are called cluster radio relics [14, and references therein]. An observed radio map of a filamentary cluster radio relic in Abell 85 is also displayed in Fig. 2 for comparison. Lower frequency observation show that the upper filament of this relic forms a closed torus [15].

Thus, sensitive observation of cluster radio relics are able to probe several properties of IGM shock waves [13]. Since the major diameter of the fossil radio plasma is approximately conserved, a rough estimate of the compression factor can be derived from relic radio maps by measuring the filament diameters. The compression factor depends only on the pressure jump in the shock and the equation of state of the radio plasma. Therefore, the shock strength is measurable for a given radio plasma equation of state. Or, if

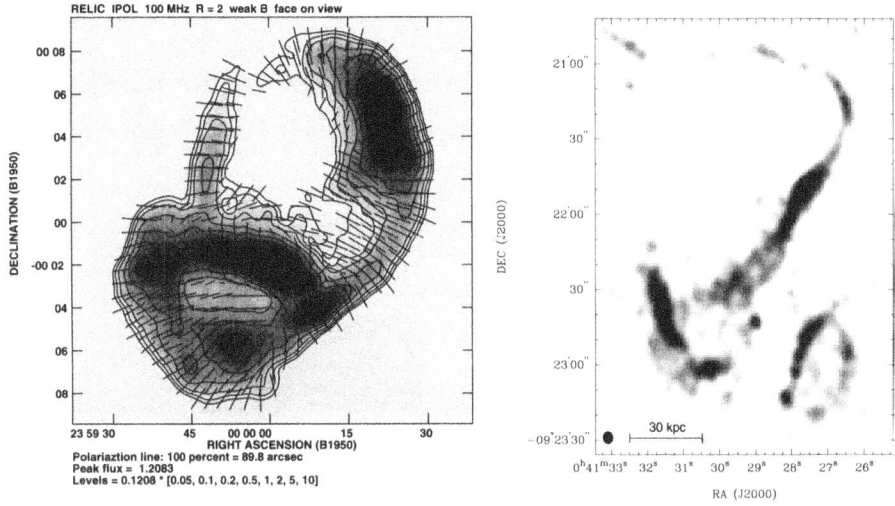

Fig. 2. Left: synthetic radio map of a shocked radio ghost [13]. Right: observed cluster radio relic in Abell 85 at 1.4 GHz [16]

detailed X-ray maps allow to estimate the shock strength independently, the equation of state of radio plasma can be measured. Furthermore, the total radio polarization of cluster radio relics (after averaging over the source) contains in principle enough information to entangle the 3-D orientation of the shock wave: The sky-projected electric vector is aligned with the projected shock normal. The polarization fraction is highly correlated with the angle between the shock normal and the line of sight [13].

3   SSC & CMB-IC

But even very old radio plasma may be detectable by its long lasting very low frequency radio emission (kHz – MHz). Even if this emission is undetectable directly for terrestrial telescopes, it can be measured indirectly due to the unavoidable inverse Compton (IC) scattering of the synchrotron photons by their source electrons [17]. These synchrotron self-Comptonized (SSC) photons have much higher energies and can therefore be in observable wavebands. The relativistic electron population also up-scatters every other present photon field. Photons of the cosmic microwave background (CMB) but also of the cosmic radio background (CRB) are removed from their original spectral location and shifted to much higher frequencies by IC encounters with the fossils radio plasma electron population. Since the frequency shift is large for IC scattering by ultra-relativistic electrons, the CMB flux is decremented within the whole typical CMB frequency range [18]. The spectral signature of all these processes are sketched in Fig. 3.

The strongest sources of such low frequency SSC emission should be radio lobes of just extinguished powerful radio galaxies (see Fig. 4 and [17] for details). If for

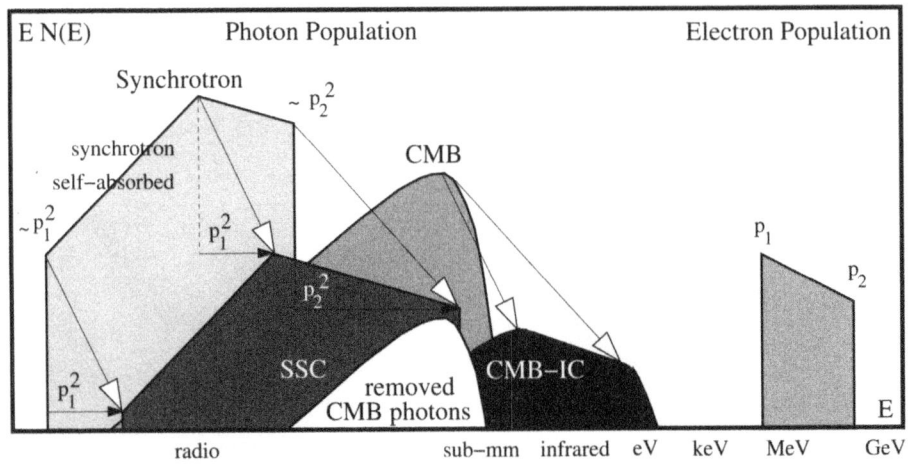

Fig. 3. Sketch of the SSC and the CMB-IC process

example the central engine of Cygnus A would decease today, its directly observable radio lobe synchrotron emission would vanish within some 10 Myr due to radiation and adiabatic losses of the (still) expanding radio plasma. But SSC and CMB-IC emission (or decrement) can remain for a few 100 Myr. Since the SSC emission is very sensitive to the compression state of the radio plasma it would decrease rapidly due to adiabatic expansion during the buoyant rise of the radio bubble in the cluster atmosphere. The CMB-IC process is much less sensitive to compression and would start to dominate the spectrum above 30 GHz after roughly 100 Myr.

Also our own galaxy might have produced radio ghosts during earlier active phases of the central black hole. Owing to the low density environment of the local group (compared to the Cygnus A galaxy cluster) such a ghost is expected to be very relaxed and to have a very low surface brightness (see Fig. 4). But it may be detectable because of its large angular scale. If further environmental compression would have increased its SSC and CMB-IC glow (in the displayed model a compression by a factor 11 in a 10 Gyr period was assumed) a future detection may be possible.

The detection of SSC from radio ghosts is an observational challenge. It would be rewarded by revealing the locations of fossil radio plasma graveyards. It would provide important information on the lower end of the relativistic electron population. This would be very valuable since the electron energy range above a few 10 keV and below 100 MeV is still an unexploited spectral regions. Further, due to the strong dependence of the SSC emission on the compression stage of radio plasma, SSC emission is also a sensitive probe of the IGM pressure.

Due to its broad frequency spectra, it can be probed with several future high sensitivity instruments, ranging from lowest radio frequency radio telescopes as GMRT and LOFAR, over microwave spacecrafts like MAP and PLANCK, balloon and ground based CMB experiments, and sub-mm/IR projects as ALMA and the HERSCHEL satellites. A multi-frequency sky survey, as will be provided by the Planck experiment,

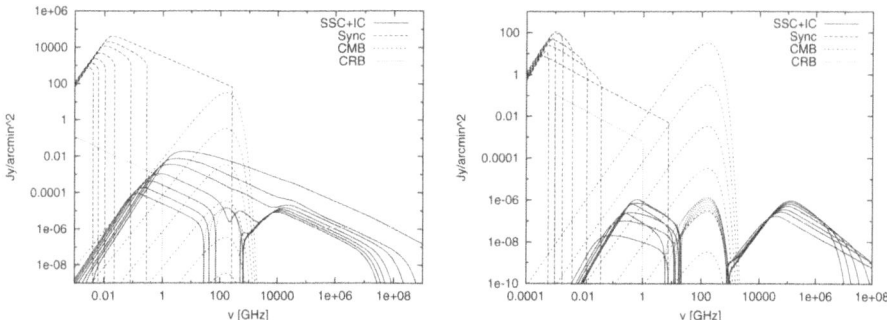

Fig. 4. Central surface brightness of a Cygnus A-like radio cocoon in a cooling and expanding phase (left) and of a possible radio ghost produced by our own galaxy during cooling and moderate compression (right). The synchrotron (long-dashed) and SSC+IC spectra (solid) are shown for the stages at the jet-power shutdown and for later stages (left: from top to bottom spectra at ages of 0, 20, 40, 80, 120, 160, and 200 Myr are displayed; right: from bottom to top spectra at ages of 0, 1, 2, 4, 6, 8, and 10 Gyr are displayed). In spectral regions, where the SSC+IC processes lead to a reduction of the brightness below the CMB brightness, the absolute value of the (negative) SSC+IC surface brightness is plotted by a dotted line. The top one of the short-dashed lines is the CMB spectrum, the short-dashed lines below this are $10^{-2}, 10^{-4}, 10^{-6}, 10^{-8}$, and 10^{-10} times the CMB spectrum for comparison of the source to the CMB brightness. The dotted power-law line at frequencies below 1 GHz is the cosmic radio background (CRB)

should allow to search for the SSC and relativistic IC spectral signature of many nearby clusters of galaxies and radio galaxies, at least in a statistical sense by co-adding the signals from similar sources [18]. In addition to this, there should be targeted observations of promising candidates, as e.g. the recently reported X-ray cluster cavities without apparent observable synchrotron emission. New and upcoming lower frequency (≤ 10 GHz) radio telescopes like LOFAR, GMRT, EVLA, and ATA should have a fairly good chance to detect such sources. E.g. a Cygnus-A like radio cocoon should be detectable for these telescopes out to a few 100 Mpc even ~ 100 Myr after the jetpower shutdown [17].

We hope that this work stimulates observational efforts to exploit the spectral landscape of the *terra incognita* of 1-100 MeV electrons residing in the intergalactic space via their combined SSC and CMB-IC emission.

References

1. T. A. Enßlin: In *Diffuse Thermal and Relativistic Plasma in Galaxy Clusters*, (1999) pp. 275, astro-ph/9906212
2. H. Böhringer, W. Voges, A. C. Fabian, A. C. Edge, D. M. Neumann: *MNRAS*, 264, L25–L28 (1993)
3. C. L. Carilli, R. A. Perley, D. E. Harris: *MNRAS* 270, pp. 173 (1994)
4. Z. Huang C. L. Sarazin: *ApJ* 496, pp. 728 (1998)

5. B. R. McNamara, M. Wise, P. E. J. Nulsen, L. P. David, C. L. Sarazin, M. Bautz, M. Marke-vitch, A. Vikhlinin, W. R. Forman, C. Jones, D. E. Harris: *ApJ Lett.* 534, pp. L135–L138 (2000)
6. A. C. Fabian, J. S. Sanders, S. Ettori, G. B. Taylor, S. W. Allen, C. S. Crawford, K. Iwasawa, R. M. Johnstone, P. M. Ogle: *MNRAS* 318, pp. L65–L68 (2000)
7. A. Finoguenov, C. Jones: *ApJ Lett.* 547, pp. L107–L110 (2001)
8. A. C. Fabian: In D. M. Neumann, editor, *XXIth Moriond Astrophysics Meeting on 'Galaxy Clusters and the High Redshift Universe Observed in X-rays'*, 2001.
9. B. R. McNamara: In F. Durret & D. Gerbal, editor, *Constructing the Universe with Clusters of Galaxies*, IAP (Paris, 2000) astro-ph/0012331.
10. S. Heinz, Y.-Y. Choi, C. S. Reynolds, M. C. Begelman: *ApJ Lett.* submitted (2001).
11. S Schindler, Castillo-Morales A, E De Filippis, A Schwope, J. Wambsganss: *A&A* submitted (2001)
12. T. A. Enßlin, Gopal-Krishna: *A&A* 366, pp. 26–34 (2001)
13. T. A. Enßlin, M. Brüggen: *MNRAS* submitted (2001) astro-ph/0104233
14. L. Feretti: In *Diffuse Thermal Relativistic Plasma in Galaxy Clusters* (1999) pp. 3
15. G. Giovannini, L. Feretti: *New Astronomy* 5, pp. 335–347 (2000)
16. O. B. Slee, A. L. Roy, M. Murgia, H. Andernach, M. Ehle: *AJ* 297, pp. 86 (2001)
17. T. A. Enßlin, R. A. Sunyaev: *A&A* submitted (2001)
18. T. A. Enßlin, C. R. Kaiser: *A&A* 360, pp. 417 (2000) astro-ph/0107432

Testing Models of the Lyα Forest

Avery Meiksin

Institute for Astronomy, University of Edinburgh,
Blackford Hill, Edinburgh EH9 3HJ, UK

Abstract. The structure of the Lyα forest is simulated for a variety of Cold Dark Matter (CDM) dominated cosmological scenarios: CHDM, OCDM, ΛCDM, SCDM, and tCDM. Synthetic spectra are constructed duplicating the resolution, wavelength and signal-to-noise ratio of a Keck HIRES spectrum of Q1937–1009. The predicted cumulative flux distribution of the ΛCDM model agrees with the measured to within 5%. This represents the most accurate agreement achieved to date between the predictions of CDM in the non-linear regime and the observed universe. Significantly larger median Doppler parameters are found in the measured spectra than predicted, indicating a possible need to introduce additional energy injection throughout the intergalactic medium, as may be provided by late He II reionization.

1 Introduction

The successful recovery of the observed statistical properties of the Lyα forest by numerical simulations in the context of various cold dark matter (CDM) dominated cosmologies suggests that the gravitational instability scenario provides a broadly accurate description of the development and evolution of the structure of the intergalactic medium (IGM) [3], [10], [5], [11], [1], [9]. The simulations show that nearly the entire IGM fragments into filaments, sheets, and fluctuations in underdense minivoids, all of which give rise to the absorption lines comprising the Lyα forest detected in the spectra of Quasi-Stellar Objects (QSOs) [3], [8], [12]. By analysing synthetic spectra drawn from the simulations, it has been demonstrated that the measured flux distributions, neutral hydrogen column density distributions, and the evolution in the number of absorption lines per unit redshift are reasonably well accounted for by the complex web of interconnecting structures found in the simulations.

For any given cosmological model, the properties of the Lyα forest are nearly completely determined. The only unknowns arise from the uncertainty in the reionization scenario. The intensity of the radiation field and, to a lesser extent, the temperature of the gas depend on the nature of the radiation sources and the manner in which they photoionized the IGM after the recombination era. The intensity of the ionizing background may be normalized by the measured average optical depth of the IGM. The determination of the temperature, however, is more difficult, since it will in general depend on the manner in which H II and He III regions grow and overlap during reionization, which is still unknown. Except for these uncertainties (and assuming galaxies do not have a large hydrodynamical impact on the IGM), the prediction of the structure of the IGM is a well-posed initial value problem, requiring only the solution to the coupled hydrodynamical and gravitational equations dictating the evolution of the structure in the non-linear regime. It should be possible for predictions of the structure of the IGM to reach a level of accuracy in the agreement with observations comparable to that of the CMB. While the results of the simulations are in generally good agreement with the measured properties of the Lyα forest, direct statistical comparisons between the models and the measured spectra have largely been lacking. In principle, with $N \simeq 10^4$ pixel elements per spectrum, it should be possible to predict the average flux distributions to an accuracy of $N^{-1/2} \approx 0.01$. Similarly, with several hundred absorption lines per spectrum, the simulations should be capable of reproducing the measured distributions of the H I column density and Doppler parameters to an accuracy of a few percent. The possibility of reaching this level of accuracy is demonstrated here. A more complete discussion is given in [7].

2 Analysis Method

The models used in the paper are described in detail in [6] and summarised in Table 1. All the models considered are in the context of CDM dominated cosmologies. We construct synthetic spectra matching the characteristics of the Keck HIRES spectrum of Q1937–1009 [2]. The redshift coverage is $3.126 < z < 3.726$, corresponding approximately to the range between Lyα and Lyβ in the QSO restframe, to avoid confusing Lyβ absorption with Lyα absorption.

Table 1. Parameters for the cosmological models. Ω_M is the total mass density parameter, Ω_Λ the cosmological constant density parameter, Ω_b the baryonic mass fraction, $h = H_0/100 \, \mathrm{km \, s^{-1} \, Mpc^{-1}}$, where H_0 is the Hubble constant at $z = 0$, n the slope of the primordial density perturbation power spectrum, $\sigma_{8h^{-1}}$ the density fluctuation normalization in a sphere of radius $8h^{-1}$ Mpc, and σ_J the density fluctuation at $z = 3$ on the scale of the cosmological Jeans length. For the CHDM model, two massive neutrino species are assumed accounting for a mass fraction of $\Omega_\nu = 0.2$

Model	Ω_M	Ω_Λ	$\Omega_b h^2$	h	n	$\sigma_{8h^{-1}}$	σ_J
CHDM	1	0	0.025	0.6	0.98	0.7	1.1
ΛCDM$_L$	0.4	0.6	0.015	0.65	1	1.0	1.7
ΛCDM$_H$	0.4	0.6	0.021	0.65	1	0.8	1.3
OCDM	0.4	0	0.015	0.65	1	1.0	2.2
SCDM	1	0	0.015	0.5	1	0.7	1.6
tCDM	1	0	0.025	0.6	0.81	0.5	0.9

There are several statistics that may be used to compare the models of the Lyα forest with the observed spectra. The most fundamental is simply the distribution of flux per pixel. Additional constraints are provided by the fluctuations in the spectra, which are related both to the underlying baryon density fluctuations and to the thermal and velocity widths of the associated absorption lines. We quantify the fluctuations using AUTOVP [4] to decompose the spectra into a set of Voigt absorption line profiles.

3 Results

3.1 Pixel Flux Distributions

We normalize the flux in the synthetic spectra by matching the mean H I optical depth $\bar{\tau}_\alpha$ to the measured intergalactic H I optical depth over the comparison redshift interval.

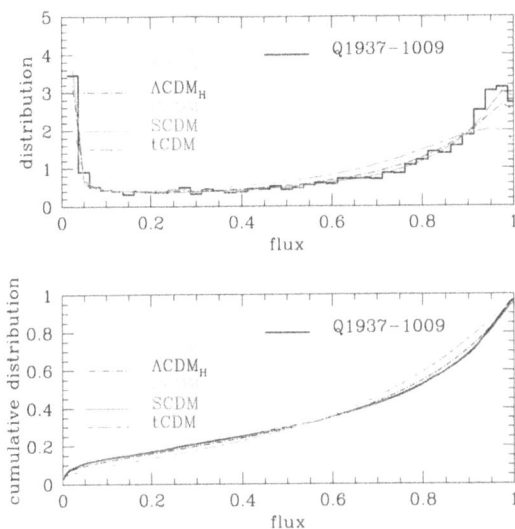

Fig. 1. Comparison between the measured flux distribution of Q1937−1009 and the model predictions. A continuum offset is applied to each model to enforce matching the cumulative distributions at flux values near unity. The SCDM model is preferred over $\Lambda\mathrm{CDM_L}$

The flux distributions for the different models, allowing for small continuum offsets, are shown in Fig. 1. The best agreement is found for the SCDM model, with a difference between the predicted and measured cumulative flux distributions of $d_{\mathrm{KS}} = 0.016$, corresponding to a formal acceptance probability by the KS test of $P_{\mathrm{KS}} = 0.011$. By comparison, the next best models are $\Lambda\mathrm{CDM_H}$ and $\Lambda\mathrm{CDM_L}$, both of which are strongly rejected by the KS test. Allowing, however, for an effective reduction in the number of independent pixels due to the finite width of the lines and finite spectral resolution, the KS probabilities for the SCDM, $\Lambda\mathrm{CDM_H}$, and $\Lambda\mathrm{CDM_L}$ models become, respectively, $P_{\mathrm{KS}} = 0.36 - 0.68$, 0.002–0.037, and 0.001–0.024, still favouring SCDM, but not excluding the ΛCDM models. The predictions of the CHDM, OCDM and tCDM models are still strongly rejected.

3.2 Absorption Line Parameter Distributions

The H I column density and Doppler parameter distributions resulting from fitting absorption lines to the spectra using AUTOVP are shown in Fig. 2. The agreement in the

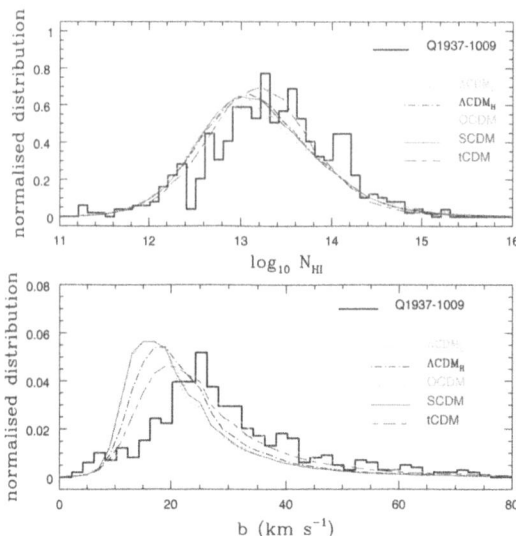

Fig. 2. Comparison between the measured absorption line parameter distributions of Q1937−1009 and the model predictions, allowing for an offset in the continuum level. While good agreement is generally found for the N_{HI} distributions, the predicted b distributions are generally skewed toward lower b-values than measured

distribution of N_{HI} between the data and all the models is reasonably good, with the prediction of CHDM providing the best match. The predicted b distributions for the remaining models peak at substantially lower values than does the measured distribution.

The measured N_{HI} distribution is best matched by the CHDM model: the maximum difference in the measured and predicted cumulative distributions is $d_{KS} = 0.083$, with the associated formal KS acceptance probability $P_{KS} = 0.002$. The CHDM model also provides the best match to the measured b distribution, with $d_{KS} = 0.071$ and $P_{KS} = 0.013$. The remaining models are very strongly rejected. All the models other than CHDM predict far too narrow lines. The narrowness of the absorption lines is reflected by the larger number of systems found in the AUTOVP analyses of the simulation results: more lines are required to recover the mean measured optical depth of the spectrum. The expected numbers of lines predicted by all the models except CHDM exceed the measured number by at least 3σ.

Absorption features with a line-centre optical depth $\tau_0 < 1$ correspond to structures that are underdense at $z = 3$, while higher optical depth systems are associated with overdense gas [12]. Because of this physical difference, we split the line samples into two subsamples at $\tau_0 = 1$ to determine if the disagreements between the measured and predicted N_{HI} and b distributions may be due to an incorrect thermal history rather than an incorrect cosmological model. The effect of this split is shown in Fig. 3. All of the models produce N_{HI} distributions for $\tau_0 > 1$ consistent with the data. The best case is CHDM, with $d_{KS} = 0.10$ and $P_{KS} = 0.041$. The ΛCDM_L and ΛCDM_H models

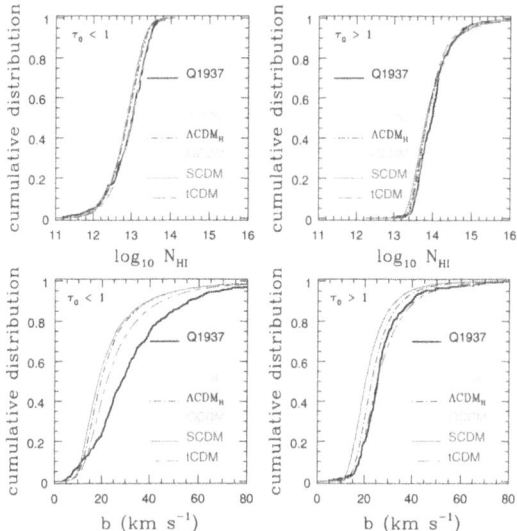

Fig. 3. Comparison between the measured absorption line parameter cumulative distributions of Q1937−1009 and the model predictions, for subsamples of lines optically thin ($\tau_0 < 1$) and optically thick ($\tau_0 > 1$) at Lyα line-centre. The agreement between the model predictions and the data is best for the optically thick systems. All the models predict much too low median b-values for the optically thin systems

give, respectively, $d_{KS} = 0.12$, $P_{KS} = 0.014$ and $d_{KS} = 0.11$, $P_{KS} = 0.018$. Only the tCDM model is able to recover the b distribution for $\tau_0 > 1$ ($d_{KS} = 0.06$, $P_{KS} = 0.50$). All the models predict a number of optically thick absorption systems consistent with the measured number.

The disagreement between the model predictions and measured distributions for the $\tau_0 < 1$ systems is more severe. None of the models are able to reproduce the measured b distribution. While the CHDM and tCDM models predict the correct shape for the N_{HI} distribution (both models give $d_{KS} = 0.06$ and $P_{KS} = 0.2$), they predict far too great a number of optically thin ($\tau_0 < 1$) lines compared with the measured number. Thus it appears the discrepancy between the model predictions and measured distributions arises primarily from the optically thin systems, although there is still a pronounced tendency for the models that agree best with the pixel flux distribution to predict too low b-values even for the optically thick ($\tau_0 > 1$) absorbers.

4   Discussion

The simulations are able to reproduce the measured flux per pixel distributions to extremely high accuracy. The best case is given by the SCDM model for Q1937−1009. The maximum difference between the measured and predicted cumulative flux distributions is $d_{KS} = 0.016$, close to the level of precision of the simulation. The flux distributions are found to vary systematically with σ_J, the density fluctuations on the scale of the cosmological Jeans length. As shown in Fig. 1, as σ_J increases, the distributions flatten (as measured between flux values of 0.2 and 0.8). Agreement between the predicted and measured flux per pixel distributions imposes the restriction $1.3 < \sigma_J < 1.7$ (better than 3σ), normalized at $z = 3$. Combining the allowed range of σ_J with recent CMB limits restricts the cosmological matter density for a flat universe to the range $0.26 < \Omega_M < 0.43$.

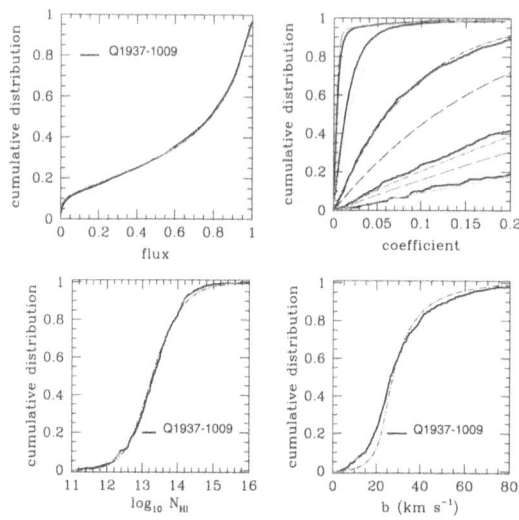

Fig. 4. The effect of additional heat input due to late He II reionization on the predictions of the ΛCDM_L model for Q1937−1009. Shown are the cumulative distributions of the flux per pixel (top left panel), wavelet coefficients (top right panel; see [7]), H I column density (bottom left panel) and Doppler parameter (bottom right panel). The effect of late reionization is mimicked by broadening the simulation optical depths by an additional $12\,\mathrm{km\,s}^{-1}$ (dot-dashed lines), corresponding to the expected increase in temperature due to the onset of sudden He II reionization. The original results without the added broadening are also shown (dashed curves), as well as the direct measurements from the spectrum of Q1937−1009 (solid curves). The extra broadening has very little effect on the flux per pixel distribution, but pronounced effects on the wavelet coefficients and fit absorption line parameters

A possible explanation for the discrepancy in absorption line widths is late reionization of He II at $z_{\mathrm{HeII}} \approx 3.5$. Late reionization will increase the temperature of the gas by an amount $\Delta T \approx 1.5 - 2 \times 10^4$ K. While the temperature of overdense gas will relax to lower values as a result of radiative losses, the lower density gas giving rise to the optically thin systems will retain a stronger memory of the boost in temperature, because it is too rarefied to radiate efficiently. We estimate the impact of the temperature boost for the ΛCDM_L model predictions for Q1937−1009 by convolving the simulation optical depths with a gaussian of velocity width $\sigma_T = 12\,\mathrm{km\,s}^{-1}$, corresponding to $\Delta T = 1.7 \times 10^4$ K. Excellent agreement is produced for the distributions of pixel flux, wavelet coefficients (on the 17–$34\,\mathrm{km\,s^{-1}}$ scale), H I column density, and the median Doppler parameter, as shown in Fig. 4. The extra broadening of the features has very little effect on the flux distribution, suggesting that the flux per pixel distribution provides a robust test of the models that is insensitive to uncertainties in the modelling of the reionization history of the IGM.

References

1. J. R. Bond, J. W. Wadsley: 'Lyα Absorption in the Cosmic Web'. In: *Structure and Evolution of the Intergalactic Medium from QSO Absorption Line Systems*, ed. by P. Petitjean, S. Charlot (Editions Frontières, Paris 1997) pp. 143–148
2. S. Burles, D. Tytler: A.J. **114**, 1330 (1997)
3. R. Cen, J. Miralda-Escudé, J. P. Ostriker, M. Rauch: Ap.J. **437**, L9 (1994)
4. R. Davé, L. Hernquist, D. H. Weinberg, N. Katz: Ap.J. **477**, 21 (1997)
5. L. Hernquist, N. Katz, D. Weinberg, J. Miralda-Escudé: Ap.J. **457**, L51 (1996)
6. M. E. Machacek, G. L. Bryan, A. Meiksin, P. Anninos, D. Thayer, M. L. Norman, Y. Zhang: Ap.J. **532**, 118 (2000)
7. A. Meiksin, G. L. Bryan, M. E. Machacek: MNRAS **327**, 296 (2001)
8. J. Miralda-Escudé, R. Cen, J. P. Ostriker, M. Rauch: Ap.J. **471**, 582 (1996)
9. T. Theuns, A. Leonard, G. Efstathiou: MNRAS **297**, L49 (1998)
10. Y. Zhang, P. Anninos, M. L. Norman: Ap.J. **453**, L57 (1995)
11. Y. Zhang, P. Anninos, M. L. Norman, A. Meiksin: Ap.J. **485**, 496 (1997)
12. Y. Zhang, A. Meiksin, P. Anninos, M. L. Norman: Ap.J. **495**, 63 (1998)

Detecting Filaments at z≈3

Johan P.U. Fynbo[1], Palle Møller[1], and Bjarne Thomsen[2]

[1] ESO, Garching, Germany
[2] IFA, Århus, Denmark

Abstract. We present the detection of a filament of Lyα emitting galaxies in front of the quasar Q1205-30 at z=3.04 based on deep narrow band imaging and follow-up spectroscopy obtained at the ESO NTT and VLT. We argue that Lyα selection of high redshift galaxies with relatively modest amounts of observing time allows the detection and redshift measurement of galaxies with sufficiently high space densities that we can start to map out the large scale structure at z≈2-3 directly. Even more interesting is it that a 3D map of the filaments will provide a new cosmological test for the value of the cosmological constant, Ω_Λ.

1 Introduction

For the past few decades computer simulations have been ahead of the observations when it comes to describing the first structures to form at high redshifts. The present consensus of the model builders is that the gas arranges itself in long string–like structures commonly referred to as filaments (see Fig. 1). Density variations along the filaments will lead to formation of lumps of cold, self–shielding HI regions and those regions are identified, in the simulations, as regions of starformation. Because of the high column density of neutral Hydrogen a sightline through such a cloud intersects, they are also identified as strong absorbers known as Damped Lyα Absorbers (DLAs). By poking random sightlines through a virtual universe one may simulate observations, and a given model universe will hence predict a specific correlation between DLA systems and the galaxies hosting the DLAs (e.g. Katz et al. 1996). Comparison to real observations of DLA galaxies (Møller & Warren 1998) has shown that there is very good agreement between observations and simulations. This agreement is encouraging, but it would be of great interest if one could observationally map out the actual filaments. Until now this his been done only at low redshifts (e.g. De Lapparent et al. 1991; Bharadwaj et al. 2000), but never at z>0.1. Knowing the distribution of scalesizes of filaments at different redshifts will help constrain the allowable parameter space of the simulations.

Unfortunately such a map cannot be constructed directly via absorption studies, because there is currently not a sufficiently tight mesh of background z>3 quasars available (e.g. Pichon et al. 2001). The best way to proceed is hence to attempt to find enough centres of starformation to be able to map out filaments by their own light. In order to identify objects to map out filaments one might at first guess that a search for Lyman Break Galaxies (LBGs, Steidel & Hamilton 1992) would be the best procedure. Unfortunately only the very brightest galaxies can be found and have their redshifts measured precisely enough with this technique, and such sparse sampling of the filamentary structure does not allow the structures to be seen. However, it has been shown

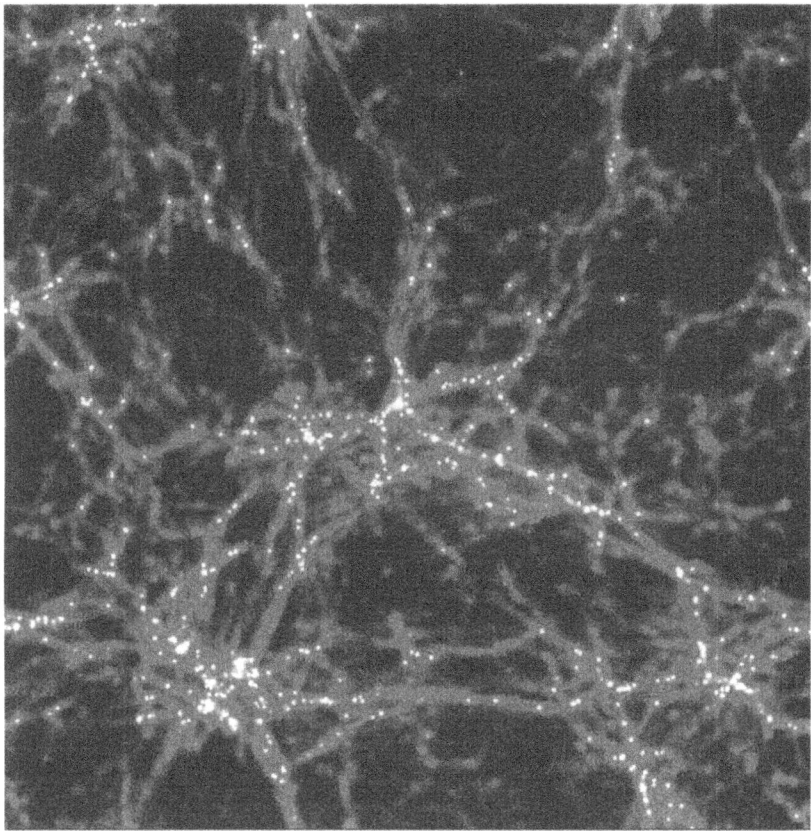

Fig. 1. A hydro simulation of a region of comoving size $12.5/h \times 12.5/h \times 1.6/h$ Mpc^3 at z=3 (courtesy of Tom Theuns, IoA, Cambridge). The image shows the density distribution. The white regions, corresponding to overdensities above 100, are distributed in filamentary structures and in the filament intersections.

that both DLA galaxies and galaxies selected for their Lyα emission, are sampling the high redshift galaxy population much further down the Luminosity function than do the LBGs, and one will therefore expect a better sampling of the high redshift structure if one uses DLA galaxies and Lyα galaxies (Fynbo, Møller & Warren 1999; Haehnelt et al. 2000). This has recently been independently confirmed, as deep narrow band Lyα imaging in a known overdensity of LBGs revealed about a factor of 10 more candidate Lyα galaxies than LBGs (Steidel et al. 2000).

2   Observations

In February through March 1998 we obtained deep narrow band imaging in a 21Å wide filter tuned to Lyα at z=3.04. The filter was tuned to the wavelength of a strong Lyα absorption line in the spectrum of the QSO (Fig. 2). The data were collected as service observing program at the 3.5-m ESO New Technology Telescope on La Silla, Chile. In total almost 18 hours of narrow band imaging was secured reaching a 5σ flux limit of 1.1×10^{-17} erg s^{-1} cm^{-2} (Fynbo, Thomsen & Møller 2000). We detected six good ($>5\sigma$) and two marginal ($\sim4\sigma$) candidate Lyα emitters in the field of the QSO as well as extended Lyα emission close to the QSO line of sight. In March 2000 we obtained Multi-Object follow-up spectroscopy at the ESO Very Large Telescope using FORS1 on the UT1 unit telescope. We also obtained deep imaging in the B and I bands (reaching B(AB)=26.7 and I(AB)=25.9 at 5σ).

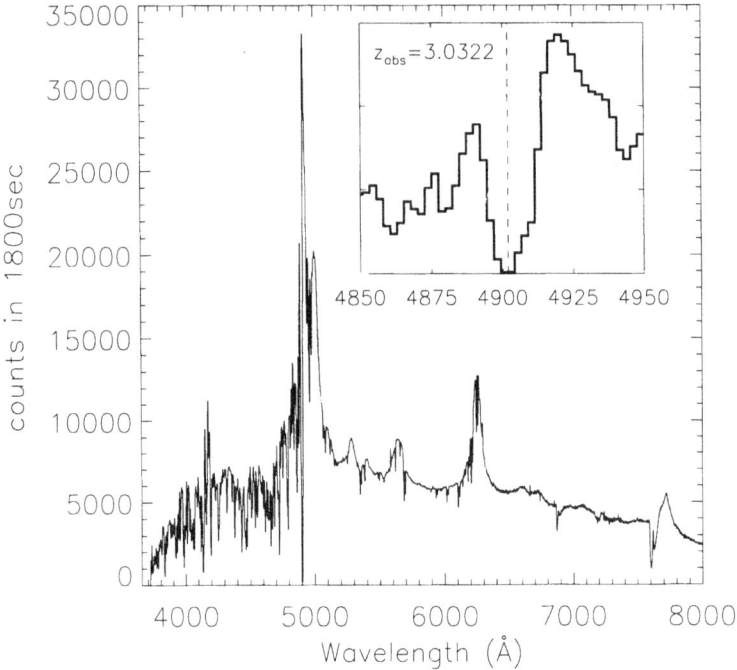

Fig. 2. The spectrum of Q1205-30 obtained at the VLT in March 2000. The insert in the uper right hand corner shows the region of the spectrum around a strong Lyα absorption line z=3.0322. The narrow filter (transmission curve overplotted) was tuned to Lyα at this redshift. The redshift of the background QSO, as measured from the low ionization OI line, is z=3.0473±0.0012.

3   Results

The VLT spectroscopy confirmed (by detecting the Lyα line and at the same time excluding the possibility of low redshift interlopers) all six good candidates and one of the two marginal candidates as Lyα emitters at z=3.04 (Fynbo, Møller & Thomsen 2001).

The spectral regions around Lyα for all confirmed candidates are shown in Fig. 3. The spectroscopy of the extended emission close to the QSO line-of-sigt will be presented in a separate paper (Weidinger et al. in preparation).

In Table 1 we present the redshifts and celestial positions for all confirmed Lyα emitters and for the absorber (from Møller & Fynbo 2001).

Table 1. Redshifts and positions of seven Lyα emitters and a Lyα absorber in the field of Q1205–30. The positions are given relative to the quasar coordinates: 12:08:12.7, -30:31:06.10 (J2000.0). The uncertainty on the redshifts is 0.0012 (1σ).

Object	ΔRA (arcsec)	Δdecl. (arcsec)	redshift
S7	-143.3±0.6	41.9±0.2	3.0402
S8	-141.5±0.6	59.7±0.2	3.0398
S9	-124.6±0.5	63.4±0.2	3.0350
S10	-119.9±0.5	59.8±0.2	3.0353
S11	-77.8±0.3	0.9±0.1	3.0312
S12	-43.9±0.2	54.4±0.2	3.0333
S13	68.3±0.3	-52.1±0.2	3.0228
abs	0.0	0.0	3.0322

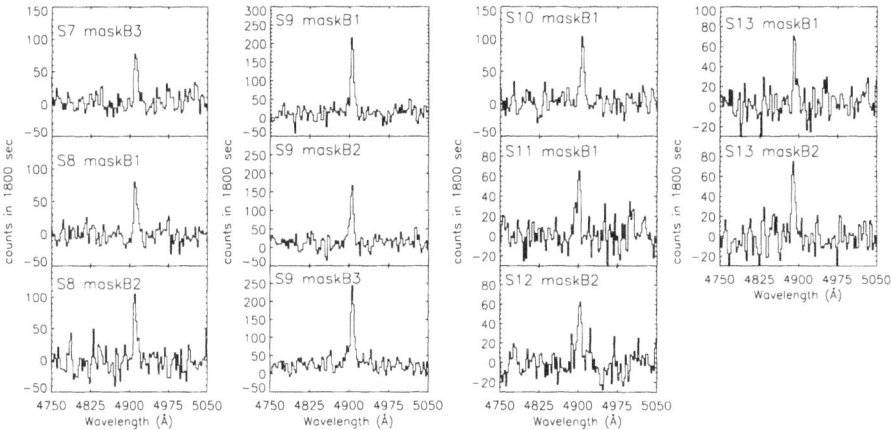

Fig. 3. The spectral regions around the Lyα emission lines for the 7 confirmed Lyα emitters (named S7–S13). For S8, S9, and S13 we show the spectra from several masks. As seen, the presence of emission lines is confirmed for all 7 candidates. By obtaining spectra in a red grism covering the spectral region from 5500Å to 7500Å we exclude the possibility of low redshift interlopers (e.g. OII emitters at z≈0.3, see Fynbo, Møller & Thomsen 2001 for details).

3.1 Filamentary Structure

In Fig. 4 we show the objects plotted in the box defined by the Field of View of the Camera and the redshift depth of the filter for Lyα at z=3. As seen the 7 Lyα emitters (marked with filled symbols) and the absorber (the open symbol) all align in this diagram. If we assume that the redshifts are all solely due to Hubble flow then this implies a real alignment in 3D space e.g. a filamentary spatial distribution of the objects. However, the measured redshift may not be due to Hubble flow alone for mainly two effects : *i)* outflows, and *ii)* peculiar velocities. We now briefly discuss the importance of each of these effects. *Outflows:* In the nearby starburst galaxy NGC1705 the outflow velocity is estimated to be around 80 km s^{-1} (Heckman et al. 2001 and references therein). Outflows of this strength will cause a shift in the redshift measurement which is of the same order as the combined uncertainty from the wavelength calibration and line centroid measurement (corresponding to 90 km s^{-1} at z=3). Outflows will either produce a systematic blueshift of the emission line redshift if the galaxies are opaque (so that we only see the gas moving towards us) or a broadening with no velocity shift of the lines (if the galaxies are transparent and we also see the gas moving away from us). The fact that the absorber, for which the redshift is detemined from the Lyα absorption line, also follows the alignment is an argument against a significant blueshift due to outflows. *Peculiar velocities:* In the local universe (v<4000 km s^{-1}) the $\sigma(v_{peculiar})$ of peculiar velocities is of the order 200 km s^{-1} (e.g. Branchini et al. 2001). We do not expect this number to be larger at z=3. Furthermore, any peculiar velocities will tend to smear out any underlying filamentary structure, so the fact that we see alignment is an argument against large peculiar velocities. We therefore conclude that the most likely interpretation of Fig. 4 is that we see a redshift z=3 filament.

3.2   Properties of the Filament

We can only determine a lower limit to the length of the filament as it seems to extend beyond the volume mapped by our instrumental setup (CCD and filter). Assuming a Hubble constant of 65 km s^{-1} Mpc^{-1}, $\Omega_m = 0.3$, and $\Omega_\Lambda = 0.7$ we find a coming length (defined as the distance between the two outhermost objects) of 4800 proper kpc. The radius of the minimum cylinder containing all objects is 400 proper kpc. Due to the effect of peculiar and outflow velocities this radius should be considered an upper limit.

The derived properties of filaments are strongly dependend on the assumed cosmology. In particular, since filaments are anchored in the Hubble flow, the observed angular distribution of a sample of filaments will be a function of the assumed cosmology. Therefore, it is in principle possible to use a sample of filaments to obtain an independend constrain on the value of the cosmological constant at z≈3 (Weidinger et al. 2001).

4   Summary and Outlook

In order to start mapping out the large scale filamentary structure suggested by numerical simulations directly at high redshift we need cosmic sources that are very numerous rather than rare, very bright light houses such as QSOs or Gamma-Ray Bursters. We have here demonstrated that by reaching flux limits below 1×10^{-17} erg s^{-1} cm^{-2} the density of Lyα emitting galaxies is sufficiently high at z=3 to allow a direct mapping of filamentary structure.

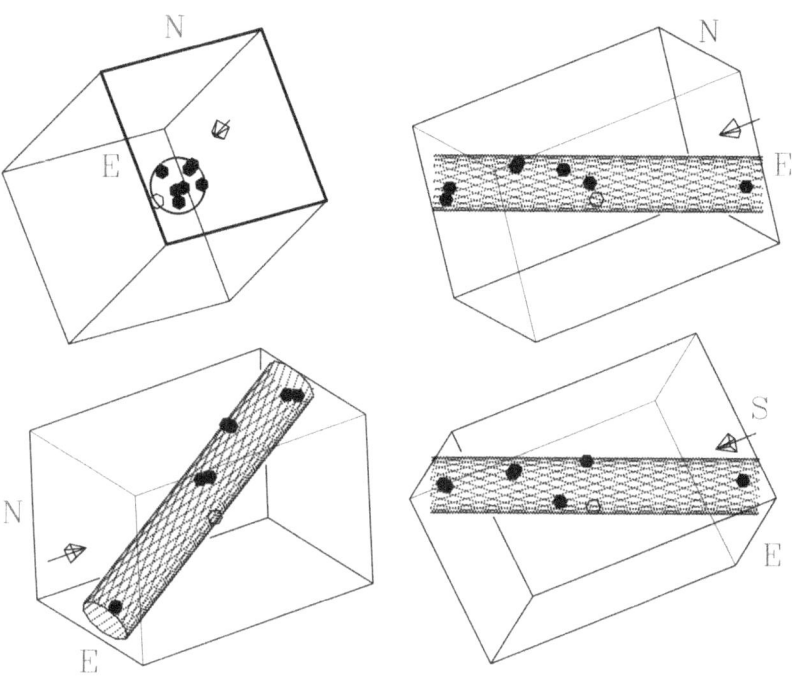

Fig. 4. 3D distribution of the eight objects seen from 4 different viewing–angles. In each of the figures the 3D arrow points in our viewing direction on the sky, and the spiral pattern maps out a cylinder with radius 400 kpc (see Table 1). The box marks the volume of space observed with our narrow–band Lyα filter. **Top left:** Here we have rotated the view to look along the filament. The thick lines mark the front "entrance window" of the box (corresponding to our CCD image). **Top right:** The box is here rotated 90 degrees to the right, hence viewing the filament from the left side compared to the end–on view. **Bottom right:** Same as top right but rotated 90 degrees around the filament to give a view of the filament as seen from "above" the view in top left. **Bottom left:** View from a random angle to give an impression of the 3D structure.

The next logical step is to try to map out larger regions of the z≈2–3 universe with Lyα emitters. Therefore we (Møller, Fynbo, Thomsen, Egholm, Weidinger, Haehnelt, Theuns) have initiated a large area survey for Lyα emitters at z=2 with the 2.56-m Nordic Optical Telescope on La Palma. Furthermore, in a pilot project conducted at the ESO VLT we (here Fynbo, Ledoux, Burud, Leibundgut, Møller and Thomsen) have obtained narrow band observations of two fields around QSO absorbers at z≈3. In the field of the z=2.85 absorber towards Q2138-4427, for which our imaging observations are complete, we reach a detection limit of about 7×10^{-18} erg s^{-1} cm^{-2} and detect 34 candidate Lyα emitters in a 45 arcmin^2 field over a redshift range of Δz=0.05. This shows that the density of z=3.04 Lyα emitters in the Q1205-30 field is not unusually high. Follow-up spectroscopic observations of the Q2138-4427 field has not yet been obtained. In the future we hope to map out a large volume with several hundred z=3 Lyα emitters with the VLT.

Acknowledgments

This paper is based on observations collected at the European Southern Observatory, La Silla and Paranal, Chile (ESO project No. 60.B-0843 and 64.O-0187).

References

1. S. Bharadwaj, V. Sahni, B.S. Sathyaprakash, S.F.
2. E. Branchini, W. Freudling. L.N. Da Costa, et al.: MNRAS **326**, 1191 (2001)
3. J.U. Fynbo, P. Møller, S.J. Warren: MNRAS **305**, 849 (1999)
4. J.U. Fynbo, B. Thomsen, P. Møller: A&A **353**, 457
5. J.U. Fynbo, P. Møller, B. Thomsen: A&A **374**, 443
6. M.G. Haehnelt, M. Steinmetz, M. Rauch: ApJ **534**, 594
7. T.M. Heckman, K.R. Sembach, G.R. Meurer, et al.: ApJ **554**, 1021 (2001)
8. N. Katz, D.H. Weinberg, L. Hernquist, J. Miralda-Escudé: ApJL **457**, L57 (1996)
9. V. De Lapparent, M.J. Geller, J.P. Huchra: ApJ **369**, 273 (1991) Shandarin, C. Yess: ApJ **528**, 21 (2000)
10. P. Møller, S.J. Warren: MNRAS **299**, 661 (1998)
11. P. Møller, J.U. Fynbo: A&AL **372**, L57 (2001)
12. C. Pichon, J.L. Vergely, E. Rollinde, S. Colombi, P. Petitjean: MNRAS **326**, 597 (2001)
13. C.C. Steidel, D. Hamilton: AJ **104**, 941
14. C.C. Steidel, K.L. Adelberger, A.E. Shapley, et al.: ApJ **532**, 170
15. M. Weidinger, P. Møller, J.U. Fynbo, B. Thomsen, M.P. Egholm: submitted to A&A (2001)

Distant Radio Galaxies in Their Environment with the 3D Spectroscopy OASIS/CFHT

Brigitte Rocca-Volmerange[1] and Emmanuel Moy[2]

[1] Institut d'Astrophysique de Paris, 98bis Bd Arago,F-75014 Paris
[2] Max-Planck-Institut für Extraterrestrische Physik, D-85748, Garching

Abstract. Increasing spatial resolution on Extended Emission Line Regions (EELRs) of high z radio sources allows to better understand the nature of the interacting processes between radio sources and their environments. In particular, it brings clues to explain the sources of gaseous emissions which are the main difference between radio galaxies and massive ellipticals. The 3D spectroscopy with the integral field units (IFUs) has an advantage on account of its capacity to map kinematics, emission line ratios and equivalent widths of extended radio sources. Recent observations of the FRII 3C171 (z = 0.284) with the IFU OASIS at CFHT are presented (Rocca-Volmerange et al, 2001). We propose the interpretation of the local emissions of this extended radio source, with a combination of AGN photoionizing and shock models(Moy & Rocca-Volmerange, 2001). The 3D distribution of emission lines confirms that the ionized gas is closely coupled to the radio emission, in particular along the radio jet propagation. Moreover far from the radio jet axis, the diffuse photoionizing radiation field is produced in the shocks generated by interactions at fringes of the ionized cocoon and on hot spots.
The number distribution of elliptical galaxies, predicted by fitting the deepest galaxy counts, is not in contradiction with the QSO distribution predicted from the recent modeling. Star formation histories in QSO and ellipticals could be similar at the highest redshifts.

1 Introduction

An essential debate on high-z ($z > 5$) radio galaxies is to relate galaxy hosts of radio sources to massive elliptical galaxies. Both show signatures of old stellar populations, are massive even at high z, and are located inside the central part of clusters. However all along the observed redshift range(z = 0 to 6), high-z radio galaxies show huge EELRs while distant elliptical galaxies rarely show emission lines. They are supposed to have ejected their gas content at early epochs (when they were 3Gyrs old), stopping any star formation activity and nebular emission (see the classical Matthews & Baker scheme).

Star formation traces are found at $z \simeq 4$ on the typical template spectrum of the radio source 4C41.17, these stellar signatures are probes of post-starburst populations (Dey et al, 1997). On the other side, the presence of a bulk of low-mass giant stars is identified in stellar continua of massive radio sources (see 3C 435A at z =0.371, Rocca-V. et al, 1994). The age of the stellar populations is from 8 to 10 Gyrs, measured with the help of the spectrophotometric models (as PÉGASE, Fioc & Rocca-Volmerange, 1997). These ages correspond to 14 Gyrs old at z=0, depending on cosmology, and are typical of elliptical galaxies.

The EELRs, extended on radial scales of 5-100 kpc from the nuclei of the host early-type galaxies (Baum et al, 1988), are not seen around ellipticals. This main difference of radio sources compared to ellipticals is due to the presence of an ionizing active nucleus. However some discussions arise when no sign of the broad cone-like emission-line distributions predicted by the unified schemes for powerful radio galaxies is detected in observations. In that case, the theory of quasar illumination reveals possible flaws. The extent of the EELRs may be interpreted by a series of other ionizing processes. Shocks may also contribute to gaseous emission as AGN illumination does. The best ways to solve this essential question is to map the emission zones of EELRs and to locally investigate various sources of ionizing emission in the close environment of radio sources. To summarize, of the many models that have been proposed to explain the alignment effect, the two that have received the most attention are the shock and AGN photoionization models, we plan to combine them with a fraction depending on the location inside the galaxy.

An important distinction of this study, compared to earlier works, is the availability for substantial subset of data of new integral field spectroscopy from OASIS/CFHT. The 3D spectroscopy observations of the radio source 3C171 (z=0.238) with OASIS (Rocca-Volmerange et al, 2001) achieve an unprecedented precision. As more sensitive probes of these two processes, line ratios in diagnostic diagrams, velocity fields and maps of FWHM and of equivalent widths can be analyzed with an increasing resolution. The balance from a process to another is followed all along the two-dimension maps, following the approach developed by Moy & Rocca-Volmerange, 2001, by fitting a series of strong classical emission line ratios.

2   Maps and Kinematics of 3C171 (z=0.238) with OASIS

The radiogalaxy 3C171 is a compact FRII radio galaxy (Baum et al, 1988), character-ized by two huge radio lobes (10" and 13"), extended perpendicularly to the radio axis (Heckman et al, 1984, Blundell, 1996). It is an excellent target to analyze the jet-cloud interaction as proposed by Clark et al, 1998. The integral field spectrograph OASIS at CFHT (Bacon et al, 1995) is the best instrument to map the spectral properties of extended objects. 3C171 was observed with OASIS during the 1999, 2000 campaigns, results are presented by Rocca-Volmerange et al, 2001. The H_α map (Fig. 1) shows a splendid cocoon structure, similar to most nearby radio galaxies, surrounded by ion-ized loops closely following the radio isophotes, even far from the central axis. The main emission lines between $H\beta$ and [SII]$\lambda\lambda6716,6731$ were observed through the two blue and red configurations, allowing to build most diagnostics to characterize pho-toionization by AGN, shocks or possibly star formation in HII regions(Robinson et al, 1987, Veilleux & Osterbrock, 1987). The map of each optical emission line is built and the distribution of line ratios are spatially plotted (Fig. 2). Velocity fields of three dif-ferent components fitted by gaussians and centred on the reference 63094km s^{-1} are distributed as following: $+250 kms^{-1} > v > -210 kms^{-1}$ (central), $-130 kms^{-1} > v > -1150 kms^{-1}$(blueshifted), $+960 kms^{-1} > v > +135 kms^{-1}$ (redshifted) with maps of FWHMs of the same components, confirming velocities of about 1000 kms^{-1} near bow shocks.

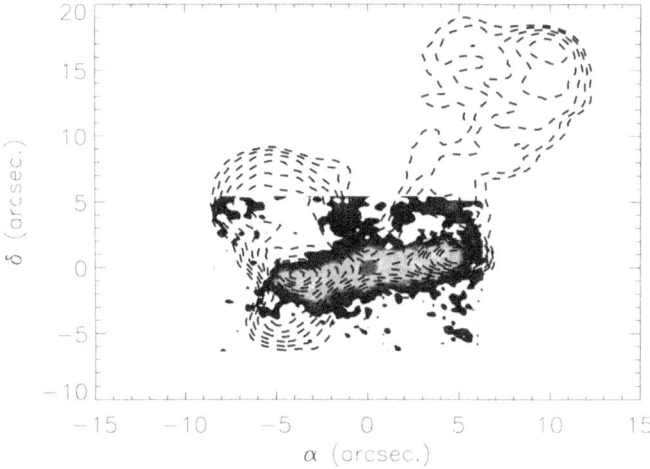

Fig. 1. Ha etendue

3 PÉGASE, CLOUDY and MAPPINGSIII to Analyze Continua and Line Ratio Maps of 3C171

An instantaneous starburst evolution model is computed with the code PÉGASE.2 (Fioc & Rocca-Volmerange, 1997 and available on the site **http://www.iap.fr/pegase**) or $ftp.iap.fr/pub/from\_users/pegase$. Continua witness a dominant stellar contribution while emission line ratios identify HII regions, LINERS, AGNs according to their location in the classical diagnostic diagrams.

To simultaneously predict continua and lines, the combination of the code PÉGASE with the photoionization code CLOUDY (Ferland, 1996)was built by Moy et al, 2001, respecting the metal enrichment and the number of Lyman continuum photons in the two codes. However in 3C171, as in most nearby radio galaxies, the ionization sources are preferentially attributed to two other dominant processes (AGN and shocks). Photoionization by AGN and ionizing shock emissions are respectively computed with the codes CLOUDY and MAPPINGSIII (Dopita & Sutherland 1996), respectively. The balance of the two processes and its main diagnostics in radio sources is analyzed by Moy and Rocca-Volmerange, 2001 in the optical. For this purpose, 7 various zones are localized on the cocoon of 3C171 (the [OIII]λ5007/$H\beta$ map), each of them is covered by several micro lenses. The selection is done in areas where emissions by shocks and AGNs have to be respectively compared (center, east center, east hot spot, west center, west hot spot, north extension). Fig. 3 shows the localization of these zones on the galaxy body and their characteristic colors are compared with model results.The adopted ionization parameter U is from 0.0001 to 0.1, the shock velocity v range extends from 100 to 1000 km s^{-1} and the density n_H from 0.1 to 10000 cm^{-3}.

4   Main Results: Combination of Photoionization by AGN + Shock Models

Maps of emission line ratios of 3C171 observed with OASIS are fitted in the line ratio diagnostic diagrams (Rocca-V. et al, 2001). The cocoon is confined in a highly dense medium. A unique model (either shock or AGN photoionisation) is unable to fit the distribution of data. Only a combination of models reproduces all data ,if the fraction locally varies. Results are then quantified, emission from the central part of the cocoon (inside 1/2 of the total length)is largely dominated by AGN illumination. As expected, the hot spot zones are strongly dominated by shocks $(200\text{-}500\text{kms}^{-1})$. The external cocoon fringes are also clearly dominated by shocks. A zone is surprinsigly identified by a high ionisation level, far from the nucleus. This South-West zone is presumed to be a knot of radiojet curvature. It requires deeper observations to be confirmed with a better S/N. These results are in excellent agreement with predictions of Moy & Rocca-Volmerange, 2001, which concluded to a sequence of ionization sources from the center to the interstellar medium: successively photoionization, shocks and photoionization. At the highest redshifts, due to the rapid fading of surface brigthness, only the central brightest parts of radio sources will be detected so that emission lines will appear dominated by photoionization processes.

5   Relation with Faint Counts of Galaxies

Because the stellar energy distributions of AGN galaxy hosts at $z \simeq 0$ present striking similarities with nearby massive ellipticals, the comparison of their respective populations at large z may bring constraints on their evolution and possible interactions. Faint galaxy counts down to the HDF (Williams et al, 1996) are fitted with classical scenarii of galaxy evolution (see figure 5 of Fioc & Rocca-Volmerange, 1999). The corresponding star formation histories were analyzed by spectral types (Rocca-Volmerange, 2001). While spiral galaxies are rapidly evolving from $z = 0$ to 1, as currently admitted, the population of elliptical galaxies dominates in number at higher $(z > 1)$ redshifts. That is due to star formation rates of ellipticals, consistent with the energy distributions of nearby templates and able to reproduce the energy distribution of the faintest galaxies of the Hubble Deep Field. The quasar counts computed with the luminosity function(Boyle et al, 2000) and a template spectrum of quasar show a similar change in the slope, evocating pure luminosity evolution scenarios for the two populations which could not to be uncorrelated. More precise data and modeling are needed before a more conclusive result.

6   Conclusions from Ionized Gas

A combination of shocks and photoionization by AGN is computed to interpret the emission lines of distant radio sources, frequently thought as the brightest lighthouses of the distant universe. From the 3D spectroscopy of 3C171, we showed that photoionization by the central AGN is dominant in the central part of the radio galaxy and in

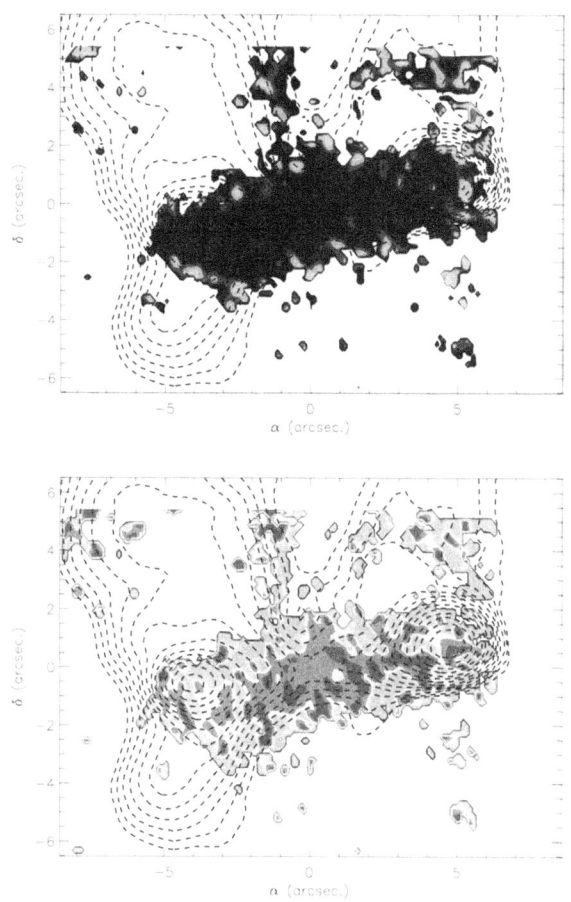

Fig. 2. Emission line ratios [OI]6300/H$_{alpha}$(up), [SII]6716/[SII]6731 (down) of the radiogalaxy 3C171 with OASIS.

zones between the center and the hot spots, even if aligned with the radio axis. However shocks are dominant near bow shocks and at the external fringes of the ionized cocoon. Moreover in this galaxy, an exemple of gas-cloud interaction is possible at the point of curvature of the radio axis. At last shocks are probable in the North extension along the bended radio isophotes. If shocks induce the gravitationnal collapse of clouds, followed by a cooling process, then stars may form in cool gas. Finally elliptical galaxy counts which dominate SFR at high redshifts could be related to radio sources counts through the jet-cloud interaction at high redshifts (Begelman & Cioffi, 1989).

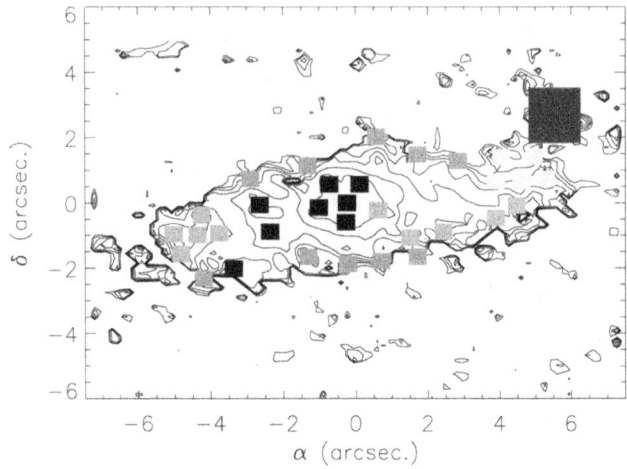

Fig. 3. Selection of preferential zones, localized on the [OIII]λ5007/Hβ map of 3C171 with OA-SIS. Various grey colors identify various zones (see Rocca-Volmerange et al, 2001 for colors)

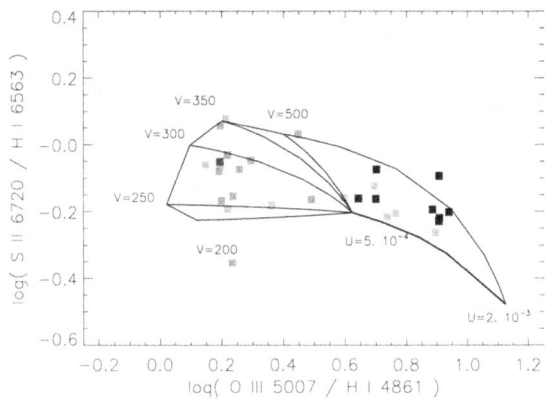

Fig. 4. Interpretation with shocks and AGN models of the [SII]λλ6716,6731/Hα vs. [OIII]λ5007/Hβ emission line ratios for the 3C171 zones plotted on fig. 3 (colors are identical). Sequences of the two component model resulting from a linear combination of shocks and AGN photoionization. The ionization parameter U is indicated at the AGN photoionization sequence extremities. The velocity V in km s^{-1} is given when the shock fraction reachs 100%. See Moy & Rocca-volmerange, 2001 for details.

References

1. Bacon, R., et al., 1995, A&A, **113**, 347
2. Baum, Stefi Alison; Heckman, Timothy M.; Bridle, Alan; van Breugel, Wil J. M.; Miley, George K., 1988, ApJS, **68**, 643
3. Begelman,M. C., Cioffi, D. F., 1989, ApJ., **345**, L21
4. Best P. N., Röttgering H. J. A., Lehnert M. D., 1999, MNRAS **310**, 223
5. Blundell, K., 1996, MNRAS,**283**, 538
6. Boyle, B. J.; Shanks, T.; Croom, S. M.; Smith, R. J.; Miller, L.; Loaring, N.; Heymans, C., 2000, MNRAS **317**, 1014
7. Clark N. E., Axon D. J., Tadhunter C. N., Robinson A., O'Brien P., 1998, ApJ **494**, 546
8. De Breuck C., Röttgering H., Miley G., van Breugel W., Best P., 2000, A&A, **362**, 519
9. Dey A., van Breugel W., Vacca W. D., Antonucci R., 1997, ApJ., **490**, 698
10. Dopita M. A., Sutherland R.S., 1996, ApJS., **102**, 161
11. Ferland G. J., 1996, University of Kentucky, Dpt Physics &Astronomy, Internal Report
12. Fioc & Rocca-Volmerange, 1999, Astronomy and Astrophysics, **344**, 393
13. Fioc & Rocca-Volmerange, 1997, A&A, **326**, 950
14. Heckman,T.M., van Breugel, W.J.M., Miley, G.K., 1984, ApJ., **286**, 509
15. Moy E., Rocca-Volmerange B., Fioc M., 2001, A&A **365**, 347
16. Moy E., Rocca-Volmerange B., 2001, in press for A & A
17. Robinson, A., Binette, L., Fosbury, R. A., Tadhunter, C., 1987, MNRAS, **227**, 97
18. Rocca-Volmerange B., 2001, in Galaxy Disks and Disk Galaxies, ASP Conferences series, vol 230, eds J. Funes & E. Corsini, p. 597
19. Rocca-Volmerange B., Moy, E., Adam G., preprint 2001
20. Rocca-Volmerange, B.; Adam, G.; Ferruit, P.; Bacon, R., 1994, A & A, **292**, 20
21. Veilleux, S., Osterbrock, D.E., 1987, ApJS **63**, 295
22. Williams R.E., Blacker B., Dickinson M., et al., 1996, AJ **112**, 1335

Spatial Distribution of Quasars in the 2QZ 10K Release and the SDSS Early Data Release

Andrei Doroshkevich[1], Douglas Tucker[2], Dan Vanden Berk[2], and Sahar Allam[3]

[1] Theoretical Astrophysics Center, Juliane Maries Vej 30,
DK-2100,Copenhagen Ø, Denmark
[2] Fermi National Accelerator Laboratory, MS 127,
P.O. Box 500, Batavia, IL 60510 USA
[3] National Research Institute for Astronomy & Geophysics,
Helwan Observatory, Cairo, Egypt

Abstract. We make use of the Minimal Spanning Tree technique to estimate some statistical characteristics of the spatial distribution of quasars in the 2QZ 10K Release and in the SDSS Early Data Release. Our investigation provides additional evidence for the small scale clustering of quasars found by Croom et al. (2001b); however, we find no evidence for clustering of quasars on very large scales.

1 Introduction

With the advent of such large redshift surveys as the Durham/UKST Galaxy Redshift Survey (Ratcliffe et al. 1996) and the Las Campanas Redshift Survey (Shectman et al. 1996), the galaxy distribution on scales of up to $\sim$300 h^{-1} Mpc could be studied. Now these investigations can be extended to even larger scales with the public data sets from the 2dF Redshift Survey (Colless et al. 1999) and the Sloan Digital Sky Survey (SDSS; York et al. 2000). In particular, both surveys include rich samples of optically selected quasars with $z \leq 3$. Thus, the 2dF QSO Redshift Survey 10K Release contains $\approx 10\,000$ QSOs with $b_J \leq 20.85$ and $z \leq 3$ (Croom et al. 2001a) while the SDSS Early Data Release (EDR) contains $\approx 4\,000$ quasars with $i^* \leq 20.5$ and $z \leq 5$ (Stoughton et al. 2001). These data sets allow us to investigate the structure in the distribution of quasars at high redshifts.

Numerical simulations (see, e.g., Governato et al. 1999) show that, at redshifts $z \sim 2 - 3$, the filamentary structure in the dark matter distribution can be clearly seen on scales up to $10 - 15h^{-1}$ Mpc ($h \equiv H_0/100$ km s^{-1} Mpc^{-1}). However, due to the small number density of quasars, any structure on these scales will be only sparsely sampled. Nonetheless, the analysis of the 2QZ survey (Croom et al. 2001b) reveals the existence of strong correlations in the distribution of quasars on scales $l \leq 60h^{-1}$ Mpc for a flat, lambda-dominated universe ($\Omega_m = 0.3$ and $\Omega_\Lambda = 0.7$), the model we use throughout this paper. It is interesting that this result is quite consistent with earlier estimates obtained for poorer samples of quasars (see, e.g., Osmer 1981; Komberg et al. 1994; more references are cited in Croom et al. 2001b).

In this paper we investigate the spatial distribution of both the 2QZ 10K and the SDSS EDR quasar samples by means of the Minimal Spanning Tree (MST) technique. In Secs. 2 and 3 we briefly describe this technique and the samples used. In Sec. 4 we present our main results, and in Sec. 5 we present our conclusions.

2   MST Technique

The MST is a *unique network* associated with a given point sample and connects all points of the sample to a *tree* in a special and unique manner which minimizes the full length of the tree. Cosmological applications of this technique are discussed in in Doroshkevich et al. (1999, 2001) and references therein.

$W_{\mathrm{MST}}(l)$, the distribution function of edge lengths l linking individual points within the MST, depends on the correlation functions (or cumulants) of all orders. For larger point separations — where correlations become small and the cumulants asymptotically approach constant values — a Poisson-like point distribution can be expected. For 1D and 2D Poissonian distributions, analytical expressions for the $W_{\mathrm{MST}}(l)$ are, respectively,

$$W_{\mathrm{MST}}(l) = \frac{1}{\langle l \rangle} e^{-l/\langle l \rangle} , \tag{1}$$

$$W_{\mathrm{MST}}(l) = 2 \frac{l}{\langle l^2 \rangle} e^{-(l^2/\langle l^2 \rangle)} . \tag{2}$$

There is no known analytical expression for 3D Poissonian distributions. Nonetheless, fits of simulated point 3D Poissonian distributions show that

$$W_{\mathrm{MST}}(l) \approx \frac{l^2}{\langle l^3 \rangle} \exp \left[-\frac{l^3}{\langle l^3 \rangle} - \beta_0 \left(\frac{l^3}{\langle l^3 \rangle} \right)^{\beta_1} \right] , \quad \beta_1 \approx 2 - 3 , \tag{3}$$

where β_0 and β_1 are constants. In this case [eq. (3)], the cutoff of $W_{\mathrm{MST}}(l)$ at $l \sim 2\langle l \rangle$ indicates the onset of percolation, the interval of linking lengths l in which all objects are suddenly connected to the MST (cf. Doroshkevich et al. 1999, 2001).

These three equations provide a reference in the analysis of $W_{\mathrm{MST}}(l)$'s constructed from real samples.

3   Observed Samples of Quasars

The 2QZ 10K Release is the first public release from the 2QZ redshift survey, a survey which covers two $75° \times 5°$ slices of sky towards the North and South Galactic Caps (Croom et al. 2001a). The SDSS EDR sample considered here is composed of two $90° \times 2.5°$ equatorial slices, also towards the North and South Galactic Caps (Stoughton et al. 2001). (The north slice of the 2QZ 10K Release overlaps somewhat with the north slice of the SDSS EDR.)

For the model under consideration the comoving distance to an object with a redshift z is defined as follows:

$$r = c \int_0^z \frac{dx}{H(x)} \approx \frac{3600}{\sqrt{\Omega_m}} \left[1 - \frac{1}{(1 + 0.78z + 0.37z^2)^{0.6}} \right] h^{-1}\mathrm{Mpc}, \tag{4}$$

where $H(z)$ is the Hubble parameter. Due to the flat geometry, the observed angular coordinates can be directly used. This approach is consistent with that was discussed by Osmer (1981) and Croom et al. (2001b).

Our analysis was performed for distances $1100h^{-1}\,\mathrm{Mpc} \leq r \leq 4100h^{-1}\,\mathrm{Mpc}$ for the 2QZ 10K Release and for distances $1500h^{-1}\,\mathrm{Mpc} \leq r \leq 4000h^{-1}\,\mathrm{Mpc}$ for the SDSS EDR, distances for which radial selection effects in the two samples are minimal. Some numerical characteristics of the samples are listed in Table 1, where N_{obj} is the number of objects in the given sample, $\langle n_{\mathrm{obj}} \rangle$ is the mean number density of objects, $\langle l_{\mathrm{MST}} \rangle$ is the mean edge length of the MST spanning the sample, $\langle b_{\mathrm{MST}} \rangle$ is the mean dimensionless edge length ($\langle b_{\mathrm{MST}} \rangle \equiv \langle l_{\mathrm{MST}} \rangle (4\pi/3\langle n_{\mathrm{obj}} \rangle)^{1/3}$), and s_0 and γ are respectively the scale length and power law index of the correlation function ($\xi(s) = (s/s_0)^{-\gamma}$; Croom et al. 2001b).

4   Main Results

On very large scales, these two quasar surveys are effectively two-dimensional, due to the small angular thicknesses of their slices. Thus, the expected characteristics of their samples could be sensitive to the influence of boundary effects. The characteristics may also be affected by strong field-to-field sampling variations, particularly for the 2QZ 10K Release. Therefore, the characteristics of the observed samples must be compared with the those obtained for random samples having the same general geometries and selection effects as the real samples.

Table 1. Properties of observed samples of quasars

sample	N_{obj}	$\langle n_{\mathrm{obj}} \rangle$ $10^{-5}h^3\mathrm{Mpc}^{-3}$	$\langle l_{\mathrm{MST}} \rangle$ $h^{-1}\mathrm{Mpc}$	$\langle b_{\mathrm{MST}} \rangle$ $h^{-1}\mathrm{Mpc}$	s_0	γ
2QZ North	3822	0.15	50	0.81	5.7 ± 0.46	1.56 ± 0.1
2QZ South	6203	0.25	44	0.90	5.7 ± 0.46	1.56 ± 0.1
SDSS North	1245	0.09	71	1.35	–	–
SDSS South	986	0.11	66	1.37	–	–

For the comparison, twenty random samples were prepared for each slice, North and South, of each survey. For the 2QZ 10K random samples, we used the the angular positions from the observed quasars but we used random distances. However, for the more homogeneous SDSS EDR, we prepared the random samples by using random coordinates for both the angular and the radial positions. For all random samples mean densities are the same as for the observed ones.

We generated the MST statistics for both the observed and random samples. The random sample results were averaged together. Main results of our analysis are listed in Table 1 and plotted in Figs. 1 – 3.

Note that variations of the mean edge lengths from sample to sample (Table 1) are directly linked to the variations of the sample richness and they remain important even for the mean dimensionless edge lengths.

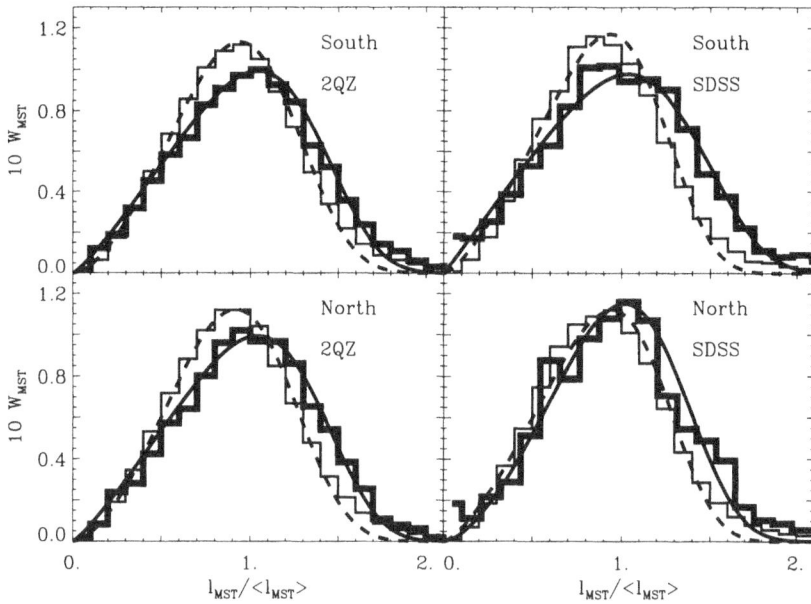

Fig. 1. Distribution functions of MST edge lengths in redshift space for the 2QZ 10K and the SDSS EDR samples of quasars near the North and South Galactic Caps (thick solid lines) and for the average of the 20 random samples (thin solid lines). Equations (5,6,7) and equations (8,9) are plotted by solid and dashed lines, respectively.

4.1 W_{MST} Approach

Fitting W_{MST}, we find for the observed samples:

$$2QZ \text{ North \& South} : W_{MST}(l) = 0.13x^{1.2} \exp(-0.1\,x^2 - 0.15x^{5.3})\,, \qquad (5)$$

$$\text{SDSS North} : \qquad W_{MST}(l) = 0.16x^{1.6} \exp(-0.33x^{4.5})\,, \qquad (6)$$

$$\text{SDSS South} : \qquad W_{MST}(l) = 0.12x \exp(-0.18x^{4.7})\,, \qquad (7)$$

and we find for the random samples:

$$2QZ \text{ North \& South} : \quad W_{MST}(l) = 0.2\,x^{1.7} \exp(-0.5\,x^{3.5} - 0.08x^{5.6})\,, \qquad (8)$$

$$\text{SDSS North \& South} : W_{MST}(l) = 0.16x^{1.4} \exp(-0.38x^5)\,, \qquad (9)$$

where $x = l_{MST}/\langle l_{MST}\rangle$.

In all these cases, the rapid drop in W_{MST} for $x > 1$ is similar to that expected for a random 3D point distribution (see Sec. 2). Considering smaller scales, note that the power law indices in eqs. (5) & (7) are close to 1, which is typical for a 2D random point distribution [eq. (2)]. These indices are smaller than those for the random samples [eqs. (8) & (9)]. However, for the observed SDSS North and the random SDSS North & South samples, the W_{MST} fits are quite similar to each other.

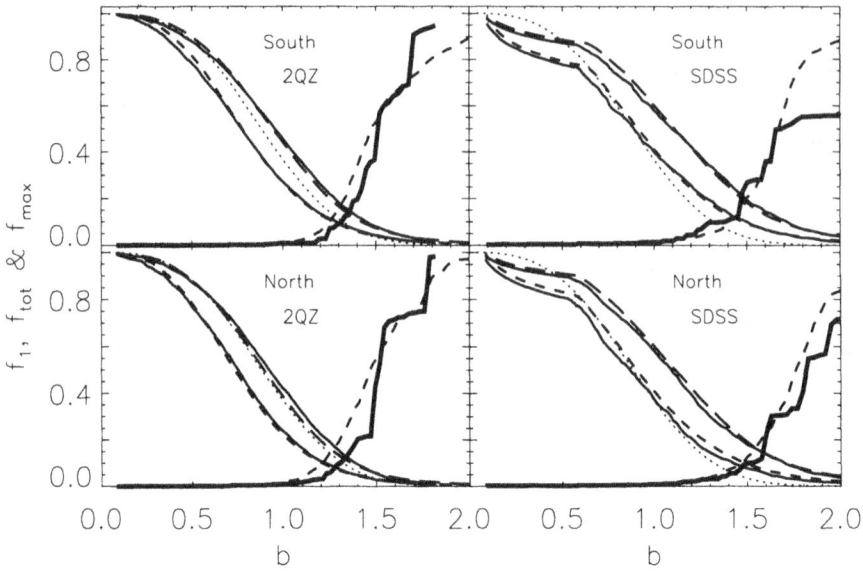

Fig. 2. Fractions of single objects, f_1 (bottom thin solid lines), of total number of clusters, f_{tot} (top thin solid lines), and of objects in the richest cluster, f_{max} (thick solid lines) are plotted vs. the dimensionless linking length, b, for the 2QZ 10K and SDSS EDR quasar samples and averaged random samples (dashed and long dashed lines). Dotted lines show fits expected for a 3D Poissonian point distribution for f_1 (eq. 10).

4.2 Fraction of Single Objects, Number of Clusters, and the Richness of the Richest Cluster

More detailed characteristics of the spatial object distribution are the variations with linking length l of the fraction of single objects (f_1), of the total number of clusters (f_{tot}), and of the number of objects in the richest cluster (f_{max}). For both the observed and the random samples, these functions (f_1, f_{tot}, f_{max}) are plotted in Fig. 2. Note that for both surveys the decreasing functions shown in Fig. 2 represent f_1 and f_{tot}, whereas the increasing function is that of f_{max}. As is clearly seen in Fig. 2, the shapes of these functions differ little between the observed and the random samples; this is true for both hemispheres and in both surveys.

More significant differences, however, can be seen between the shapes of f_1 for the observed and random samples and the shape of f_1 expected for a 3D Poissonian distribution given by

$$f_1(< b) = \exp(-b^3) , \quad b \equiv l(4\pi/3\langle n_{obj}\rangle)^{1/3} , \tag{10}$$

where b is again the dimensionless linking length. These differences can be attributed to boundary effects and field-to-field variations in sample completeness.

4.3 Fraction of Doublets and Triplets

For the small scale clustering of quasars, a more sensitive test is the variations in the cumulative fractions of doublets and triplets, $f_2(< b)$ and $f_3(< b)$, versus the dimen-

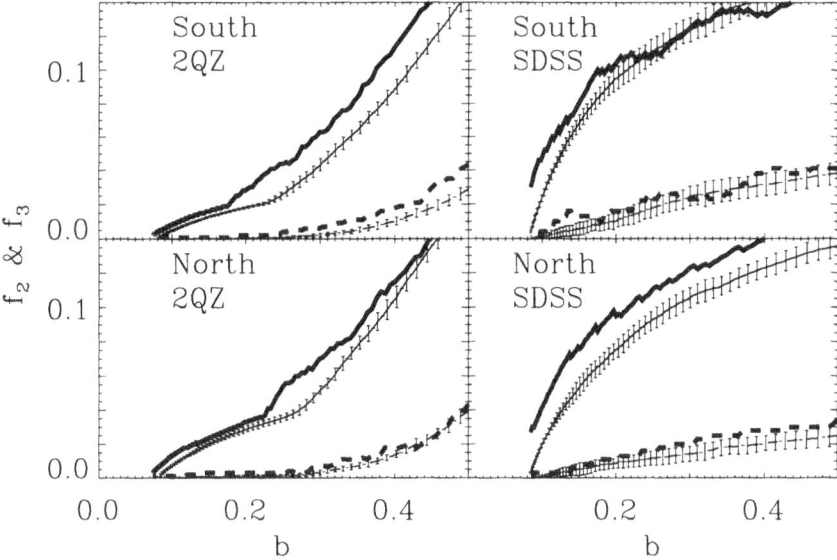

Fig. 3. Fraction of doublets $f_2(< b)$ (thick solid lines) and triplets $f_3(< b)$ (thick dashed lines) vs. the dimensionless linking length b for the same samples of quasars and averaged random samples (thin solid and dashed lines) as in Fig. 2.

sionless linking length, b. These functions are plotted in Fig. 3 for all observed and random samples. For the random samples, 1σ rms error bars were found by averaging the twenty random samples prepared as described in Sec. 3.

For all samples at $b \leq 0.4$, $l_{link} \leq 20h^{-1}$ Mpc, the fraction of doublets is much greater than the fraction of triplets, $f_2(< b) \gg f_3(< b)$. For the richer samples, 2QZ South and SDSS North, the fraction of observed doublets exceeds that for the random samples; this again demonstrates the small scale clustering of observed quasars. However, even for these samples, the fraction of doublets does not exceed $\sim$10%, implying an important role for cosmic variance.

5   Summary

Our results show that on larger scales, $l \geq 20h^{-1}$ Mpc, the observed quasar distribution is quite similar to a 3D random distribution. On smaller scales, we see some evidence for the clustering of quasars. However, the strong sample dependence of the results indicates that cosmic variance might still play an important role. In particular, for the 2QZ 10K Release, the small scale clustering we see may be enhanced by the strong field-to-field sampling variations.

Acknowledgments

The 2dF QSO Redshift Survey (2QZ) was compiled by the 2QZ survey team from observations made with the 2-degree Field on the Anglo-Australian Telescope.

Funding for the creation and distribution of the SDSS Archive has been provided by the Alfred P. Sloan Foundation, the Participating Institutions, the National Aeronautics and Space Administration, the National Science Foundation, the U.S. Department of Energy, the Japanese Monbukagakusho, and the Max Planck Society. The SDSS Web site is http://www.sdss.org/.
The Participating Institutions are The University of Chicago, Fermilab, the Institute for Advanced Study, the Japan Participation Group, The Johns Hopkins University, the Max-Planck-Institute for Astronomy (MPIA), the Max-Planck-Institute for Astrophysics (MPA), New Mexico State University, Princeton University, the United States Naval Observatory, and the University of Washington.
This paper was supported in part by Denmark's Grundforskningsfond through its support for an establishment of the Theoretical Astrophysics Center.
We would like to thank Brian C. Lee (FNAL) for useful comments in the final stages of preparing this text.

References

1. M. Colless et al.: 'The 2dF Galaxy Redshift Survey'. In: *Looking Deep in the Southern Sky*, ed. by F. Morgani, W. Couch, (Springer-Verlag, 1999), pp. 9-14.
2. S. Croom et al.: MNRAS, 322, L29 (2001a)
3. S. Croom et al.: MNRAS, 325, 483 (2001b)
4. A. Doroshkevich et al.: MNRAS, 306, 575 (1999)
5. A. Doroshkevich et al.: MNRAS, 322, 369 (2001)
6. F. Governato et al.: Nature, 392, 389 (1998)
7. B. Komberg et al.: Astron.Astrophys, 286, L19 (1994)
8. P.S. Osmer: ApJ, 247, 762 (1981)
9. A. Ratcliffe et al.: MNRAS, 281, L47 (1996)
10. S. Shectman et al.: ApJ, 470, 172 (1996)
11. S. Stoughton et al.: AJ, submitted (2001)
12. D. York et al: AJ, 120, 1579 (2000)

Galaxy Distributions in Tsallis and Levy-Stable Statistical Mechanics

Akika Nakamichi[1] and Masahiro Morikawa[2]

[1] Gunma Astronomical Observatory, Takayama, Agatsuma, Gunma 377-0702, JAPAN
[2] Department of Physics, Ochanomizu University, 2-1-1 Otsuka, Bunkyo, Tokyo 112-8610, JAPAN

1 Introduction

Well-developed large scale structure of the Universe is mainly described as a statistical distribution of galaxies. In order to describe this distribution and the evolution of it, we cannot simply apply the ordinary extensive statistical mechanics due to the non-extensive property of the force gravity. Firstly, in order to describe the large scale structure, we apply Tsallis statistical mechanics as one of non-extensive generalizations of the ordinary statistical mechanics. Secondly, we point out theoretically unsatisfactory issues in Tsallis version, and propose new statistical mechanics based on the Levy-Stable distributions or the generalized central limit theorem. We demonstrate that this new statistical mechanics and the Tsallis version are remarkably identical with each other despite the difference in theoretical constructions.

2 Galaxy Distribution and Tsallis Statistical Mechanics

Tsallis statistical mechanics [1] is characterized by the special form of the entropy functional: $S_q(P) = [(\sum_i P_i^{1/q})^{-q} - 1]/(1 - q)$, and the power-law behavior of the distribution functional: $P_i = p_i{}^q / \sum_{j=1}^{W} p_j{}^q \equiv p_i{}^q / C$, where q is a parameter.

We first suppose that the galaxy distribution obeys the Tsallis statistics with grand canonical ensemble and see how well it works. The void probability $f(0)$, the probability of finding no galaxies in randomly positioned volume V, is given by $f(0) \equiv \sum_E P_{0,E}$. Using the properties of Euler relation etc., it reduces to

$$f(0) \equiv \sum_E P_{0,E} = P_{0,0} = (1 + (1 - q) S)^{\frac{-1}{1-q}} \left(1 + (1 - q) \tilde{\beta} \left(\bar{E} - \mu \bar{N} \right) \right)^{\frac{q}{1-q}} \quad (1)$$

In the above, we have succeeded to avoid summing up all the N and E levels. Furthermore, the probability of finding exactly N galaxies in randomly positioned volume V, $f(N)$, is calculated as: $f(N) = (-n)^N (N!)^{-1} d^N f(0)/dn^N$, where the parameter n is the number density of the galaxies. [3] We fit the above theoretical result with the data of CfA II South redshift observations. [4]

We reduce the data to the uniform 870 galaxies sample: absolute luminosity is brighter than 19.1 mag., distance is between 4000 and 8000 km/s, exclude edge of the observed region, compensate reddening by K-correction.

As is seen in Fig. 1, the results are remarkably well; Tsallis statistics is not excluded by the CfAII South observations.

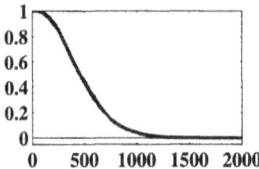

Fig. 1. The void probability f (0). The horizontal line represents the linear size of the region in the unit km/sec. The dots with error bars represent the CfAII data and the solid line represents our calculation with the best fitting parameter $q = -5.7^{+0.6}_{-0.3}$. [2]

3   Levy Stable Distribution

Though we have seen that the Tsallis statistical mechanics is phenomenologically successful, we observe several unsatisfactory points in the theoretical formalism. For example, a) we have no principle to select the form of entropy; Tsallis entropy is not unique, and b) sum of many Tsallis distributions is NOT a Tsallis distribution.

Therefore, we would like to construct better statistical mechanics that is
(i) based on a certain principle and
(ii) retains phenomenological advantage of the Tsallis statistical mechanics.

For the first point (i), we focus our attention to the generalized central limit theorem: Sums of independent distributions, which may or may not have finite dispersion, approaches to the Levy stable distributions [5]. The stable distribution is characterized by $N^{-1/\alpha} \sum_{i=1}^{N} x_i \approx x$, where α is an index.

In order to construct a statistical theory, we start with the Langevin eq. including friction and fluctuation terms:

$$\frac{dv(t)}{dt} = \xi(t) - \beta v(t) \tag{2}$$

We consider the noise that has power-law distribution for large variable. Then we obtain time-dependent distribution function as a solution of the Langevin equation as follows.

$$P(v,t) = \int \frac{d\lambda_1}{2\pi} \exp\left[iv\lambda_1 - \frac{s}{\beta\alpha}\left(1 - e^{-\beta\alpha(t-t_0)}\right)|\lambda_1|^{\alpha}\right] \tag{3}$$

In order to examine the point (ii) above, we compare this stable distribution with Tsallis distribution. As a result, these two kinds of distributions are remarkably identical with each other. The following figure 2 shows a typical example of the resemblance.

The resemblance guarantees the phenomenological success of our formalism as Tsallis statistical mechanics.

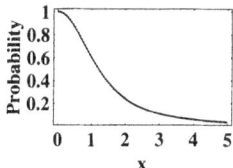

Fig. 2. For $\alpha = 1.2$, the corresponding parameter becomes $q = 3.53643$. The dotted line represents the Levy-stable distribution. The solid line the Tsallis escote dsitribtion.

4   Divergence of the Dispersion

Next, we study the Langevin equation that includes potential term as well as friction and fluctuation terms:

$$\ddot{x} + \beta\dot{x} + \mu^2 x = \xi \tag{4}$$

Similarly, we obtain time-dependent distribution function as follows:

$$P(x, p, t) = \int \frac{d\lambda_1}{2\pi} \int \frac{d\lambda_2}{2\pi} \exp\left[ix\lambda_1 + ip\lambda_2 - s \int_{t_0}^t dt' \left|\begin{matrix} \lambda_1 Q_1(t - t') \\ + \lambda_2 Q_2(t - t') \end{matrix}\right|^\alpha\right] \tag{5}$$

Here $Q_1(t) = \theta(t)\frac{2\pi}{\omega_R}e^{-\beta t/2}\sin(\omega_R t)$, $\omega_R \equiv \sqrt{\mu^2 - \frac{\beta^2}{4}}$, $Q_2(t) = \frac{\partial Q_1(t)}{\partial t}$. Formally, dispersion of the stable distribution is infinite except $\alpha = 2$. However, since actual observation time is finite, we have to discard the very small distribution regions. We introduce cutoff, because, in finite observed time t, probability $P(x, t) < t^{-1}$ does not contribute. Then we obtain observable time-dependent dispersion:

$$\left\langle x(t)^2 \right\rangle = \int_{-x_*}^{x_*} dx P(x, t) x^2 = \frac{1}{2 - \alpha}\left(\frac{s}{\mu\alpha}\right)^{\frac{3}{1+\alpha}} t^{\frac{2-\alpha}{1+\alpha}} \tag{6}$$

The statistical mechanics based on the Levy stable distribution has many interesting properties. They will be discussed in separate publications.

References

1. Tsallis C, J Stat. Phys. **52** 479 (1988).
2. A. Nakamichi, I. Joichi, O. Iguchi and M. Morikawa: Chaos, Solitons & Fractals, **13 (3)**, 595 (2001).
3. S. D. M. White., MNRAS 186 145 (1979).
4. J. Hucra, et al. ApJS 121 287 (1999).
5. Gnedenko B. V. and A. N. kolmogorov *Limit Distribution for Sums of Independent Random Variables* Addison-Wesley (1954).

Cold Dark Matter Halos Must Burn

Paolo Salucci[1] and Annamaria Borriello[1]

(1) International School for Advanced Studies SISSA Trieste, Italy

Abstract. High–quality optical rotation curves for a sample of low–luminousty spirals evidence that the dark halos around galaxies are inconsistent with the output of proper CDM simulations. In fact, dark halos enveloping stellar disks are structures with approximately a constant density out to the optical edges. This is in strong disagreement with the characteristic $\rho(r) \propto r^{-1.5}$ CDM regime and severely challenges the "standard" CDM theory also because the halo density appears to be heated up, at gross variance with the hierarchical evolution of collision–free particles.

1 CDM and Galaxy Halos

Dark matter (DM) halos in the Cold Dark Matter scenario are formed via dissipation-less hierarchical merging and harbor the infall/cooling of the primordial gas that leads to the formation of the present–day galaxies. Several studies have investigated the de-tailed structure of these halos by means of N-body simulations at a progressively higher spatial/mass resolution (e.g. [1], [2]). The outcome is well known: CDM halos have an "universal" profile in their density which includes a steep central cusp. In detail, on galactic scales $\simeq 1 - 10$ kpc: $\rho(r) \propto r^{-\gamma}$. Simulations at the highest resolution indi-cate $\gamma = -1.5$ [3].

Name	M_I	R_{opt}	β	ρ_s	r_s
116-G12	−20.0	5.4	$0.29^{+0.03}_{-0.1}$	$2.7^{+3}_{-0.7}$	10^{+10}_{-5}
531-G22	−21.4	10.5	$0.11^{+0.07}_{-0.06}$	$2.1^{+0.5}_{-0.4}$	12^{+4}_{-2}
533-G4	−20.7	8.4	$0.07^{+0.05}_{-0.02}$	$4.3^{+0.5}_{-0.9}$	6^{+2}_{-1}
545-G5	−20.4	7.5	$0.21^{+0.03}_{-0.03}$	$1.0^{+0.3}_{-0.3}$	22^{+300}_{-4}
563-G14	−20.5	6.3	$0.25^{+0.02}_{-0.07}$	$4.3^{+1}_{-0.9}$	6^{+2}_{-2}
79-G14	−21.4	12.3	$0.19^{+0.03}_{-0.07}$	$1.3^{+0.3}_{-0.3}$	14^{+4}_{-2}
M-3-1042	−20.1	4.7	$0.45^{+0.06}_{-0.04}$	$2.2^{+3}_{-0.8}$	15^{+300}_{-10}
N7339	−20.6	4.9	$0.53^{+0.05}_{-0.03}$	$2.1^{+0.3}_{-0.5}$	22^{+300}_{-7}
N755	−20.1	4.8	$0.05^{+0.02}_{-0.04}$	$4.1^{+1}_{-0.7}$	5^{+3}_{-1}

Let us represent generic halos, including the CDM ones, by means of:

$$\rho(r) = \frac{\rho_s}{[c + (r/r_s)^\gamma] \, [1 + (r/r_s)^\alpha]^{(\delta-\gamma)/\alpha}} \tag{1}$$

where γ is the density inner slope, δ is the outer slope and α settles the turnover point between the inner and outer regimes. The parameter c indicates the presence of a "con-stant density" (inner) region (CDR). If $c = 0$, as in collisionless CDM, the density

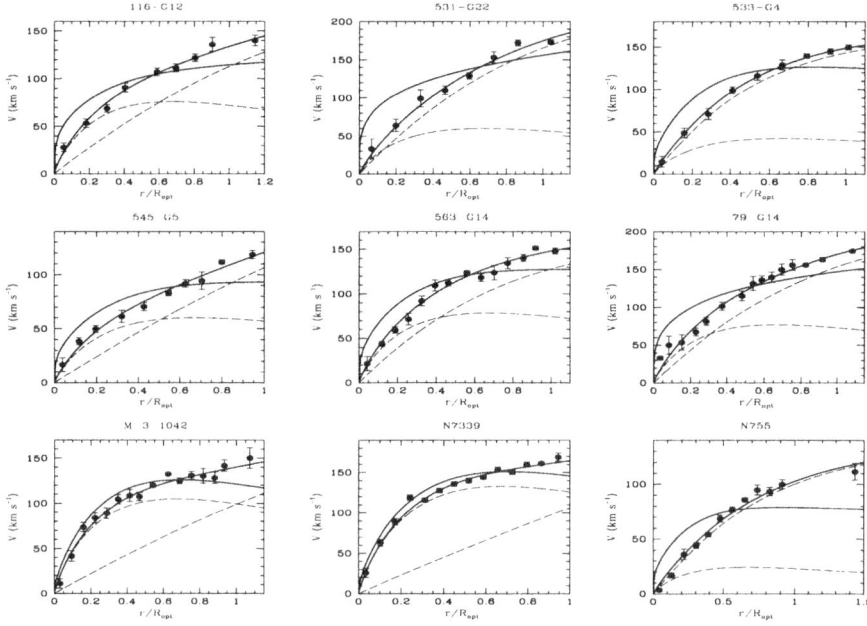

Fig. 1. RC's of the sample (*points*) *vs* CDM model (*discrepant continuous line*) and *vs* BS model, of which the disk and halo contributions are shown as (*dotted lines*).

diverges for $r \to 0$. In the case of $c = 1$, r_s is the size of the CDR, whose value is ρ_s. According to the above notation, the Moore density profile corresponds to $(c, \alpha, \delta, \gamma) = (0, 1.5, 3, 1.5)$ and the Burkert–Salucci one to $(1, 2, 3, 1)$ [4].

We test these mass models by means of the dark halos detected around galaxies of a sample of low luminosity late–type spirals. In these objects the DM distribution can be easily derived from the available high-quality *optical* rotation curves because the stellar and HI disks contribute very little to the gravitational potential so as the beam smearing effect, plaguing the HI RC's, is obviously absent. The sample includes 9 high–quality *optical* RC's that are smooth and symmetric and extend at least out to the optical edge R_{opt}. The characteristics of merit are the following: the spatial resolution is better than $1/20\, R_{opt}$ and the velocity *rms* is $< 3\%$. The mass modeling is furthermore simplified in that in these spirals the gas contribution to $V(r)$ (for $R < R_{opt}$) is modest [5] and the stellar surface brightness is very well fit by a Freeman disk.

2   The Structure of Dark Halos

The mass model includes *i)* a stellar exponential thin disk with free parameter $\beta \equiv (V_D/V)^2_{R_{opt}}$ i.e. the fractional disk contribution to the total circular velocity at R_{opt}, and *ii)* a dark spherical halo, whose contribution to $V(r)$ is obtained from (1) in that: $V_h^2(r) \equiv \frac{G}{r} \int_0^r 4\pi r^2 \rho(r) dr$. Once we choose the scenario (i.e. $c = 1$ or $c = 0$), we are

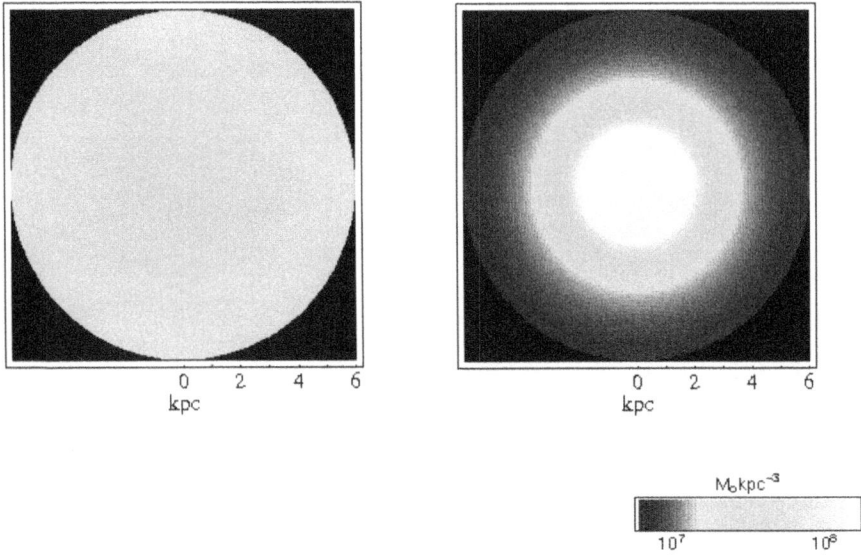

Fig. 2. *left*) Dark halo around 116–G12, *right*) CDM prediction for a halo of the same mass (inside $R_{\rm opt}$).

left with two free parameters, ρ_s and r_s, the characteristic density and radius. Both get determined by adjusting the model to data as in [5].

The CDM halo profiles fail: in no case they can reproduce (with or without an exponential disk) the observed RC's. This is shown in Fig. 1: the discrepancy with data is very high at any radius. The CDM halos, even allowed to take any value for r_s (or for the concentration parameter), have definitely a too steep density profile in the innermost region and show a too flat profile in the outer regions.

Next, we fit the BS profile to the data: the results are shown in Fig. 1 and Table 1. This model fits perfectly, indicating the values of parameters which are reported in Table 1. Each of 9 halos has a central density ρ_s of about $1 - 4 \times 10^{-24}$ g/cm^3 that keeps constant out the edge of the stellar distribution.

3   You Don't Need More Evidence

We definitely conclude, as in [5], that the dark halos embedding the stellar disks show a density distribution inconsistent with that predicted by CDM (see Fig.2). Real halos in the Universe rather resemble, in the regions where the stars reside, some homogeneous spheres. Crucially, the sample and the method employed here level off to zero the criticisms raised to previous claims for DM core radii in galaxies. Furthermore, since the DM radial distribution out to R_{opt} is featureless, we cannot link together the local and the global properties of the dark halos; such a connection, which is a main consequence of the bottom–up merging scenario appears to be just missing in Nature. The CDM scenario, then, must find a way to cut off any signature of the gravitational collisionless collapse. For instance, the "temperature" of the CDM particles, which initially $\to 0$ for $r \to 0$, must be largely heated up and kept constant out to the disk edge (see [6]). Solving this deep mystery will be the guideline of our (and others) future research.

References

1. Navarro J.F. , Frenk C.S. , White S.D.M. , ApJ , **462**, 563 (1996)
2. Moore B. , Quinn T., Governato F. , Stadel J. , Lake G., MNRAS , **310**, 1147 (1999)
3. Ghigna S. , Moore B. , Governato F. , Lake G. , Quinn T. , Stadel J., ApJ , **544**, 616 (2000)
4. Burkert A. , ApJ, **477**, 25 (1995)
5. Borriello A. , Salucci P. , MNRAS, **323**, 285 (2001)
6. Burkert A. , ApJ, **534**, 143 (2000)

Can X-Ray Jets Be Cosmic Beacons?

Dan Schwartz

Smithsonian Astrophysical Observatory, Cambridge MA 02138, USA

Abstract. If X-rays observed from *any* extragalactic radio jets are due to inverse Compton scattering on the cosmic microwave background (CMB) radiation, then such a source will be detectable with the same surface brightness anywhere in the more distant universe. Chandra observations imply that such systems do exist, and will therefore serve as Cosmic Beacons out to the redshift at which they form.

1 Introduction

PKS 0637–752, the first celestial X-ray target of the Chandra X-ray Observatory, revealed a surprisingly strong X-ray jet coincident with the previously known radio jet [12]. The great difficulty in explaining the X-ray emission *and* allowing the magnetic field and relativistic electrons to be near equipartition, led to suggestions [1], [14] that the extended jet structure was in bulk relativistic motion with $\Gamma \sim 10$ on scales greater than 100 kpc. With Chandra detection of X-ray jets soon to be in the tens of objects, models for at least 3C 273 [9], and 3C 371 [8] also call for large scale relativistic motions.

We point out here that if relativistic beaming is common, we should see X-ray jets from those sources pointed toward us *anywhere in the universe*. Even powerful unbeamed sources may be seen at all redshifts, if their intrinsic magnetic fields are somewhat less than the typical values $\sim 2 \times 10^{-4}$. This follows from the standard relation that the ratio of IC power to synchrotron power is the ratio of the energy density of target photons to the magnetic field energy density [3]. The CMB energy density increases with redshift according to $(1 + z)^4$, (Fig. 1). This factor compensates for the decrease of surface brightness due to the expanding universe. X-ray jets are resolved by Chandra and will always be resolved (on-axis) anywhere in the universe (Fig. 2).

As the jets maintain their apparent surface brightness, they can even outshine the quasar cores. Such a case might be speculated for PKS 2215+020, at z=3.572, where ROSAT detects an unusually high X-ray to optical flux ratio [13]. Other cases should occur among the distant radio emitting Sloan Survey quasars [2].

2 Observational Implications

Jets may be displaced by $5''$ to $10''$ from the optical object which emits them. They may therefore be blank fields within the best location capability of ROSAT, Chandra, and XMM/Newton. They might also be incorrect identifications of chance coincidences with optical objects. Jets require $\leq 1''$ angular resolution for clear identification. If

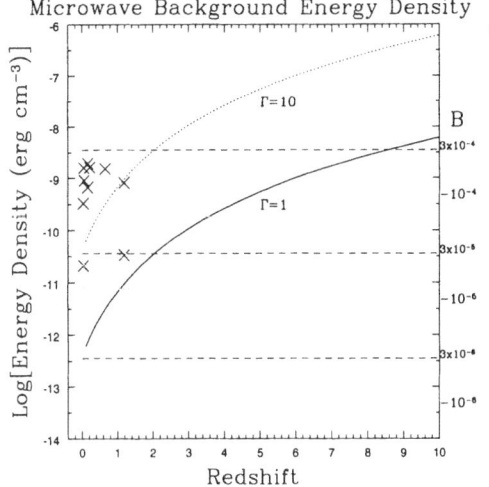

Microwave Background Energy Density

Apparent Angular Size of a Proper Length

Fig. 1. The continuous curves give the apparent CMB energy density (on the left vertical scale) vs. redshift as the solid line (in the rest frame of the CMB) or the dotted line (in a rest frame moving with bulk Lorentz factor $\Gamma = 10$ with respect to the CMB). The crosses and diamonds plot the redshift and inferred magnetic field (on the right vertical scale) for observed knots and hotspots in X-ray jets. The crosses assume non-relativistic bulk motion, while the diamonds allow the jet to have the necessary bulk Lorentz factor to produce the X-ray emission while maintaining equipartition in its rest frame. (Data from [4], [5], [8], [15], [16])

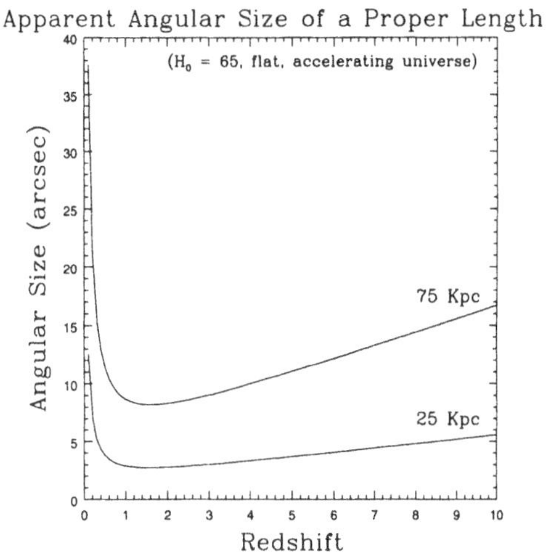

Fig. 2. The X-ray jets in PKS 0637-752 [12], and 3C 273 [6] brighten about 20 to 30 kpc from the quasar core, and extend 50 to 100 kpc (all distances projected in the plane of the sky). We plot the apparent angular size of projected distances of 25 and 75 kpc as a function of redshift. With the $1''$ resolution of Chandra, 10 kpc can be resolved anywhere in the universe. Throughout this paper we assume a flat, accelerating cosmology, with $H_0=65$, $\Omega_0 = 0.3$, and $\Omega_\Lambda = 0.7$, and use the formulas given by Pen [7]

resolved, the jet should point toward the AGN. Algorithms to detect extended sources for telescopes with poorer resolution, e.g. when more than $5'$ offaxis in Chandra, must fit to a linear shape, *not* to a circularly symmetric shape, for optimum efficiency.

3   What Can We Learn?

- Cosmogony and Jet Physics
  - Discover distant, early activity in the universe.
  - Verify the common occurrence of relativistic beaming on 100 kpc scales. If this occurs, X-ray jets from such objects are certain to be seen at any redshift at which they exist.
  - Is the CMB the limiting dynamical factor in radio lobe and jet formation? The X-rays are produced with electrons of $\gamma \sim 1000/\Gamma$, which will have lifetimes decreasing from $\sim 10^8\,\Gamma$ to $10^5\,\Gamma$ years, for redshifts z=1 to 10, where Γ is the bulk Lorentz factor of the jet.

– Are jets established before the accretion luminosity in the core becomes significant? (In this case there might be no optical counterpart to an X-ray jet.)
– Are there radio quiet X-ray jets? This could happen if "radio" electrons at higher γ have lost their energy to gamma rays, or if larger jet structures can transport upwards of 10^{46} erg s^{-1} while $B \leq 10^{-5}$ Gauss.
– Do jets produce an appreciable component of the X-ray background? Although the diffuse X-ray background clearly arises from AGN, there are significant implications for producing the observed spectral shape, and for the mix of absorbed and unabsorbed AGN.

- **Cosmology**
  – Measure the temperature of the cosmic microwave background *in situ*.
  – Verify the decrease of surface brightness according to $(1 + z)^{-4}$ (Tolman effect; [10], [11]).

4 Conclusions

We should vigorously search for distant X-ray jets in both wide field and deep field surveys. Wide field surveys will discover the relatively rare but bright objects such as PKS 0637-752, if the explanation of relativistic beaming with $\Gamma \sim 10$ occurs in Nature. If relativistic beaming is not common, then deep surveys will detect weak X-ray jets from *all* radio jet sources at redshifts 3 to 10, *if* they exist!

This research was supported by NASA contract NAS8-39073.

References

1. Celotti, A., Ghisellini, G., and Chiaberge, M. 2001, MNRAS, **321**, L1
2. Fan, X. et al. 2001, A.J. **122**, in press
3. Felten, J. E. and Morrison, P. 1966, Ap.J., **146**, 686
4. Hardcastle, M.J., Birkinshaw, M. and Worrall, D.M. 2001, MNRAS, **323**, L17
5. Harris, D. E., and Krawczynski, H. 2002, Ap.J., **565**, in press
6. Marshall, H.L., et al. 2001, Ap.J., **549**, L167
7. Pen, U-I. 1999, Ap.J.Suppl., **120**, 49
8. Pesce, J.E., et al. 2001, Ap.J., **556**, L79
9. Sambruna, et al. 2001, Ap.J., **549**, L161
10. Sandage, A. 1961, Ap.J., **133**, 355
11. Sandage, A. and Lubin, L.M. 2001, A.J., **121**, 2271
12. Schwartz, D. A., et al. 2000, Ap.J., **540**, L69
13. Siebert, J. and Brinkmann, W. 1998, Astron. Astrophys., **333**, 63
14. Tavecchio, F., Maraschi, L., Sambruna, R. M., and Urry, C. M. 2000, Ap.J., **544**, L23
15. Wilson, A.S., Young, A.J., and Shopbell, P.L., 2000, Ap.J., **544**, L27
16. Wilson, A.S., Young, A.J., and Shopbell, P.L., 2001, Ap.J., **547**, 740

Cosmic Acceleration From Effective Forces?

Dominik J. Schwarz[1], Winfried Zimdahl[2], Alexander B. Balakin[3], and Diego Pavón[4]

[1] Institut für Theoretische Physik, Technische Universität Wien,
   Wiedner Hauptstraße 8-10, A-1040 Wien, Austria
[2] Fachbereich Physik, Universität Konstanz,
   Postfach M678, D-78457 Konstanz, Germany
[3] Department of General Relativity and Gravitation, Kazan State University,
   RU-420008 Kazan, Russia
[4] Departamento de Física, Universidad Autónoma de Barcelona,
   E-08193 Bellaterra (Barcelona), Spain

Abstract. Accelerated expansion of the Universe may result from an anti-frictional force that is self-consistently exerted on cold dark matter (CDM). Cosmic anti-friction is shown to give rise to an effective negative pressure of the cosmic medium. While other models introduce a component of dark energy besides "standard" CDM, we resort to a phenomenological one-component model of CDM with internal self-interactions. We demonstrate how the dynamics of the ΛCDM model may be recovered as a special case of cosmic anti-friction [1].

Measurements of supernovae of type Ia at high redshifts provide evidence that the expansion of the Universe accelerates [2–4] and there are first observational indications that this acceleration sets in at a redshift below $z \sim 1$ [5]. Observational aspects and some cosmological consequences of high-z supernovae have been discussed by R. Kirshner at this conference [6]. In our contribution we point out that other explanations besides introducing a cosmological constant or another scalar ("quintessence") field are possible.

Recently, the spatial geometry of the Universe has been measured by means of the temperature fluctuations in the cosmic microwave background (CMB) [7,8]. Based on a seven parameter fit the Boomerang team finds for the spatial curvature $\Omega_k = 0.03\pm0.06$ (value for weak priors) [7], which is consistent with a spatially flat Universe.

The combination of CMB and high-z supernova data shows that dust (vanishing pressure P) alone cannot make up all of the mass in the Universe. An obvious solution to this problem is provided by a cosmological constant, parameterised by Ω_Λ. However, this solution is plagued by heavy theoretical problems: Why is Ω_Λ so small? Why can we see it just know? While the first question is probably one of fundamental physics, an answer to the second question may come from cosmology. To avoid this "coincidence problem", models with a new scalar field ("quintessence" [9]) have been introduced. These models can be characterised by the density of the scalar field, Ω_ϕ, and its equation of state $w_\phi \equiv P_\phi/\rho_\phi$.

In the following we present an alternative explanation of CMB and high-z supernova data in which one does not introduce another dark component besides CDM (see [1] for more details). Instead we add an extra force that acts on CDM. For non-relativistic CDM particles the most general force which is compatible with the cosmological principle was shown to reduce to

$$F = -B(z)mv,$$ (1)

where m is the mass of the particles and v their peculiar velocity. This force is directed parallel or anti-parallel to the motion of the individual CDM particles. The quantity $B(z)$, which has the dimension of an inverse time, obviously plays the role of a coefficient of (anti-)friction. Macroscopically, the action of a force of this type generates an effective viscous pressure. Thus it is not allowed to describe CDM by a perfect fluid in that case. The exact expression for the dynamic pressure in the presence of (anti-)frictional forces reads

$$P = \frac{B}{H}\rho,$$ (2)

which gives rise to an accelerated expansion of the Universe if $B < -H/3$, thus in the case of anti-friction. In our calculation we have assumed that the CDM particle distribution is invariant with respect to elastic collisions (the collision term in the kinetic equation vanishes), which implies that a negative pressure is related to the phenomenon of particle production.

Even without having a microscopic model for cosmic anti-friction at hand we can check this idea by a comparison with the data. We need to make an ansatz for $B(z)$. It is natural to assume $B = -\mathcal{O}(H_0)$, since H_0 is the only rate that is distinguished by cosmology. In [1] we investigated three different models, which all fit the supernova data. Figure 1 shows the prediction for two different models. With the help of the CMB data we can rule out one of our models. The two other models seem to be consistent, as is shown in figure 2.

It turns out that our third ansatz $B = -\nu H_0^2/H$, where ν is a constant, is dynamically equivalent to the ΛCDM model, since $P = -\nu\rho_0$ in that case. This ansatz fits both the CMB and the supernova data and predicts an onset of acceleration at $z_{\mathrm{acc}} \approx 0.7$, but with the important difference that there is no separate dark energy component. This degeneration with ΛCDM may be resolved by investigating large scale structure or clusters. Naively, one might think that cluster data [10] immediately rule out $\Omega_{\mathrm{M}} = 1$.

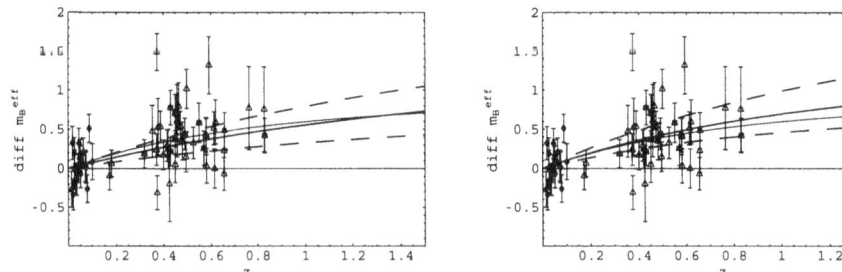

Fig. 1. Differences of the magnitudes with respect to a $(\Omega_{\mathrm{M}}, \Omega_{\Lambda}) = (1, 0)$ universe versus redshift. The data points are taken from [3]. The thin line denotes the $(\Omega_{\mathrm{M}}, \Omega_{\Lambda}) = (0.3, 0.7)$ universe. For the lhs figure we assume $B = -\nu H$ and plot the predictions for $\nu = 0.7, 0.5, 0.3$ (thick lines from top to bottom). For the figure on the rhs we assume $B = -\nu H_0$ with $\nu = 0.9, 0.7, 0.5$ (thick lines from top to bottom).

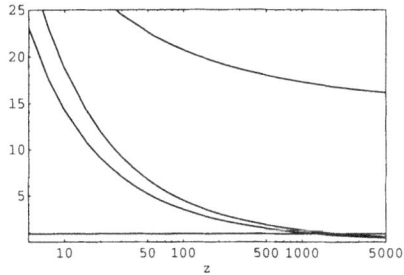

Fig. 2. The angular scale under which the Hubble horizon at redshift z would be observed today. The horizontal line is the angular scale, $0.9°$, of the first acoustic peak in the CMB temperature spectrum. The curves show our three models $-B \propto H, H_0, 1/H$ (from top to bottom), respectively. The first model is excluded by the CMB data, the last one is dynamically equivalent to ΛCDM.

However, one should keep in mind that $\Omega_{cluster}$ typically is obtained under the assumption that CDM is a perfect fluid. In general relativity the gravitating mass is given by $\rho + 3P$, thus if there were a similar negative dynamic pressure at cluster scales, the matter densities from clusters could be underestimated.

We have shown that effective anti-frictional forces could explain the accelerated expansion of the Universe. An interesting hint for finding a microscopic model of cosmic anti-friction might be that it is the cosmological scale $1/H_0$ that gives rise to the correct order of magnitude, which seems to suggest a gravitational mechanism.

D.J.S. thanks the Austrian Academy of Sciences for financial support. This work was also supported by the Deutsche Forschungsgemeinschaft and NATO.

References

1. W. Zimdahl, D. J. Schwarz, A. B. Balakin and D. Pavón, Phys. Rev. D **64**, 063501 (2001)
2. A. G. Riess et al., Astron. J. **116**, 1009 (1998)
3. S. Perlmutter et al., Astrophys. J. **517**, 565 (1999)
4. A. G. Riess, PASP **112**, 1284 (2000)
5. A. G. Riess et al., astro-ph/0104455
6. R. Kirshner, this volume
7. C. B. Netterfield et al., astro-ph/0104460
8. A. T. Lee et al., astro-ph/0104459; N. W. Halverson et al., astro-ph/0104489
9. R. R. Caldwell, R. Dave and P. J. Steinhardt, Phys. Rev. Lett. **82**, 896 (1998)
10. R. Carlberg et al., Astrophys. J. **478**, 462 (1997); N. Bahcall et al., Astrophys. J. **541**, 1 (2000)

STIS Spectroscopy of the Lyα Forest Toward 3C 273 and PKS 0405–123

Gerard M. Williger[1,2], Sara R. Heap[1], Ray J. Weymann[3], Romeel Davé[4], and Robert F. Carswell[5]

[1] Code 681, NASA's Goddard Space Flight Center, Greenbelt MD 20771, USA
[2] present address: Dept. of Physics & Astronomy, Johns Hopkins U., Baltimore MD 21218 USA
[3] Obs. Carnegie Inst. Washington, 813 Santa Barbara St., Pasadena CA 91101, USA
[4] Steward Obs., Univ. Arizona, Tucson AZ 85721, USA
[5] Inst. of Astronomy, Madingley Rd, Cambridge CB3 0HA, England

Abstract. We present on-going results from ~ 7 km s^{-1} resolution STIS spectra for two bright low z QSOs, 3C 273 ($z = 0.158$) and PKS 0405–123 ($z = 0.574$), as part of a STIS Investigation Definition Team (IDT) key project to determine the redshift density, H I column density and Doppler parameter distributions of weak Lyα forest systems at low redshift. In a preliminary analysis of features at the 4σ level, our samples are complete to log $N_{HI} \sim 12.5, 13.2$ for 3C 273 and PKS 0405–123 respectively, with sample sizes of 36 and 169 Lyα lines at velocities $\Delta v \geq 5000$ km s^{-1} from their background QSOs. Our ultimate goal is to compare results with simulations of cosmological evolution, in order to make constraints on the physical characteristics of the intergalactic medium and UV background flux. We also find a variety of Galactic and intergalactic metal absorption lines, including O VI from intervening systems, which will be useful in studies of gas from the halo of the Galaxy and galaxies out to $z \sim 0.6$.

1 Introduction

Observations of the Lyα forest at low redshift allow us to follow the evolution of class of objects from redshifts of nearly 6 to the present. However, the same density perturbations seen in optical spectra at high redshift are predicted to produce only weak absorption at $z < 1$, which requires us to probe to low H I column densities if we are to compare systems of similar density/structure regimes at higher redshift [2]. We therefore have embarked upon a project to measure the statistical properties of the low z Lyα forest toward a set of relatively bright quasars, and present results for two of them here.

2 Observations and Reductions

3C 273: We obtained data from two 5 orbit visits with HST using the E140M echelle grating on STIS and the 0.2 × 0.2 arcsec slit.
PKS 0405–123: We obtained data from HST visits of 2 and 5 orbits' duration, using the E140M echelle grating on STIS and the 0.2 × 0.06 arcsec slit.
Correction for scattered light, spectral extraction and summation of the exposures for both objects were performed with STIS IDT team software at NASA/GSFC.

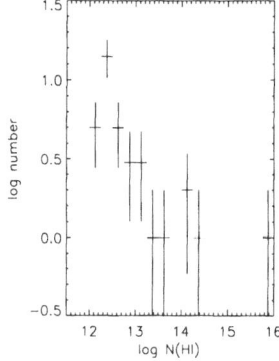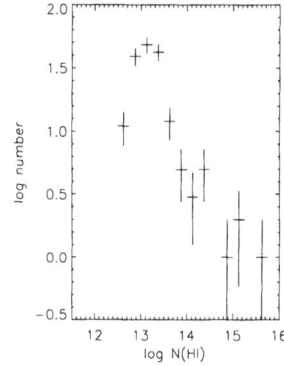

Fig. 1. H I column density distribution for 3C 273 (left) and PKS 0405–123 (right). The binsize in log N_{HI} is shown, and a range corresponding to the square root of the number within each bin is given as a guide to the error

3   Feature Selection and Profile Fitting

Absorption features were selected with a Gaussian filter, with half-widths of 10, 15, 20 pixels (in which 1 pixel is $\sim$ 3 km s^{-1}). We selected all features above a detection threshold of 4.0σ for profile fitting. Williger and Carswell interactively fitted the spectra for PKS 0405–123 independently with VPFIT [4], with largely consistent results; this provided metal line and Lyman series identifications. Williger also profile-fitted the 3C 273 data. Davé is analysing the Lyα forest in both sets of data with the automated routine AUTOVP [3], in anticipation of comparing results to mock QSO spectra drawn from cosmological hydrodynamical simulations.

Simulated spectra of various signal to noise (s/n) ratios are being analysed with both VPFIT and AUTOVP in order to determine the completeness limits for both objects. We include in our analysis all Lyα systems at velocities $\Delta v \geq$ 5000 km s^{-1} from their background QSOs, with Doppler parameters $b \geq$ 10 km s^{-1} .

The 3C 273 data provide the highest signal to noise ratio, high resolution spectrum of a low z QSO, comparable in sensitivity to 8-10m class telescope observations of $z > 1.6$ objects. The completeness limit is $\sim$ log $N_{HI} \sim$ 12.5, though more analysis of simulated spectra is necessary to determine accurately the completeness fraction as a function of Doppler parameter, H I column density and s/n ratio. The mean Doppler parameter $\langle b \rangle = 31.3 \pm 12.5$. Lyβ information from FUSE archival spectra have been used to constrain N_{HI}, b for two of the systems. We find Galactic absorption from Al II; C I, II, II*, IV; Fe II; Mg II; Mn II; N I, N V; Ni II; O I; P II; S II, III; Si II, III, IV. We also find intervening metal absorption at $z = 0.0053$ (C II, Si II, III), which is in the Virgo Cluster, and $z = 0.1200$ (O VI).

PKS 0405–123 provides a long redshift baseline, crucial for studies of redshift evolution. The completeness limit is log $N_{HI} \sim$ 13.2; simulations will provide a more accurate measure. Many Lyα lines with $b > 40$ are likely unresolved blends, due to the lower s/n ratio: $\langle b \rangle = 37.8 \pm 19.5$. A sample selected at 4.5σ, fitted independently by Carswell, has $\langle b \rangle = 45.4 \pm 32.3$ km s^{-1}. Archival FOS, GHRS and STIS G230M

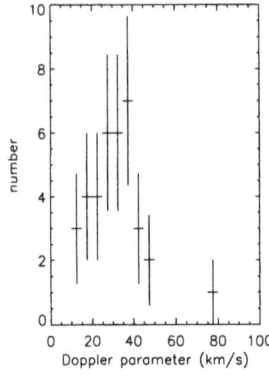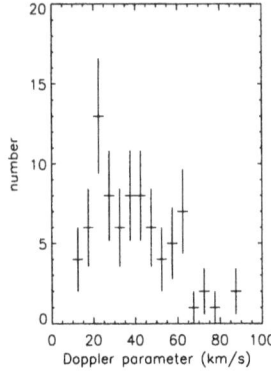

Fig. 2. Doppler parameter distribution for 3C 273 (left) and PKS 0405–123 (right). The binsize in Doppler parameter is shown, and a range corresponding to the square root of the number within each bin is given as a guide to the error

data aided line identifications. We find many higher order Lyman lines at $\lambda < 1614$ Å and metal lines, especially from a partial Lyman limit system at $z = 0.167$ [1]. We find Galactic absorption from Al II; C I, II, II*, IV; Fe II; N I; Ni II; O I; P II; S II, III; Si II, II*, III, IV. We also find intervening metal absorption at $z = 0.167$ (C II; Fe II, Fe III; N II, V; O I, VI; Si II, III, IV); $z = 0.1829$ (O VI); $z = 0.3381$ (Si IV); $z = 0.3608$ (Si III; C IV); $z = 0.3633$ (O VI); $z = 0.4951$ (C III; O VI). H I column density and Doppler parameter distributions for the two objects are shown in Figs. 1 and 2.

References

1. Chen, H.-W. & Prochaska, J. X. 2000, ApJ, 543, L9
2. Davé, R., Hernquist, L., Katz, N., Weinberg, D. H., 1999, ApJ, 511, 521
3. Davé, R., Hernquist, L., Weinberg, D. H., Katz, N. 1997, ApJ, 477, 21
4. Webb, J. K. 1987, PhD thesis, Cambridge University

Cosmological Constraint
from QSO Spatial Power Spectrum

Kazuhiro Yamamoto

Department of Physics, Hiroshima University, Higashi-hiroshima, 739-8526, JAPAN

Abstract. In this paper we consider constraints on the cosmological density parameters from the spatial power spectrum of QSOs. We first review an analytic approach to the spatial power spectrum of QSOs, then we compare the result of the analytic approach with a preliminary result of the power spectrum from the two-degree Field QSO redshift (2QZ) survey. From a simple χ^2 test, we show that the finite baryon fraction better explains observation of the QSO power spectrum, which might suggest a possible detection of the baryonic oscillations in the QSO power spectrum.

1 Introduction

The 2QZ group has recently reported their preliminary results of the spatial correlation function and the power spectrum. These analyses are based on an initial sample of 10,000 QSOs [1,4]. Numerical simulation is the most common technique through which observational results and theoretical predictions can be compared. Indeed, the 2QZ group has utilized the Hubble Volume simulation, which is a huge N-body simulation of horizon box size, containing 1 billion mass particles run by the Virgo consortium [3]. However, a simple, semi-analytic formula that reproduces numerical results would be useful. The purpose of this paper is to report on the development of such a semi-analytic formula and to apply it to recent observational results. As a demonstration of the usefulness of our approach, we use the formula to place constraints on the cosmological density parameters by comparing the theoretical predictions to the 2QZ power spectrum reported by Hoyle et al. [4].

2 Theoretical Formula and a Simple Application

In the clustering statistics of high-redshift objects in a redshift survey, several observational effects must be incorporated for careful comparison between theoretical predictions and an observational result. A useful theoretical formula for the two-point statistics has been developed incorporating the redshift distortions due to peculiar motion of sources and the light-cone effect simultaneously, as well as the geometric distortion [7]. According to the result, the power spectrum is obtained by averaging the local power spectrum $P_0^a(k, z)$ over the redshift,

$$P_0^{\mathrm{LC}}(k) = \frac{\int dz W(z) P_0^a(k, z)}{\int dz W(z)}, \tag{1}$$

with the weight factor $W(z) = (dN/dz)^2 (s^2 ds/dz)^{-1}$, where dN/dz denotes the number count of the objects per unit redshift and per unit solid angle, and $s = s(z)$ denotes the distance-redshift relation of the radial coordinate that we chose to plot a map of the objects. In the expression (1), z-integration arises due to the light-cone effect within the small-angle approximation, and the power spectrum $P_0^a(k, z)$ is given by

$$P_0^a(k, z) = \frac{1}{c_\perp^2 c_\parallel} \int_0^1 d\mu P_{\rm QSO}\left(q_\parallel \to \frac{k\mu}{c_\parallel}, |\mathbf{q}_\perp| \to \frac{k\sqrt{1-\mu^2}}{c_\perp}, z\right), \qquad (2)$$

where $P_{\rm QSO}(q_\parallel, |\mathbf{q}_\perp|, z)$ is the QSO power spectrum, $q_\parallel$ ($\mathbf{q}_\perp$) is the wave number component parallel (perpendicular) to the line-of-sight direction in the real space. In equation (2), with the comoving distance in the real space $r(z)$, we defined $c_\perp = r(z)/s(z)$ and $c_\parallel = dr(z)/ds(z)$. We model the power spectrum of Q.SO distribution by introducing the bias factor $b(z)$,

$$P_{\rm QSO}(q_\parallel, |\mathbf{q}_\perp|, z) = b(z)^2 \left\{1 + \beta(z)\left(\frac{q_\parallel}{q}\right)^2\right\}^2 P_{\rm mass}(q, z), \qquad (3)$$

where $q = \sqrt{q_\parallel^2 + |\mathbf{q}_\perp|^2}$ and we model the CDM mass power spectrum $P_{\rm mass}(q, z) \propto q^n T(q, \Omega_m, \Omega_B, h)^2 D_1(z)^2$ with the transfer function T and the linear growth rate $D_1(z)$. We adopt the fitting formula of the transfer function by Eisenstein & Hu [2], which is useful when the baryon fraction is large.

As a simple application, we consider a cosmological implication comparing with the power spectrum from a preliminary result of the 2QZ survey. We simply introduce χ^2 defined by

$$\chi^2 = \sum_{i=1}^{17} \frac{\left[P_0^{\rm LC}(k_i) - P^{\rm obs}(k_i)\right]^2}{\Delta P(k_i)^2}, \qquad (4)$$

where $P^{\rm obs}(k_i)$ is the observational value at k_i and $\Delta P(k_i)^2$ is the variance of observational errors, for which we adopt the 17 data points in the range $0.012\ h{\rm Mpc}^{-1} < k < 0.2\ h{\rm Mpc}^{-1}$ on Figure 13 in the paper by Hoyle et al. [4]. Figure 1 displays the contours of the χ^2 for various cosmological models on the $\Omega_m - \Omega_b/\Omega_m$ plane. For the clustering bias, we assumed the form $b(z) = b_0/D_1(z)$, where b_0 is a constant, which we determined to minimize the value of χ^2. Alternation of this assumption does not alter our conclusion qualitatively. From Figure 1 it is clear that the QSO power spectrum favors the low density universe rather than the standard CDM model with $\Omega_m = 1$. The minimum of the χ^2 is located at $\Omega_m \simeq 0.2 \sim 0.3$ and $\Omega_b/\Omega_m \simeq 0.2 \sim 0.3$. An interesting fact is that the QSO power spectrum is better explained with the finite baryonic component, though the peak of χ^2 is broad and thus the constraint is not that tight [5,6].

3 Conclusion

By performing a simple χ^2 test we have shown that the QSO power spectrum can be consistent with a simply biased mass power spectrum based on the familiar CDM cosmology with a cosmological constant. We have also shown that the finite baryon fraction

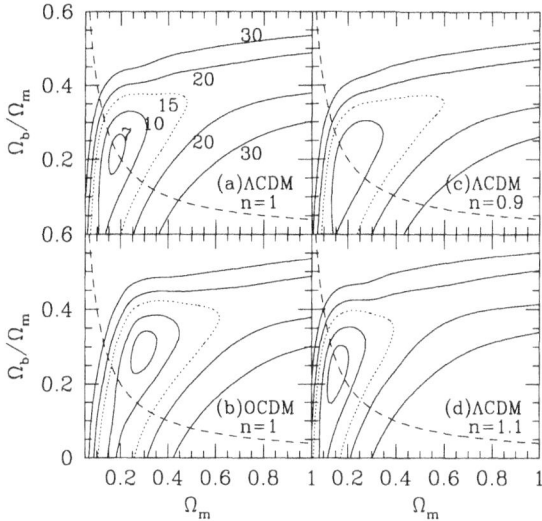

Fig. 1. Contours of χ^2 on the $\Omega_m - \Omega_b/\Omega_m$ plane for various cosmological models: (a) The ΛCDM model with the initial power spectrum with $n = 1$; (b) The open CDM model with $n = 1$; (c) The ΛCDM model with $n = 0.9$; (d) The same as (c) but with $n = 1.1$. Here the 17 data points in Figure 13 in ref.[4] are used and b_0 is determined to minimize χ^2. Levels of the contour curves are $\chi^2 = 7$, 10, 15, 20, 30. The dotted line is the contour of the level $\chi^2 = 15$, and $\Omega_b = 0.04$ on the dashed line.

better explains observation of the 2QZ power spectrum, which might suggest a possible detection of the baryonic oscillations in the QSO power spectrum.

Acknowledgements: The author thanks Fiona Hoyle for providing his results and also for useful discussions and comments. He also thanks Prof. S. D. M. White and the people at Max-Planck-Institute for Astrophysics (MPA) for their hospitality and useful discussions, where parts of this work were done. He acknowledges financial supports from the DAAD and the JSPS for visiting programs to MPA.

References

1. S. M. Croom, et al., MNRAS 325, 483 (2001)
2. D. J. Eisenstein & W. Hu, ApJ 496, 605 (1998)
3. C. S. Frenk, et al., astro-ph/0007362
4. F. Hoyle, et al., submitted to MNRAS, astro-ph/0102163
5. C. J. Miller, R. C. Nichol & D. J. Batuski, Science 292, 2302 (2001)
6. J. A. Peacock, et al., astro-ph/0105450
7. Y. Suto, H. Magira & K. Yamamoto, PASJ 52, 249 (2000)

Constraining Dark Matter with the Long-Term Variability of Quasars

Erik Zackrisson[1], Nils Bergvall[1], and Phillip Helbig[2]

[1] Department of Astronomy and Space Physics, Box 515, SE-75120 Uppsala, Sweden
[2] Institut für theoretische Physik, Johann Wolfgang Goethe-Universität
   Postfach 11 19 32, 60054 Frankfurt am Main, Germany

Abstract. By comparing the results of numerical microlensing simulations to the observed long-term variability of quasars, strong upper limits on the cosmological density of compact objects in the mass range 10^{-2}–10^{-4} $M_{\odot}$ may be imposed. Using recently developed methods to better approximate the amplification of large sources, we investigate in what way the constraints are affected by assumptions concerning the size of the optical continuum-emitting region of quasars in the currently favored ($\Omega_{\mathrm{M}} = 0.3$, $\Omega_{\Lambda} = 0.7$) cosmology.

1 Introduction

Although the optical variability of quasars on time scales of a few years could be both intrinsic in nature (e.g. due to accretion disk instabilities or supernova explosions) and caused by microlensing along the line of sight, adding the effects of both mechanisms can only increase the probability of variations. By assuming that *all* variability is due to microlensing and comparing the predictions from microlensing scenarios to the observed amplitudes of quasar light curves, upper limits to the cosmological density of compact object may therefore be imposed. This technique was first implemented in [1] to constrain compact dark matter populations in the mass range 1–10^{-4} $M_{\odot}$ for an Einstein–de Sitter universe. Here, we extend the analysis to the $\Omega_{\mathrm{M}} = 0.3$, $\Omega_{\Lambda} = 0.7$ cosmology and the full range of quasar sizes allowed.

2 Analysis

In our model of quasar microlensing, the multiplicative magnification assumption outlined in [2] is assumed to adequately reproduce the statistical probability of variability. The results from the numerical simulations are compared to the observational sample of [3], containing 117 quasars in the redshift range $z = 0.29$–3.23, monitored for ten years. The method of comparing observed and synthetic samples closely follows that of [1], which may be consulted for further details.

The constraints derived assume that the luminosities of these objects are independent of the radii of their optical continuum-emitting regions, R_{QSO}. Under this condition, the most conservative upper limits are inferred from volume-limited synthetic samples. Our analysis is therefore restricted to this case only.

3   Upper Limits on the Cosmological Density of Compact Objects

The size of the optical continuum-emitting region of quasars is not a well-determined quantity, but typically assumed to lie somewhere in the range $R_{QSO} = 10^{12}–10^{14}$ m. Whereas only the $R_{QSO} = 10^{13}$ m case was considered when deriving the constraints in [1], we here investigate the impact that smaller and larger quasar sizes may have on the upper limits inferred.

Figure 1 indicates the constraints on the cosmological density of compact objects, $\Omega_{compact}$, for different lens masses, $M_{compact}$, and R_{QSO} in both Einstein-de Sitter and Ω_M=0.3, Ω_Λ=0.7 cosmologies. Poisson probabilities of 10% (see [1]) have been chosen to define the upper limits. A Hubble constant of $H_0 = 65$ km/s/Mpc has also been assumed.

Even though the transition to the Λ-dominated universe significantly strengthens the constraints, the limits on $\Omega_{compact}$ are seen to be very sensitive to the value of R_{QSO} used.

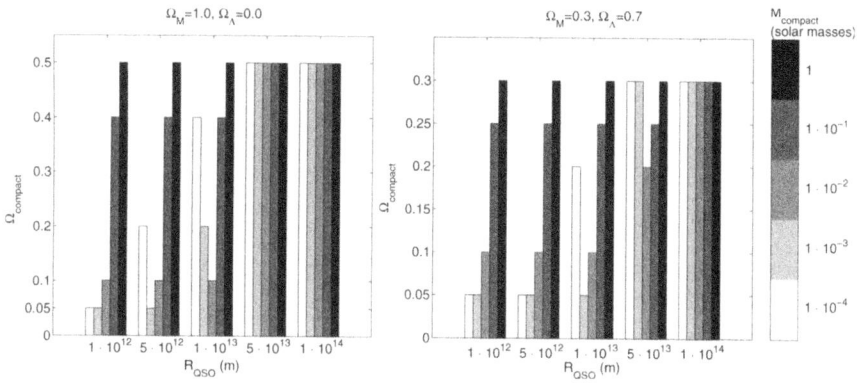

Fig. 1. Allowed regions of the ($M_{compact}$, R_{QSO}, $\Omega_{compact}$) parameter space in the Einstein–de Sitter (**Left**) and $\Omega_M = 0.3$, $\Omega_\Lambda = 0.7$ (**Right**) cosmologies. Lower values of $\Omega_{compact}$ indicate stronger constraints

4   The Impact of Large-Source Microlensing

In [4], the amplification formula derived in [1] was shown to underpredict the amplification of large sources and replaced by a modified expression. As predicted in [4], the improved amplification formula should therefore result in tighter limits on $\Omega_{compact}$. In Fig. 2, however, we show that the implementation of the large-source formalism merely has a modest effect on the upper limits, and only alters the constraints at the lowest mass (10^{-4} $M_\odot$) considered.

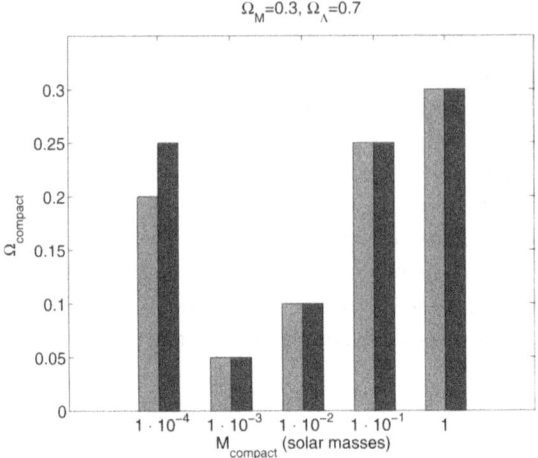

Fig. 2. The allowed region of the (M_{compact}, Ω_{compact}) parameter space for a source size of $R_{\text{QSO}} = 1 \cdot 10^{13}$ m in the $\Omega_{\text{M}} = 0.3$, $\Omega_{\Lambda} = 0.7$ cosmology, with (*light gray*) and without (*dark gray*) the use of the amplification formula for large sources ([4])

5 Conclusions

By comparing the results of numerical microlensing simulations to the observed long-term variability of quasars, strong upper limits may be placed on the cosmological density of dark matter in the form of compact objects. In the currently favored Λ-dominated universe ($\Omega_{\text{M}} = 0.3$, $\Omega_{\Lambda} = 0.7$), compact objects with masses $\sim 10^{-2} \, M_{\odot}$ and $\sim 10^{-3} \, M_{\odot}$ cannot contribute more to the cosmological density than $\Omega_{\text{compact}} = 0.1$ and $\Omega_{\text{compact}} = 0.05$ respectively, provided that the continuum-emitting region of quasars is not significantly larger than $1 \cdot 10^{13}$ m. We do however note that the method used is sensitive to the assumptions concerning the size of the continuum-emitting region of quasars, and becomes effectively useless for $R_{\text{QSO}} \geq 5 \cdot 10^{13}$ m.

The large source amplification formula developed in [4] is shown not to have any significant impact in the part of parameter space for which the long-term variability method provides the most interesting constraints.

A more detailed analysis of the uncertainties present in this technique to constrain the cosmological density of compact objects is currently underway ([5]).

References

1. P. Schneider: A&A **279**, 1 (1993)
2. J.P. Ostriker, M. Vietri: ApJ **267**, 488 (1983)
3. M.R.S. Hawkins, P. Véron: MNRAS **260**, 202 (1993)
4. G. Surpi, S. Refsdal, P. Helbig: submitted to A&A
5. E. Zackrisson, N. Bergvall: in preparation

Part XI

Lighthouses and the Cosmic Background Radiation

Evolution of X-Ray Sources at High Redshift

Günther Hasinger[1,2]

[1] Max-Planck Institut für extraterrestrische Physik, Giessenbachstrasse, Garching, D-85741, Germany

[2] Astrophysikalisches Institut Potsdam, An der Sternwarte 16, D-14482 Potsdam, Germany

Abstract. Deep X-ray surveys have shown that the cosmic X-ray background (XRB) is largely due to the accretion onto supermassive black holes, integrated over the cosmic time. The *ROSAT*, *Chandra* and *XMM-Newton* satellites have resolved more than 80 % of the 0.1-10 keV X-ray background into discrete sources. Optical spectroscopic identifications are about 90% and 60% complete, for the deepest ROSAT and *Chandra/XMM-Newton* surveys, respectively, and show that the sources producing the bulk of the X-ray background are a mixture of obscured (type-1) and unobscured (type-2) AGNs, as predicted by the XRB population synthesis models, following the unified AGN scenarios. The characteristic hard spectrum of the XRB can be explained if most of the AGN are heavily absorbed, and in particular a class of highly luminous type-2 AGN, so called QSO-2s exist. The deep *Chandra* and *XMM-Newton* have recently detected several examples of QSO-2s. The space density of the X-ray selected AGN, as determined from ROSAT surveys does not seem to decline as rapidly as that of optically selected QSO, however, the statistics of the high-redshift samples is still rather poor. The new *Chandra* and *XMM-Newton* surveys at significantly fainter fluxes are starting to provide additional constraints here, but the preliminary observed redshift distribution peaks at much lower redshifts (z=0.5-1) than the predictions based on the ROSAT data.

1 Introduction

Deep X-ray surveys indicate that the cosmic X-ray background (XRB) is largely due to accretion onto supermassive black holes, integrated over cosmic time. In the soft (0.5-2 keV) band more than 90% of the XRB flux has been resolved using 1.4 Msec observations with *ROSAT* [18] and recently 1 Msec *Chandra* observations [33,4] and 100 ksec observations with *XMM-Newton* [20] (see Fig. 1). In the harder (2-10 keV) band a similar fraction of the background has been resolved with the above *Chandra* and *XMM-Newton surveys*, reaching source densities of about 4000 deg^{-2}. Surveys in the very hard (5- 10 keV) band have been pioneered using *BeppoSAX*, which resolved about 30% of the XRB [8]. *XMM-Newton* and *Chandra* have now also resolved the majority (60-70%) of the very hard X-ray background. The Log N–Log S distribution shows a significant cosmological flattening in the softer bands (see Fig. 2), while in the very hard band it is still relatively steep, indicating that those surveys have not yet sampled the redshifts where the strong cosmological evolution of the sources saturates.

Optical follow-up programs with 8-10m telescopes have been completed for the ROSAT deep surveys and find predominantly Active Galactic Nuclei (AGN) as counterparts of the faint X-ray source population [35,43,24] mainly X-ray and optically unobscured AGN (type-1 Seyferts and QSOs) and a smaller fraction of obscured AGN (type-2 Seyferts). Optical identifications for the deepest *Chandra* and *XMM-Newton* fields are still far from complete, however a mixture of obscured and unobscured AGN with an increasing fraction of obscuration seems to be the dominant population in these samples, too [9,1,42,33] (see below). Interestingly, first examples of the long-sought class of high-redshift, high-luminosity, heavily obscured active galactic nuclei (type-2 QSO) have been detected in deep *Chandra* fields [32,39] and in the *XMM-Newton* Deep survey in the *Lockman Hole* field [21].

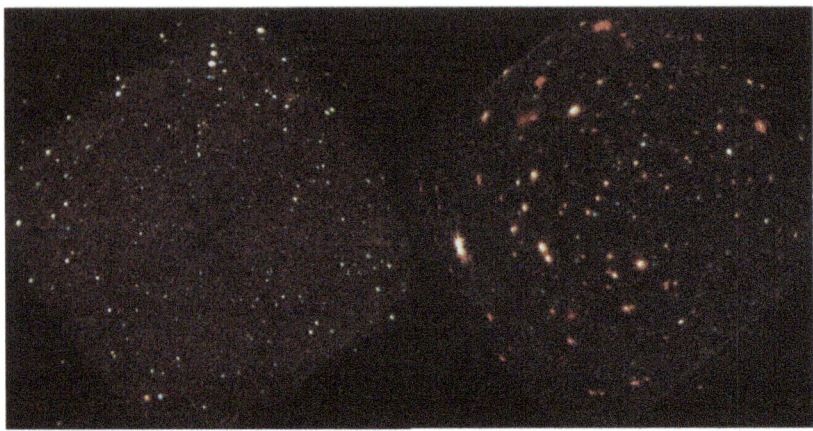

Fig. 1. X-ray colour images *XMM-Newton* PN and MOS images of the *Lockman Hole* field (left, [20]) and the *Chandra* ACIS-I image of the *Chandra Deep Field South* (right, [33]). The field sizes are about 30 x 30 arcmin and 20 x 20 arcmin, respectively. The colours refer to sources in different energy bands: red, green and blue correspond to the soft, medium and hard X-ray range, respectively.

After having understood the basic contributions to the X-ray background, the general interest is now focussing on understanding the physical nature of these sources, the cosmological evolution of their properties, and their role in models of galaxy evolution. We know that basically every galaxy with a spheroidal component in the local universe has a supermassive black hole in its center [10]. The luminosity function of X-ray selected AGN shows strong cosmological density evolution at redshifts up to 2, which goes hand in hand with the cosmic star formation history [28,29]. At the redshift peak of optically selected QSO around z=2.5 the AGN space density is several hundred times higher than locally, which is in line with the assumption that most galaxies have been active in the past and that the feeding of their black holes is reflected in the X-ray background. While the comoving space density of optically and radio-selected QSO declines significantly beyond a redshift of 3 [34,7,38], a similar decline has not yet been observed in the X-ray selected AGN population [28], although the statistical quality of the high-redshift AGN samples needs to be improved. The new *Chandra* and *XMM-Newton* surveys are bound to give additional constraints here.

The X-ray observations have so far been about consistent with population synthesis models based on unified AGN schemes [5,13], which explain the hard spectrum of the X-ray background by a mixture of absorbed and unabsorbed AGN, folded with the corresponding luminosity function and its cosmological evolution. According to these models, most AGN spectra are heavily absorbed and about 80% of the light produced by accretion will be absorbed by gas and dust [6]. However, these models are far from unique and contain a number of hidden assumptions, so that their predictive power remains limited until complete samples of spectroscopically classified hard X-ray sources are available. In particular they require a substantial contribution of high-luminosity obscured X-ray sources (type-2 QSOs), which so far have only scarcely been detected. The cosmic history of obscuration and its potential dependence on intrinsic source luminosity remain completely unknown. Gilli, Salvati & Hasinger [13] e.g. assumed strong evolution of the obscuration fraction (ratio of type-2/type-1 AGN) from 4:1 in the local universe to much larger covering fractions (10:1) at high redshifts (see also [6]). The gas to dust ratio in high-redshift, high-luminosity AGN could be completely different from the usually assumed galactic value due to sputtering of the dust particles in the strong radiation field [14]. This might provide objects which are heavily absorbed at X-rays and unobscured at optical wavelengths.

In this paper I shortly discuss the current status of the optical identification work in the *ROSAT/XMM-Newton/Chandra* deep survey in the *Lockman Hole*, which is largely based on optical work with the Keck telescope led by Maarten Schmidt (see [35] and [21] for more detail). I also present preliminary results of optical identifications in the *Chandra Deep Field South*, obtained with the ESO VLT, which will be formally published in [40] (see also [42,33]). I then discuss the results and try to come to some tentative conclusions about the evolution of X-ray sources at high redshifts.

2 Optical Identifications of Deep X-Ray Surveys

2.1 The *Lockman Hole* Field

The Lockman Hole field has been observed with the *XMM-Newton* observatory during the performance verification phase (see Fig. 1a). About 100 ksec good data, centered on the same sky position as the *ROSAT HRI* pointing, have been accumulated with the European Photon Imaging Camera (EPIC) reaching minimum fluxes of 0.31, 1.4 and 2.4 · 10^{-15} erg cm^{-2} s^{-1} in the 0.5-2, 2-10 and 5-10 keV energy bands. Within an off-axis angle of 10 arcmin 148, 112 and 61 sources, respectively, have been detected. In the 5-10 keV energy band a somewhat lower sensitivity compared to the 1Msec *Chandra Deep Field South* observation has been reached (see [33]), resolving ~60 % of the very hard X-ray background [20]. This is about a factor of 20 more sensitive than the previous *BeppoSAX* observations. A total of 300 ksec observation of the Lockman Hole has been accumulated with the *Chandra* HRC in the 0.5-7.0 keV band, reaching a similar flux limit compared to the *XMM-Newton* pointing [30]. The *Chandra* HRC data provide very accurate source positions, whereas *XMM-Newton* allows spectrophotometry of very faint intrinsically absorbed X-ray sources due to its unprecedented sensitivity in the hard band.

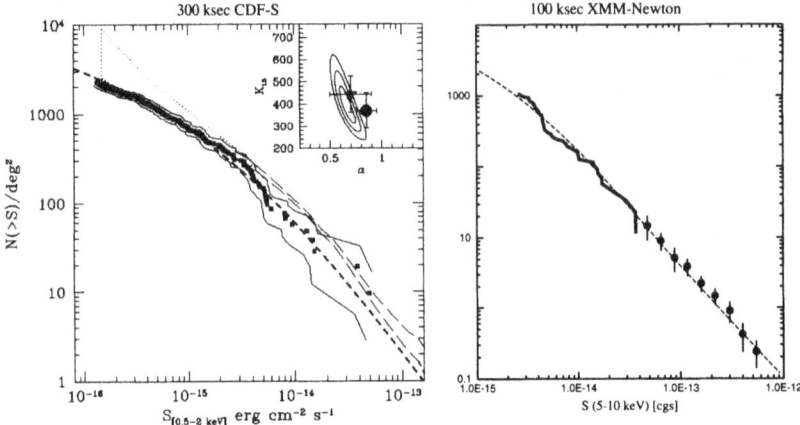

Fig. 2. Left: Log N–Log S in the 0.5-2.0 keV band from the 300 ksec *Chandra Deep Field South* observation (filled squares) [42]. The long-dashed lines are source counts from the *Lockman Hole* field and the dotted contour is from the *ROSAT* fluctuation analysis [17]. Right: Log N–Log S in the 5-10 keV band from the 100 ksec *XMM-Newton* observation (thick line) in the *Lockman Hole* field [20] compared to the *BeppoSAX* source counts (data points with error bars) [8]. The short dashed line in both figures gives the predicton based on the most recent background synthesis models [13].

The optical counterparts for ~60 X-ray sources are already known from the spectroscopic identification of the UDS sample [24]. Among them are one of the most distant X-ray selected quasars at $z = 4.45$ [36] and one of the highest redshift, probably merging cluster of galaxies at $z = 1.26$ [19,41,16]. We have identified 25 new *XMM-Newton* sources using low-resolution multi-slit mask spectra taken with the LRIS instrument at the Keck II telescope in March 2001 (PI: M. Schmidt, [25]). Among the new XMM sources we have found only a few new broad emission line AGNs (type-1), while the optical spectra of most new sources show narrow emission lines and/or only galaxy-like continuum emission at redshifts $z < 1.0$. In several cases high ionisation emission lines like [Ne V] $\lambda 3426$ are absent and thus we see no sign for AGN activity in the optical spectrum, however, their high X ray luminosity ($L_X > 10^{43}$ erg s^{-1}) and/or the strong intrinsic absorption (log $N_H > 22.0$ cm^{-2}) reveal a type-2 AGN in these sources. Three new sources showing typical galaxy spectra, have been detected only in the 0.5-2.0 keV band. Due to their relatively low X-ray luminosities (log $L_X < 42.0$) and their soft X-ray spectra (no indication for intrinsic absorption) we classify them as normal galaxies. Several sources with X-ray luminosities in the range of 42.0 $<L_X < 43.0$, which show galaxy-like optical spectra, are hard to classify due to the small number of photons in their X-ray spectra. We preliminarily classify them as type-2 AGN/galaxy. The completeness of the identification ranges from 61% in the soft sample to 79% in the ultra hard sample (5-10 keV energy band) [25]. The majority of the so far spectroscopically identified sources are type-1 and type-2 AGNs. Although we have no complete identification so far we find a strong indication for a larger fraction of type-2 AGNs, especially in the ultra hard sample, compared to that (~20%) of the UDS. Nearly all spectroscopically identified type-2 AGNs are at moderate redshift ($z < 1$). One type-2 QSO candidate (X174A) at $z = 3.240$ has been identified in the Lockman Hole region so far [21].

Most of the unidentified faint *XMM-Newton* sources have very faint optical counterparts ($R > 24.0$) and at least half of them are extremely red objects (EROs, $R - K' > 5.0$). The new *XMM-Newton* sources with EROs as optical counterparts are similar to those objects in the UDS with photometric redshifts suggesting obscured AGNs at redshifts $1 < z < 3$. The photometric redshift technique is probably the only tool to identify such faint optical objects. The *XMM-Newton* source population at faint fluxes is therefore likely dominated by obscured AGNs (type-2), as predicted by the AGN population synthesis models for the X-ray background.

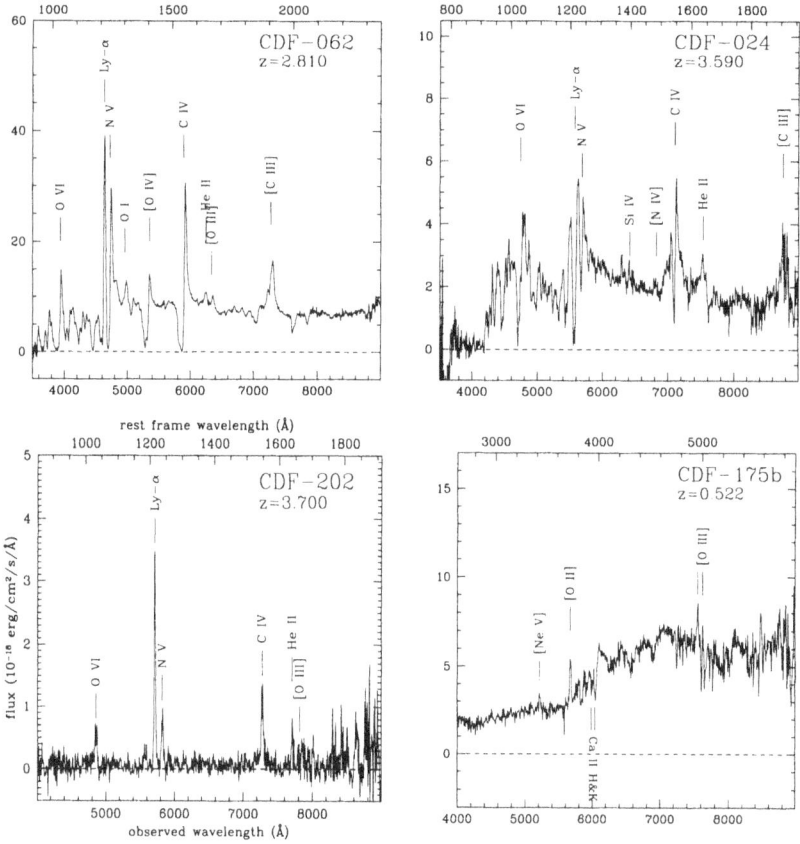

Fig. 3. Optical spectra of some selected CDFS sources obtained using multiobject-spectroscopy with FORS at the VLT [40]. a: broad absorption line (BAL) QSO CDFS-062 at z=2.822(see also [11]; b: high-redshift QSO CDFS-024 at z=3.605, showing strong absorption lines; c: QSO-2 CDFS-202 at z=3.700 (see [32]); d: Seyfert-2 CDFS-175b at z=0.522, showing a weak high excitation line of [NeV].

2.2   The *Chandra Deep Field South* (CDFS)

The *Chandra X–ray Observatory* has performed deep X-ray surveys in a number of fields with ever increasing exposure times [31,22,11,42,3] and has recently completed two 1 Megasec *Chandra* exposures, in the *Chandra Deep Field South* (CDFS, [12,33]) and in the *Hubble Deep Field North* (HDF-N, [4]), the latter exposure is currently being increased to 2 Megasec.

Here I discuss results from the 940 ksec CDFS observation The source counts (see figure 2) have been extended to 5.5×10^{-17} erg cm^{-2} s^{-1} in the soft 0.5–2 keV band and 4.5×10^{-16} erg cm^{-2} s^{-1} in the hard 2–10 keV band, reaching a space density of almost 4000 deg^{-2}, resolving > 80% of the background in both bands. A total of about 360 sources has been detected.

Optical imaging and spectroscopy has been performed in $\sim$ 10 nights with the ESO Very Large Telescope (VLT) in the time frame April 2000 - December 2001, using deep optical imaging and low resolution multiobject spectroscopy with the FORS instruments with individual exposure times ranging from 1-5 hours. Some preliminary results including the VLT optical spectroscopy have already been presented [32,42,33]. The complete optical spectroscopy will be published in [40]. Figure 3 shows examples of 4 VLT spectra. The upper two spectra show high-redshift QSOs with restframe-UV absorption features (BAL or -mini-BAL QSOs), which both have some indication of intrinsic absorption in their X-ray spectra. The object in the lower left is the famous, highest redshift type-2 QSO detected in the CDFS with heavy X-ray absorption in the QSO rest frame [32]. The spectrum in the lower right shows a Seyfert-2 galaxy with heavy X-ray absorption and an AGN-type luminosity. The latter spectrum is characteristic for the bulk of the detected galaxies, which show either no or very faint high excitation lines indicating the AGN nature of the object, so that we have to resort to a combination of optical and X-ray diagnostics to classify them as AGN (see below). Redshifts could be obtained so far for 169 of the 360 sources in the CDFS, of which 123 are very reliable (high quality spectra with 2 or more spectral features), while the remaining optical spectra contain only a single emission line, or are too noisy. For objects fainter than R=24 reliable redshifts can be obtained (see also Figure 5), if the spectra contain strong emission lines. for the remaining optically faint objects we have to resort to photometric redshift techniques. About 11% of the CDFS sources have no counterpart even in deep VLT optical images ($R < 27.5$) or near–IR imaging (15% at $K < 22$) [33]. Nevertheless, for a subsection of the sample at off-axis angles smaller than 8 arcmin we obtain a spectroscopic completeness of about 60%.

2.3   Optical/X-ray Classification

Type-1 AGN (Seyfert-1 and QSOs) can be often readily identified by the broad permitted emission lines in their optical spectra. Luminous Seyfert-2 galaxies show strong forbidden emission lines and high-excitation lines indicating photoionization by a hard continuum source. However, already in the spectroscopic identifications of the *ROSAT* Deep Surveys it became apparent, that an increasing fraction of faint X-ray selected AGN shows a significant, sometimes dominant contribution of stellar light from the

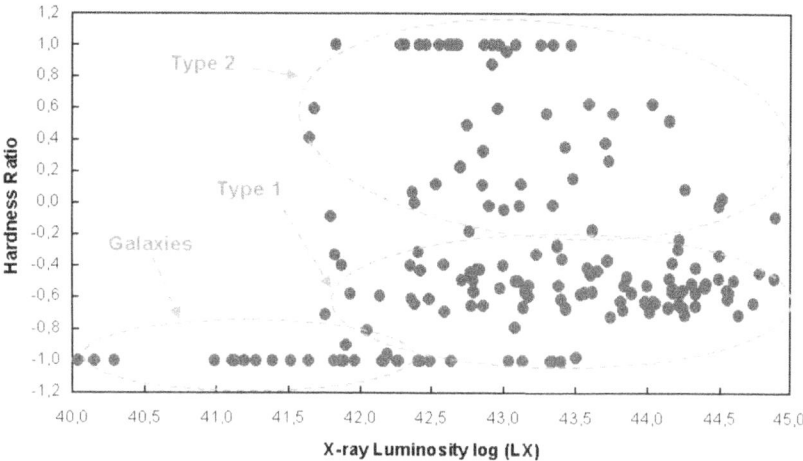

Fig. 4. Hardness ratio of CDFS and Lockman Hole Sources with redshifts as a function of X-ray luminosity. The dashed ellipses indicate the regions expected for type-1 and type-2 AGN as well as for galaxies whose X-ray emission is dominated by thermal processes.

host galaxy in their optical spectra, depending on the ratio of optical luminosity between nuclear and galaxy light [23,24]. If an AGN is outshined by its host galaxy it is not possible to detect it optically. Many of the counterparts of the faint X-ray sources detected by *Chandra* and *XMM-Newton* show optical spectra which are dominated by their host galaxy and only a minority has clear indications of an AGN nature (see also [1,2]). In these cases, the X-ray emission could still be dominated by the active galactic nucleus, while a contribution from stellar and thermal processes (hot gas from supernova remnants, starbursts and thermal halos, or a population of X-ray binaries) can be important as well.

In these cases X-ray diagnostics in addition to the optical spectroscopy can be crucial to classify the source of the X-ray emission. AGN have typically (but not always!) X-ray luminosities above 10^{42} erg s^{-1} and power law spectra, often with significant intrinsic absorption. Local, well-studied starburst galaxies have X-ray luminosities typically below 10^{42} erg s^{-1} and very soft X-ray spectra. Thermal haloes of galaxies and the intergalactic gas in groups can have higher X-ray luminosities, but has soft spectra as well. The redshift effect in addition helps the X-ray diagnostic, because soft X-ray spectra appear even softer already at moderate redshift, while the typical AGN power law spectra appear harder over a very wide range of redshifts.

In Figure 4 the X-ray hardness ratio is shown as a function of the X-ray luminosity (in the 0.5–2 keV, 2–10 keV, or 0.5–10 keV band, depending on in which band the object was detected) for 170 sources for which we have optical spectra and reliable redshifts in the CDFS [40](see also [33]) and the *Lockman Hole* [24,25] for X-ray sources detected by *Chandra* and *XMM-Newton*, respectively. The hardness ratio is defined as $HR \equiv (H - S)/(H + S)$ where H and S are the net count rates in the hard (2–7 keV for Chandra and 2-4.5 keV for XMM-Newton) and the soft band (0.5–2 keV), respectively. The X–ray luminosities are not corrected for internal absorption and are computed assuming $H_0 = 50$ km s^{-1} Mpc^{-1} and $q_0 = 0.5$.

Although this diagram is for illustration purposes only and a a correct treatment would have to properly take into account the different instrument characteristics and detection bands, it clearly shows a segregation of the different X-ray emitters (indicated by the dashed elliptical outlines. Type-1 AGNs have luminosities above 10^{42} erg s^{-1} and hardness ratios scattered around $HR \simeq -0.6$, corresponding to a power law photon index around $\Gamma = 1.8$, typical for the intrinsic continuum of AGN. Type-2 AGN also have luminosities above 10^{42} erg s^{-1}, but are scattered to much higher hardness ratios ($HR > -0.2$). Direct spectral fits of the XMM-Newton and (some) Chandra spectra clearly indicate, that these harder spectra are due to neutral gas absorption and not due to a flatter intrinsic slope [32,21,26]. It is interesting to note that no high-luminosity, very hard sources exist in this diagram. This is due to a selection effect of the pencil beam surveys: due to the small solid angle, the rare high luminosity sources are only sampled at high redshifts, where the absorption cutoff of type-2 AGN is redshifted to softer X-ray energies. Indeed, the type-2 QSOs in this sample [32,25] are the objects at $L_X > 10^{44}$ erg s^{-1} and $HR > 0$.

About 10% of the sources have optical spectra of normal galaxies, X-ray luminosities below 10^{42} erg s^{-1} and very soft spectra ($HR \sim -1$), typical for starburst galaxies or hot gas halos. The deep *Chandra* and *XMM-Newton* surveys therefore for the first time detect the population of normal starburst galaxies out to intermediate redshifts [31,11,25],for which a significant contribution to the XRB had been claimed for a long time [27], however, at much lower fluxes and therefore almost negligible contribution to the background. These galaxies might become an important means to study the star formation history in the universe completely independently from optical/UV, sub-mm or radio observations. However, in the X-ray luminosity range around 10^{42} erg s^{-1}, where the emission from star forming processes and the central AGN may be comparable, there will always remain ambiguities.

The joint optical/X-ray diagnostics scheme can also be applied to the spectroscopically identified X-ray sources in other deep Chandra fields in order to obtain an as complete sample of faint X-ray source classification as possible. Table 1 gives a summary of optical identifications and X-ray source types in the two deep fields discussed here, as well as in the Hawaii 13hr field, the Abell 370 cluster field and the Hubble Deep Field North which all have spectroscopically identified samples in the literature [1,2].

Table 1. *Chandra* and *XMM-Newton* Survey Identifications

Field	Type-1	Type-2	Galaxies	Total	Reference
Lockman Hole	41	26	7	74	[24,25]
CDFS	47	73	49	169	[40]
Abell 370	9	5	6	20	[2]
13hr[a]	5	7	1	20	[1]
HDF-N[a]	10	10	0	20	[2]

[a] only 2-7 keV band detections considered

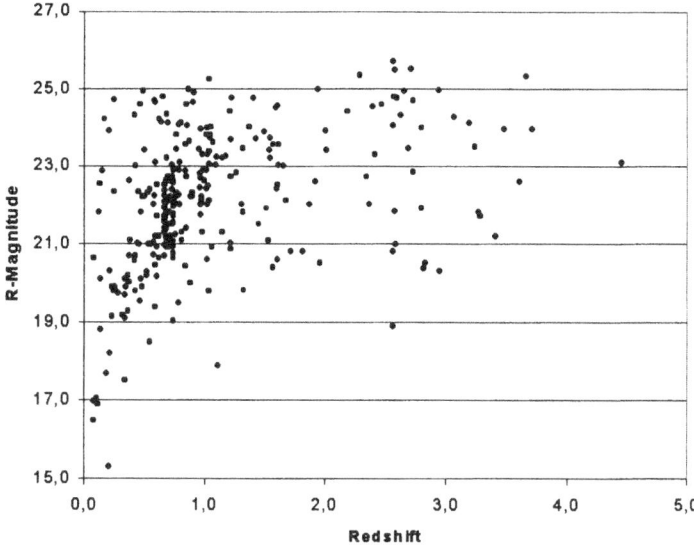

Fig. 5. Optical magnitudes of AGN and galaxies from all the samples in table 1 as a function of redshift. R-magnitudes are taken from the Lockman Hole [24,25] and the CDFS samples [12]. For the Hawaii 13hr [1], Abell 370, and HDF-N samples [2], where only I magnitudes are given, a colour of R-I=1 has been assumed. An accumulation of objects in two redshift bins around z=0.7 is due to a large scale structure in the CDFS.

3 The Chandra/XMM-Newton Redshift Distribution

All the above samples have a spectroscopic completeness of about 60%, which is mainly caused by the fact that about 40% of the counterparts are optically too faint to obtain reliable spectra. This incompleteness is probably also reflecting some redshift bias, most likely higher redshift objects are missing, as well as faint emission line objects, where the strongest emission lines ([OII], Ly_α) fall outside the optical bands. On the other hand, the optically faintest identified sources (R=24-25) are distributed throughout the whole redshift range z=0-4 (see Figure 5), therefore there is reason to believe that a substantial fraction of the so far unidentified sources follows the same redshift distribution as the identified sources. The completeness of 60% therefore allows to compare the redshift distribution with predictions from X-ray background population synthesis models [13], based on the AGN X-ray luminosity function and its evolution as determined from the ROSAT surveys [28], which predict a maximum at redshifts around z=1.5. Figure 6 shows two predictions of the redshift distribution from the Gilli et al. model [13] for a flux limit of 2.3×10^{-16} erg cm^{-2} s^{-1} in the 0.5-2 keV band with different assumptions for the high-redshift evolution of the QSO space density. The two models from [13] have been normalized at the peak of the distribution.

The actually observed redshift distribution does not vary significantly within the flux limit range covered by the samples in table 1, therefore the total observed redshift distribution is shown in figure 6 for the total number of ~ 300 sources in all samples. The observed redshift distribution, arbitrarily normalized to roughly fit the population synthesis models in the redshift range 1.5 – 3 keV is radically different from the prediction, with a peak at a redshift in the range 0.5-0.7. This is still the case, if the objects belonging to the large scale structures around z=0.7 in the CDFS are removed. The total number of objects at redshift less than 1 is significantly higher than the model predictions, even ignoring the 40% spectroscopic incompleteness. The peak at redshifts below 1 is also significant, if the normal star forming galaxies in the sample are removed. This clearly demonstrates that the population synthesis models will have to be modified to incorporate different different luminosity functions and evolutionary scenarios for intermediate-redshift, low-luminosity AGN.

Chandra/XMM-Newton Redshift Distribution

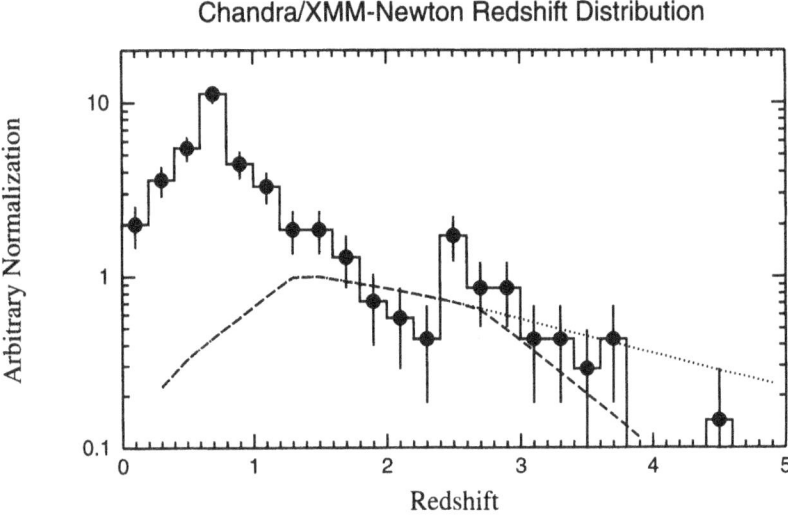

Fig. 6. Redshift distribution of ~ 300 X-ray selected AGN and galaxies in the deep Chandra and XMM-Newton survey samples given in table 1 (solid circles and histogram), compared to model predictions from population synthesis models [13]. The dashed line shows the prediction for a model, where the commoving space density of high-redshift QSO follows the decline above z=2.7 observed in optical samples [34,7]. The dotted line shows a prediction with a constant space density for $z > 1.5$. The two model curves have been normalized to their peak at z=1, while the observed distribution has been normalized to roughly fit the models in the redshift range 1.7-3.

4   The AGN Evolution at High Redshift

The comparison between the observed and predicted N(z) distributions at high redshifts is complicated by the possible existence of large-scale structure in the pencil beam survey (there is e.g. a possibly significant excess of objects around z=2.5 in the CDFS), but also by redshift-dependent selection effects and in general by the still relatively small volume sampled and therefore poor counting statistics in the number of objects. In addition, the overall normalization of the curves is uncertain because of the significant mismatch of the distribution at low z. Nevertheless, the observed distribution is roughly consistent with both predictions in the redshift range z=1.6-3.8. There is, however, a significant discrepancy between the observed distribution and the constant space density model (dotted line) at redshifts above 4, where only one object was detected, while about 8 objects would be predicted. From Figure 5 it becomes apparent, that the dearth of X-ray selected AGN is probably not due to optical spectroscopic selection effects. The one object detected at z=4.45 already in the ROSAT data of the Lockman Hole [36] has an optical magnitude of R=23 and is therefore not at the spectroscopic limit of the samples. Also the Ly_α and CIV lines for QSOs in the redshift range 4-5 fall well into the optical range. The observed redshift distribution therefore gives a strong indication for a decline of the QSO space density beyond a redshift of 3.8.

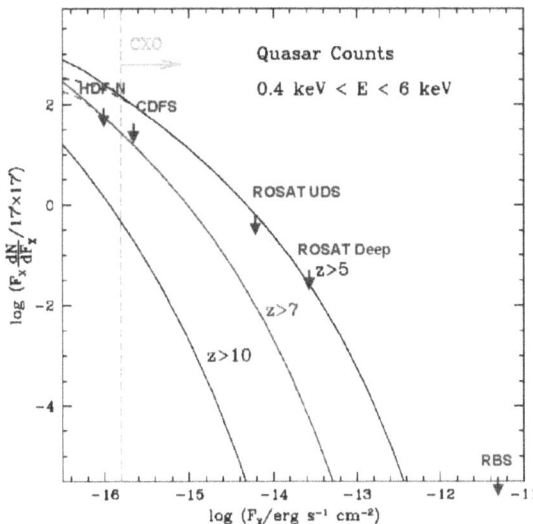

Fig. 7. Prediction of the number density of AGN with redshits larger than 5, 7 and 10, respectively as a function of flux in a typical 17 × 17 arcmin Chandra field of view from Haiman & Loeb [15]. Upper limits measured in X-ray surveys at various flux limits are indicated.

A similar conclusion about a decline of the X-ray selected AGN space density at high redshifts can be obtained from the absence of QSOs with $z > 5$ in all X-ray survey samples so far. (There was a recent announcement of a QSO at z=5.2 in the Chandra observation of the HDF-N, but this does not change the conclusions discussed below). Figure 7 shows a prediction of number counts for high-redshift QSO from Haiman & Loeb [15], according to which a large number of $z > 5$ AGN should be detected in any deep survey with Chandra. This theoretical model assumes the X-ray luminosity function at z=3.5 determined from the ROSAT surveys and extrapolates it backwards in time assuming a simple hierarchical CDM model. The figure also shows limits for the number counts of $z > 5$ AGN from X-ray surveys at varying flux limits. The most distant QSO among ~ 2000 objects in the ROSAT Bright Survey (RBS, [37] has a redshift of 2.8, the lack of higher redshift objects is, however, not constraining given the high flux limit of this survey. The lack of $z > 5$ AGN in the ROSAT Deep and Ultradeep Surveys [35,23,24] is still just consistent with the Haiman & Loeb predictions, the highest-redshift object in the UDS is RX J105225.9+571905 at z=4.45 [36]. The Chandra Deep survey, while only about 60% spectroscopically identified, still provides an upper limit for the number counts of $z > 5$ AGN significantly lower than the prediction, using the conservative assumption that less than half of the unidentified objects are at redshifts larger than 5. Finally, the 400 ksec Chandra observation in the Hubble Deep Field proper, providing 100% identifications for 12 sources in the field and their highest redshift object at z=4.42 just outside the HDF-N also gives an upper limit about a factor of three lower than the Haiman & Loeb prediction.

The information about the space density of X-ray selected AGN is still limited by the small number statistics in the deep X-ray surveys which cover too small a solid angle. More and wider fields have been surveyed by both Chandra and XMM-Newton. As soon as the tedious and time consuming optical follow-up work in these fields is completed, we will be able to learn more about the decline of the X-ray AGN and therefore their formation at early redshifts. The possible discrepancy between a declining space density of optical and radio-selected QSOs above a redshift of 2.7 and an apparently constant space density of X-ray selected AGN with a decline beyond a redshift of ~ 4 could still be understood in terms of the different luminosity and therefore different black hole mass of the objects involved. The optical and radio surveys cover a large solid angle to a modest flux limit and therefore pick up only the most luminous and therefore most massive objects at high redshift. The deep pencil beam surveys, on the other hand, sample a much smaller volume to much fainter flux limits and therefore select high-redshift AGN which are intrinsically more than a factor of 10 less luminous and therefore probably less massive than the objects selected in wide-angle surveys. In the hierarchical large scale structure formation the smaller cold dark matter halos collapse earlier than the larger ones. Given the correlation between black hole mass and galaxy mass (and presumably dark matter mass), it is expected that the lower mass black holes are formed earlier than the most massive objects and thus that lower luminosity AGN appear earlier than the most luminous QSOs. This concept can be tested with more optical identifications of Chandra and XMM-Newton surveys and with future, even more sensitive X-ray telescopes, like the ESA/ISAS XEUS mission.

5   Acknowledgements

I thank the organisers of the conference "Lighthouses in the Universe" for the invitation for this review. I thank my co-workers in the Chandra Deep Field South and Lockman Hole identification teams for the fruitful collaboration and the permission to show data in advance of publication.

References

1. Barger, A. J., Cowie, L. L., Mushotzky, R. F., Richards, E. A., 2001, AJ 121, 662
2. Barger, A.J., Cowie, L.L., Bautz M.W., et al., 2001, AJ 122, 2177
3. Brandt, W.N., Hornschemeier A.E., Alexander D.M., et al., 2001, AJ 122, 1
4. Brandt, W.N., Alexander D.M., Hornschemeier A.E., et al., 2001, AJ 122, 2810
5. Comastri A., et al., 1995, A&A 296, 1
6. Fabian A.C., Barcons X., Almaini O., Iwasawa K., 1998, MNRAS 297, L11
7. Fan X., et al., 2001, AJ 121 54
8. Fiore F., LaFranca F., Giommi P., et al., 1999, MNRAS 306, 55
9. Fiore F., LaFranca F., Vignali C., et al., 2000, NewA 5, 143
10. Gebhardt K., Bender R., Bower G., et al., 2000, ApJ 539, 13
11. Giacconi, R., Rosati P., Tozzi P., et al., 2001, ApJ 551, 624
12. Giacconi, R., Zirm A., Wang P., et al., 2002, ApJS (in press), astro-ph/0112184
13. Gilli, R., Salvati, M., Hasinger, G., 2001, A&A 366, 407
14. Granato G.L., Danese L., Franceschini A., 1997, ApJ 486, 147
15. Haiman, Z. & Loeb A., 1999, ApJ 519, 479
16. Hashimoto, Y., Hasinger, G., Arnaud, M., et al., 2002, A&A, in press
17. Hasinger, G., Burg, R., Giacconi, R., et al., 1993, A&A 275, 1
18. Hasinger, G., Burg, R., Giacconi, R., et al., 1998, A&A 329, 482
19. Hasinger, G., Giacconi, R., Gunn, J.E., et al., 1999, A&A 340, 27
20. Hasinger, G., Altieri, B., Arnaud, M., et al., 2001, A&A 365, 45
21. Hasinger, G. & Lehmann, I. 2001, Proceedings for "Where's the Matter?", Marseille, France, 25-29 June 2001, eds. L. Tresse & M. Treyer, in press.
22. Hornschemeier, A.E., Brandt, W.N., Garmire, G.P., et al., 2000, ApJ 541, 49
23. Lehmann, I., Hasinger, G., Schmidt, M., et al., 2000, A&A 354, 35
24. Lehmann, I., Hasinger, G., Schmidt, M., et al., 2001, A&A 371, 833
25. Lehmann, I., Hasinger, G., Murray, S.S., Schmidt, M., 2002, High Energy Universe at Sharp Focus: Chandra Science, proceedings, ASP Conference Series, eds. S. Vrtilek, E.M. Schlegel, L. Kuhi (astro-ph/0109172)
26. Mainieri V., et al., 2002, in prep.
27. McHardy I., Jones L.R., Merrifield M.R., et al., 1998, MNRAS 295, 641
28. Miyaji, T., Hasinger, G., Schmidt, M., 2000, A&A 353, 25
29. Miyaji, T., Hasinger, G., Schmidt, M., 2001, A&A XXX, YY
30. Murray, S.S., et al., 2002, in preparation
31. Mushotzky, R.F., Cowie L.L., Barger, A.J., Arnaud, K.A., 2000, Nature 404, 459
32. Norman C., Hasinger G., Giacconi R., et al. 2001, submitted to ApJ (astro-ph/0103198)
33. Rosati P., Tozzi P., Giacconi R., et al., 2002, ApJ, in press (astro-ph/0110452)
34. Schmidt, M., Schneider, D.P. & Gunn J.E., 1995, AJ 114, 36
35. Schmidt, M., Hasinger, G., Gunn, J.E., et al., 1998, A&A 329, 495
36. Schneider, D.P., Schmidt, M., Hasinger, G., et al., 1998, AJ 115, 1230
37. Schwope A., Hasinger G., Lehmann I., et al., 2000, AN 321, 1
38. Shaver P.A. et al., 1996, Nature 384, 439
39. Stern D., et al., 2002, ApJ (in press), astro-ph/0111513
40. Szokoly, G., Hasinger G., Rosati, P. et al. 2002 (in prep.)
41. Thompson D., Pozzetti, L., Hasinger, G., et al., 2001, A&A 377, 778
42. Tozzi, P., Rosati, P., Nonino, M., et al., 2001, ApJ, in press (astro-ph/0103014)
43. Zamorani G., Mignoli M., Hasinger G., et al., 1999, A&A 346,731

Resolving the X-ray Spectral Paradox: XMM-Newton Spectra of Faint Sources in the Lockman Hole

Richard E. Griffiths[1], Andrew Ptak[1], and Takamitsu Miyaji[1]

Carnegie Mellon University, Pittsburgh, PA 15213, USA

Abstract. We report on the X-ray spectra of the faint sources detected by using the XMM-Newton Observatory in a 100ks exposure on the Lockman Hole, a region previously studied exceptionally well using ROSAT and other observatories. When the X-ray spectra of optically identified and unidentified X-ray sources are stacked together by optical type, we find that the summed spectra are best fit by (i) a spectral photon index of 1.9 for Type I AGN, (ii) a spectral index of 1.2 for Type II AGN, and (iii) a harder index of 1.1 for (non-ROSAT) unidentified sources selected in the 5 – 8 keV band. We conclude that the composition of the X-ray background between 2 and 10 keV is largely comprised of the latter two types of source.

Comparison between a CXO (1 Ms) and XMM (160ks) exposure in the HDF-N indicates agreement in spectral results, but also shows some confusion in the XMM sources at flux levels below about 5×10^{-15} ergs cm^{-2} s^{-1}. Although such deep exposures with CXO will be very limited in number, XMM exposures at such depths will be more common and the latter will eventually be used to study the cosmic variance in the deep survey sources and to study the evolution in the luminosity function of the multiple classes of galaxies and AGN which contribute to the faint source counts.

1 Introduction

CXO and XMM-Newton have enabled us to make rapid progress in deciphering the long standing problem of the origin of the X-ray background. A key factor in this problem has been the mismatch between the spectra of AGN and that of the XRB, a problem known as the 'spectral paradox', and we focus in this paper on the resolution of this paradox in broad terms. The XRB spectrum, approximated as a power law, has an effective photon index of 1.4 between 2 and 20 keV, and an energy density which reaches a peak near 30 keV (see the 1992 review by Fabian & Barcons [1]. One of the major goals of the current generation of X-ray observatories has been to identify those source populations which are primarily responsible for the bulk of the overall energy density. In particular, we would like to identify those sources contributing to the source counts in the range 2 - 10 kev, and to identify those with the hardest spectral indices, since they are the sources which make the greatest contribution to the total XRB energy.

In the photon energy range between 0.5 and 2 keV (effectively, at 1 kev), the XRB has been resolved into point sources by the Einstein and ROSAT X-ray observatories (the deepest such survey was that of Hasinger et al. [2] with ROSAT). The fraction of the XRB resolved into discrete sources identified as AGN has increased from about 30% with Einstein observations (Griffiths et al. [3]) to more than 60% with ROSAT (Hasinger et al. [2], Schmidt et al.[4]). When evolution of the AGN population is taken into account, the AGN contribution may be as high as $\sim 80 - 100\%$ below 2 kev (Miyaji et al. [5]).

In the era ending in 1999 with the launch of CXO and XMM-Newton, the best estimates of the AGN contribution to the 2 - 10 keV XRB were made using data from ASCA and Beppo-SAX. Both of these instruments had been used to show that AGN probably contribute at least 60% of the 2 - 10 kev XRB, when AGN luminosity functions were extrapolated to high redshift. But they had also been used to show that the faintest extant sources, near 10^{-15} cgs, seemed to be associated with galaxies rather than AGN (Georgantopoulos et al. [6]). This result was confirmed by Ueda et al. [7] and then by observations with Beppo-SAX (Comastri et al.[8]). In this 2 – 10 keV energy range the XRB has now been resolved into point sources at the $\sim 80 - 90\%$ level by using CXO ([9], [10], [11], [12], [13] and refs. therein) and by using XMM-Newton [14]. These observations have effectively been made at a typical photon energy of about 3 kev, since most of the detected photons are at the soft end of the specrum. The fraction of the XRB resolved by the CXO at 8 - 10 kev is about 20% in a few hundred ks.

XMM-Newton has shown great effectiveness in terms of collecting area for measurement of deep survey source spectra. In the Lockman Hole deep survey area, the optical identifications have largely been completed as a result of the deep ROSAT surveys, and we are thus able to obtain X-ray spectra of the various classes of source contributing to the X-ray background. Rather than commenting on the percentage make-up of sources to the XRB within a particular energy range, we are able to comment on their summed spectra so as to be able to extrapolate to energies above 10 kev where the bulk of the energy of the XRB resides.

2   Observations and Results

2.1   The Lockman Hole

The Lockman Hole is a region in which the Galactic hydrogen column density N_H is only 6×10^{19} cm^{-2}, and it has therefore been the subject of intense study at soft X-ray energies using the ROSAT observatory [2]. As many as 1000 sources deg^{-2} were detected in this field (0.5 – 2 keV) and a considerable amount of follow-up optical spectroscopy has been performed using the Keck and other telescopes [4], [15].

The XMM-Newton observatory was pointed at the Lockman Hole in Apr./May 2000, resulting in a total of $\sim$100 ks. of useful data. A total of $\sim$ 150 sources was detected, as reported in Hasinger et al. 2000 [14]. Of these sources, about 50 were previously detected with ROSAT, so that about 100 are new detections with XMM-Newton. The resulting log(N) - log(S) analysis and hardness ratios of the sources are also presented in ref. [14].

2.2   Spectra of Optically Identified Sources

In Table 1 and Figure 1, we show the summed x-ray spectra of sources for which the redshift has been measured via optical spectroscopy (from ref.[15]), together with the spectra of the unidentified sources. In summary, the best-fit power-law spectral indices for these summed spectra are 1.9 for the Type I objects, 1.2 for the Type II's, and 1.1 for unidentified sources (hard-selected), respectively.

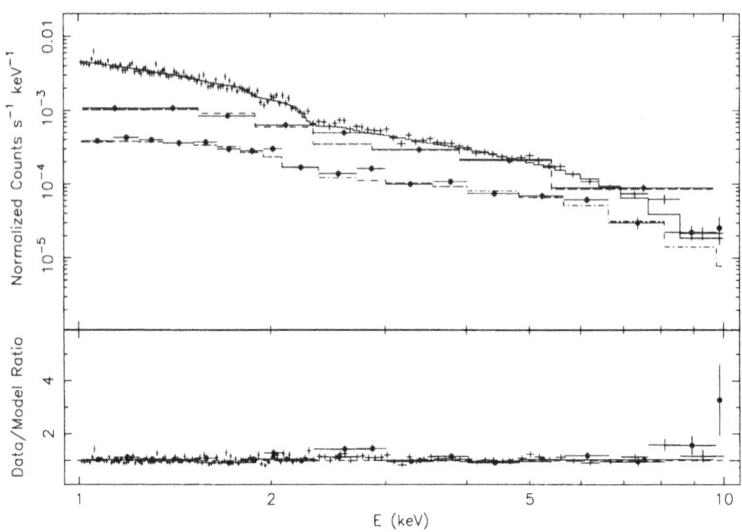

Fig. 1. Spectra Summed by AGN Type
top AGN Type I, *middle* AGN Type II, *bottom* unidentified sources, hard-selected

Table 1. X-ray Spectral Indices

Fit	N	Nh 10^{21} cm^{-2}	Gamma	F(2-10)	Chisq/dof
AGN I	42	0.5 (< 1.1)	1.87 (1.80-1.94)	2.32e-14	632/606
AGN I*	27	0.4 (< 1.0)	1.85 (1.78-1.92)	3.55e-14	600/562
AGN II	5	2.6 (< 5.4)	1.19 (0.93-1.45)	2.68e-15	71/71
No ROSAT ID	99	0.0 (< 1.4)	1.49 (1.39-1.68)	3.58e-15	711/635
No ROSAT ID*	40	1.8 (< 3.7)	1.11 (0.93-1.31)	7.06e-15	284/309

Notes to Table 1:
* = selected to have a > 90% confidence detection in the 5-8 keV bandpass.
F(2-10) is the mean flux/galaxy in cgs units. All AGN II galaxies have significant detection in the hard band.
N = number of galaxy spectra summed

AGN I = sources in Lehmann et al. [15] with class = a-c

AGN II = sources in Lehmann et al. [15] with class = d

2.3 Hubble Deep Field - North

A 160ks exposure of the HDF-N field was taken with XMM in May 2001 and completely overlaps the CXO 1 Ms archived exposure. AGN spectra extracted from corresponding sources show excellent agreement between the two experiments.

At flux levels at or below 5×10^{-15} ergs cm^{-2} s^{-1}, XMM sources start to become confused, i.e. the CXO observation shows that some of the sources detected as single sources in the XMM data are shown to consist of two or more fainter sources in the CXO data. The log(N)/log(S) for XMM source counts is therefore unreliable below this level. In the 5 – 10 kev band, confusion has not yet been reached at the 160ks exposuree level.

3 Conclusions

- From a deep XMM-Newton observation of the Lockman Hole, and using established optical identifications from the deep ROSAT survey:
  (i) AGN of type I have an average photon spectral index of 1.9 over the energy range 2 – 10 keV, as found by many previous experiments
  (ii) AGN of type II have an average spectral energy index of 1.2 (2 - 10 keV)
  (iii) unidentified sources have an energy spectral index of 1.4 (2 –10 keV) and 1.1 (hard selected 5 - 8 keV).
- For sources identified with galaxies, the centroid source position as measured using XMM may not be sufficient to distinguish between an active nucleus and an off-nuclear transient or ultraluminous object of stellar or intermediate mass.
- Extrapolating to higher energies, it is evident that AGNs of Type II and the unidentified sources are likely to make up the bulk of the energy in the XRB.
- XMM deep surveys start to become confused at flux levels of 5×10^{-15} ergs cm^{-2} s^{-1} (2 - 10 kev), reached in exposure times of $\sim$ 150 ks.
- The multiple deep surveys with XMM will allow the study of cosmic variance and evolution of the source populations.

4 Acknowledgments

These observations were made using the XMM-Newton X-ray observatory, built, launched and operated by the European Space Agency with NASA participation. These results rely heavily on the analysis performed and reported by Guenther Hasinger and colleagues [14].

References

1. A. C. Fabian and X. Barcons: Ann. Rev. Astron. Astroph., **30**, 429 (1992)
2. G. Hasinger et al.: Astron. & Astroph., **329**, 482 (1998)
3. R. E. Griffiths et al.: Mon. Not. R. Astr. Soc., **255**, 545 (1992)
4. M. X. Schmidt et al.: Astron. & Astroph. **329**, 495 (2001)
5. T. Miyaji et al.: Astron. & Astroph., **369**, 49 (2001)

 6. I. Georgantopoulos et al.: Mon. Not. R. Astr. Soc., **291**, 203 (1997)
 7. Y. Ueda et al.; Nature, **391**, 866 (1998)
 8. A. Comastri et al.: Adv. Sp. Res., **25**, 833, (2000)
 9. R. F. Mushotzky, L. L. Cowie, A. J. Barger, K. A. Arnaud: Nature **404**, 459 (2000)
10. G. P. Garmire et al.: Ap. J., submitted (2002)
11. A.E. Hornschemeier et al.: Ap. J. **554**, 742 (2001)
12. W. N. Brandt et al.: Astron. J. **122**, 1 (2001)
13. R. Giacconi et al.: Ap. J. **551**, 624 (2001)
14. G. Hasinger et al.: Astron. & Astroph. **365**, L45 (2001)
15. Lehmann et al.: Astron. & Astroph. **371**, 833 (2001)

Resolving the Hard X-Ray Background in the Chandra Deep Field South

Roberto Gilli[1,2], Riccardo Giacconi[3], Paolo Tozzi[4], Piero Rosati[5], Günther Hasinger[6,7], Lisa Kewley[8], JunXian Wang[1], Andrew Zirm[1], Vincenzo Mainieri[5], Stefano Borgani[9], Jacqueline Bergeron[5], Roberto Gilmozzi[5], Ethan Schreier[10], Anton Koekemoer[10], Norman Grogin[10], Mario Nonino[4], and Colin Norman[1,10]

[1] Dept. of Physics and Astronomy, The Johns Hopkins University, 3400 N Charles St., Baltimore, MD 21218
[2] Osservatorio Astrofisico di Arcetri, Largo E. Fermi 5, I-50125 Firenze, Italy
[3] Associated Univ. Inc., 1400 16th Street, NW, Suite 730, Washington, DC 20036
[4] Osservatorio Astronomico di Trieste, Via G. Tiepolo 11, I-34131 Trieste, Italy
[5] ESO, Karl-Schwarzschild-Strasse 2, D-85748 Garching, Germany
[6] Astroph. Institut Potsdam, An der Sternwarte 16, D-14482 Potsdam, Germany
[7] Max-Planck-Institut für extraterrestrische Physik, Giessenbachstraße, D-85748 Garching, Germany
[8] Harvard-Smithsonian Center for Astrophysics, Cambridge, MA 02138
[9] INFN, Sezione di Trieste, c/o Dipartimento di Astronomia dell'Università, via G. Tiepolo 11, I-34131 Trieste, Italy
[10] Space Telescope Science Institute, 3700 San Martin Drive, Baltimore, MD 21218

Abstract. The 1Msec observation of the Chandra Deep Field South (CDFS) has almost entirely resolved into single sources the hard X-ray background as measured by the HEAO-1 A2 experiment. The main population of sources, as determined by the first optical identifications, seems to be constituted by an admixture of type 1 and type 2 AGNs. We will consider the X-ray properties of the sources in the CDFS and test their consistency with the predictions from the AGN synthesis models of the X-ray background. An apparent excess of sources at $z < 0.8$ is found with respect to the model predictions. The X-ray background flux produced by these sources is also exceeding the predictions, suggesting that the bulk of the XRB is produced closer than expected.

1 Introduction

The origin of the hard X-ray background ([6]) seems to be finally understood. X-ray surveys with increasing sensitivity have been resolving an increasing fraction of the hard XRB into single sources ([11]). The extremely deep surveys in the Chandra Deep Field North ([2]) and South ([7]) have now resolved between 70 and 100% of the 2-10 keV XRB ([16]), the main uncertainty on the resolved fraction laying in the absolute sky flux (see e.g. [18] and [5] for a comparison among different measures). As predicted by population synthesis models ([14], [4], [8]) the sources producing the hard XRB are an admixture of unobscured and obscured AGN, the latter being more numerous by a factor $\sim 4 - 10$. Optical identifications have confirmed this view, although the relative fraction of obscured AGN is still uncertain. The Chandra Deep Field South (CDFS) is a unique opportunity to study the nature of the sources making the XRB and to verify the AGN population scenario predicted by the models. So far all the main observational constraints like the XRB spectrum, the logN-logS relations in different bands, the AGN redshift and absorption distributions, which were derived in previous surveys, were consistent with the model expectations. Also, the XRB spectrum and the logN-logS relations derived in the CDFS are consistent with the expectations. However, in the last Section we will present further constraints from the CDFS showing that our view of the XRB synthesis might change.

2   The CDFS Survey

The 1 Msec observation of the CDFS is a combination of 11 single pointings of the Chandra ACIS-I array on a southern sky field with a very low Galactic column density ($\sim 8 \ 10^{19}$ cm^{-2}). The combined pointings give a total exposure of 942 ksec (1 Msec) covering a 0.1 deg^2 area. Due to vignetting effects and different roll angles in the single pointings the limiting flux varies by two orders of magnitude from the center to the outskirts of the field. The limiting fluxes reached in the center are 5.5 10^{-17} erg cm^{-2} s^{-1} (hereafter cgs) in the 0.5-2 keV band and 4.5 10^{-16} cgs in the 2-10 keV band. We detected about 340 X-ray sources in the whole field. The details of the detection process are presented in [17] and [7].

2.1   Resolved XRB Spectrum and logN-logS Relations

By means of the "stacking analysis" presented by [17] the source spectra have been summed and their integrated contribution to the 2-10 keV XRB as a function of flux has been calculated, as shown in Fig.1 (*left*). The XRB flux measured by HEAO-1 A2 ([10]) is almost entirely resolved, while 30% of the BeppoSAX flux ([18]) is still unresolved. A naive look at the curve flattening shown in Fig. 1 (*left*) would suggest that only a small integrated flux will be added by sources below 4.5 10^{-16} cgs. The stacked spectrum of all the CDFS sources is also shown in Fig. 1 (*right*), where it is compared with the XRB measurements from different satellites and with the curve obtained from model B by [8] rescaled to fit the HEAO-1 A2 sky flux.

The logN-logS relations in the 0.5-2 keV, 2-10 keV and 5-10 keV bands have been calculated taking into account the corrections for the sky coverage. The 0.5-2 keV and 2-10 keV logN logS are shown in Fig. 2 (*left* and *right*, respectively), along with the predictions of the rescaled model B by [8].

3   Optical Data

We obtained optical spectroscopy with the FORS instrument at the Very Large Telescope for a subsample of ~ 100 sources which were already present in the first 300 ksec exposure of the CDFS at a limiting flux of 1 10^{-16} cgs in the 0.5-2 keV band and 1 10^{-15} cgs in the 2-10 keV band. The optical spectroscopy work is still in progress and we consider about 80 sources to have a secure redshift determination. Also, optical (U,B,V,R,I) and near-IR (J,H,K) photometry have been obtained from the ESO Imaging Survey ([1]). Only a fraction of the CDFS field, however, was observed in the full band range.

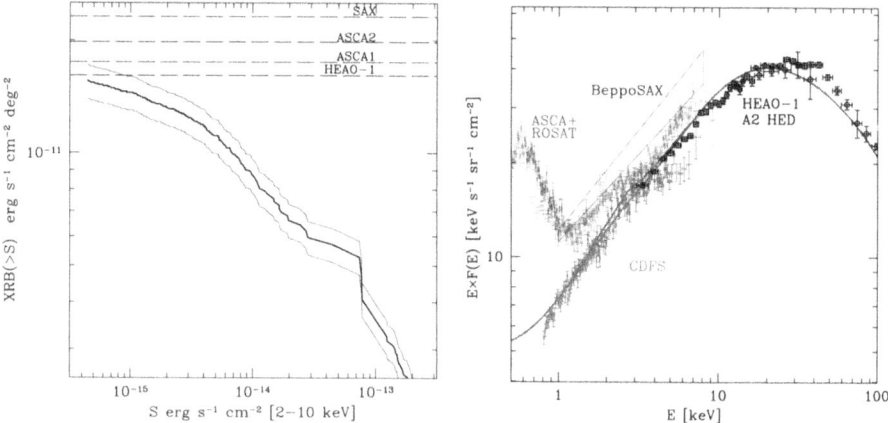

Fig. 1. *Left*: Source contribution to the 2-10 keV XRB emission as a function of limiting flux. Different XRB measurements are also plotted as dashed lines. *Right*: Spectrum of the resolved XRB in the CDFS compared with the XRB spectrum measured by previous missions. The curve from the rescaled model B by [8] is also plotted.

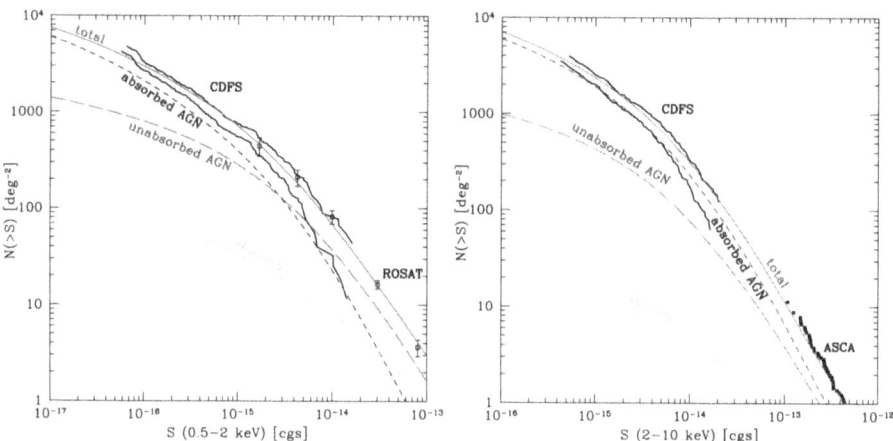

Fig. 2. *Left*: Soft logN-logS relation obtained in the CDFS compared with ROSAT measurements ([9]). The curves from the rescaled model B of [8] are also plotted. The total curve is the sum of the other labeled curves. *Right*: Hard logN-logS relation obtained in the CDFS compared with ASCA measurements ([3]). Curves are analogous to those plotted in the *left* panel.

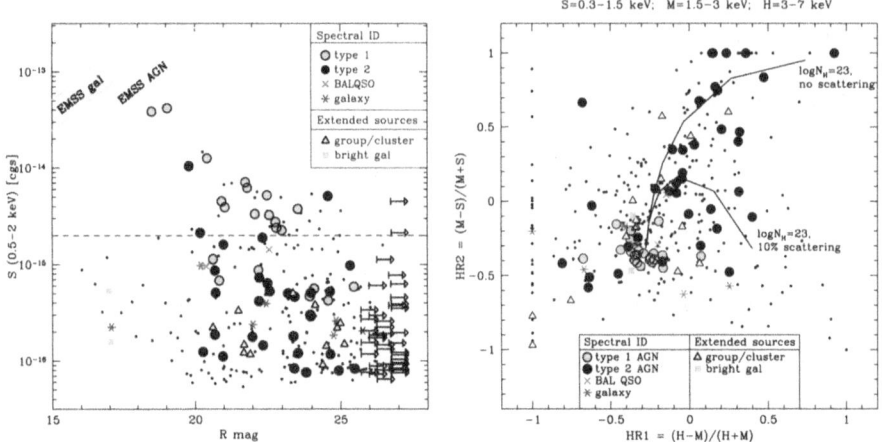

Fig. 3. Soft X-ray flux to R magnitude plot for the CDFS sources. Different colors/symbols refer to different source classes as shown in the inset. Unidentified sources are plotted as dots. Sources with no optical counterparts are plotted as arrows. In the following Figures we will made use of the same color/symbol code. The shaded areas show the range of X-ray-to-optical flux ratios spanned galaxies and AGN in the EMSS ([15]). The dashed line shows the flux at which type 2, absorbed AGN start to dominate over type 1s (compare with Fig. 2 (*left*)). *Right*: X-ray color-color diagram for the CDFS sources. Tracks have been calculated for absorbed powerlaw spectra (with $\Gamma = 1.8$ and $\log N_{\mathrm{H}} = 23$) assuming different source redshifts. From right to left the source is moved from z=0 to z=4 in steps of z=0.5. Two different scattered components have also been considered, as labeled in the Figure.

In Fig. 3 (*left*) we plot the soft X-ray vs optical flux for our sources, where different colors/symbols refer to different classifications. Most of the sources fall in the range already spanned by the EMSS and ROSAT sources ([15], [13])). However, some sources with very low X-ray to optical flux ratio have been observed, which are identified as bright normal galaxies. Nevertheless, the majority of the identified sources are AGN. Also, it seems that there is a change in the AGN population at a flux of $\sim 2\ 10^{-14}$ cgs, where obscured sources start to dominate over unobscured ones. This is in good agreement with what is shown in Fig. 2 (*left*), where the logN-logS of obscured and unobscured AGN cross at a flux of $\sim 2\ 10^{-14}$ cgs.

In order to have a sense on what the unidentified sources are, we divided the source counts into three energy bands to achieve two X-ray colors (hardness ratios). The X-ray color-color plot is shown in Fig. 3 (*right*), where the energy bands and the X-ray colors are also defined. The same color/symbol code introduced in Fig. 3 (*left*) has been adopted. Also, absorbed AGN tracks have been calculated and overplotted. The details on the AGN tracks are given in the caption. Inspection of Fig. 3 (*right*) reveals that most of the unidentified sources are likely to be absorbed AGN having different scattered components.

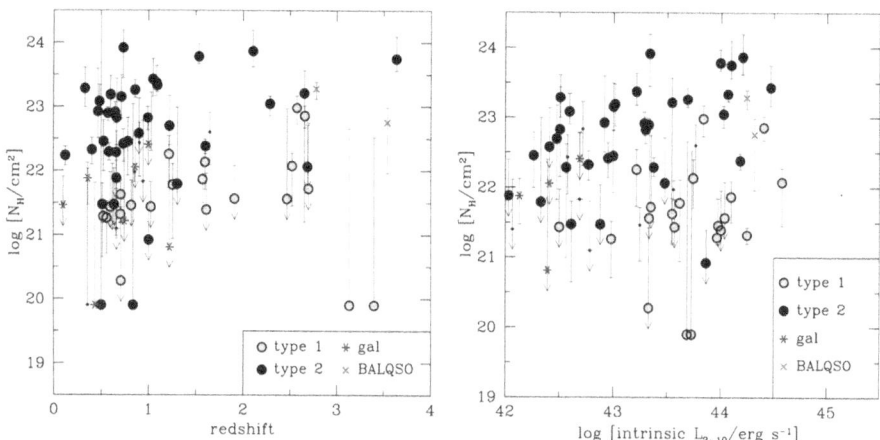

Fig. 4. *Left*: Absorption vs. redshift plot for the CDFS sources. Sources without intrinsic absorption have been plotted at the Galatic value. Dots are sources without secure optical classification. *Right*: Intrinsic luminosity in the 2-10 keV rest frame band vs. absorption for the CDFS sources. Sources with luminosities above 10^{44} erg s^{-1} and absorbing columns above 10^{22} cm^{-2} are considered type 2 QSOs.

4 X-Ray Source Spectra

For those sources with measured redshift we performed X-ray spectroscopy with XSPEC simply fitting the data with an absorbed powerlaw. For those sources with a few X-ray counts, the photon index was fixed to $\Gamma = 1.8$. In Fig. 4 (*left*) we plot the redshift of the sources against the measured column density. Sources with no intrinsic absorption have been plotted at the Galactic value. The color/symbol code is the same adopted in the previous Figures. Absorbed sources are found at all the redshifts and no evidence for a decreasing average obscuration with increasing redshift is found. We then calculated the 2-10 keV intrinsic (de-absorbed) and rest frame luminosities for our sources. They are plotted against the absorbing column densities in Fig. 4 (*right*). We note the existence of an abundant population of sources with luminosities above 10^{44} erg s^{-1} and absorbing columns above 10^{22} cm^{-2}, which can well be considered type 2 QSOs. We also note that the lack of very luminous type 2 QSOs ($L_{2-10} > 10^{45}$ erg s^{-1}) in the CDFS is not a strong argument against their existence, since very luminous AGN are rare objects and can be picked up efficiently only with wide area surveys. Type 1 QSOs with $L_{2-10} > 10^{45}$ erg s^{-1} are also lacking in the CDFS (which supports the above statement), therefore we find no evidence of a decreasing average column density with increasing AGN luminosity.

5 The "Local Excess"

We compared the redshift distribution of our sources with those predicted by the rescaled model B of [8]. Our optical identifications are largely incomplete, since large regions of the CDFS have not been observed spectroscopically yet. To enhance completeness we considered only a subregion of the CDFS (a circle with a 0.047 deg^2 area) where optically identified sources are 50% of the total X-ray sources. We limited our analysis to sources selected in the 2-10 keV band. Two extreme hyphotesis are then made on the unidentified sources: i) they have the same redshift distribution as the identified ones, i.e. they peak around $z = 0.7$; ii) they are all at high redshift, e.g. $z > 1.4$. The real distribution should be bracketed by these two, since the unidentified sources, which have mean X-ray and optical fluxes lower than those of the identified ones, are likely to be on the average at higher redshifts. We perform a KS test to verify the consistency between the two redshift distributions with that predicted by the rescaled model B. It is found that the model is completely ruled out if distribution i) is assumed, while it cannot be ruled out at more than 98% confidence level if distribution ii) is assumed. Overall, these results are suggesting that some model refinements have to be introduced.

A further test can be done by summing the flux of our identified sources in different redshift bins and then dividing the flux by the area covered by the sources. In this way we are calculating how much XRB flux is produced by our sources at different redshifts. If we compare our calculations with the predictions of model B, we found that the integrated flux of the CDFS sources at $z < 1$ is already exceeding that predicted by the XRB model in the same redshift range. Since the redshifts of many CDFS sources still have to be measured, the observed curve is indeed a lower limit, which enchance the significance of the result. These comparisons suggest that the XRB is produced closer than expected, and therefore the AGN XLF and evolution have to be reconsidered. It could be objected that the sources in the $z < 1$ bin are not all AGN. However, although the optical classification is uncertain for many of them, their 2-10 keV luminosities generally exceed 10^{42} erg s^{-1}, which are not easily explained with mechanisms other than AGN.

References

1. S. Arnouts et al.: A&A in press, astro-ph/0103071 (2001)
2. W.N. Brandt et al.: AJ in press, astro-ph/0108404 (2001)
3. I. Cagnoni et al.: ApJ **493**, 54 (1998)
4. A. Comastri, G. Setti, G. Zamorani, G. Hasinger: A&A **296**, 1 (1995)
5. A. Comastri: Astroph. and Lett. in press, astro-ph/0003137 (1999)
6. R. Giacconi, H. Gursky, F.R. Paolini, B.B. Rossi: Phys. Rev. Lett. **9**, 439 (1962)
7. R. Giacconi, et al.: ApJ, submitted (2001)
8. R. Gilli, M. Salvati, G. Hasinger: A&A **366**, 407 (2001)
9. G. Hasinger et al.: A&A **329**, 482 (1998)
10. F. Marshall et al.: ApJ **235**, 4 (1980)
11. R. Mushotzky et al.: Nature **404**, 459 (2000)
12. C.A. Norman et al.: ApJ submitted, astro-ph/0103198 (2001)
13. M. Schmidt et al.: A&A **329**, 495 (1998)
14. G. Setti, L. Woltjer: A&A **224**, L21 (1989)
15. J.T. Stocke et al.: ApJS **76**, 813 (1991)
16. P. Rosati et al.: ApJ submitted (2001)
17. P. Tozzi, et al.: ApJ **560**, 407 (2001)
18. A. Vecchi et al.: A&A **349**, L73 (1999)

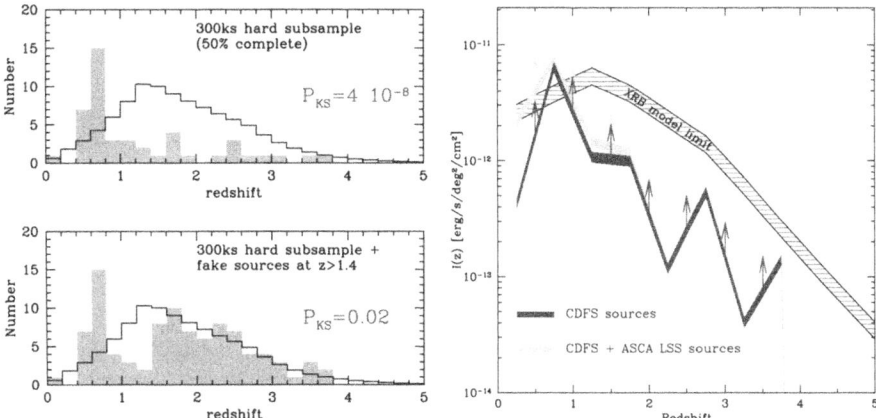

Fig. 5. *Left*: Redshift distribution of the CDFS sources compared with the predictions of the rescaled model B of [8]. In the upper panel we show the redshift distribution of a subsample of sources in the 2-10 keV band with 50% spectroscopic completeness. As shown by the KS test probablity P_{KS}, the observed and predicted distributions disagree. Then, if the still unidentified sources would have the same distribution of the identified ones (hypothesis i)) the model would be ruled out at a high confidence. In the lower panel we consider a fake redshift distribution for the unidentified sources, assuming they are all at $z > 1.4$ (hypothesis ii)). In this extreme case the model cannot be ruled out at more than 98% confidence level. *Right*: Comparison between the 2-10 keV XRB production history observed in the CDFS and that predicted by model B. The CFDS curve has been obtained by summing up the 2-10 keV source flux in different redshift bins and then dividing it by the covered area. The upper and lower boundaries of the model predictions have been obtained fitting the SAX and the HEAO-1 XRB flux, respectively. The observed curve is already exceeding the model predictions at $z \sim 0.7$. Since 50% of the sources have still to be observed, this curve is a lower limit, enhancing the significance of the result. If we combine the CDFS sources, which are all detected below $1\ 10^{-13}$ cgs with those observed in the ASCA LSS, all at fluxes above $1\ 10^{-13}$ cgs, the discrepancy increases further. The big fluctuations at high redshift are due to low source statistics.

Distribution of the Brightest X-Ray Sources

Andrzej M. Sołtan

Nicolaus Copernicus Astronomical Center, Bartycka 18, 00-716 Warsaw, Poland

Abstract. Brightest sources in the *ROSAT* pointings as well as in the All-Sky Survey generate disproportionately high fraction of the total X-ray background (XRB) fluctuations. The autocorrelation function of the soft XRB indicates that substantial fraction of the XRB variations over a wide range of angular scales is associated with relatively few bright sources. Possible explanation of this effect are listed. It is likely that most of the fluctuations of the XRB are not produced by source clustering represented by the standard power law autocorrelation function with the canonical index of -1.8.

1 Introduction

Deep X-ray exposures explicitly demonstrate that the soft X-ray background (XRB) is dominated by discrete extragalactic sources (e.g. [6], [4] and references therein). However, both observational data and theoretical considerations indicate that some fraction of the XRB flux is related to diffuse emission by hot gas. The *ROSAT* All-Sky Survey (RASS) as well as pointing observations reveal a presence of substantial thermal component produced by galactic plasma [11], [5]. Diffuse emission of extragalactic origin is also predicted. Hydrodynamic simulations [2] show that large fraction of baryonic matter in the Universe is accumulated in potential wells formed by galaxies and clusters. Models describing physical parameters of this gas postulate that thermal emission should be generated in filaments and halos surrounding galaxy groups ([1], [3], [18], [7]). Significant amount of diffuse baryonic matter is indicated by quasar absorption spectra in far-UV [17]. Despite strong theoretical arguments the extended extragalactic XRB component has not yet been directly detected. Deep X-ray source counts impose strong constraints on the diffuse XRB and apparently its surface brightness is so low that it is difficult to separate it from instrumental and particle background of the present day X-ray experiments.

Diffuse component generates specific variations of the XRB. Thus, investigation of the XRB fluctuations could provide indirect information on the extended X-ray sources. Anisotropy of the soft XRB has been investigated using the *ROSAT* observations and fluctuations have been detected over a wide range of angular scales ([15], [13]). The signal is produced by various effects. In order to extract from the total amplitude, fluctuations generated by diffuse component, careful examination of other sources contributing to the XRB variations is needed.

In a series of papers we have investigated relations between distinct classes of X-ray sources and the XRB fluctuations. It was shown in [15] that rich clusters of galaxies are surrounded by large low surface brightness halos of the X-ray emission. The enhanced signal extending up to $\sim 10\,\mathrm{Mpc}$ most likely results from a positive, extending over several Mpc, correlation between Abell clusters and a general galaxy population [9]. Effects of a nonuniform distribution of galaxies in the Lick counts [10] on the XRB fluctuations were analyzed in [16]. It was demonstrated that relatively small but measurable fraction of the XRB fluctuations is generated by a local population of "normal" galaxies. In [12] we have shown that a substantial fraction of the XRB variations could be attributed to the clustering of AGNs. But large amplitude of the XRB autocorrelation function (ACF) is only marginally consistent with the present models of the quasar clustering.

The fact that the XRB is produced by several classes of extragalactic objects makes the interpretation of the XRB fluctuations difficult and ambiguous. A comparison of the ACF estimates based on the RASS [12] with our earlier fluctuation analysis [14] of the deep *ROSAT* exposures seem to indicate that removal of detected sources substantially reduces the fluctuation amplitude. In the present work a question on the relation between the bright X-ray sources and the amplitude of the XRB fluctuations is addressed. Our analysis is based on a large material extracted from the archived *ROSAT* pointing observations. An analogous analysis of the RASS data is also presented.

2   *ROSAT* Pointings

In our study of the small angular scale fluctuations of the XRB [13] we have used a large sample of the *ROSAT* archive observations. Criteria applied in the data selection are listed in that paper. Here I describe the final sample which seems to satisfy requirements of statistical homogeneity. The sample consists of 141 PSPC pointings accumulated between April 1991 and April 1994. All the observations were carried out with the PSPC detector B. Pointings are distributed fairly uniformly within an area of approximately 1 sr with $70° < l < 250°$ and $b > 40°$. This area is apparently free from galactic contamination and the ACF signal is generated mostly by the extragalactic component of the XRB. Our investigation of the large scale XRB fluctuations based on the RASS were based on the same region of the sky.

However, the sample is not fully representative for the overall XRB distribution. The *ROSAT* archive observations are strongly biased towards bright and extended sources. In the selection process we have eliminated all the observations of known extended sources (supernova remnants, nearby normal galaxies, clusters and groups of galaxies). This procedure in effect removed many known nearby X-ray clusters and introduced deficiency of extended sources in the sample.

Although the sample contains smaller number of nearby clusters than expected for the random selection of pointings, the known extended serendipitous sources generate substantial signal in the ACF at scales below $\sim 0.°1$. This is shown in Fig. 1. A solid curve denotes the ACF derived from the sample of 140 pointings[1] and open symbols

---

[1] The strongest source in the sample gives roughly 8 times higher count rate than the second brightest source. To avoid the situation that our results are dominated by just one prominent source, the pointing with the source RXJ1341.0+3959 has not been used in most of our calculations. This source coincides with the BL Lac object and the cluster of galaxies A 1774 and it produces high amplitude of the ACF for separations up to $7'$.

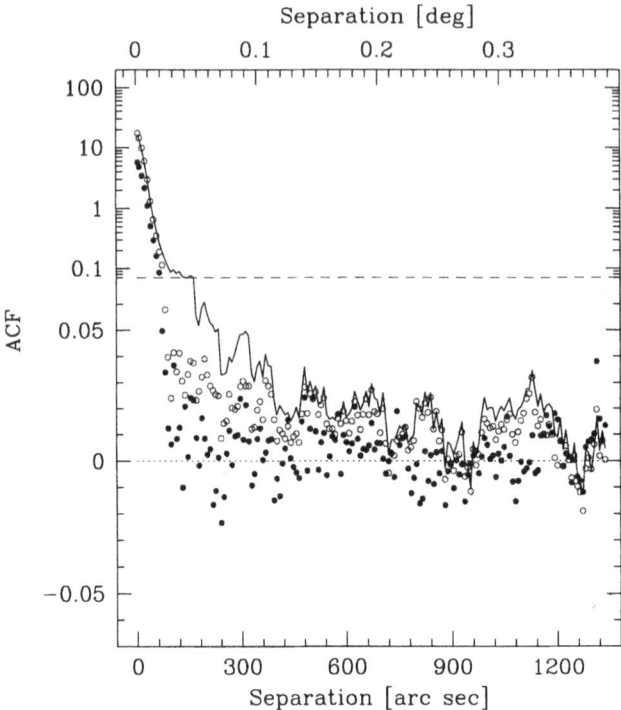

Fig. 1. The autocorrelation functions of three analyzed samples: solid curve – set of 140 pointings, open circles – 129 pointing without known extended sources (clusters of galaxies), full symbols - subset of 70 pointings with the least prominent discrete sources (see text)

– the ACF after a removal of 11 pointings containing known clusters of galaxies. The ACF signal produced by known clusters is conspicuous to separations below $\sim 0.°1$ and drops below noise at large separations.

The sample of 129 pointings is free from known extended sources. Nevertheless, the distribution of count rates in this sample exhibits fluctuations over a wide range of angular scales. Although statistical fluctuations and possible systematic effects introduce large uncertainties of the XRB variations [13], the ACF signal is clearly detected with the amplitude declining from ~ 0.035 at $2'$ to ~ 0.02 at $\sim 12'$.

To examine the relationship between the ACF and prominent X-ray sources, I have selected a subsample of 70 pointings with a relatively low content of bright sources. To separate fields without prominent point-like sources from the entire sample, the shape of the ACF at small separations has been used. Pointings deficient in strong sources generate the ACF smooth over all the separations with relatively weak single maximum at low separations. Strong sources in the area produce substantial fluctuations of the ACF at separations corresponding to positions of sources and a strong maximum at separations below $2'$. The low separation peak is generated by discrete nature of the XRB. The shape of this peak is defined by the point spread function of the X-ray

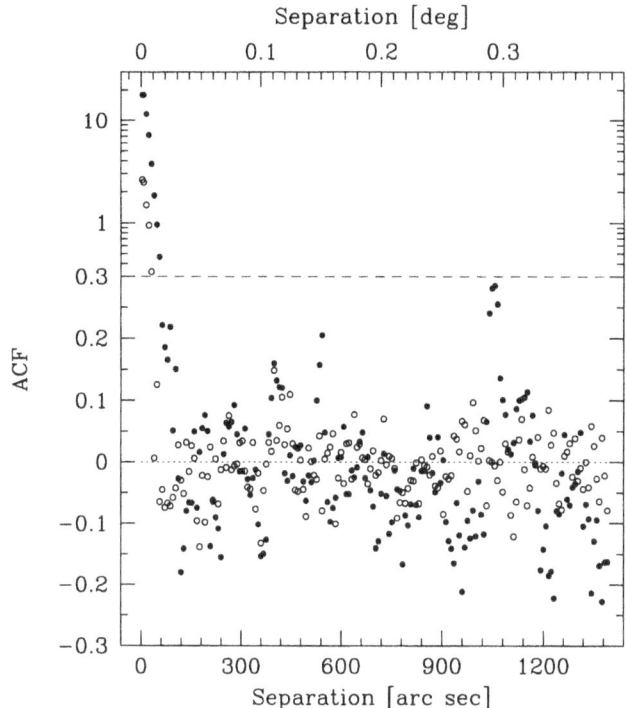

Fig. 2. Examples of the autocorrelation functions for two pointings of similar exposure times but distinctly different in respect of bright source contents: open circles - no obvious sources in the field of view (exposure time - 8500 s), full symbols - several conspicuous point-like sources (exposure time - 11100 s)

telescope/detector combination, while its amplitude is related to the brightest sources within each field. These ACF features are exemplified in Fig. 2 where the ACFs are plotted for two pointings. Although exposure times of both observations are similar, amplitude of the ACF fluctuations are distinctly different. One of the pointings shows no conspicuous sources and the corresponding ACF exhibits only moderate variations as compared to the second ACF representing the field with several strong sources.

The sample of 70 pointings 'without bright sources' is defined as the half of all the observations with the lowest ACF signal below 1'. Obviously, this criterion does not define pointings 'without sources' in an absolute sense, but it constitutes a convenient way to separate pointings deficient in bright sources from those dominated by such sources. The ACF for the subsample 'without bright sources' is shown with Fig. 1 with full symbols. Despite large statistical scatter, a conspicuous reduction of the ACF amplitude in relation to the entire sample is evident. Apparently bright sources seem to produce a large fraction of the XRB fluctuations and generate a substantial ACF signal at scales of several arc min. It is not clear, however, what physical mechanism is responsible for this effect. One potential explanation is that bright sources are on the

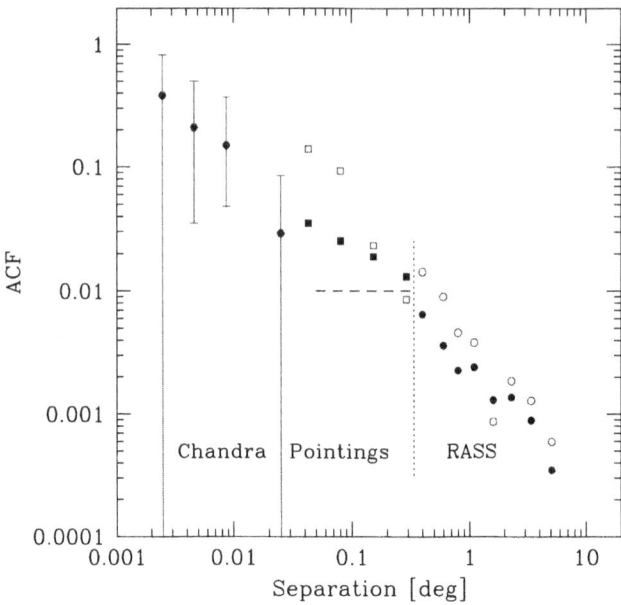

Fig. 3. The ACF of the XRB from the *ROSAT* All-Sky Survey and pointings combined with the ACF of the *Chandra* sources (see text for details)

average more clustered than the rest of source population responsible for the remaining, unresolved fraction of the XRB. It is also possible that strong sources are surrounded by areas of slightly enhanced X-ray emission. Even a minute increase of the XRB flux around the bright source generates relatively strong signal in the ACF. This effect is in fact observed at larger angular scale in the RASS [15] (see below).

To resolve the question of the XRB structure in the vicinity of strong sources, a comprehensive analysis of the XRB distribution is required. In particular, one should study distribution of counts around known galactic and extragalactic point sources. To explore how the present results correspond to the ACF measurements at different angular scales, I have compared the effect revealed in the *ROSAT* pointings with both larger (RASS) and smaller scales (*Chandra* observation).

3   Comparison with *ROSAT* All-Sky Survey and *Chandra*

In Fig. 3 the composite ACF from three experiments is plotted. The *ROSAT* pointing results are shown with squares. The sample without known clusters is represented by full symbols, while the ACF computed from the entire sample of 141 pointings is shown with the open squares. The ACF measurements for the sample 'without bright sources' based on 70 pointings (full symbols in Fig. 1) are dominated by statistical noise. A horizontal dashed line in Fig. 3 indicates approximate upper limit of the ACF signal for this sample.

Fluctuations of the XRB at the degree scale detected in the RASS [15], [12] are represented by two ACFs shown with circles. The open symbols denote the ACF computed using all the pixels in the investigated area. The ACF plotted with full circles denotes the ACF after a removal of 1 % of pixels with the highest count rates. Although the average background level in the sample devoid of those pixels is lower just by by 6 % as compared to the full sample, the ACF amplitude is reduced by a factor of $\sim$ 2 over all the separations below roughly $1°$. Because of a relatively large size of the RASS pixels ($12'$ a side), the reduction of the ACF amplitude at small separations is expected. Unfortunately, low signal-to-noise ratio and potential systematic errors of our ACF estimates at larger separations do not allow for more quantitative investigation of this effect.

At separations around $1'$ and below the *Chandra X-ray Observatory* is expected to provide definite data on the XRB fluctuations. In Fig. 3 results based on a single deep pointing taken from [4] and shown. The ACF amplitude and error bars have been calculated using the integral ACF from the original paper. Although the area and number of sources in the individual observation are too small to give restrictive estimates on the XRB variations, a potential power of *Chandra* in the analysis of the XRB distribution is evident. A couple dozen of *Chandra* fields will allow to calculate the ACF of discrete sources with an unprecedented accuracy.

4   Discussion

The available data on the XRB fluctuations are reviewed. It is shown that the ACF amplitude of the XRB is strongly related to the distribution of the brightest sources. This effect is detected both at small angular scales using *ROSAT* pointings as well as at larger scales using the RASS. At the degree scale it is likely that the ACF signal is partially generated by large halos of the X-ray emission surrounding X-ray luminous rich cluster of galaxies which constitute a substantial fraction of the brightest sources in the RASS [15]. At the arc min scale the effect could be produced by the scattering of background photons by dust in foreground objects (e.g. [8]). However, quantitative analysis is problematical because detailed models of this effect depend on several unknown parameters. Although the clustering of bright sources in the *ROSAT* pointings could also contribute to the ACF signal, strong dependence of the ACF amplitude on the apparent source brightness would imply significant differences in the clustering of bright and dim sources.

The main conclusion of the present study is following. Noticeable fraction of the XRB fluctuations is related to the bright sources and it is highly probable that the observed surface XRB fluctuations measured by the 2D ACF are generated by some other effects than spatial clustering of discrete sources described by a universal 3D correlation function of the power law shape $\xi \propto r^{-\gamma}$, where r is spatial distance and $\gamma \approx 1.8$.

First results on the discrete source distribution obtained from *Chandra* indicate that the moderate enlargement of the observational material will provide a unique opportunity to compare the fluctuations of the integrated background as measured by *ROSAT* with the variations generated by the discrete sources. Careful evaluation of both fluctuation measurements would help to estimate the amplitude of the truly diffuse component of the XRB.

The ROSAT project is supported by the German Bundesministerium für Bildung und Forschung/Deutsches Zentrum für Luft- und Raumfahrt (BMBF/DLR) and by the Max-Planck-Gesellschaft (MPG). This paper was supported by the Polish KBN grant 5 P03 D 022 20.

References

1. G. L. Bryan, G. M. Voit: ApJ **556**, 590 (2001)
2. R. Cen, J. P. Ostriker: ApJ **514**, 1 (1999)
3. R. A. C. Croft, T. di Matteo, R. Davé et al.: ApJ **557**, 67 (2001)
4. R. Giacconi, P. Rosati, P. Tozzi et al.: ApJ **551**, 624 (2001)
5. Hasinger, G.: '*ROSAT* Deep Surveys'. In: *Highlights of Astronomy, 9, 199,* ed. by J. Bergeron (1992)
6. I. Lehmann, G. Hasinger, M. Schmidt et al.: *astro-ph/0103368* (2001)
7. L. A. Phillips, J. P. Ostriker, R. Cen: ApJ **554**, L9 (2001)
8. B. Rudak., P. Meszaros: ApJ **371**, 29
9. M. Seldner, P. J. E. Peebles: ApJ **215**, 703 (1997)
10. C. D. Shane, C. A. Wirtanen: Publ. Lick Obs. **22**, Part 1 (1967)
11. S. L. Snowden, M. J. Freyberg, P. P. Plucinsky et al.: ApJ **454**, 643 (1995)
12. A. M. Sołtan, M. J. Freyberg, G. Hasinger et al.: A&A **349**, 354 (1999)
13. A. Sołtan, M. J. Freyberg, J. Trümper: A&A (in press) (2001)
14. A. Sołtan, G. Hasinger: A&A **288**, 77 (1994)
15. A. M. Sołtan, G. Hasinger, R. Egger et al.: A&A **305**, 17 (1996)
16. A. M. Sołtan, G. Hasinger, R. Egger et al.: A&A **320**, 705 (1997)
17. T. M. Tripp, B. D. Savage, E. B. Jenkins: ApJ **534**, L1 (2000)
18. K. K. S. Wu, A. C. Fabian, P. E. J. Nulsen: MN **324**, 95 (2001)

Mining the EIS Fields: Quasar Candidates in CDF-S Point-Source Catalogue

Evanthia Hatziminaoglou and The EIS team

European Southern Observatory
Karl-Schwartzschild-Str. 2, 85748 Garching bei München, Germany

Abstract. The χ^2-technique, a standard fitting procedure between the observed SED and a template library, is applied on the point-source catalogue of the Chandra Deep Field South (CDF-S), with the aim of identifying quasar candidates. The CDF-S, a part of the EIS Deep Public Survey (DPS), has been observed in five broad band filters U, B, V, R and I. Its central part has also been observed in the near infrared passbands J and K_s. From this catalogue a total of 253 quasar candidates have been selected. The number of candidates is consistent with both theoretical predictions, derived from a two-slope luminosity function, as well as empirical predictions based on previous spectroscopic surveys. The photometric redshifts of the candidates span from 0 to ~ 5, reproducing remarkably well the expected absolute and apparent magnitude distributions versus redshift. Within the CDF-S point-source catalogue, there are 7 quasar candidates with photometric redshifts higher than 3.5. Their spectroscopic confirmation would mean that over the 12 EIS pointings that will be publicly available, a total number of over a hundred high-redshift quasars will be detected.

1 χ^2-technique, Data and Spectral Library

The χ^2-technique (presented in detail in [4]) consists of a standard fitting procedure, where the observed spectral energy distributions are compared to a template library. Thus, the traditional multi-dimensional method ($2 \times N$ dimensions, with N the number of the colour-colour diagrams) is reduced to a one-dimensional technique.

The EIS Deep Public Survey (DPS) consists of 12 pointings, covering ~ 3 square degrees in the south hemisphere. The χ^2-technique is applied on the point-source multi-colour data of the Chandra Deep Field South (CDF-S), as a part of the DPS. This field (0.25 square degrees) has been covered in U, B, V, R, and I, while its central region of 0.1 square degrees has been also observed in J and K_s ([1]; [11]). The combination of the multi-passband information and the survey depth offers an excellent opportunity for tracking down quasars at all redshifts. The point-source catalogues examined comprise 1494 sources in five passbands and 605 in seven passbands with $R < 23.2$, the morphological classification limit.

The spectral library in use consists of: i) a series of template quasar spectra, assumed to have three different power-law continua (0, 0.5 and 1), Gaussian emission line profiles and typical relative intensities (e.g. [9]) with redshifts spanning from $z = 0$ to $z = 6$, in steps of $dz = 0.1$.; ii) a set of theoretical white dwarf spectra ([3]; [6]) as well as three observed spectra of very cool white dwarves ([7]; [8]); iii) a series of theoretical low-mass star and brown dwarf spectra [2] and iv) the stellar library of Pickles (1998) [10], which contains 131 spectra for main sequence stars, giants and sub-giants.

2   Comparison with the Model

For the calculation of the estimated number counts and the construction of the simulated catalogues a single representation of the luminosity function has been used

$$\Phi(M_B, z) = \frac{\Phi^*}{10^{0.4[M_B - M_B(z)](\beta_1 + 1)} + 10^{0.4[M_B - M_B(z)](\beta_2 + 1)}},$$

yielding an estimated quasar number of 202^{+131}_{-78}, down to $B \sim 25$. An extended discussion on the mock catalogues will be presented in [5]. Figure 1 shows the location of quasar candidates on a $(B - R) \times (R - I)$ diagram (left panel) and their photometric redshift distributions in five (blue) and seven (red) passbands, versus the model prediction (black) (right panel). Figure 2 presents the absolute magnitude versus redshift distribution for the candidates (left) and the model (right).

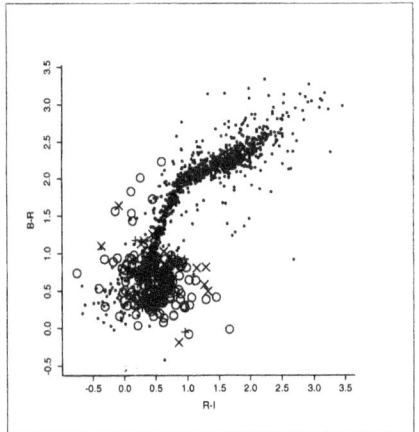 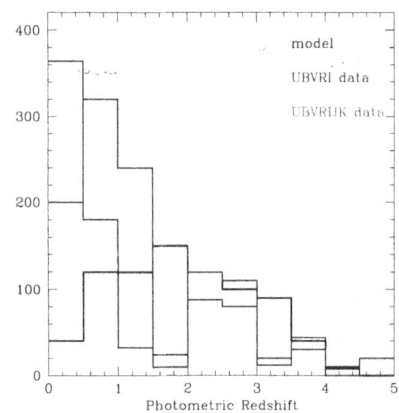

Fig. 1. Left panel: location of quasar candidates on a $(B - R) \times (R - I)$ diagram; right panel: and photometric redshift distributions of quasar candidates in five (blue) and seven (red) passbands, versus the model prediction (black).

3   Perspectives

Combining the data from the five and seven passband catalogues, one finds a total of 253 quasar candidates with estimated photometric redshifts up to $z \sim 5$, among which seven have $z \sim 3.5$. If the classifications are confirmed, samples comprising over 100 high-redshift quasars ($z > 3.5$), will become available at the end of the survey, expected to cover 3 square degrees, only within the (bright) point-source catalogues. Taking into account that the faint source catalogues go as much as 2 magnitudes deeper than the morphological classification limits, one would expect to find a total of more than 500 high-redshift quasars, or a factor of two more than the number already known. It is

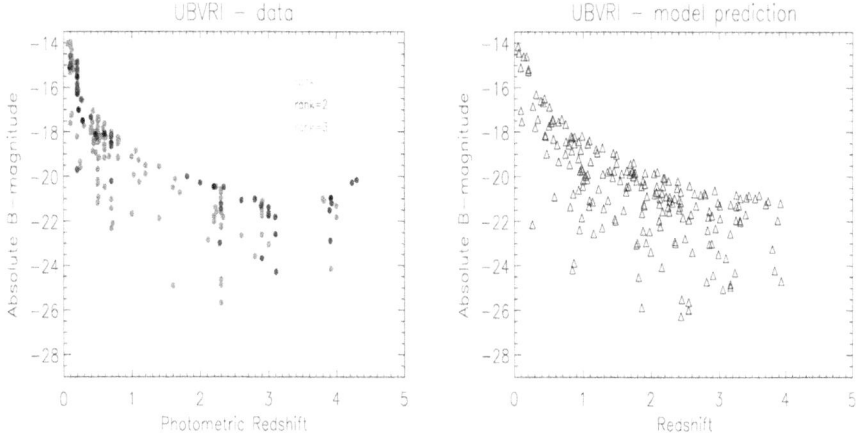

Fig. 2. Absolute magnitude versus redshift distribution for the candidates (left) and the model (right).

worth emphasizing the **contribution of the near-infrared data:** it increases the accuracy of the determination of the photometric redshifts and significantly increases the number of quasar candidates in the redshift interval $2.5 < z < 3.5$. Furthermore, it allows for selecting very high redshift candidates that would be dropouts in U, B and/or V and therefore difficult to pick up by their optical colours only.

References

1. Arnouts, S. et al., 2001, accepted for publication in A&A, astro-ph/0103071
2. Chabrier G., Baraffe I., Allard F., Hauschild P., 2000, ApJ, 542, 464
3. Finley, D. S., Koester, D., Basri, G., 1997, ApJ, 488, 375
4. Hatziminaoglou, E. 2001, A&A, submitted
5. Hatziminaoglou, E. et al., in preparation
6. Homeier D., Koester, D., Hagen, H.-J. et al., 1998, A&A, 338, 563
7. Ibata, R., Irwin, M., Bienaymé, O. et al., 2000, ApJ, 532, L41
8. Oppenheimer, B. R., Saumon, D., Hodgkin, S. T. et al., 2001, ApJ, 550, 448
9. Peterson, B. M., 1997, in *An introduction to active galactic nuclei*
10. Pickles, A. J., 1998, PASP, 110, 863
11. Vandame, B. et al., 2001, accepted for publication in A&A, astro-ph/0102300

A Faint Quasar Survey in the Marano Field

Marco Mignoli[1], Gianni Zamorani[1], and Bruno Marano[1]

Osservatorio Astronomico di Bologna, via Ranzani 1, I-40127 Bologna, ITALY

1 Introduction

The "Marano Field" is a southern sky region extensively surveyed in the optical (by means of multi-color imaging, slit-less and slit spectroscopy and variability), in the X-rays (with ROSAT in the band 0.2÷2 keV) and in the radio band (with the ATCA radio telescope at 1.4 and 2.4 GHz).

The existing observations have already provided 70 confirmed AGNs with B_J ≤22.5 (Zitelli et al., 1992) and a complete sample of 50 X-ray sources with S_J ≥3.7×10^{-15} erg cm^{-2} s^{-1} (Zamorani et al., 1999). About 86% of these X-ray sources have been optically identified and AGNs are by far the dominant class of counterparts, representing more than 80% of the optical identifications obtained so far. Finally, we obtained optical photometric identifications for 63% of the radio sources down to $S_{1.4GHz}$ =0.2 mJy, and spectroscopic for more than 50% up to R_{lim} =23 (Gruppioni et al., 1999); these percentages are among the highest identification fractions available in literature for sub-mJy radio samples.

2 The Deep Multicolour Survey

In order to select a fainter sample of quasar candidates, we performed a deep survey using EMMI at ESO/NTT. The relevant characteristics of the survey are:

- $UBVR$ filters in the standard Johnson-Kron-Cousins system
- the area covered by all the filters is 0.132 square degrees
- the average measured seeing in the U B V R is respectively
  $1\rlap{.}''14 \pm 0\rlap{.}''19$, $0\rlap{.}''94 \pm 0\rlap{.}''17$, $0\rlap{.}''89 \pm 0\rlap{.}''11$, $0\rlap{.}''88 \pm 0\rlap{.}''15$
- the typical limits are U_{lim} ~23.7, B_{lim} ~25.0, V_{lim} ~24.3, R_{lim} ~24.0
- the photometric errors are <0.1 at m~22 and ≤0.2 at m~24.

For the data analysis (i.e. detection, photometry and star/galaxy separation) we used the SExtractor program (Bertin & Arnouts, 1996).

2.1 The Photometric Catalogue and AGN-Candidate Selection

We extracted from the whole multi-color catalogue all the **point-like** objects brighter than B=23.5. Figure 1 shows the distribution of the selected objects with 18.5<B≤22.5 (left panel), where almost all the AGN candidates outside the star locus (defined by the dashed line) were already spectroscopically identified. In the right panel the fainter

objects (22.5<B≤23.5) are shown: in this case most of the spectroscopic identifications have been obtained in a recent FORS/VLT run (only 5 AGNs and 1 NELG had
previously been observed as X-ray counterparts).

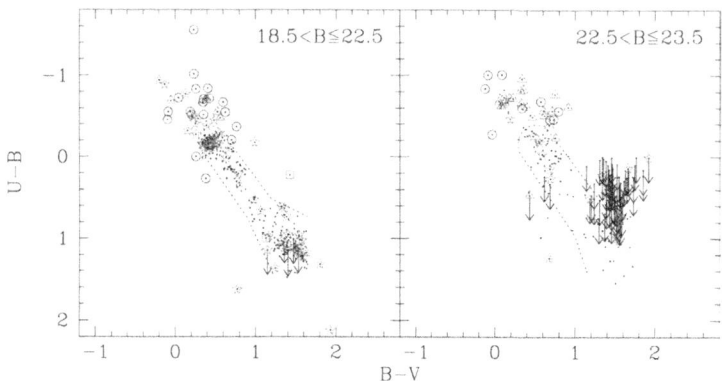

Fig. 1. U-B/B-V color diagrams for stellar objects: spectroscopically confirmed AGNs, CNELGs,
Early Galaxies and stars are plotted as circles, triangles, squares and stars

3   The Spectroscopic Follow-Up

The spectroscopy of the AGN-candidates up to B=23.5 was carried out with FORS1 at
VLT-Antu in the MOS mode. The observations have been performed in service mode
and due to geometric and schedule constraints we obtained useful spectra of only 28 out
of the 40 new candidates. A few random stellar objects within the stellar locus have been
also observed (in order to fill up all the slitlets in each MOS): all these targets turned
out to be stars. **The new identified candidates are 5 quasars, 2 stars and 21 compact
galaxies.** The CNELG's contamination becomes important in the faintest bin: it grows
up from ∼10% for the candidates with 18.5<B≤22.5 to ≳50% for 22.5<B≤23.5.

4   The Final Spectroscopic Sample

The total number of broad-line AGNs detected so far in the Marano Field is 85. Seventy-
nine of these can be used to define three magnitude limited samples over three different
areas, which are summarized in table 1. In the faintest sample two identified AGNs
were classified extended and have been discovered as counterparts of X-ray sources.

5   Conclusions

- The *measured* AGN surface density at B=23.5 is ∼250 per square degree; correcting for the spectroscopic incompleteness, our estimate for the *total* surface density
  of broad line AGNs is ∼290 per square degree at B=23.50.

Table 1. The spectroscopic sub-samples

Sample	mag.range [a]	Sky area	completeness	# AGNs
BRIGHT PLATE	$J \leq 20.87$	0.69	100%	24
FAINT PLATE	$20.87 < J \leq 22.00$	0.34	100%	34
	$22.00 < J \leq 22.40$	0.34	~70%	9
DEEP CCD [b]	$22.50 < B \leq 23.50$	0.13	~75%	12

[a] We note that $B \simeq J + 0.1$

- More than 50% of the objects in the sample have $M_B > -23$, i.e. in the magnitude range typical of Seyfert galaxies (see Fig.2): in combination with larger and brighter surveys (e.g. 2dF QSO survey), these data will allow a better determination of the faint end slope of the Luminosity Function (LF).
- Although with relatively poor statistics, these data suggest a dearth of faint, high redshift quasars (i.e. $z > 2.2$): none is found in the fainter subsample (stars in Fig.2). Optical identifications of deep XMM data, currently in progress in the same field, will confirm whether this effect is significant.

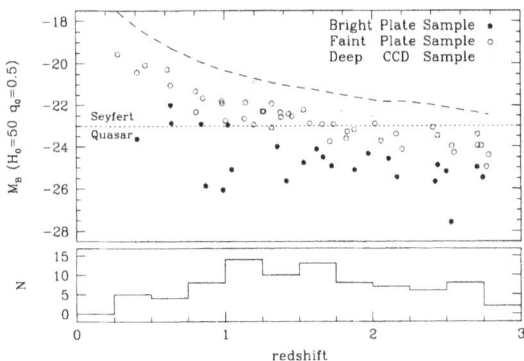

Fig. 2. Luminosity vs. redshift for the 3 sub-samples and the corresponding histogram of the redshift distribution. The dashed line indicates our deep CCD magnitude limit

References

1. C. Gruppioni, M. Mignoli, G. Zamorani: MNRAS **304**, 199 (1999)
2. V. Zitelli, M. Mignoli, G. Zamorani, B. Marano et al.: MNRAS **256**, 349 (1992)
3. G. Zamorani, M. Mignoli, G. Hasinger et al.: Astron.Astrophys. **346**, 731 (1999)

The BTC40 Survey for High-z Quasars

Patrick S. Osmer[1], Eric Monier[1], Julia Kennefick[2], Patrick B. Hall[3], Malcolm G. Smith[4], and Richard F. Green[5]

[1] Department of Astronomy, The Ohio State University, 140 W. 18th Ave., Columbus, OH 43210, USA
[2] Oxford University, NAPL, Keble Rd., Oxford OX1 3RH, England, UK
[3] Princeton University Observatory, Princeton, NJ 08544-1001, and Pontificia Universidad Católica de Chile, Departamento de Astronomía y Astrofísica, Facultad de Física, Casilla 306, Santiago 22, Chile
[4] Cerro Tololo Inter-American Observatory, Casilla 603, La Serena, Chile
[5] National Optical Astronomy Observatory, 950 N. Cherry Ave., AZ 85719, USA

Abstract. The BTC40 Survey is a multicolor survey using images obtained with the BTC camera on the CTIO 4-m telescope in B, V, I, & Z filters to search for high-redshift quasars. The survey covers 40 deg^2 in B, V, & I and 36 deg^2 in V, I, and Z. Limiting magnitudes reach to 25.8 in V and 23.8 in I. We used the $V - I$ vs. $I - Z$ two-color diagram to select high-redshift quasar candidates from the objects classified as stellar in the imaging data. Follow-up spectroscopy with the AAT and CTIO 4-m telescopes on candidates with $I < 21.5$ has yielded two quasars with redshifts of 4.6 and 4.8 as well as four emission-line galaxies with redshifts of about 0.6.

1 Introduction

Surveys for quasars at high redshift are important for many reasons: 1) to determine when they first became active in the universe; 2) to establish their relation to the formation and evolution of galaxies; 3) to estimate their contribution to the ionizing background; and 4) provide beacons for the study of the high-redshift universe via absorption-line studies. The Sloan Digital Sky Survey (SDSS) is being remarkably successful at finding $z > 4$ quasars, having discovered more than 100 such objects, including one at $z = 5.8$, the most distant published (see, e.g., [1], [2]). However, SDSS is limited to high-redshift quasars with $M_B < -26$ and misses the bulk of the population (at lower luminosities). Accordingly, we began the BTC40 survey to carry out a deeper survey using a 4-m telescope and large format camera to complement and extend the work of the SDSS.

2 Observations

We have carried out a deep survey with the BTC camera on the CTIO 4-m telescope to cover 40 deg^2 in B, V, & I and 36 deg^2 in V, I, and Z; limiting magnitudes are 25.8 in V and 23.8 in I. As we show below in Fig. 1, the Z filter is necessary to separate high-redshift quasars from cool stars, in accord with the SDSS findings. The survey imaging data were reduced using standard IRAF techniques and processed using the

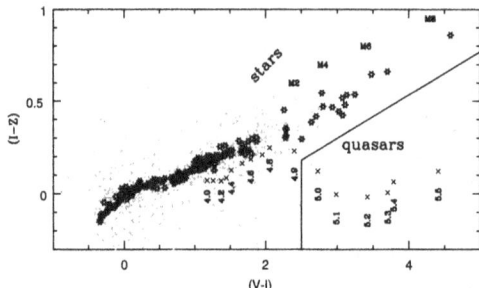

Fig. 1. $(V - I)$ vs. $(I - z)$ diagram of stellar objects from 0.25 deg^2 of the VIZ data set (dots) along with stars synthesized from the Bruzual Persson Gunn Stryker atlas (open stars) and average synthetic quasar colors at high redshift (crosses). Quasars with $z > 5$ are distinguished from stars by having bluer $I - Z$ colors, as indicated, because of the presence of Ly α in the I band

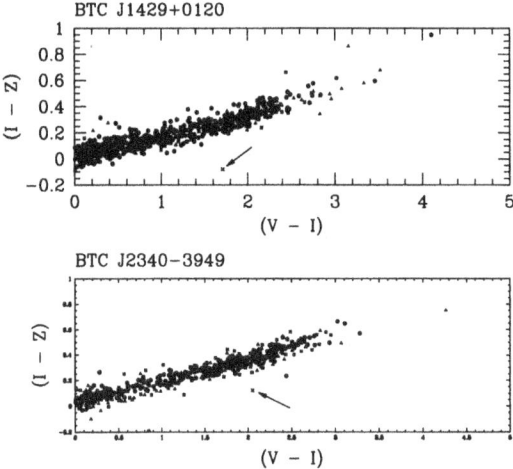

Fig. 2. Color-color diagrams showing the $z = 4.6$ and 4.8 quasars (crosses below the stellar locus) that we confirmed with slit spectra

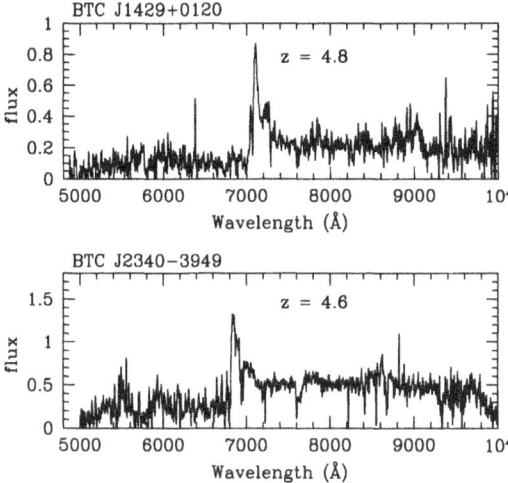

Fig. 3. CTIO 4-m spectra of the two quasars with $z = 4.6$ and 4.8. Their I magnitudes are both 19.4

SKICAT package to produce catalogs of objects matched in the different filters. Color-color diagrams such as shown in Fig. 2 were produced and used to select candidates for follow-up slit spectroscopy.

Using the CTIO 4-m and AAT, we made spectroscopic observations of 92 candidates with $I < 21.5$ over six observing runs in 1999 and 2000 and confirmed two quasars with redshifts of 4.6 and 4.8, as shown in Fig. 3. In addition, we found four emission-line galaxies with redshifts of about 0.6; the remaining objects either were stars or were unclassifiable at the low signal-to-noise ratio of the data.

3 Discussion

The finding of two $z > 4.5$ quasars with $I < 20$ in 36 deg^2 is consistent with the SDSS results [3], thereby providing validation for our approach.

To go fainter with the spectroscopy will require 8 - 10-m telescopes (not surprisingly, since the imaging was done with a 4-m telescope). Also, deeper Z data are needed to extend the limiting magnitude for the detection of $z > 5$ quasars.

The BTC40 data, in conjunction with the Big Faint Quasar Survey (BFQS) survey (see, e.g., [4]), which goes about 1 mag. deeper over 7 deg^2, will be valuable for studies of $3.5 < z < 5$ quasars and of the faint galaxies and stars in the fields.

Acknowledgement. Support by NSF grant AST-9802658 is gratefully acknowledged.

References

1. Fan, X. et al.: AJ **120**, 1167 (2000)
2. Anderson, S. et al.: AJ **122**, 503 (2001)
3. Fan, X. et al.: AJ, **121**, 54 (2001)
4. Hall, P.: astro-ph/0010501 (2000)

The Contribution of AGN to the Far Infrared Background

Guido Risaliti[1,2], Martin Elvis[1], and Roberto Gilli[2,3]

[1] Harvard-Smithsonian Center for Astrophysics, 60 Garden Street, Cambridge, MA 02138, USA
[2] Osservatorio Astrofisico di Arcetri, Largo E. Fermi 5, I-50125, Firenze, Italy
[3] Johns Hopkins University, Baltimore, MY ,USA

Abstract. We use synthesis models for the X-ray background, together with quasar Spectral Energy Distributions, to estimate the contribution of Active Galactic Nuclei to the Far Infrared Background and to the total luminosity of the Universe. We find that the AGN contribution to the energy output of the Universe is between 7% and 15%, and comparing this value with the mass of supermassive black holes, we find that the average efficiency of accretion in converting mass to energy must be at least 15%. The contribution of AGN to the overall FIR background is on average low, but it can be dominant at some wavelengths.

1 The Total Energy from Quasars

We can estimate the total energy released by quasars with the following two steps: (1) Using a synthesis model for the X-ray background (Gilli et al. 2001), we estimate the *intrinsic* emission of AGNs at 2 keV; (2) we compute the total energy output from quasars, Ψ_T, using an average quasar SED. The first step involves uncertainties of order 30%, the second step, by contrast, has a correction of a factor 1.6-3 to be made.

1) AGN emission at 2 KeV: We used the code of Gilli et al. 2001, which reproduces the X-ray background spectrum and source counts. To estimate the total intrinsic luminosity emitted by AGNs, we assumed an unabsorbed SED for all AGNs. The result we obtain is $\Psi_X = EI(E) = 48 \, \mathrm{keV \, s^{-1} \, cm^{-2} \, sr^{-1}}$.

2) 2 keV-to-bolometric correction: we use Ψ_X, together with a bolometric correction, to estimate the total integrated emission of quasars, Ψ_T. At wavelengths from the radio to the UV we can assume that the SED by Elvis et al. 1994 is representative for the average emission of unobscured quasars. The SED of Elvis et al. is obtained from a sample of quasars that gave a large enough number of photons in the Einstein IPC to construct a well-constrained spectrum. Therefore, this SED is biased toward X-ray brighter quasars. We corrected for this bias using an optically selected sample.

Using the 2 keV intensity and the X-ray to bolometric correction estimated above, and taking into account for all the uncertainties, we find that the total intensity produced by AGNs is: $\Psi_T \geq (3.6 - 8) \, \mathrm{nW \, m^{-2} \, sr^{-1}}$. Since the total intensity in the cosmic background from the far infrared to the X-rays is $\sim 55 \, \mathrm{nW \, m^{-2} \, sr^{-1}}$, we conclude that the contribution of AGNs to the energy output of the Universe is between 7% and 15%.

Converted into an energy density this gives: $U_T = \frac{4\pi\Psi}{c} \geq (1.5 - 3.4) \times 10^{-15} \, \mathrm{erg \, cm^{-3}}$. If we assume that this energy comes from accretion with efficiency ϵ, the mass density in black holes is: $\rho_\bullet \geq (7.5 - 16.8) \times 10^4 \, \epsilon^{-1} M_\odot \, \mathrm{Mpc^{-3}}$, where we

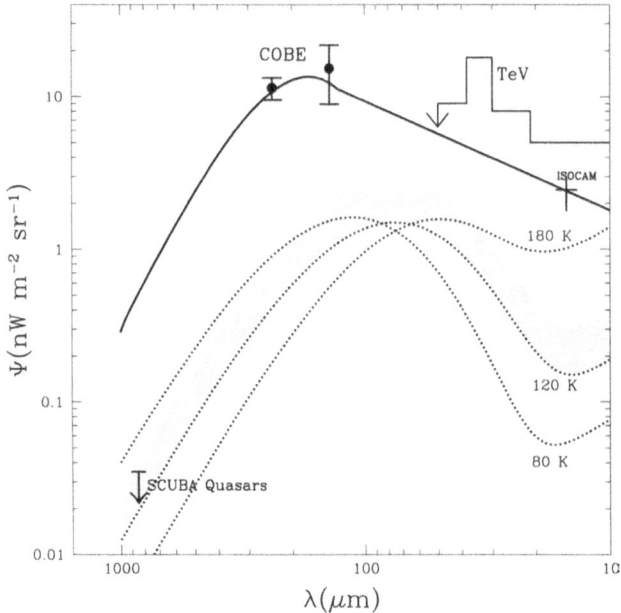

Fig. 1. The contribution of AGN to the FIR background.

assume that the peak of AGN emission is at z=2, in agreement with the XRB synthesis models.

The mass density in central black holes has been estimated by Salucci et al. (1999) to be $\rho_\bullet \sim 4 \times 10^5\ M_\odot\ \mathrm{Mpc}^{-3}$ using a quasar evolution model and starting from X-ray counts. These results are in agreement with the recent work of Merritt & Ferrarese (2001) on the relation between central black hole mass and the velocity dispersion of the bulge (best estimate $\rho_\bullet \sim 5 \times 10^5\ M_\odot\ \mathrm{Mpc}^{-3}$). Combining Equation 1 with the estimates of the mass density of supermassive black holes in the nearby Universe ($\rho_\bullet \sim 4 \times 10^5\ M_\odot\ \mathrm{Mpc}^{-3}$, Salucci et al. 1999, Merritt & Ferrarese 2000) we can obtain a value for the average accretion efficiency, ϵ. We find that the average efficiency in converting accreting mass in radiated energy is $\epsilon > 0.15$. This rules out the possibility that a "silent" process like ADAF accretion gives a significant contribution (in terms of accreting mass and radiated energy) to the total accretion in the Universe.

2   The FIR Background

We used the Gilli et al. 2001 model to estimate the contribution of AGNs to the far infrared background, substituting FIR SEDs in place of the X-ray SEDs. For type 1 quasars we assumed the SED of Elvis et al. (1994) (the normalization was re-scaled from the 2 keV value using the bolometric corrections described above). For type 2 quasars we assumed a FIR SED with several shapes, in order to be consistent with the observational data and to investigate how the final results of our analysis depend on the details of the SED. We discuss these issues below.

We first assumed a simple, single temperature, blackbody in the 10-1000 μm band. The results are plotted in Fig. 1 for three different temperatures. The normalizations are determined only by the X-ray data and the bolometric correction discussed above, not by the FIR data. The allowed temperature range is constrained by the SCUBA data of Severgnini et al. (2000): average temperatures lower than 120 K overproduce the AGN contribution to the FIRB at 850μm. The shaded region in Fig. 1 gives an idea of the uncertainties related to the bolometric correction. To investigate the effect of multi-temperature FIR SEDs we held the lowest temperature at 80 K and added two component with temperatures of 120 K and 160 K. The result of the model is a curve very similar from that obtained with a single temperature SED of the appropriate temperature ($\sim$ 110 K). We also tried to add strong emission lines to the SED, and obtained very similar results. This weak dependence of the final result on the details of the single SEDs is due to a redshift effect: the final contribution of AGN to the FIRB is made by AGNs at different redshifts, therefore narrow spectral features are diluted.

This model will be tested and constrained in the near future with the identification of the ISOCAM sources at 15μ, and, in a few years, with the observations of SIRTF, which is expected to resolve most of the far infrared background at 70μm.

References

1. Elvis, M., et al., 1994, ApJS, 95, 1
2. Fabian, A.C., & Iwasawa, K., 1999, MNRAS 303, L34
3. Gilli, R., Savati, M., & Hasinger, G. 2001, A&A, 366, 407
4. Merritt, D., & Ferrarese, L. 2001, MNRAS, 320, L30
5. Salucci, p., Szuszkiewicz, E., Monaco, P., & Danese, L. 1999, MNRAS, 307, 637
6. Severgnini, P., et al. 2000, A&A 360, 457

List of Participants

Name	Affiliation and E-Mail
Tom Abel	Harvard, CfA tabel@cfa.harvard.edu
Immo Appenzeller	Landessternwarte Heidelberg I.Appenzeller@lsw.uni-heidelberg.de
Halton Arp	MPA, Garching arp@mpa-garching.mpg.de
Katsuaki Asano	Osaka University asano@vega.ess.sci.osaka-u.ac.jp
Anthony Banday	MPA, Garching banday@MPA-Garching.MPG.DE
Thomas Beckert	Max-Planck-Institut fuer Radioastronomie tbeckert@mpifr-bonn.mpg.de
Thomas Behrens	MPA, Garching tbehrens@mpa-garching.mpg.de
Ralf Bender	Universitaets-Sternwarte Muenchen bender@usm.uni-muenchen.de
Narciso Benitez	Department of Physics and Astronomy, JHU txitxo@pha.jhu.edu
Ilfan Bikmaev	Astronomy Department, Kazan State University Ilfan.Bikmaev@ksu.ru
Luc Binette	Instituto de Astronomia, UNAM Binette@astroscu.unam.mx

Roger Blandford

Caltech
rdb@caltech.edu

Joshua Bloom

Caltech
jsb@astro.caltech.edu

Katherine Blundell

Oxford University
kmb@astro.ox.ac.uk

Hans Boehringer

MPE, Garching
hxb@mpe.mpg.de

Thomas Boller

MPE, Garching
bol@mpe.mpg.de

Stefano Borgani

INFN - Trieste, Dipartimento di Astronomia
dell'Universita'
borgani@ts.astro.it

Luis Borgonovo

Stockholm Observatory
luis@astro.su.se

Volker Bromm

University of Cambridge
volker@ast.cam.ac.uk

Marcus Brueggen

Churchill College, Cambridge
marcus@MPA-Garching.MPG.DE

Rodion Burenin

CfA/IKI
rodion@head-cfa.harvard.edu

Vadim Burwitz

MPE, Garching
burwitz@mpe.mpg.de

Max Camenzind

Landessternwarte Koenigstuhl
M.Camenzind@lsw.uni-heidelberg.de

Alberto Cappi

Osservatorio Astronomico di Bologna
cappi@astbo3.bo.astro.it

Nicoletta Carangelo

Universita' dell'Insubria, Como, Italy
nicoletta.carangelo@mib.infn.it

Alfonso Cavaliere

Dip. Fisica, Univ. Roma Tor Vergata
alfonso.cavaliere@roma2.infn.it

Annalisa Celotti	S.I.S.S.A.	`celotti@sissa.it`
Catherine Cesarsky	ESO, Garching	`ccesarsk@eso.org`
Symeon Charalabides	NUI Galway	`symeon@itc.nuigalway.ie`
Eugene Churazov	MPA, Garching	`churazov@mpa-garching.mpg.de`
Benedetta Ciardi	MPA, Garching	`ciardi@mpa-garching.mpg.de`
Douglas Clowe	IAEF, Universitaet Bonn	`clowe@astro.uni-bonn.de`
Werner Collmar	MPE, Garching	`wec@mpe.mpg.de`
Andrea Comastri	Osservatorio Astronomico di Bologna	`comastri@anastasia.bo.astro.it`
Conrad Cramphorn	MPA, Garching	`conrad@mpa-garching.mpg.de`
Stefano Cristiani	ST European Coordinating Facility	`Stefano.Cristiani@eso.org`
Maria Cruz	University of Oxford	`mjc@astro.ox.ac.uk`
Luiz Da Costa	ESO, Garching	`ldacosta@eso.org`
Frederic Daigne	MPA, Garching	`daigne@mpa-garching.mpg.de`
Matthias Dietrich	University of Florida	`dietrich@astro.ufl.edu`
Tiziana Di Matteo	Harvard University	`tdimatteo@cfa.harvard.edu`

Andrei Doroshkevich   Theoretical Astrophysics Center
                             `dorr@tac.dk`

Georg Drenkhahn   MPA, Garching
                             `georg@mpa-garching.mpg.de`

Kees Dullemond   MPA, Garching
                             `dullemon@mpa-garching.mpg.de`

Torsten Ensslin   MPA, Garching
                             `ensslin@mpa-garching.mpg.de`

Andrew Fabian   IoA, Cambridge University
                             `acf@ast.cam.ac.uk`

Heino Falcke   Max-Planck-Institut fuer Radioastronomie
                             `hfalcke@mpifr-bonn.mpg.de`

Xiaohui Fan   School of Natural Sciences,
                             Institute for Advanced Study
                             `fan@sns.ias.edu`

Pascal Favre   INTEGRAL Science Data Center (ISDC) and
                             Geneva Observatory
                             `Pascal.Favre@obs.unige.ch`

William Forman   Harvard-Smithsonian Center for Astrophysics
                             `wrf@head-cfa.harvard.edu`

Steven Furlanetto   Harvard University/CfA
                             `sfurlanetto@cfa.harvard.edu`

Johan Fynbo   ESO, Garching
                             `jfynbo@eso.org`

Titus Galama   California Institute of Technology
                             `tjg@astro.caltech.edu`

Lisa Germany   ESO, Garching
                             `lgermany@eso.org`

Marat Gilfanov   MPA, Garching
                             `gilfanov@mpa-garching.mpg.de`

Roberto Gilli   Johns Hopkins University
                                 rgilli@pha.jhu.edu

Alvaro Gimenez   European Space Agency
                                 Alvaro.Gimenez-Canete@esa.int

Marcello Giroletti   Istituto di Radioastronomia CNR
                                 giroletti@ira.bo.cnr.it

Mario Gliozzi   MPE, Garching
                                 mgliozzi@mpe.mpg.de

Robert Goodrich   The W. M. Keck Observatory
                                 rgoodrich@keck.hawaii.edu

Federica Govoni   Istituto di Radioastronomia CNR
                                 fgovoni@ira.bo.cnr.it

Sergey Grebenev   IKI, Moscow
                                 grebenev@hea.iki.rssi.ru

Richard Griffiths   Carnegie Mellon University
                                 griffith@astro.phys.cmu.edu

Hans Grimm   MPA, Garching
                                 grimm@mpa-garching.mpg.de

Rainer Gruber   MPE, Garching
                                 gru@mpe.mpg.de

Dirk Grupe   MPE, Garching
                                 dgrupe@mpe.mpg.de

Takashi Hamana   National Astronomical Observatory of Japan
                                 hamana@yukawa.kyoto-u.ac.jp

Guenter Hasinger   AIP, Potsdam; MPE, Garching
                                 ghasinger@mpe.mpg.de

Evanthia Hatziminaoglou   ESO, Garching
                                 ehatzimi@eso.org

Christian Haydn   MPA, Garching
                                 ch@mpa-garching.mpg.de

Alexander Heger

University of California, Santa Cruz
alex@ucolick.org

Sebastian Heinz

MPA, Garching
heinzs@mpa-garching.mpg.de

Lars Hernquist

Harvard University
lars@cfa.harvard.edu

Wolfgang Hillebrandt

MPA, Garching
wfh@mpa-garching.mpg.de

Kevin Hurley

UC Berkeley Space Sciences Lab
khurley@sunspot.ssl.berkeley.edu

Jose Ibanez

Departamento de Astronomia y Astrofisica,
Universidad de Valencia, Spain
jose.m.ibanez@uv.es

Yasushi Ikebe

MPE, Garching
ikebe@mpe.mpg.de

Leopoldo Infante

P.Universidad Catolica de Chile
linfante@astro.puc.cl

Hajime Inoue

The Institute of Space and Astronautical Science
inoue@astro.isas.ac.jp

Kunihito Ioka

Department of Earth and Space Science,
Osaka University
ioka@vega.ess.sci.osaka-u.ac.jp

Anatoli Iyudin

MPE, Garching
ani@mpe.mpg.de

Hans-Thomas Janka

MPA, Garching
thj@mpa-garching.mpg.de

Sebastian Jester

MPI fuer Astronomie
jester@mpia.de

Guinevere Kauffmann

MPA, Garching
gamk@mpa-garching.mpg.de

Lucyna Kedziora-Chudczer	AAO/ATNF lkedzior@atnf.csiro.au
Ralf Keil	MPE, Garching rkeil@mpe.mpg.de
Irek Khamitov	TUBITAK National Observatory irekk@tug.tug.tubitak.gov.tr
Robert Kirshner	Harvard-Smithsonian Center for Astrophysics kirshner@cfa.harvard.edu
Shinji Koide	Toyama University, Faculty of Engineering koidesin@ecs.toyama-u.ac.jp
Stefanie Komossa	MPE, Garching skomossa@mpe.mpg.de
Shri Kulkarni	Caltech srk@astro.caltech.edu
Don Lamb	University of Chicago lamb@oddjob.uchicago.edu
Walter Lewin	MIT lewin@mit.edu
Richard Lieu	Dept. of Physics, University of Alabama lieur@cspar.uah.edu
Bifang Liu	Yukawa Institute for Theoretical Physics bfliu@yukawa.kyoto-u.ac.jp
Abraham Loeb	Harvard University aloeb@cfa.harvard.edu
Malcolm Longair	Cavendish Laboratory msl@mrao.cam.ac.uk
Dirk H. Lorenzen	Wissenschaftsjournalist, Deutschlandfunk DLorenzen@CompuServe.com
Kazuo Makishima	Department of Physics, University of Tokyo maxima@phys.s.u-tokyo.ac.jp

Jon Marcaide	Univ. Valencia `J.M.Marcaide@uv.es`
Sabino Matarrese	Dipartimento di Fisica "Galileo Galilei" `matarrese@pd.infn.it`
Kyoko Matsushita	MPE, Garching `matusita@xray.mpe.mpg.de`
Pasquale Mazzotta	European Space Agency and Smithsonian Astrophysical Observatory `mazzotta@head-cfa.harvard.edu`
Avery Meiksin	University of Edinburgh `aam@roe.ac.uk`
Doerte Mehlert	Landessternwarte Heidelberg `dmehlert@lsw.uni-heidelberg.de`
Mariano Mendez	SRON `M.Mendez@sron.nl`
Emmi Meyer-Hofmeister	MPA, Garching `emm@mpa-garching.mpg.de`
Friedrich Meyer	MPA, Garching `frm@mpa-garching.mpg.de`
Marco Mignoli	Bologna Observatory `mignoli@bo.astro.it`
Shin Mineshige	Yukawa Institute, Kyoto University `minesige@yukawa.kyoto-u.ac.jp`
Takeo Minezaki	University of Tokyo `minezaki@mtk.ioa.s.u-tokyo.ac.jp`
Francesco Miniati	MPA, Garching `fm@mpa-garching.mpg.de`
Felix Mirabel	SAP. CE-Saclay `mirabel@discovery.saclay.cea.fr`
Houjun Mo	MPA, Garching `hom@mpa-garching.mpg.de`

Palle Moller

ESO, Garching
pmoller@eso.org

Masahiro Morikawa

Dept. of Physics, Ochanomizu University
hiro@phys.ocha.ac.jp

Emmanuel Moy

MPE, Garching
moy@mpe.mpg.de

Ewald Mueller

MPA, Garching
emueller@mpa-garching.mpg.de

Akika Nakamichi

Gunma Astronomical Observatory
akika@astron.pref.gunma.jp

Ramesh Narayan

Center for Astrophysics
rnarayan@cfa.harvard.edu

Sergei Nayakshin

GSFC/NASA and MPA, Garching
serg@milkyway.gsfc.nasa.gov

Ryoichi Nishi

Department of Physics, Kyoto Univ.
nishi@tap.scphys.kyoto-u.ac.jp

S.Peng Oh

Caltech
peng@tapir.caltech.edu

Ken Ohsuga

Yukawa Institute for Theoretical Physics,
Kyoto University
ohsuga@yukawa.kyoto-u.ac.jp

Patrick Osmer

The Ohio State University
posmer@astronomy.ohio-state.edu

Alin Panaitescu

Princeton University
adp@astro.princeton.edu

Laura Pentericci

Max Planck Institut fuer Astronomie
laura@mpia-hd.mpg.de

Frank Pfefferkorn

MPE, Garching
pfefferk@mpe.mpg.de

Christoph Pfrommer	MPA, Garching pfrommer@mpa-garching.mpg.de
Steven Phillipps	University of Bristol s.phillipps@bristol.ac.uk
Marguerite Pierre	Service d'Astrophysique CEA/Saclay mpierre@cea.fr
Fulvio Pompilio	SISSA (ISAS) pompilio@sissa.it
Alexei Pozanenko	IKI, Moscow apozanen@iki.rssi.ru
Almudena Prieto	ESO, Garching aprieto@eso.org
Peter Predehl	MPE, Garching predehl@mpe.mpg.de
Jesper Rasmussen	Copenhagen University Observatory jr@astro.ku.dk
Martin Rees	Cambridge University mjr@ast.cam.ac.uk
Thomas H. Reiprich	MPE, Garching reiprich@mpe.mpg.de
Mike Revnivtsev	IKI, Moscow mikej@hea.iki.rssi.ru
Guido Risaliti	University of Florence risaliti@arcetri.astro.it
Brigitte Rocca-Volmerange	Institut d'Astrophysique de Paris rocca@iap.fr
Albrecht Ruediger	MPI f Gravitationsphysik atr@mpq.mpg.de
Remo Ruffini	ICRA and University of Rome Ruffini@icra.it

Pilar Ruiz-Lapuente	University of Barcelona & MPA pilar@am.ub.es
Nail Sakhibullin	Astronomy Department, Kazan State University Nail.Sakhibullin@ksu.ru
Isabel Salamanca	Anton Pannekoek Institute (UvA) isabel@science.uva.nl
Paolo Salucci	SISSA salucci@sissa.it
Norbert Schartel	XMM-Newton Observatory, ESA, Villafranca del Castillo nscharte@xmm.vilspa.esa.es
Paul Schechter	MIT schech@mit.edu
Hermann-Ulrich Schmidt	MPA, Garching schmidt@mpa-garching.mpg.de
Maarten Schmidt	Caltech mxs@astro.caltech.edu
Peter Schneider	Inst. f. Astrophysik, Univ. Bonn peter@astro.uni-bonn.de
Bernard Schutz	MPI fuer Gravitationphysik schutz@aei-potsdam.mpg.de
Dan Schwartz	Harvard-Smithsonian Center for Astrophysics (SAO) das@cfa.harvard.edu
Dominik Schwarz	Institut fuer Theoretische Physik, TU Wien dschwarz@hep.itp.tuwien.ac.at
Peter Shaver	ESO, Garching pshaver@eso.org
Chenggang Shu	Shanghai Astronomical Observatory cgshu@center.shao.ac.cn

Nail Sibgatullin	Faculty of Mechanics and Mathematics, Moscow State University `nail@mpa-garching.mpg.de`
Andrzej Soltan	Nicolaus Copernicus Astronomical Center `soltan@camk.edu.pl`
Joerg Sommer	MPA, Garching `joerg@mpa-garching.mpg.de`
Hendrik Spruit	MPA, Garching `henk@mpa-garching.mpg.de`
Rashid Sunyaev	MPA/IKI `sunyaev@mpa-garching.mpg.de`
Ewa Szuszkiewicz	Institute of Physics, University of Szczecin `szusz@univ.szczecin.pl`
Yasuo Tanaka	MPE, Garching `ytanaka@mpe.mpg.de`
Joachim Truemper	MPE, Garching `jtrumper@mpe.mpg.de`
Sachiko Tsuruta	Montana State University `uphst@gemini.oscs.montana.edu`
Ed van den Heuvel	Astronomical Institute "Anton Pannekoek" `edvdh@science.uva.nl`
Matteo Viel	MPA, Garching `viel@mpa-garching.mpg.de`
Alexey Vikhlinin	Harvard-Smithsonian Center for Astrophysics `alexey@head-cfa.harvard.edu`
Wolfgang Voges	MPE, Garching `wvoges@mpe.mpg.de`
Stefan Wagner	LSW Heidelberg `swagner@lsw.uni-heidelberg.de`
Joachim Wambsganss	Potsdam University `jkw@astro.physik.uni-potsdam.de`

Ken Watarai

Department of Astronomy, Graduate School of Science,
Kyoto University
`watarai@kusastro.kyoto-u.ac.jp`

Achim Weiss

MPA, Garching
`aweiss@mpa-garching.mpg.de`

Simon White

MPA, Garching
`swhite@mpa-garching.mpg.de`

Gerard Williger

NASA/GSFC
`williger@tejut.gsfc.nasa.gov`

Kazuhiro Yamamoto

Hiroshima University
`kazuhiro@hiroshima-u.ac.jp`

Feng Yuan

MPIfR, Bonn
`fyuan@mpifr-bonn.mpg.de`

Erik Zackrisson

Uppsala Astronomical Observatory, Sweden
`Erik.Zackrisson@astro.uu.se`

Saleem Zaroubi

MPA, Garching
`saleem@mpa-garching.mpg.de`

Anton Zensus

Max-Planck-Institut f. Radioastronomie
`azensus@mpifr-bonn.mpg.de`

Shu Zhang

MPE, Garching
`shz@mpe.mpg.de`

You Hong Zhang

Dipartimento di Scienze, Universita' dell'Insubria
`youhong.zhang@uninsubria.it`

Wei Zheng

Johns Hopkins University
`zheng@pha.jhu.edu`

Hans-Ulrich Zimmermann

MPE, Garching
`zim@mpe.mpg.de`

Author Index

Printing: Mercedes-Druck, Berlin
Binding: Stein+Lehmann, Berlin

GPSR Compliance

The European Union's (EU) General Product Safety Regulation (GPSR)
is a set of rules that requires consumer products to be safe and our
obligations to ensure this.

If you have any concerns about our products, you can contact us on
ProductSafety@springernature.com

In case Publisher is established outside the EU, the EU authorized
representative is:

Springer Nature Customer Service Center GmbH
Europaplatz 3
69115 Heidelberg, Germany

Batch number: 09625320

Printed by Printforce, the Netherlands